T0073764

Metaheuristic Applications in Structures and Infrastructures

Metaheuristic Applications in Structures and Infrastructures

Edited by

Amir Hossein Gandomi
Civil Engineering, The University of Akron, OH, USA

Xin-She Yang
School of Science and Technology, Middlesex University, London, UK

Siamak Talatahari
Marand Faculty of Engineering, University of Tabriz, Tabriz, Iran

Amir Hossein Alavi
Civil Engineering, Iran University of Science and Technology, Tehran, Iran

ELSEVIER

AMSTERDAM • BOSTON • HEIDELBERG • LONDON • NEW YORK • OXFORD
PARIS • SAN DIEGO • SAN FRANCISCO • SINGAPORE • SYDNEY • TOKYO

Elsevier
32 Jamestown Road, London NW1 7BY
225 Wyman Street, Waltham, MA 02451, USA

First edition 2013

Notices
Knowledge and best practice in this field are constantly changing. As new research and experience broaden our understanding, changes in research methods, professional practices, or medical treatment may become necessary.

Practitioners and researchers must always rely on their own experience and knowledge in evaluating and using any information, methods, compounds, or experiments described herein. In using such information or methods they should be mindful of their own safety and the safety of others, including parties for whom they have a professional responsibility.

To the fullest extent of the law, neither the Publisher nor the authors, contributors, or editors, assume any liability for any injury and/or damage to persons or property as a matter of products liability, negligence or otherwise, or from any use or operation of any methods, products, instructions, or ideas contained in the material herein.

British Library Cataloguing-in-Publication Data
A catalogue record for this book is available from the British Library

Library of Congress Cataloging-in-Publication Data
A catalog record for this book is available from the Library of Congress

ISBN: 978-0-12-398364-0

This book has been manufactured using Print On Demand technology. Each copy is produced to order and is limited to black ink. The online version of this book will show color figures where appropriate.

Contents

List of Contributors

Amir Hossein Alavi School of Civil Engineering, Iran University of Science and Technology, Tehran, Iran

Jennifer Avakian Department of Civil Engineering and Architecture, Technical University of Bari, Bari, Italy

Alireza Azarbakht Department of Civil Engineering, Faculty of Engineering, Arak University, Arak, Iran

Gebrail Bekdaş Department of Civil Engineering, Istanbul University, Avcılar, Istanbul, Turkey

André Borrmann Chair for Computational Modelling and Simulation, Technische Universität München

Giuseppe Carlo Marano Department of Civil Engineering and Architecture, Technical University of Bari, Bari, Italy

Benoît Descamps Building, Architecture & Town planning (BATir) Department, Université libre de Bruxelles, Brussels, Belgium

Emad E. Elbeltagi Structural Engineering Department, Mansoura University, Mansoura, Egypt

Rajan Filomeno Coelho Building, Architecture & Town planning (BATir) Department, Université libre de Bruxelles, Brussels, Belgium

Yiannis Fourkiotis Institute of Structural Analysis and Seismic Research, National Technical University of Athens, Zografou Campus, Athens, Greece

Kohei Fujita Department of Architecture and Architectural Engineering Kyoto University, Kyoto, Japan

Amir Hossein Gandomi Department of Civil Engineering, The University of Akron, Akron, OH, USA

Shobeir Pirayeh Gar Zachry Department of Civil Engineering, Texas A&M University, College Station, TX, USA

Saeed Gholizadeh Department of Civil Engineering, Urmia University, Urmia, Iran

Kasthurirangan Gopalakrishnan Department of Civil, Construction and Environmental Engineering, Iowa State University, Ames, IA, USA

Muhammad N.S. Hadi School of Civil, Mining and Environmental Engineering, University of Wollongong, NSW, Australia

Monique Head Department of Civil Engineering, Morgan State University, Baltimore, MD, USA

Manoj K. Jha Department of Civil Engineering, Morgan State University, Baltimore, MD, USA

Matthew G. Karlaftis Department of Transportation Planning and Engineering, National Technical University of Athens, Zografou Campus, Athens, Greece

Ali Kaveh Centre of Excellence for Fundamental Studies in Structural Engineering, Iran University of Science and Technology, Narmak, Tehran, Iran

Sunghoon Kim Department of Civil and Environmental Engineering, Korea Advanced Institute of Science and Technology, Daejeon, Republic of Korea

Chan Ghee Koh Department of Civil and Environmental Engineering, National University of Singapore, Singapore

Nikos D. Lagaros Institute of Structural Analysis and Seismic Research, National Technical University of Athens, Zografou Campus, Athens, Greece

Luciano Lamberti Politecnico di Bari, Dipartimento di Meccanica, Matematica e Management, Viale Japigia, Bari, Italy

Katharina C. Lukas Chair for Computational Modelling and Simulation, Technische Universität München

Chara Ch. Mitropoulou Institute of Structural Analysis and Seismic Research, National Technical University of Athens, Zografou Campus, Athens, Greece

Parviz Mohammad Zadeh Faculty of New Sciences and Technology, University of Tehran, Tehran, Iran

Mehdi Mousavi Department of Civil Engineering, Faculty of Engineering, Arak University, Arak, Iran

Pruettha Nanakorn Sirindhorn International Institute of Technology, Thammasat University, Thailand

Sinan Melih Nigdeli Department of Civil Engineering, Istanbul University, Avcılar, Istanbul, Turkey

Anan Nimtawat Faculty of Technology, Udon Thani Rajabhat University, Thailand

Alessandro Palmeri School of Civil and Building Engineering, Loughborough University, Loughborough, UK

Carmine Pappalettere Politecnico di Bari, Dipartimento di Meccanica, Matematica e Management, Viale Japigia, Bari, Italy

Giuseppe Quaranta Department of Civil and Environmental Engineering, University of California Davis, Davis, CA, USA

Masoud Rais-Rohani Mississippi State University, Starkville, MS, USA

Mohammad Rouhi Mississippi State University, Starkville, MS, USA

Mohammed Ghasem Sahab Department of Civil and Environmental Engineering, Tafresh University, Tafresh, Iran

Pejman Sharafi School of Civil, Mining and Environmental Engineering, University of Wollongong, NSW, Australia

Mohadeseh Alsadat Sadat Shirazi Faculty of Aerospace, K.N.T. University, Tehran, Iran

Sehyun Tak Department of Civil and Environmental Engineering, Korea Advanced Institute of Science and Technology, Daejeon, Republic of Korea

Izuru Takewaki Department of Architecture and Architectural Engineering Kyoto University, Kyoto, Japan

Siamak Talatahari Marand Faculty of Engineering, University of Tabriz, Tabriz, Iran

Lip H. Teh School of Civil, Mining and Environmental Engineering, University of Wollongong, NSW, Australia

Vassili V. Toropov School of Civil Engineering, University of Leeds, Leeds, UK

Thanh N. Trinh Applied Computing and Mechanics Laboratory, Structural Engineering Institute, Swiss Federal Institute of Technology in Lausanne (EPFL), Lausanne, Switzerland

Xin-She Yang School of Science and Technology, Middlesex University, London, UK

Hwasoo Yeo Department of Civil and Environmental Engineering, Korea Advanced Institute of Science and Technology, Daejeon, Republic of Korea

1 Metaheuristic Algorithms in Modeling and Optimization

Amir Hossein Gandomi[1], *Xin-She Yang*[2], *Siamak Talatahari*[3] and *Amir Hossein Alavi*[4]

[1]Department of Civil Engineering, The University of Akron, Akron, OH, USA, [2]School of Science and Technology, Middlesex University, London, UK, [3]Marand Faculty of Engineering, University of Tabriz, Tabriz, Iran, [4]School of Civil Engineering, Iran University of Science and Technology, Tehran, Iran

1.1 Introduction

In metaheuristic algorithms, meta- means "beyond" or "higher level." They generally perform better than simple heuristics. All metaheuristic algorithms use some trade-off of local search and global exploration. The variety of solutions is often realized via randomization. Despite the popularity of metaheuristics, there is no agreed upon definition of heuristics and metaheuristics in the literature. Some researchers use "heuristics" and "metaheuristics" interchangeably. However, the recent trend tends to name all stochastic algorithms with randomization and global exploration as metaheuristic. Randomization provides a good way to move away from local search to the search on the global scale. Therefore, almost all metaheuristic algorithms are usually suitable for nonlinear modeling and global optimization.

Metaheuristics can be an efficient way to use trial and error to produce acceptable solutions to a complex problem in a reasonably practical time. The complexity of the problem of interest makes it impossible to search every possible solution or combination; the aim is to find good, feasible solutions in an acceptable timescale. There is no guarantee that the best solutions can be found; we do not even know whether an algorithm will work, and if it does work, why (Yang, 2008, 2010a). The idea is to have an efficient and practical algorithm that will work most of the time and is able to produce good-quality solutions. Among the quality solutions found, it can be expected that some of them are nearly optimal, though there is no guarantee for such optimality.

The main components of any metaheuristic algorithm are: intensification and diversification, or exploitation and exploration (Blum and Roli, 2003). Diversification means generating diverse solutions so as to explore the search space on the global scale, while intensification means focusing on the search in a local

Metaheuristic Applications in Structures and Infrastructures. DOI: http://dx.doi.org/10.1016/B978-0-12-398364-0.00001-2

region by exploiting the information that a current good solution is found in this region. This is in combination with the selection of the best solutions (Yang, 2011a). The selection of the best ensures that the solutions will converge to the optimality. On the other hand, the diversification via randomization increases the diversity of the solutions while keeping the solutions from being trapped at local optima. The good combination of these two major components will usually ensure that the global solution is achievable.

Metaheuristic algorithms can be classified in many ways. One way is to classify them as either population-based or trajectory-based (Yang, 2010a). For example, genetic algorithms (GAs) and genetic programming (GP) are population-based as they use a set of strings; the particle swarm optimization (PSO) algorithm, which uses multiple agents or particles, is also population-based (Kennedy and Eberhart, 1995). On the other hand, simulated annealing (SA) (Kirkpatrick et al., 1983) uses a single solution that moves through the design space or search space, while artificial neural networks (ANNs) use a different approach.

Modeling and optimization may have a different emphasis, but for solving real-world problems, we often have to use both modeling and optimization because modeling makes sure that the objective functions are evaluated using the correct mathematical/numerical model for the problem of interest, while optimization can achieve the optimal settings of the design parameters. For optimization, the essential parts are the optimization algorithms. For this reason, we will focus on the algorithms, especially metaheuristic algorithms.

1.2 Metaheuristic Algorithms

1.2.1 Characteristics of Metaheuristics

Throughout history, especially during the early periods of human history, the main approach to problem-solving has always been heuristic or metaheuristic—by trial and error. Many important discoveries were done by "thinking outside the box," and often by accident; that is heuristics. Archimedes' Eureka moment was a heuristic triumph. In fact, our daily learning experiences (at least as children) are dominantly heuristic (Yang, 2010a). There are many reasons for the popularity and success of metaheuristics, and one of the main reasons is that these algorithms have been developed by mimicking the most successful processes in nature, including biological systems, and physical and chemical processes. For most algorithms, we know their fundamental components, but how exactly these components interact to achieve efficiency still remains largely a mystery, which inspires more active studies. Convergence analysis of a few algorithms shows some insight, but in general the mathematical analysis of metaheuristic algorithms still has many open questions and is still an ongoing active research topic (Yang, 2011a,c).

The notable performance of metaheuristic algorithms is due to how they imitate the best features in nature. Intensification and diversification are two main features of metaheuristic algorithms (Blum and Roli, 2003; Gandomi and Alavi, 2012a; Yang, 2010a, 2011c). The intensification phase, also called the exploitation phase, searches the current best solutions and selects the best candidates or solutions. The diversification phase, also called the exploration phase, ensures that the algorithm explores the search space more efficiently. The overall efficiency of an algorithm is mainly influenced by a fine balance between these two components. The system may be trapped in local optima if there is too little exploration or too much exploitation. In this case, it would be very difficult or even impossible to find the global optimum. On the other hand, if there is too much exploration but too little exploitation, it may be difficult for the system to converge. In this case, the overall search performance decelerates. Balancing these two components is itself a major optimization problem (Yang, 2011c). Evidently, simple exploitation and exploration are a part of the search. During the search, a proper mechanism or criterion should be considered to select the best solutions. "Survival of the fittest" is a common criterion. It is based on repeatedly updating the current best solution found so far. Moreover, certain elitism should be used. This is to verify that the best or fittest solutions are not lost and are passed onto the next generations.

Each algorithm and its variants use different ways to obtain a balance between exploration and exploitation. Certain randomization in combination with a deterministic procedure can be considered an efficient way to achieve exploration or diversification. This makes sure that the newly generated solutions distribute as diversely as possible in the feasible search space. From the implementation viewpoint, the actual way of implementing the algorithm does affect the performance to some degree. Hence, validation and testing of implementation of any algorithm are important (Talbi, 2009).

1.2.2 No Free Lunch Theorems

There are the so-called "no free lunch theorems," which can have significant implications in the field of optimization (Wolpert and Macready, 1997). One of the theorems states that if algorithm A outperforms algorithm B for some optimization functions, then B will be superior to A for other functions. In other words, if averaged over all possible function space, both algorithms A and B will perform, on average, equally well. That is to say, there are no universally better algorithms. An alternative viewpoint is that there is no need to find the average over all the possible functions for a given optimization problem. In this case, the major task is to find the best solutions, which has nothing to do with the average over all possible function space. Other researchers believe that there is no universal tool and, based on experiences, some algorithms outperform others for given types of optimization problems. Thus, the main objective would be either to choose the most suitable algorithm for a given problem or to design better algorithms for most types of problems, not necessarily for all the problems.

1.3 Metaheuristic Algorithms in Modeling

Various methodologies can be employed for nonlinear system modeling. Each method has its own advantages or drawbacks. The need to determine both the structure and the parameters of the engineering systems makes the modeling of these systems a difficult task. In general, models are classified into two main groups: (i) phenomenological and (ii) behavioral (Metenidis et al., 2004). The first class is established by taking into account the physical relationships governing a system. The structure of a phenomenological model is chosen on the basis of *a priori* knowledge about the system. To cope with the design complexity of phenomenological models, behavioral models are usually used. The behavioral models capture the relationships between the inputs and the outputs on the basis of a measured set of data. Thus, there is no need for *a priori* knowledge about the mechanisms that produced the experimental data. Such models are beneficial because they can provide very good results with minimal effort (Gandomi and Alavi, 2011, 2012a,b; Metenidis et al., 2004). Statistical regression techniques are widely used in behavioral modeling approaches.

Several alternative metaheuristic approaches have been developed for behavioral modeling. Developments in computer hardware during the last two decades have made it much easier for these techniques to grow into more efficient frameworks. In addition, various metaheuristics may be used as efficient tools in problems where conventional approaches fail or perform poorly. Two well-known classes of metaheuristic algorithms used in nonlinear modeling are ANNs (Haykin, 1999) and GP (Koza, 1992). ANNs have been used for a wide range of structural engineering problems (Alavi and Gandomi, 2011a; Sakla and Ashour, 2005). In spite of the successful performance of ANNs, they usually do not give a deep insight into the process for which they obtain a solution. GP, as an extension of GAs, possess completely new characteristics. GP is essentially a supervised machine-learning approach that searches a program space instead of a data space and automatically generates computer programs that are represented as tree structures and expressed using a functional programming language (Gandomi and Alavi, 2011; Koza, 1992). The ability to generate prediction models without assuming the form of the existing relationships is surely a main advantage of GP over regression and ANN techniques. GP and its variants are widely used for solving real-world problems (Alavi and Gandomi, 2011b; Gandomi et al., 2011a,b). There are some other metaheuristic algorithms have been described in the literature for modeling; these include, fuzzy logic (FL) and support vector machine (SVM). These algorithms (ANNs, GP, FL, and SVM) are explained in the following sections.

1.3.1 Artificial Neural Networks

ANNs emerged as a result of simulating a biological nervous system. The ANN method was developed in the early 1940s by McCulloch and coworkers (Perlovsky, 2001). The first studies were focused on building simple neural networks to model simple logic functions. At present, ANNs have been applied to problems that do not have algorithmic solutions or problems with complex solutions. In this study, the approximation ability of two of the most well-known ANN architectures, multilayer perceptron (MLP) and radial basis function (RBF), are investigated.

1.3.1.1 Multilayer Perceptron Network

MLP networks are a class of ANN structures using feed-forward architecture. They are among the most widely used metaheuristics for modeling complex systems in real-world applications (Alavi et al., 2010a). The MLP networks are usually applied to perform supervised learning tasks, which involve iterative training methods to adjust the connection weights within the neural network. MLPs are universal approximators; that is, they are capable of approximating essentially any continuous function to an arbitrary degree of accuracy. They are often trained using back propagation (BP) (Rumelhart et al., 1986) algorithms. MLPs consist of an input layer, at least one hidden layer of neurons, and an output layer. Each of these layers has several processing units, and each unit is fully interconnected with weighted connections to units in the subsequent layer. Each layer contains a number of nodes. Every input is multiplied by the interconnection weights of the nodes. The output (h_j) is obtained by passing the sum of the product through an activation function. Further details of MLPs can be found in Cybenko (1989) and Haykin (1999).

1.3.1.2 Radial Basis Function

RBFs have feed-forward architectures. Compared with other ANN structures such as MLPs, the RBF procedure for finding complex relationships is generally faster, and their training is much less computationally intensive. The structure of the RBF network consists of an input layer, a hidden layer with a nonlinear RBF activation function, and a linear output layer. Input vectors are transformed into RBFs by means of the hidden layer (Alavi et al., 2009).

The transformation functions used are based on a Gaussian distribution as an activation function. The center and width are two important parameters that are related to the Gaussian basis function. As the distance, usually Euclidean distance, between the input vector and its center increases, as the output given by the activation function decays to zero. The rate of decrease in the output is controlled by the width of RBF. The RBF networks with Gaussian basis functions have been shown to be universal function approximators with high point-wise convergence (Girosi and Poggio, 1990).

1.3.2 Genetic Programming

GP is a symbolic optimization technique that creates computer programs to solve a problem using the principle of Darwinian natural selection (Koza, 1992). Friedberg (1958) left the first footprints in the area of GP by using a learning algorithm to stepwise improve a program in a stepwise manner. Much later, Cramer (1985) applied GAs and tree-like structures to evolve programs. The breakthrough in GP then came in the late 1980s with the experiments of Koza (1992) on symbolic regression. GP was introduced by Koza (1992) as an extension of GA. The main difference between GP and GA is the representation of the solution. The GP solutions are computer programs that are represented as tree structures and expressed in a functional programming language (like LISP) (Koza, 1992). GA first creates a string of numbers that represent the solutions. In GP, the evolving programs (individuals) are parse trees that, unlike fixed-length binary strings, can vary in length throughout the run. Essentially, this was the beginning of computer programs that could program themselves (Koza, 1992). Since GP often evolves computer programs, the solutions can be executed without post-processing, while coded binary strings typically evolved by GA require post-processing. The optimization techniques, like GA, are generally used in parameter optimization to evolve so as to find the best values for a given set of model parameters. GP, on the other hand, provides the basic structure of the approximation model, together with the values of its parameters (Javadi and Rezania, 2009). GP optimizes a population of computer programs according to a fitness landscape determined by a program's ability to perform a given computational task. The fitness of each program in the population is evaluated using a predefined fitness function. Thus, the fitness function is the objective function GP aims to optimize (Torres et al., 2009).

This classical GP approach is referred to as tree-based GP. In addition to the traditional tree-based GP, there are other types of GP where programs are represented in different ways (Figure 1.1). These are linear GP and graph-based GP (Alavi et al., 2012; Banzhaf et al., 1998). The emphasis of the present study is on the linear-based GP techniques.

1.3.2.1 Linear-Based GP

There are a number of reasons for using linear GP. Basic computer architectures are fundamentally the same now as they were 20 years ago when GP began. Almost all the architectures represent computer programs in a linear fashion. Also, computers do not naturally run tree-shaped programs. Hence, slow interpreters have to be used as part of tree-based GP. Conversely, by evolving the binary bit patterns, in fact, used by computers, the use of an expensive interpreter (or compiler) is avoided and GP can run several orders of magnitude faster (Poli et al., 2007). Several linear variants of GP have been recently proposed. Some of them are (Oltean and Groşan, 2003a): linear genetic programming (LGP) (Brameier and Banzhaf, 2007), gene expression programming (GEP) (Ferreira, 2001), multiexpression programming (MEP) (Oltean and Dumitrescu, 2002), Cartesian genetic programming (CGP) (Miller and Thomson, 2002), genetic algorithm for deriving software (GADS) (Patterson, 2002), and infix form genetic programming (IFGP) (Oltean and Groşan, 2003b). LGP, GEP, and MEP are the most common linear-based GP methods. These variants make a clear distinction between the genotype

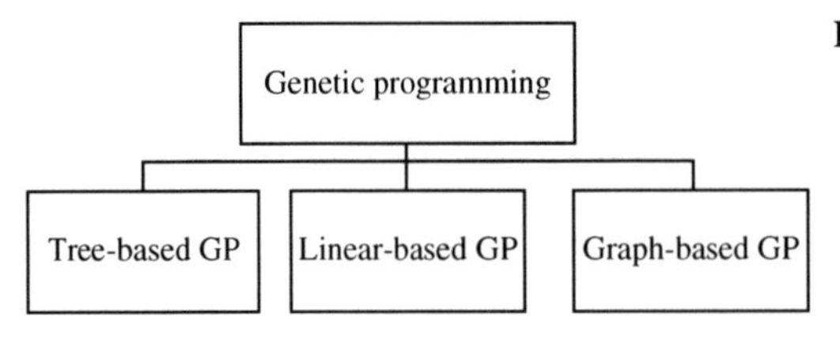

Figure 1.1 Different types of GP.

and the phenotype of an individual. The individuals in these variants are represented as linear strings (Oltean and Grosşan, 2003a).

1.3.2.1.1 Linear Genetic Programming

LGP is a subset of GP with a linear representation of individuals. There are several main differences between LGP and the traditional tree-based GP. Figure 1.2 presents a comparison of program structures in LGP and tree-based GP. Linear genetic programs can be seen as a data flow graph generated by multiple usage of register content. LGP operates on genetic programs that are represented as linear sequences of instructions of an imperative programming language (like C/C++) (see Figure 1.2A). As shown in Figure 1.2B, the data flow in tree-based GP is more rigidly determined by the tree structure of the program (Brameier and Banzhaf, 2001; Gandomi et al., 2010).

In the LGP system described here, an individual program is interpreted as a variable-length sequence of simple C instructions. The instruction set or function set of LGP consists of arithmetic operations, conditional branches, and function calls. The terminal set of the system is composed of variables and constants. The instructions are restricted to operations that accept a minimum number of constants or memory variables, called registers (r), and assign the result to a destination register, e.g., $r_0 = r_1 + 1$. LGPs can be converted into a functional representation by successive replacements of variables, starting with the last effective instruction (Oltean and Grosşan, 2003a). Automatic induction of machine code by genetic programming (AIMGP) is a particular form of LGP. In AIMGP, evolved programs are stored as linear strings of native binary machine code and are directly executed by the processor during fitness calculation. The absence of an interpreter and complex memory handling results in a significant speedup in the AIMGP execution compared to tree-based GP. This machine-code-based LGP approach searches for the computer program and the constants at the same time. Comprehensive descriptions of the basic parameters used to direct a search for a linear genetic program can be found in Brameier and Banzhaf (2007).

1.3.2.1.2 Gene Expression Programming

GEP is a natural development of GP. It was first presented by Ferreira (2001). GEP consists of five main components: function set, terminal set, fitness function, control parameters, and termination condition. Unlike the parse-tree representation in the conventional GP, GEP uses a fixed length of character strings to represent solutions

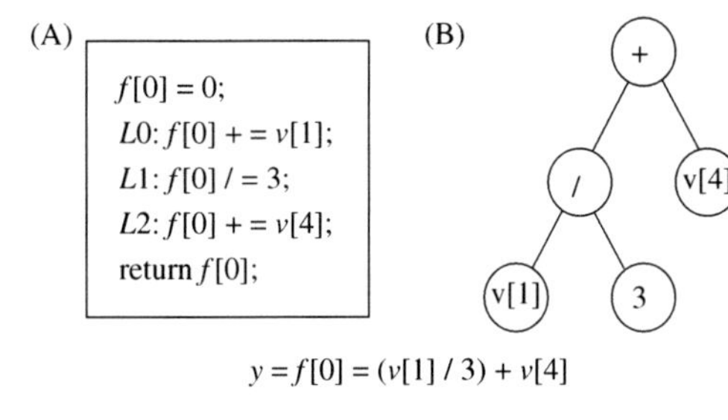

Figure 1.2 Comparison of the GP structures. (A) LGP and (B) Tree-based GP.
Source: After Alavi et al. (2010).

to the problems, which are afterward expressed as parse trees of different sizes and shapes. These trees are called GEP expression trees (ETs). One advantage of the GEP technique is that the creation of genetic diversity is extremely simplified as genetic operators work at the chromosome level. Another strength of GEP is that it refers to its unique; the multigenic nature allows the evolution of more complex programs composed of several subprograms (Gandomi and Alavi, 2011b, 2012c).

GEP genes have a fixed length, which is predetermined for a given problem. Thus, what varies in GEP is not the length of genes but the size of the corresponding ETs.

1.3.2.1.3 Multiexpression Programming

MEP is a subarea of GP developed by Oltean and Dumitrescu (2002). MEP uses linear chromosomes for solution encoding. It has a special ability to encode multiple solutions (computer programs) of a problem in a single chromosome. Based on the fitness values of the individuals, the best encoded solution is chosen to represent the chromosome. There is no increase in the complexity of the MEP decoding process, compared with the other GP variants that store a single solution in a chromosome. The exception is for the situations where the set of training data is not known (Oltean and Groşan, 2003a,c). The evolutionary steady-state MEP algorithm typically starts by the creation of a random population of individuals.

MEP is represented in a similar way to that of C and Pascal compilers translating mathematical expressions into machine code. The number of MEP genes per chromosome is constant, which specifies the length of the chromosome. A terminal (an element in the terminal set T) or a function symbol (an element in the function set F) is encoded by each gene. A gene that encodes a function includes pointers toward the function arguments. Function parameters always have indices of lower values than the position of that function itself in the chromosome. The first symbol in a chromosome must be a terminal symbol as stated by the proposed representation scheme.

The fitness of each expression in an MEP chromosome is calculated to designate the best encoded expression in that chromosome (Alavi et al., 2010b).

1.3.3 Fuzzy Logic

FL is a process of mapping an input space onto an output space using membership functions and linguistically specified rules (Ceven and Ozdemir, 2007). The concept of "fuzzy set" was preliminarily introduced by Zadeh (1965). The fuzzy approach is more in line with human thought as it provides possible rules relating input variables to the output variable. FL is well suited to implementing control rules that can only be expressed verbally. It can also be used for the modeling of systems that cannot be modeled with linear differential equations (Afandizadeh-Zargari et al., 2012).

The essential idea in FL is the concept of partial belongings of any object to different subsets of the universal set instead of full belonging to a single set. Partial belonging to a set can be described numerically by a membership function

(Topcu and Sarıdemir, 2008). A membership function is a curve, mapping an input element to a value between 0 and 1, showing the degree to which it belongs to a fuzzy set. Membership degree is the value of every element, varying between 0 and 1. A membership function can have different shapes for different kinds of fuzzy sets, such as bell, sigmoid, triangle, and trapezoid (Ceven and Ozdemir, 2007). In FL, rules and membership sets are used to make a decision. The idea of a fuzzy set is basic and simple: an object is allowed to have a gradual membership of a set. It means the degree of truth of a statement can range between 0 and 1, which is not limited to just two logic values {true, false}.

When linguistic variables are used, these degrees may be managed by specific functions. A fuzzy system consists of output and input variables. For each variable, fuzzy sets that characterize those variables are formulated, and for each fuzzy set a membership function can be defined. After that, the rules that relate the output and input variables to their fuzzy sets are defined. Figure 1.3 depicts a typical FL system where a general fuzzy inference system has basically four components: fuzzification, fuzzy rule base, fuzzy inference engine, and defuzzification (Topcu and Sarıdemir, 2008).

1.3.4 Support Vector Machines

SVM is a well-known machine-learning method, based on statistical learning theory (Boser et al., 1992; Vapnik, 1995, 1998). Similar to ANNs, the SVM procedure involves a training phase in which a series of input and target output values are fed into the model. A trained algorithm is then employed to evaluate a separate set of testing data. Two fundamental concepts underlying the SVM are (Goh and Goh, 2007):

1. An optimum margin classifier. This is a linear classifier that constructs a separating hyperplane (decision surface) such that the distance between the positive and the negative examples is maximized.
2. Use of kernel functions. A kernel function is a function that calculates the dot product of two vectors. A suitable nonlinear kernel can map the original example data onto a new data set that become linearly separable in a high-dimensional feature space, even though they are nonseparable in the original input space (Goh and Goh, 2007; Vapnik, 1995, 1998).

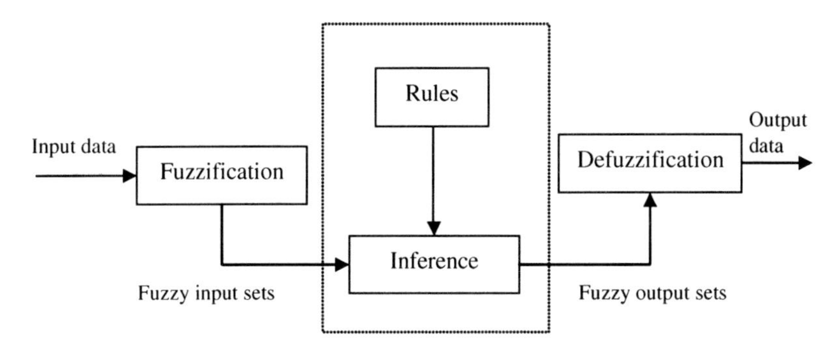

Figure 1.3 FL system.

The SVM procedure can be outlined as follows (Goh and Goh, 2007):

a. Choosing a kernel function with related kernel parameters.
b. Solving a Lagrange cost function and obtaining the Lagrange multipliers.
c. Carrying out the binary classification task, with training input data points.

Comprehensive descriptions of SVM can be found in more advanced literature (Goh and Goh, 2007; Vapnik, 1995).

1.4 Metaheuristic Algorithms in Optimization

To find an optimal solution to an optimization problem is often a very challenging task, depending on the choice and the correct use of the right algorithm. The choice of an algorithm may depend on the type of problem, the available of algorithms, computational resources, and time constraints. For large-scale, nonlinear, global optimization problems, there is often no agreed guideline for algorithm choice, and in many cases, there is no efficient algorithm. For hard optimization problems, especially for nondeterministic polynomial-time hard, or NP-hard, optimization problems, there is no efficient algorithm at all. In most applications, an optimization problem can be commonly expressed in the following generic form (Yang, 2010a, 2011e):

$$\text{minimize}\, x \in \Re^n f_i(x), \quad (i = 1, 2, \ldots, M), \tag{1.1}$$

$$\text{subject to} \quad h_j(x) = 0, \quad (j = 1, 2, \ldots, J), \tag{1.2}$$

$$g_k(x) \leq 0, \quad (k = 1, 2, \ldots, K), \tag{1.3}$$

where $f_i(x)$, $h_j(x)$, and $g_k(x)$ are functions of the design vector $x = (x_1, x_2, \ldots, x_n)^T$. Here, the components x_i of x are called design or decision variables, and they can be real continuous, discrete, or a mix of these two. The functions $f_i(x)$, where $i = 1, 2, \ldots, M$ are called the objective functions, or simply cost functions, and in the case of $M = 1$, there is only a single objective. The space spanned by the decision variables is called the design space or search space. The equalities for h_j and inequalities for g_k are called constraints. It is worth pointing out that we can also write the inequalities in the other way ≥ 0, and we can also formulate the objectives as a maximization problem.

Various algorithms may be used for solving optimization problems. The conventional or classic algorithms are mostly deterministic. As an instance, the simplex method in linear programming is deterministic. Some other deterministic optimization algorithms, such as Newton–Raphson algorithm, use the gradient information and are called gradient-based algorithms. Nongradient-based, or gradient-free/derivative-free, algorithms only use the function values, not any derivative (Yang, 2011b).

Heuristic and metaheuristic are the main types of the stochastic algorithms. The difference between heuristic and metaheuristic algorithms is negligible. Heuristic means "to find" or "to discover by trial and error." Quality solutions to a tough optimization problem can be found in a reasonable amount of time, but there is no guarantee that optimal solutions are reached. This is useful when good solutions, but not necessarily the best solutions, are needed within a reasonable amount of time (Koziel and Yang, 2011; Yang, 2010a).

As discussed earlier in this chapter, metaheuristic optimization algorithms are often inspired by nature. These metaheuristic algorithms can be classified into different categories based on the source of inspiration as shown in Figure 1.4. The main category is the biology-inspired algorithms, which generally use biological evolution and/or collective behavior of animals as their models. Science is another source of inspiration for metaheuristic algorithms. These algorithms are usually inspired by physics and chemistry. Moreover, art-inspired algorithms have been successful for global optimization. These are generally inspired by the creative behavior of artists such as musicians and architects. Social behavior is another source of inspiration and the socially inspired algorithms simulate social behavior to solve optimization.

Although there are different sources of inspiration for the metaheuristic optimization algorithms, they have similarities in their structures. Therefore, they can also be classified into two main categories: evolutionary algorithms and swarm algorithms.

1.4.1 Evolutionary Algorithms

The evolutionary algorithms generally use an iterative procedure based on a biological evolution progress to solve optimization problems. Some of the evolutionary algorithms are described below.

1.4.1.1 Genetic Algorithm

GAs are a powerful optimization method based on the principles of genetics and natural selection (Holland, 1975). Holland (1975) was the first to use the crossover and recombination, mutation, and selection in the study of adaptive and artificial systems. These genetic operators form the essential part of GA for problem-solving. Up to now, many variants of GA have been developed and applied to a wide range of optimization problems (Nikjoofar and Zarghami, 2013; Rani et al., 2012).

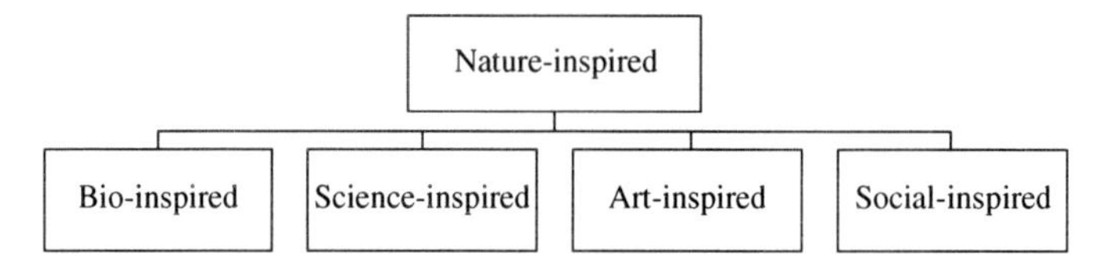

Figure 1.4 Source of inspiration in metaheuristic optimization algorithms.

One of the main advantages of GA is that it is a gradient-free method with the flexibility to deal with various types of optimization, whether the objective function is stationary or nonstationary, linear or nonlinear, continuous or discontinuous, or with random noise. In GA, a population can simultaneously find the search space in many directions because multiple offsprings in the population act like independent agents. This feature idealizes the parallelization of the algorithms for implementation. Moreover, different parameters and groups of encoded strings can be manipulated at the same time. Despite several advantages, Gas have some disadvantages pertaining to the formulation of fitness function, the usage of population size, the choice of the important parameters, and the selection criteria of a new population. The convergence of GA can be seriously dependent on the appropriate choice of these parameters.

1.4.1.2 *Differential Evolution*

Differential evolution (DE) was developed by Storn and Price (1997). It is a vector-based evolutionary algorithm and can be considered as a further development to GAs. It is a stochastic search algorithm with a self-organizing tendency and does not use the information of derivatives. DE carries out operations over each component (or each dimension of the solution). Solutions are represented in terms of vectors, and then mutation and crossover are carried out using these vectors (Gandomi et al., 2012a). For example, in GAs, mutation is carried out at one site or multiple sites of a chromosome; but in DE, a difference vector of two randomly chosen vectors is used to perturb an existing vector. Such vectorized mutation can be viewed as a self-organizing search, directed toward optimality (Yang, 2008, 2010a). This kind of perturbation is carried out over each population vector, and thus can be expected to be more efficient. Similarly, crossover is also a vector-based component-wise exchange of chromosomes or vector segments.

1.4.1.3 *Harmony Search*

Harmony search (HS) algorithm is a music-inspired algorithm, based on the improvisation process of a musician (Geem et al. 2001). Previous reviews of the HS literature have focused on applications in civil engineering such as engineering optimization (Lee and Geem, 2005), design of structures (Lee et al., 2005), design of water-distribution networks (Geem, 2006), geometry design of geodesic domes (Saka, 2007), design of steel frames (Degertekin, 2008a,b), groundwater management problems (Ayvaz and Elci, 2012), and geotechnical engineering problems (Cheng and Geem, 2012).

HS algorithms include a number of optimization operators, such as the harmony memory (HM), the harmony memory size (HMS), the harmony memory considering rate (HMCR), and the pitch adjusting rate (PAR). In the HS algorithm, the HM stores the feasible vectors, which are all in the feasible space. The HMR determines the number of vectors to be stored.

During the optimization process, a new harmony vector is generated from the HM, based on memory considerations, pitch adjustments, and randomization. After generating a new harmony vector, if it is better than the worst harmony in the HM, judged in terms of the objective function value, the new harmony is included in the HM and the existing worst harmony is excluded from the HM. Pitch adjustment is similar to the mutation operator in GAs. Although adjusting pitch has a similar role, it is limited to a certain local pitch adjustment and thus corresponds to a local search. The use of randomization can drive the system further to explore various regions with high solution diversity so as to find the global optimality.

1.4.2 Swarm-Intelligence-Based Algorithms

Swarm-intelligence-based algorithms use the collective behavior of animals such as birds, insects, or fishes. Here, we introduce briefly some of the most widely used swarm algorithms.

1.4.2.1 Particle Swarm Optimization

The PSO algorithm, inspired by social behavior simulation, was initially proposed by Kennedy and Eberhart (1995). PSO used the idea that social sharing of information among members may have some evolutionary advantage (Kennedy et al., 2001). PSO has been applied to many real-world problems (Talatahari et al., 2012a; Yang, 2008, 2010a). A standard PSO algorithm is initialized with a population (swarm) of random potential solutions (particles). Each particle iteratively moves across the search space and is attracted to the position of the best fitness historically achieved by the particle itself (local best) and by the best among the neighbors of the particle (global best) (Kaveh and Talatahari, 2009a). In fact, in the PSO, instead of using more traditional genetic operators, each particle adjusts its flying according to its own flying experience and its companions' flying experience (Hadidi et al., 2011; Kaveh and Talatahari, 2008, 2009b). Chaos theory can also improve the performance of the PSO by tuning its main constants or random variables (Gandomi et al., 2013a).

The original PSO, developed by Kennedy and Eberhart (1995), used an equation to calculate the velocity of each particle according to previous velocity, direction of the best position of each particle itself and direction of the best swarm, and then update the particle position.

After many numerical simulations, Shi (1998) added a weighting/inertia factor to the velocity equation to control the trade-off between the global exploration and the local exploitation abilities of the flying particles.

A well-chosen weight can stabilize the swarm as well as speed up the convergence. By using the linearly decreasing inertia weight, the PSO lacks global search ability at the end of run even when the global search ability is required to jump out of the local minimum in some cases. Nevertheless, the results shown in literature illustrate that by using a linearly decreasing inertia weight the performance of the PSO can be improved greatly and have better results than that of both a simple

PSO and an evolutionary programming as reported in Angeline (1998) and Shi and Eberhart (1999). Eusuff and Lansey (2003) combined the benefits of the local search tool of the PSO and the idea of mixing information from parallel local searches (Duan et al., 1993) to solve global optimization problems. They called this algorithm as shuffled frog-leaping (SFL) algorithm.

1.4.2.2 Ant Colony Optimization

In 1992, Dorigo developed a paradigm known as the ant colony optimization (ACO), a cooperative search technique that mimics the foraging behavior of real-life ant colonies (Dorigo, 1992; Dorigo et al., 1996). The ant algorithms mimic the characteristics of real ants that can rapidly establish the shortest route from food source to their nest and vice versa. Ants start searching the area surrounding their nest in a random manner. Ethologists observed that ants can construct the shortest path from their colony to the feed source and back using pheromone trails (Deneubourg and Goss, 1989; Goss et al., 1990). When ants encounter an obstacle, at first, there is an equal probability for all ants to move right or left, but after a while, the number of ants choosing the shorter path increases because of the increase in the amount of the pheromone on that path. With the increase in the number of ants and pheromone on the shorter path, all of the ants will choose and move along the shorter one (Kaveh and Talatahari, 2010a, Talatahari et al., 2012b).

In fact, real ants use their pheromone trails as a medium for communication of information among them. When an isolated ant comes across some food source in its random sojourn, it deposits a quantity of pheromone on that location. Other randomly moving ants in the neighborhood can detect this marked pheromone trail. Furthermore, these ants can follow this trail with a very high degree of probability and simultaneously enhance the trail by depositing their own pheromone. More and more ants follow the pheromone-rich trail and the probability of the trail being followed by other ants is further enhanced by the increased trail deposition. This is an autocatalytic (positive feedback) process that favors the path along which more ants previously traversed. The ant algorithms are based on the indirect communication capabilities of the ants. In ACO algorithms, virtual ants are deputed to generate rules by using heuristic information or visibility and the principle of indirect pheromone communication capabilities for iterative improvement of rules.

The general procedure of the ACO algorithm manages the scheduling of three steps: initialization, solution construction, and pheromone updating. The initialization of the ACO includes two parts: the first part is initialization of the pheromone trail. In the second part, a number of ants are arbitrarily placed on the nodes chosen randomly. Then each of the distributed ants will perform a tour on the graph by constructing a path according to the node transition rule described next.

For generation of a solution, each ant constructs a complete solution to the problem according to a probabilistic state transition rule. The state transition rule depends mainly on the state of the pheromone and visibility of ants. Visibility is an additional ability used to make this method more efficient. When every ant has constructed a solution, the intensity of pheromone trails on each edge is updated by the pheromone updating rule, which is applied in two phases: first, an evaporation phase where a fraction of the pheromone evaporates, and then a reinforcement phase, where the elitist ant, which has the best solution among the others, deposits an amount of pheromone. At the end of each movement, local pheromone update reduces the level of pheromone trail on paths selected by the ant colony during the preceding iteration.

1.4.2.3 Bee Algorithms

Bee algorithms are another class of metaheuristic algorithms that mimic the behavior of bees (Karaboga, 2005; Yang, 2005, 2008). Different variants of bee algorithms use slightly different characteristics of the behavior of bees. For example, in the honeybee-based algorithms, forager bees are allocated to different food sources (or flower patches) so as to maximize the total nectar intake (Karaboga, 2005; Nakrani and Tovey, 2004; Pham et al., 2006; Yang, 2005). In the virtual bee algorithm (VBA), developed by Yang (2005), pheromone concentrations can be linked with the objective functions more directly. On the other hand, the artificial bee colony (ABC) optimization algorithm was first developed by Karaboga (2005). In the ABC algorithm, the bees in a colony are divided into three groups. Unlike the honey bee algorithm, which has two groups of the bees (forager bees and observer bees), bees in ABC are more specialized (Afshar et al., 2007; Karaboga, 2005).

In the ABC algorithm, the colony of the artificial honey bees contains three groups of bees including employed bees (forager bees), onlooker bees (observer bees), and scouts. The first half of the colony consists of the employed artificial bees and the second half includes the onlookers. The position of a food source represents a possible solution to the considered optimization problem and the amount of nectar at the food source corresponds to the quality or fitness of the associated solution. At first, the ABC algorithm generates a randomly distributed, predefined number of initial population. After initialization, the population of the positions (solutions) is subjected to repeated cycles of the search process of the employed bees, onlooker bees, and scout bees. An employed bee produces a modification on the position (solution) in its memory depending on the local information (visual information) and tests the nectar amount (fitness value) of the new food source (new solution). Provided that the nectar amount of the new source is higher than that of the previous one, the bee memorizes the new position and forgets the old one. Otherwise, it keeps the position of the previous source in its memory. When all the employed bees complete the search process, they share the nectar information of the food sources and their position information with the onlooker bees in the dance area.

1.4.2.4 Firefly Algorithm

The firefly algorithm (FA) was first developed by Yang (2008, 2009), and was based on the flashing patterns and behavior of fireflies. In essence, FA uses the following three idealized rules: (i) fireflies are unisexual so that one firefly will be attracted to other fireflies regardless of their sex; (ii) The attractiveness is proportional to the brightness and they both decrease as their distance increases. Thus, for any two flashing fireflies, the less bright one will move toward the brighter one. If neither is brighter, they will each move randomly. (iii) The brightness of a firefly is determined by the landscape of the objective function.

A demo version of FA implementation, without Lévy flights, can be found at Mathworks file exchange website.[1] FA has attracted much attention (Apostolopoulos and Vlachos, 2011; Gandomi et al., 2011c; Sayadi et al., 2010; Talatahari et al., 2012c; Yang et al., 2012). A discrete version of FA can efficiently solve NP-hard scheduling problems (Sayadi et al., 2010), while a detailed analysis has demonstrated the efficiency of FA over a wide range of test problems, including multiobjective load dispatch problems (Apostolopoulos and Vlachos, 2011). A chaos-enhanced FA with a basic method for automatic parameter tuning is also developed (Yang, 2011b), and the use of various chaotic maps can significantly improve the performance of FA, though different chaotic maps may have different effects (Gandomi et al., 2013b).

1.4.2.5 Cuckoo Search

Cuckoo search (CS) is one of the latest nature-inspired metaheuristic algorithms, developed by Yang and Deb (2009). CS is based on the brood parasitism of some cuckoo species. In addition, this algorithm is enhanced by the so-called Lévy flights (Pavlyukevich, 2007), rather than by simple isotropic random walks. Recent studies show that CS is potentially far more efficient than PSO and GAs (Yang and Deb, 2010). For simplicity in describing the standard CS, we now use the following three idealized rules: (i) each cuckoo lays one egg at a time, and dumps it in a randomly chosen nest; (ii) the best nests with high-quality eggs will be carried over to the next generations; and (iii) the number of available host nests is fixed, and the egg laid by a cuckoo is discovered by the host bird with a probability between 0 and 1.

A Matlab implementation is given by the author, and can be downloaded.[2] CS is very efficient in solving engineering optimization problems (Gandomi et al., 2011d, 2012c).

1.4.2.6 Bat Algorithm

Bat algorithm (BA) is a relatively new metaheuristic, developed by Yang (2010b). It was inspired by the echolocation behavior of microbats. Microbats use a type of sonar, called echolocation, to detect prey, avoid obstacles, and locate their roosting crevices in the dark. These bats emit a very loud sound pulse and listen for the echo that bounces back from the surrounding objects. Their pulses vary in properties and can be correlated with their hunting strategies, depending on the species. Most bats use short, frequency-modulated signals to sweep through about an octave, while others more often use constant-frequency signals for echolocation.

[1] http://www.mathworks.com/matlabcentral/fileexchange/29693-firefly-algorithm

[2] www.mathworks.com/matlabcentral/fileexchange/29809-cuckoo-search-cs-algorithm

Their signal bandwidth varies depends on the species and is often increased by using more harmonics.

BA has been extended to multiobjective bat algorithm (MOBA) by Yang (2011d), and preliminary results suggest that it is very efficient (Yang and Gandomi, 2012; Gandomi et al., 2012d).

1.4.2.7 Charged System Search

The charged system search (CSS) is another of the more recently introduced metaheuristic algorithms (Kaveh and Talatahari, 2010b). This algorithm has been used to solve different types of optimization problems such as the design of skeletal structures (Kaveh and Talatahari, 2010c), the design of grillage systems (Kaveh and Talatahari, 2010d), parameter identification of MR dampers (Talatahari et al., 2012d), and the design of composite open channels (Kaveh et al., 2012).This algorithm was inspired by the governing laws of charged systems. Like the swarm algorithms, CSS uses multiple agents/charged particles, and each agent can be considered as a charged sphere. However, these agents are treated as charged particles (CP) they can affect each other according to the Coulomb and Gauss laws of electrostatics.

The governing laws of motion is from the Newtonian mechanics. CPs can impose electrical forces on the others, and the forces vary with the separation distance between the CPs, and for a CP located outside the sphere is inversely proportional to the square of the separation distance between the particles. At each iteration, each CP moves toward its new position considering the resultant forces and its previous velocity. If each CP exits from the allowable search space, its position is corrected using the HS-based handling approach as described by Kaveh and Talatahari (2009c). In addition, to store the best design, a charged memory (CM) is considered containing the CM number of positions for the so far best agents.

1.4.2.8 Krill Herd

Krill herd (KH) is another novel biologically inspired algorithm, proposed by Gandomi and Alavi (2012a). The KH algorithm is based on simulating the herding behavior of krill individuals. The minimum distances of each individual krill from food and from highest density of the herd are considered as the objectives for the Lagrangian movement. The time-dependent positions of the krill individuals are updated by three main components/factors:

1. movement induced by the presence of other individuals,
2. foraging activity, and
3. random diffusion.

This algorithm is also a gradient-free method because derivatives are not needed in the KH algorithm, and KH is also a metaheuristic algorithm because it uses a stochastic/random search in addition to some deterministic components.

For any metaheuristic algorithm, it is important to tune its related parameters. One of the interesting parts of the KH algorithm is that it can carefully simulate the krill behavior, and the values of these coefficients are based on empirical studies of the real-world krill systems. For this reason, only time interval needs fine-tuning in the KH algorithm. This can be considered a first attempt to use a real-world system to derive algorithm-dependent parameters, which can be advantageous. The preliminary results indicate that the KH method is very encouraging for its further application to optimization tasks.

1.5 Challenges in Metaheuristics

As we have seen from this review, metaheuristic algorithms have been used successfully for solving a variety of real-world problems. However, there remain some challenging issues concerning metaheuristics. First, theoretical analysis of these algorithms still lacks a unified framework, and there are many open problems as outlined by Yang in a recent review (Yang, 2011c). For example, in what ways do algorithm-dependent parameters affect the efficiency of an algorithm? What is the optimal balance between exploration and exploitation for metaheuristic algorithms so that they can perform most efficiently? How can memory in algorithm help to improve the performance of an algorithm?

Another important issue is the gap between theory and practice because metaheuristic applications are expanding rapidly, far more rapidly than mathematical analysis. At the same time, most applications are concerned with small-scale problems. Future applications and studies should focus on larger-scale applications.

On the other hand, there are many new algorithms, but more algorithms make it even harder to understand the working mechanisms of metaheuristics in general. We may need a unified approach to analyze algorithms, and ideally to classify these algorithms, so that we can understand all metaheuristics in a more insightful way (Yang, 2011c). These challenges also provide some timely and hot research opportunities for researchers so that important progress can be made in the near future.

References

Afandizadeh-Zargari, S., Zabihi, S., Alavi, A.H., Gandomi, A.H., 2012. A computational intelligence based approach for short-term traffic flow prediction. Expert Syst. 29 (2), 124–142.

Afshar, A., Haddad, O.B., Marino, M.A., Adams, B.J., 2007. Honey-bee mating optimization (HBMO) algorithm for optimal reservoir operation. J. Franklin Inst. 344, 452–462.

Alavi, A.H., Gandomi, A.H., Heshmat, A.A.R., 2010. Discussion on "Soft computing approach for real-time estimation of missing wave heights" by S.N. Londhe [Ocean Engineering 35 (2008) 1080–1089]. Ocean Eng. 37 (13), 1239–1240.

Alavi, A.H., Gandomi, A.H., 2011a. Prediction of principal ground-motion parameters using a hybrid method coupling artificial neural networks and simulated annealing. Comput. Struct. 89 (23–24), 2176–2194.

Alavi, A.H., Gandomi, A.H., 2011b. A robust data mining approach for formulation of geotechnical engineering systems. Int. J. Comput. Aided Meth. Eng. Eng. Comput. 28 (3), 242–274.

Alavi, A.H., Gandomi, A.H., Gandomi, M., Sadat Hosseini, S.S., 2009. Prediction of maximum dry density and optimum moisture content of stabilized soil using RBF neural networks. IES J. A Civ. Struct. Eng. 2 (2), 98–106.

Alavi, A.H., Gandomi, A.H., Mollahasani, A., Heshmati, A.A.R., Rashed, A., 2010a. Modeling of maximum dry density and optimum moisture content of stabilized soil using artificial neural networks. J. Plant Nutr. Soil Sci. 173 (3), 368–379.

Alavi, A.H., Gandomi, A.H., Sahab, M.G., Gandomi, M., 2010b. Multi expression programming: a new approach to formulation of soil classification. Eng. Comput. 26 (2), 111–118.

Alavi, A.H., Gandomi, A.H., Bolury, J., Mollahasani, A., 2012. Linear and tree-based genetic programming for solving geotechnical engineering problems. In: Yang, X.S., et al., (Eds.), Metaheuristics in Water Resources, Geotechnical and Transportation Engineering. Elsevier, Waltham, MA, pp. 289–310. (Chapter 12).

Angeline P.J., 1998. Evolutionary optimization versus particle swarm optimization: philosophy and performance difference. Proceedings of Annual Conference on Evolutionary Programming, San Diego, 1998, pp. 601 – 610.

Apostolopoulos, T., Vlachos, A., 2011. Application of the firefly algorithm for solving the economic emissions load dispatch problem. Int. J. Combin. (Volume 2011, Article ID 523806. <http://www.hindawi.com/journals/ijct/2011/523806.html> (accessed 14.01.2012.).

Ayvaz, M.T., Elci, A., 2012. Application of the hybrid HS solver algorithm to the solution of groundwater management problems. In: Yang, X.S., et al., (Eds.), Metaheuristics in Water Resources, Geotechnical and Transportation Engineering. Elsevier, pp. 79–97. (Chapter 4).

Banzhaf, W., Nordin, P., Keller, R., Francone, F., 1998. Genetic Programming—An Introduction on the Automatic Evolution of Computer Programs and its Application. dpunkt/Morgan Kaufmann, Heidelberg/San Francisco, CA.

Blum, C., Roli, A., 2003. Metaheuristics in combinatorial optimization: overview and conceptual comparison. ACM Comput. Surv. 35, 268–308.

Boser B.E., Guyon I.M., Vapnik V.N., 1992. A training algorithm for optimal margin classifiers. In: Proceedings of the Fifth Annual ACM Workshop on Computational Learning Theory, vol. 5. Pittsburgh, pp. 144–152.

Brameier, M., Banzhaf, W., 2001. A comparison of linear genetic programming and neural networks in medical data mining. IEEE Trans. Evol. Comput. 5 (1), 17–26.

Brameier, M., Banzhaf, W., 2007. Linear Genetic Programming. Springer Science + Business Media LLC, New York, NY.

Ceven, E.K., Ozdemir, O., 2007. Using fuzzy logic to evaluate and predict Chenille Yarn's shrinkage behaviour. Fibres Textiles East. Europe. 15 (3), 55–59.

Cheng, Y.M., Geem, Z.W., 2012. Hybrid heuristic optimization methods in geotechnical engineering. In: Yang, X.S., et al., (Eds.), Metaheuristics in Water Resources, Geotechnical and Transportation Engineering. Elsevier, Waltham, MA, pp. 205–229. (Chapter 9).

Cortes, C., Vapnik, V., 1995. Support vector networks. Mach. Learn. 20, 273–297.

Cramer, N.L., 1985. A representation for the adaptive generation of simple sequential programs. In: Genetic Algorithms and Their Applications, Davis, L., Ed. Pittsburgh, PA. pp. 183 – 187.

Cybenko, J., 1989. Approximations by superpositions of a sigmoidal function. Math. Cont. Sign. Syst. 2, 303–314.
Degertekin, S.O., 2008a. Optimum design of steel frames using harmony search algorithm. Struct. Multidiscip. Optim. 36, 393–401.
Degertekin, S.O., 2008b. Harmony search algorithm for optimum design of steel frame structures: a comparative study with other optimization methods. Struct. Eng. Mech. 29, 391–410.
Deneubourg, J.L., Goss, S., 1989. Collective patterns and decision-making. Ethnol. Ecol. Evol. 1, 295–311.
Dorigo, M., 1992. Optimization, Learning and Natural Algorithms. Dip. Elettronica e Informazione, Politecnico di Milano, Milano, PhD Thesis.
Dorigo, M., Maniezzo, V., Colorni, A., 1996. The ant system: optimization by a colony of cooperating agents. IEEE Trans. Syst. Man, Cybern. B Cybern. 26 (1), 29–41.
Duan, Q.Y., Gupta, V.K., Sorooshian, S., 1993. Shuffled complex evolution approach for effective and efficient global minimization. J. Optim. Theor. Appl. 76, 502–521.
Eberhart R.C., Kennedy J., 1995. A new optimizer using particle swarm theory. In: Proceedings of the Sixth International Symposium on Micro Machine and Human Science, Nagoya, Japan, 1995.
Eusuff, M.M., Lansey, K.E., 2003. Optimization of water distribution network design using the shuffled frog leaping algorithm. J. Water. Res. Plan. Manage. 29 (3), 10–25.
Ferreira, C., 2001. Gene expression programming: a new adaptive algorithm for solving problems. Complex Syst. 13 (2), 87–129.
Ferreira, C., 2006. Gene Expression Programming: Mathematical Modeling by an Artificial Intelligence. second ed. Springer, Germany.
Francone, F.D., Deschaine, L.M., 2004. Extending the boundaries of design optimization by integrating fast optimization techniques with machine-code-based, linear genetic programming. Inf. Sci. 161, 99–120.
Friedberg, R.M., 1958. A learning machine: Part I. IBM J. Res. Dev. 2, 2–13.
Gandomi, A.H., Alavi, A.H., 2011. Multi-stage genetic programming: a new strategy to nonlinear system modeling. Inf. Sci. 23, 5227–5239.
Gandomi, A.H., Alavi, A.H., 2012a. Krill herd: a new bio-inspired optimization algorithm. Commun. Nonlinear Sci. Numer. Simul. 17 (12), 4831–4845.
Gandomi, A.H., Alavi, A.H., 2012b. A new multi-gene genetic programming approach to nonlinear system modeling. Part I: materials and structural engineering problems. Neural Comput. Appl. 21 (1), 171–187.
Gandomi, A.H., Alavi, A.H., 2012c. A new multi-gene genetic programming approach to nonlinear system modeling. Part II: geotechnical and earthquake engineering problems. Neural Comput. Appl. 21 (1), 189–201.
Gandomi, A.H., Alavi, A.H., Sahab, M.G., 2010. New formulation for compressive strength of CFRP confined concrete cylinders using linear genetic programming. Mater. Struct. 43 (7), 963–983.
Gandomi, A.H., Alavi, A.H., Mirzahosseini, M.R., Moqaddas Nejad, F., 2011a. Nonlinear genetic-based models for prediction of flow number of Asphalt mixtures. J. Mater. Civil Eng. ASCE. 23 (3), 248–263.
Gandomi, A.H., Alavi, A.H., Yun, G.J., 2011b. Nonlinear modeling of shear strength of SFRC beams using linear genetic programming. Struct. Eng. Mech. 38 (1), 1–25.
Gandomi, A.H., Yang, X.S., Alavi, A.H., 2011c. Mixed variable structural optimization using firefly algorithm. Comput. Struct. 89 (23–24), 2325–2336.

Gandomi, A.H., Yang, X.S., Alavi, A.H., 2011d. Cuckoo search algorithm: a metaheuristic approach to solve structural optimization problems. Eng. Comput. doi: 10.1007/s00366-011-0241-y (in press).

Gandomi, A.H., Yang, X.S., Talatahari, S., Deb, S., 2012a. Coupled eagle strategy and differential evolution for unconstrained and constrained global optimization. Comput. Math. Appl. 63 (1), 191–200.

Gandomi, A.H., Babanajad, S.K., Alavi, A.H., Farnam, Y., 2012b. A novel approach to strength modeling of concrete under triaxial compression. J. Mater. Civ. Eng. 24 (9), 1132–1143.

Gandomi, A.H., Talatahari, S., Yang, X.S., Deb, S., 2012c. Design optimization of truss structures using cuckoo search algorithm. Struct. Des. Tall Spec. Buildings. 10.1002/tal.1033.

Gandomi, A.H., Yang, X.S., Talatahari, S., Alavi, A.H., 2012d. Bat algorithm for constrained optimization tasks. Neural Comput. Appl. doi: 10.1007/s00521-012-1028-9.

Gandomi, A.H., Yun, G.J., Yang, X.S., Talatahari, S., 2013a. Chaos-enhanced accelerated particle swarm algorithm. Commun. Nonlinear Sci. Numer. Simul. 18 (2), 327–340.

Gandomi, A.H., Yang, X.S., Talatahari, S., Alavi, A.H., 2013b. Firefly algorithm with chaos. Commun. Nonlinear Sci. Numer. Simul. 18 (1), 89–98.

Geem, Z.W., 2006. Optimal cost design of water distribution networks using harmony search. Eng. Optim. 38, 259–277.

Geem, Z.W., Kim, J.H., Loganathan, G.V., 2001. A new heuristic optimization algorithm; harmony search. Simulation. 76, 60–68.

Girosi, F., Poggio, T., 1990. Networks and the best approximation property. Biol. Cybern. 63 (3), 169–176.

Goh, A.T.C., Goh, S.H., 2007. Support vector machines: their use in geotechnical engineering as illustrated using seismic liquefaction data. Comput. Geotech. 34, 410–421.

Goss, S., Beckers, R., Deneubourg, J.L., Aron, S., Pasteels, J.M., 1990. How trail laying and trail following can solve foraging problems for ant colonies. In: Hughes, R.N. (Ed.), Behavioural Mechanisms in Food Selection, NATO-ASI Series, vol. G 20, Berlin.

Hadidi, A., Kaveh, A., Farahmand Azar, B., Talatahari, S., Farahmandpour, C., 2011. An efficient optimization algorithm based on particle swarm and simulated annealing for space trusses. Int. J. Optim. Civ. Eng. 1 (3), 375–395.

Haykin, S., 1999. Neural Networks—A Comprehensive Foundation. second ed. Prentice Hall, Englewood Cliffs, NJ.

Holland, J., 1975. Adaptation in Natural and Artificial Systems. University of Michigan Press, Ann Anbor, MI.

Javadi, A.A., Rezania, M., 2009. Applications of artificial intelligence and data mining techniques in soil modeling. Geomech. Eng. 1 (1), 53–74.

Karaboga, D., 2005. An Idea Based on Honey Bee Swarm for Numerical Optimization. Erciyes University, Computer Engineering Department (Technical Report-TR06).

Kaveh, A., Talatahari, S., 2008. A discrete particle swarm ant colony optimization for design of steel frames. Asian J. Civ. Eng. 9 (6), 563–575.

Kaveh, A., Talatahari, S., 2009a. Hybrid algorithm of harmony search, particle swarm and ant colony for structural design optimization, studies in computational intelligence, Harmony Search Algorithms for Structural Design Optimization, vol. 239. Springer, Berlin, Heidelberg (pp. 159–198).

Kaveh, A., Talatahari, S., 2009b. A particle swarm ant colony optimization algorithm for truss structures with discrete variables. J. Construct. Steel Res. 65 (8–9), 1558–1568.

Kaveh, A., Talatahari, S., 2009c. Particle swarm optimizer, ant colony strategy and harmony search scheme hybridized for optimization of truss structures. Comput. Struct. 87 (5–6), 267–283.

Kaveh, A., Talatahari, S., 2010a. An improved ant colony optimization for constrained engineering design problems, engineering computations. Int. J. Comput. Aided Eng. Softw. 27 (1), 155–182.

Kaveh, A., Talatahari, S., 2010b. A novel heuristic optimization method: charged system search. Acta Mech. 213 (3–4), 267–289.

Kaveh, A., Talatahari, S., 2010c. Optimal design of skeletal structures via the charged system search algorithm. Struct. Multidiscip. Optim. 41 (6), 893–911.

Kaveh, A., Talatahari, S., 2010d. Charged system search for optimum grillage systems design using the LRFD-AISC code. J. Construct. Steel Res. 66 (6), 767–771.

Kaveh, A., Talatahari, S., Farahmand Azar, B., 2012. Optimum Design of Composite Open Channels Using Charged System Search Algorithm. Iranian Journal of Science & Technology, Transaction B: Engineering. 36 (C1), 67–77.

Kennedy, J., Eberhart, R., 1995. Particle swarm optimization. In: Proceedings of the IEEE International Conference on Neural Networks, Piscataway, NJ, pp. 1942 – 1948.

Kennedy, J., Eberhart, R.C., Shi, Y., 2001. Swarm Intelligence. Morgan Kaufman Publishers, San Francisco, CA.

Kirkpatrick, S., Gellat, C.D., Vecchi, M.P., 1983. Optimization by simulated annealing. Science. 220, 671–680.

Koza, J.R., 1992. Genetic Programming: on the Programming of Computers by means of Natural Selection. MIT Press, Cambridge, MA.

Koziel, S., Yang, X.S., 2011. Computational Optimization, Methods and Algorithms, Studies in Computational Intelligence, 356. Springer, Berlin, Germany.

Lee, K.S., Geem, Z.W., 2005. A new meta-heuristic algorithm for continuous engineering optimization: harmony search theory and practice. Comput. Method Appl. Mech. Eng. 194, 3902–3933.

Lee, K.S., Geem, Z.W., Lee, S.-H., et al., 2005. The harmony search heuristic algorithm for discrete structural optimization. Eng. Optim. 37, 663–684.

Metenidis, M.F., Witczak, M., Korbicz, J., 2004. A novel genetic programming approach to nonlinear system modelling: application to the DAMADICS benchmark problem. Eng. Appl. Artif. Intell. 17, 363–370.

Miller, J., Thomson, P., 2002. Cartesian genetic programming. In: Poli, R., Banzhaf, W., Langdon, B., Miller, J., Nordin, P., Fogarty, T.C. (Eds.), Genetic Programming. Springer, Berlin.

Nakrani, S., Tovey, C., 2004. On Honey Bees and Dynamic Server Allocation in Internet Hosting Centers. Adaptive Behaviour. 12 (3–4), 223–240.

Nikjoofar, A., Zarghami, M., 2013. Water distribution networks designing by the multiobjective genetic algorithm and game theory. In: Yang, X.S., et al., (Eds.), Metaheuristics in Water Resources, Geotechnical and Transportation Engineering. Elsevier, Waltham, MA, pp. 43–77. (Chapter 5).

Oltean, M., Dumitrescu, D., 2002. Multi Expression Programming. Babeş-Bolyai University, Cluj-Napoca, Romania (Technical Report, UBB-01-2002).

Oltean, M., Groşşan, C., 2003a. A comparison of several linear genetic programming techniques. Adv. Complex Syst. 14 (4), 1–29.

Oltean, M., Groşşan, C., 2003b. Solving classification problems using infix form genetic programming. In: Berthold, M. (Ed.), Intelligent Data Analysis. Springer, Berlin, pp. 242–252. (LNCS 2810).

Oltean, M., Groşan, C., 2003c. Evolving evolutionary algorithms using multi expression programming. In Artificial life, LNAI, vol. 2801. Springer, pp. 651–658.

Patterson, N., 2002. Genetic Programming with Context-Sensitive Grammars. School of Computer Science, University of Scotland, Scotland, UK (Ph.D. Thesis).

Pavlyukevich, I., (2007). Lévy flights, non-local search and simulated annealing, J. Computational Physics, vol. 226 , 1830–1844.

Perlovsky, L.I., 2001. Neural Networks and Intellect. Oxford University Press, Oxford, UK.

Pham, D.T., Ghanbarzadeh, A., Koc, E., Otri, S., Rahim, S., Zaidi, M., 2006. The bees algorithm: a novel tool for complex optimisation problems. Proceedings of IPROMS 2006 Conference, pp.454 – 461.

Poli, R., Langdon, W.B., McPhee, N.F., Koza, J.R., 2007. Genetic programming: an introductory tutorial and a survey of techniques and applications. University of Essex, UK (Technical report [CES-475], 2007).

Rani, D., Jain, S.K., Srivastava, D.K., Perumal, M., 2012. Genetic algorithms and their applications to water resources systems. In: Yang, X.S., et al., (Eds.), Metaheuristics in Water Resources, Geotechnical and Transportation Engineering. Elsevier, Waltham, MA, pp. 43–77. (Chapter 3).

Rumelhart, D.E., Hinton, G.E., Williams, R.J., 1986. Learning internal representations by error propagation. Proceedings Parallel Distributed Processing. MIT Press, Cambridge, MA.

Saka, M.P., 2007. Optimum geometry design of geodesic domes using harmony search algorithm. Adv. Struct. Eng. 10, 595–606.

Sakla, S.S., Ashour, A.F., 2005. Prediction of tensile capacity of single adhesive anchors using neural networks. Comput. Struct. 83 (21–22), 1792–1803.

Sayadi, M.K., Ramezanian, R., Ghaffari-Nasab, N., 2010. A discrete firefly meta-heuristic with local search for make span minimization in permutation flow shop scheduling problems. Int. J. Ind. Eng. Comput. 1, 1–10.

Shi Y., Eberhart R.C., 1998. A modified particle swarm optimizer. Proceedings of IEEE International Conference on Evolutionary Computation, Alaska, 1998, pp. 69–73.

Shi Y., Eberhart R.C., 1999. Empirical study of particle swarm optimization. Proceedings of the 1999 IEEE Congress on Evolutionary Computation 1999. vol. 3, pp. 1945 – 1950.

Storn, R., Price, K.V., 1997. Differential evolution—a simple and efficient heuristic for global optimization over continuous spaces. J. Global Opt. 11 (4), 341–359.

Talatahari, S., Kheirollahi, M., Farahmandpour, C., Gandomi, A.H., 2012a. A multi-stage particle swarm for optimum design of truss structures. Neural Comput. Appl. doi: 10.1007/s00521-012-1072-5.

Talatahari, S., Singh, V.P., Hassanzadeh, Y., 2012b. Ant colony optimization for estimating parameters of flood frequency distributions. In: Yang, X.S., et al., (Eds.), Metaheuristics in Water Resources, Geotechnical and Transportation Engineering. Elsevier, Waltham, MA, pp. 121–146. (Chapter 6).

Talatahari, S., Gandomi, A.H., Yun, G.Y., 2012c. Optimum design of tower structures by firefly algorithms. Struct. Des. Tall Spec. Build. doi: 10.1002/tal.1043.

Talatahari, S., Kaveh, A., Mohajer Rahbari, N., 2012d. Parameter identification of Bouc-Wen model for MR fluid dampers using adaptive charged system search optimization. J. Mech. Sci. Technol. 26 (8), 2523–2534.

Talbi, E.G., 2009. Metahueristics: From Design to Implementation. Wiley, Hoboken, NJ.

Topcu, I.B., Sarı, M., 2008. Prediction of compressive strength of concrete containing fly ash using artificial neural networks and fuzzy logic. Comput. Mater. Sci. 41, 305–311.

Torres, R.S., Falcão, A.X., Gonçalves, M.A., Papa, J.P., Zhang, B., Fan, W., et al., 2009. A genetic programming framework for content-based image retrieval. Pattern Recognit. 42 (2), 283–292.

Vapnik, V., 1995. The Nature of Statistical Learning. Springer, New York, NY.

Vapnik, V., 1998. Statistical Learning Theory. Wiley, New York, NY.

Wolpert, D.H., Macready, W.G., 1997. No free lunch theorems for optimization. IEEE Trans. Evol. Comput. 1 (1), 67–82.

Yang, X.S., 2005. Engineering optimization via nature-inspired virtual bee algorithms. Artificial Intelligence and Knowledge Engineering Applications: A Bioinspired Approach. Springer, Berlin, Germany (Lecture Notes in Computer Science, **3562**, pp. 317 – 323).

Yang, X.S., 2008. Nature-Inspired Metaheuristic Algorithms. first ed. Luniver Press, Frome.

Yang, X.S., 2009. Firefly algorithms for multimodal optimization. In: Watanabe, O., Zeugmann, T. (Eds.), Fifth Symposium on Stochastic Algorithms, Foundation and Applications (SAGA 2009), LNCS, 5792, pp. 169–178.

Yang, X.S., 2010a. Engineering Optimization: An Introduction with Metaheuristic Applications. John Wiley and Sons, Hoboken, NJ.

Yang, X.S., 2010b. A new metaheuristic bat-inspired algorithm. In: Gonzalez, J.R., et al., (Eds.), *Nature-Inspired Cooperative Strategies for Optimization* (NICSO 2010). Springer, Berlin, pp. 65–74. (**SCI 284**).

Yang, X.S., 2011a. Review of metaheuristics and generalized evolutionary walk algorithm. Int. J. Bio-Inspired Comput. 3 (2), 77–84.

Yang, X.S., 2011b. Chaos-enhanced firefly algorithm with automatic parameter tuning. Int. J. Swarm Intell. Res. 2 (4), 1–11.

Yang, X.S., 2011c. Metaheuristic optimization: algorithm analysis and open problems. In: Pardalos, P.M., Rebennack, S. (Eds.), Proceedings of the Tenth Symposium of Experimental Algorithms (SEA2011), vol. 6630. Springer (Lecture Notes in Computer Science, pp. 21 – 32).

Yang, X.S., 2011d. Bat algorithm for multi-objective optimisation. Int. J. Bio-Inspired Comput. 3 (5), 267–274.

Yang, X.S., 2011e. Metaheuristic optimization. Scholarpedia. 6 (8), 11472, http://www.scholarpedia.org/article/Metaheuristic_Optimization. doi:10.4249/scholarpedia.11472.

Yang, X.S., Deb, S., 2009. Cuckoo search via Lévy flights. In: Proceedings of World Congress on Nature and Biologically Inspired Computing (NaBic 2009), IEEE Publications, USA, pp. 210 – 214. doi:10.1109/NABIC.2009.5393690.

Yang, X.S., Deb, S., 2010. Engineering optimization by cuckoo search. *Int. J. Math. Modelling Num. Optimisation.* 1 (4), 330–343.

Yang, X.S., Gandomi, A.H., 2012. Bat algorithm: a novel approach for global engineering optimization. Eng. Comput. 29 (5), 464–483.

Yang, X.S., Sadat, H.S.S., Gandomi, A.H., 2012. Firefly algorithm for solving non-convex economic dispatch problems with valve loading effect. Appl. Soft Comput. 12 (3), 1180–1186.

Zadeh, L.A., 1965. Fuzzy sets. Inf. Control. 8, 338–353.

2 A Review on Traditional and Modern Structural Optimization: Problems and Techniques

Mohammed Ghasem Sahab[1], *Vassili V. Toropov*[2] *and Amir Hossein Gandomi*[3]

[1]*Department of Civil and Environmental Engineering, Tafresh University, Tafresh, Iran,* [2]*School of Civil Engineering, University of Leeds, Leeds, UK,* [3]*Department of Civil Engineering, The University of Akron, Akron, OH, USA*

2.1 Optimization Problems

Optimization is a branch of mathematics concerned with obtaining the conditions that give the extreme value of a function, or many functions, under given circumstances. In other words, optimization entails engaging in an action to find the best solution. Optimization plays a key role in finding feasible solutions to real-life problems, from mathematical programming to operations research, economics, management science, business, medicine, life sciences, and artificial intelligence, to mention only few. An optimization problem can be mathematically stated as follows:

$$\text{Find } \boldsymbol{X} = (x_1, x_2, \ldots, x_n) \text{ which minimizes } f_i(\boldsymbol{X}) \quad i = 1, 2, \ldots, n_o \tag{2.1}$$

$$\begin{aligned} &\text{subject to:} \quad \text{(or maximizes)} \\ &g_{j(X)} \le 0, \quad j = 1,\ 2, \ldots, n_g \end{aligned} \tag{2.2}$$

$$h_k(\boldsymbol{X}) = 0, \quad k = 1, 2, \ldots, n_e \tag{2.3}$$

$$x_m^l \le x_m \le x_m^u \quad m = 1, 2, \ldots, n_s \tag{2.4}$$

where $\boldsymbol{X}$ is the vector of n design variables, $f_i(\boldsymbol{X})$ is an objective or merit function, $g_j(\boldsymbol{X})$ and $h_k(\boldsymbol{X})$ are the inequality and the equality constraints, respectively. These constraints represent limitations on the behavior or performance of the system. Therefore, they are called behavioral or functional constraints. Side constraints (2.4) restrict the acceptable range of potential solutions of the problem based on

Metaheuristic Applications in Structures and Infrastructures. DOI: http://dx.doi.org/10.1016/B978-0-12-398364-0.00002-4

nonbehavioral constraints. In this expression, x_m^l and x_m^u are the lower and upper limits on the design variable x_m, respectively. In the above expressions, n_o, n_g, n_e, and n_s are the number of objective functions, number of inequality, equality, and side constraints, respectively. Depending on the specific choice of design variables, objective functions, and constraints, various types of optimization problems may be created. Table 2.1 presents a classification of optimization problems that have been collected from various reviews (Arora, 1989; Foulds, 1981; Haftka and Gurdal, 1992; Kirsch, 1993; Rao, 2009; Sarma and Adeli, 1998, 2000; Vanderplaats, 1999).

2.2 Optimization Techniques

Optimization techniques can be classified in several ways. In a very general way, they can be divided into two broad categories, namely, function and parameter optimization techniques. In function optimization, an object under consideration is described by a number of unknown functions and through the optimization process, the optimum form of these functions will be found. The techniques for function optimization use differential calculus, the calculus of variations, etc., to find the optimum function. The *brachistochrone* problem is a traditional functional optimization problem posed and solved by Johann Bernoulli in 1696. The problem consists of finding the plane curve connecting two given points A and B, which allow a bead to slide down under gravity from A to B in minimum time. The term *brachistochrone* derives from the Greek *brachistos* (shortest) and *chronos* (time). The solution to this problem is a cycloid curve, $y(x)$, lying between A and B, which minimizes the following integral (functional) as:

$$t = \int_{\mathrm{A}}^{\mathrm{B}} \sqrt{\frac{1 + (y')^2}{2gy}}\, \mathrm{d}x$$

where y' is the derivation of $y(x)$ and g is the gravitational acceleration.

In parameter optimization, instead of searching for an optimum continuous function, the optimum values of design variables for a specific problem are obtained. Mathematical programming, optimality criteria (OC), and metaheuristic methods are some subsets of parameter optimization techniques. Figure 2.1 shows a classification of numerical optimization techniques. This diagram has been produced on the basis of these subsequent references (Arora, 1989; Cohn, 1994; Foulds, 1981; Haftka and Gurdal, 1992; Kirsch, 1993; Rao, 2009; Vanderplaats, 1999). In the following sections, the basic concepts of some of these methods as applied to structural optimization problems will be explained.

2.3 Optimization History

Mathematicians have been involved with optimization problems from the beginning of mathematics and evidence of the existence of optimization techniques may be

Table 2.1 Classification of Optimization Problems

Base of Classification	Category	Specifications
Number of design variables	Single variable	The vector of design variables includes only one variable.
	Multivariable	The vector of design variables includes more than one variable.
Number of objective functions	Single objective	There is one criterion expressed as an objective function.
	Multiobjective	There are many criteria that are considered together to determine an optimum solution.
Presence of constraints	Unconstrained	A minimum or maximum of objective function without any limitation is attempted.
	Constrained	Some constraints define the set of feasible solutions.
Features of constraints and objective functions	Linear programming	Objective functions and constraints are linear.
	Nonlinear programming	Some of the objective functions and constraints can be nonlinear. Quadratic programming and geometric programming problems are two specific kinds of nonlinear optimization problems.
Nature of design variables	Static	Design variables are independent. They are not functions of other parameters.
	Dynamic	Design variables are functions of other parameters, e.g., time.
Type of design variables	Discrete	Design variables can only take integer or discrete values.
	Continuous	Design variables take any real value.
	Mixed	Some design variables take integer values and others take real values.
Nature of design variables and design input data	Deterministic	All design variables or preassigned parameters such as loads acting on a structure are assumed to be deterministic.
	Probabilistic	All or some of design variables or preassigned parameters are described by random or probabilistic variables within a given range.
Nature of objective functions and design constraints	Crisp	The constraints and objective functions are expressed by nonfuzzy and unambiguous expressions or responses.
	Fuzzy	Some of the constraints and objective functions are described by fuzzy expressions or responses.

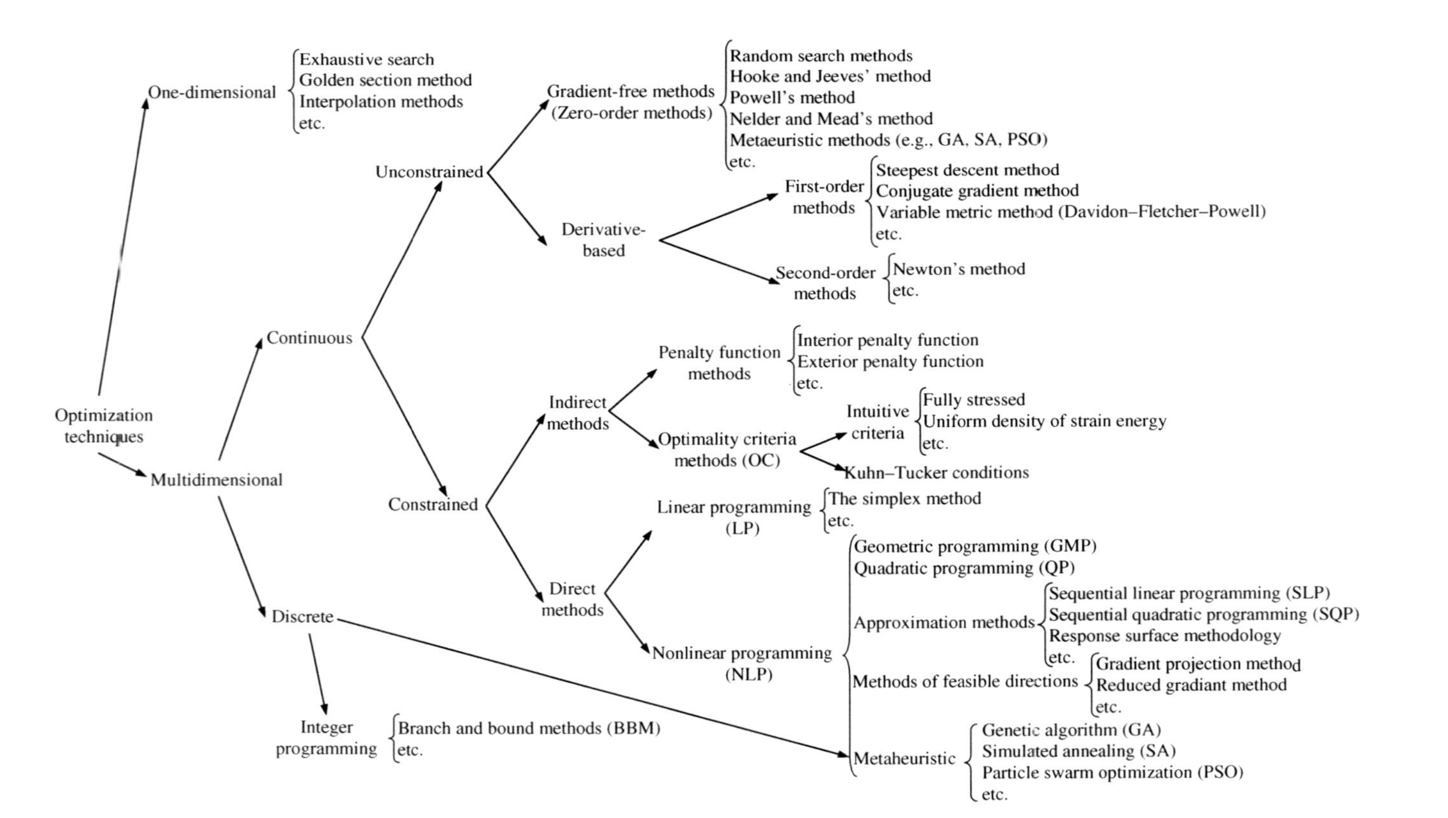

Figure 2.1 Classification of numerical optimization techniques.

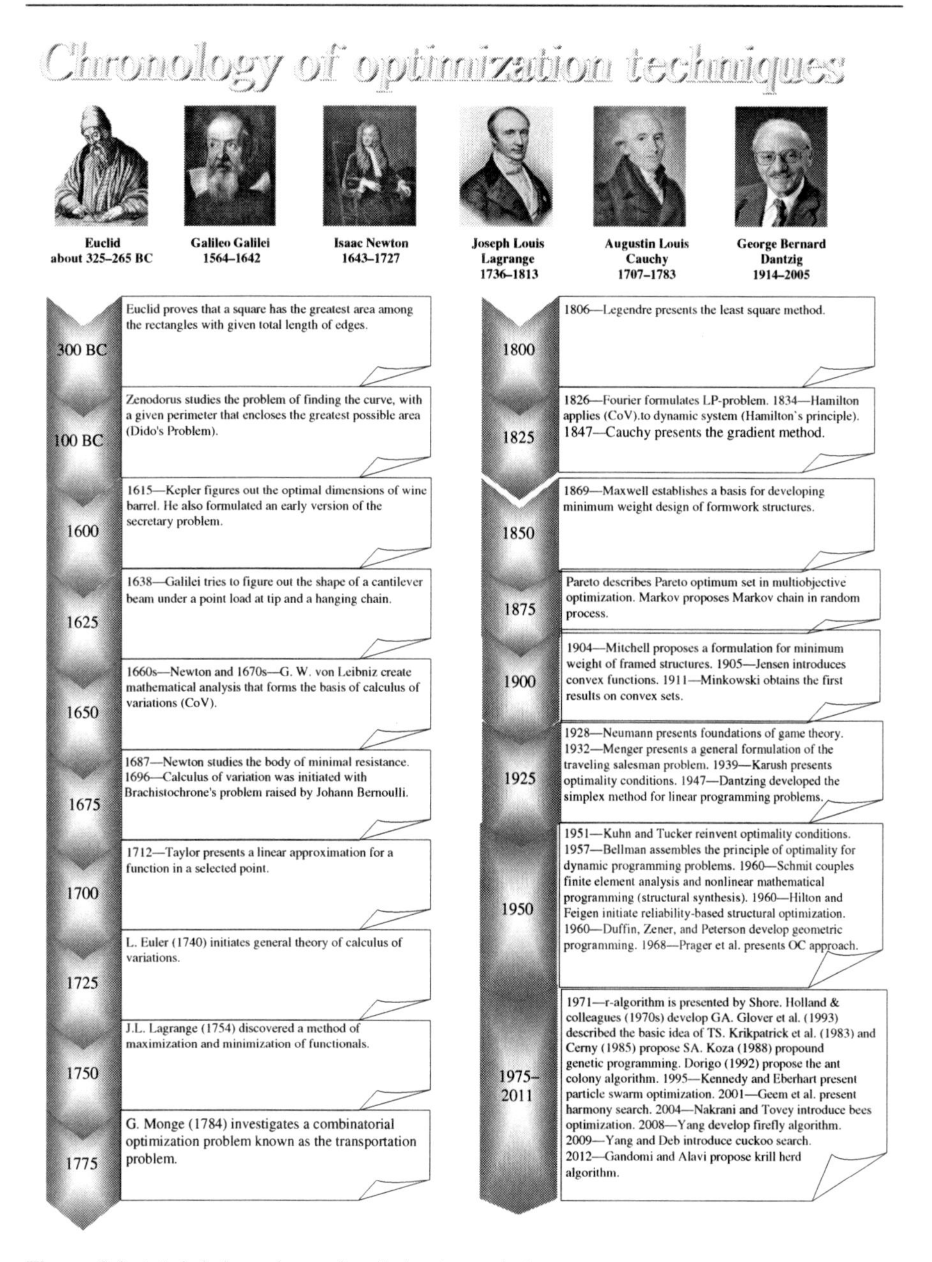

Figure 2.2 A brief chronology of optimization technique.

traced to the days of Galileo and Newton. Figure 2.2 shows a brief review of the development of optimization techniques (History of optimization, 2012; Index of history, 2012; Rao, 2009).

The purpose here is to briefly review the development of optimization techniques up to recent advances. Despite some contributions by several mathematicians, among them Lagrange, Legendre, Hamilton, and Cauchy, very little progress was made until the middle of the twentieth century, when high-speed digital computers made it possible to implement optimization techniques for the design optimization of large-scale structures. Modern optimization techniques like genetic algorithms (GAs), simulated annealing (SA), particle swarm optimization (PSO), ant colony optimization (ACO), harmony search (HS) algorithm, firefly algorithm (FA), cuckoo search (CS), bat algorithm (BA), Big Bang–Big Crunch (BB–BC), charged system search (CSS), krill herd (KH) algorithm, and others have emerged as powerful and popular methods for solving complex engineering optimization problems in recent years.

This history also shows the noticeable role of practicing engineers in the development of optimization techniques. However, optimization is a branch of mathematics; it has always been attractive for engineers to employ optimization techniques in their practical designs. This can be one reason why engineers have been so involved in the development of optimization techniques.

We follow this review by taking a glance through the development of structural optimization. The distinction is that general optimization provides the actual optimization algorithm while structural optimization offers advanced methods for making the best use of these algorithms.

2.4 Structural Optimization

2.4.1 General Concept

Optimization has applications in various branches of science and engineering, including structural engineering. In structural optimization, design objectives are structural criteria used to evaluate the merit of a design such as minimum construction cost, minimum life-cycle cost, minimum weight, and maximum stiffness. Building code provisions, which provide safety and serviceability requirements to the structure, usually appear as the design constraints. Moreover, some nonbehavioral constraints, like the type and size of available structural elements, may limit the acceptable designs. Design variables may describe shape, topology, and geometry of the structure, or they may define the size or properties of structural elements.

2.4.2 Major Advances in Structural Optimization

It is acknowledged that Galileo initiated the simple bending theory of beams in 1638 (Barnett, 1966). He also considered the uniform strength criterion to find the optimum form of a cantilever beam with constant width and under tip loading. He proved that to have equal resistance at all sections of the beam, it should be parabolic in shape. This can be considered as a basic model for a trend in design optimization that is called fully stressed design (FSD). Since in statically determinate structures internal forces are independent of member dimensions, it could be easily proved that the uniform strength conditions lead to minimum weight. But in statically indeterminate structures, internal forces are influenced by member dimensions; therefore, in these cases such a condition does not necessarily result in minimum weight. Also, it is not evident that there is a unique solution for member dimensions, which provides uniform strength condition. Many other researchers, including Newton, Bernoulli, Young, Saint Venant, Morin, Cullman, Levy, and others, developed upon Galileo's idea for different cases of loading, states of forces, and types of structures (Wasiutynski and Brandt, 1963).

Maxwell (1869) formulated a theory that established a basis for developing a minimum weight design of framework structures. This theory states that, of all fully stressed frameworks, which equilibrate a given system of applied forces, the lightest has the least volume of compression members or, alternatively, the least volume of tension members. He proved that for a given system of applied forces, if a fully stressed structure with only compression or tension members can be found, this structure has the smallest possible weight. He also formulated a way of finding the total required volume of material to resist the tensile and compressive forces in a framework.

On the basis of Maxwell's theorem, Michell (1904) proposed a formulation for minimum weight of framed structures. His work provided a basis for many researchers in the field of topology optimization of structures. The main result of his theory is that members of the optimum structure must lie along lines of principal strain. Thus, tension and compression members cross each other orthogonally and they form an orthogonal mesh in optimum framework. Figure 2.3 shows two of Michell's examples of a minimum weight structure. Eschenauer and Olhoff (2001) provided a review on topology optimization of continuum structures. Sturler et al. (2008) presents a robust method for topology optimization with adaptive mesh refinement (AMR) and derefinement. Shimoda et al. (2010) apply the shape optimization method for the out-of-plane shape variation to a stiffness design problem and a vibration eigenvalue design problem of shell structures.

Schmit (1960) coupled finite element structural analysis and nonlinear mathematical programming that formed a new approach to structural optimization. By example of a simple three-bar truss, he explained the concept of structural synthesis. In this

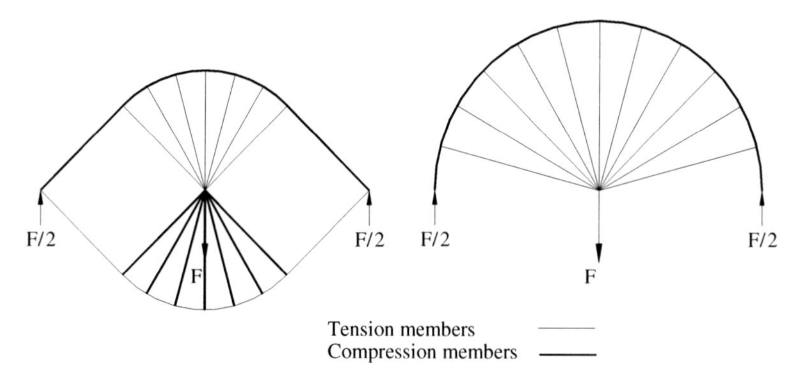

Figure 2.3 Two alternatives of Michell optimum structure for a centrally loaded beam.

example, he showed, contrary to the traditional belief, that a fully stressed structure is not the optimal structure. Later, the approximation concepts were employed in structural synthesis to decrease the number of accurate finite element analysis of structures during the optimization process (Schmit and Farshi, 1974; Schmit and Miura, 1976). In this modified approach, after the initial analysis, the gradients of some critical constraints are evaluated (the sensitivity analysis), and an approximation of the main problem is established. This problem is solved by general-purpose optimization code and a solution is obtained. Using the new set of data, an accurate finite element analysis is carried out and again an approximate optimization problem is solved. This process is repeated until convergence is achieved (Vanderplaats, 1993).

Griffith and Stewart (1961) and Kellog (1960) introduced the idea of treating a nonlinear optimization problem as a sequence of linear programming (sequential linear programming, SLP). Moses (1964) applied SLP to structural optimization. He presented a simple planar truss and a planar frame example to illustrate the method employed. In this method, the objective function and constraints can be linearized using a Taylor series expansion about an initial guess, X_0. This approximate linear optimization problem can be solved by a linear programming technique. Then, X_0 is replaced by the obtained solution and linearization is repeated about this new obtained point. This process is continued until the difference between two sequential solutions is less than a predefined value. It has been found that the initial guess plays a crucial role not only in terms of convergence to an unfeasible solution or local optimum but also in terms of increasing the computing time (Haftka and Gurdal, 1992). SLP has been used in the field of reinforced concrete (RC) structures, mainly applied to simplified problem formulations (Chung and Sun, 1994; Kanagasundaram and Karihaloo, 1990). Zienkiewics and Campbell (1973) and Campbell and Kelliher (2000) applied SLP to structural shape optimization.

Using a penalty function formulation, the constrained minimization problem may be transformed into a sequence of unconstrained minimization problems (Fiacco and McCormick, 1968). Since in these methods a sequence of unconstrained optimization problems with different values of imposed penalty are solved to find the solution of a constrained optimization problem, these methods are called sequential unconstrained minimization techniques (SUMT). One of the first applications of this technique to structural optimization was presented by Schmit and Fox (1965). Krishnamoorthy and Mosi (1981) used SUMT for optimization of two-dimensional (2D) frames with rectangular cross-sections. Dinno and Mekha (1993) applied SUMT to cost optimization of one- and two-story RC frames.

2.4.3 OC Methods

OC and mathematical programming (MP) methods are two broad categories of numerical optimization techniques. Using MP methods, the objective function is minimized directly; but with OC methods, a criterion or some criteria are defined so that when this criterion or these criteria are satisfied, the optimum is found. In 1968, the OC approach was presented in analytical form by Prager and his colleagues (Prager and Shield, 1968; Prager and Taylor, 1968) and in numerical form by

Venkayya and his colleagues (Venkayya et al., 1968). The essence of these methods is to first establish the criteria that define an optimum and then create a recursive algorithm that meets the criteria. These methods can be very efficient when the number of constraints is small compared to the number of design variables (Haftka and Gurdal, 1992). OC methods are problem dependent because they use the known or assumed features of the optimum or special behavior of the structure. The OC can be rigorous mathematical statements or an intuitive one such as the stipulation that the strain energy density in the optimum structure is uniform. FSD, uniform density of the strain energy, maximum stiffness, and simultaneous failure mode are some intuitive criteria that have been considered as the OC for minimum weight structures in the past. Wasiutynski and Brandt (1963) and Barnett (1966) presented a comprehensive review of progress in these fields.

During the 1970s and 1980s, these methods were developed for the purpose of large-scale optimization (Kirsch, 1993). Moharrami and Grierson (1993) applied an OC method to optimization of RC frameworks. Fadee and Grierson (1996, 1998) used an OC method for three-dimensional (3D) RC structures including shear walls. Adamu and Karihaloo (1995a,b) presented the application of a discretized continuum-type optimality criteria (DCOC) to design of RC plane frames.

2.4.4 Reliability-Based Optimization Approach

Reliability-based structural optimization was initiated by Hilton and Feigen (1960). Since the strength properties of the structural elements and the magnitude of loads are in reality probabilistic, they can be treated as random variables with a specified statistical distribution. In reliability-based structural optimization, the optimum values of the design variables are obtained so that the objective function is minimized considering the overall probability of failure (P_f), while loads acting on structure and structural strengths are random variables. Overall probability of failure is calculated considering different failure modes under different load conditions. There are two approaches to incorporate the probability of failure in design optimization of a structure. In the first approach, the objective function (total cost, C) is considered as the sum of initial cost (C_i) and the expected failure cost (C_{fa}) as follows:

$$C = C_i + P_f C_{fa} \tag{2.5}$$

The expected failure cost may include reconstruction cost, damage to properties, casualties, etc. In the second approach, the objective function is the initial cost of the structure such as the objective function in the deterministic optimization approach but, a new constraint is added to the design constraints such as:

$$P_f \le P_{fa} \tag{2.6}$$

where P_{fa} is the allowable probability of failure.

Moses and Stevenson (1970) investigated the reliability-based minimum weight design of planar frames based on plastic collapse analysis. A comprehensive review on the progresses and concepts of reliability-based optimization has been carried out by Moses (1971). Lin and Frangopol (1996) studied minimum cost design of RC girders based on reliability theory. In their chapter, the initial cost included only the costs of concrete and reinforcement.

In some recent publications (Koskisto and Ellingwood, 1997; Wen and Kang, 2000a,b), the life-cycle cost of a structure has been minimized using the reliability theory. The total life-cycle cost of a structure in those papers can include that of construction, maintenance and operation, repair, damage, failure consequence (loss of revenue, deaths, and injuries, etc.), and discounting of cost over time.

The computation of the probability of failure, as a function of random variables of loading and member resistances, is a substantial subject in reliability-based optimization. Sarma and Adeli (1998, 2000) cited that the probability of failure and expected failure costs cannot be computed consistently due to insufficient statistical data; they have to be chosen somewhat arbitrarily. Moreover, the existence of a large number of possible failure modes, especially for large structures, makes the evaluation of system reliability practically an impossible task. They concluded that all the papers published on reliability-based cost optimization of structures take an academic, theoretical, and idealistic approach to the problem.

2.4.5 Fuzzy Optimization

In a traditional design optimization approach, it is assumed that all the design input data are known precisely; but in reality, some of these data are imprecise. In cost optimization of a structure, three major sources of fuzziness and imprecision can be identified. The first one is in the evaluation of the structural behavior and resistance, the second one is in determining the loads acting on the structure, and the third one is in the formulation and evaluation of the cost function. This means that the relationships and statements, which are used for the description of the problem, cannot be presented in a crisp and precise form. In the fuzzy optimization approach, an optimum solution is found considering fuzzy expressions or responses for some of the constraints and objective functions. Fuzzy optimization is based on the theory of fuzzy sets initiated by Zadeh (1965). In this approach, the boundary between feasible and infeasible domains or the objective functions are represented in a fuzzy form. On the other hand, the criteria for acceptance, rejection, and evaluation of a solution are not sharp and strict. In the fuzzy optimization approach, a solution may be acceptable with a degree of its compatibility with fuzzy constraints and objective functions. The degree of compatibility of a solution, X, with a constraint, $g_j(\boldsymbol{X})$, or objective function, $f_i(\boldsymbol{X})$, is evaluated by a membership function, $\mu_{gi}(\boldsymbol{X})$ and $\mu_{fi}(\boldsymbol{X})$, with a value in the interval [0, 1], respectively. For each constraint or objective function, a specific linear or nonlinear membership function can be defined based on a presumed general form for membership function and the range of predicted tolerances for the constraints or the objective function. A fuzzy design is characterized by its membership function, which can be evaluated as the intersection of the fuzzy goal and fuzzy constraints. In a very common approach, the optimum design is chosen in such a way that the membership function for the optimum design becomes maximum.

Brown and Yao (1983) investigated the application of fuzzy set theory in structural engineering. They outlined the elements of fuzzy set theory and contrasted them with crisp sets. Wang and Wang (1985) used a simplified fuzzy optimization by considering the fuzzy constraints and nonfuzzy cost function. Yeh and Hsu (1990) introduced a procedure for cost optimization of structures with fuzzy allowable strength and fuzzy loads. They considered exponential functions for the fuzzy allowable strength and fuzzy loads. A simple three-bar truss and a one-story one-bay frame have been optimized using this procedure.

2.5 Metaheuristic Optimization Techniques

In the past three decades, from about 1980 to 2012 metaheuristic methods have been rapidly developed to solve optimization problems. These methods are principally intuitive and do not have theoretical support. Metaheuristic methods such as GAs, SA, and TS provide general ways to search for a good but not necessarily the best solution. Since GAs and SA use statistical rules in their search process, they may be classified as stochastic search methods.

Manoharan and Shanmuganathan (1999) applied the four different search mechanisms of TS, SA, GA, and branch and bound technique to the design optimization of three different truss problems and compared them. They concluded that all three metaheuristic search methods, TS, SA, and GA, work well and produce an acceptable solution within a reasonable amount of time. Among these three methods, GA has the added advantage that it does not need an initial guess to search for the optimum. Branch and bound technique consumes an enormous amount of computing time to find an optimum solution. Also, it is not viable for structures of any reasonable size (e.g., a truss structure with more than six bars).

2.5.1 Genetic Algorithm

GAs were introduced by Holland in the 1960s. With the aid of his colleagues and students, he further developed these algorithms during the 1970s at the University of Michigan. He summarized the results of his research in the book, *Adaptation in Natural and Artificial Systems* (Holland, 1975). GAs are numerical optimization techniques inspired by the natural laws of evolution. A GA starts searching the design space with a population of individuals (designs), which are initially created over the design space at random. In the basic GA, every individual of the population (design) is described by a binary string (encoded form). GA uses four main operators, namely, selection, creation of the mating pool, crossover, and mutation, to direct the population toward the optimum design. In the selection process, some designs of a population are selected by randomized methods for GA operations; for example, to create the mating pool, some good designs in the population are selected and copied to form a mating pool. The better (fitter) designs have a greater chance of being selected. Crossover allows the characteristics of the designs to be altered, depending on the crossover probability for creation of a better generation of designs. In this process, different digits of binary strings of each parent are transferred to their children (new designs produced by the crossover operation). Mutation is an occasional random change of the value of some randomly selected design variables. The mutation operation changes each bit of string from 0 to 1 or vice versa in a design's binary code depending on the mutation probability. Mutation can be one factor used to prevent premature convergence.

A GA uses a discrete set of design variables in the optimization process. However, by defining the number of decimal digits for representation of continuous variables or step size between the sequential values of design variables, this method can be applied to continuous problems as well. Lin and Hajela (1992) implemented GAs in the optimal design of structural systems with a mix of continuous, integer, and discrete design variables.

Goldberg and Samtani (1986) used GAs to minimize the weight of a 10-bar truss. Other researchers such as Adeli and Cheng (1993, 1994), Adeli and Kumar (1995), Camp et al. (1998), Coello and Christiansen (2000), Erbatur et al. (2000), Ghasemi et al. (1999), Hajela (1990), Jenkins (1991a,b, 1992, 1997), Kaveh and Rahami (2006), Koumousis and Arsenis (1994), Krishnapillai and Jones (2009), Lin and Hajela (1994), Rahami et al. (2008), Rajeev and Krishnamoorthy (1992, 1997), Sahab et al. (2005a,b), Toğan and Daloğlu (2006), and Wang and Tai (2005) reported on various applications of GAs to structural optimization. Also, several papers have been published on modification of the basic GA and different methods of encoding, selection, crossover, and constraint handling. For example, Janikow and Michalewicz (1991), Jingui et al. (1996), Lemonge and Barbosa (2004), Lim et al. (2000), Nanakorn and Meesomklin (2001), and Yang and Soh (1997) are a few to be mentioned. Sahab et al. (2005a,b) propose a hybrid GA and apply it to column layout and size optimization of RC flat slab buildings.

2.5.2 Simulated Annealing

The SA algorithm was proposed by Kirkpatrick et al. (1983) and Cerny (1985) independently. SA is based on the analogy between the way in which the crystalline structure of a metal achieves near global minimum energy states during the process of annealing and the way in which a function may reach minimum during a statistical search of the design space. The objective function corresponds to the energy state and moving to any new set of design variables corresponds to a change of the energy state. Although the method has been basically developed for discrete problems, it can be used in continuous problems in the same way as GAs are used.

Elperin (1988) used SA for design optimization of a 10-bar truss. Design optimization of steel frames using this algorithm was studied by Balling (1991). In another paper, May and Balling (1992) applied a SA algorithm for 3D steel frameworks. Bennage and Dhingra (1995a,b) proposed SA for single and multiobjective structural optimization in discrete–continuous variables. Shim and Manoochehri (1997) used SA to find optimal configuration of structural members. Leite and Topping (1999) proposed a parallel SA for structural optimization. Ceranic (2000) applied a constrained SA algorithm to the minimum cost design of RC cantilever retaining walls. During the decade of the 2000s, other researchers, such as Bureerat and Limtragool (2008), Hasançebi and Erbatur (2002), Lamberti (2008), and Park and Sung (2002), have applied SA to different kinds of structural engineering problems.

2.5.3 Tabu Search

TS is an iterative procedure appropriate for discrete optimization problems. However, this method, like SA and GA, can be applied to continuous problems. The basic idea of the method has been described by Glover et al. (1993). The method consists of a sequence of exploration movements over the design search space, which is started from a feasible design toward the optimum. To move from one solution to another, the algorithm explores around the point and chooses the best available solution. To avoid retracing the steps used before, the method records recent moves in one particular iteration as a forbidden or tabu list. Tabu moves are based on the short-term and long-term history of the sequence of moves.

The applications of TS in structural optimization are quite new. TS has been applied to the design of a 10-bar cantilever truss problem by Bland (1994). Dhingra and Bennage (1995) investigated the applicability of TS by solving structural topology optimization problems. Kargahi et al. (2006) applied TS for optimization of frame structures.

2.5.4 Ant Colony Optimization

ACO was developed by Dorigo and his associates in the early 1990s (Colorni et al., 1991, Dorigo, 1992, Dorigo et al., 1996). The main idea of ACO is to imitate the cooperative behavior of an ant colony, which finds the shortest path to a food source. In this method, a combinatorial optimization problem with n design variables ($x_1 - x_n$) is modeled as a multilayered graph as shown in Figure 2.1. The number of layers is equal to the number of design variables and the number of nodes in each layer is equal to the number of discrete values permitted for the corresponding design variable. Thus, each node in a specific position of the graph is associated with a permissible discrete value of a design variable. Artificial ants walk through this graph, looking for good paths. An ant colony consists of N ants. The ants start at the nest node, travel through the various layers from the first layer to the last layer, and end at the destination node in each cycle or iteration. Each ant can select only one node in each layer in accordance with the state transition rule given by metaheuristic information. The nodes selected along the path visited by an ant represent a candidate solution. A typical path visited by an ant is shown by thick lines in Figure 2.4. Once the path is complete, the ant deposits some pheromone on the path based on a local updating rule. In the beginning of the optimization process (i.e., in iteration 1), all the edges or rays are initialized with an equal amount of pheromone. As such, in iteration 1, all the ants start from the home node and end at the food node by randomly selecting a node in each layer. Small quantities of pheromone are deposited during the construction phase, while larger amounts are deposited at the end of each iteration in proportion to solution quality. The optimization process is terminated if any of the defined termination conditions are satisfied. The values of the design variables denoted by the nodes on the path with the largest amount of pheromone are considered as the components of the optimum solution vector. In general, at the optimum solution, all ants travel along the same best (converged) path.

One of the first applications of ACO in structural engineering was presented by Bland (2001) for design optimization of a 25-bar space truss. Camp and Bichon (2004) also employed ACO for design optimization of space trusses. Later, they extended this work to optimize rigid steel frames (Camp et al., 2004). Kaveh and Shojaee (2007) employed ACO for optimal design of skeletal structures. A hybridization of ACO and other metaheuristic techniques, like PSO, have been applied for design optimization of steel frames and truss structures (Kaveh and Talatahari, 2007; Kaveh and Talatahari, 2009). ACO has been employed to determine the optimum design of 3D irregular steel frames, taking into account warping deformations of thin-walled sections (Aydoğdu and Saka, 2009). ACO has been used for structural topology optimization (Luh and Lin, 2009). Zhang and Li (2011) presented a two-level optimization algorithm based on ACO to design the shape of a transmission tower. Majumdar et al. (2012) utilized ACO to identify damages in truss structures.

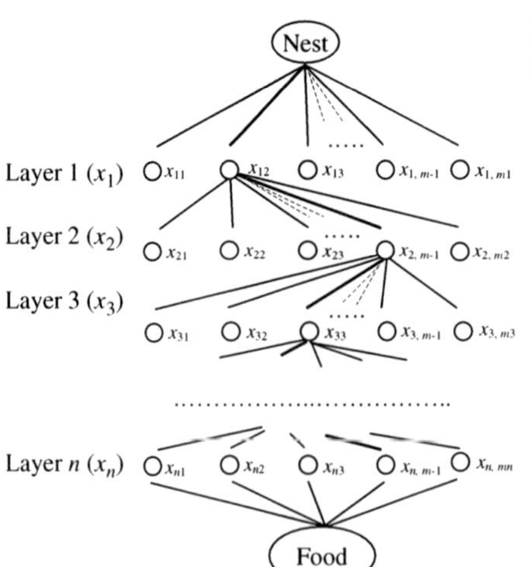

Figure 2.4 The ACO process in the form of a multilayered graph.

2.5.5 Particle Swarm Optimization

PSO is one of the most well-known metaheuristics; it was proposed by Kennedy and Eberhart (1995a,b). This algorithm is inspired from swarm behavior such as bird flocking and schooling in nature. PSO has been widely used and it is the inspiration for a new research area called swarm intelligence (Yang, 2008). The standard PSO is very simple and easy to implement. In standard PSO, the new location of each particle is determined by a velocity term, which reflects the attraction of global best (g_b) and its own best (o_b) during the history of the particle and random coefficients (Figure 2.5).

PSO has been improved a lot, and its parameters have been tuned during the past years. Their range of random coefficients is also determined using an analytical convergence study (Carlisle and Dozier, 2001). PSO has been used for many real-world engineering cases, especially structural engineering problems. Fourie and Groenwold (2002) proposed PSO for structural size and shape optimization. PSO has been successfully used for discrete truss optimization (Li et al., 2009; Luh and Lin, 2011; Talatahari et al., 2012a). Dimou and Koumousis (2009) have proposed PSO for reliability-based optimal design of truss structures. A hybrid PSO is also utilized by Kaveh and Talatahari (2007) for frame optimization.

2.5.6 Harmony Search

HS is a metaheuristic algorithm proposed by Geem et al. (2001). HS has a different source of inspiration from other well-known metaheuristics; it is inspired by the behavior of musicians to find a better harmony. Lee and Geem (2004) introduced it as a new structural optimization technique and verified it using several benchmark structural engineering problems. Many applications of HS in structural optimization can be found in a book recently edited by Geem (2009). Hasançebi et al. (2010) have proposed an adaptive HS for structural optimization. A new review of the HS applications in structural optimization has also been provided by Saka et al. (2011).

2.5.7 Big Bang–Big Crunch

BB–BC optimization is a two-phase metaheuristic optimization algorithm proposed by Erol and Eksin (2006). BB–BC is inspired from the famous physical cosmology theory. The first phase is Big Bang when candidate solutions are randomly distributed over the search space. In the Big Crunch phase, a contraction operation

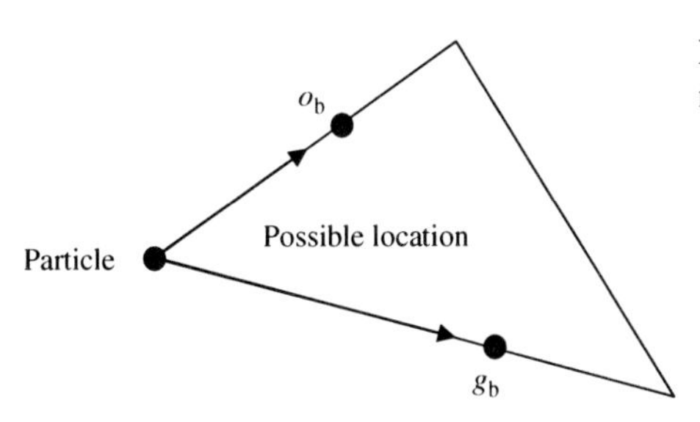

Figure 2.5 Schematic representation of the motion of a particle in PSO.

estimates a center of mass for the candidates. Then, a sequential Big Bang is randomly distributed about the mass center. Camp (2007) first proposed BB–BC for space truss optimization with discrete and continuous variables. Kaveh and Talatahari (2009, 2010) used a hybrid BB–BC for truss optimization with linear and nonlinear responses. Tang et al. (2010) estimate parameters in structural systems using BB–BC. Camp and Akin (2012) have recently proposed BB–BC for optimization of retaining walls.

2.5.8 Firefly Algorithm

FA is a population-based optimization algorithm and mimics a firefly's attraction to flashing light. This algorithm has been proposed by Yang (2008) at the University of Cambridge. Gandomi et al. (2012a) have recently improved FA by using chaotic maps. This algorithm is naturally a multimodal algorithm (Yang 2009). Therefore, it can be suitable for structural engineering problems, especially when we need to prepare some engineering alternatives in multimodal problems. This algorithm was first verified by Gandomi et al. (2011) for different structural optimization problems with continuous–discrete variables. Talatahari et al. (2012b) also used FA method for design optimization of tower structures.

2.5.9 Cuckoo Search

CS is inspired from brood parasitism of cuckoos. CS is proposed by Yang and Deb (2009) as a robust optimization algorithm. Gandomi et al. (2011b) have introduced CS as a new metaheuristic for structural optimization problems by solving several types of structural engineering problems. CS has been successfully applied for frame optimization by Kaveh and Bakhshpoori (2012). Gandomi et al. (2012b) have also used CS to solve different truss structures and found that it can outperform the most well-known algorithm.

2.5.10 Other Metaheuristics

There are several other algorithms in the literature and these can also be useful in structural optimization. Kaveh and Talatahari (2010a) developed the CSS based on electrostatic and Newtonian mechanics laws and have successfully applied it to structural optimization problems such as skeletal structures (Kaveh and Talatahari, 2010b). Sonmez (2011a,b) proposed bee colony optimization for continuous and discrete truss optimization problems. Bat algorithm (BA) has been successfully used to solve benchmark structural optimization problems (Gandomi et al. (2012c); Yang and Gandomi (2012)). Krill herd (KH) algorithm is also a new metaheuristic that may be able to outperform the other well-known metaheuristics (Gandomi and Alavi (2012)) and can be successful for structural optimization.

References

Adamu, A., Karihaloo, B.L., 1995a. Minimum cost design of RC frames using the DCOC method, Part I: columns under uniaxial bending actions. Struct. Optim. 10 (1), 16–32.

Adamu, A., Karihaloo, B.L., 1995b. Minimum cost design of RC frames using the DCOC method, Part II: columns under biaxial bending actions. Struct. Optim. 10 (1), 33–39.

Adeli, H., Cheng, N.T., 1993. Integrated genetic algorithms for optimization of space structures. J. Aerosp. Eng. 6 (4), 315–328.

Adeli, H., Cheng, N.T., 1994. Augmented Lagrangian genetic algorithm for structural optimization. J. Aerosp. Eng. 7 (1), 104–118.

Adeli, H., Kumar, S., 1995. Distributed genetic algorithm for structural optimization. J. Aerosp. Eng. 8 (3), 156–163.

Arora, J.S., 1989. Introduction to Optimum Design. McGraw-Hill, Singapore.

Aydoğdu, I., Saka, M.P., 2009. Ant colony optimization of irregular steel frames including effect of warping. In: Topping, B.H.V., Costa Neves, L.F., Barros, R.C. (Eds.), Civil-Comp 09, Proceedings of the Twelfth International Conference on Civil. Struct Environ Eng Comput, 1-4 September, Madeira, Portugal, Paper ID: 69.

Balling, R.J., 1991. Optimal steel frames design by simulated annealing. J. Struct. Eng. 123 (2), 193–202.

Barnett, R.L., 1966. Survey of optimum structural design. Exp. Mech. 6 (12), 19A–26A.

Bennage, W.A., Dhingra, A.K., 1995a. Single and multiobjective structural optimization in discrete–continuous variables using simulated annealing. Int. J. Numer. Methods Eng. 38 (16), 2753–2773.

Bennage, W.A., Dhingra, A.K., 1995b. Optimization of truss topology using tabu search. Int. J. Numer. Methods Eng. 38 (23), 4035–4052.

Bland, J.A., 1994. A tabu search approach to engineering optimisation. In: Proceedings of Ninth International Conference on Applications of Artificial Intelligence in Engineering. Computational Mechanics Publishers, University Park, PA, pp. 423–430.

Bland, J.A., 2001. Optimal structural design by ant colony optimization. Eng. Optim. 33, 425–443.

Brown, C.B., Yao, J.T.P., 1983. Fuzzy sets and structural engineering. J. Struct. Eng. 109 (5), 1211–1225.

Bureerat, S., Limtragool, J., 2008. Structural topology optimisation using simulated annealing with multiresolution design variables. Finite Elem. Anal. Des. 44 (12–13), 738–747.

Camp, C.V., 2007. Design of space trusses using Big Bang–Big Crunch optimization. J. Struct. Eng. 133, 999–1008.

Camp, C.V., Akin, A., 2012. Design of retaining walls using Big Bang–Big Crunch optimization. J. Struct. Eng. 137 (3), 438–448.

Camp, C.V., Bichon, B.J., Stovall, S.P., 2004. Design of steel frames using ant colony optimization. J. Struct. Eng. ASCE. 131 (3), 369–379.

Camp, C.V., Bichon, B.J., 2004. Design of space trusses using ant colony optimization. J. Struct. Eng. ASCE. 130 (5), 741–751.

Camp, C.V., Pezeshk, S., Cao, G., 1998. Optimized design of two-dimensional structures using a genetic algorithm. J. Struct. Eng. 124 (5), 551–559.

Campbell, J.S., Kelliher, D., 2000. Structural shape optimization of elastic continua: a study on nonlinearity and a 2D benchmark. In: Sienz, J. (Ed.), The Second ASMO UK/ISSMO Engineering Design Optimization Conference. University of Wales, Swansea, pp. 65–72.

Carlisle, A., Dozier, G., 2001. An off-the-shelf PSO. In: Proceedings of the Particle Swarm Optimization Workshop. pp. 1–6.
Ceranic, B., 2000. Optimum Design of Reinforced Concrete Skeletal Systems Using Non-Linear Programming Techniques. School of Engineering, University of Derby, UK, PhD Thesis.
Cerny, V., 1985. Thermodynamical approach to the travelling salesman problem: an efficient simulation algorithm. J. Optim. Theor. Appl. 45 (1), 41–51.
Chung, T.T., Sun, T.C., 1994. Weight optimization for flexural reinforced concrete beams with static nonlinear response. Struct. Optim. 8 (2–3), 174–180.
Coello, C.A., Christiansen, A.D., 2000. Multiobjective optimization of trusses using genetic algorithms. Comput. Struct. 75 (6), 647–660.
Cohn, M.Z., 1994. Theory and practice of structural optimization. Struct. Optim. 7 (1–2), 20–31.
Colorni, A., Dorigo, M., Maniezzo, V., 1991. Distributed optimization by ant colonies. In: Varela, F., Bourgine, P. (Eds.), Proceedings ECAL91 - European Conference Artificial Life. Elsevier, Amsterdam, pp. 134–142.
Dhingra, A.K., Bennage, W.A., 1995. Discrete and continuous variable structural optimization using tabu search. Eng. Optim. 24 (3), 177–196.
Dimou, C.K., Koumousis, V.K., 2009. Reliability-based optimal design of truss structures using particle swarm optimization. J. Comput. Civil Eng. 23 (2), 100–109.
Dinno, K.S., Mekha, B.B., 1993. Optimal design of reinforced concrete frames based on inelastic analysis. Comput. Struct. 47 (2), 245–252.
Dorigo, M., 1992. Optimization, learning, and natural algorithms. PhD Thesis, Dip. Electronica e Informazione Politecn. Milano. (In Italian).
Elperin, T., 1988. Monte Carlo structural optimization in discrete variables with anncaling algorithm. Int. J. Numer. Methods Eng. 26, 815–821.
Erbatur, F., Hasancebi, O., Tutuncu, I., Kilic, H., 2000. Optimal design of planar and space structures with genetic algorithms. Comput. Struct. 75 (2), 209–224.
Erol, O.K., Eksin, I., 2006. New optimization method: Big Bang–Big Crunch. Adv. Eng. Software. 37, 106–111.
Eschenauer, H.A., Olhoff, N., 2001. Topology optimization of continuum structures: a review. Appl. Mech. Rev. 54 (4), 331–389.
Fadaee, M.J., Grierson, D.E., 1996. Design optimization of 3D reinforced concrete structures. Struct. Optim. 12 (2–3), 127–134.
Fadaee, M.J., Grierson, D.E., 1998. Design optimization of 3D reinforced concrete structures having shear walls. Eng. Comput. 14 (2), 139–145.
Fiacco, A.V., McCormick, G.P., 1968. Nonlinear Programming: Sequential Unconstrained Minimization Techniques. John Wiley & Sons, New York, NY.
Foulds, L.R., 1981. Optimization Techniques. Springer-Verlag, New York, NY.
Fourie, P.C., Groenwold, A.A., 2002. The particle swarm optimization algorithm in size and shape optimization. Struct. Multidiscip. Optim. 23 (4), 259–267.
Gandomi, A.H., Alavi, A.H., 2012. Krill herd: a new bio-inspired optimization algorithm. Commun. Nonlinear Sci. Numer. Simulat. 17 (12), 4831–4845.
Gandomi, A.H., Talatahari, S., Yang, X.S., Deb, S., 2012b. Design optimization of truss structures using cuckoo search algorithm. Struct. Design Tall Spec. Build., in press. doi: 10.1002/tal.1033.
Gandomi, A.H., Yang, X.S., Alavi, A.H., 2011a. Mixed variable structural optimization using firefly algorithm. Comput. Struct. 89 (23–24), 2325–2336.

Gandomi, A.H., Yang, X.S., Alavi, A.H., 2011b. Cuckoo search algorithm: a metaheuristic approach to solve structural optimization problems. Eng. Comput., in press. doi: 10.1007/s00366011-0241-y.

Gandomi, A.H., Yang, X.S., Talatahari, S., Alavi, A.H., 2012a. Firefly algorithm with chaos. Commun. Nonlinear Sci. Numer. Simulat., in press. doi: 10.1016/j.cnsns.2012.06.009.

Gandomi, A.H., Yang, X.S., Talatahari, S., Alavi, A.H., 2012c. Bat algorithm for constrained optimization tasks. Neural Comput. Appl., in press. doi: 10.1007/s00521012-1028-9.

Geem, Z.W., 2009. Harmony Search Algorithms for Structural Design Optimization. Springer, Berlin.

Geem, Z.W., Kim, J.H., Loganathan, G.V., 2001. A new heuristic optimization algorithm: harmony search. Simulation. 76, 60–68.

Ghasemi, M.R., Hinton, E., Wood, R.D., 1999. Optimization of trusses using genetic algorithms for discrete and continuous variables. Eng. Comput. 16 (3), 272–301.

Glover, F., Taillard, E., de Werra, D., 1993. A user's guide to tabu search. Ann. Oper. Res. 41, 3–28.

Goldberg, D.E., Samtani, M.P., 1986. Engineering optimization via genetic algorithm. Proceedings of Ninth Conference on Electronic Computation. ASCE, New York, NY, pp. 471–482.

Griffith, R.E., Stewart, R.A., 1961. A nonlinear programming technique for the optimization of continuous processing systems. Manage. Sci. 7, 379–392.

Haftka, R.T., Gurdal, Z., 1992. Elements of Structural Optimization. third ed. Kluwer Academic Publishers, London.

Hajela, P., 1990. Genetic search—an approach to nonconvex optimization problem. AIAA J. 28 (7), 1205–1210.

Hasançebi, O., Erbatur, F., 2002. Layout optimisation of trusses using simulated annealing. Adv. Eng. Software. 33 (7–10), 681–696.

Hasançebi, O., Erdal, F., Saka, M.P., 2010. An adaptive harmony search method for structural optimization. J. Struct. Eng. 136 (4), 419–431.

Hilton, H.H., Feigen, M., 1960. Minimum weight analysis based on structural reliability. J. Aerosp. Sci. 27 (9), 641–652.

History of optimization, 2012, April 13. Retrieved April 16, 2012, from <www.mitrikitti.fi/opthist.html>.

Holland, J.H., 1975. Adaptation in Natural and Artificial Systems. University of Michigan, Ann Arbor, MI.

Index of history, 2012. Retrieved April 15, 2012, from Index of History/Mathematicians: <http://turnbull.mcs.st-and.ac.uk/~history/Mathematicians/>.

Janikow, C.Z., Michalewics, Z., 1991. An experimental comparison of binary and floating point representations in genetic algorithms. In: Belew, R.K., Booker, L.B. (Eds.), Proceedings of the Fourth International Conference on Genetic Algorithms. Morgan Kaufmann, San Mateo, CA, pp. 31–36.

Jenkins, W.M., 1991a. Towards structural optimization via the genetic algorithm. Comput. Struct. 40 (5), 1321–1327.

Jenkins, W.M., 1991b. Structural optimization with the genetic algorithm. Struct. Eng. 69 (24), 418–422.

Jenkins, W.M., 1992. Plane frame optimum design environment based on genetic algorithm. J. Struct. Eng. 118 (11), 3103–3112.

Jenkins, W.M., 1997. On the application of natural algorithms to structural design optimization. Eng. Struct. 19 (4), 302–308.

Jingui, L., Yunliang, D., Bin, W., Shide, X., 1996. An improved strategy for GAs in structural optimization. Comput. Struct. 61 (6), 1185–1196.
Kanagasundaram, S., Karihaloo, B.L., 1990. Minimum cost design of reinforced concrete structures. Struct. Optim. 2 (3), 173–184.
Kargahi, M., Anderson, J.C., Dessouky, M.M., 2006. Structural weight optimization of frames using tabu search. I: optimization procedure. J. Struct. Eng. 132 (12), 1858–1868.
Kaveh, A., Bakhshpoori, T., 2012. Optimum design of steel frames using cuckoo search algorithm with Levy flights. Struct. Design Tall Spec. Build. doi: 10.1002/tal.754.
Kaveh, A., Rahami, H., 2006. Nonlinear analysis and optimal design of structures via force method and genetic algorithm. Comput. Struct. 84 (12), 770–778.
Kaveh, A., Shojaee, S., 2007. Optimal design of skeletal structures using ant colony optimization. Int. J. Numer. Meth. Eng. 70, 563–581.
Kaveh, A., Talatahari, S., 2007. A discrete particle swarm ant colony optimization for design of steel frames. Asian J. Civil. Eng. 9 (6), 563–575.
Kaveh, A., Talatahari, S., 2009. Size optimization of space trusses using Big Bang–Big Crunch algorithm. Comput. Struct. 87 (17–18), 1129–1140.
Kaveh, A., Talatahari, S., 2010a. A novel heuristic optimization method: charged system search. Acta Mech. 213 (3–4), 267–289.
Kaveh, A., Talatahari, S., 2010b. Optimal design of skeletal structures via the charged system search algorithm. Struct. Multidiscip. Optim. 41 (6), 893–911.
Kellog, H.J., 1960. The cutting plane method for solving complex programs. SIAM J. 8, 703–712.
Kennedy, J., Eberhart, R., 1995a. Particle swarm optimization. Proceedings of IEEE International Conference on Neural Networks. IEEE Press, Piscataway, NJ, pp. 1942–1948.
Kennedy, J., Eberhart, R.C., 1995b. Particle swarm optimization. In: IEEE International Conference on Neural Networks, Perth, Australia, pp. 1942–1948.
Kirkpatrick, S., Gelatt Jr., C.D., Vecchi, M.P., 1983. Optimization by simulated annealing. Science. 220 (4598), 671–680.
Kirsch, U., 1993. Structural Optimization: Fundamental and Application. Springer-Verlag, London.
Koskisto, O.J., Ellingwood, B.R., 1997. Reliability-based optimization of plant precast concrete structures. J. Struct. Eng. 123 (3), 298–304.
Koumousis, V.K., Arsenis, S.J., 1994. Genetic algorithms in a multi-criterion optimal detailing of reinforced concrete members. In: Topping, B.H.V., Papadrakakis, M. (Eds.), Advances in Structural Optimization. CIVIL-COMP Ltd., Edinburgh, Scotland, pp. 233–240.
Krishnamoorthy, C.S., Mosi, D.R., 1981. Optimal design of reinforced concrete frames based on inelastic analysis. Eng. Optim. 5, 151–167.
Krishnapillai, K., Jones, R., 2009. Three-dimensional structural design optimisation based on fatigue implementing a genetic algorithm and a non-similitude crack growth law. Finite Elem. Anal. Design. 45 (2), 132–146.
Lamberti, L., 2008. An efficient simulated annealing algorithm for design optimization of truss structures. Comput. Struct. 86 (19–20), 1936–1953.
Lee, K.S., Geem, Z.W., 2004. A new structural optimization method based on the harmony search algorithm. Comput. Struct. 82 (9–10), 781–798.
Leite, J.P.B., Topping, B.H.V., 1999. Parallel simulated annealing for structural optimization. Comput. Struct. 73 (1–5), 545–564.

Lemonge, A.C.C., Barbosa, H.J.C., 2004. An adaptive penalty scheme for genetic algorithms in structural optimization. Int. J. Numer. Methods Eng. 59 (5), 703–736.
Li, L.J., Huang, Z.B., Liu, F., 2009. A heuristic particle swarm optimization method for truss structures with discrete variables. Comput. Struct. 87 (7–8), 435–443.
Lim, M.H., Yuan, Y., Omatu, S., 2000. Efficient genetic algorithms using simple genes exchange local search policy for the quadratic assignment problem. Comput. Optim. Appl. 15, 249–268.
Lin, K., Frangopol, D.M., 1996. Reliability-based optimum design of reinforced concrete girders. Struct. Saf. 18 (2–3), 239–258, Amsterdam, The Netherlands.
Lin, C.Y., Hajela, P., 1992. Genetic algorithms in optimization problems with discrete and integer design variables. Eng. Optim. 19, 309–327.
Lin, C.Y., Hajela, P., 1994. Design optimization with advanced genetic search strategies. Adv. Eng. Software. 21 (3), 179–189.
Luh, G.C., Lin, C.Y., 2009. Structural topology optimization using ant colony optimization algorithm. Appl. soft comput. 9 (4), 1343–1353.
Luh, G.C., Lin, C.Y., 2011. Optimal design of truss-structures using particle swarm optimization. Comput. Struct. 89 (23–24), 2221–2232.
Majumdar, A., Maiti, D.K., Maity, D., 2012. Damage assessment of truss structures from changes in natural frequencies using ant colony optimization. Appl. Math. Comput. 218 (19), 9759–9772.
Manoharan, S., Shanmuganathan, S., 1999. A comparison of search mechanisms for structural optimization. Comput. Struct. 73 (1–5), 363–372.
Maxwell, C., 1869. Scientific Papers II. Dover Publications, New York, NY, pp. 175–177.
May, S.A., Balling, R.J., 1992. A filtered simulated annealing strategy for discrete optimization of 3D steel frameworks. Struct. Optim. 4, 142–146.
Michell, A., 1904. The limit of economy of material in frame-structures. Philos. Mag. 8 (47), 589–597.
Moharrami, H., Grierson, D.E., 1993. Computer-automated design of reinforced concrete frameworks. J. Struct. Eng. 119 (7), 2036–2058.
Moses, F., 1964. Optimum structural design using linear programming. J. Struct. Div. 90 (ST6), 89–104.
Moses, F., 1971. Optimization of structures with reliability constraints. In: Pope, G.G., Schmit, L.A. (Eds.), Structural Design Applications of Mathematical Programming Techniques. Technical Editing and Reproduction Ltd., London, pp. 126–143. AGARDograph 149, Chapter 10.
Moses, F., Stevenson, J.D., 1970. Reliability based structural design. J. Struct. Div. 96 (ST2), 221–244.
Nanakorn, P., Meesomklin, K., 2001. An adaptive penalty function in genetic algorithms for structural design optimization. Comput. Struct. 79 (29–30), 2527–2539.
Park, H.S., Sung, C.W., 2002. Optimization of steel structures using distributed simulated annealing algorithm on a cluster of personal computers. Comput. Struct. 80 (14–15), 1305–1316.
Prager, W., Shield, R.T., 1968. Optimal design of multi-purpose structures. Int. J. Solids Struct. 4, 469–475.
Prager, W., Taylor, J.E., 1968. Problems of optimal structural design. J. Appl. Mech. 35 (1), 102–106.
Rahami, H., Kaveh, A., Gholipour, Y., 2008. Sizing, geometry and topology optimization of trusses via force method and genetic algorithm. Eng. Struct. 30 (9), 2360–2369.

Rajeev, S., Krishnamoorthy, C.S., 1992. Discrete optimization of structures using genetic algorithms. J. Struct. Eng. 118 (5), 1233–1250.
Rajeev, S., Krishnamoorthy, C.S., 1997. Genetic algorithms-based methodologies for design optimization of trusses. J. Struct. Eng. 123 (3), 350–358.
Rao, S.S., 2009. Optimization Theory and Application. fourth ed. John Wiley & Sons, Hoboken, NJ.
Sahab, M.G., Ashour, A.F., Toropov, V.V., 2005a. A hybrid genetic algorithm for reinforced concrete flat slab buildings. Comput. Struct. 83, 551–559.
Sahab, M.G., Ashour, A.F., Toropov, V.V., 2005b. Cost optimisation of reinforced concrete flat slab buildings. Eng. Struct. 27, 313–322.
Saka, M.P., Aydogdu, I., Hasancebi, O., Geem, Z.W., 2011. Harmony search algorithms in structural engineering. In: Yang, X.-S., Koziel, S. (Eds.), Computational Optimization and Applications in Engineering and Industry, Studies in Computational Intelligence, 359. pp. 145–182.
Sarma, K.C., Adeli, H., 1998. Cost optimization of concrete structures. J. Struct. Eng. 124 (5), 570–578.
Sarma, K.C., Adeli, H., 2000. Cost optimization of steel structures. Eng. Optim. 32, 777–802.
Schmit, L.A., 1960. Structural design by systematic synthesis. Proceedings of the Second National Conference on Electronic Computation. ASCE, pp. 105–122.
Schmit, L.A., Farshi, B., 1974. Some approximation concepts for structural synthesis. AIAA J.,. 12 (5), 692–699.
Schmit, L.A., Fox, R.L., 1965. An integrated approach to structural synthesis and analysis. AIAA J. 3 (6), 1104–1112.
Schmit, L.A., Miura, H., 1976. Approximation Concepts for Efficient Structural Synthesis. NASA.
Shim, P.Y., Manoochehri, S., 1997. Generating optimal configurations in structural design using simulated annealing. Int. J. Numer. Methods Eng. 40, 1053–1069.
Shimoda, M., Iwasa, K., Tsukada, S., 2010. Optimal shape design of shell structures, In: Proceedings of IV European Conference on Computational Mechanics, ECCM 2010, May 16–21, 2010, Paris, France.
Sonmez, M., 2011a. Artificial bee colony algorithm for optimization of truss structures. Appl. Soft. Comput. 11 (2), 2406–2418.
Sonmez, M., 2011b. Discrete optimum design of truss structures using artificial bee colony algorithm. Struct. Multidiscip. Optim. 43 (1), 85–97.
Sturler, E.D., Paulino, G.H., Wang, S., 2008. Topology optimization with adaptive mesh refinement, In: Abel, J.F., Cooke, J.R. (Eds.), Proceedings of the Sixth International Conference on Computation of Shell and Spatial Structures, IASS-IACM 2008, "Spanning Nano to Mega" May 28–31, 2008, Cornell University, Ithaca, NY, pp. 1–4.
Talatahari, S., Kheirollahi, M., Farahmandpour, C., Gandomi, A.H., 2012a. A multi-stage particle swarm for optimum design of truss structures. Neural Comput. Appl. doi: 10.1007/s00521-012-1072-5.
Talatahari, S., Gandomi, A.H., Yun, G.J., 2012b. Optimum design of tower structures by firefly algorithm. Struct. Design Tall Spec. Build. doi: 10.1002/tal.1043.
Tang, H., Zhou, J., Xue, S., Xie, L., 2010. Big Bang–Big Crunch optimization for parameter estimation. Mech. Syst. Signal Process. 24 (8), 2888–2897.
Toğan, V., Daloğlu, A.T., 2006. Optimization of 3D trusses with adaptive approach in genetic algorithms. Eng. Struct. 28 (7), 1019–1027.

Vanderplaats, G.N., 1993. Thirty years of modern structural optimization. Adv. Eng. Software. 16 (2), 81–88.

Vanderplaats, G.N., 1999. Numerical Optimization Techniques for Engineering Design. third ed. Vanderplaats Research Inc., Colorado Springs, CO.

Venkayya, V.B., Khot, N.S., Reddy, V.S., 1968. Energy distribution in an optimum structural design. AFFDL-TR.68–150.

Wang, S.Y., Tai, K., 2005. Structural topology design optimization using genetic algorithms with a bit-array representation. Comput. Method Appl. Mech. Eng. 194 (36–38), 3749–3770.

Wang, G., Wang, W., 1985. Fuzzy optimum design of structures. Eng. Optim. 8, 291–300.

Wasiutynski, Z., Brandt, A., 1963. The present state of knowledge in the field of optimum design of structures. Appl. Mech. Rev. 16 (5), 341–350.

Wen, Y.K., Kang, Y.J., 2000a. Minimum building life-cycle cost design criteria. I: Methodology. J. Struct. Eng. 127 (3), 330–337.

Wen, Y.K., Kang, Y.J., 2000b. Minimum building life-cycle cost design criteria. II: applications. J. Struct. Eng. 127 (3), 338–346.

Yang, X.S., 2008. Nature-Inspired Metaheuristic Algorithms. Luniver Press, Frome.

Yang, X.S., 2009. Firefly algorithms for multimodal optimization. Stochastic algorithms: foundations and applications, SAGA 2009. Lect. Notes Comput. Sci. 5792, 169–178.

Yang, X.S., Deb, S., 2009. Cuckoo search via Lévy flights. World congress on nature and biologically inspired computing (NaBIC 2009). IEEE Publ.210–214.

Yang, X.S., Gandomi, A.H., 2012. Bat algorithm: a novel approach for global engineering optimization. Eng. Comput. 29 (5), 464–483.

Yang, J., Soh, C.K., 1997. Structural optimization by genetic algorithms with tournament selection. J. Comput. Civil Eng. 11 (3), 195–200.

Yeh, Y., Hsu, D., 1990. Structural optimization with fuzzy parameters. Comput. Struct. 37 (6), 917–924.

Zadeh, L.A., 1965. Fuzzy sets. Inf. Control. 8 (3), 338–353.

Zhang, Z., Li, H., 2011. Two level optimization method of transmission tower structure based on ant colony algorithm. Adv. Mater. Res. 243-249, 5849–5853.

Zienkiewics, O.C., Campbell, J.S., 1973. Shape optimization and sequential linear programming. In: Zienkiewics, O.A., Zienkiewics, O.C. (Eds.), Optimum Structural Design. Wiley, New York, NY, pp. 109–126.

3 Particle Swarm Optimization in Civil Infrastructure Systems: State-of-the-Art Review

Kasthurirangan Gopalakrishnan

Department of Civil, Construction and Environmental Engineering, Iowa State University, Ames, IA, USA

3.1 Introduction

Applications of metaheuristic search algorithms inspired by natural phenomena are rapidly growing in diverse scientific fields to solve tough optimization problems. Some popular stochastic optimization algorithms include genetic algorithms (GA) inspired by Darwinian evolution and natural selection of biological systems (Goldberg, 1989; Holland, 1975), simulated annealing (SA) inspired by the annealing process of metals (Kirkpatrick, 1984), ant colony optimization (ACO) modeled on the actions of an ant colony (Dorigo and Maniezzo, 1996), particle swarm optimization (PSO) inspired by bird flocking behavior (Eberhart and Kennedy, 1995), and harmony search (HS) inspired by the improvisation process of musicians (Geem et al., 2001). These are approximate algorithms that efficiently explore the search space in order to find near-optimal solutions in a reasonable time.

Nature-inspired metaheuristic search algorithms are also steadily attracting much attention from the civil engineering research community owing to their ability to solve a wide range of continuous, discrete, and combinatorial optimization problems and at the same time to be easily implemented (Gandomi and Alavi, 2012; Yang and Koziel, 2011). Since they are not problem-specific, do not require the objective function to be continuous or differentiable (unlike gradient-based optimization algorithms like the quasi-Newton method), can incorporate constraints, can search very large spaces of candidate solutions, and in general be parallelizable, researchers have successfully applied them to a wide-variety of civil engineering optimization problems. The objective of this paper is to provide a state-of-the-art survey of PSO applications reported in civil engineering literature from 1995 to 2010.

PSO is a population-based global optimization tool, belonging to the class of swarm intelligence (SI) techniques, that was developed in 1995 and is based on swarming behaviors observed in flocks of birds or schools of fish (Eberhart and

Metaheuristic Applications in Structures and Infrastructures. DOI: http://dx.doi.org/10.1016/B978-0-12-398364-0.00003-6

Kennedy, 1995; Kennedy and Eberhart, 1995; Kennedy et al., 2001). PSO has very few parameters to adjust, can be implemented in a few lines of computer code, and with no or slight variations works well in a wide variety of applications including areas such as signal processing, graphics, robotics, and so on (Eberhart and Shi, 2001).

Poli (2008) reported over 1100 publications on the topic of PSO stored in the IEEE Xplore database. A recent search in the same IEEE Xplore database with the terms "particle swarm optimization" retrieved over 6700 records as of January 2011! There is an exponential growth of publications reporting on PSO applications in different fields including civil engineering (Alrashidi and El-Hawary, 2006; Poli, 2008; Song and Gu, 2004; Teodorovic, 2008). This chapter provides an up-to-date review of PSO applications to solve civil engineering problems and generate a greater awareness among civil engineering researchers and practitioners in this topic of growing interest.

3.2 Particle Swarm Optimization

The PSO concept introduced by Eberhart and Kennedy (1995) draws its inspiration from artificial life, bird flocking, fish schooling, swarming theory, as well as GA and evolutionary programming. Although it was originally introduced for optimization of nonlinear continuous functions, many advances in PSO development have enabled it to handle a wide class of complex engineering and science optimization problems. Similar to GA, PSO uses a population of potential solutions to the problem under consideration to probe the search space. However, each individual of the population in PSO has an adaptable *velocity* (position change), according to which it moves in the search space. Moreover, each individual has a *memory*, remembering the best position of the search space it has ever visited (Eberhart and Shi, 2001). The movement of the individual is thus an aggregated acceleration toward its best previously visited position and toward the best individual of a topological neighborhood. Since the "acceleration" term was mainly used for particle systems in particle physics (Reeves, 1983) and the term "swarm" for describing population, this algorithm was named PSO. In essence, PSO employs a swarm of particles or possible solutions that fly through the feasible solution space to explore optimal solutions.

Two variants of the PSO algorithm were originally developed: one with a global neighborhood (Gbest model) and one with a local neighborhood. Each particle moves toward its best previous position and toward the best particle in the whole swarm, according to the global variant. In contrast, each particle moves toward its best previous position and toward the best particle in its restricted neighborhood, according to the local variant (Eberhart and Kennedy, 1995).

The global variant PSO algorithm, which is the most standard one, is described as follows. Suppose that the search space is D-dimensional, then the ith particle of the swarm can be represented by a D-dimensional vector, $X_i = (x_{i1}, x_{i2}, \ldots, x_{iD})^T$.

The *velocity* (position change) of this particle can be represented by another D-dimensional vector $V_i = (v_{i1}, v_{i2},\ldots, v_{iD})^T$. The best previously visited position of the ith particle is denoted as $P_i = (p_{i1}, p_{i2}, \ldots, p_{iD})^T$. Defining g as the index of the best particle in the swarm (i.e., the gth particle is the best), and let the superscripts denote the iteration number. Each particle updates its position based on its own best exploration, best swarm overall experience, and its previous velocity vector according to the following two equations (Eberhart and Kennedy, 1995) which define the initial version of the PSO algorithm:

$$v_{id}^{n+1} = v_{id}^n + cr_1^n(p_{id}^n - x_{id}^n) + cr_2^n(p_{gd}^n - x_{id}^n),\ v_{id}^{n+1} = v_{id}^n + cr_1^n(p_{id}^n - x_{id}^n) + cr_2^n(p_{gd}^n - x_{id}^n), \tag{3.1}$$

$$x_{id}^{n+1} = x_i^d + v_{id}^{n+1} \tag{3.2}$$

where $d = 1, 2,\ldots,D$; $i = 1, 2,\ldots,N$ and N is the size of the swarm; c is a positive constant called *acceleration constant*; r_1 and r_2 are random numbers, uniformly distributed in [0, 1]; and $n = 1, 2,\ldots,$ determines the iteration number. The performance of each particle is measured according to a predefined fitness function or objective function, which is related to the problem under consideration.

Later on, a maximum velocity parameter, V_{max}, was imposed to improve the efficiency of PSO in the region of the optimum by allowing its velocity step-size to continue the search at a finer grain. Thus, in the later versions of the PSO, Eqs. (1) and (2) were modified as follows by incorporating a weight parameter for the previous velocity of the particle (Eberhart and Shi, 1998):

$$v_{id}^{n+1} = \chi(wv_{id}^n + c_1r_1^n(p_{id}^n - x_{id}^n) + c_2r_2^n(p_{gd}^n - x_{id}^n)), \tag{3.3}$$

$$x_{id}^{n+1} = x_i^d + v_{id}^{n+1} \tag{3.4}$$

where w is called *inertia weight*; c_1 and c_2 are two positive constants called *cognitive* and *social* parameters, respectively; and χ is a *constriction factor*, which is used, alternatively to w to limit velocity.

The termination criteria for PSO is typically one of the following: (i) to terminate the optimization process after a fixed number of iterations even if convergence is not achieved, (ii) to terminate the optimization after some fixed number of iterations without any improvement in the solution, and (iii) to predefine an error tolerance and terminate the optimization if the error between the objective function value and the best fitness value is less than the error tolerance (Abraham et al., 2006).

The swarm behavior in conventional PSO is influenced by the number of particles, inertia weight, maximum velocity, and acceleration coefficients to modify the velocity. The choice of the PSO algorithm's parameters is considered to be of utmost importance for the speed, convergence, and efficiency of the algorithm. The impact of the previous history of velocities on the current velocity is controlled by employing the inertia weight, *w*, which influences the trade-off between global (wide-ranging) and local (nearby) exploration abilities of the "flying points." The global exploration (searching new areas) capability is facilitated by a larger inertia weight, while a smaller inertia weight tends to facilitate local exploration to fine-tune the current search area. Thus, a good balance between global and local exploration abilities requiring fewer iterations on average to find the optimum solution can be achieved by suitable selection of the inertia weight (Shi and Eberhart, 1998).

PSO requires only primitive mathematical operators as it involves only two model equations and is computationally inexpensive in terms of both memory requirements and speed. It has several advantages over other optimization techniques including (Alrashidi and El-Hawary, 2006) its derivative-free nature unlike many conventional techniques, flexibility to be integrated with other optimization and soft-computing techniques to form hybrid tools, fewer parameters to adjust, ability to handle objective functions with stochastic nature, ease of implementation, lack of requirement of seed solution to start its iteration process, and so on.

Since its original development in 1995, there have been more than 20 different variants of PSO derived mainly by modifying the way the velocity of a particle is updated to improve the PSO performance; these variants include discrete PSO (DPSO), unified PSO (UPSO), chaotic PSO (C-PSO), and so on (Song and Gu, 2004).

PSO applications reported in different branches of civil engineering are summarized in the following sections after stating a brief note of caution. According to the no-free-lunch theorem (NFLT) of Wolpert and Macready (2002), all optimization methods perform equally well when averaged over all problems. The NFLT was further interpreted by Ho and Pepyne (2002) to mean that a general-purpose, universal optimization strategy is impossible. Also, it should be kept in mind that metaheuristic search algorithms are not guaranteed to find the optimum solution. Therefore, PSO may not work well for all classes of problems and it rests on the practitioner to find the best and most efficient algorithm for a given problem or a problem domain.

3.3 Structural Engineering

3.3.1 Shape and Size Optimization Problems in Structural Design

PSO has been used for addressing shape and size optimization problems in structural design and its performance has been shown superior to GA and comparable to gradient-based algorithms (Fourie and Groenwold, 2000, 2002). Schutte and Groenwold (2003) studied the optimal sizing design of truss structures using PSO. Camp et al. (2004) applied PSO to the low-weight design of trusses. PSO has also been applied for solving structural optimization problems for postbuckling behavior (Bochenek and Foryś, 2005, 2006). Dimou and Koumousis (2009) employed PSO for the reliability-based optimal design of statically determinate truss structures. Gomes (2011) applied PSO for structural truss mass optimization on size and shape taking into account frequency constraints.

Perez and Behdinan (2007) showed the suitability of the PSO approach for constraint truss optimization problems. Li et al. (2007a) presented an improved PSO (IPSO) algorithm, which is based on the PSO with passive congregation (PSOPC) scheme and a HS scheme, for solving truss structure optimization problems. Heinisuo et al. (2007) attempted to optimize the design of welded steel beams for a typical structure that contains both primary and secondary beams using PSO, but without much success.

Kaveh and Talatahari (2009b) implemented a heuristic particle swarm ACO (HPSACO) methodology, which is an efficient hybridized approach based on the HS scheme, PSO, and ACO, to find an optimum design of different types of structures. The authors demonstrated the effectiveness of HPSACO methodology in finding an optimum design of truss structures with continuous or discrete search domains and for frame structures with a discrete search domain. The discrete version of HPSACO, namely DHPSACO, was employed by Kaveh and Talatahari (2009a) to optimize truss structures with discrete variables.

A heuristic PSO (HPSO), based on HS schemes and standard PSO, was applied by Li and Liu (2009) to pin connected space structures with different variable types including continuous and discrete variables. To illustrate the effectiveness of HPSO, an optimal result of a complex practical double-layer grid shell structure was presented by Li and Liu (2009). The HPSO was also tested on several truss structures with discrete variables (Li et al., 2009a).

Seyedpoor et al. (2010, 2011) used a combination of simultaneous perturbation stochastic approximation and PSO algorithms for structural design optimization and to find the optimal shapes of arch dams considering fluid–structure interaction subject to earthquake loading. Plevris and Papadrakakis (2011) implemented an enhanced PSO algorithm combined with a gradient-based quasi-Newton sequential quadratic programming method for handling structural (global) optimization problems.

3.3.2 Structural Condition Assessment and Health Monitoring

Rao and Anandakumar (2007) proposed a new hybrid PSO (HPSO) algorithm by combining a self-configurable PSO with the Nelder–Mead algorithm for optimal placement of sensors for structural system identification and health monitoring. Abdalla (2009) examined the PSO approach in solving the inverse structural damage detection problem using a simulated cantilever beam model. Begambre and Laier (2009) proposed a HPSO-simplex (PSOS) algorithm for structural damage identification using frequency domain data. The authors demonstrated that the PSOS algorithm performs better than SA and standard PSO in different benchmark functions and applied it to two structural damage identification problems taking into consideration the effects of noisy and incomplete data. Koh et al. (2010) presented an experimental verification of the structural damage detection process using Latin hypercube sampling and PSO algorithm.

Muller and Chang (2009) used a PSO-based finite element modeling (FEM) scheme for real-time inverse impact identification and monitoring of composite structures as part of the structural health management approach. Liang and Tian (2010) used PSO for training a neural network (NN) to detect the depth and range of defect in concrete structures based on infrared thermal imaging. A gray model based PSO algorithm was developed by Liu and Dong (2011) for fatigue strength prognosis of concrete.

A cooperative PSO (CPSO) algorithm to optimize the weights of the NN was proposed by Lei and Ziyang (2009) for dam safety monitoring. Zhen-zhong et al. (2010) presented a deformation prediction model of a concrete arch dam based on an IPSO algorithm that combines the crossover and mutation operators of GA with PSO. PSO has been applied while searching the most critical slip circle in analyzing the effect of nonlinear strength of rockfill material on a dam's stability (Yi et al., 2010).

3.3.3 Structural Material Characterization and Modeling

Hideaki et al. (2006) applied PSO with FEM to estimate the thermal properties of massive concrete structures through inverse analysis. Travassos et al. (2008) employed a two-phase algorithm that combines matched-filter-based reverse-time migration algorithm with the PSO for the characterization of inclusions in concrete structures. McCluskey and McCarthy (2009) utilized a PSO approach for the cost optimum design of reinforced-concrete beams with multiple constraints according to Australian Standard (AS) 3600.

Ahmadi-Nedushan and Varee (2009) used PSO for the optimal design of reinforced-concrete earth-retaining walls. Tsai (2010) designed a center UPSO approach for learning of hybrid multilayer perceptron networks to predict strength in concrete cylinders, reinforced-concrete deep beams, and reinforced-concrete squat walls.

3.3.4 Other PSO Applications in Structural Engineering

PSO has been used for developing a novel hysteretic model for magnetorheological fluid dampers (a semi-active building control device) and parameter identification (Kwok et al., 2006). Schmidt (2010) used a modified PSO method for the design of an active structural vibration reduction system. PSO has also been used for the design of an active seismic control system for a building (Schmidt and Lewandowski, 2010).

3.4 Transportation and Traffic Engineering

3.4.1 Transportation Network Design

SI techniques, including PSO, have been shown to be very promising for solving complex transportation and traffic engineering problems (Teodorovic, 2008). Hu (2009) proposed a PSO-based approach for solving a typical model of bilevel programming (BLP) models in urban transportation equilibrium network design. PSO has been proposed for solving continuous network design problems in transportation, which seeks to optimize the capacity expansion of existing links in a given transportation network by minimizing the total system cost and at the same time considering the route choice behavior of individual users (Xu et al., 2010).

Previous studies on urban traffic network design mainly accounted for automobile exhaust emission as an environmental constraint without paying attention to traffic noise pollution, which has become one of the main pollution sources in cities. Huang et al. (2009) applied PSO for establishing a feasible BLP model of an urban traffic network considering noise pollution control. Lv et al. (2010) used PSO in BLP-based contra flow optimization to reduce traffic congestion of transportation networks during evacuation events in the face of disasters.

In highway transportation networks with private sector participants, there is always a competition in the provision of services. Koh (2009) explored the possibilities of employing a coevolutionary PSO algorithm to solve the resulting bilevel variational inequalities and determine the optimal choice of strategic variables of these private firms. The UPSO, which combined both the local and global search elements, was successfully utilized in this application.

3.4.2 Traffic Flow Forecasting

Zhao et al. (2006) used PSO to optimize the hidden and output layer weights of a radial basis function NN-based urban traffic flow forecasting model. A similar application of PSO was presented by Xiaobin (2009) for forecasting urban traffic flow. Zhao et al. (2008a) also presented a traffic flow forecasting model for a typical multi-intersection for urban trunk road using a dynamic recursion NN. The initial inputs of the context unit and all the weights of the dynamic recursion NN model were optimized by a dissimilation PSO algorithm.

Chengtao and Jianmin (2007) used an IPSO algorithm for optimizing the support vector regression (SVR) parameters for short-term traffic flow prediction. Compared to the classical PSO, the IPSO uses dynamic best inertia weight and acceleration coefficient, which makes it converge faster and avoids getting trapped in local optima.

3.4.3 Traffic Control

A fuzzy-based PSO methodology was presented by Chen and Xu (2006) for solving the traffic signal timing optimization problem. Dong et al. (2006) proposed a chaos-PSO (C-PSO) algorithm, developed by introducing a logistic map in PSO, for the optimization of signal timing in urban area traffic control. Wang et al. (2007) employed PSO for optimal coordination of the traffic signals of an artery system. Peng et al. (2009) proposed an isolation niches embedded PSO algorithm for traffic lights control. Lianyu and Jianfu (2009) proposed the use of quantum behaved PSO algorithm for origin-destination matrix calculation used in urban traffic management and control.

Kachroudi and Bhouri (2009) employed multiobjective PSO (MOPSO) for developing a traffic responsive strategy for multimodal urban traffic regulation for private vehicle traffic and public transport. Goccia (2010) recommended the use of

PSO over GA to speed up the optimization of traffic light phasing for particular traffic conditions. Cao et al. (2010) proposed a two-direction green wave control algorithm of traffic signal based on PSO.

3.4.4 Traffic Accident Forecasting

PSO has been successfully used in place of the backpropagation (BP) algorithm to train the weights of multilayer feed-forward NNs for automatic incident detection on traffic highways (Cheu et al., 2004; Srinivasan et al., 2003). Jiang et al. (2007) used PSO to tune the parameters of a support vector machine (SVM) model to develop a road traffic accident forecasting model. Similarly, Qing-Wei et al. (2009) employed PSO for choosing the parameters of SVR for improving the highway traffic accident forecasting performance. It was shown that the proposed PSO−SVR method performed better compared to BP NN in traffic accident forecasting.

3.4.5 Vehicle Routing Problem

A number of research studies in recent years have reported on the successful application of PSO in solving the vehicle routing problem (VRP) (Potvin, 2009).

Zhu et al. (2006) presented a computationally efficient PSO algorithm for solving VRP with time windows (VRPTW). Wu et al. (2008) presented a decision support model based on a hybrid algorithm of the PSO and ACO for solving the VRPTW.

Chen et al. (2006) employed a HPSO algorithm that combines DPSO and SA for solving large-scale capacitated VRP (CVRP). Jian (2009) proposed a modified genetic PSO to solve the CVRP. Shao et al. (2009) presented a HPSO algorithm to solve the VRP with stochastic travel time in transportation logistics distribution.

Ai and Kachitvichyanukul (2009) applied a real-value version of a PSO algorithm for solving VRP with simultaneous pickup and delivery. Liu et al. (2009) proposed a HPSO algorithm that utilizes the crossover operator of the GA for solving the VRP. Yin and Liu (2009) presented a similar HPSO algorithm for solving the single-depot complex VRP. Shanmugam et al. (2010) employed a HPSO with a genetic operator for solving the VRP.

Marinakis and Marinaki (2010) enhanced the ability of PSO to explore more effectively the solution space by hybridizing it with GA and applied it for solving the VRP. In a similar study, a HPSO that combines a PSO algorithm, the multiple phase neighborhood search−greedy randomized adaptive search procedure algorithm, the expanding neighborhood search strategy, and a path relinking strategy was proposed for solving very large-scale VRP and other more difficult combinatorial optimization problems (Marinakis et al., 2010).

3.4.6 Other PSO Application in Transportation and Traffic Engineering

Wai and Chuang (2008, 2010) designed a backstepping PSO control (BSPSOC) for the on-line levitated balancing and propulsive positioning of a magnetic levitation transportation system. The superiority of the BSPSOC scheme over total sliding-mode control and backstepping control strategies was also demonstrated in their study.

Hu et al. (2006, 2008) utilized PSO for optimal deployment of wireless sensor nodes in topology planning of the Urban Traffic Surveillance System, a subsystem of Intelligent Transportation Systems playing a vital role in advanced traffic management.

A HPSO–SA algorithm was proposed by Ali et al. (2008) for aircraft departure scheduling. Similarly, Xiujuan et al. (2008) proposed a second-order oscillating PSO for solving the aircraft sequencing problem. PSO has also been applied to optimizing the recovery scheduling of large-scale flight delays (Liu et al., 2008).

Shengwen et al. (2008) presented a method that integrates geographic information systems with an IPSO algorithm for simultaneously optimizing three-dimensional highway alignments. Ma et al. (2009) applied PSO in the decision-making process of hazardous materials transportation routing.

Lin et al. (2009) used an IPSO algorithm for solving the problem of public traffic transfer hub location selection. Situ et al. (2010) applied PSO for creating a combination bus timetable for a multiroute bus network with the goal of minimizing the waiting times of nontransfer and transfer passengers. Meng et al. (2010) proposed an IPSO for train timetable optimization and rescheduling.

3.5 Hydraulics and Hydrology

3.5.1 River Stage Prediction

Accurate prediction of river stages is an important research topic in the field of hydrology as it enables the concerned agency to issue forewarning of the impending flood as well as implement early evacuation measures. Chau (2004, 2006a) employed PSO as an alternative training algorithm for NN (in place of BP) to predict water levels in the Shing Mun River of Hong Kong with different lead times. As a further improvement, Chau (2007a) presented a split-step PSO algorithm coupled with the Levenberg–Marquardt (LM) algorithm to train the weights of NNs for forecasting real-time water levels at Fo Tan in Shing Mun River of Hong Kong with different lead times. The proposed split-step PSO algorithm combines the global search capability of PSO algorithm in the first step and local fast convergence of LM algorithm in the second step.

3.5.2 Design Optimization of Water/Wastewater Distribution Networks

Water supply systems (WSS) design optimization is one of the heavily researched areas in hydraulics. Several research studies have reported on the use of evolutionary methods for solving optimal design problems for WSS.

Montalvo et al. (2008a) demonstrated the use of DPSO for the design optimization of the Hanoi water-distribution network and the New York City water supply tunnel system. To further improve the robustness and performance of PSO for optimal design of WSS, Montalvo et al. (2008b, 2010a) proposed a self-adaptive framework with no a priori PSO parameter tuning.

A multiobjective variant of the PSO algorithm allowing better development of certain zones of the Pareto front through real-time human interaction was presented by Montalvo et al. (2010b) for multiobjective optimal design of water-distribution systems. Montalvo et al. (2010c) also proposed agent swarm optimization, a generalization of PSO oriented toward distributed artificial intelligence, for optimal design of water-distribution systems using a multiobjective approach.

Yang and Zhai (2009) employed PSO for solving an optimal operation model with multiobjective mixed discrete variables for large-scale WSS. Filion and Jung (2008, 2010) presented a PSO-based water-distribution network optimization framework that accounts for the cost of pipes as well as the potential economic damages sustained in fires in its objective function.

Izquierdo et al. (2008a,b, 2009a, 2010) illustrated the usefulness of PSO for handling different problems of industrial interest in water management, namely, the optimal design of water-distribution and wastewater collection networks, the calibration of a water-distribution network, and the identification and detection of leaks. In their study, the authors developed a variant of PSO that considers both discrete and continuous decision variables, has increased population diversity, and manages its parameters self-adaptively.

Hul et al. (2007) proposed a modified PSO for solving mixed integer nonlinear programming models for the synthesis of industrial water reuse/recycle networks. Bansal and Deep (2009) utilized PSO for optimal design of both serial and branched water-distribution networks. Geem (2009) proposed a modified HS algorithm incorporating PSO concept for optimal design of water-distribution networks.

In the area of water-distribution systems, PSO has also been used for optimizing pump operations (Wegley et al., 2000), developing a least-cost WSS design methodology (Vairavamoorthy and Shen, 2004), exploring the uncertainty and convergence of inverse transient calibration techniques (Jung and Karney, 2005), optimal pipe sizing design (Suribabu and Neelakantan, 2006), stormwater network design (Afshar, 2008), and developing an optimal surge-protection strategy for worst-case transient loadings (Jung and Karney, 2009).

3.5.3 Reservoir Operation Problems

Reservoir operations are characterized by multiple objectives that often compete with one another, and therefore involve generating optimal trade-offs among multiple objectives. Reddy and Nagesh Kumar (2007) presented a MOPSO approach for generating Pareto-optimal solutions for reservoir operation problems by integrating Pareto dominance principles into a PSO algorithm that also included a variable-size external repository and an efficient elitist-mutation (EM) operator. Reddy and Kumar (2009) also showed that their EM MOPSO algorithm outperforms the nondominated sorting GA (NSGA-II) model in solving multiobjective water resource management problems.

The MOPSO solver has also been applied for solving multipurpose reservoir operation problems with up to four objectives and the problem of selective withdrawal from a thermally stratified reservoir with three objectives (Baltar and Fontane, 2008). Fallah-Mehdipour et al. (2009) presented another application of MOPSO in multipurpose operation of a single-reservoir system. Azadnia and Zahraie (2010a) applied MOPSO in operation management of reservoirs with sedimentation problems.

3.5.4 Parameter Estimation/Calibration of Hydrological Models

Gill et al. (2006) applied the MOPSO algorithm for parameter estimation of a well-known conceptual rainfall-runoff model, namely the Sacramento soil moisture accounting model with 13 parameters as well as tested it to calibrate a three-parameter SVR model for soil moisture prediction. Wang et al. (2008) presented a PSO-based calibration tool for parameter estimation of a one-dimensional steady state stream water quality model. Scheerlinck et al. (2009) evaluated the use of PSO for hydrological parameter estimation of a simplified water and energy balance model. PSO has also been used in the automatic calibration of HEC-1 lumped conceptual rainfall-runoff model (Zakermoshfegh et al., 2008).

Zhang and Cui (2009) presented a framework for automatic calibration of a hydrological model (namely the Xinanjiang model) with multiobjectives that combines an MOPSO algorithm and an entropy-based TOPSIS ranking method. Liu (2009) proposed a new nondominated sorting PSO algorithm that draws its ideas from the multiobjective GA NSGA-II, for automatic calibration of a rainfall-runoff model.

Chu and Chang (2009) applied PSO to the parameter estimation of the nonlinear Muskingum model, the most widely used method for flood routing in hydrologic engineering. Azadnia and Zahraie (2010b) presented an application of MOPSO for optimization of the extended Muskingum model with variable parameters. Jiang et al. (2010) presented a new method, namely the MSSE-PSO (master–slave swarms shuffling evolution algorithm based on PSO), which combines the ideas of PSO with hierarchical evolution, for calibrating the Xinjiang conceptual hydrological model with 15 parameters.

3.5.5 Other PSO Applications in Hydraulics and Hydrology

Other miscellaneous PSO applications in the field of water resources include estimation of water table elevation from streaming potentials (Fernandez-Martinez et al., 2010; Naudet et al., 2008), retrieval of water quality parameters based on optical remote sensing data (Campbell and Phinn, 2009), optimal synthesis of an integrated water system (Luo and Uan, 2008), optimized design and operation of river basin (Shourian et al., 2008), optimal design of cascade stilling basins (Daraeikhah et al., 2009), and optimal design and operation of irrigation water pumping systems (Afshar and Rajab, 2007; Reddy and Adarsh, 2010).

In the field of hydraulics and hydrology, PSO has also been used to train the weights of a NN that mimics the outputs of the soil and water assessment tool hydrological model (Bekele and Nicklow, 2005), in NN-based forecasting of real-time algal bloom dynamics (a source of water pollution) (Chau, 2005), in optimal design and calibration of large and complex urban stormwater management models (Muleta et al., 2006), and in optimization of the preliminary selection, sizing, and placement of hydraulic devices in a pipeline system in order to control its transient response (Jung and Karney, 2006).

3.6 Construction Engineering

3.6.1 Construction Planning and Management

Construction scheduling problems have remained an active area of research in the field of construction engineering. PSO has been applied for solving the resource-constrained project scheduling problem (RCPSP) in construction with minimization of project duration as the objective (Zhang et al., 2005, 2006a; Zhao and Ru, 2008). A permutation-based PSO algorithm based on a hybrid particle-updating mechanism that incorporates a partially mapped crossover operator of GA and a definition of an activity-move-range, was developed by Zhang et al. (2006b) for solving the RCPSP. A PSO-based methodology was also introduced by Zhang et al. (2006c) to implement preemptive scheduling under break and resource-constraints for construction projects. A justification PSO scheme that includes other designed mechanisms has been recently proposed for solving the RCPSP (Chen, 2011).

Zhang et al. (2006d) introduced a methodology based on PSO for solving a multimode RCPSP (MRCPSP) that considers both renewable and nonrenewable resources. Jarboui et al. (2008) used a combinatorial PSO (CPSO) algorithm to solve a MRCPSP and showed that the CPSO algorithm outperforms the SA algorithm and has a similar performance as the classical PSO algorithm.

Lu et al. (2008) combined the advantages of the simplified discrete event simulation approach and the PSO technique to automate the formulation of a resource-constrained critical path method (CPM)-based construction schedule with the shortest total project duration. The authors applied the proposed method to a real drainage project in Hong Kong and achieved its results better than those produced by the current CPM software.

A PSO-supported construction-simulation approach was proposed by Zhang et al. (2006e) for efficient determination of optimal resource combination for a construction operation. Yang (2007a) developed and tested a PSO algorithm to aid in the construction project crashing analysis, which is to minimize the required cost while meeting a specified deadline.

A time–cost tradeoff (TCT) analysis is typically used in construction management to solve two types of problems, namely the budget problem (which is to find the shortest project duration without exceeding a given budget) and the deadline problem (which is to minimize the total cost while meeting a specific deadline).

Various multiobjective optimization methods have been applied to study the TCT issue with the objectives of minimizing project duration and total cost in determining an optimal combination of construction activity methods. Yang (2007b) used an elitist PSO algorithm to find the complete time–cost profile (Pareto front) over a set of feasible project durations, thereby solving the generalized bicriterion optimization problem that incorporates both the budget and deadline problems. Zhang and Li (2010) presented a combined scheme-based MOPSO method for solving the TCT problem.

More recently, construction planners are faced with the time–cost–quality tradeoff (TCQT) problem, where the goal is to determine an optimal set of construction methods consisting of suitable construction technologies and resource utilization plans to minimize cost and time, but at the same time maximize the quality of construction for long-term performance. Wang and Feng (2008) presented a hierarchical subpopulation PSO algorithm for solving the TCQT problem in construction management. Since the construction quality or performance may be recorded in terms of vague data rather than precise numbers, Zhang and Xing (2010) presented a fuzzy-MOPSO algorithm which utilizes a fuzzy multiattribute utility methodology for solving the TCQT problem.

3.6.2 Construction Litigation

Because of very high costs involved in construction claims and their complicated nature, it would be advantageous to know with some certainty the outcome of a dispute resolution if it were taken to a court. In a study demonstrating the use of NNs for predicting the outcome of such construction claims in Hong Kong, Chau (2007b, 2006b) employed the PSO algorithm to train the weights of the feedforward MLP NN.

3.6.3 Construction Cost Estimation and Prediction

Shi and Li (2010a,b) presented a novel method integrating fuzzy logic, rough sets theory, and NN whose weights are trained by a PSO algorithm to estimate conceptual construction costs more precisely during the early stage of a project based on historical construction data from the World Wide Web. Yang (2008) presented a PSO-based distribution-free Monte Carlo simulation approach for construction cost estimation and reliability analysis.

3.6.4 Other PSO Applications in Construction Engineering

Rui and Xiaoya (2008) proposed an IPSO algorithm for selecting the optimal combination of construction contractors to achieve the minimum cost within the required duration in completing large-scale construction projects. PSO has been utilized for solving the construction site unequal-area layout problem where the goal is to optimize the layout of temporary facilities on a construction site to enhance productivity and safety (Zhang and Wang, 2008).

A HPSO algorithm combining PSO with the simplex method has been used for developing a BLP model of earthwork allocation in construction projects (Xianjia et al., 2009). Xu and Zeng (2010) applied PSO for solving a dynamic equipment allocation problem encountered during the implementation of a concrete-faced rockfill dam construction project.

3.7 Geotechnical Engineering

3.7.1 Inverse Parameter Identification and Geotechnical Model Calibration

Du et al. (2006) used an IPSO algorithm for back-analysis of geotechnical parameters of concrete face rockfill dams. Knabe (2009) employed PSO for estimating parameters of advanced constitutive models for clays using an iterative approach. Zhang et al. (2009a) proposed a novel parallel hybrid moving PSO (hmPSO) algorithm and used it for selecting parameter values in constitutive models for unsaturated soils. Zhang et al. (2009b) applied the hmPSO for the calibration of a water infiltration model for unsaturated soils and suggested its use for general back-analysis of geotechnical case studies.

Meier et al. (2008) used PSO in the calibration processes of geomechanical constitutive models used in soil mechanics and engineering geology. Chen and Feng (2007) proposed an IPSO (CSV-PSO), in which flight velocity limit and flight space of particles are constricted dynamically with flying of particles, for inverse identification of rheological parameters of soft and weak rock mass. Zimmerer and Schanz (2006) utilized PSO for inverse soil stiffness parameter identification for modeling of deep excavations.

PSO has been applied to inverse parameter identification in desaturation of a field column (Schanz et al., 2006), solving inverse optimization problems in geomechanics (Meier et al., 2008), and to tune the parameters of SVM in displacement back-analysis for establishing geomechanical parameters in rock mechanics and engineering (Zhao and Yin, 2009). Xing et al. (2010) studied the adaptive control of tunnel excavation based on numerical simulation and PSO in which PSO was used to back-analyze the rock mechanics parameters.

3.7.2 Slope Stability Analysis

Cheng et al. (2007) used a modified PSO with a termination criterion for the location of the critical noncircular failure surface in two-dimensional slope stability analysis. Zhao et al. (2008b) incorporated chaos mapping into PSO and applied the updated PSO method for identifying the noncircular critical slip surface in slope stability analysis. Li et al. (2009a,b) proposed a mixed search procedure integrating both a HS scheme and PSO to determine the minimum factor of safety of soil slopes. In studying the same problem of slope stability analysis, Li et al. (2010) developed and applied a new discontinuous flying PSO algorithm. Another research study reported on a similar application of classical PSO to reliability evaluation of earth slope and locating the critical probabilistic slip surface (Khajehzadeh et al., 2011).

3.8 Pavement Engineering

Shen et al. (2009) applied a chaos-PSO (CPSO) algorithm in pavement maintenance decision-making. Gopalakrishnan (2010) applied PSO for inverse modeling and real-time nondestructive evaluation of pavement systems. Tayebi et al. (2010) applied PSO to programming of pavement maintenance activities in a pavement management system. Le (2010) proposed the application of a MOPSO for sustainable flexible pavement design.

3.9 PSO Applications in Other Civil Engineering Fields

Elbeltagi et al. (2005) applied PSO to optimize maintenance and repair decisions in bridge deck rehabilitation. Yang and Hsu (2010) applied a ε-constrained PSO algorithm in developing a risk-based multiobjective optimization model for bridge maintenance planning. Khajehzadeh et al. (2010) used the PSOPC in the economic design of retaining walls.

A PSO-based dynamic scheduling algorithm has been proposed by Li et al. (2007b) for vertical traffic in buildings. Izquierdo et al. (2009b) applied PSO for modeling pedestrian evacuation simulation and estimating pedestrian evacuation times from high-traffic buildings or buildings of cultural, governmental, or industrial importance under critical circumstances. Wang et al. (2010) applied PSO to optimize the queuing of aircrafts on multirunway airports during their departures or arrivals. PSO has been used to train the weights of NN in predicting air pollutant levels (Lu et al., 2003).

3.10 Concluding Remarks

PSO is a type of artificial intelligence method based on the collective behavior of decentralized, self-organized systems, and has proved to be an efficient method for many global optimization problems, and in some cases it does not suffer the difficulties encountered by other evolutionary computation techniques. This chapter provided an up-to-date review of PSO applications in different branches of civil engineering including structural, traffic and transportation, hydraulics and hydrology, construction, geotechnical, and pavement engineering. More than 190 application papers were reviewed; most of these were published in the last few years, indicating a rapidly increasing interest in this topic among the civil engineering community. A case study was also presented to demonstrate the successful application of PSO in solving a practical pavement engineering problem.

Researchers and practitioners find PSO attractive, owing to its simplicity in having fewer parameters to adjust, its flexibility to be integrated with other optimization and soft-computing techniques to form hybrid tools, its ability to handle objective functions with a stochastic nature, and above all its ease of implementation. The application areas reported in this chapter attest to PSO's versatility in dealing with diverse optimization problems in civil engineering. It is expected that PSO will continue to see similar success in the future, especially in combination with other soft-computing techniques, in solving tough civil engineering optimization problems.

References

Abdalla, M.O., 2009. Particle swarm optimization (PSO) for structural damage detection. In: Proceedings of the Third International Conference on Applied Mathematics, Simulation, Modelling, Circuits, Systems and Signals. ASMCSS'09. Stevens Point, World Scientific and Engineering Academy and Society (WSEAS), WI, USA, pp. 43–48. Available from: <http://portal.acm.org/citation.cfm?id = 1737863.1737873/>.

Abraham, A., Guo, H., Liu, H., 2006. Swarm intelligence: foundations, perspectives and applications. In: Nedjah, N., Mourelle, L. (Eds.), Swarm Intelligent Systems. Studies in Computational Intelligence. Springer, Berlin/ Heidelberg, pp. 3–25. Available from: <http://dx.doi.org/10.1007/978-3-540-33869-7_1/>.

Afshar, M.H., 2008. Rebirthing particle swarm optimization algorithm: application to storm water network design. Can. J. Civil Eng. 35 (10), 1120–1127.

Afshar, M.H., Rajab, P.R., 2007. Optimal design and operation of irrigation pumping systems using particle swarm optimization algorithm. Int. J. Civil Eng. 5 (4), 302–311.

Ahmadi-Nedushan, B., Varee, H., 2009. A particle swarm optimisation approach to reinforced concrete beam design according to AS3600. In: Topping, B.H.V., Tsompanakis, Y. (Eds.), Proceedings of the First International Conference on Soft Computing Technology in Civil, Structural and Environmental Engineering. Civil-Comp Press, Madeira, Portugal.

Ai, T.J., Kachitvichyanukul, V., 2009. A particle swarm optimization for the vehicle routing problem with simultaneous pickup and delivery. Comput. Oper. Res. 36 (5), 1693–1702.

Ali, F., Xiujuan, L., Xiao, X., 2008. The aircraft departure scheduling based on particle swarm optimization combined with simulated annealing algorithm. In: Proceedings of the IEEE Congress on Evolutionary Computation, CEC 2008. IEEE World Congress on Computational Intelligence (CEC 2008). Hong Kong, pp. 1393–1398.

Alrashidi, M.R., El-Hawary, M.E., 2006. A survey of particle swarm optimization applications in power system operations. Elec. Power Compon. Syst. 34 (12), 1349–1357.

Azadnia, A., Zahraie, B., 2010a. Application of multi-objective particle swarm optimization in operation management of reservoirs with sedimentation problems. In: Palmer, R.N. (Ed.), Proceedings of the World Environmental and Water Resources Congress 2010. ASCE, Providence, Rhode Island. Available from: <http://link.aip.org/link/?ASC/371/233/1/> (accessed 07.01.2011).

Azadnia, A., Zahraie, B., 2010b. Optimization of nonlinear muskingum method with variable parameters using multi-objective particle swarm optimization. In: Palmer, R.N. (Ed.),

Proceedings of the World Environmental and Water Resources Congress 2010. ASCE, Providence, RI, p. 235. Available from: <http://link.aip.org/link/?ASC/371/235/1/> (accessed 07.01.2011).

Baltar, A.M., Fontane, D.G., 2008. Use of multiobjective particle swarm optimization in water resources management. J. Water Resour. Plann. Manage. 134 (3), 257–265.

Bansal, J.C., Deep, K., 2009. Optimal design of water distribution networks via particle swarm optimization. In: 2009 IEEE International Advance Computing Conference (IACC). Patiala, India, pp. 1314–1316.

Begambre, O., Laier, J.E., 2009. A hybrid particle swarm optimization—simplex algorithm (PSOS) for structural damage identification. Adv. Eng. Softw. 40 (9), 883–891.

Bekele, E.G., Nicklow, J.W., 2005. Hybrid evolutionary search methods for training an artificial neural network. In: Walton, R. (Ed.), ASCE Conference Proc. Proceedings of World Water and Environmental Resources Congress 2005. ASCE, Anchorage, AK. Available from: <http://link.aip.org/link/?ASC/173/342/1/> (accessed 07.01.2011).

Bochenek, B., Foryś, P., 2005. Structural optimization against instability using particle swarms. In: Proceedings of the Sixth World Congress on Structural and Multidisciplinary Optimization.

Bochenek, B., Foryś, P., 2006. Structural optimization for post-buckling behavior using particle swarms. Struct. Multidiscip. Optim. 32 (6), 521–531.

Camp, C.V., Meyer, B.J., Palazolo, P.J., 2004. Particle swarm optimization for the design of trusses. In: Blandford, G.E. (Ed.), Proceedings of Structures Congress 2004. ASCE, Nashville, TN, pp. 1–10. Available from: <http://link.aip.org/link/?ASC/137/160/1/> (accessed 07.01.2011).

Campbell, G., Phinn, S., 2009. Accuracy and precisions of water quality parameters retrieved from particle swarm optimisation in a sub-tropical lake. In: Remote Sensing of the Ocean, Sea Ice, and Large Water Regions 2009: SPIE—The International Society for Optical Engineering 2009. Berlin, Germany.

Cao, C.T., Cui, F., Guo, G.Q., 2010. Two-direction green wave control of traffic signal based on particle swarm optimization. Appl. Mech. Mater. 26, 507–511.

Chau, K., 2004. River stage forecasting with particle swarm optimization. In: Orchard, B., Yang, C., Ali, M. (Eds.), Innovations in Applied Artificial Intelligence. Lecture Notes in Computer Science. Springer, Berlin/Heidelberg, pp. 1166–1173. Available from: <http://dx.doi.org/10.1007/978-3-540-24677-0_119/>.

Chau, K., 2005. A split-step PSO algorithm in prediction of water quality pollution. In: Wang, J., Liao, X., Yi, Z. (Eds.), Advances in Neural Networks – ISNN 2005. Lecture Notes in Computer Science. Springer, Berlin/Heidelberg, pp. 1034–1039. Available from: <http://dx.doi.org/10.1007/11427469_164/>.

Chau, K.-W., 2006a. A split-step PSO algorithm in predicting construction litigation outcome. In: Yang, Q., Webb, G. (Eds.), PRICAI 2006: Trends in Artificial Intelligence. Lecture Notes in Computer Science. Springer, Berlin/Heidelberg, pp. 1211–1215. Available from: <http://dx.doi.org/10.1007/978-3-540-36668-3_163/>.

Chau, K.W., 2006b. Particle swarm optimization training algorithm for ANNs in stage prediction of Shing Mun River. J. Hydrol. 329 (3-4), 363–367.

Chau, K.W., 2007a. A split-step particle swarm optimization algorithm in river stage forecasting. J. Hydrol. 346 (3–4), 131–135.

Chau, K.W., 2007b. Application of a PSO-based neural network in analysis of outcomes of construction claims. Autom. Constr. 16 (5), 642–646.

Chen, A., Yang, G., Wu, Z., 2006. Hybrid discrete particle swarm optimization algorithm for capacitated vehicle routing problem. J. Zhejiang Univ. Sci. A. 7 (4), 607–614.

Chen, B.R., Feng, X.T., 2007. CSV-PSO and its application in geotechnical engineering. In: Chan, F.T., Tiwari, M.K. (Eds.), Swarm Intelligence, Focus on Ant and Particle Swarm Optimization. I-Tech Education and Publishing.

Chen, J., Xu, L., 2006. Road-junction traffic signal timing optimization by an adaptive particle swarm algorithm. In: Proceedings of the Ninth International Conference on Control, Automation, Robotics and Vision (ICARCV'06), pp. 1–7.

Chen, R.-M., 2011. Particle swarm optimization with justification and designed mechanisms for resource-constrained project scheduling problem. Expert Syst. Appl. (In Press. Available from: <http://www.sciencedirect.com/science/article/B6V03-51S25CW-8/2/3576438aa5e85da720b6fbf05d936c8f/> (accessed 06.01.11)).

Cheng, Y.M., Li, L., Chi, S.C., Wei, W.B., 2007. Particle swarm optimization algorithm for the location of the critical non-circular failure surface in two-dimensional slope stability analysis. Comput. Geotech. 34 (2), 92–103.

Chengtao, C., Jianmin, X., 2007. Improved particle swarm optimized SVM for short-term traffic flow predication. In: Chinese Control Conference (CCC). Hunan, China, pp. 6–9.

Cheu, R.L., Srinivasan, D., Loo, W.H., 2004. Training neural networks to detect freeway incidents by using particle swarm optimization. Trans. Res. Rec. 1867, 11–18.

Chu, H.-J., Chang, L.-C., 2009. Applying particle swarm optimization to parameter estimation of the nonlinear muskingum model. J. Hydrol. Eng. 14 (9), 1024–1027.

Daraeikhah, M., Meraji, S.H., Afshar, M.H., 2009. Application of particle swarm optimization to optimal design of cascade stilling basins. Sci. IranicaTrans. A Civil Eng. 16 (1), 50–57.

Dimou, C.K., Koumousis, V.K., 2009. Reliability-based optimal design of truss structures using particle swarm optimization. J. Comput. Civil Eng. 23 (2), 100–109.

Dong, C., Liu, Z., Liu, X., 2006. Chaos-particle swarm optimization algorithm and its application to urban traffic control. Int. J. Comput. Sci. Network Security. 6 (1), 97–101.

Dorigo, M., Maniezzo, V., 1996. The ant system: optimization by a colony of cooperating agents. IEEE Trans. Syst. Man Cybern. B. 26 (1), 29–41.

Du, H., Chi, S.C., Wang, F., 2006. Using an improved particle swarm optimization for back analysis of geotechnical parameters of concrete face rock-fill dams. In: Proceedings of the Sixth International Conference on Hybrid Intelligent Systems. HIS'06, IEEE Computer Society, Washington, DC, USA. Available from: <http://dx.doi.org/10.1109/HIS.2006.75/>.

Eberhart, R., Kennedy, J., 1995. A new optimizer using particle swarm theory. In: Proceedings of the Sixth International Symposium on Micro Machine and Human Science (MHS'95). Nagoya, Japan, pp. 39–43.

Eberhart, R.C., Shi, Y., 1998. Evolving artificial neural networks. In: Proceedings of the International Conference on Neural Networks and Brain, pp. 5–13.

Eberhart, R., Shi, Y., 2001. Particle swarm optimization: developments, applications and resources. In: Proceedings of the 2001 Congress on Evolutionary Computation. Seoul, South Korea, pp. 81–86.

Elbeltagi, E., Elbehairy, H., Hegazy, T., Grierson, D., 2005. Evolutionary algorithms for optimizing bridge deck rehabilitation. In: Soibelman, L., Pena-Mora, F. (Eds.), Proceedings of the 2005 ASCE International Conference on Computing in Civil Engineering. ASCE, Cancun, Mexico. Available from: <http://link.aip.org/link/?ASC/179/59/1/> (accessed 07.01.2011).

Fallah-Mehdipour, E., Haddad, O.B., Marino, M.A., 2009. MOPSO in multipurpose operation of single-reservoir system. In: Starrett, S. (Ed.), Proceedings of World Environmental and Water Resources Congress 2009. ASCE Press, Kansas City, MO.

Fernandez-Martinez, J.L., Garcia-Gonzalo, Esperanza, Naudet, V., 2010. Particle swarm optimization applied to solving and appraising the streaming-potential inverse problem. Geophysics. 75 (4), 3–15.

Filion, Y.R., Jung, B.S., 2008. Particle swarm optimization of water distribution networks with economic damages. In: Van Zyl, K. (Ed.), Proceedings of Water Distribution Systems Analysis 2008. ASCE Press, Kruger National Park, South Africa, pp. 1–8.

Filion, Y.R., Jung, B.S., 2010. Least-cost design of water distribution networks including fire damages. J. Water Resour. Plan. Manage. 136 (6), 658–668.

Fourie, P.C., Groenwold, A.A., 2000. Particle swarms in size and shape optimization. In: Proceedings of the International Workshop on Multidisciplinary Design Optimization. Pretoria, South Africa, pp. 97–106.

Fourie, P.C., Groenwold, A.A., 2002. The particle swarm optimization algorithm in size and shape optimization. Struct. Multidiscip. Optim. 23 (4), 259–267.

Gandomi, A.H., Alavi, A.H., 2012. Krill herd: a new bio-inspired optimization algorithm. Commun. Nonlinear. Sci. Numer. Si. 17 (12), 4831–4845.

Geem, Z.W., 2009. Particle-swarm harmony search for water network design. Eng. Optimiz. 41 (4), 297–311.

Geem, Z.W., Kim, J.H., Loganathan, G.V., 2001. Harmony search. Simulation. 76 (2), 60–68.

Gill, M.K., Kaheil, Y.H., Khalil, A., Mckee, M., Bastidas, L., 2006. Multiobjective particle swarm optimization for parameter estimation in hydrology. Water Resour. Res. 42, p.14.

Goccia, M., 2010. Traffic optimisation: speed up the flow. Available from: <http://www.thalesresearch.com/Publications/WhitePapers/Documents/EDE100101.pdf/>.

Goldberg, D.E., 1989. *Genetic Algorithmsin Search, Optimization, and Machine Learning*. Addison-Wesley.

Gomes, H.M., 2011. Truss optimization with dynamic constraints using a particle swarm algorithm. Expert Syst. Appl. 38 (1), 957–968.

Gopalakrishnan, K., 2010. Neural network–swarm intelligence hybrid nonlinear optimization algorithm for pavement moduli back-calculation. J. Trans. Eng. 136 (6), 528–536.

Heinisuo, M. et al., 2007. Welded steel beam design using particle swarm analysis. In: Proceedings of the Fourteenth EG-ICE Workshop. Maribor, Slovenia, pp. 675–680.

Hideaki, N., et al., 2006. Application of the particle swarm optimization to the inverse thermal analysis for the mass concrete structures. Proc. JSCE. 809, 41–52.

Ho, Y.C., Pepyne, D.L., 2002. Simple explanation of the no-free-lunch theorem and its implications. J. Optimiz. Theor. Appl. 115 (3), 549–570.

Holland, J.H., 1975. Adaptation in Natural and Artificial Systems. University of Michigan Press, Ann Arbor, MI.

Hu, H., 2009. A particle swarm optimization algorithm for bilevel programming models in urban traffic equilibrium network design. In: Wang, Y., et al., (Eds.), Proceedings of the Ninth International Conference on Chinese Transportation Professionals. ASCE, Harbin, China.

Hu, J. et al., 2006. A study of particle swarm optimization in urban traffic surveillance system. In: International Multiconference on Computational Engineering in Systems Applications (IMACS). Beijing, China, pp. 2056–2061.

Hu, J., et al., 2008. Topology optimization for urban traffic sensor network. Tsinghua Sci. Technol. 13 (2), 229–236.

Huang, K., et al., 2009. Bi-level programming model of urban traffic network considering noise pollution control. In: Peng, Q., et al., (Eds.), Proceedings of the Second International Conference on Transportation Engineering. ASCE, Chengdu, China, pp. 3399–3404.

Hul, S., et al., 2007. Water network synthesis using mutation-enhanced particle swarm optimization. Process Saf. Environ. Protect. 85 (6), 507–514.

Izquierdo, J., Montalvo, I., Pérez, R., Fuertes, V.S., 2008a. Design optimization of wastewater collection networks by PSO. Comput. Math. Appl. 56 (3), 777–784.

Izquierdo, J., Montalvo, I., Pérez, R., Tavera, M., 2008b. Optimization in water systems: a PSO approach. In: Proceedings of the 2008 Spring Simulation Multiconference. SpringSim'08. Society for Computer Simulation International, San Diego, CA, USA, pp. 239–246. Available from: <http://portal.acm.org/citation.cfm?id = 1400549.1400581/>.

Izquierdo, J., et al., 2009a. Forecasting pedestrian evacuation times by using swarm intelligence. Physica A Stat. Mech. Appl. 388 (7), 1213–1220.

Izquierdo, J., et al., 2009b. Robust design of water supply systems through evolutionary optimization. In: Bru, R., Romero-Vivó, S. (Eds.), Positive Systems. Lecture Notes in Control and Information Sciences. Springer, Berlin/Heidelberg, pp. 321–330. Available from: <http://dx.doi.org/10.1007/978-3-642-02894-6_31/>.

Izquierdo, J. et al., 2010. Distributed particle swarm intelligence for optimization in the water industry. In: Progress in Industrial Mathematics at ECMI 2008. Mathematics in Industry. pp. 893–898.

Jarboui, B., et al., 2008. A combinatorial particle swarm optimization for solving multimode resource-constrained project scheduling problems. Appl. Math. Comput. 195 (1), 299–308.

Jian, L., 2009. Solving capacitated vehicle routing problems via genetic particle swarm optimization. Third International Symposium on Intelligent Information Technology Application, 2009. IITA 2009. pp. 528–531.

Jiang, A., Liu, L., Zhang, J., 2007. The evolutionary support vector machine forecasting model of road traffic accident. In: Peng, Q., et al., (Eds.), Proceedings of the First International Conference on Transportation Engineering 2007 (ICTE 2007). ASCE, Chengdu, China. Available from: <http://link.aip.org/link/?ASC/246/131/1/> (accessed 07.01.11).

Jiang, Y., et al., 2010. Improved particle swarm algorithm for hydrological parameter optimization. Appl. Math. Comput. 217 (7), 3207–3215.

Jung, B.S., Karney, B.W., 2005. A systematic exploration of uncertainty and convergence of inverse transient calibration for WDSs. In: Walton, R. (Ed.), Proceedings of World Water and Environmental Resources Congress 2005. ASCE, Anchorage, AL. Available from: <http://link.aip.org/link/?ASC/173/50/1/> (accessed 07.01.2011).

Jung, B.S., Karney, B.W., 2006. Hydraulic optimization of transient protection devices using GA and PSO approaches. J. Water Resour. Plann. Manage. 132 (1), 44–52.

Jung, B.S., Karney, B.W., 2009. Systematic surge protection for worst-case transient loadings in water distribution systems. J. Hydraulic Eng. 135 (3), 218–223.

Kachroudi, S., Bhouri, N., 2009. A multimodal traffic responsive strategy using particle swarm optimization. In: Chassiakos, A.G. (Ed.), Twelfth IFAC Symposium on Control in Transportation Systems. Available from: <http://www.ifac-papersonline.net/Detailed/40476.html/> (accessed 04.01.11).

Kaveh, A., Talatahari, S., 2009a. A particle swarm ant colony optimization for truss structures with discrete variables. J. Constr. Steel. Res. 65 (8-9), 1558–1568.

Kaveh, A., Talatahari, S., 2009b. Hybrid algorithm of harmony search, particle swarm and ant colony for structural design optimization. In: Geem, Z. (Ed.), Harmony Search

Algorithms for Structural Design Optimization. Studies in Computational Intelligence. Springer, Berlin/Heidelberg, pp. 159–198. Available from: <http://dx.doi.org/10.1007/978-3-642-03450-3_5/>.

Kennedy, J., Eberhart, R.C., 1995. Particle swarm optimization. In: Proceedings of IEEE International Conference on Neural Networks. pp. 1942–1948.

Kennedy, J., Eberhart, R.C., Shi, Y., 2001. Swarm Intelligence. Morgan Kaufmann.

Khajehzadeh, M., et al., 2010. Economic design of retaining wall using particle swarm optimization with passive congregation. Aust. J. Basic. Appl. Sci. 4 (11), 5500–5507.

Khajehzadeh, M., Taha, M.R., El-Shafie, A., 2011. Reliability evaluation of slopes using particle swarm optimization. In: Proceedings of the International Conference on Advanced Science, Engineering and Information Technology 2011. Malaysia, pp. 613–617.

Kirkpatrick, S., 1984. Optimization by simulated annealing: quantitative studies. J. Statist. Phys. 34 (5-6), 975–986.

Knabe, K., 2009. Constitutive models for subsoil in the context of structural analysis in construction engineering. In: Gurlebeck, K., Konke, C. (Eds.), Proceedings of the 18th International Conference on the Application of Computer Science and Mathematics in Architecture and Civil Engineering. Weimar, Germany.

Koh, A., 2009. A coevolutionary particle swarm algorithm for bi-level variational inequalities: applications to competition in highway transportation networks. In: Chiong, R., Dhakal, S. (Eds.), Natural Intelligence for Scheduling, Planning and Packing Problems. Studies in Computational Intelligence. Springer, Berlin/Heidelberg, pp. 195–217. Available from: <http://dx.doi.org/10.1007/978-3-642-04039-9_8/>.

Koh, B.H., Choi, J.H., Jeong, M.J., 2010. Damage detection through genetic and swarm-based optimization algorithms. In: Song, G., Malla, R.B. (Eds.), Proceedings of the Twelveth International Conference on Engineering, Science, Construction, and Operations in Challenging Environments. ASCE, Honolulu, HI. Available from: <http://link.aip.org/link/?ASC/366/215/1/> (accessed 07.01.11).

Kwok, N.M., et al., 2006. A novel hysteretic model for magnetorheological fluid dampers and parameter identification using particle swarm optimization. Sensors Actuators A. 132 (2), 441–451.

Le, T.S., 2010. Application of Stochastic Multi-Objective Particle Swarm Optimization for Sustainable Flexible Pavement Design. National Taiwan University of Science and Technology, Taipei, Taiwan.

Lei, Z., Ziyang, L., 2009. Dam safety monitoring forecast model based on CPSO. In: International Workshop on Intelligent Systems and Applications (ISA). Wuhan, China, pp. 1–4.

Li, L., Liu, F., 2009. Harmony particle swarm algorithm for structural design optimization. In: Geem, Z. (Ed.), Harmony Search Algorithms for Structural Design Optimization. Studies in Computational Intelligence. Springer, Berlin/Heidelberg, pp. 121–157. Available from: <http://dx.doi.org/10.1007/978-3-642-03450-3_4/>.

Li, L., Huang, Z., Liu, F., 2007a. An improved particle swarm optimizer for truss structure optimization. In: Wang, Y., Cheung, Y., Liu, H. (Eds.), Computational Intelligence and Security. Lecture Notes in Computer Science. Springer, Berlin/Heidelberg, pp. 1–10. Available from: <http://dx.doi.org/10.1007/978-3-540-74377-4_1/>.

Li, Z., Tan, H.-Z., Zhang, Y. 2007b. Particle swarm optimization applied to vertical traffic scheduling in buildings. In Proceedings of the Eleventh International conference, KES 2007 and XVII Italian Workshop on Neural Networks Conference on Knowledge-based Intelligent Information and Engineering Systems: Part I. KES'07/WIRN'07. Springer-Verlag, Berlin, Heidelberg, pp. 831–838. Available from: <http://portal.acm.org/citation.cfm?id = 1771110.1771223/>.

Li, L.J., Huang, Z.B., Liu, F., 2009a. A heuristic particle swarm optimization method for truss structures with discrete variables. Comput. Struct. 87 (7-8), 435–443.

Li, L et al., 2009b. The harmony search algorithm in combination with particle swarm optimization and its application in the slope stability analysis. In: International Conference on Computational Intelligence and Security (CIS'09). Beijing, China, pp. 133–136.

Li, L., et al., 2010. Discontinuous flying particle swarm optimization algorithm and its application to slope stability analysis. J. Cent. S. Univ. Technol. 17 (4), 852–856.

Liang, B.L., Tian, Y., 2010. A novel defect evaluation method for concrete structures in infrared based on ANN and PSO algorithm. Key Eng. Mat. 439, 552–557.

Lianyu, W.E.I., Jianfu, D.U., 2009. Research on OD matrix calculation based on quantum behaved particle swarm optimization algorithm. J. Softw. Eng. Appl. 2, 344–349.

Lin, L., et al., 2009. A study on the location selection of public traffic transfer hub based on improved particle swarm algorithm. In: Wang, Y., et al., (Eds.), Proceedings of the Ninth International Conference on Chinese Transportation Professionals. ASCE, Harbin, China, pp. 1–7.

Liu, Y., 2009. Automatic calibration of a rainfall-runoff model using a fast and elitist multi-objective particle swarm algorithm. Expert Syst. Appl. 36 (5), 9533–9538.

Liu, Q.M., Dong, M., 2011. Grey model based particle swarm optimization algorithm for fatigue strength prognosis of concrete. Adv Mater. Res. 148, 420–424.

Liu, W., Zhu, M., Wang, X., 2008. Hybrid particle swarm optimization applied to recovery scheduling of large-scale flight delays. In: Proceedings of the Fourth International Conference on Natural Computation (ICNC'08). Jinan, China, pp. 634–639.

Liu, X., Jiang, W., Xie, J., 2009. Vehicle routing problem with time windows: a hybrid particle swarm optimization approach. Fifth International Conference on Natural Computation, 2009. ICNC'09. pp. 502–506.

Lu, W.Z., Fan, H.Y., Lo, S.M., 2003. Application of evolutionary neural network method in predicting pollutant levels in downtown area of Hong Kong. Neurocomputing. 51, 387–400.

Lu, M., Lam, H.-C., Dai, F., 2008. Resource-constrained critical path analysis based on discrete event simulation and particle swarm optimization. Autom. Constr. 17 (6), 670–681.

Luo, Y., Uan, X., 2008. Global optimization for the synthesis of integrated water systems with particle swarm optimization algorithm. Chin. J. Chem. Eng. 16 (1), 11–15.

Lv, N., et al., 2010. Bi-level programming based contra flow optimization for evacuation events. Kybernetes. 39 (8), 1227–1234.

Ma, C., Wu, F., Lu, R., 2009. Decision-making method of hazardous material transportation route based on particle swarm optimization algorithm and neural network. In: Proceedings of the Pacific-Asia Workshop on Computational Intelligence and Industrial Application (PACIIA'08). Wuhan, China, pp. 1023–1027.

Marinakis, Y., Marinaki, M., 2010. A hybrid genetic—particle swarm optimization algorithm for the vehicle routing problem. Expert Syst. Appl. 37 (2), 1446–1455.

Marinakis, Y., Marinaki, M., Dounias, G., 2010. A hybrid particle swarm optimization algorithm for the vehicle routing problem. Eng. Appl. Artif. Intell. 23 (4), 463–472.

McCluskey, S., McCarthy, T.J., 2009. A particle swarm optimisation approach to reinforced concrete beam design according to AS3600. In: Topping, B.H.V., Tsompanakis, Y. (Eds.), Proceedings of ehe First International Conference on Soft Computing Technology in Civil, Structural and Environmental Engineering. Civil-Comp Press, Madeira, Portugal, pp. 1–14.

Meier, J., et al., 2008. Inverse parameter identification technique using PSO algorithm applied to geotechnical modeling. J. Artif. Evol. Appl. 3, 1–14.

Meng, X., Jia, L., Qin, Y., 2010. Train timetable optimizing and re-scheduling based on improved particle swarm algorithm. Proceedings of Transportation Research Board 89th Annual Meeting. Transportation Research Board, Washington, DC, p. 30.

Montalvo, I., et al., 2008a. A diversity-enriched variant of discrete PSO applied to the design of water distribution networks. Eng. Optimiz. 40 (7), 655–668.

Montalvo, I., et al., 2008b. Particle swarm optimization applied to the design of water supply systems. Comput Math. Appl. 56 (3), 769–776.

Montalvo, I., Izquierdo, J., Schwarze, S., et al., 2010a. Agent swarm optimization: a paradigm to tackle complex problems. Application to water distribution system design. In: Swayne, D.A., et al., (Eds.), 2010 International Congress on Environmental Modelling and Software Modelling for Environment's Sake. International Environmental Modelling and Software Society, Ottawa, Canada (iEMSs).

Montalvo, I., Izquierdo, J., Pérez-García, R., et al., 2010b. Improved performance of PSO with self-adaptive parameters for computing the optimal design of water supply systems. Eng. Appl. Artif. Intell. 23 (5), 727–735.

Montalvo, I., Izquierdo, J., Schwarze, S., et al., 2010c. Multi-objective particle swarm optimization applied to water distribution systems design: an approach with human interaction. Math. Comput. Model. 52 (7-8), 1219–1227.

Muleta, M.K., et al., 2006. Using genetic algorithms and particle swarm optimization for optimal design and calibration of large and complex urban stormwater management models. In: Graham, R. (Ed.), Proceedings of the World Environmental and Water Resources Congress 2006. ASCE, Omaha, NE. Available from: <http://link.aip.org/link/?ASC/200/113/1/> (accessed 07.01.2011).

Muller, I., Chang, F.K., 2009. Model-based impact monitoring by inverse methods using particle swarm optimization. Proceedings of the IMAC-XXVII. Society for Experimental Mechanics (SEM) Inc., Orlando, FL.

Naudet, V., et al., 2008. Estimation of water table from self-potential data using particle swarm optimization (PSO). SEG Tech. Program Expanded Abstr. 27 (*1*), 1203–1207.

Peng, L. et al., 2009. Isolation niches particle swarm optimization applied to traffic lights controlling. In: Proceedings of the 48th IEEE Conference on Decision and Control (CDC) held jointly with the 2009 28th Chinese Control Conference (CCC). Shanghai, China, pp. 3318–3322.

Perez, R.E., Behdinan, K., 2007. Particle swarm approach for structural design optimization. Comput. Struct. 85 (19), 1579–1588.

Plevris, V., Papadrakakis, M., 2011. A hybrid particle swarm—gradient algorithm for global structural optimization. Comput-Aided Civil Infrastr. Eng. 26 (1), 48–68.

Poli, R., 2008. Analysis of the publications on the applications of particle swarm optimisation. J. Artif. Evol. Appl.10p (Article ID 685175).

Potvin, J.-Y., 2009. A review of bio-inspired algorithms for vehicle routing. In: Pereira, F., Tavares, J. (Eds.), Bio-Inspired Algorithms for the Vehicle Routing Problem. Studies in Computational Intelligence. Springer, Berlin/ Heidelberg, pp. 1–34. Available from: <http://dx.doi.org/10.1007/978-3-540-85152-3_1/>.

Qing-Wei, Z., Ai-Ying, F., Zhi-Hai, X., 2009. Application of support vector regression and particle swarm optimization in traffic accident forecasting. 2009 International Conference on Information Management, Innovation Management and Industrial Engineering. pp. 188–191.

Rao, A., Anandakumar, G., 2007. Optimal placement of sensors for structural system identification and health monitoring using a hybrid swarm intelligence technique. Smart Mater. Struct. 16 (6), 2658–2672.

Reddy, M.J., Adarsh, S., 2010. Overtopping probability constrained optimal design of composite channels using swarm intelligence technique. J. Irrig. Drain. Eng. 136 (8), 532–542.

Reddy, M.J., Kumar, D.N., 2009. Performance evaluation of elitist-mutated multi-objective particle swarm optimization for integrated water resources management. J. Hydroinform. 11 (1), 79–88.

Reddy, M.J., Nagesh Kumar, D., 2007. Multi-objective particle swarm optimization for generating optimal trade-offs in reservoir operation. Hydrol. Process. 21 (21), 2897–2909.

Reeves, W.T., 1983. Particle systems—a technique for modeling a class of fuzzy objects. ACM Trans. Graphics. 2 (2), 91–108.

Rui, L., Xiaoya, W., 2008. Application of improved particle swarm optimization in construction contractors' selection and optimization. In: Proceedings of the Fourth International Conference on Wireless Communications, Networking and Mobile Computing (WiCOM'08). Dalian, China, pp. 1–4.

Schanz, T., et al., 2006. Identification of constitutive parameters for numerical models via inverse approach. Felsbau Rock Soil Eng. 24 (2), 11–21.

Scheerlinck, K., et al., 2009. Calibration of a water and energy balance model: recursive parameter estimation versus particle swarm optimization. Water Resour. Res. 45, 22.

Schmidt, A., 2010. The design of an active structural vibration reduction system using a modified particle swarm optimization. In: Dorigo, M., Birattari, M., Di Caro, G. (Eds.), Swarm Intelligence. Lecture Notes in Computer Science. Springer, Berlin/Heidelberg, pp. 544–551. Available from: <http://dx.doi.org/10.1007/978-3-642-15461-4_55/>.

Schmidt, A., Lewandowski, R., 2010. The design of an active seismic control system for a building using the particle swarm optimization. In: Rutkowski, L., et al., (Eds.), Artificial Intelligence and Soft Computing. Lecture Notes in Computer Science. Springer, Berlin/Heidelberg, pp. 651–658. Available from: <http://dx.doi.org/10.1007/978-3-642-13232-2_80/>.

Schutte, J.F., Groenwold, A.A., 2003. Sizing design of truss structures using particle swarms. Struct. Multidiscip. Optim. 25 (4), 261–269.

Seyedpoor, S.M., Gholizadeh, S., Talebian, S.R., 2010. An efficient structural optimisation algorithm using a hybrid version of particle swarm optimisation with simultaneous perturbation stochastic approximation. Civ. Eng. Environ. Syst. 27 (4), 295–313.

Seyedpoor, S.M., et al., 2011. Optimal design of arch dams subjected to earthquake loading by a combination of simultaneous perturbation stochastic approximation and particle swarm algorithms. Appl. Soft. Comput. 11 (1), 39–48.

Shanmugam, G., Ganesan, P., Vanathi, P.T., 2010. A hybrid particle swarm optimization with genetic operators for vehicle routing problem. J. Adv. Inf. Technol. 1 (4), 181–188.

Shao, Z.J., Gao, S.P., Wang, S.S., 2009. A hybrid particle swarm optimization algorithm for vehicle routing problem with stochastic travel time. In: Cao, B.Y., Zhang, C.Y., Li, T.F. (Eds.), Fuzzy Information and Engineering. Advances in Soft Computing. Springer, Berlin/Heidelberg, pp. 566–574. Available from: <http://dx.doi.org/10.1007/978-3-540-88914-4_70/>.

Shen, Y., Bu, Y., Yuan, M., 2009. A novel chaos particle swarm optimization (PSO) and its application in pavement maintenance decision. In: Proceedings of the Fourth IEEE Conference on Industrial Electronics and Applications (ICIEA 2009). Xi'an, China, pp. 3521–3526.

Shengwen, T., Xiucheng, G., Shengwu, T., 2008. Optimizing highway alignments based on improved particle swarm optimization and ArcGIS. In: Cohn, L.F. (Ed.), Proceedings of the First International Symposium on Transportation and Development Innovative Best Practices 2008. ASCE, Beijing, China, pp. 419–425. Available from: <http://link.aip.org/link/?ASC/319/69/1/> (accessed 07.01.11).

Shi, H., Li, W., 2010a. Integration of fuzzy logic, particle swarm optimization and neural networks in quality assessment of construction project. J. Softw. 5 (7), 737–744.

Shi, H., Li, W., 2010b. A web-based integrated system for construction project cost prediction. In: Luo, Q. (Ed.), Advancing Computing, Communication, Control and Management. Lecture Notes in Electrical Engineering. Springer, Berlin Heidelberg, pp. 31–38. Available from: < http://dx.doi.org/10.1007/978-3-642-05173-9_5/>.

Shi, Y., Eberhart, R., 1998. Parameter selection in particle swarm optimization. In: Evolutionary Programming VII: Lecture Notes in Computer Science. pp. 591–600.

Shourian, M., Mousavi, S., Tahershamsi, A., 2008. Basin-wide water resources planning by integrating PSO algorithm and MODSIM. Water Resour. Manage. 22 (10), 1347–1366.

Situ, B., Jin, W., Wei, M., 2010. Creating multi-routes combination bus timetable. In: Mao, B., et al., (Eds.), Proceedings of the Seventh International Conference on Traffic and Transportation Studies. ASCE, Kunming, China. Available from: <http://link.aip.org/link/?ASC/383/68/1/> (accessed 07.01.11).

Song, M.-P., Gu, G.-C., 2004. Research on particle swarm optimization: a review. In: Proceedings of 2004 International Conference on Machine Learning and Cybernetics. pp. 2236–2241.

Srinivasan, D., Loo, W.H., Cheu, R.L. 2003. Traffic incident detection using particle swarm optimization. In: Proceedings of the 2003 IEEE Swarm Intelligence Symposium (SIS'03). pp. 144–151.

Suribabu, C.R., Neelakantan, T.R., 2006. Particle swarm optimization compared to other heuristic search techniques for pipe sizing. J. Environ. Inform. 8 (1), 1–9.

Tayebi, N., Nejhad, F.M., Hassani, A., 2010. Analysis of pavement management activities programming by particle swarm optimization. In: Proceedings of the Fifth Civil Engineering Conference in the Asian Region and Australasian Structural Engineering Conference 2010. Sydney, Australia. Available from: <http://www.cecar5.com/abstract/452.asp/> (accessed 08.01.11).

Teodorovic, D., 2008. Swarm intelligence systems for transportation engineering: principles and applications. Trans. Res. Part C Emerg. Technol. 16 (6), 651–667.

Travassos, X.L., et al., 2008. Inverse algorithms for the GPR assessment of concrete structures. IEEE Trans. Magn. 44 (6), 994–997.

Tsai, H.-C., 2010. Predicting strengths of concrete-type specimens using hybrid multilayer perceptrons with center-unified particle swarm optimization. Expert Syst. Appl. 37 (2), 1104–1112.

Vairavamoorthy, K., Shen, Y., 2004. Least cost design of water distribution network using particle swarm optimization. In: Liong, S.-Y., Phoon, K.-K., Babovic, V. (Eds.), Proceedings of the Sixth International Conference on Hydroinformatics. pp. 834–841.

Wai, R.-J., Chuang, K.-L., 2008. On-line particle-swarm-optimization control for maglev transportation system via backstepping design procedure. IEEE International Conference on Systems, Man and Cybernetics (SMC). IEEE Press, Singapore, pp. 2355–2360.

Wai, R.J., Chuang, K.L., 2010. Design of backstepping particle-swarmoptimisation control for maglev transportation system. Control. Theory Appl. 4 (4), 625–645.

Wang, W., Feng, Q., 2008. Multi-objective optimization in construction project based on a hierarchical subpopulation particle swarm optimization algorithm. In: Proceedings of the Second International Symposium on Intelligent Information Technology Application (IITA'08). Shanghai, China, pp. 746–750.

Wang, Z., Luo, D., Huang, Z., 2007. Optimal coordination of artery system based on modified cell transmission model and particle swarm algorithm. In: Liu, R., Yang, D., Lu, J.J. (Eds.), Proceedings of the Seventh International Conference of Chinese Transportation Professionals Congress 2007. ASCE, Shanghai, China, pp. 500–511. Available from: <http://link.aip.org/link/?ASC/317/49/1/> (accessed 07.01.11).

Wang, K. et al., 2008. Particle swarm optimization for calibrating stream water quality model. In: Proceedings of the 2008 Second International Symposium on Intelligent Information Technology Application. IITA'08. IEEE Computer Society, Washington, DC, USA, pp. 682–686. Available from: <http://dx.doi.org/10.1109/IITA.2008.555/>.

Wang, J.-N., et al., 2010. Aircraft's assignment on multi-runway airport based on PSO. In: Wei, H., et al., (Eds.), Proceedings of the Tenth International Conference of Chinese Transportation Professionals. ASCE, Beijing, China. Available from: <http://link.aip.org/link/?ASC/382/129/1/> (accessed 07.01.11).

Wegley, C., Eusuff, M., Lansey, K., 2000. Determining pump operations using particle swarm optimization. In: Hotchkiss, R.H., Glade, M. (Eds.), Proceedings of Joint Conference on Water Resources Engineering and Water Resources Planning and Management. ASCE, Minneapolis, MN, p. 206. Available from: <http://link.aip.org/link/?ASC/104/206/1/> (accessed 07.01.11).

Wolpert, D.H., Macready, W.G., 2002. No free lunch theorems for optimization. IEEE Trans. Evol. Computat. 1 (1), 67–82.

Wu, J, et al., 2008. A PSO-based ant colony optimization for VRPTW. In: Liu, R., Zhang, J., Guan, C. (Eds.), Proceedings of the Eighth International Conference of Chinese Logistics and Transportation Professionals. ASCE, Chengdu, China, pp. 2082–2088. Available from: <http://link.aip.org/link/?ASC/330/306/1/> (accessed 07.01.11).

Xianjia, W., Yuan, H., Wuyue, Z., 2009. The bilevel programming model of earthwork allocation system. In: Shi, Y., et al., (Eds.), Cutting-Edge Research Topics on Multiple Criteria Decision Making. Communications in Computer and Information Science. Springer, pp. 275–281. Available from: <http://dx.doi.org/10.1007/978-3-642-02298-2_42/>.

Xiaobin, L., 2009. RBF neural network optimized by particle swarm optimization for forecasting urban traffic flow. In: Third International Symposium on Intelligent Information Technology Application (IITA 2009). Nanchang, China, pp. 124–127.

Xing, J., Jiang, A., Qiu, J., 2010. Studying the adaptive control of tunnel excavation based on numerical simulation and particle swarm optimization. In: Wei, H., et al., (Eds.), Proceedings of the Tenth International Conference of Chinese Transportation Professionals. ASCE, Beijing, China, p. 334. Available from: <http://link.aip.org/link/?ASC/382/334/1/> (accessed 07.01.11).

Xiujuan, L., Ali, F., Zhongke, S., 2008. The aircraft departure scheduling based on second-order oscillating particle swarm optimization algorithm. IEEE Congress on Evolutionary Computation, 2008. CEC 2008. (IEEE World Congress on Computational Intelligence). pp. 1399–1403.

Xu, J., Zeng, Z., 2010. Applying optimal control model to dynamic equipment allocation problem: a case study of concrete-faced rockfill dam construction project. J. Constr. Eng. Manage.

Xu, M., Yang, J., Gao, Z., 2010. Particle swarm optimization algorithm in transport continuous network design problems. In: Proceedings of the 2010 Third International Joint

Conference on Computational Science and Optimization, vol. 02. CSO'10. IEEE Computer Society, Washington, DC, USA, pp. 513–517. Available from: <http://dx.doi.org/10.1109/CSO.2010.53/>.

Yang, I.-T., 2007a. Performing complex project crashing analysis with aid of particle swarm optimization algorithm. Int. J. Proj. Manage. 25 (6), 637–646.

Yang, I.-T., 2007b. Using elitist particle swarm optimization to facilitate bicriterion time-cost trade-off analysis. J. Constr. Eng. Manage. 133 (7), 498–505.

Yang, I.-T., 2008. Distribution-free monte carlo simulation: premise and refinement. J. Constr. Eng. Manage. 134 (5), 352–360.

Yang, I.-T., Hsu, Y.-S., 2010. Risk-based multiobjective optimization model for bridge maintenance planning. In: Lu, J.W.Z., et al. (Eds.), Proceedings of the Second International Symposium on Computational Mechanics and the Twelfth International Conference on the Enhancement and Promotion of Computational Methods in Engineering and Science. pp. 477–482. Available from: <http://link.aip.org/link/?APC/1233/477/1/> (accessed 08.01.11).

Yang, K., Zhai, J., 2009. Particle swarm optimization algorithms for optimal scheduling of water supply systems. In: 2009 Second International Symposium on Computational Intelligence and Design (ISCID'09). pp. 509–512.

Yang, X.S., Koziel, S. (Eds.), 2011. Computational Optimization and Applications in Engineering and Industry. Springer, Germany.

Yi, P., Xu, C., Liu, J., 2010. The effect of nonlinear strength of rockfill material on dam's stability. In: Song, G., Malla, R.B. (Eds.), Proceedings of the Twelfth International Conference on Engineering, Science, Construction, and Operations in Challenging Environments. ASCE, Honolulu, HI. Available from: <http://link.aip.org/link/?ASC/366/294/1/> (accessed 07.01.11).

Yin, L. & Liu, X., 2009. A single–depot complex vehicle routing problem and its PSO solution. In: Proceedings of the Second Symposium International Computer Science and Computational Technology (ISCSCT'09). Huangshan, China, pp. 266–269.

Zakermoshfegh, M., Neyshabouri, S., Lucas, C., 2008. Automatic calibration of lumped conceptual rainfall-runoff model using particle swarm optimization. J. Appl. Sci. 8 (20), 3703–3708.

Zhang, H., Li, H., 2010. Multi-objective particle swarm optimization for construction time-cost tradeoff problems. Constr. Manage. Econ. 28 (1), 75–88.

Zhang, H., Wang, J.Y., 2008. Particle swarm optimization for construction site unequal-area layout. J. Constr. Eng. Manage. 134 (9), 739–748.

Zhang, H., Xing, F., 2010. Fuzzy-multi-objective particle swarm optimization for time-cost-quality tradeoff in construction. Autom. Constr. 19 (8), 1067–1075.

Zhang, L., Cui, G., 2009. Automatic calibration of a hydrological model using multiobjective particle swarm optimization and TOPSIS. Proceedings of the 2009 WRI World Congress on Computer Science and Information Engineering. IEEE Computer Society, Washington, DC, pp. 617–621. Available from: <http://portal.acm.org/citation.cfm?id = 1579194.1579725/>.

Zhang, H., et al., 2005. Particle swarm optimization-based schemes for resource-constrained project scheduling. Autom. Constr. 14 (3), 393–404.

Zhang, H., Li, H., Tam, C.M., 2006a. Particle swarm optimization for preemptive scheduling under break and resource-constraints. J. Constr. Eng. Manage. 132 (3), 259–267.

Zhang, H., Li, H., Tam, C.M., 2006b. Particle swarm optimization for resource-constrained project scheduling. Int. J. Project Manage. 24 (1), 83–92.

Zhang, H., et al., 2006c. Particle swarm optimization-supported simulation for construction operations. J. Constr. Eng. Manage. 132 (12), 1267–1274.

Zhang, H., Li, H., Tam, C.M., 2006d. Permutation-based particle swarm optimization for resource-constrained project scheduling. J. Comput. Civil Eng. 20 (2), 141–149.

Zhang, H., Tam, C.M., Li, H., 2006e. Multimode project scheduling based on particle swarm optimization. Comput-Aided Civil Infrastruct. Eng. 21 (2), 93–103.

Zhang, Y., Gallipoli, D., Augarde, C., 2009a. Parallel hybrid particle swarm optimization and applications in geotechnical engineering. In: Cai, Z., et al., (Eds.), Advances in Computation and Intelligence. Lecture Notes in Computer Science. Springer, Berlin/Heidelberg, pp. 466–475. Available from: <http://dx.doi.org/10.1007/978-3-642-04843-2_49/>.

Zhang, Y., Gallipoli, D., Augarde, C.E., 2009b. Simulation-based calibration of geotechnical parameters using parallel hybrid moving boundary particle swarm optimization. Comput. Geotech. 36 (4), 604–615.

Zhao, H., Ru, Z., 2008. Construction schedule optimization using particle swarm optimization. In: Proceedings of the Fourth International Conference on Wireless Communications, Networking and Mobile Computing (WiCOM'08). pp. 1–4.

Zhao, H.-B., Yin, S., 2009. Geomechanical parameters identification by particle swarm optimization and support vector machine. Appl. Math. Model. 33 (10), 3997–4012.

Zhao, J. et al., 2006. Urban traffic flow forecasting model of double rbf neural network based on PSO. Sixth International Conference on Intelligent Systems Design and Applications, 2006. ISDA'06. pp. 892–896.

Zhao, H.B., Zou, Z.S., Ru, Z.L., 2008a. Chaotic particle swarm optimization for non-circular critical slip surface identification in slope stability analysis. In: Cai, M., Wang, J. (Eds.), Boundaries of Rock Mechanics: Recent Advances and Challenges for the 21st Century. Taylor & Francis, pp. 585–588.

Zhao, J., Gao, H., Jia, L., 2008b. Traffic flow forecasting model of typical multi-intersection for urban trunk road based on dissimilation particle swarm optimization. Proceedings of the 2008 Congress on Image and Signal Processing. IEEE Computer Society, Washington, DC, pp. 127–131. Available from: <http://portal.acm.org/citation.cfm?id = 1446300.1446945/>.

Zhen-zhong, S., et al., 2010. Deformation prediction model of concrete arch dam based on improved particle swarm optimization algorithm. In: Song, G., Malla, R.B. (Eds.), Proceedings of the Twelfth International Conference on Engineering, Science, Construction, and Operations in Challenging Environments. ASCE, Honolulu, HI, p. 45. Available from: <http://link.aip.org/link/?ASC/366/45/1/> (accessed 06.01.2011).

Zhu, Q. et al., 2006. An improved particle swarm optimization algorithm for vehicle routing problem with time windows. In: IEE Congress on Evolutionary Computation (CEC). Vancouver, BC, pp. 1386–1390.

Zimmerer, M.M., Schanz, T., 2006. Determination of soil parameters for modeling deep excavations utilizing an inverse approach. In: Triantafyllidis, T. (Ed.), Numerical Modelling of Construction Processes in Geotechnical Engineering for Urban Environment: Proceedings of the International Conference on Numerical Simulation of Construction Processes in Geotechnical Engineering. Taylor & Francis, pp. 21–28.

Part One

Structural Design

4 Evolution Strategies-Based Metaheuristics in Structural Design Optimization

Chara Ch. Mitropoulou[1], *Yiannis Fourkiotis*[1], *Nikos D. Lagaros*[1] *and Matthew G. Karlaftis*[2]

[1]Institute of Structural Analysis and Seismic Research, National Technical University of Athens, Zografou Campus, Athens, Greece, [2]Department of Transportation Planning and Engineering, National Technical University of Athens, Zografou Campus, Athens, Greece

4.1 Introduction

During the last three decades, many numerical methods have been developed to meet the demands of structural design optimization. These methods can be classified in two categories, the deterministic methods and the probabilistic methods. Mathematical programming methods are the most popular methods of the first category and in particular the gradient-based optimizers. These methods make use of local curvature information, derived from linearization of the objective and constraint functions by using their derivatives with respect to the design variables at points obtained in the process of optimization, to construct an approximate model of the initial problem. Heuristic and metaheuristic algorithms are nature-inspired or bioinspired as they have been developed based on the successful evolutionary behavior of natural systems by learning from nature. These methods belong to the probabilistic category of methods. Modern metaheuristic algorithms for engineering optimization include genetic algorithms (GAs) (Holland, 1975), simulated annealing (SA) (Kirkpatrick et al., 1983), particle swarm optimization (PSO) (Kennedy and Eberhart, 1995), ant colony optimization (ACO) algorithm (Dorigo and Stützle, 2004), artificial bee colony (ABC) algorithm (Bozorg Haddad et al., 2005), harmony search (HS) algorithm (Geem et al., 2001), cuckoo search (CS) algorithm (Yang and Deb, 2010), firefly algorithm (FA) (Yang, 2008), bat algorithm (BA) (Yang and Gandomi, 2012), krill herd (KH) algorithm (Gandomi and Alavi, 2012), and many others. Evolutionary algorithms (EAs) are the most widely used class of metaheuristic algorithms and include evolutionary programming (EP) (Fogel, 1992), GAs (Goldberg, 1989; Holland, 1975), evolution strategies (ES) (Rechenberg, 1973; Schwefel, 1981), and genetic programming (GP) (Koza, 1992).

Metaheuristic Applications in Structures and Infrastructures. DOI: http://dx.doi.org/10.1016/B978-0-12-398364-0.00004-8

Gradient-based optimizers can quickly capture the right path to the nearest optimum, irrespective of whether it is a local or a global optimum, but it cannot ensure that the global optimum can be found. On the other hand, metaheuristics, due to their random search, are being considered more robust in terms of global convergence; they may suffer, however, from a slow rate of convergence toward the global optimum. When metaheuristics are adopted to perform the optimization, the solution of the finite element equations is of paramount importance since more than 95% of the total computing time is spent for the solution of the finite element equilibrium equations (Papadrakakis et al., 1999). A second characteristic is that in place of a single design point, metaheuristics work simultaneously with a population of design points in the space of design variables. This allows for a natural implementation of the evolution procedure in parallel computer environments.

In single-objective optimization problems, the optimal solution is usually clearly defined since it is the minimum or maximum value of the objective function. This does not hold in real-world problems where multiple and conflicting objectives frequently exist. Instead of a single optimal solution, there is usually a set of alternative solutions, generally denoted as the set of Pareto optimal solutions. These solutions are optimal in the wider sense since no other solution in the search space is superior to them when all objectives are considered. In the absence of preference information, none of the corresponding trade-offs can be said to be better than the others. On the other hand, the search space can be too large and too complex, which is the usual case with real-world problems, hence the implementation of gradient-based optimizers for these types of problems becomes even more cumbersome. Thus, efficient optimization strategies are required to deal with the presence of multiple objectives and the complexity of the search space. Metaheuristics, in particular EA, have several characteristics that are desirable for these kinds of problems and most frequently outperform the deterministic optimizers such as gradient-based optimization algorithms. In sizing optimization, the aim is to minimize the objective function, which is usually the weight or the cost of the structure under certain restrictions imposed by the design codes when the characteristics of the cross sections of the members are under investigation.

The objective of this study is to assess the performance of metaheuristic optimization when implemented for the design of 3D steel structures having single and multiple objectives. These problems are formulated as sizing design optimization, where the cross-sectional dimensions of the structural elements constitute the sizing design variables. In particular, this chapter is concerned with the structural optimization of skeletal truss structures under static loading conditions with single and multiple objectives. Combinatorial optimization methods, and in particular algorithms based on ES, are implemented for the solution of both type of problems.

4.2 Literature Survey

In the past, a number of studies have been published where structural optimization problems with single and multiple objectives are solved using metaheuristics.

A small survey can be found in the studies presented in this section. Perez and Behdinan (2007) presented the background and implementation of a PSO algorithm suitable for constraint structural optimization problems, while improvements that affect the setting parameters and functionality of the algorithm were shown. Hasançebi (2008) investigated the computational performance of adaptive ES in large-scale structural optimization. Bureerat and Limtragool (2008) presented the application of SA for solving structural topology optimization, while a numerical technique, known as multiresolution design variables, was proposed as a numerical tool to enhance the searching performance. Hansen et al. (2008) introduced an optimization approach based on an ES that incorporates multiple criteria by using nonlinear finite element analyses for stability and a set of linear analyses for damage-tolerance evaluation; the applicability of the approach was presented for the window area of a generic aircraft fuselage. Kaveh and Shahrouzi (2008) proposed a hybrid strategy combining indirect information share in ant systems with direct constructive genetic search; for this purpose, some proper coding techniques were employed to enable testing the method with various sets of control parameters. Farhat et al. (2009) proposed a systematic methodology for determining the optimal cross-sectional areas of buckling restrained braces used for the seismic upgrading of structures against severe earthquakes; for this purpose, single and multiobjective optimization problems were formulated. Chen and Chen (2009) proposed modified ES for solving mixed-discrete optimization problems; in particular, three approaches were proposed for handling discrete variables. Gholizadeh and Salajegheh (2009) proposed a new meta-modeling framework that reduces the computational burden of the structural optimization against the time history loading; for this purpose, a meta-model consisting of an adaptive neuro-fuzzy inference system, subtractive algorithm, self-organizing map, and a set of radial basis function networks were used to accurately predict the time history responses of structures. Wang et al. (2010) studied an optimal cost base isolation design or retrofit design method for bridges subject to transient earthquake loads. Hasançebi et al. (2010) utilized metaheuristic techniques like GAs, SA, ES, PSO, tabu search, ant colony optimization, and HS in order to develop seven optimum design algorithms for real-size rigidly connected steel frames. Manan et al. (2010) employed four different biologically inspired optimization algorithms (binary GA, continuous GA, PSO, and ant colony optimization) and a simple meta-modeling approach on the same problem set. Gandomi and Yang (2011) provided an overview of structural optimization problems of both truss and nontruss cases. Martínez et al. (2011) described a methodology for the analysis and design of reinforced concrete tall bridge piers with hollow rectangular sections, which are typically used in deep valley bridge viaducts. Kripakaran et al. (2011) presented computational approaches that can be implemented in a decision support system for the design of moment-resisting steel frames, while trade-off studies were performed using GAs to evaluate the savings due to the inclusion of the cost of connections in the optimization model. Gandomi et al. (2011a) used the CS method for solving structural optimization problems; furthermore, for the validation against structural engineering optimization problems, the CS method was applied to 13 design problems taken from the literature. Kunakote and Bureerat (2011) dealt with the comparative performance of some established multiobjective EAs for structural topology optimization; four multiobjective problems, having design objectives like structural compliance, natural frequency and mass, and subjected to constraints on stress, were used for performance testing. Su et al. (2011) used a GA to handle topology and sizing optimization of truss structures, in which a sparse node matrix encoding approach is used and individual identification technique is employed to avoid duplicate structural analysis to save computation time. Gandomi et al. (2011b) used an FA for solving mixed continuous/discrete structural optimization problems; the results of a trade study carried out on six classical structural optimization problems taken from the literature confirm the validity of the proposed algorithm. Degertekin (2012) proposed two improved HS algorithms for sizing optimization of truss structures, while four truss structure weight minimization problems were presented to demonstrate the robustness of the proposed algorithms. The main part of the work by Muc and Muc-Wierzgoń (2012) was devoted to the definition of design variables and the forms of objective functions for multilayered plated and shell structures, while the ES method was used as the optimization algorithm. Comparative studies of metaheuristics on engineering problems can be found in two recent studies by the authors (Lagaros and Karlaftis, 2011; Lagaros and Papadrakakis, 2012) and in the edited book by Yang and Koziel (2011).

4.3 The Structural Optimization Problem

Structural optimization problems are characterized by various objective and constraint functions that are generally nonlinear functions of the design variables. These functions are usually implicit, discontinuous, and nonconvex. For the mathematical formulation of structural optimization problems with respect to the design variables, the objective and constraint functions depend on the type of the application. However, all optimization problems can be expressed in standard mathematical terms as a nonlinear programming problem (NLP), which in general form can be stated as follows:

$$\begin{aligned} &\min && F(\mathbf{s}) \\ &\text{subject to} && g_j(\mathbf{s}) \leq \quad j=1,\ldots,k \\ & && s_i^{\text{low}} \leq s_i \leq s_i^{\text{up}}, \quad i=1,\ldots,n \end{aligned} \tag{4.1}$$

where $\mathbf{s}$ is the vector of design variables, $F(\mathbf{s})$ is the objective function to be minimized, $g_j(s)$ are the behavioral constraints, and s_i^{low} and s_i^{up} are the lower and the upper bounds of the ith design variable.

There are mainly three classes of structural optimization problems: sizing, shape and topology, and layout. Initially, structural optimization was focused on sizing optimization, such as optimizing cross-sectional areas of truss and frame structures, or the thickness of plates and shells. The next step was to consider finding optimum boundaries of a structure, and therefore to optimize its shape. In the former case, the structural domain is fixed, while in the latter case, it is not fixed but it has a predefined topology. In both cases, a nonoptimal starting topology can lead to suboptimal results. To overcome this deficiency, structural topology optimization needs to be employed, which allows the designer to optimize the layout or the topology of a structure by detecting and removing the low-stressed material in the structure which is not used effectively.

4.3.1 Sizing Optimization

In sizing optimization problems, the aim is usually to minimize the weight of the structure under certain behavioral constraints on stresses and displacements. The design variables are most frequently chosen to be dimensions of the cross-sectional areas of the members of the structure. Due to engineering practice demands, the members are divided into groups having the same design variables. This linking of elements results in a trade-off between the use of more material and the need for symmetry and uniformity of structures due to practical considerations. Furthermore, it has to be taken into account that due to fabrication limitations, the design variables are not continuous but discrete since cross sections belong to a certain set. A discrete structural optimization problem can be formulated in the following form:

$$\begin{array}{ll} \min & F(\mathbf{s}) \\ \text{subject to} & g_j(\mathbf{s}) \leq 0 \quad j = 1, \ldots, k \\ & s_i \in R^d \quad i = 1, \ldots, n \end{array} \tag{4.2}$$

where R^d is a given set of discrete values representing the available structural member cross sections and design variables s_i ($i = 1, \ldots, n$) can take values only from this set.

The sizing optimization methodology proceeds with the following steps:

1. At the outset of the optimization, the geometry, the boundaries, and the loads of the structure under investigation, have to be defined.
2. The design variables, which may or may not be independent to each other, are also properly selected. Furthermore, the constraints are also defined in this stage in order to formulate the optimization problem as in Eq. (4.2).
3. A finite element analysis is then carried out and the displacements and stresses are evaluated.
4. If a gradient-based optimizer is used, then the sensitivities of the constraints and the objective function to small changes of the design variables are computed.
5. The design variables are being optimized. If the convergence criteria for the optimization algorithm are satisfied, then the optimum solution has been found and the process is terminated, or else the optimizer updates the design variable values and the whole process is repeated from Step 3.

4.3.2 Shape Optimization

In structural shape optimization problems, the aim is to improve the performance of the structure by modifying its boundaries. This can be numerically achieved by minimizing an objective function subjected to certain constraints (Hinton and Sienz, 1994; Ramm et al., 1994). All functions are related to the design variables, which are some of the coordinates of the key points in the boundary of the structure. The shape optimization methodology proceeds with the following steps:

1. At the outset of the optimization, the geometry of the structure under investigation has to be defined. The boundaries of the structure are modeled using cubic B-splines that, in turn, are defined by a set of key points. Some of the coordinates of these key points will be the design variables, which may or may not be independent to each other.
2. An automatic mesh generator is used to create a valid and complete finite element model. A finite element analysis is then carried out and the displacements and stresses are evaluated. In order to increase the accuracy of the analysis, an h-type adaptivity analysis may be incorporated in this stage.
3. If a gradient-based optimizer is used, then the sensitivities of the constraints and the objective function to small changes of the design variables are computed either with the finite difference or with the semianalytical method.
4. The optimization problem is solved; the design variables are being optimized and the new shape of the structure is defined. If the convergence criteria for the optimization algorithm are satisfied, then the optimum solution has been found and the process is terminated, or else a new geometry is defined and the whole process is repeated from Step 2.

4.3.3 Topology Optimization

Structural topology optimization assists the designer in defining the type of structure, which is best suited to satisfying the operating conditions for the problem in question. It can be seen as a procedure of optimizing the rational arrangement of the available material in the design space and eliminating the material that is not needed. Topology optimization is usually employed in order to achieve an acceptable initial layout of the structure, which is then refined with a shape optimization tool. The topology optimization procedure proceeds step by step with a gradual "removal" of small portions of low-stressed material, which are being used inefficiently.

Many researchers have presented solutions for structural topology optimization problems. Topological or layout optimization can be undertaken by employing one of the following main approaches (Hinton and Sienz, 1993):

1. Ground structure approach (Shieh, 1994)
2. Homogenization method (Bendsoe and Kikuchi, 1988; Suzuki and Kikuchi, 1993)
3. Bubble method (Eschenauer et al., 1993)
4. Fully stressed design technique (Xie and Steven, 1993).

The first three approaches have several things in common. They are optimization techniques with an objective function, design variables, and constraints, and they solve the optimization problem by using an algorithm based either on a sequential quadratic programming (approach 1) or on an optimality criterion concept (approaches 2 and 3). The fully stressed design technique, on the other hand, although not an optimization algorithm in the conventional sense, proceeds by removing inefficient material, and therefore optimizes the use of the remaining material in the structure, in an evolutionary process.

The domain of the structure, which is called the reference domain, can be divided into the design domain and the nondesign domain. The nondesign domain covers regions with stress concentrations, such as supports and areas where loads are applied, and therefore it cannot be modified throughout the whole topology optimization process. After the generation of the finite element mesh, the evolutionary fully stressed design cycle is activated, where a linear elastic finite element analysis is carried out. The maximum principal stress for each element can be computed, which for convenience is called stress level and is denoted as σ_{evo}. The maximum stress level σ_{max} of the elements in the structure at the current optimization step is defined, and all elements that fulfill the condition:

$$\sigma_{evo} < \text{ratre} \times \sigma_{max} \tag{4.3}$$

are removed, or *switched-off*, where ratre is the rejection rate parameter (Xie and Steven, 1994). The elements are removed by assigning them a relatively small elastic modulus, which is typically:

$$E_{off} = 10^{-5} \times E_{on} \tag{4.4}$$

In this way, the elements switched-off virtually do not carry any load and their stress levels are accordingly small in subsequent analyses. This strategy is called "hard kill," since the low-stressed elements are immediately removed, in contrast with the "soft kill" method where the elastic modulus varies linearly and the elements are removed more gradually. The remaining elements are considered *active* and they are sorted in ascending order according to their stress levels before a subsequent analysis is performed. The iterative process of element removal and addition, if element growth is allowed, is continued until one of several specified convergence criteria are met:

1. All stress levels are larger than a certain percentage value of the maximum stress. This criterion assumes that a fully stressed design has been achieved and the material is used efficiently.
2. The number of active elements is smaller than a specified percentage of the total number of elements. For uniform meshes, which are commonly used in topology optimization problems, this criterion is equivalent to an area or volume fraction of the initial design, which will be in use in the final layout.
3. When element growth is allowed, the evolutionary process is completed when more elements are switched-on than they are switched-off.

4.4 Problem Formulations

The general formulations of the single and the multiobjective problems are presented in this part of the study.

4.4.1 Single-Objective Structural Optimization

In sizing optimization problems, the aim is to minimize a single-objective function, usually the weight or the cost of the structure, under certain behavioral constraints on stress and displacements. The design variables are most frequently chosen to be dimensions of the cross-sectional areas of the members of the structure. Due to fabrication limitations, the design variables are not continuous but discrete since cross sections belong to a certain set. A discrete structural optimization problem can be formulated in the form of Eq. (4.2). In the optimal design of frames, the constraints are the member stresses and nodal displacements or interstory drifts, while in the optimal design of trusses, the constraints are the member stresses, nodal displacements, or frequencies.

4.4.2 Multiobjective Structural Optimization

In practical applications of structural optimization, the material weight or the structural cost rarely give a representative measure of the performance of the structure. In fact, several conflicting and incommensurable criteria usually exist in real-life design problems that have to be dealt with simultaneously. This situation forces the designer to look for a good compromise design between the conflicting requirements. These kinds of problems are called optimization problems with many objectives. The consideration of multiobjective optimization in its present sense originated toward the end of the nineteenth century when Pareto presented the optimality concept in economic problems with several competing criteria (Pareto, 1897). Since then, although many techniques have been developed in order to deal with multiobjective optimization problems, the corresponding applications were confined strictly to mathematical functions. The first applications in the field of structural optimization with multiple objectives appeared at the end of the 1970s.

4.4.2.1 Criteria and Conflict

The designer looking for the optimum design of a structure is faced with the question of selecting the most suitable criteria for measuring the economy, the strength, the serviceability, or any other factor that affects the performance of a structure. Any quantity that has a direct influence on the performance of the structure can be considered as a criterion. On the other hand, those quantities that must satisfy some imposed-only requirements are not criteria but they can be treated as constraints. Most of the structural optimization problems are treated with one single objective, usually the weight of the structure, subjected to some strength constraints. These constraints are set as equality or inequality constraints using some upper and lower limits. When there is a difficulty in selecting these limits, then these parameters are better treated as criteria.

One important basic property in the multicriterion formulation is the conflict that may or may not exist between the criteria. Only those quantities that are competing should be treated as independent criteria, whereas the others can be combined into a single criterion to represent the whole group. The local conflict between two criteria can be defined as follows: The functions f_i and f_j are called locally collinear with no conflict at point s if there is $c > 0$ such that $\nabla f_i(s) = c\nabla f_j(s)$. Otherwise, the functions are called locally conflicting at s. According to this definition, any two criteria are locally conflicting at a point of the design space if their maximum improvement is achieved in different directions. The global conflict between two criteria can be defined as follows:

> The functions f_i and f_j are called globally conflicting in the feasible region F of the design space when the two optimization problems $\min_{s \in F} f_i(s)$ and $\min_{s \in F} f_j(s)$ have different optimal solutions.

4.4.2.2 Formulation of a Multiple Objective Optimization Problem

In formulating an optimization problem, the choice of the design variables, criteria, and constraints represents undoubtedly the most important decision made by the engineer. In general, the mathematical formulation of a multiobjective problem includes a set of n design variables, a set of m objective functions, and a set of k constraint functions and can be defined as follows:

$$\begin{aligned} \min_{s \in F} \quad & [f_1(\mathbf{s}), f_2(\mathbf{s}), \ldots, f_m(\mathbf{s})]^{\mathrm{T}} \\ \text{subject to} \quad & g_j(\mathbf{s}) \leq 0 \quad j = 1, \ldots, k \\ & s_i \in R^d \quad i = 1, \ldots, n \end{aligned} \tag{4.5}$$

where the vector $\mathbf{s} = [s_1, s_2, \ldots, s_n]^{\mathrm{T}}$ represents a design variable vector and F is the feasible set in design space R^n. It is defined as the set of design variables that satisfy the constraint functions $g(s)$ in the form:

$$F = \{\mathbf{s} \in R^n | g(\mathbf{s}) \leq 0\} \tag{4.6}$$

Usually, there exists no unique point that would give an optimum for all m criteria simultaneously. Thus, the common optimality condition used in single-objective optimization must be replaced by a new concept, the so-called Pareto optimum.

A design vector $\mathbf{s}^* \in F$ is Pareto optimal for the problem of Eq. (4.5) if and only if there exists no other design vector $\mathbf{s} \in F$ such that:

$$f_i(\mathbf{s}) \leq f_i(\mathbf{s}^*) \quad \text{for } i = 1, 2, \ldots, m \tag{4.7}$$

with $f_j(s) < f_j(s^*)$ for at least one objective j. The solutions of optimization problems with multiple objectives constitute the set of the Pareto optimum solutions. The problem of Eq. (4.5) can be regarded as being solved after the set of Pareto optimal solutions has been determined. In practical applications, however, the designer seeks for a unique final solution. Thus, a compromise should be made among the available Pareto optimal solutions.

4.5 Metaheuristics

Nature has been solving various problems over millions or even billions of years. Only the best and the most robust solutions remain based on the principle of the survival of the fittest. Similarly, heuristic algorithms use the trial and error, learning, and adaptation to solve problems. Modern metaheuristic algorithms are almost guaranteed an efficient performance for a wide range of combinatorial optimization problems. The main aim of research in optimization and algorithm development is to design and/or choose the most suitable and efficient algorithms for a given optimization problem. In this section, metaheuristics based on ES implemented for solving single and multiobjective optimization problems are presented.

4.5.1 *Solving the Single-Objective Optimization Problems*

Three metaheuristic optimization algorithms are tested in this chapter, and they appear to be very promising as they have been implemented in various challenging problems with success. We present here a short description of the three algorithms used.

4.5.1.1 Evolution Strategies

ES are population-based probabilistic direct search optimization algorithms gleaned from principles of Darwinian evolution. Starting with an initial population of μ candidate designs, an offspring population of λ designs is created from the parents using variation operators. Depending on the manner in which the variation and selection operators are designed and the spaces in which they act, different classes of ES have been proposed. In the ES algorithm employed in this study (Rechenberg, 1973; Schwefel, 1981), each member of the population is equipped with a set of parameters:

$$\begin{aligned} \mathbf{a} &= [(\mathbf{s}_d, \boldsymbol{\gamma}), (\mathbf{s}_c, \boldsymbol{\sigma}, \boldsymbol{\alpha})] \in (I_d, I_c) \\ I_d &= D^{n_d} \times R_+^{n_\gamma} \\ I_c &= R^{n_c} \times R_+^{n_\sigma} \times [-\pi, \pi]^{n_a} \end{aligned} \tag{4.8}$$

where $\mathbf{s}_d$ and $\mathbf{s}_c$ are the vectors of discrete and continuous design variables defined in the discrete and continuous design sets D^{n_d} and R^{n_c}, respectively. Vectors $\boldsymbol{\gamma}$, $\boldsymbol{\sigma}$, and $\boldsymbol{\alpha}$ are the distribution parameter vectors taking values in $R_+^{n_\gamma}$, $R_+^{n_\sigma}$, and $[-\pi, \pi]^{n_a}$, respectively. Vector $\boldsymbol{\gamma}$ corresponds to the variances of the Poisson distribution. Vector $\boldsymbol{\sigma} \in R_+^{n_\sigma}$ corresponds to the standard deviations $(1 \leq n_\sigma \leq n_c)$ of the normal distribution. Vector $\boldsymbol{\alpha} \in [-\pi, \pi]^{n_a}$ is related to the inclination angles $(n_\alpha = (n_c - n_\sigma/2)(n_\sigma - 1))$ defining linearly correlated mutations of the continuous design variables $\mathbf{s}_c$, where $n = n_d + n_c$ is the total number of design variables.

Let $P(t) = \{a_1,\ldots,a_\mu\}$ denote a population of individuals at the tth generation. The genetic operators used in the ES method are denoted by the following mappings:

$$\begin{aligned} &\text{rec}\ :(I_d,I_c)^\mu \rightarrow (I_d,I_c)^\lambda \ \text{(recombination)} \\ &\text{mut}:(I_d,I_c)^\lambda \rightarrow (I_d,I_c)^\lambda \ \text{(mutation)} \\ &\text{sel}_\mu^k:(I_d,I_c)^k \rightarrow (I_d,I_c)^\mu \ (\text{selection}, k\in\{\lambda,\mu+\lambda\}) \end{aligned} \tag{4.9}$$

A single iteration of the ES, which is a step from the population $P_p^{(t)}$ to the next parent population $P_p^{(t+1)}$ is modeled by the mapping:

$$\text{opt}_{\text{EA}}: (I_d,\ I_c)_t^\mu \rightarrow (I_d,\ I_c)_{t+1}^\mu \tag{4.10}$$

4.5.1.2 *Covariance Matrix Adaptation*

The covariance matrix adaptation (CMA), proposed by Hansen and Ostermeier (2001), is a completely derandomized self-adaptation scheme. First, the covariance matrix of the mutation distribution is changed in order to increase the probability of producing the selected mutation step again. Second, the rate of change is adjusted according to the number of strategy parameters to be adapted. Third, under random selection the expectation of the covariance matrix is stationary. Further, the adaptation mechanism is inherently independent of the given coordinate system. The transition from generation g to $g+1$, given in the following steps, completely defines the algorithm (Figure 4.1).

4.5.1.2.1 Generation of Offsprings

Creation of λ new offsprings as follows:

$$s_k^{(g+1)} \sim N(\mathbf{m}^{(g)}, {\sigma^{(g)}}^2\mathbf{C}^{(g)}) \sim \mathbf{m}^{(g)} + \sigma^{(g)}N(\mathbf{0},\mathbf{C}^{(g)}) \tag{4.11}$$

where $s_k^{(g+1)}\in\Re^n$ is the design vector of the kth offspring in generation $g+1$ $(k=1,\ldots,\lambda)$, $N(\mathbf{m}^{(g)},\mathbf{C}^{(g)})$ are normally distributed random numbers where $\mathbf{m}^{(g)}\in\Re^n$ is the mean value vector and $\mathbf{C}^{(g)}$ is the covariance matrix while $\sigma^{(g)}\in\Re_+$ is the global step size. To define a generation step, the new mean value vector $\mathbf{m}^{(g+1)}$, global step size $\sigma^{(g+1)}$, and covariance matrix $\mathbf{C}^{(g+1)}$ have to be defined.

4.5.1.2.2 New Mean Value Vector

After selection scheme $(\mu,\ \lambda)$ operates over the λ offsprings, the new mean value vector $\mathbf{m}^{(g+1)}$ is calculated according to the following expression:

$$\mathbf{m}^{(g+1)} = \sum_{i=1}^{\mu} w_i s_{i:\lambda}^{(g+1)} \tag{4.12}$$

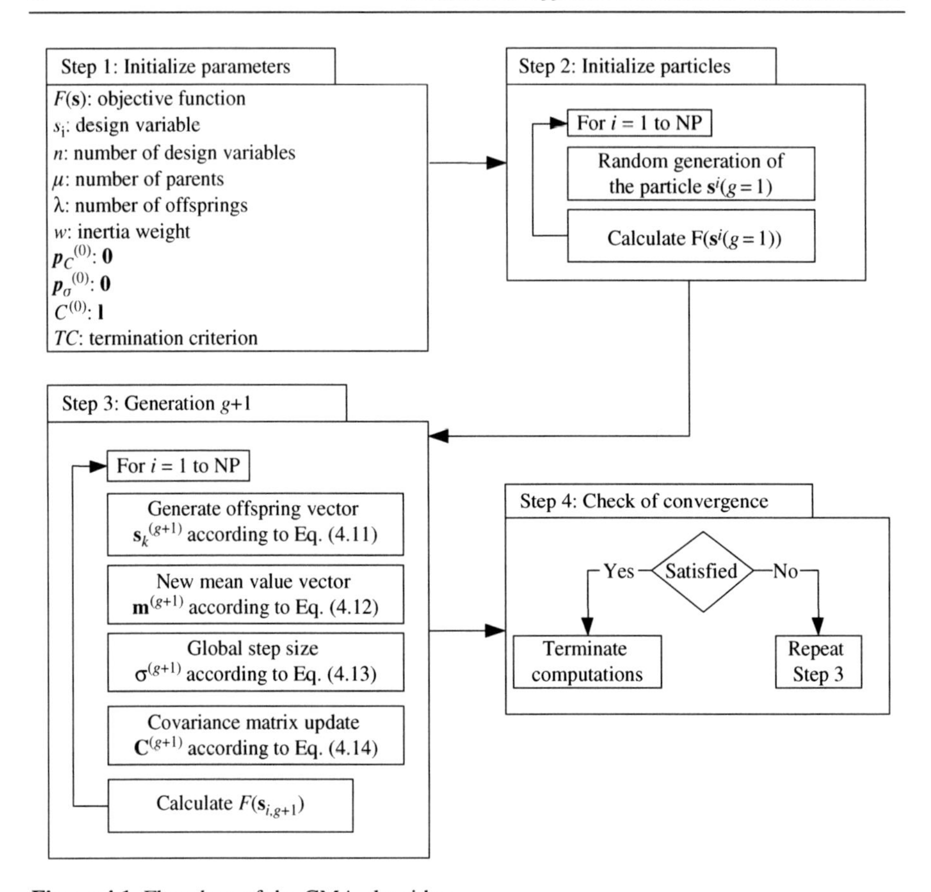

Figure 4.1 Flowchart of the CMA algorithm.

where $s_{i:\lambda}^{(g+1)}$ is the ith best offspring and w_i are the weight coefficients.

4.5.1.2.3 Global Step Size

The new global step size is calculated according to the following expression:

$$\sigma^{(g+1)} = \sigma^{(g)} \exp\left(\frac{c_\sigma}{d_\sigma}\left(\frac{\left\|\mathbf{p}_\sigma^{(g+1)}\right\|}{E\left\|N(\mathbf{0},\mathbf{I})\right\|} - 1\right)\right) \tag{4.13}$$

while the matrix $\mathbf{C}^{(g)^{-\frac{1}{2}}}$ is given by:

$$\mathbf{C}^{(g)^{-\frac{1}{2}}} = \mathbf{B}^{(g)}\mathbf{D}^{(g)^{-1}}\mathbf{B}^{(g)^{\mathrm{T}}} \tag{4.14}$$

where the columns of $\mathbf{B}^{(g)}$ are an orthogonal basis of the eigenvectors of $\mathbf{C}^{(g)}$ and the diagonal elements of $\mathbf{D}^{(g)}$ are the square roots of the corresponding positive eigenvalues.

4.5.1.2.4 Covariance Matrix Update

The new covariance matrix $\mathbf{C}^{(g+1)}$ is calculated from the following equation:

$$\mathbf{C}^{(g+1)} = (1 - c_{\text{cov}})\mathbf{C}^{(g)} + \frac{c_{\text{cov}}}{\mu_{\text{cov}}}\mathbf{p}_c^{(g+1)}\mathbf{p}_c^{(g+1)^{\mathrm{T}}} + c_{\text{cov}}\left(1 - \frac{1}{\mu_{\text{cov}}}\right)\sum_{i=1}^{\mu} w_i \mathrm{OP}\left(\frac{s_{i:\lambda}^{(g+1)} - \mathbf{m}^{(g)}}{\sigma^{(g)}}\right) \tag{4.15}$$

OP denotes the outer product of a vector with itself and $\mathbf{p}_c^{(g)} \in \Re^n$ is the evolution path ($\mathbf{p}_c^{(0)} = 0$).

4.5.1.3 *Elitist CMA*

The elitist CMA (ECMA) ES is a combination of the well-known $(1 + \lambda)$ selection scheme of ES (Rechenberg, 1973), with CMA (Igel et al., 2007). The original update rule for the covariance matrix can be reasonably applied in the $(1 + \lambda)$ selection. The cumulative step size adaptation (path length control) of the CMA $(\mu/\mu, \lambda)$ is replaced by a success rule-based step size control. Every individual a of the ECMA algorithm is comprised by five components:

$$a = \{\mathbf{s}, \overline{p}_{\text{succ}}, \sigma, \mathbf{p}_c, C\} \tag{4.16}$$

where $\mathbf{s}$ is the design vector, $\overline{p}_{\text{succ}}$ is a parameter that states the success rate during the evolution process, σ is the step size, $\mathbf{p}_c$ is the evolution path, and $\mathbf{C}$ is the covariance matrix of the mutation strengths. Contrary to the CMA algorithm, each individual has its own step size σ, evolution path $\mathbf{p}_c$ and covariance matrix $\mathbf{C}$. A pseudocode of the ECMA algorithm is shown in Figure 4.2A. In line 1, a new parent $a_{\text{parent}}^{(g)}$ is generated. In lines 4–6, λ new offsprings are generated from the parent vector $a_{\text{parent}}^{(g)}$. The new offsprings are sampled according to Eq. (4.8), with variable $\mathbf{m}^{(g)}$ being replaced by the design vector $\mathbf{s}_{\text{parent}}^{(g)}$ of the parent individual. After the λ new offsprings are sampled, the parent's step size is updated by means of UpdateStepSize subroutine (see Figure 4.2B). The arguments of the subroutine are the parent $a_{\text{parent}}^{(g)}$ and the success rate$\lambda_{\text{succ}}^{(g+1)}/\lambda$, where $\lambda_{\text{succ}}^{(g+1)}$ is the number of offsprings having better fitness function than the parent. The step size update is based on the 1/5 success rule. When the ratio $\lambda_{\text{succ}}^{(g+1)}/\lambda$ is larger than 1/5, the step size increases, and when it is smaller the step size decreases. If the best offspring has a better fitness value than the parent, it becomes the parent of the next generation (see lines 8–9). If the inequality of line #8 is satisfied, then the covariance matrix of the new parent is updated by means of UpdateCovariance subroutine

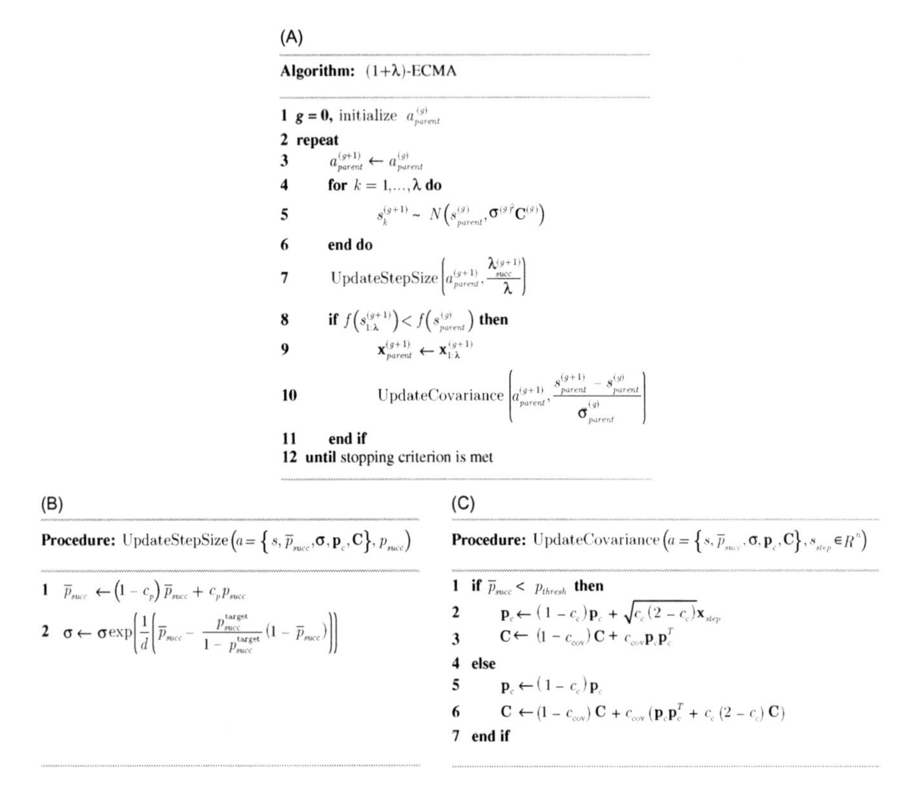

Figure 4.2 ECMA: (A) pseudocode, (B) update the step size subroutine, and (C) update the covariance matrix subroutine.

(see Figure 4.2C). The arguments of the subroutine are the current parent and the step change:

$$\frac{\mathbf{s}_{\text{parent}}^{(g+1)} - \mathbf{s}_{\text{parent}}^{(g)}}{\sigma_{\text{parent}}^{(g)}} \tag{4.17}$$

The update of the evolution path and the covariance matrix depends on the success rate:

$$\bar{p}_{\text{succ}} = \frac{\lambda_{\text{succ}}}{\lambda} \tag{4.18}$$

If the success rate is below a given threshold value p_{thresh}, then the step size is taken into account and the evolution path and the covariance matrix is updated (see lines #2–3 of Figure 4.2C). If the success rate is above the given threshold p_{thresh}, the step change is not taken into account and the update of the evolution path and the covariance matrix happens (see lines #5–6).

4.5.2 Solving the Multiobjective Optimization Problems

Several methods have been proposed in the past for treating structural multiobjective optimization problems (Coello Coello, 2000; Marler and Arora, 2004). In this chapter, three algorithms are used in order to handle the two objective optimization problems at hand. The first one is based on the nondomination sort genetic algorithm (NSGA) developed by Deb et al. (2002); the second one is based on the strength Pareto evolutionary algorithm (SPEA) developed by Zitzler et al. (2001); while the third one is the multiobjective elitist covariance matrix adaptation (MOECMA) method presented in Hansen et al. (2005). The ES method has been proved very efficient for solving single-objective structural optimization problems (Bäck and Schwefel, 1993; Lagaros et al., 2004); therefore, the ES method is combined with the philosophies of the first two multiobjective optimization methods (NSGA and SPEA). The resulting multiobjective optimization algorithms are denoted as NSES($\mu + /,\lambda$) and SPES($\mu + /,\lambda$).

4.5.2.1 Nondominated Sorting Evolution Strategies

The main part of the nondominated sorting evolution strategies (NSES) algorithm is the fast-nondomination-sort procedure according to which a population is sorted in nondominated fronts; it is based on the work by Deb et al. (2002). This algorithm identifies nondominated individuals in the population, at each generation, to form Pareto fronts, based on the concept of nondominance. After this step, the basic operators of ES are implemented. In the ranking procedure, the nondominated individuals in the current population are first identified. Then, these individuals are assumed to constitute the first nondominated front and a large dummy fitness value is assigned to each one. All these solutions have an equal reproductive potential. In order to maintain population diversity, these nondominated solutions are then shared with their dummy fitness value. Afterward, the individuals of the first front are ignored temporarily, and the rest of the population is processed in the same way to identify individuals for the second nondominated front. They are also assigned a dummy fitness value, which is a little smaller than the worst shared fitness value observed in the solutions of the first nondominated front. This process continues until the whole population is classified into nondominated fronts. Since the nondominated fronts are defined, the population is then reproduced according to the dummy fitness value.

4.5.2.2 Strength Pareto Evolution Strategies

The basic option of the strength Pareto evolution strategies (SPES($\mu + /,\lambda$)) algorithm was proposed by Zitzler et al. (2001) as an approach that incorporates several of the desirable features of other well-known multiobjective EAs. SPES($\mu + /,\lambda$) implements elitism through the maintenance of an external set of best solutions found during the whole iteration loop. Elitism, when applied by an EA, guarantees that the solutions with higher fitness will not be eliminated during the execution of the optimization algorithm. The nondominated solutions in the external set are used to determine the fitness of the current population (set of solutions) and also take part in the selection process for reproduction. In SPES(μ+ /,λ), the fitness of a solution in the population depends on the best solutions in the external set but is independent of the number of solutions. This solution dominates, or is dominated, within the population. The most important aspects of this algorithm are the fitness assignments and the clustering procedure. In each iteration, a population of individuals $\mathbf{B}_p^{(g)}$ is obtained, and the nondominated solutions of this population are copied to $\mathbf{A}^{(g)}$ (external population). Next, the solutions of $\mathbf{A}^{(g)}$ that are dominated by other solutions are eliminated, obtaining the front of Pareto of $\mathbf{A}^{(g)}$. In SPES(μ+ /,λ), the number of externally stored nondominated solutions is limited to λ. If the number of solutions of the Pareto front is greater than λ, it is necessary to reduce the external population by some means of clustering.

4.5.2.3 Multiobjective Elitist Covariance Matrix Adaptation

The MOECMA EA (Hansen et al., 2005) is a multiobjective optimization algorithm that is based on ES and was proposed by Hansen et al. (2005). The algorithm combines the ECMA(1 + λ) ES with the population sorting and diversity preservation mechanisms of NSES. According to the MOECMA algorithm, λ_{MO} parallel ECMA (1 + λ) ES are performed. An MOECMA design is a set that comprises the design vector $\mathbf{s}$, the parameter $\overline{p}_{\mathrm{succ}}$, the step size σ, the evolution path vector $\mathbf{p}_c$, and the covariance matrix C. The pseudocode of the MOECMA is shown in Figure 4.3. Initially, λ_{MO} parents are created (line 1). In lines 3–8, λ offsprings are sampled for every one of the λ_{MO} parallel ES. The jth offspring that is created from the kth parent is written as a_{kj} and its design vector as $\mathbf{s}_{kj}$. All designs are combined in one population $\mathbf{Q}^{(g)}$ in line 8. The population $\mathbf{Q}^{(g)}$ is sorted in nondomination fronts in line 10. Afterwards, the new step size of each parent is calculated (line 14) by the subroutine UpdateStepsize. The evolution path vectors and the new covariance matrices of every offspring are then calculated (lines 15–18). The step size calculation is based on the number of successful offsprings $\lambda_{\mathrm{succ},\,k}$ of the kth parent. The number $\lambda_{\mathrm{succ},k}$ is equal to the number of offsprings that were created by the kth parent and that are in a better nondomination front or if they are in the same nondomination front, they are in a better position (less crowded). Finally, in lines 20–22, the new λ_{MO} parents are chosen as the first λ_{MO} designs of population $\mathbf{Q}^{(g)}$.

4.6 39-Bar Truss—Test Example

A three-dimensional 39-bar truss shown in Figure 4.4 is considered for testing the efficiency of the ES-based metaheuristics employed for solving single and multiobjective optimization problems. The height of the structure is 16 m (Figure 4.4B), while its basis is an equilateral triangle of side 6.93 m (Figure 4.4C). In the single-objective optimization problem, the objective function to be minimized is the weight

1 $g = 0$, initialize $a_k^{(g)}$ for $k = 1,\ldots,\lambda_{MO}$
2 **repeat**
3 **for** $k = 1, \lambda_{MO}$ **do**
4 **for** $j = 1, \lambda$ **do**
5 $a_{kj}^{(g)} = a_k^{(g)}$
6 $s_{kj}^{(g+1)} \sim N\ s_k^{(g)}, \left(\sigma_k^{(g)^2}\ \mathrm{C}_k^{(g)}\right)$
7 **end**
8 **end**
9 $\mathbf{Q}^{(g)}$ = whole population
10 $\mathbf{F}^{(g)}$ = fast–nondomination–sort $\left(\mathbf{Q}^{(g)}\right)$
11 crowding–distance–assignment $\left(\mathbf{F}^{(g)}\right)$
12 crowded–comparison–operator $\left(\mathbf{F}^{(g)}\right)$
13 **for** $k = 1, \lambda_{MO}$ **do**
14 UpdateStepSize $\left(a_k^{(g)}, \frac{\lambda_{succ,k}^{(g+1)}}{\lambda}\right)$
15 **for** $j = 1, \lambda$ do
16 UpdateStepSize $\left(a_{kj}^{(g)}, \frac{\lambda_{succ,k}^{(g+1)}}{\lambda}\right)$
17 UpdateCovariance $\left(a_{kj}^{(g+1)}, \frac{s_{kj}^{(g+1)} - s_k^{(g)}}{\sigma_k^{(g)}}\right)$
18 **end**
19 **end**
20 **for** $k = 1, \lambda_{MO}$ **do**
21 $a_k^{(g+1)} = \mathrm{Q}^{(g)}\ [k]$
22 **end**
23 $g = g+1$
24 Until **termination_condition**

Figure 4.3 Pseudocode of the MOECMA method.

of the structure, while for the multiobjective problem the two objectives considered are the weight of the structure and the displacement of a characteristic node representing the response of the structure; in particular, the top displacement in x-direction is selected. The design variables considered are the dimensions of the members of the structure, four groups in total, taken from the circular hollow section (CHS) table of the Eurocode. A vertical load $V = 2$ kN is applied to all nodes, while a horizontal load F of 8 kN is also applied to the top three nodes at along the x-direction.

For both formulations, three types of constraints are considered: (i) stress, (ii) compression force (for buckling), and (iii) displacement constraints, as imposed by

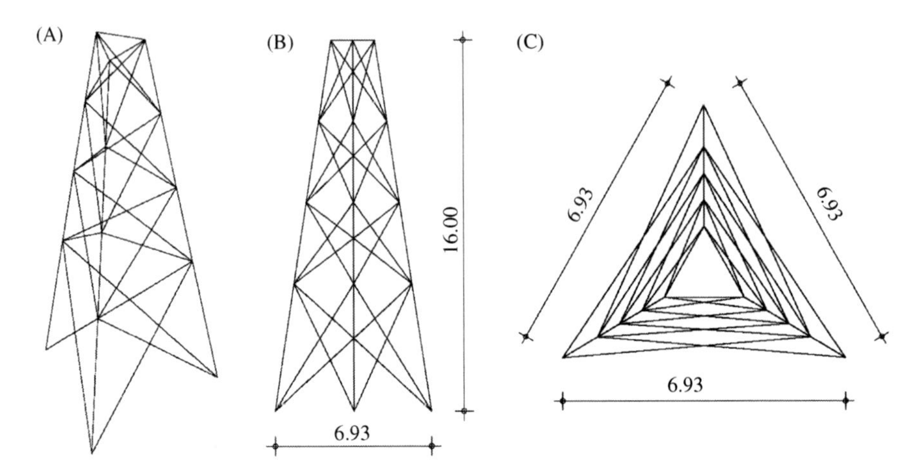

Figure 4.4 39-bar truss example: (A) three-dimensional view, (B) side view, and (C) top view.

the European design code EC3 (2005). The stress constraints considered can be written as follows:

$$\begin{aligned}\sigma_{\max} &\leq \sigma_a \\ \sigma_a &= \frac{\sigma_y}{1.10}\end{aligned} \tag{4.19}$$

where $\sigma_{\max}$ is the maximum axial stress in each element group for all loading cases, σ_a is the allowable axial stress according to Eurocode 3 (2005), and σ_y is the yield stress. For members under compression, an additional buckling constraint is implemented:

$$\begin{aligned}\left|P_{c,\max}\right| &\leq P_{cc} \\ P_{cc} &= \frac{P_e}{1.05} \\ P &= \frac{\pi^2 EI}{L_{\text{eff}}^2}\end{aligned} \tag{4.20}$$

where $P_{c,\max}$ is the maximum axial compression force for all loading cases, P_e is the critical Euler buckling force in compression, taken as the first buckling mode of a pin-connected member, and L_{eff} is the effective length. The effective length is

Table 4.1 Efficiency of the Procedures

Algorithm	Weight (kN)				Design Vector			
	Mean	COV	Min	Max	Sec_1	Sec_2	Sec_3	Sec_4
ES(10 + 10)	23.788	7.39×10^{-2}	22.617	33.568	88.9 × 8	101.0 × 5	88.9 × 5	88.9 × 3.2
CMA(4,8)	26.164	9.39×10^{-2}	22.762	33.415	88.9 × 8	101.0 × 5	88.9 × 5	76.1 × 4
ECMA(1 + 19)	31.559	1.63×10^{-1}	23.039	45.781	139.7 × 5	101.0 × 5	88.9 × 5	88.9 × 3.2

taken equal to the actual length of the member. Similarly, the displacement constraints can be written as:

$$|d| \leq d_a \tag{4.21}$$

where d_a is the limit value of the displacement at a certain node or the maximum nodal displacement, taken as 200 mm.

For the single-objective optimization problem, the simple yet effective, multiple linear segment penalty function is used in this study for handling the constraints. According to this technique, if no violation is detected, then no penalty is imposed on the objective function. If any of the constraints is violated, a penalty, relative to the maximum degree of constraints' violation, is applied to the objective function. On the other hand, the optimization procedure is terminated when the best value of the objective function in the last 10 generations remains unchanged. For assessing the efficiency and the robustness of the metaheuristic optimization algorithms, 100 independent runs have been performed for each algorithm resulting in 300 independent optimization runs. The robustness of the three metaheuristic algorithms is demonstrated in Table 4.1, showing the sensitivity of the results with reference to the best objective function achieved. ES and CMA seem to be more robust, since a coefficient of variation (COV) of the optimum objective function value is below 10%. For the multiobjective optimization methods, the death penalty method is implemented, while the optimization schemes used are NSES(10 + 10), SPES (10 + 10), and MOECMA(1 + 19). The resultant Pareto front curve is depicted in Figure 4.5, with the weight of the structure and the standard deviation of the horizontal displacement on the horizontal and vertical axis, respectively. The Pareto front curve shows a strong conflict between the two objective functions in question. In order to assess the three methods, the Pareto front curves obtained after 10, 50, and 500 generations are compared (see Figure 4.5A–C, respectively). As it can be seen, a good quality Pareto front curve is obtained for all three methods. As can be seen, 50 generations are sufficient to obtain a good quality Pareto front curve for all three methods adopted.

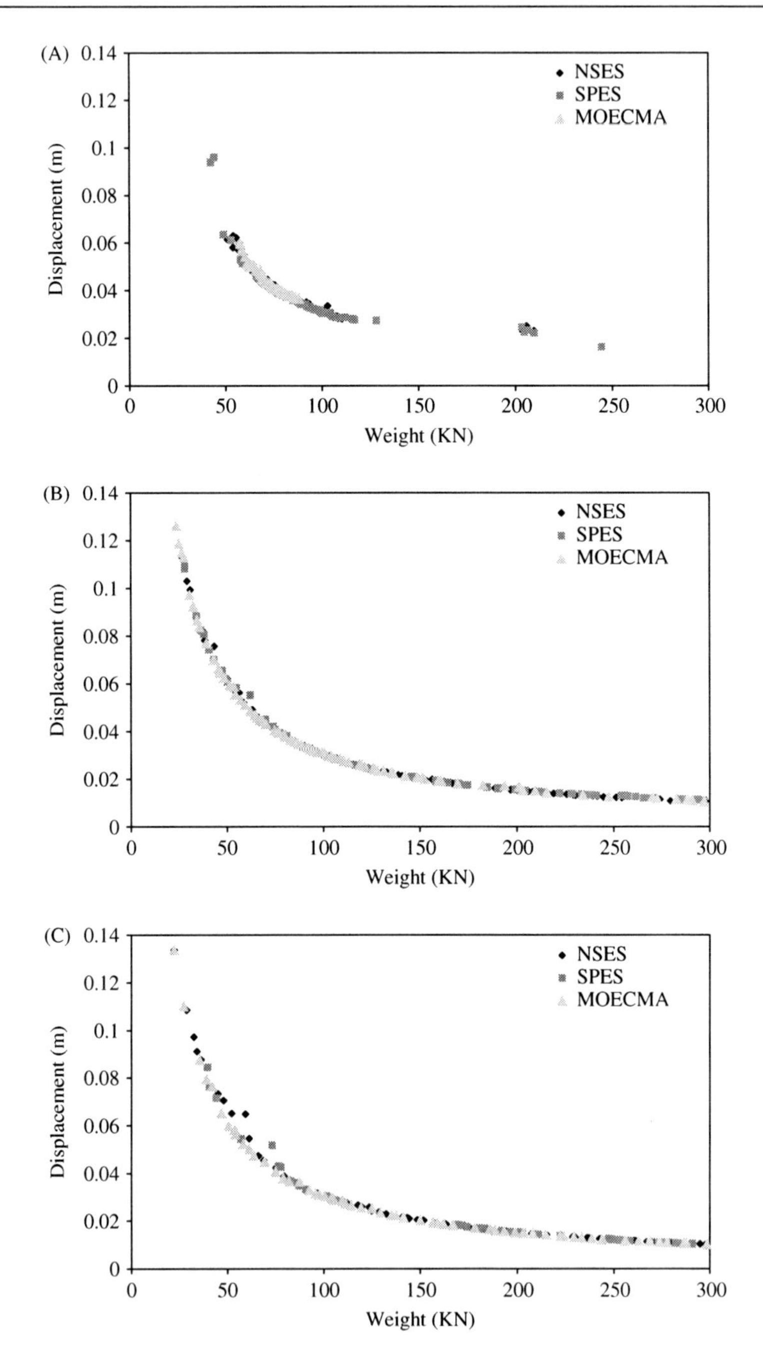

Figure 4.5 39-bar truss example Pareto front curves after: (A) 10, (B) 50, and (C) 500 generations.

4.7 Conclusions

In this chapter, successful implementation of metaheuristics is presented for solving single and multiobjective structural design optimization problems. In particular, sizing structural optimization problems of a truss structure are formulated, where the cross-sectional dimensions of the structural elements constitute the sizing design variables while constraints imposed by Eurocode 3 are implemented. Both single and multiobjective ES-based optimization methods were used. For the single-objective problems, the ES, the CMA, and the ECMA methods are used. For the multiobjective problems, the NSES, the SPES, and the MOECMA methods are used.

Comparing the three algorithms, it can be said that algorithms based on ES can be considered as efficient tools for both single and multiobjective design optimization of structural problems such as space trusses. In terms of computational efficiency, it appears that all three methods considered require similar computational effort with approximately the same number of generation steps. In both the problems, a large number of solutions need to be found and evaluated in search of the optimum one. The metaheuristics employed in this study have been found efficient in finding optimized solutions, overcoming excessive the computational effort and the local optima; while they are capable dealing with discrete variables when needed.

References

Bäck, T., Schwefel, H.-P., 1993. An overview of evolutionary algorithms for parameter optimization. J. Evol. Comput. 1 (1), 1–23.

Bendsoe, M.P., Kikuchi, N., 1988. Generating optimal topologies in structural design using a homogenization method. Comput. Method Appl. Mech. Eng. 71, 197–224.

Bozorg Haddad, O., Afshar, A., Mariño, M.A., 2005. Honey bees mating optimization algorithm: a new heuristic approach for engineering optimization. In: Proceeding of the First International Conference on Modelling, Simulation and Applied Optimization (ICMSA0/05). February 1–3 2005, Sharjah, United Arab Emirates.

Bureerat, S., Limtragool, J., 2008. Structural topology optimisation using simulated annealing with multiresolution design variables. Finite Elem. Anal. Des. 44 (12-13), 738–747.

Chen, T.Y., Chen, H.C., 2009. Mixed-discrete structural optimization using a rank-niche evolution strategy. Eng. Optim. 41 (1), 39–58.

Coello Coello, C.A., 2000. An updated survey of GA-based multi-objective optimization techniques. ACM Comput. Surv. 32 (2), 109–143.

Deb, K., Pratap, A., Agarwal, S., Meyarivan, T., 2002. A fast and elitist multi-objective genetic algorithm: NSGA-II. IEEE Trans. Evol. Comput. 6 (2), 182–197.

Degertekin, S.O., 2012. Improved harmony search algorithms for sizing optimization of truss structures. Comput. Struct. 92–93, 229–241.

Dorigo, M., Stützle, T., 2004. Ant Colony Optimization. The MIT Press, Cambridge, MA, USA.

EC3, 2005. Eurocode 3: Design of Steel Structures, Part 1.1: General Rules and Rules for Buildings. European Committee for Standardisation, Brussels, Belgium (The European Standard EN 1993-1-1: 2005).

Eschenauer, H.A., Schumacher, A., Vietor, T., 1993. Decision makings for initial designs made of advanced materials. In: Bendsoe, M.P., Soares, C.A.M. (Eds.), Topology Design of Structures. Kluwer Academic Publishers, Sesimbra, Portugal, Dordrecht, the Netherlands, pp. 469–480.

Farhat, F., Nakamura, S., Takahashi, K., 2009. Application of genetic algorithm to optimization of buckling restrained braces for seismic upgrading of existing structures. Comput. Struct. 87 (1–2), 110–119.

Fogel, D.B., 1992. Evolving Artificial Intelligence. University of California, San Diego, CA, PhD Thesis.

Gandomi, A.H., Alavi, A.H., 2012. Krill herd: a new bio-inspired optimization algorithm. Commun. Nonlinear Sci. Numer. Simul. 17 (12), 4831–4845.

Gandomi, A.H., Yang, X.S., 2011. Benchmark problems in structural optimization. In: Koziel, S., Yang, X.S. (Eds.), *Computational Optimization, Methods and Algorithms*, 356. Springer Berlin, Heidelberg, pp. 259–281. (Studies in Computational Intelligence).

Gandomi, A.H., Yang, X.-S., Alavi, A.H., 2011a. Cuckoo search algorithm: a metaheuristic approach to solve structural optimization problems. Eng. Comput. doi: 10.1007/s00366-011-0241-y (in press).

Gandomi, A.H., Yang, X.-S., Alavi, A.H., 2011b. Mixed variable structural optimization using firefly algorithm. Comput. Struct. 89 (23–24), 2325–2336.

Geem, Z.W., Kim, J.H., Loganathan, G.V., 2001. A new heuristic optimization algorithm: harmony search. Simulation. 76, 60–68.

Gholizadeh, S., Salajegheh, E., 2009. Optimal design of structures subjected to time history loading by swarm intelligence and an advanced metamodel. Comput. Method Appl. Mech. Eng. 198 (37–40), 2936–2949.

Goldberg, D.E., 1989. *GeneticAlgorithms in Search Optimization and Machine Learning*. Addison-Wesley, Boston, MA, USA.

Hansen, L.U., Häusler, S.M., Horst, P., 2008. Evolutionary multicriteria design optimization of integrally stiffened airframe structures. J. Aircr. 45 (6), 1881–1889.

Hansen, N., Ostermeier, A., 2001. Completely derandomized self-adaptation in evolution strategies. Evol. Comput. 9 (2), 159–195.

Hansen, N., Igel, C., Roth, S., 2005. The multi-objective variable metric evolution strategy, Part I. Technical Report, IR-INI 2005-04, ISSN 0943-2752, Institut für Neuroinformatik.

Hasançebi, O., 2008. Adaptive evolution strategies in structural optimization: enhancing their computational performance with applications to large-scale structures. Comput. Struct. 86 (1–2), 119–132.

Hasançebi, O., Çarbaş, S., Doğan, E., Erdal, F., Saka, M.P., 2010. Comparison of non-deterministic search techniques in the optimum design of real size steel frames. Comput. Struct. 88 (17–18), 1033–1048.

Hinton, E., Sienz, J., 1993. Fully stressed topological design of structures using an evolutionary procedure. Eng. Comput. 12, 229–244.

Hinton, E., Sienz, J., 1994. Aspects of adaptive finite element analysis and structural optimization. In: Topping, B.H.V., Papadrakakis, M. (Eds.), Advances in Structural Optimization. Civil-Comp Press, Edinburgh, pp. 1–26.

Holland, J., 1975. Adaptation in Natural and Artificial Systems. University of Michigan Press, Ann Arbor, MI.

Igel, C., Hansen, N., Roth, S., 2007. Covariance matrix adaptation for multi-objective optimization. Evol. Comput. 15 (1), 1–28.

Kaveh, A., Shahrouzi, M., 2008. Dynamic selective pressure using hybrid evolutionary and ant system strategies for structural optimization. Int. J. Numer. Methods Eng. 73 (4), 544–563.

Kennedy, J., Eberhart, R., 1995. Particle swarm optimization. 4th IEEE International Conference on Neural Networks, 1942-1948, IEEE Press, 27 November-1 December 1995, Piscataway, Australia.

Kirkpatrick, S., Gelatt Jr., C.D., Vecchi, M.P., 1983. Optimization by simulated annealing. Science. 220, 671–680.

Koza, J.R., 1992. *Genetic Programming/ On the Programming of Computers by means of Natural Selection*. The MIT Press, Cambridge, MA, USA.

Kripakaran, P., Hall, B., Gupta, A., 2011. A genetic algorithm for design of moment-resisting steel frames. Struct. Multidiscip. Optim. 44 (4), 559–574.

Kunakote, T., Bureerat, S., 2011. Multi-objective topology optimization using evolutionary algorithms. Eng. Optim. 43 (5), 541–557.

Lagaros, N.D., Karlaftis, M.G., 2011. A critical assessment of metaheuristics for scheduling emergency infrastructure inspections. Swarm Evol. Comput. 1 (3), 147–163.

Lagaros, N.D., Papadrakakis, M., 2012. Applied soft computing for optimum design of structures. Struct. Multidiscip. Optim. 45 (6), 787–799.

Lagaros, N.D., Fragiadakis, M., Papadrakakis, M., 2004. Optimum design of shell structures with stiffening beams. AIAA J. 42 (1), 175–184.

Manan, A., Vio, G.A., Harmin, M.Y., Cooper, J.E., 2010. Optimization of aeroelastic composite structures using evolutionary algorithms. Eng. Optim. 42 (2), 171–184.

Marler, R.T., Arora, J.S., 2004. Survey of multi-objective optimization methods for engineering. Struct. Multidiscip. Optim. 26 (6), 369–395.

Martínez, F.J., González-Vidosa, F., Hospitaler, A., Alcalá, J., 2011. Design of tall bridge piers by ant colony optimization. Eng. Struct. 33 (8), 2320–2329.

Muc, A., Muc-Wierzgoń, M., 2012. An evolution strategy in structural optimization problems for plates and shells. Compos. Struct. 94 (4), 1461–1470.

Papadrakakis, M., Tsompanakis, Y., Lagaros, N.D., 1999. Structural shape optimization using evolution strategies. Eng. Optim. J. 31, 515–540.

Pareto, V., 1897. Cours d' économique politique, 1&2. Rouge, Lausanne.

Perez, R.E., Behdinan, K., 2007. Particle swarm approach for structural design optimization. Comput. Struct. 85 (19–20), 1579–1588.

Ramm, E., Bletzinger, K.-U., Reitinger, R., Maute, K., 1994. The challenge of structural optimization. In: Topping, B.H.V., Papadrakakis, M. (Eds.), Advances in Structural Optimization. Civil-Comp Press, Edinburgh, pp. 27–52.

Rechenberg, I., 1973. Evolution Strategy: Optimization of Technical Systems According to the Principles of Biological Evolution. Frommann-Holzboog, Stuttgart (in German).

Schwefel, H.P., 1981. Numerical Optimization for Computer Models. Wiley & Sons, Chichester.

Shieh, R.C., 1994. Massively parallel structural design using stochastic optimization and mixed neural net/finite element analysis methods. Comput. Syst. Eng. 5 (4–6), 455–467.

Su, R., Wang, X., Gui, L., Fan, Z., 2011. Multi-objective topology and sizing optimization of truss structures based on adaptive multi-island search strategy. Struct. Multidiscip. Optim. 43 (2), 275–286.

Suzuki, K., Kikuchi, N., 1993. Layout optimization using the homogenization method. In: Rozvany, G.I.N. (Ed.), Optimization of Large Structural Systems. Kluwer Academic Publishers, Berchtesgaden, Germany, Dordrecht, the Netherlands, pp. 157–175.

Wang, Q., Fang, H., Zou, X., 2010. Application of Micro-GA for optimal cost base isolation design of bridges subject to transient earthquake loads. Struct. Multidiscip. Optim. 41 (5), 765–777.

Xie, Y.M., Steven, G.P., 1993. A simple evolutionary procedure for structural optimization. Comput. Struct. 49, 885–896.

Xie, Y.M., Steven, G.P., 1994. Optimal design of multiple load case structures using an evolutionary procedure. Eng. Comput. 11, 295–302.

Yang, X.S., 2008. Nature-Inspired Metaheuristic Algorithms. Luniver Press, Frome.

Yang, X.S., Deb, S., 2010. Engineering optimization by cuckoo search. Int. J. Math. Model. Numer. Optim. 1 (4), 330–343.

Yang, X.S., Gandomi, A.H., 2012. Bat algorithm: a novel approach for global engineering optimization. Eng. Comput. 29 (5), 464–483.

Yang, X.S., Koziel, S., 2011. Computational Optimization and Applications in Engineering and Industry. Springer, Germany.

Zitzler, E., Laumanns, M., Thiele, L., 2001. SPEA 2: improving the strength Pareto evolutionary algorithm. In: Giannakoglou, K., Tsahalis, D., Periaux, J., Papailou, P., Fogarty, T. (Eds.), EUROGEN 2001, Evolutionary Methods for Design, Optimization and Control with Applications to Industrial Problems, Athens, Greece, pp. 95 – 100.

5 Multidisciplinary Design and Optimization Methods

Parviz Mohammad Zadeh[1] *and Mohadeseh Alsadat Sadat Shirazi*[2]

[1]Faculty of New Sciences and Technology, University of Tehran, Tehran, Iran, [2]Faculty of Aerospace, K.N.T University, Tehran, Iran

5.1 Introduction

Optimization has evolved from a methodology of academic interest into a technology that has made, and continues to make, a significant impact in industry. The use of optimization methods in structural design in the early 1960s started a new phase in engineering design. In the mid-1970s, Schmit and Miura, 1976 investigated applications of optimization methods with approximations in structural design. In addition to advances in theoretical aspects of structural optimization during the past two decades, particularly the development of efficient and robust optimization algorithms, there have been significant advances made in the application of optimization methods to various engineering problems. The use of such methods has proven to be the most successful tool for treating simulation-based design problems, such as finite element and computational fluid dynamic based design (for more detail see Alexandrov and Lewis, 2003; Jameson, 1997; Newman et al., 1999; Tatting and Gürdal, 2000).

The tremendous progress in computer technology, optimization techniques, and numerical analysis tools has significantly increased the possibilities of integrating these tools into the design process. Combined with a growing demand for improved designs within shorter design cycle times over the years, this has led to the development of multidisciplinary design optimization (MDO) methodologies (Balling and Sobieszczanski-Sobieski, 1994a; Sobieszczanski-Sobieski and Haftka, 1996).

During the past two decades, MDO methodology has emerged as an engineering discipline that focuses on the development of new design and optimization strategies for complex systems such as aerospace design problems. In recent years, MDO methodology has broken into other fields such as automotive, mechanical, civil, naval, and off-shore engineering. MDO is a broad area that encompasses design optimization methods, sensitivity analysis, approximation concepts, information processing, and management methods and strategies, all in the context of integrated design dealing with multiple disciplines and subsystem interactions. This

Metaheuristic Applications in Structures and Infrastructures. DOI: http://dx.doi.org/10.1016/B978-0-12-398364-0.00005-X

methodology is formally defined as "a systematic methodology for design and analysis of complex engineering systems and subsystems, which coherently exploits the synergism of mutually interacting phenomena." The emphasis in this definition is on the systematic mathematical methods rather than ad hoc approaches (Balling and Sobieszczanski-Sobieski, 1994b).

The main MDO concept pertains to the decomposition of complex engineering systems into a set of smaller and less complex subsystems together with formal approaches of accounting for system interactions and couplings. The aim is to improve the iterative design of complex systems by making the process systematic and based upon a set of mathematical concepts. An additional advantage of partitioning may be the ability to solve the subsystems concurrently.

A decomposition is usually based upon engineering disciplines or mathematical models governing the system. The challenge of solving such distributed optimum design problems lies in addressing the interactions (or coupling) among the individual disciplines. Coupling variables (also termed linking variables or shared variables) are common to more than one subsystem or are shared by the system level with at least one subsystem. Equality constraints for coupling variables are included in the decoupled problem to ensure compatibility of the subsystem solutions. The shared design variables must achieve the same value in the final solution, while the equality constraints enforce system-level feasibility.

Development of decomposition and coordination methods is an active area of research. There have been several MDO decomposition and system-level coordination strategies proposed over the past two decades (Kodiyalam and Sobieszczanski-Sobieski, 2001). Important features that characterize these solution strategies are the allowable level of disciplinary autonomy in the analysis and optimization process, the ease of implementation, and robustness and computational efficiency. Concurrent subspace optimization (CSSO) (Bloebaum et al., 1992; Renaud and Tappeta, 1997; Sobieszczanski-Sobieski, 1988) and collaborative optimization (CO) (Balling and Sobieski, 1994a; Braun et al., 1996; Kroo et al., 1994) have been recognized as two main MDO methods enabling disciplinary autonomy in optimization processes. However, CO differs from CSSO in the issue of control of coupled variables. Each discipline in CO is given control of its own set of local design variables and is responsible for satisfying its own set of local constraints, and a system-level optimizer is used to coordinate this process while minimizing the overall objective. This high degree of disciplinary autonomy in CO provides designers with significant potential benefits in terms of design freedom and makes it an attractive method for solving complex multidisciplinary design problems. However, there are a number of problems associated with this method that limit its application based on high-fidelity simulation models as described later in more detail.

5.2 Coupled Multidisciplinary System

Prior to describing the mathematical formulation of MDO, it is necessary to review some of the mathematical terms and concepts defined in Balling and

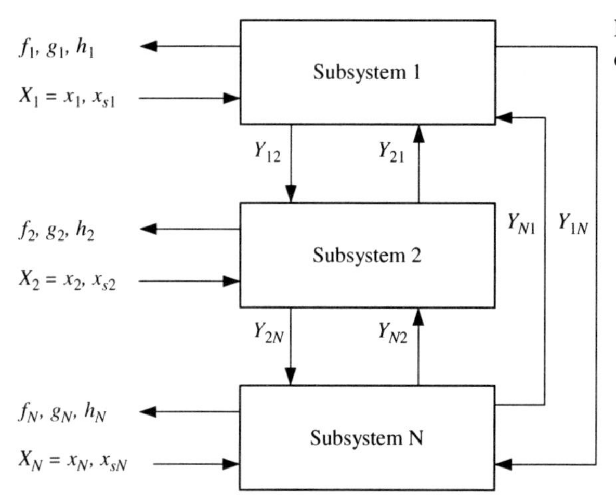

Figure 5.1 Multidisciplinary coupled system.

Sobieszczanski-Sobieski (1994b) that characterize MDO frameworks. For illustrative purposes, we consider the N-discipline system. Each box represents a module that utilizes a set of inputs to compute its outputs. These modules are referred to as disciplines and exchange information (couplings) in order to evaluate system states. The interdisciplinary consistency constraints are introduced to drive to zero the discrepancy among the disciplinary inputs and outputs. The values of constraints at the system level are computed by solving disciplinary optimization problems.

Each subsystem is based on a disciplinary analysis, depicted as an input–output relation in Figure 5.1. $\mathbf{X} = \{x_1, \ldots, x_N\}$ is the vector of design variables and $\mathbf{x_s} = \{x_{s1}, \ldots, x_{sN}\}$ are the design variables (parameters) appearing in more than one subsystem (shared design variables). X_i $(i = 1,2, \ldots,N)$ are the local design variables of discipline i. $Y_{ij}(i \neq j)$ are linking (coupling) variables evaluated as outputs of discipline i at the same time as being inputs to discipline j.

5.3 Classifications of MDO Formulations

MDO problems are characterized as having two or more disciplinary analyses. Each disciplinary analysis may be a simulation tool or a collection of simulation models that are coupled through performance inputs and outputs of the individual disciplines. Couplings of such systems take the form of shared design variables and both input and output state performances. An analysis of these systems for a given set of values of the design variables requires iteration between individual disciplines until convergence is reached. The general MDO optimization problem can be stated in the following form:

$$\text{Minimize: } f(x, \mathbf{u}(\mathbf{x})) \tag{5.1}$$

$$\text{subject to: } g(\mathbf{x}, \mathbf{u}(\mathbf{x})) \leq 0 \tag{5.2}$$

$$h(\mathbf{x}, \mathbf{u}(\mathbf{x})) = 0 \tag{5.3}$$

$\mathbf{x} = \{x_1, x_2, x_3, \ldots, x_n\}$ is the vector of design variables and $\mathbf{u}(\mathbf{x}) = \{u_1(x), u_2(x), u_3(x), \ldots, u_n(x)\}$ is a vector of behavior variables.

To solve the MDO problem Eqs. (5.1)–(5.3), several solution methods can be found in the literature. Classifications of these solution methods are presented in Balling and Sobieszczanski-Sobieski (1996), Cramer et al. (1994), and Alexandrov and Lewis (1999). The key element in these classifications is that the way feasibility of the constraints is satisfied. Cramer et al. (1994) classified MDO methods into all-at-once (AAO), individual-discipline feasible (IDF), and multiple-discipline feasible (MDF). Balling and Sobieszczanski-Sobieski (1996) distinguished between single-level and multilevel approaches. Single-level refers to an approach where only the system optimization problem determines the design variable values. In the multi-level case, disciplinary optimizations are introduced to determine the independent discipline design variables, while the system optimizer determines the shared design variables. From the system optimizer's point of view, the disciplinary constraints $g_i(\mathbf{x}, \mathbf{u}(\mathbf{x})) \leq 0$ will always be satisfied. This means that they are closed constraints according to the classification in Alexandrov and Lewis (1999). In this chapter, the classification of MDO solutions is based on single-level and multilevel approaches and these are described in the following sections.

5.4 Single-Level Optimization

5.4.1 Multiple-Discipline Feasible

The MDF method (Cramer et al., 1994) can be seen as a cycle of full multidisciplinary analysis (MDA) followed by design updates, which is the most common way of solving MDO problem Eqs. (5.1)–(5.3). It is an example of the variable reduction method to nonlinear programming where only the design variables $\mathbf{x}$ are used and not state (behavior) variables as independent optimization variables. The basic idea is to insert a multiple-discipline analyzer between the optimizer and the disciplines. In this method, the vector of design variables $\mathbf{x}$ is provided to the coupled system of disciplinary analysis and a complete MDA is carried out by a fixed-point iteration with that value of $\mathbf{x}$ to obtain the output variable $\mathbf{u}(\mathbf{x})$ of the system MDA, which is then used in evaluating the objective function $f(\mathbf{x}, \mathbf{u}(\mathbf{x}))$ and the constraints $g(\mathbf{x}, \mathbf{u}(\mathbf{x}))$.

MDF has been used since nonlinear programming techniques were first applied to engineering design optimization problems and hence, it is a well-established method for solving MDO problems. The primary disadvantage of MDF is its high computational cost because a complete MDA must be carried out not only at every iteration but also for computing derivatives. Hence, achieving multidisciplinary compatibility can be prohibitively expensive in realistic applications (Kodiyalam and Sobieszczanski-Sobieski, 2001). In addition, optimization algorithms applied to MDF are sensitive to the robustness of the method and its speed (Alexandrov and Kodiyalam, 1998).

5.4.2 All-At-Once Method

The AAO method, known as simultaneous analysis and design (Sobieszczanski-Sobieski and Haftka, 1997) or optimizer-based decomposition, (Kroo, 1995), treats the entire multidisciplinary design as a single optimization problem in which each analysis-block is executed in parallel. The basic idea behind AAO is that iterations should not be wasted determining a feasible design when the current design iteration is far from optimum. This is achieved by converting the system analysis equations into equality constraints and by treating both system-design variables and subsystem outputs (state variables) as optimization variables. In AAO, these is no explicit coupling between the analyses. Instead, the optimizer enforces the coupling by imposing constraints on the input and output variables.

The AAO approach has proven to be the most computationally efficient of the MDO approaches in several test problems (Balling and Wilkinson, 1997). It has the advantage of eliminating the iterative design cycle for achieving an optimum design but does retain some of the single-level optimization limitations. The main limitation is that a single optimizer must control all the design variables and constraints, and, hence, imposes a greater computational burden. An additional disadvantage of this approach is that disciplinary feasibility is only achieved at a relative or at an absolute extremum. Hence, if the optimizer is unsuccessful in reaching the global optimum solution, the likelihood of completing the solution cycle with a feasible design solution is reduced (Hulme, 2000).

5.4.3 Individual-Discipline Feasible

The IDF formulation is another popular alternative to the classical MDF approach and exhibits characteristics that lie between the extremes of MDF and AAO (Cramer et al., 1994; Lewis, 1997). IDF maintains individual-discipline feasibility by allowing the optimizer to drive the individual disciplines to multidisciplinary feasibility by controlling interdisciplinary coupling variables. The phrase *individual-discipline feasible* refers to maintaining disciplinary feasibility at each iteration, but not multidisciplinary feasibility until a solution is reached, which provides a way to avoid a complete MDA at optimization. IDF has a similar property to CO in terms of coupling and problem decomposition. IDF maintains autonomy with respect to analyses but lacks disciplinary optimization autonomy of CO. Another limitation of IDF is its treatment of the disciplinary constraints, which are transferred to the system, despite the need to handle disciplinary constraints at the discipline level (Alexandrov and Kodiyalam, 1998).

5.4.4 *Comparative Characteristics of Single-Level Optimization*

The general characteristics of the single-level optimization approaches (MDF, AAO, and IDF) are compared in Table 5.1 (Braun, 1996). In multidisciplinary design, most optimization studies performed are based on a MDF approach. While this approach may be simplest to implement, it has several performance limitations as shown in Table 5.1.

On the other hand, while AAO may provide the most efficient means toward an optimum, the solution strategy is difficult to implement and has potential communication requirements using a single optimizer. The MDF approach also inherits the limitations of the single-level optimization solution method.

The high computational cost of single-level optimization approaches and their potential convergence difficulties (due to the use of a single optimizer to handle all the design variables) coupled with the increasing requirement of addressing the organizational challenges led to the development of multilevel optimization approaches. These are reviewed in the sections below.

5.5 Multilevel Optimization

The multilevel optimization (also known as distributed design optimization or disciplinary feasible constrained method) involves the use of a system-level optimizer to coordinate the overall system performance to satisfy disciplinary constraints and to ensure multidisciplinary feasibility of the converged solution. These approaches retain several advantages of AAO while incorporating disciplinary optimization to augment the role of the disciplinary expert. They offer a practical solution to large-scale coupled design problems and allow for solving MDO problems concurrently (hence, reducing design cycle time and maintaining autonomy of disciplinary optimization in the design process). The next sections briefly analyze three of the main multilevel optimization approaches.

5.5.1 *Concurrent Subspace Optimization*

The CSSO formulation, Sobieszczanski-Sobieski (1988), is a decomposition-based strategy that enables each discipline to run independently with a system-level

Table 5.1 Single-Level Optimization Comparison

Characteristics	**AAO**	**IDF**	**MDF**
Satisfaction of governing equations	Only at convergence	At each optimization iteration	At each optimization iteration
System convergence	Only at convergence	Only at convergence	At each optimization iteration
Expected speed	Fast	Medium	Slow
Robustness	Unknown	High	Medium

optimizer to provide overall coordination. This corresponds to common design practice where individual teams optimize their local component design and compromises are made at the system level. CSSO provides MDA feasibility at each cycle and deals with all design variables simultaneously at the system level. This approach has been successfully applied to numerous design problems (Sellar et al., 1996). However, convergence difficulties are reported by Shankar et al. (1993). The main deficiencies of CSSO are as follows:

- The formulation of the coordination problem is based on an optimum-sensitivity analysis procedure that cannot be regarded as robust.
- Coupling in CSSO is solved via the use of approximations, which require move limits to be placed on the design variables.
- The use of heuristics in solving the coordination problem has shown to be of limited benefit, often introducing convergence problems.
- Sensitivity information is unavailable when dealing with discrete and integer design variables.

5.5.2 Bilevel Integrated System Synthesis

The bilevel integrated system synthesis (BLISS) formulation Sobieszczanski-Sobieski (1988) is based on decompositions that separate system-level optimization, having a relatively small number of design variables, from the subspace optimizations that may have a large number of local design variables.

The subspace optimizations are autonomous and may be carried out concurrently. The subspace and system-level optimizations alternate, linked by sensitivity data producing a design improvement at each iteration of the optimization process. Starting from a best guess initial design, BLISS improves the design in iterative cycles. Each cycle involves two steps. In the first step, the system-level variables are frozen and the design improvement is achieved by separate, concurrent disciplinary optimization. In the second step, additional design improvement is carried out using system-level variables. Optimum-sensitivity data link these two steps. This method has the advantage of using a relatively small number of system-level variables in comparison with other MDO methods. However, the use of sensitivities is critical for the efficiency of the method, and traditional optimization formulation has to change to derivative-based formulation.

5.5.3 Collaborative Optimization

CO is a popular MDO approach that provides design flexibility by using a system-level optimizer to act as an overall design objective subject to disciplinary compatibility constraints (Braun et al., 1997; Gu and Renaud, 2001; Sobieski and Kroo, 1996; Tappeta and Renaud, 1997). In this formulation, different optimizers can be used in different disciplinary optimizations. The system-level optimizer only provides system-level and coupling variables, and local design variables are treated exclusively in the discipline level. At the discipline level, the optimization problem is not a conventional discipline problem, but rather an optimization aiming at

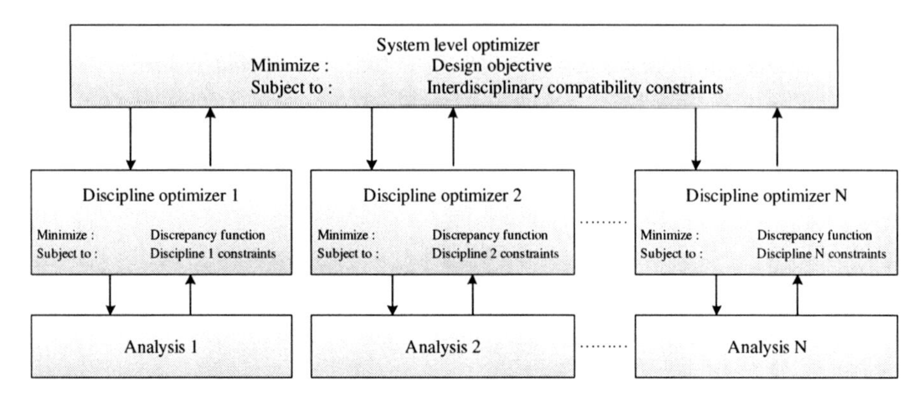

Figure 5.2 Collaborative optimization.

reduction of incompatibility between design variables shared by the disciplines. This formulation permits considerable disciplinary autonomy while achieving interdisciplinary compatibility and is intended for solving design problems with loosely coupled analyses of large-scale multidisciplinary design problems. The main disadvantage of the approach is high computational cost, which limits its application to practical design problems. The main advantages of CO can be summarized as follows:

- System-level optimization generally has a small number of design variables.
- Each discipline is allowed to function autonomously.
- CO is robust in comparison to other multilevel optimization approaches.

5.5.3.1 Decomposition of Coupled Systems into CO

The MDO problem, Eqs. (5.1)–(5.3), can be restated in terms of CO. It can be seen in Figure 5.2 that the problem is hierarchically decomposed along disciplinary analysis boundaries into N disciplinary optimization problems. A system-level optimizer coordinates coupling among the disciplinary optimization problems in CO. Compatibility among the disciplinary optimization problems is measured through the system-level design variables (target values of discipline level). The role of each disciplinary optimizer is to minimize, in a least-squares sense, the discrepancy between the disciplinary design variables and target values provided by the system-level optimizer. To enforce interdisciplinary compatibility in the solution, the optimum disciplinary objectives appear as constraints at the system-level optimizer.

5.6 Optimization Algorithms

There have been many optimization techniques proposed in the literature (e.g., Goldberg, 1989; Haftka and Gürdal, 1992; Luenberger, 1984) each having its own advantages and disadvantages. There are many ways to classify these techniques, starting with classification into the categories of function and parameter techniques. In function optimization, a design problem is described by a number of unknown functions and optimization techniques, such as differential calculus and the calculus of variations. By contrast, in parameter optimization, the optimum values of design variables for a specific problem are obtained using, e.g., mathematical programming, optimality criteria, and heuristic methods. In the following sections, three main categories of optimization techniques are briefly discussed: direct search, gradient search, and heuristic optimization.

5.6.1 Direct Search Methods

Direct search methods, developed in the 1960s and 1970s, are based on well-defined search paradigms that rely only upon the objective function values. The optimization literature contains examples of methods that do not require derivatives and approximate the objective function without recourse to Taylor's expansion. These methods are not considered "direct search methods," but are referred to as *derivative-free* methods. For example, Hooke and Jeeves (1961) considered direct search to involve the comparison of each trial solution with the best previous solution. Hence, a distinguishing characteristic of direct search methods, at least in the case of unconstrained optimization, is that they do not require numerical function values: the relative rank of objective values is sufficient (Lewis, 2000). They accept new iterations that produce a simple decrease in the objective function value.

Historically, most optimization techniques are based on a Taylor series expansion of the objective function and are classified in terms of number of terms used in the expansion. For example, Newton's method, which requires first and second derivatives and uses the second-order Taylor polynomial to build a local quadratic approximation to the objective function, is a second-order method. Steepest descent, which requires the availability of first-order derivatives and uses a first-order Taylor polynomial to build a local linear approximation to the objective function, is a first- order method. Based on this classification, *zero-order* methods (which do not require derivatives and approximation construction) are direct search methods and remain popular as they are easy to use and are less sensitive to numerical noise. Methods that fall into this class include the conjugate direction method of Powell (1964), simplex search methods (in particular Nelder and Mead (1965)), and adaptive random search methods (Luus and Jaakola, 1973).

In recent years, due to an increasing need to incorporate complex simulation models in optimization, a number of researchers have considered direct search methods. Dennis and Torczon (1991) developed a multidimensional search algorithm that extends the simplex method of Nelder and Mead. They concluded that the Nelder and Mead approach fails as the number of design variables increases, even for a simple problem. To address this, their multidimensional simplex method combines reflection, expansion, and contraction steps that act as line search algorithms for a number of linearly independent search directions. This approach is referred to in the literature as a derivative-free optimization technique, and it stimulated research interest in the analysis and development of code for derivative-free optimization methods. For example, Conn et al. (1997) developed a multivariable derivative-free optimization algorithm that uses a surrogate model for the objective function within a trust region method. A number of trust region methods that rely on this method are reviewed in Conn et al. (1997). In addition, a number of derivative-free optimization codes have been developed for large complex models, including the DAKOTA system at Sandia National Lab (Eldred, 2002), and FOCUS, developed at Boeing Corporation (Booker et al., 1998).

The above derivative-free optimization techniques have the advantage of being easy to use for a wide variety of optimization problems and that their termination criteria are not based on gradient information. These methods are more suitable for unconstrained problems or problems with simple bounds.

5.6.2 Gradient-Based Optimization Techniques

There have been numerous optimization techniques making use of first- and higher-order derivatives of the function (with respect to the design variables) that have been developed, tested, and successfully applied to many design optimization problems. The most popular of these methods are conjugate gradient and Newton's method for unconstrained optimization, the penalty function, gradient projection, augmented Lagrangian, and sequential quadratic programming (SQP) methods for constrained optimization.

First-order and higher-order gradient-based techniques make use of function derivatives with respect to the design variables to provide better convergence characteristics of the optimization process. In spite of the widespread application of gradient-based optimization methods, there is to date no universal method for solving all problems. In general, many gradient-based methods are tailored to a particular type of optimization problem. Certain difficulties have prevented the general application of these methods to design optimization problems. These are outlined as follows:

- Functions are often very expensive to evaluate (e.g., finite element analysis of structures or computational fluid dynamics for aerodynamic analysis);
- The existence of noise in the objective and constraint functions, and the presence of discontinuities in the functions;
- Possibility of multiple local minima requiring a global optimization technique;
- The presence of regions in the design space where functions are not defined; and
- Existence of a very large number of design variables.

To address these problems, significant research effort continues to be expended on the metaheuristic optimization techniques described in the section below.

5.6.3 Metaheuristic Optimization Techniques

During the last decades, metaheuristic methods have received considerable attention and have experienced rapid development. Their popularity lies in their ease of use and their ability to locate globally optimum designs. Gradient-based optimization operates on a single potential solution and seeks improvements in its neighborhood. In contrast, metaheuristic methods maintain large sets (populations) of potential solutions and apply recombination operators on them to reach an optimum solution. Their ability to search an entire design space makes them more suitable for handling optimization problems with highly nonlinear objective functions with many local optima. Moreover, they are more suited to optimization problems with discrete design variables. The main disadvantage of metaheuristic methods is high computational costs, which make them difficult to apply to a wide range of optimization problems. Metaheuristic methods include genetic algorithms (GA) (Adeli and Cheng, 1993; Konaka et al., 2006), simulated annealing (SA) (Alrefaei and Diabat, 2009; Bertsimas and Nohadani, 2009; Cemy, 1985; Kirpatrick et al., 1983), particle swarm optimization (PSO) (Kennedy and Eberhart, 1995; Settles, 2005), and ant colony optimization (Dorigo and Stützle, 2004). In addition, variations of the above methods such as bat algorithm (Yang and Gandomi, 2012), cuckoo search algorithm (Gandomi et al., 2011), and mixed variable structural optimization (Gandomi et al., 2012) have been introduced. (For more detail on metaheuristic algorithms, see Chapters 1 and 2.)

5.7 High-Fidelity MDO Using Metaheuristic Algorithms

The computational cost and complexity of typical multidisciplinary systems based on high-fidelity simulation models hinders the application of formal optimization algorithms to this class of problem. The use of approximations to represent the system-design objectives and constraints has become essential for reducing computational expense and achieving good performance for solving MDO problems based on high-fidelity simulation models. Because of this, it has long been recognized that in the field of design optimization, approximations are an effective tool for reducing the computational effort. The basic approach is to replace the computationally expensive simulation model with a simplified mathematical approximation, which is then used in optimization runs. Approximation models act as a *surrogate* for the original model; they are often referred to as a surrogate approximation, an approximation model, or a metamodel ("model of a model").

The use of approximation or response surface methodology in single-discipline optimization in general and in MDO in particular has become popular for reducing computational costs of high-fidelity models in the optimization process. The main benefits of using approximations are summarized as follows:

- The use of low-order polynomials in place of high-fidelity simulation models reduces the number of expensive high-fidelity analyses and can smooth out numerical noise.
- Separation of the analysis code from the optimization routines is enabled and the integration of codes from various disciplines for MDO is facilitated.
- Utilization of parallel computer architecture.

Computational issues of using high-fidelity models in engineering design optimization have been addressed over the past three decades. It was first demonstrated by Schmit and his coworkers that approximations can significantly reduce computational cost in optimization processes (Schmit and Farshi, 1974; Schmit and Miura, 1976). They used approximation to replace the original nonlinear programming problem by a sequence of approximate optimizations in which design constraints were replaced by Taylor series approximations. As the use of approximation models, or surrogates, has grown in popularity, a variety of modeling methods has been employed. A survey of the use of approximations in structural optimization has been carried out by Barthelemy and Haftka (1993). Recently, Toropov and Mahfouz (2001) reviewed some of the modeling and approximation strategies in design optimization.

In this chapter, two levels of metamodel building techniques are introduced in order to significantly reduce the computational effort while handling high-fidelity discipline-level simulation models in CO. Metamodels in the disciplinary optimization are based on multi-fidelity modeling; for system-level optimization, global metamodels are introduced using the moving least-squares method (MLSM) combined with a trust region strategy. The multi-fidelity modeling consists of computationally efficient simplified numerical models (low-fidelity) and expensive detailed (high-fidelity) models. The main advantage of using simplified numerical models is that they reflect the most prominent features of the original model while remaining computationally inexpensive (can be used repeatedly in the optimization process). (Due to space limitations, the main concept of metamodeling is not discussed in this chapter. For more detail on this topic, see reference by Zadeh et al. (2005). The implementation of metamodeling and metaheuristic algorithms for solving an MDO problem using high-fidelity simulation models is illustrated through a test problem in the below section.

5.8 Test Problem

This section describes the application of the metahueristic algorithm and metamodel-based CO framework for solving MDO problems with high-fidelity models used in individual disciplines. The test problem deals with the weight minimization of a cantilevered composite beam subject to a parabolic distributed load (q) as shown in Figure 5.3. The objective function (to be minimized) is the weight of the beam. The design constraints include maximum deflection at the free end of the beam ($\delta_{\max}$), maximum bending stress in the beam ($\sigma_{\max}$), and the geometric requirement that the depth of the beam does not exceed ten times the width (to avoid torsional lateral buckling), as well as the side constraints on the design variables. The design data for this problem are listed in Table 5.2.

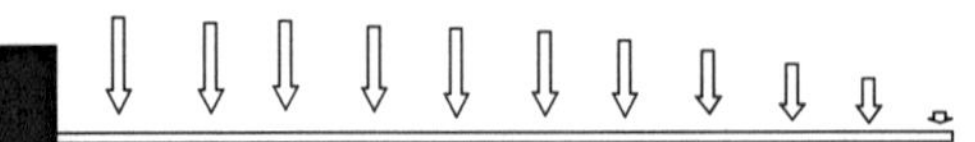

Figure 5.3 Beam subjected to a parabolic distributed load.

The maximum stress and deflection of the beam can be calculated as follows:

$$\sigma_{\max} = \frac{M_{\max} h}{2I} = \frac{3}{2}\frac{q_0 L^2}{wh^2} = \frac{q_0 L^2 h}{8I} \tag{5.4}$$

$$\delta_{\max} = \frac{19 q_0 L^4}{360\, EI} \tag{5.5}$$

where the parabolic load $q(x)$ over the beam is defined as:

$$q(x) = q_0\left(1 - \frac{x^2}{L^2}\right) \tag{5.6}$$

The geometry of the fiber packing in a unit volume that corresponds to the maximum volume of v_f used in this test problem is shown in Figure 5.4. Based on the rule of mixtures for a continuous fiber-reinforced composite material with a fiber volume fraction v_f and a matrix volume fraction v_m, the following relationship must be satisfied:

$$v_f + v_m = 1 \tag{5.7}$$

Table 5.2 Design Data

Description (notation)	Unit	Value
Parabolic distributed load (q_0)	N/mm	1
Length of the beam (L)	mm	1000
Elastic modulus graphite fiber (E_f)	N/mm^2	2.3×0^5
Elastic modulus epoxy resin (E_m)	N/mm^2	3.45×10^3
Weight density graphite fiber (ρ_f)	N/mm^3	1.72×10^{-5}
Weight density epoxy resin (ρ_m)	N/mm^3	1.2×10^{-5}
Stress limit (σ)	N/mm^2	166.667
Displacement limit (δ)	mm	12.9387

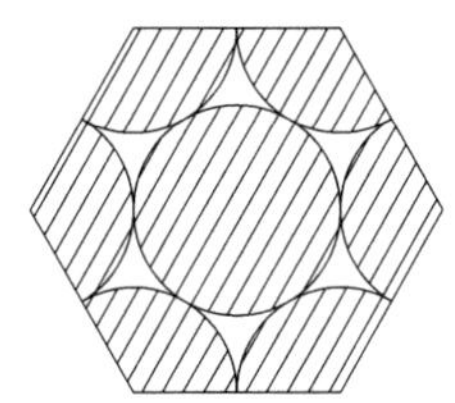

Figure 5.4 Geometry of fiber packing in a unit volume.

The longitudinal (fiber direction) Young's modulus E_f and the composite weight density, ρ, in terms of the fiber volume fraction using the rule of mixtures can be expressed as:

$$E = E_f v_f + E_m(1 - v_f) \tag{5.8}$$

$$\rho = \rho_f v_f + \rho_m(1 - v_f) \tag{5.9}$$

where E_f and E_m are the elastic moduli for graphite fiber and epoxy resin, respectively and ρ_f and ρ_m are the weight density values of the graphite fiber and epoxy resin, respectively.

The fiber volume fraction can vary from zero (no fiber) to the maximum value of v_f^{max}. The maximum possible fiber volume fraction using the packing geometry provided in Figure 5.4 is calculated as follows:

$$v_f^{max} = \frac{A_{fiber}}{A_{total}} = \frac{3\pi R^2}{6\sqrt{3}R^2} = \frac{\pi}{2\sqrt{3}} = 0.9069 \tag{5.10}$$

5.8.1 Conventional Optimization Problem Formulation

The objective is to minimize the weight of a cantilevered composite beam. The design variables are shown in Table 5.3. The function to be minimized is the weight of the composite beam:

Objective: minimize weight $= AL\,\rho$

where

$$A = \frac{12I}{h^2}\ \text{mm}^2; \quad L = 1000\ \text{mm} \quad \text{and} \quad \rho = \rho_m + v_f(\rho_f - \rho_m)$$

The constraints are:

$$g_1(w, h) = \frac{\sigma_{max}}{\sigma} = \frac{q_0 L^2 h}{8I\sigma} = \frac{3q_0 L^2}{2wh^2\sigma} \leq 1 \tag{5.11}$$

Table 5.3 Design Variables

Description (notation)	Unit	Baseline Design	Lower Limit	Upper Limit
Second moment of area (I)	mm^4	2.25×10^4	0.333×10^4	20.833×10^4
Height of the beam (h)	mm	30	20	50
Fiber volume fraction (v_f)		0.785	0.4	0.9069

$$g_2(w,h,v_f) = \frac{\delta_{max}}{\delta} = \frac{19q_0L^4}{36EI\delta} = \frac{19q_0L^4}{30wh^3[E_m + v_f(E_f - E_m)]\delta} \le 1 \quad (5.12)$$

$$g_3(w,h) = \frac{h}{10w} \le 1 \quad (5.13)$$

The inequalities that Eqs. (5.11)–(5.13) represent, respectively, are the constraints on the maximum stress, maximum deflection at the end of the beam, and the geometric requirements that the depth of the beam must be equal to or greater than ten times the width of the composite beam. The limiting values of stress and displacement are shown in Table 5.2, and they correspond to the values for the baseline design. The side constraints on the design variables are obtained by the lower and upper limits shown in Table 5.4.

Conventional (AAO) optimization is carried out using SQP. Results are shown in Table 5.4.

5.8.2 CO Formulation

The test problem (composite beam) described in the above sections is now posed to suit the CO framework. The test problem is decomposed into two disciplines (the stress-constrained and the deflection-constrained problems) and a system-level optimizer to coordinate the overall optimization procedure. The design variables shown in Table 5.4 are grouped into two disciplines, 1 and 2, shown in Table 5.5.

The system-level design variables s_1, s_2, and s_3 represent the second moment of area, depth of the beam, and the fiber volume fraction, respectively. These are treated as system-level target values corresponding to discipline-level design variables

Table 5.4 Results of Optimization (AAO) Using SQP

Design Variables	Unit	Baseline Design	Lower Limit	Upper Limit	Optimum
$I(x_1)$	mm^4	2.25×10^4	0.333×10^4	20.833×10^4	3.361×10^4
$h(x_2)$	mm	30	20	50	44.814
$v_f\ (x_3)$	–	0.785	0.4	0.9069	0.5205
Objective $F(\mathbf{x})$	N	4.8246	–	–	2.9535
$C_1(\mathbf{x})$	1.0	–	–	–	1.0
$C_2(\mathbf{x})$	1.0	–	–	–	1.0
$C_3(\mathbf{x})$	0.3	–	–	–	1.0

Table 5.5 Design Variables and Constraints of the Test Problem Using CO

Discipline Levels	Design Variables	Shared Design Variables	Constraints
Discipline 1	l_1, s_1	s_1	g_1, g_2
Discipline 2	l_2, s_1	s_1	g_3
System level	s_1, s_2, s_3	–	C_1, C_2

s_1, s_2, and s_3, respectively. The design variables l_2 and l_3 are the local design variables for disciplines 1 and 2, respectively. The design variable s_1 is a shared design variable in both disciplines 1 and 2. The optimization objective at the discipline level is to find optimum values of the design variables, which satisfy the discipline's own constraints and minimize the discrepancy from the target values generated by the system-level optimizer. The constraints at the system level are:

$$C_1(\psi_1, s_1) = 0, \quad C_2(\psi_2, s_1) = 0 \tag{5.14}$$

where $C_1(\psi_1, s_1)$ and $C_2(\psi_2, s_1)$ are the discrepancies between the actual and target values returned from discipline 1 and discipline 2, respectively. The system level must satisfy these consistency requirements among disciplines 1 and 2 by enforcing the equality constraints in Eq. (5.14).

5.8.3 Discipline-Level Optimization

This section demonstrates discipline-level optimization of the test problem using a multi-fidelity modeling strategy, focusing on the use of a tuned low-fidelity model in place of a high-fidelity model in the optimization process. The models used here involve three levels of fidelity: a coarse FE model consisting of two elements as a low-fidelity model, a fine-meshed FE model consisting of 100 elements as a high-fidelity model, and a tuned low-fidelity model. The discrepancy function to be minimized in the first discipline is:

$$\text{Minimize: } ||\psi_1 - s_1||^2 \tag{5.15}$$

where s_1 and ψ_1 are the shared design variable (system level) and its local copy (discipline level), respectively. The constraints in discipline 1 are:

$$g_1 = \frac{\delta_{\max}}{[\delta]} \le 1 \tag{5.16}$$

$$g_2 = \frac{l_2^4}{1.2. \times 10^6} \le 1 \tag{5.17}$$

The discrepancy function of discipline 2 can be expressed as:

$$\text{Minimize: } ||\psi_2 - s_1||^2 \tag{5.18}$$

where s_1 and ψ_2 are the shared (system level) and its local copy (discipline level) variable, respectively.

The second discipline has one constraint on the maximum deflection of the beam:

$$g_3 = \frac{\sigma_{\max}}{[\sigma]} \le 1 \tag{5.19}$$

5.8.4 Implementation of Multi-Fidelity Modeling Methodology in CO

The main steps in multi-fidelity modeling methodology applied to discipline level optimization in CO are described below:

Step 1. Choice of a design of experiments: In this step, the scheme suggested by Audze and Eglais (1997) is used to plan P points in the design variable space. In order to use a minimum number of points for the construction of high-quality meta-model models, five separate *designs of experiments* involving a 10-point plan, a 5-point plan and, 4-, 3-, and 2-point plans were studied on the test problem (Zadeh and Toropov, 2002).

Step 2. Run low- and high-fidelity simulation models: The designs of experiments established in step 1 are used to compute response values at the selected plan points using a low-fidelity simulation model, which requires at most a few computer processing unit (CPU) seconds per evaluation. Similarly, a high-fidelity model is used to compute response values at the selected plan points.

Step 3. Choice of metamodel function: In this step, the original high-fidelity model, used for the response analysis, is replaced by a low-fidelity model. The analysis is conducted using a coarser finite element mesh. The low-fidelity model is used to build the metamodel. In this study, several types of metamodel function (linear and multiplicative of types 1 and 2, quadratic and cubic) were examined on a test problem. The selection of the best metamodel function was based on the error of the corrected low-fidelity model as compared to the high-fidelity model at points of a selected plan of experiments, referred to as the verification plan. The low-fidelity model is tuned using a small number of runs of the high-fidelity model (Zadeh and Toropov, 2002).

The performance of six types of metamodel (linear type 1 and 2, multiplicative type 1 and 2, quadratic and cubic polynomials) has been studied for the constraints in both disciplines of the test problem. Each model was verified against an additional verification plan of experiments. The quality of these metamodels is measured by the root mean square (RSM) error and the maximum deviation between the high-fidelity model and the corrected low-fidelity model.

The five-point plan was selected as an appropriate one for building metamodels for constraints in both disciplines. This was verified against the four-point plan. Based on the results, three types of metamodel have been selected, namely the linear and multiplicative type 1, and multiplicative type 2; the linear metamodel type 1 was selected as the appropriate metamodel to be used in the CO framework.

Step 4. Tuning low-fidelity model: Since the metamodel of the original high-fidelity model $F(\mathbf{x})$ by the simplified numerical model $f(\mathbf{x})$ is not exact, the discrepancy between the original and metamodels is to be minimized by using the tuning parameter a. Hence, in building metamodels, $\tilde{F}(x)$, to express them as a function of design variables $\mathbf{x}$ and tuning parameters a.

Step 5. Run optimization using corrected low-fidelity model: In this step, the corrected (tuned) low-fidelity model is used in place of the original high-fidelity model in the optimization process. This model has the same accuracy as the high-fidelity model, and the discrepancy of the optimization solution between the models is negligible.

5.8.5 System-Level Optimization Using MLSM

This section focuses upon the system-level optimization of the test problem, which is derived by:

$$\text{Minimize: } f = \frac{s_1(1440 + 624s_3)}{s_2^2} \tag{5.20}$$

$$\text{subject to: } C_1(\psi_1, s_1) = 0,\ C_2(\psi_2, s_1) = 0 \tag{5.21}$$

$$0.333.10^4 \leq s_1 \leq 20.833.10^4,\ \ 20.0 \leq s_2 \leq 50.0,\ \ 0.4 \leq s_3 \leq 0.9069 \tag{5.22}$$

where s_1, s_2, and s_3 are system-level design variables and $C_1(\psi_1,s_1)$ and C_2 (ψ_2,s_1) are the system-level equality compatibility constraints. The functions $C_1(\psi_1,s_1)$ and $C_2(\psi_2,s_1)$ can be expensive to evaluate and are replaced by inexpensive metamodels. The applications of the steps to the test problem are outlined below:

Step 1. Choice of a design of experiments: The selection of points in the design variable space is based on a uniform Latin hypercube for the case shown in Figure 5.2A and B for disciplines 1 and 2, respectively.

Step 2. Compute corrected low-fidelity model response values: The Design of Experiment (DoE) established in step 1 (Figure 5.2) for disciplines 1 and 2 is used to compute the corrected low-fidelity response values at the selected 50 plan points.

Step 3. Construct metamodel: The corrected low-fidelity model response values calculated in step 2 are used to build global metamodels for disciplines 1 and 2.

Step 4. Study of "closeness of fit" parameter on the test problem: There are several parameters that can be used, such as the size of the domain of influence and the weight decay function to control the quality of the metamodel. In this study, the weighting parameter and additional points are examined for best fit. These are described below:

- *Weighting parameter*: Selection of a suitable expression for the weight decay function plays an important role in the construction of a high-quality metamodel. In this study, the Gaussian function, $w_i = \exp\ (-\theta r_i^2)$, is used as a suitable function to examine the "closeness of fit" (Zadeh, 2005).
- *Additional points*: In order to provide a sufficient number of points for metamodel building, it is necessary to ensure that the sphere of influence has at least $K+N$ points. K is the number of coefficients in the base polynomial (linear, quadratic, or cubic) and N is the number of additional points to provide the necessary amount of redundant information for the least-squares model fitting. One and three additional points were studied on the test problem using a polynomial function (linear, quadratic, and cubic) (Zadeh et al., 2005). In this study, it was found that the quadratic function with one additional point (Figure 5.5) provides an accurate metamodel for the test problem.

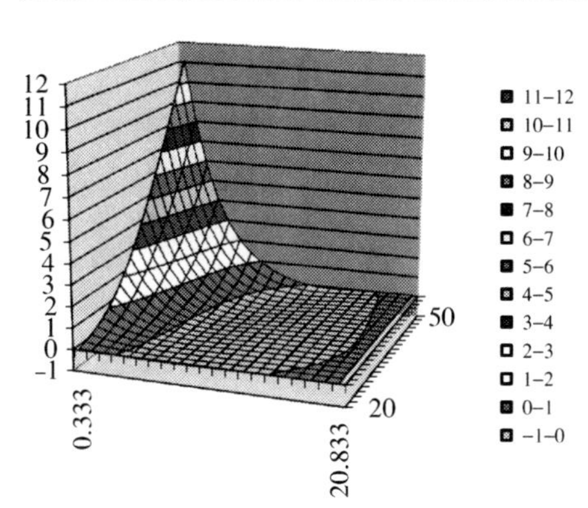

Figure 5.5 $\theta = 10$ with one additional sampling point.

Step 5. Solve the system-level metamodel optimization problem using a GA: Metamodels constructed in step 3 are used in the system-level optimization run. The process treats response values below 0.0002 as zero (due to metamodel error, which is corrected during the optimization process).

Step 6. Trust region: Construct a new sub-region design space, add new plan points, and return to step 2. This step focuses on the localized search for an optimum solution. In this process, a new sub-region of the design space is constructed. This new sub-region is centered on the new design point obtained in step 5, which is resized to 25% of the original size of the design variable space. The new plan points are generated in such a way as to ensure the homogeneous distribution of the points inside the first search sub-region. The metamodel is implemented in the optimization process using a GA and is checked for convergence (stop if convergence is obtained, otherwise the trust region process is continued until the optimum solution is reached).

The trust region is an important part of the methodology described here. If the solution has not converged, the trust region determines the next iteration and next design space search. The acceptance criteria for convergence are based on constraint violation and objective function improvement.

5.8.6 Evaluation of Predictive Capabilities of the Metamodels

The construction of highly accurate metamodels is a requirement for system-level optimization within a CO framework and it is therefore important to evaluate the predictive capabilities of such models. An accuracy estimation using several statistical criteria was used in this work. These include RMS, R^2, relative average absolute error (RAAE), relative maximum absolute error (RMAE), over plan points. The larger R^2 and smaller RMS and RAAE values indicate a more accurate metamodel. (Note the parameters of the MLSM-based metamodels were fixed in all runs.)

In Tables 5.6 and 5.7, R^2 values for disciplines 1 and 2 increase from 0.9021 to 0.998 and from 0.7984 to 0.9909, respectively. This is indicative of the fact that as the design space reduces, the accuracy of the metamodel increases. From

Table 5.6 Evaluation of Predictive Capabilities of Metamodels Constructed During CO Runs for the System Level (Discipline 1)

Iterations	RMS	R^2	RMAE	RAAE
Global metamodel	0.5355	0.9021	1.3235	0.1402
Trust region 1	0.2895	0.9267	0.7582	0.1590
Trust region 2	0.3207	0.9914	0.2255	0.0711
Trust region 3	0.0903	0.9980	0.1173	0.0323

Table 5.7 Evaluation of Predictive Capabilities of Metamodels Constructed During CO Runs for the System Level (Discipline 2)

Iterations	RMS	R^2	RMAE	RAAE
Global metamodel	0.1924	0.7984	2.1062	0.1801
Trust region 1	0.0147	0.9300	0.6985	0.1716
Trust region 2	0.1591	0.9789	0.3612	0.1215
Trust region 3	0.0571	0.9909	0.2858	0.0611

Tables 5.6 and 5.7, it can be seen that values for RMS, RAAE, and RMAE become smaller as the size of the design space reduces. Hence, accuracy of the constructed metamodel increases with the reduction of the size of design space using the move limit strategy.

5.8.7 Optimization Algorithms

In a CO, a single optimization algorithm may be used for both discipline and system-level optimization of the framework. There are several classes of gradient-based method available, including SQP and penalty methods, which can be used in CO. However, the use of equality constraints at the system level to represent the disciplinary feasible regions introduces numerical and computational difficulties that hinder the application of these optimization algorithms at the system level. In this study, three different optimization algorithms (SQP, Nelder and Mead (1965), and GA) were used for the system-level optimization of the test problem in CO. Due to the special features of the system-level optimization, both SQP and Nelder and Mead failed to provide a converged solution and, consequently, a more robust optimization algorithm (GA) was used for solving the CO system-level optimization. SQP was used for solving the optimization problems in disciplines 1 and 2. It can be seen from Table 5.3 that the numerical values of the design variables are different by several orders of magnitude. To avoid numerical difficulties, all design variables at the system level are scaled to lie between 1 and 11. The result of CO based on multi-fidelity modeling on the test problem is shown in Table 5.8.

Table 5.8 Results of Metamodel-Based Collaborative Optimization and Comparison with AAO Method

Iteration Number	Number of Plan Points Used in Disciplines 1 and 2		System Level						Discipline Level				
			Design Variables			Constraints		Objective	Discipline 1			Discipline 2	
	1	2	s_1	s_2	s_3	C_1	C_2	f	Objective	g_1	g_2	Objective	g_3
	50	50	6.771	42.889	0.401	0	0	6.221	0	0.53	0.58	0	0.36
Trust region 1	35	30	2.620	36.577	0.402	0	0	3.312	0	0	0.45	0.07	0
Trust region 2	39	28	3.314	43.350	0.497	0	0	3.085	0	0.02	0.11	0	0
Trust region 3	25	24	3.359	44.819	0.524	0	0	2.955	0	0	0	0	0
AAO			3.361	44.814	0.521	–	–	2.954	–	0	0	–	0

5.9 Conclusions

This chapter presented conventional and MDO problem formulation and application of metahueristic algorithms for solving high-fidelity MDO problems. Multilevel optimization approaches provide an effective decomposition strategy for solving complex MDO problems. The attempt to preserve disciplinary autonomy and reduce system-level complexity are the two most important features of these approaches. CO has been identified in this chapter as one of the most effective multilevel optimization approaches for solving MDO problems. Despite its many advantages, CO has not been widely used due to several difficulties. One of the most prominent difficulties is its extreme computational cost, which hinders the use of high-fidelity simulation models for realistic engineering design problems.

This is due to the fact that CO uses equality constraints at the system level. The values of these system-level constraints are obtained by optimization and hence they are non-smooth at the transition from a plateau of zero values to a region of non-zero values. This feature causes slow convergence of the CO system-level optimization. These characteristics of system-level optimization in CO make it difficult to directly employ conventional approximations; on the other hand, use of metaheuristic algorithms can be computationally prohibitive. To address these issues, a combination of a metaheuristic algorithm (GA) and metamodeling techniques is introduced in this chapter for solving MDO using high-fidelity simulation models. The use of approximations to represent the design objectives and constraints has become vital for reducing computational costs and achieving good performance in most MDO algorithms in general and CO in particular.

Conventional optimization techniques used for solving MDO problems, including direct search, gradient-based, and evolutionary methods each has significant advantages and disadvantages. Despite widespread use of optimization in engineering design, there is still no universal optimization technique that can be reliably and efficiently used in design in general and MDO in particular. However, in recent years, metaheuristic methods (e.g., GA, SA and PSO) have become increasingly popular because of their robustness for solving optimization problems. Application of these optimization techniques in MDO is prohibitively expensive in realistic applications. In order to overcome this, a combination of approximation methods and metaheuristic algorithms appears useful in terms of solving system-level optimization in CO while reducing computational cost.

References

Adeli, H., Cheng, N.T., 1993. Integrated genetic algorithms for optimization of space structures. J. Aerosp. Eng. 6 (4), 315–328.

Alexandrov, N.M., Kodiyalam, S., 1998. Initial results of an MDO Method Evaluation Study. Seventh AIAA/NASA/USAF/ISSMO Symposium on Multidisciplinary Analysis and Optimization. AIAA 98-4884, St. Louis, MO.

Alexandrov, N.M., Lewis, R.M., 1999. Comparative properties of collaborative optimization and other approaches to MDO. Eng. Design. Optim. MCB University Press.
Alexandrov, N.M., Lewis, R.M., 2003. Dynamically reconfigurable approach to multidisciplinary problems. AIAA Conference Paper 2003-3431, Orlando, Florida.
Alrefaei, M.H., Diabat, A.H., 2009. A simulated annealing technique for multi-objective simulation optimization. Appl. Math. Comput. 215, 3029–3035.
Audze, P., Eglais, V., 1997. New approach for planning out of experiments, Problems of Dynamics and Strengths 35, 104–107 (in Russian).
Balling, R.J., Sobieszczanski-Sobieski, J., 1994a. Optimization of coupled systems: a critical overview of approaches. AIAA Paper 94-4330-CP, pp. 753–773.
Balling, R.J., Sobieszczanski-Sobieski, J., 1994b. An algorithm for solving the system level problem in multilevel optimization. Proceedings of the Fifth AIAA/NASA/USAF/ISSMO Symposium on Multidisciplinary Analysis and Optimization. AIAA-94-4333- CP, 7–9 September, Panama City, FL, pp. 494–809.
Balling, R.J., Sobieszczanski-Sobieski, J., 1996. Optimization of coupled systems: a critical overview of approaches. AIAA J. 34 (1), 6–17.
Balling, R.J., Wilkinson, C.A., 1997. Execution of multidisciplinary design optimization approaches on common test problems. AIAA J. 35 (1), 178–186.
Barthelemy, J.F. M., Haftka, R. T., 1993. Approximation Concepts for optimum structural design – A Review, Structural Optimization, 5 (3), 129–144.
Bertsimas, D., Nohadani, O., 2009. Robust Optimization with Simulated Annealing Journal of Global Optimization. Springer Science + Business Media LLC.
Bloebaum, C.L., Hajela, P., Sobieszczanski-Sobieski, J., 1992. Non-hierarchic system decomposition in structural optimization. Eng. Optim. 19, 171–186.
Booker, A.J., Dennis Jr., J.E., Frank, P.D., Serafini, D.B., Torczon, V., Trosset, M.W., 1998. A rigorous framework for optimization of expensive functions by surrogates, CRPC Technical Report 98739, Rice University,Texas, USA.
Braun, R.D. 1996. Collaborative optimization: an architecture for large-scale distributed design. Ph.D. thesis. Stanford University, Department of Aeronautics and Astronautics, California, USA.
Braun, R.D., Kroo, I.M., Moore, A.A., 1996. Use of the collaborative optimization architecture for launch vehicle design. Sixth AIAA/NASA/USAF/ISSMO Symposium on Multidisciplinary Analysis and Optimization. Bellevue, WA. AIAA 96-4018.
Braun, R.D., Moore, A.A., Kroo, I.M., 1997. Collaborative architecture for launch vehicle design. J. Spacecr. Rocket. 34 (4), 478–486.
Cemy, V., 1985. Thermodynamical approach to the travelling salesman problem: an efficient simulation algorithm. J. Optim. Theor. Appl. 45 (1), 41–45.
Conn, A.R., Scheinberg, K., Toint, P., 1997. Recent progress in unconstrained nonlinear optimization without derivatives. Math. Program. 79 (3), 397.
Cramer, E.J., Dennis, J.E., Frank, P.D., Lewis, R.M., Shubin, G.R., 1994. Problem formulation for multidisciplinary design optimization. SIAM J. Optim. 4 (4), 754–776.
Dennis, J.E., Torczon, V., 1991. Direct search methods on parallel machines. SIAM J. Optics. 1, 448.
Dorigo, M., Stützle, T., 2004. Ant colony optimization. first ed. MIT Press Publisher, USA.
Eldred, M., 2002. DAKOTA: A Multilevel Parallel Object-Oriented Framework for Design Optimization, Parameter Estimation, Uncertainty Quantification, and Sensitivity Analysis. Available from: <http://endo.sandia.gov/DAKOTA/software.html/>.
Gandomi, A.H., Yang, X.S., Alavi, A.H., 2011. Mixed variable structural optimization using firefly algorithm. Comput. Struct. 89 (23-24), 2325–2336.

Gandomi, A.H., Yang, X.S., Alavi, A.H., 2012. Cuckoo search algorithm: a metaheuristic approach to solve structural optimization problems. Eng. Comput. doi: 10.1007/s00366-011-0241.

Goldberg, D.E., 1989. Genetic Algorithms in Search, Optimization and Machine Learning. Addison -Wesley, Reading, MA.

Gu, X., Renaud, J.E., 2001. Implicit uncertainty propagation for robust collaborative optimization. Twenty-seventh ASME Design Automation Conference. DAC-21118, 9–12 September, 2001,Pittsburgh, PA.

Haftka, R.T., Gürdal, Z., 1992. Elements of structural optimization. third ed. Kluwer Academic Publishers, London.

Hulme, K.F., 2000. The Design of a simulation-based framework for the development of solution approaches in multidisciplinary design optimization. State University of New York at Buffalo (Ph. D. Dissertation, January, 2000).

Hooke, R., Jeeves, T.A., 1961. Direct search solutions of numerical and statistical problems. J. Assoc. Comput. Mach. 8, 641–652.

Jameson, A., 1997. Essential Elements of Computational Algorithms for Aerodynamic Analysis and Design. Institute for Computer Applications in Science and Engineering, NASA Langley Research Center, Hampton, VA, Tech. Rep. 97–68.

Kennedy, J., Eberhart, R., 1995. Particle swarm optimization. IEEE J. 4, 1942–1948.

Kirpatrick, S., Gellatt Jr., C.D., Vecchi, M.P., 1983. Optimization by simulated annealing. Science. 220 (4598), 671–680.

Kodiyalam, S., Sobieszczanski-Sobieski, J., 2001. Multidisciplinary design optimization: some forma methods, framework requirements, and application to vehicle design. Int. J. Vehicle Design. 25 (1–2), 3–32 (Special Issue).

Konaka, A., Coitb, D.W., Smith, A.E., 2006. Multi-objective optimization using genetic algorithms: a tutorial. Reliab. Eng. Syst. Safety. 91, 992–1007.

Kroo, I.M., 1995. Decomposition and collaborative optimization for large-scale aerospace design programs. Accepted for publication in Multidisciplinary Design Optimization: State of the Art, N. Alexandrov and M.Y. Hussaini, editors, SIAM.

Kroo, I., Altus, S., Braun, R., Gage, P., Sobieski, I., 1994. Multidisciplinary optimization method for aircraft preliminary designProceedings of the Fifth AIAA/NASA/USAF/ ISSMO Symposium on Multidisciplinary Analysis and Optimization. AIAA-94-4333-CP, 7–9 September, Panama City, FL, pp. 697–707.

Lewis, R.M., 1997. Practical aspects of variable reduction formulations and reduced basis algorithms in multidisciplinary design optimization. In: Alexandrov, N.M., Hussaini, M.Y. (Eds.), Multidisciplinary Design Optimization: State-of-the-Art. SIAM, Philadelphia, PA.

Lewis, R.M., Torczon, V., Trosset, M.W., 2000. Direct search methods: then and now. J. Comput. Appl. Math. 124 (1–2), 191–207. <http://rmlewi.people.wm.edu/pubs/LeToTr00a/LeToTr00a.pdf>.

Luenberger, D.G., 1984. Introduction to Linear and Nonlinear Programming. *Addison-Wesley, Reading, MA.*

Luus, R., Jaakola, T.H.I., 1973. Direct search for complex systems. AIChE J. 19, 645–646.

Nelder, J.A., Mead, A., 1965. A simplex method for function minimization. Comput. J. 7, 308–313.

Newman III, J.C., Taylor III, A.C., Barnwell, R.W., Newman, P.A., Hou, G.W., 1999. Overview of sensitivity analysis and shape optimization for complex aerodynamic configurations. J. Aircr. 36 (1), 87–96.

Powell, M.J.D., 1964. An efficient method for finding the minimum of a function of several variables without calculating derivatives. Comput. J. 7, 155–162.

Renaud, J.E., Tappeta, R.V., 1997. Multiobjective collaborative optimization. ASME J. Mech Eng. Des. 119 (3), 403–411.
Schmit, L.A., Farshi, B., 1974. Some approximation concepts for structural synthesis. AIAA J. 11, 489–494.
Schmit, L.A., Miura, H. 1976. Approximation concepts for efficient structural synthesis. NASA CR-2552.
Sellar, R.S., Batill, S.M., Renaud, J.E., 1996. Response surface based, concurrent subspace optimization for multidisciplinary design optimization. Thirty-fourth Aerospace Science Meeting and Exhibit. AIAA 96-0714, 15–18 January 1995, Reno, NV.
Settles, M., 2005. An introduction to particle swarm optimization. GECCO.185–192.
Shankar, J., Ribbens, C., Haftka, R., Watson, L., 1993. Computational study of non-hierarchical decomposition algorithm. Comput. Optim. Appl. 2, 273–293.
Sobieski, I., Kroo, I.M., 1996. Aircraft design using collaborative optimization. Presented at the 34th AIAA Aerospace Science Meeting. AIAA Paper 96-0715, 15–18 January 1996, Reno, NV.
Sobieszczanski-Sobieski, J., 1988. Optimization by decomposition: a step from hierarchic to non-hierarchic systems. Second NASA/Air Force Symposium on Recent Advances in Multidisciplinary Analysis and Optimization. NASA-CP-3031, Hampton, VA.
Sobieszczanski-Sobieski, J., Haftka, R.T., 1996. Multidisciplinary aerospace design optimization: a survey of recent advances. Aerospace Science Conference. AIAA 96-0711, Reno, NV.
Sobieszczanski-Sobieski, J., Haftka, R.T., 1997. Multidisciplinary aerospace design optimization: survey of recent developments. Struct. Optim. 14 (1), 1–23.
Tappeta, R.V., Renaud, J.E., 1997. Multiobjective collaborative optimization. J. Mech. Design. 119, 403–411.
Tatting, B., Gurdal, Z., 2000. Cellular automata for design of two-dimensional structures. Ninth AIAA/USAF/NASA/ISSMO on Multidisciplinary Analysis and Optimization. AIAA-2000-4832.
Toropov, V.V., Mahfouz, S.Y., 2001. Design optimization of structural steelwork using a genetic algorithm, fem and a system of design rules. Eng. Comput. 18 (3/4), 437–459.
Yang, X.S., Gandomi, A.H., 2012. Bat algorithm: a novel approach for global engineering optimization. Eng. Comput. 29 (5), 464–483.
Zadeh P.M., Toropov V.V., 2002. Multi-fidelity multidisciplinary design optimization based on collaborative optimization framework. Ninth AIAA/ISSMO Symposium. Multidisciplinary Analysis & Optimization, Atlanta, Georgia, AIAA 2002 – 5503, 4–6 September.
Zadeh P.M., Toropov V.V., Wood A.S., 2005. Use of moving least squares method in collaborative optimization. Sixth World Congress of Structural and Multidisciplinary Optimization, Rio de Janeiro, Brazil.

6 Cost Optimization of Column Layout Design of Reinforced Concrete Buildings

P. Sharafi, Muhammad N.S. Hadi and Lip H. Teh

School of Civil, Mining and Environmental Engineering, University of Wollongong, NSW, Australia

6.1 Introduction

For the design of structural systems, it is often of interest to determine the general geometric layout of the system that most efficiently supports the expected design actions. Such a design can be done by optimizing the overall layout of the structural system as well as the topology of the structure. Layout optimization is probably the most difficult class of problems in structural optimization. It is also a very important one, because it results in much higher material savings than sizing optimization of cross sections (Rozvany, 1992). In a comprehensive structural optimization process, selecting an appropriate preliminary geometric layout is of enormous importance, as it influences all the succeeding stages of the design procedure.

One of the challenging issues that researchers who study the layout optimization problem are facing is how to efficiently represent the structure's layout so it covers all the related cost elements. In fact, cost variables in combination with layout variables for a large structure may make the optimization problem quite complicated. In the literature, the layout optimization of structures is mostly confined to the weight minimization of structures and related cost components are not taken into account (Bendsøe and Sigmund, 2003; Liu and Qiao, 2011; Rozvany and Olhoff, 2000; Surhone et al., 2010). On the other hand, cost optimization of reinforced concrete (RC) structures is mostly considered for structures with predefined shapes (Adeli and Sarma, 2006; Burns, 2002). In practice, a minimum weight design may not lead to a minimum cost design in RC structures. Ideally, the optimization problem should be considered in terms of costs of materials, fabrication, erection, maintenance, and disassembling the structure at the end of its life cycle.

In designing a building, employing cost optimization methods in the phase of detailed design without considering the effect of cost elements on the preliminary layout design, or employing preliminary layout optimization methods without

Metaheuristic Applications in Structures and Infrastructures. DOI: http://dx.doi.org/10.1016/B978-0-12-398364-0.00006-1

considering the cost of elements, would result in a suboptimal solution. In the present methodology, the preliminary layout design is optimized such that the cost optimization procedure starts from an optimal point and is more likely to get a global optimization compared to other methods. In fact, in a comprehensive structural optimization process, cost optimization of structures with predefined shapes may result in suboptimal solutions as they suffer from the problem of not necessarily having an optimum starting point, and layout optimization of structures without considering all involved cost factors may not lead to globally optimal results either. Therefore, a layout design approach, considering the cost elements, along with a cost optimization method in the phase of detailed design can result in a comprehensive optimization approach.

The reason why the reciprocal effects of shape and the cost elements have not been adequately taken into consideration is that most objective functions make use of cross-sectional variables and relative prices that are indeterminate during the layout planning of a structure. Using such design variables and cost functions in the preliminary geometric layout design leads to a complex formulation for large RC structures (Burns, 2002). Therefore, there is a need to shift from the traditional cost optimization approach to methods that are easily applicable to layout optimization of large structures.

In the preliminary geometric layout design of buildings, the span lengths, as geometric layout variables, may be determined based only on architectural requirements and constraints. In this case, selecting the optimum among the possible layouts can result in a substantial cost saving, as the primary layout design will influence the entire design process. Therefore, an optimization procedure for this phase of design that takes the related cost elements into account, along with a cost optimization method in the phase of detailed design, can lead to a comprehensive optimal design procedure.

In recent decades, new classes of stochastic metaheuristics have been developed that are generic, population-based, and use bio-inspired mechanisms like mutation, crossover, natural selection, and survival of the fittest. These algorithms have found many applications in structural engineering as well (Gandomi et al., 2012).

In this chapter, a methodology is proposed to deal with the cost optimization problem of column layout design of RC frame buildings using a max–min ant system (MMAS) algorithm (Stutzle and Hoos, 1997). The outcome can be employed as an optimum starting point for the rest of the design process, say the phase of the detailed design, in a comprehensive optimization procedure.

Sharafi et al. (2012a,b) have presented such a methodology for column layout optimization of RC structures without considering the concrete floor slabs. However, concrete floor slabs constitute a significant portion of the total cost of an RC building. Therefore, the effect of floor slabs should ideally be taken into account together with other structural members. In the present work, all structural members including beams, columns, and slabs are considered as part of an integrated system.

6.2 Statement of the Problem

In general, structural designers are faced with four classes of design variables: material design variables such as the type of concrete and the grade of steel, topological variables such as the number of members in a structure, geometric layout variables such as the lengths of members, and cross-sectional variables such as the dimensions of sections. By exploiting the mathematical relationships between design parameters, one may be able to shift from one set of variables to another. In an optimization problem, such a transition leads to a new definition for the objective function and constraints. In other words, depending on the nature of the optimization problem, the process of achieving an optimum solution may be much quicker, than shifting from one design space to another by changing design variables as the space dimensions.

In an optimization procedure, the definition of the cost function may be considered the most important decision, which represents the aim of the problem. Therefore, it is essential to define a cost function that represents the most influential cost components and, more importantly, is applicable to a variety of optimization problems. Furthermore, it must be capable of matching the explicit constraints of structures, which are often given by formulae in design standards. A cost function generally can include the cost of materials, transportation, fabrication, and even maintenance costs, in addition to repair and insurance costs, which can be represented by a weighted sum of a number of factors. The effects of these factors on the total cost can be imposed on the weighted coefficients of the cost function.

In concrete structures, at least three different cost items should be considered in an optimization problem: costs of concrete, steel, and the formwork. In the earlier studies, the general cost function for RC slabs is expressed in the following form (Adeli and Sarma, 2006):

$$C = c_c A_c + c_{sl} A_{sl} + c_{sv} A_{sv} + c_f P_f \tag{6.1}$$

in which c_c, c_{sl}, c_{sv}, and c_f are, respectively, the unit costs for concrete, longitudinal steel, shear steel, and formwork, and A_c, A_{sl}, A_{sv}, and P_f are their corresponding quantities.

Equation (6.1) represents a cost function that only deals with cross-sectional variables that is mainly appropriate for structures with a small number of members and a predefined geometric layout. In fact, in the layout design of structures, the cross-sectional variables are functions of design action effects that are indeterminate and vary as the layout changes. Therefore, in an iterative procedure of solving an optimization problem, each step includes dealing with both the structural analysis and the structural design variables. In such cases, unless alternative design variables are selected for the cost function, the optimization procedure might be too unwieldy. Since cross-sectional variables are obtained from implicit functions of structural analysis outputs based on the relations and constraints in the design standards, which do not uniquely provide the exact values for these cross-sectional parameters, they are not obtained from an explicit mathematical procedure.

If the reciprocal relationships between the cross-sectional design variables and the design action effects are established, the cost function can be represented by a function of design action effects. In the design of a structural member, design action effects are only determined for the critical sections of the member and cross-sectional variables are calculated for each critical section. In other words, knowing the design action effects in critical sections, all the cross-sectional variables and consequently, the total cost defined by Eq. (6.1) can be calculated.

Using structural analysis outputs, say internal actions of a member, as design variables has some advantages over using structural design outcomes, such as cross-sectional variables of a beam. First, design action effects of each section can be easily obtained from structural analysis, and in an iterative mathematical procedure, reanalyzing a structure is considerably less time-consuming and more precise than redesigning the structure. Second, using action effects, the cost function will be considered as a section rather than a member. It enables the designer to select a number of sections for each member and in the whole structure to control the cost, and there is no necessity to conduct the optimization process over the entire member.

The present work finds the relations between the variations of action effects of RC members with the variations of the cross-sectional parameters and determines how these two types of variables affect each other. A new cost function is then defined in a way that it is a function of action effects rather than cross-sectional parameters. The formulation is carried out based on the Australian Standards for RC structures (AS3600, 2009).

6.3 Formulation in a New Space

In the present formulation of the optimization problem of column layout design of RC buildings, three types of structural members are considered: slabs, columns, and beams.

6.3.1 Slabs

Consider a beam-slab floor with N_x and N_y spans in the x and y directions, respectively. The floor is divided by $NS = NS_x + NS_y$ critical sections: NS_x sections in the x direction and NS_y sections in the y direction as shown in Figure 6.1. The critical sections are mainly selected near the supports and mid-spans. Each section contains N_b beam sections and N_s slab sections. The total cost of the floor can be represented by the sum of the individual costs of the critical sections. Now, if the relationship between cross-sectional parameters and the design action effects are established, the cost function can be defined in terms of action effects.

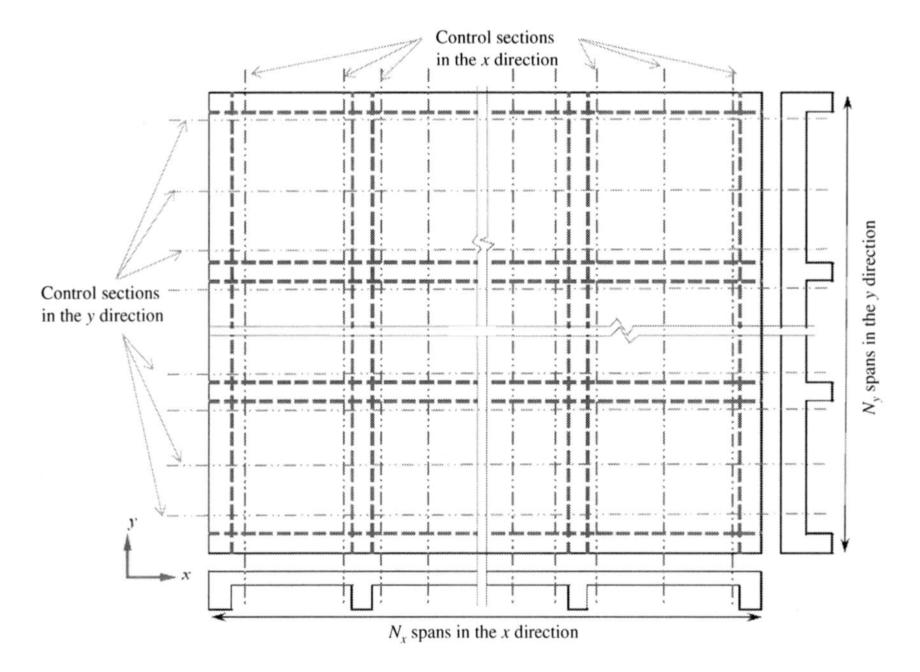

Figure 6.1 Spans and control sections in the x and y directions of slabs.

Based on the simplified design method for slabs in the Australian Standard for concrete structures (AS3600, 2009), the only action effect that needs to be considered is the bending moment in strips spanning in the x and y directions. These strips can be formed by taking the control sections along the slabs in both directions, and the moment distribution along the strips edges is determined accordingly.

Consider Eq. (6.2) as a potential alternative cost function to Eq. (6.1) in a slab

$$C_i^{(s)} = c_1 M_{u_i}^{(s)} \tag{6.2}$$

where $C^{(s)}$ is the cost of each slab section and $M_{u_i}^{(s)}$ is the bending moment capacity of each slab at a critical section i and c_1 is a coefficient. For the entire slab floor including N_s sections, variations in section capacities would change the total cost of the sections as follows:

$$\Delta C^{(s)} = \sum_{1}^{N_s} \Delta C_i^{(s)} = \sum_{1}^{N_s} c_1 \Delta M_{u_i}^{(s)} \tag{6.3}$$

Alternatively, according to Eq. (6.1), if any of the cross-sectional parameters changes, the cost function of slabs varies as follows:

$$\Delta C^{(s)} = c_c \Delta A_c^{(s)} + c_{sl} \Delta A_{sl}^{(s)} \tag{6.4}$$

In layout optimization of slab floors, the shear reinforcement and the formwork costs can be removed from the calculation process. The reasons are, first, slabs are considered not to be reinforced for shear, and second, the total area of slabs is constant, while floor layout and span lengths have no effect on the final amount of formwork.

In order to shift from Eq. (6.1) to Eq. (6.2) and come up with the weighted coefficient c_1, the first step is to determine how the variations of A_c and A_{sl} affect $M_u^{(s)}$ and vice versa. That is, we need to know how the varying amount of each cross-sectional parameter influences the section strength capacities.

For slabs, both the positive and the negative bending moment capacities are obtained from the ultimate strengths of the section in flexure $M_u^{(s)}$, which can be calculated from Eq. (6.5) (Loo and Chowdhury, 2010).

$$\begin{cases} M_u^{(s)} \cong A_s^{(s)} f_{yl}(D^{(s)} - c^{(s)} - d_c{}^{(s)}) \\ d_c^{(s)} = 0.5\gamma k_u(D^{(s)} - c^{(s)}) \end{cases} \tag{6.5}$$

where $D^{(s)}$ is the thickness of the slab. Other parameters are defined in Figure 6.1.

The variation of bending moment capacity of the slabs with respect to $A_s^{(s)}$ is as follows:

$$\begin{aligned} \frac{\Delta M_u^{(s)}}{\Delta A_s^{(s)}} &\cong f_{yl}(D^{(s)} - c^{(s)} - d_c^{(s)}) \rightarrow \Delta A_s^{(s)} \\ &= (f_{yl}(D^{(s)} - c^{(s)})(1 - 0.5\gamma k_u))^{-1} \Delta M_u^{(s)} = K_1 \Delta M_u^{(s)} \end{aligned} \tag{6.6}$$

The only parameter that affects the variation of the volume of concrete in slabs is the thickness of the slab. That is,

$$\Delta A_c^{(s)} \cong L_s \cdot \Delta D^{(s)} \tag{6.7}$$

where L_s is the width of the slab. Any changes in the thickness of the slab results in variation of the bending moment capacity of the slab as follows:

$$\begin{aligned} \frac{\Delta M_u^{(s)}}{\Delta A_c^{(s)}} \cong \frac{\Delta M_u^{(s)}}{L_s \Delta D^{(s)}} &\cong \frac{f_y A_s^{(s)}}{L_s}(1 - 0.5\gamma k_u) \rightarrow \Delta A_c^{(s)} \\ &\cong \left[\frac{f_y A_s}{L_s}(1 - 0.5\gamma k_u)\right]^{-1} \Delta M_u^{(s)} = K_2 \Delta M_u^{(s)} \end{aligned} \tag{6.8}$$

Multiplying both sides of Eq. (6.8) by c_c and comparing it with Eqs. (6.3) and (6.4) results in:

$$c_1 = c_{sl} K_1 + c_c K_2 \tag{6.9}$$

The coefficient c_1 determines how the parameter $M_{u_i}^{(s)}$ contributes to the slab's cost function, as shown in Eq. (6.2).

6.3.2 Beams

For a beam to be designed, three action effects are taken into account: positive bending moment, negative bending moment, and shear forces. For an arbitrary section of a rectangular RC beam as shown in Figure 6.2, the dimensions of the section are b and $h^{(b)}$, the areas of tension and compression reinforcement steel are $A_{st}^{(b)}$ and $A_{sc}^{(b)}$, and the area of shear reinforcement steel in a unit length of beam is $A_{sv}^{(b)}/s$. The capacity or the ultimate strength of the section in negative and positive flexure and shear for section i are $M_{u_i}^{-(b)}$, $M_{u_i}^{+(b)}$, and $V_{u_i}^{(b)}$, respectively.

Consider Eq. (6.10) as a potential alternative cost function to Eq. (6.1) in the ith RC beam section.

$$C_i^{(b)} = c_2 M_{u_i}^{+(b)} + c_3 M_{u_i}^{-(b)} + c_4 V_{u_i}^{(b)} \tag{6.10}$$

The variations in section capacities would change the cost function as follows:

$$\Delta C^{(b)} = \sum_{1}^{N_b} \Delta C_i^{(b)} = \sum_{1}^{N_b} \left(c_2 \Delta M_{u_i}^{+(b)} + c_3 \Delta M_{u_i}^{-(b)} + c_4 \Delta V_{u_i}^{(b)} \right) \tag{6.11}$$

where N_b is the number of beams' control sections. On the other hand, using Eq. (6.1), if any of the cross-sectional parameters change, the cost function varies as follows:

$$\Delta C^{(b)} = c_c \Delta A_c^{(b)} + c_{sl} \Delta A_{sl}^{(b)} + c_{sv} \Delta A_{sv}^{(b)} + c_f \Delta P_f^{(b)} \tag{6.12}$$

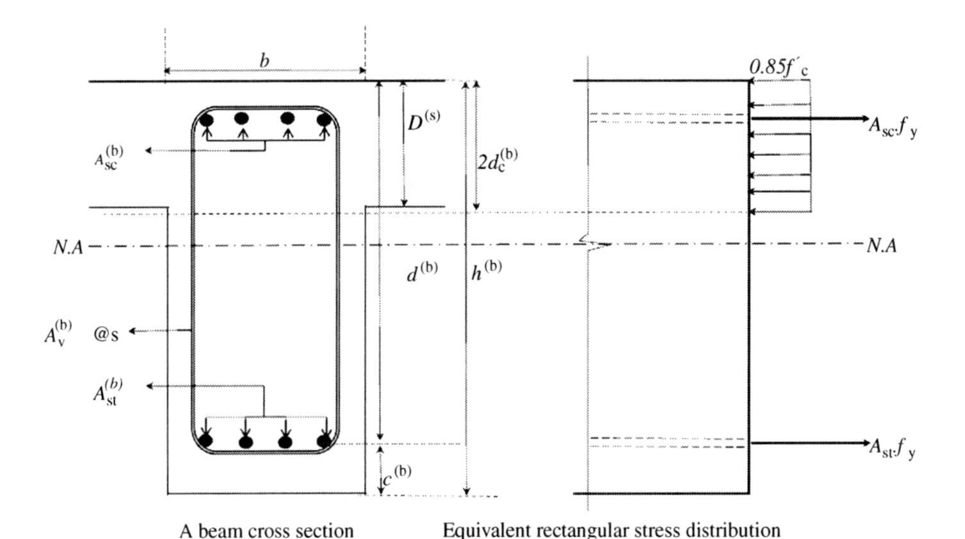

Figure 6.2 A general beam cross section and the related parameters.

Equations (6.11) and (6.12) show the contribution of each factor to cost changes and sensitivity of the cost to each term. For example, changing a unit of $A_{\mathrm{c}}^{(\mathrm{b})}$ causes a change of c_c units in cost. Therefore, if the effects of variations of $A_{\mathrm{c}}^{(\mathrm{b})}$, $A_{\mathrm{sl}}^{(\mathrm{b})}$, $A_{\mathrm{sv}}^{(\mathrm{b})}$, and $P_{\mathrm{f}}^{(\mathrm{b})}$ on the variations of $M_u^{+(\mathrm{b})}$, $M_u^{-(\mathrm{b})}$, and $V_u^{(\mathrm{b})}$ are established, the contribution of each section capacity to cost changes; that is, the set of $\{c_2, c_3, c_4\}$ in Eq. (6.10), can be determined. Sharafi et al. (2012b) established the set of $\{c_2, c_3, c_4\}$ based on the Australian Standard for concrete structures (AS3600, 2009), as shown in Eq. (6.13)

$$
\begin{cases}
c_2 = \dfrac{1}{3}c_{\mathrm{c}}K_6 + \dfrac{1}{2}c_{\mathrm{sl}}K_3 + \dfrac{2}{3}c_{\mathrm{f}}K_9 \\[2ex]
c_3 = \dfrac{1}{3}c_{\mathrm{c}}K_5 + \dfrac{1}{2}c_{\mathrm{sl}}K_3 + \dfrac{2}{3}c_{\mathrm{f}}K_8 \\[2ex]
c_4 = \dfrac{1}{3}c_{\mathrm{c}}K_7 + c_{\mathrm{sv}}K_4 + \dfrac{2}{3}c_{\mathrm{f}}K_{10} \\[2ex]
-- -- -- -- -- -- -- \\
K_3 = \left(f_{\mathrm{yl}}d^{(\mathrm{b})}(1-0.5\gamma k_u)\right)^{-1} \\
K_4 = (f_{\mathrm{yv}}d)^{-1}\Delta V_u^{(\mathrm{b})} \\
K_5 = \left[f_{\mathrm{yl}}\dfrac{A_{\mathrm{sc}}^{(\mathrm{b})}}{\mathrm{b}}(1-0.5\gamma k_u)\right]^{-1} \\[2ex]
K_6 = \left[f_{\mathrm{yl}}\dfrac{A_{\mathrm{st}}^{(\mathrm{b})}}{\mathrm{b}}(1-0.5\gamma k_u)\right]^{-1} \\[2ex]
K_7 = \left[f_{yv}\dfrac{A_{sv}^{(\mathrm{b})}}{bs} + \beta\left(f_c^{\prime}\right)^{0.5}\right]^{-1} \\[2ex]
K_8 = 2\left[f_{\mathrm{yl}}A_{\mathrm{sc}}^{(\mathrm{b})}(1-0.5\gamma k_u)\right]^{-1} \\
K_9 = 2\left[f_{\mathrm{yl}}A_{\mathrm{st}}^{(\mathrm{b})}(1-0.5\gamma k_u)\right]^{-1} \\
K_{10} = \left[\dfrac{2}{\dfrac{f_{\mathrm{yv}}A_{\mathrm{sv}}^{(\mathrm{b})}}{s} + \beta b(f_{\mathrm{c}}^{\prime})^{\frac{1}{2}}} + \dfrac{1}{\beta d(f_{\mathrm{c}}^{\prime})^{\frac{1}{2}}}\right]
\end{cases}
\tag{6.13}
$$

where f_{yl} is the yield strength of the longitudinal reinforcement, f_{yv} is the yield strength of the shear reinforcement, $f_{\mathrm{c}}^{\prime}$ is the characteristic compressive cylinder strength of concrete at 28 days, $d_c^{(b)}$ is the distance from the extreme compression fiber of the concrete to the compressive force. The coefficients γ and k_u are calculated based on the characteristic strength of the concrete and reinforcing steel, β is a coefficient based on the standard, and s is the center-to-center spacing of shear reinforcement. Other parameters are defined in Figure 6.2.

The obtained coefficients c_2, c_3, and c_4 determine how the variations of $M_{u_i}^{+(b)}$, $M_{u_i}^{-(b)}$, and $V_{u_i}^{(b)}$ contribute to the variation of cost function for a beam.

6.3.3 Columns

For an arbitrary section of a square RC column as shown in Figure 6.3, the dimensions of the section are $h^{(c)}$, the area of the longitudinal reinforcement is $A_{sl}^{(c)}$, and the area of shear reinforcement in a unit length of column is $A_{sv}^{(c)}/s$.

Consider Eq. (6.14) as a potential alternative cost function to Eq. (6.1) in an arbitrary RC column section

$$C_i^{(c)} = c_5 N_{u_i}^{(c)} + c_6 M_{u_i}^{(c)} + c_7 V_{u_i}^{(c)} \tag{6.14}$$

in which $N_{u_i}^{(c)}$, $M_{u_i}^{(c)}$, and $V_{u_i}^{(c)}$ are the axial load-carrying capacity, bending moment capacity, and the shear capacity of the ith column section, respectively. The variations in section capacities would change the cost function as follows:

$$\Delta C^{(c)} = \sum_{1}^{N_c} \Delta C_i^{(c)} = \sum_{1}^{N_c} \left(c_5 \Delta N_{u_i}^{(c)} + c_6 \Delta M_{u_i}^{(c)} + c_7 \Delta V_{u_i}^{(c)} \right) \tag{6.15}$$

where N_c is the number of columns' control sections. On the other hand, using Eq. (6.1), if any of the cross-sectional parameters changes, the cost function varies as follows:

$$\Delta C^{(c)} = c_c \Delta A_c^{(c)} + c_{sl} \Delta A_{sl}^{(c)} + c_{sv} \Delta A_{sv}^{(c)} + c_f \Delta P_f^{(c)} \tag{6.16}$$

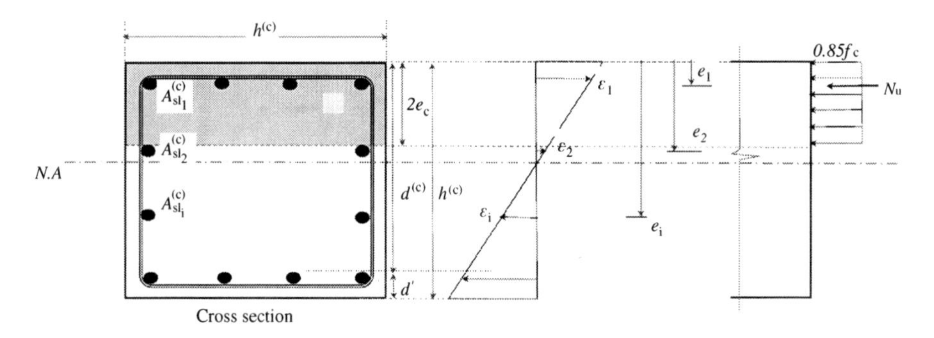

Figure 6.3 A general column cross section and the related parameters.

Equations (6.15) and (6.16) show the contribution of each factor to cost changes and sensitivity of the cost to each term. If the effect of variations of $A_{\mathrm{c}}^{(\mathrm{c})}$, $A_{\mathrm{sl}}^{(\mathrm{c})}$, $A_{\mathrm{sv}}^{(\mathrm{c})}$, and $P_{\mathrm{f}}^{(\mathrm{c})}$ on the variations of $N_{u}^{(\mathrm{c})}$, $M_{u}^{(\mathrm{c})}$, and $V_{u}^{(\mathrm{c})}$ are established, the contribution of each section capacity to cost changes; that is, the set of $\{c_5, c_6, c_7\}$, can be found. Sharafi et al. (2012b) established the set of $\{c_5, c_6, c_7\}$ based on the Australian Standard for concrete structures (AS3600, 2009), as shown in Eq. (6.17).

$$\begin{cases} c_5 = \dfrac{1}{3}c_{\mathrm{c}}K_{14} + \dfrac{1}{2}c_{\mathrm{sl}}K_{11} + \dfrac{1}{3}c_{\mathrm{f}}K_{17} \\ c_6 = \dfrac{1}{3}c_{\mathrm{c}}K_{15} + \dfrac{1}{2}c_{\mathrm{sl}}K_{12} + \dfrac{2}{3}c_{\mathrm{f}}K_{18} \\ c_7 = \dfrac{1}{3}c_{\mathrm{c}}K_{16} + c_{\mathrm{sv}}K_{13} + \dfrac{2}{3}c_{\mathrm{f}}K_{19} \\ K_{11} = \left(E_{\mathrm{s}}\displaystyle\sum_{i}^{n}\varepsilon_i\right)^{-1} \\ K_{12} = \left(E_{\mathrm{s}}\displaystyle\sum_{i}^{n}\varepsilon_i e_i\right)^{-1} \\ K_{13} = (f_{\mathrm{yv}}d^{(\mathrm{c})})^{-1} \\ K_{14} = (0.85\gamma k_u f'_{\mathrm{c}})^{-1} \\ K_{15} = (0.85\gamma k_u f'_{\mathrm{c}}e_{\mathrm{c}})^{-1} \\ K_{16} = \left(\dfrac{f_{\mathrm{yv}}A_{\mathrm{sv}}^{(\mathrm{c})}}{2hs} + \beta(f'_{\mathrm{c}})^{0.5}\right)^{-1} \\ K_{17} = \left(0.85\gamma k_u f'_{\mathrm{c}}\dfrac{h^{(\mathrm{c})}}{2}\right)^{-1} \\ K_{18} = \left(0.85\gamma k_u f'_{\mathrm{c}}\dfrac{h^{(\mathrm{c})}}{2}e_{\mathrm{c}}\right)^{-1} \\ K_{19} = \left(\dfrac{f_{\mathrm{yv}}A_{\mathrm{sv}}^{(\mathrm{c})}}{4s} + \dfrac{1}{2}\beta h^{(\mathrm{c})}(f'_{\mathrm{c}})^{0.5}\right)^{-1} \end{cases} \tag{6.17}$$

The obtained coefficients c_5, c_6, and c_7 determine how the variations of bearing capacities $N_{u_i}^{(\mathrm{c})}$, $M_{u_i}^{(\mathrm{c})}$, and $V_{u_i}^{(\mathrm{c})}$ contribute to the variation of cost function.

6.4 The Optimization Problem

Consider a multispan RC building with N_x and N_y spans and total lengths of L_x and L_y in the x and y directions under an arbitrary loading system $f(x)$. The aim is to redesign the preliminary column layout to determine the optimum span lengths in each direction in order to minimize the cost. The final cost will be a function of two sets of variables. According to Eqs. (6.9), (6.13), and (6.17), the total cost is a function of the sections' action effects under the loading system, which in turn are functions of the span lengths that can be represented as follows:

$$C = \sum_{1}^{N_s} c_1 \Delta M_{u_i}^{(s)} + \sum_{1}^{N_b} \left(c_2 \Delta M_{u_i}^{+(b)} + c_3 \Delta M_{u_i}^{-(b)} + c_4 \Delta V_{u_i}^{(b)} \right) + \sum_{1}^{N_c} \left(c_5 \Delta N_{u_i}^{(c)} + c_6 \Delta M_{u_i}^{(c)} + c_7 \Delta V_{u_i}^{(c)} \right) \tag{6.18}$$

Having the relevant coefficients, one can use Eq. (6.18) instead of Eq. (6.1). In this case, the cost is the sum of the cost functions of all selected sections in the structure based on Eq. (6.18). Using this function as alternative cost functions to Eq. (6.1), the design variables can be shifted from cross-sectional variables to action effects. For the structural optimization problem, the strength constraints on each selected section i, whether for the beams or the columns, under a load case may be written as:

$$\begin{cases} \varnothing N_{u_i} \geq N_i^* \\ \varnothing M_{u_i} \geq M_i^* \\ \varnothing V_{u_i} \geq \left| V_i^* \right| \\ U_{\max_i} \leq \Delta_{\max_i} \end{cases} \tag{6.19}$$

in which, N_i^*, M_i^*, and $\left| V_i^* \right|$ are the axial force, the flexure, and shear action effects of section i, in either columns, beams, or slabs, $\varnothing$ is the strength reduction factor, and $\Delta_{\max}$ is the maximum deflection limitation $U_{\max}$ on the entire member under the serviceability load case.

For an RC building with N_x spans in the x direction and N_y spans in the y direction, under an arbitrary loading system f to be optimally designed for preliminary geometric layout, the general formulation for the structural optimization problem can be written as follows:

$$\begin{cases} \min\limits_{l_1, l_2, \ldots, l_{\mathrm{NSP}}} \quad \mathrm{Cost}(l_1, l_2, \ldots, l_{\mathrm{NSP}}) = \sum_{1}^{N_b} C_i^{(b)} + \sum_{1}^{N_c} C_i^{(c)} + \sum_{1}^{N_s} C_i^{(s)} \\ \text{------------------------------} \\ s.t \begin{cases} \text{strength constraints and serviceability requirements based on Eq. (6.19)} \\ \text{Architectural constraints like } \{l_{\min}\} \leq \{l_i\} \leq \{l_{\max}\} \text{ : for all spans} \\ \text{other constraints based on design standards} \end{cases} \end{cases} \tag{6.20}$$

As a structural optimization problem, Eq. (6.20) can be dealt with employing various optimization methods. The nature of the geometric layout optimization of structures is a continuous problem. However, it can be dealt with as a discrete problem by discretizing the domain. In fact, in the design process, we usually deal with the dimensions as discrete sizes, and it is often the case that design variables must be chosen within a discrete manufacturer's inventory. The number and size of reinforcing bars and dimensions of concrete sections are usually treated discretely, since in practice these parameters are rounded to the nearest whole inch (or 10 mm), or the existing manufacturing products that causes the optimization problem to be treated as a discrete problem. Therefore, one can define the optimization problem of RC sections as a discrete optimization problem rather than a continuous one. In the following section, an ant colony optimization (ACO) algorithm, as a discrete optimization tool, is proposed to solve the problem.

In traditional cost optimization of RC structures, say using Eq. (6.1) as the objective function, the design variables are the cross section and the lengths of members. Displacements and stresses, as the responses of the structure, form the state variables. Therefore, the constraints stated by Eq. (6.19) are state constraints, which act on state variables. Using the objective function, stated in Eq. (6.20) as an alternative function, causes the state variables of traditional method to change to design variables, and consequently all state constraints change to design constraints. The only design variable, which is common in both methods, is the span lengths. Using the alternative objective function has no effect on behavioral constraints, which are mainly governed by structural analysis principles. However, the behavioral constraints are imposed on sections' bending moments, shear forces, and displacements of control sections, rather than the entire structure. Moreover, removing the designed section variables, such as the sections' area or reinforcement details, from the design variables, helps the iterative optimization procedure not to deal with the design parameters.

6.5 ACO Algorithm for Column Layout Optimization

The ACO algorithms mimic the characteristics of real ants that can rapidly establish the shortest route from a food source to their nest and vice versa (Gandomi et al., 2013). The ACO method can be categorized as a stochastic method. There are various types of ACO algorithms with a variety of implementations (Dorigo and Stützle, 2004). Most of the present ACO algorithms, such as MMAS, are direct successors to the ant system (AS) algorithm. The main phases of the AS algorithm are the ants' solution construction and the pheromone update.

MMAS introduces some main modifications with respect to AS. First, it strongly exploits the best solution found. Only the iteration-best ant or the best-so-far ant is allowed to deposit the pheromone. Such a strategy may result in a stagnation in which all the ants follow the same solution, due to the excessive growth of pheromone trails on arcs of a good, although suboptimal, tour. To counteract this effect, a second modification introduced by MMAS limits the possible range of pheromone trail values to the interval $[\tau_{min}, \tau_{max}]$. Moreover, the pheromone trails are initialized to the upper pheromone trail limit, which, together with a small pheromone evaporation rate, increases the exploration of solutions at the start of the search. Finally, in MMAS, pheromone trails are reinitialized each time the system approaches stagnation or when no improved solution has been generated for a certain number of consecutive iterations.

MMAS achieves a strongly improved performance, compared to AS and to other improved versions of AS, for a wide range of combinatorial optimization problems. One feature MMAS has in common with other improved AS algorithms is the fact that the best solutions found during the search are strongly exploited to direct the ants' search. MMAS has related this feature to recent results of the analysis of search space characteristics for combinatorial optimization problems. Earlier research has shown that there exists a strong correlation between the solution quality and the distance to a global optimum for the travelling salesman problem (TSP) and for some other problems (Dorigo and Stützle, 2004). MMAS can provide an effective guidance mechanism to direct the search toward the best solutions. Yet, exploitation of the best solutions is not the only remedy to achieve very high-performing ACO algorithms. To avoid premature convergence, the exploitation of the best solutions has to be combined with effective mechanisms for performing search space exploration. MMAS explicitly addresses this aspect, which is possibly the main reason why it is currently one of the best performing ACO algorithms (Stutzle and Hoos, 1997). The main ideas introduced by MMAS, the utilization of pheromone trail limits to prevent premature convergence, can also be applied in a different way, which can be interpreted as a hybrid between MMAS and the action choice rule.

The principles of the ACO algorithm are presented by Sahab et al. (2012). The present ACO algorithm for column layout optimization of RC buildings is an MMAS. The MMAS algorithm comprises three phases: initializing data, constructing ant solution, and updating pheromone. The aim of these phases is to determine a set of integers as the spans' lengths in both x and y directions to minimize the cost function. The algorithm deals with the discrete form of the problem. The domain must be discretized by forming the construction graph as a multilayered graph (Sahab et al., 2012). In the column layout optimization of frames, the number of layers represents the number of design variables, which are spans' lengths in the x and y directions, and the number of nodes in a particular layer represents the number of discrete probable values permitted for the corresponding design variable. Thus, each node on the graph is associated with an allowable discrete value of a design variable, which results in a graph with $N_x + N_y$ layers. Knowing that each span length is bounded in $[L_{e\ max}, L_{e\ min}]$, the permissible values for each span length, which are represented by the nodes on the graph, can be discretized with intervals (accuracy) equal to ε. That is, each L_e rests in the set of $\{L_{min}, L_{min} + \varepsilon, L_{min} + 2\varepsilon, \ldots, L_{max} - \varepsilon, L_{max}\}$. Each member of this set corresponds to a node on the graph. So, the number of nodes for each layer is $(L_{e\ max} - L_{e\ min})/\varepsilon$. The smaller ε is chosen, the more accurate the results will be and the more running time the algorithm needs.

In the phase of initial design, some parameters need to be assigned values. The heuristic values are specified according to the designers' preferences. In the present case, if there are any preferences for a certain span, say due to architectural constraints, higher values are assigned to the heuristic arrays corresponding to that span. Such an assignment will cause the desired lengths to be more likely chosen by ants for the corresponding spans. By choosing appropriate values for α and β, the degree of influence of the heuristic values relative to the pheromone trail is set. In order to save time, some criteria such as the symmetry of the plan are considered when the heuristic matrix is formed. Using such a heuristic matrix, or defining the initial pheromone matrix using the above-mentioned structural rules, helps the algorithm converge sooner.

The initial magnitude of pheromone on the construction graph in MMAS is initialized to the upper pheromone trail limit, which is a value slightly higher than the expected amount of pheromone deposited by the ants in one iteration. A rough estimate of the values can be obtained by setting $\forall(i,j)\tau_{ij} = \tau_0 = N_A/C^{nn}$, where N_A is the number of ants and C^{nn} is the cost of a tour generated by the nearest neighbor heuristic (Kaveh and Sharafi, 2008a,b, 2009). Since boundary conditions, constraints, and the loading condition can affect the length of the span, the entries of the heuristic matrix and the initial pheromone matrix might be organized in a way that considers such parameters according to a designer's experience. Each value (i,j) of the choice information matrix, which is obtained by multiplying the corresponding arrays of the heuristic matrix by those of the pheromone matrix, shows the tendency or desirability of the ant located on node i of the construction graph to choose edge j to move toward node $i+1$.

N ants are located on the home node and construct their solution by selecting only one node in each layer in accordance with the random proportional rule (also known as action choice rule) given by Eq. (6.21).

$$p_{ij}^k = \begin{cases} \dfrac{[\tau_{ij}]^\alpha [\eta_{ij}]^\beta}{\sum_{l \in N_i^k} [\tau_{il}]^\alpha [\eta_{il}]^\beta} & \text{if } j \in N_i^k \\ 0 & \text{if } j \notin N_i^k \end{cases} \tag{6.21}$$

In each iteration, there are as many as $(L_{max} - L_{min})/\varepsilon$ possible options (nodes) to be selected. This relationship forms the basis of the AS algorithm and shows that if ant k is positioned on node i, it will move to the next node j with the probability of p_{ij}^k. In this relationship, τ_{ij} is the magnitude of pheromone on the trails and η_{ij} is the heuristic value. The parameters α and β determine the relative influence of the pheromone trail and the heuristic information, respectively. N_i^k is the feasible neighborhood of ant k when being at node i; that is, the set of edges that ant k is allowed to choose as its next destination and is decided depending on the problem condition. The nodes selected along the path visited by an ant represent a candidate solution.

After all ants have constructed a solution, pheromones are updated by applying evaporation (lowering the pheromone value on all edges) by a constant factor ρ

according to Eq. (6.22), followed by the deposit of new pheromone (adding pheromone on selected edges) according to Eq. (6.23).

$$\tau_{ij} \leftarrow (1-\rho)\tau_{ij} \quad \forall(i,j)\in A \tag{6.22}$$

$$\tau_{ij} \leftarrow \tau_{ij} + \Delta\tau_{ij}^{\text{best}} \quad \forall(i,j)\in A \tag{6.23}$$

where $\Delta\tau_{ij}^{\text{best}}$ is $1/C^{\text{best}}$. The ant that is allowed to add pheromone may be either the best-so-far, in which case $\Delta\tau_{ij}^{\text{best}}$ is $1/C^{\text{bs}}$, or the iteration-best, in which case $\Delta\tau_{ij}^{\text{best}}$ is $1/C^{\text{ib}}$, where C^{ib} is the length of the iteration-best tour.

Pheromone deposit increases the probability that other ants will follow the same path. That is, by letting ants deposit a higher amount of pheromone on short paths, the ants' path searching is more quickly biased toward the best solutions. Pheromone evaporation can be seen as an exploration mechanism that avoids quick convergence of all the ants toward a suboptimal path. In fact, the decrease in pheromone intensity favors the exploration of different paths during the whole search process. Making pheromone update a function of the generated solution quality can help in directing future ants more strongly toward better solutions.

In general, in MMAS, both the iteration-best and the best-so-far update rules are alternatively used. Obviously, the choice of the relative frequency with which the two pheromone update rules are applied has an influence on how greedy the search is. In MMAS, lower and upper limits $\tau_{\min}$ and $\tau_{\max}$ on the possible pheromone values on all edges are imposed in order to avoid search stagnation. MMAS uses $1/C^{\text{bs}}$, to define $\tau_{\max}$: Each time a new best-so-far tour is found, the value of $\tau_{\max}$ is updated. The lower pheromone trail limit is set to $\tau_{\min}=\tau_{\max}/a$, where a is a parameter (Stutzle and Hoos, 1997).

In each iteration, all the ants, in parallel, start from the home node and end at the destination node by randomly selecting a node in each layer. The optimization process is terminated if no better solution is found in a prespecified number of successive iterations. The above procedure continues until the termination criterion is satisfied; then, the optimum spans will be given by the best-so-far solution.

In each step, a basic structural analysis is required to calculate the section action. Making use of a structural analysis approach, say finite element method or simplified method, based on the design standards, such a calculation takes a small amount of time for each section. Using Eq. (6.1) as the objective function would be significantly more time-consuming than using Eq. (6.18). Due to the variation of lengths in each step, the estimation of design variables in each step would be considerably encumbered, apart from the fact that, in each step, the algorithm would iteratively need to deal with the design formulas. Equation (6.18) makes the optimization process much more tangible and greatly usable in the layout optimization of large structures.

6.5.1 Numerical Example

A four-story RC building with a plan as shown in Figure 6.4 is optimized. The live load on intermediate floors is 5.0 kN/m^2 and on the roof is 1.5 kN/m^2. The dead

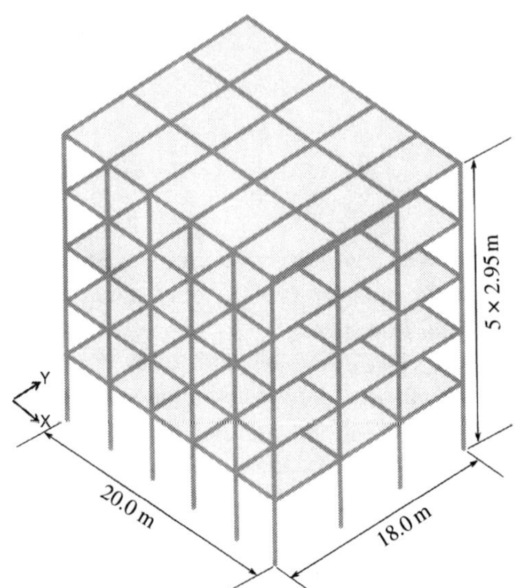

Figure 6.4 Numerical example: a four-story RC building.

loads comprise self weight plus an imposed load of 1.5 kN/m^2. The average unit price for concrete is assumed to be 54 units/m^3 and 3140 units/m^3 for reinforcing steel. The average unit price for formwork is 19 units/m^2. The characteristic strength of the main reinforcement f_y is 460 N/mm^2; the characteristic strength of the shear reinforcement f_{yv} is 250 N/mm^2; the characteristic strength of concrete f'_c is 35 N/mm^2; the top and bottom covers of steel bars are 20 and 25 mm for slabs, respectively; and the cover of bars in columns is 40 mm.

This example was analyzed in a report on comparative costs of concrete framed buildings (Goodchild et al., 2009) that has been recommended as a benchmark for future studies. The conventional design of this example has been carried out and optimized by a team of professional structural engineers. The total length and width of the building is 20 m and 18 m, respectively, and the height of each story is 2.95 m. The permissible spans are defined within the bounds of $L_{max} = 8.5$ m and $L_{min} = 5$ m. Goodchild et al. (2009) suggested the cost of 239,575 units for $L_{x1} = L_{x2} = L_{x3} = L_{x4} = 5.0$ m and $L_{y1} = L_{y2} = L_{y3} = 6.0$ m.

Three control sections are selected for every member. The values of K_1 through K_{19}, and consequently the values of c_1 through c_7 for all selected slab, column, and beam sections were obtained. Having the necessary coefficients, the optimization problem can be formulated based on Eq. (6.20). Now, considering the above design as a primary design for the proposed ACO algorithm, the ACO algorithm searches for the optimum spans based on the objective function and observing other relevant constraints.

Computations were performed on P9700 at 2.80 GHz computer running MATLAB R2009b. The termination criterion for the ACO algorithms is defined as the number of iterations, when the improvement in the solution quality was less than 0.02% after 10 consecutive iterations. After 110 iterations, and at CPU time of

56.6 s, the optimum span lengths of $L_{x1} = 4550$, $L_{x2} = 5450$, $L_{x3} = 5450$, and $L_{x4} = 4550$ mm in the x direction, and $L_{y1} = 5850$, $L_{y2} = 6300$, and $L_{y3} = 5850$ mm in the y direction were obtained, resulting in a total cost of 217,534 units, which equals a 9.2% cost saving compared to the initial design.

6.6 Conclusions

The main objective of this study is to propose a new cost optimization model for column layout design of RC buildings. Using cross-sectional action effects in critical sections as design variables in lieu of cross-sectional design parameters is the distinct advantage of the present method. The cost function presented in this study simplifies the process of cost and layout optimization and is capable of being easily employed in topology optimization of RC structures. The proposed cost function can act more efficiently in an iterative optimization procedure than its traditional counterpart, as it takes advantage of structural analysis variables instead of design variables. The proposed algorithm for solving the optimization problem is an MMAS algorithm, which can be easily employed for optimizing the span lengths and is capable of working with different cost functions. The presented example shows that the proposed algorithm using the new cost optimization function provides satisfactory results.

References

Adeli, H., Sarma, K., 2006. Cost Optimization of Structures: Fuzzy Logic, Genetic Algorithms, and Parallel Computing. Wiley, England.

AS3600, 3600. Concrete Structures. Standards Association of Australia, Sydney.

Bendsøe, M.P., Sigmund, O., 2003. Topology Optimization: Theory, Methods, and Applications. Springer, New York.

Burns, S.A., 2002. Recent Advances in Optimal Structural Design, American Society of Civil Engineers. Structural Engineering Institute. Technical Committee on Optimal Structural Design.

Dorigo, M., Stützle, T., 2004. Ant Colony Optimization. MIT Press, Cambridge, Massachusetts.

Gandomi, A.H., Yang, X.S., Talatahari, S., Alavi, A.H., 2013. Metaheuristics in modeling and optimization. In: Gandomi, et al., (Eds.), Metaheuristic Applications in Structures and Infrastructures. Elsevier, Waltham, MA (Chapter 1).

Goodchild, C.H., Webster, R.M., Elliott, K.S., 2009. Economic concrete frame elements to Eurocode 2: a pre-scheme handbook for the rapid sizing and selection of reinforced concrete frame elements in multi-storey buildings designed to Eurocode 2, The Concrete Center, Michael Burbridge Ltd, Maidenhead, UK.

Kaveh, A., Sharafi, P., 2008a. Ant colony optimization for finding medians of weighted graphs. Eng. Comput. 25, 102–120.

Kaveh, A., Sharafi, P., 2008b. Optimal priority functions for profile reduction using ant colony optimization. Finite Elem. Anal. Des. 44, 131–138.

Kaveh, A., Sharafi, P., 2009. Nodal ordering for bandwidth reduction using ant system algorithm. Eng. Comput. 26, 313–323.

Liu, S., Qiao, H., 2011. Topology optimization of continuum structures with different tensile and compressive properties in bridge layout design. Struct. Multidiscip. Optim. 43, 369–380.

Loo, Y.C., Chowdhury, S.H., 2010. Reinforced and Prestressed Concrete: Analysis and Design with Emphasis on Application of AS3600-2009. Cambridge University Press, New York.

Rozvany, G.I.N., 1992. Shape and Layout Optimization of Structural Systems and Optimality Criteria Methods. Springer, New York.

Rozvany, G.I.N., Olhoff, N., 2000. Topology Optimization of Structures and Composite Continua. Kluwer Academic Publishers, Netherlands.

Sahab, M.G., Toropov, V.V., Gandomi, A.H., 2012. A review on traditional and modern structural optimization: problems and techniques. In: Gandomi, et al., (Eds.), Metaheuristic Applications in Structures and Infrastructures. Elsevier, London (Chapter 2).

Sharafi, P., Hadi, M.N.S., Teh, L.H., 2012a. Geometric design optimization for dynamic response problem of continuous reinforced concrete beams. J. Comput. Civ. Eng. doi: 10.1061/(ASCE)CP.1943-5487.0000263 (In Press.)

Sharafi, P., Hadi, M.N.S., Teh, L.H., 2012b. Heuristic approach for optimum cost and layout design of 3D reinforced concrete frames. J. Struct. Eng. 138, 853–863.

Stutzle, T., Hoos, H., 1997. Max–min ant system and local search for the traveling salesman problem. In: IEEE International Conference on Evolutionary Computation, 13–16 April 1997, pp. 309–314.

Surhone, L.M., Tennoe, M.T., Henssonow, S.F., 2010. Topology Optimization. VDM Verlag Dr. Mueller AG & Co. Kg.

Weise, T., 2009. Global Optimization Algorithms—Theory and Application (Germany: it-weise.de (self-published), 2009). Available from: <http://www.it-weise.de/>.

7 Layout Design of Beam–Slab Floors by a Genetic Algorithm

Pruettha Nanakorn[1] *and Anan Nimtawat*[2]

[1]Sirindhorn International Institute of Technology, Thammasat University, Thailand, [2]Faculty of Technology, Udon Thani Rajabhat University, Thailand

7.1 Introduction

This chapter presents how layout design of beam–slab floors can be performed by a genetic algorithm (GA). Among various types of floor systems used in buildings, the beam–slab floor system is the most versatile one. This is because this floor system can effectively be used in most types of buildings. Design of a beam–slab floor begins with layout design of beams, slabs, and columns. Architects often fix the locations of columns during the architectural design process. Consequently, structural designers have to determine the layouts of beams and slabs that efficiently utilize the prescribed columns. Most of the time, structural designers have no difficulty in creating beam–slab layouts and can come up with good layouts without any design calculations. In fact, they use their own experience and good engineering sense to perform the task. With these characteristics, beam–slab layout design is clearly a heuristic task, and it is therefore difficult to automate this task using computers. Consequently, this simple task prevents the whole design process from being completely automated.

7.1.1 Heuristic Versus Algorithmic Design Tasks

For practical purposes, the whole structural design process of a building is usually broken into several intermediate design tasks. Certainly, an entire structural design process must have its design goals, which include design objectives and satisfaction of design constraints. Each intermediate design task within the whole design process possesses its own intermediate design goals. The intermediate design goals of each intermediate design task must move the design toward the main design goals. Designers have to make sure that when a design acceptably satisfies all intermediate design goals, it must also acceptably satisfy all the main design goals. An obvious difficulty that arises from the subdivision of an entire structural design process into several intermediate tasks is that the values of design variables in different intermediate design tasks are mostly dependent on each other due to design

Metaheuristic Applications in Structures and Infrastructures. DOI: http://dx.doi.org/10.1016/B978-0-12-398364-0.00007-3

constraints. Changing the value of one design variable in an intermediate task will definitely affect the appropriate values of other design variables in other intermediate tasks. In theory, all design variables should therefore be considered together.

Figure 7.1 schematically shows a general structural design process. Some intermediate design tasks are heuristic design tasks while some are algorithmic design tasks. We may define a heuristic design task as a task that does not have explicit relationships between its design goals and design variables. We may define an algorithmic design task as a task that has explicit relationships between its design goals and design variables in the form of algorithms or equations. There can be various reasons that a heuristic intermediate design task does not have explicit relationships between its design goals and design variables. For instance, it may be possible to express the objectives of an intermediate design task only qualitatively, not quantitatively. Beam–slab layout design is a good example of this particular shortcoming. Designers know very well that their objective in designing beam–slab layouts is to create layouts that result in safe and economical beam–slab floors. In the scope of the beam–slab layout design process, the design results are only beam–slab layouts. The design results will not include the detailed structural designs of beams, slabs, and columns. Thus, within the beam–slab layout design task, it is virtually not possible to write the objective of the task quantitatively. As a result, the objective is expressed only in a qualitative form. Consequently, an explicit relationship between this objective and the design variables, which are the positions of beams, slabs, and columns, practically does not exist. When designers perform an intermediate design task that is heuristic, they mainly use their experience and intuition to perform the task. It is not possible to spell out how the final design is obtained, except perhaps in hindsight. On the contrary, when designers perform an intermediate design task that is algorithmic, they often employ a trial-and-error design method, which can be called the conventional design method. The designers will use their experience and intuition to guess the initial design. After that, the design is checked against all intermediate design constraints as well as all intermediate design objectives. If the design is satisfactory, then the intermediate design task is completed. If the design is not satisfactory, it is incrementally improved by the designers by using the information of the employed design algorithms or equations as well as their experience and intuition. The updated design is checked against the intermediate constraints and objectives again and will be improved repeatedly until it is satisfactory. During the checking process, it is also possible that, in order to obtain a usable design, the designers must alter the design significantly by redoing some earlier design tasks. A structural design process usually starts with heuristic design tasks because, in the early design steps, there is not much information about the design itself. After more details of the design become available in the later design steps, algorithmic design tasks start to appear.

Theoretically, any algorithmic design process can be formulated as an optimization problem. After the formulation, we can select an appropriate optimization method to solve the representative optimization problem. On the contrary, heuristic design processes cannot be formulated as optimization problems simply because there are no explicit relationships between their design goals and design variables.

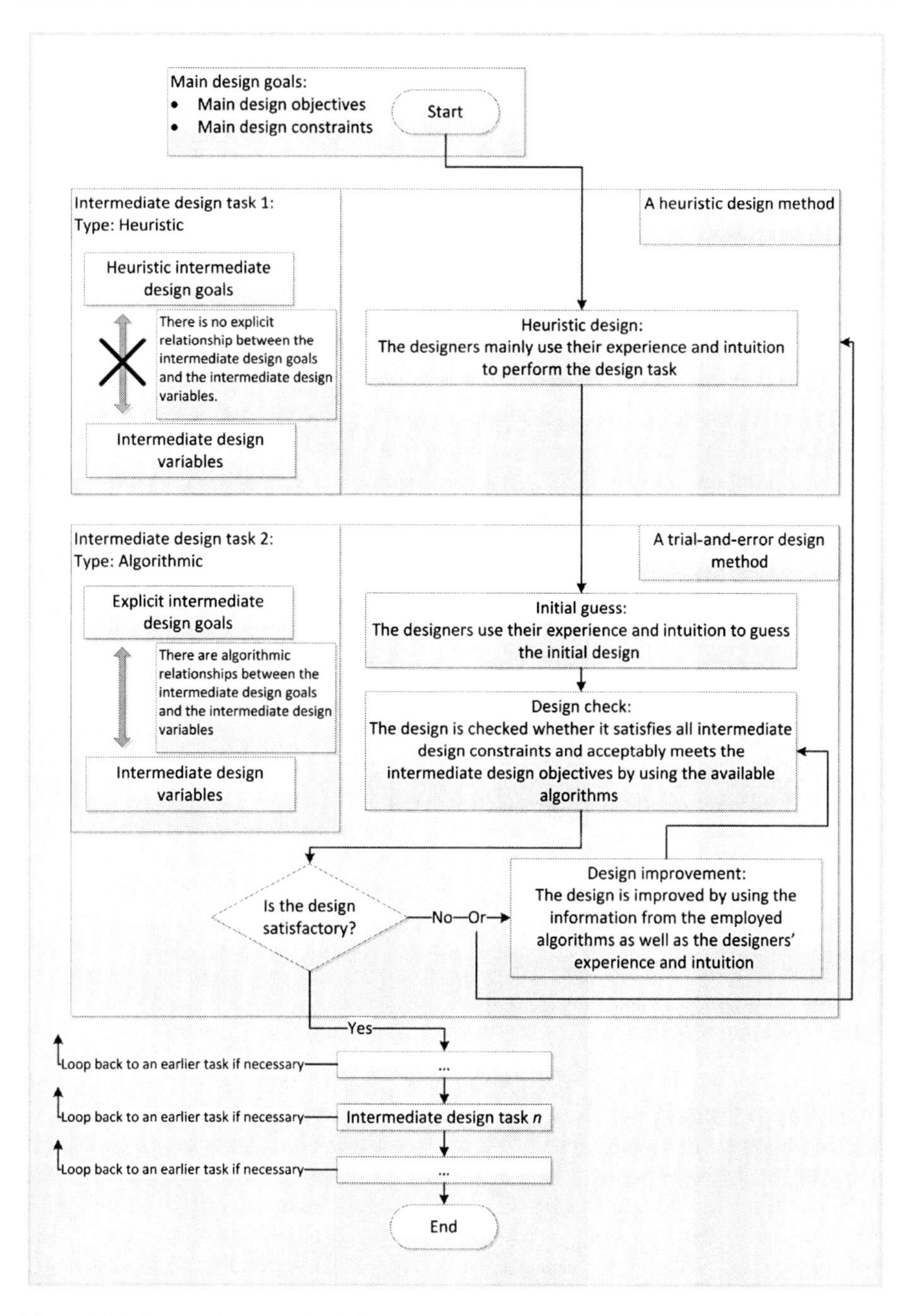

Figure 7.1 A general structural design process.

Formulating an algorithmic design process as an optimization problem and employing an efficient optimization method to solve the obtained problem certainly improves the design solution when compared with the solution obtained from the conventional trial-and-error method. In addition, using an optimization method allows the process to be solved by computers with minimum human involvement. As a result, the process can be automated. It is therefore desirable to remove heuristic intermediate design tasks from the whole design process in order that we can completely automate the whole process.

7.1.2 Conversion of Heuristic to Algorithmic Tasks

One way to remove heuristic intermediate design tasks is to consider the whole structural design task as one design task. Automatically, the task becomes an algorithmic design task. This is because, in any valid structural design process, there must be algorithms for checking any design that comes out of the complete design process against all of the design goals. Considering the whole design process as one task means that all the design variables of the whole process have to be considered together at once. Unfortunately, the number of the design variables to be considered together will be so large that it will be virtually impossible to construct this one design problem that includes all the design variables. The amount of information required for constructing one big design problem that includes all the design variables will simply be too large. However, it is quite likely that, in the future, when the required information is available in manageable forms, this the design approach will be doable and become a practical approach. Another way to remove heuristic intermediate design tasks is to convert these tasks into algorithmic ones. Since the tasks are heuristic, a question of how they can be converted into algorithmic tasks immediately arises. In fact, the representative algorithmic tasks of heuristic tasks must be heuristically created. In other words, the explicit design goals of a heuristic design task and their algorithmic relationships with the design variables must be established by means of explicit rules of thumb or estimated facts.

7.1.3 Beam–Slab Layout Design as an Optimization Problem

In this chapter, we discuss how the beam–slab layout design problem can be converted from a heuristic design task to an algorithmic one. After the conversion, the obtained algorithmic task can then be solved by any appropriate optimization method. However, we will discuss only how a GA can be developed to solve this particular problem. There exist many heuristic optimization methods (Dorigo et al., 1996; Gandomi et al., 2011, 2013; Goldberg, 1989; Kennedy and Eberhart, 1995; Kirkpatrick et al., 1983; Sahab et al., 2013). The GA is considered to be one of the best. The scope of the discussion is limited to those cases where the positions of columns and walls are given. The idea of performing floor layout design by computers has been explored by many researchers. The most popular computing methods used in this problem include knowledge-based expert systems (KBESs), case-based reasoning (CBR), and GAs. Examples of these works are those by Bailey and Smith (1994),

Maher (1984), Nimtawat and Nanakorn (2009, 2010), Park and Grierson (1999), Rafiq et al. (2003), Shaw et al. (2008), Soibelman and Pena-Mora (2000), Syrmakezis and Mikroudis (1997), and Tsakalias (1994). KBESs and CBR utilize knowledge and past experience to obtain the design. GAs, however, treat the layout design problem as an optimization problem. In many cases, a representative optimization problem is constructed by basing the objective function on the estimated cost or profit of the building project as well as the space utilization (Park and Grierson, 1999; Rafiq et al., 2003). Nimtawat and Nanakorn (2009, 2010) proposed a representative optimization problem that is based on heuristic engineering considerations. In their study, the objective function mimics real structural designers' reasoning by heuristically considering how well slabs are supported by columns. The representative optimization problem is solved by using a GA. The primary input of their algorithm is an architectural floorplan with given positions of columns and walls. The methodology used by Nimtawat and Nanakorn (2009, 2010) will form the main discussion in this chapter.

In order to solve the beam–slab layout design problem by using an optimization method, two important steps must be taken:

1. A good representation of beam–slab layouts must be created.

   The beam–slab layout design problem can be formulated as an optimization problem only after a data structure that can efficiently represent spatial configurations of beam–slab layouts is created. For example, if a GA is to be used, a binary coding system to create binary codes that represent different beam–slab layouts must be devised. The representation of beam–slab layouts, in fact, plays a very important role in the success of any employed optimization method.
2. A representative optimization problem for the beam–slab layout design problem must be developed.

   Since the beam–slab layout design problem is a heuristic design problem, a representative optimization problem will have to be heuristically developed. This is done by establishing an appropriate objective function and appropriate constraints for the problem.

These two steps will be now discussed.

7.2 A Representation of Beam–Slab Layouts

Often the difficulty in formulating an optimization problem for a design problem begins with the representation or coding of design solutions. This is true also for the beam–slab layout design problem. To create an optimization problem for the beam–slab layout design problem, it is necessary to develop an efficient way to represent beam–slab layouts. It is usually sufficient to limit the scope of the problem to rectilinear floors. A rectilinear floor is defined as a polygonal floor, all of whose edges meet at right angles. The development of a representation for beam–slab layouts can start from the definition of slabs. It is practical to define a slab as a rectangular area that is completely surrounded by beams. Thus, a beam–slab layout can be considered as a spatial arrangement of rectangular subareas that completely fill the entire floor area without having any nonrectangular subareas. Even with this simple definition of a layout, it is quite difficult to create a coding system that can efficiently represent all the possible layouts. The causes of the difficulty are quite obvious. The number of slabs, their dimensions, and their arrangement can be varied almost arbitrarily within the geometry constraints of the floor. An acceptable representation has to be able to take care of all of these variations. There are similar classes of problems called rectangle packing and floorplanning (Balasa, 2000; Huang et al., 2007; Murata et al., 1996; Yao et al., 2003). In a rectangle packing or floorplanning problem, a set of rectangles is given. The rectangles are to be placed in a bounding rectangular box. One of the common objectives of this type of problem is to find the smallest possible bounding box that can accommodate all the given rectangles. The smallest bounding box is found by finding the optimal arrangement of the given rectangles. The representation of floorplans has been intensively studied. The concepts of graphs and permutations are often used to create representations of floorplans (Balasa, 2000; Murata et al., 1996; Yao et al., 2003), which are appropriate since the rectangles are given from the beginning in this type of problem. Nevertheless, floorplan representations are still considered complex.

To establish a representation for beam–slab layouts, the following aspects have to be considered:

1. The representation has to be able to represent all possible layouts.
2. If the representation allows patterns that are not considered as layouts, such as patterns with nonrectangular subareas or patterns with unsupported beams, it is desirable that the number of these invalid patterns is not large.
3. It is possible that several representation codes correspond to the same layout. This problem is described as the redundancy of the representation. A one-to-one mapping between the representation code and the corresponding layout without any redundancy is obviously preferred.
4. Low time complexity in the translation of representation codes into corresponding layouts is desirable.

It is difficult to establish a representation for beam–slab layouts that fulfills all the above demands without any flaws. The difficulty can be largely removed if only floors with no beam intersections that form T-shaped interior junctions are considered (Maher, 1984; Rafiq et al., 2003; Sisk et al., 2003). Figure 7.2 shows examples of layouts with and without T-shaped interior junctions. Unfortunately, most practical beam–slab layouts utilize T-shaped interior junctions of beams, and therefore a practical representation cannot avoid this type of layout.

7.2.1 A Representation of Beam Locations

To develop a representation for beam–slab layouts, it is natural to start from defining possible locations of beams. Then, coding that designates whether beams exist or not at these predefined possible locations can be created. This can be done quite easily by superimposing a grid on the architectural floorplan under consideration.

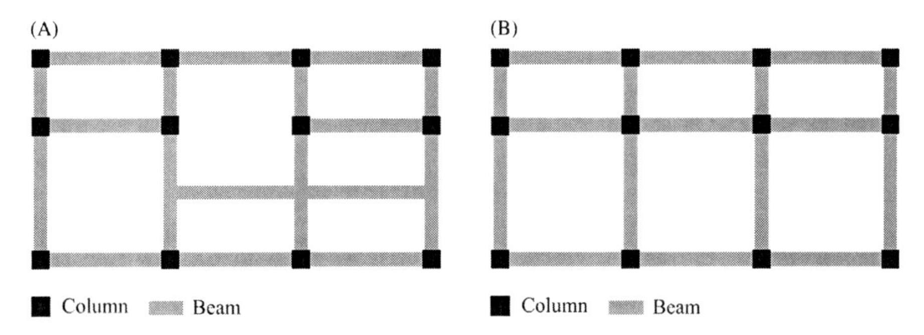

Figure 7.2 (A) A layout with T-shaped interior junctions. (B) A layout without T-shaped interior junctions.

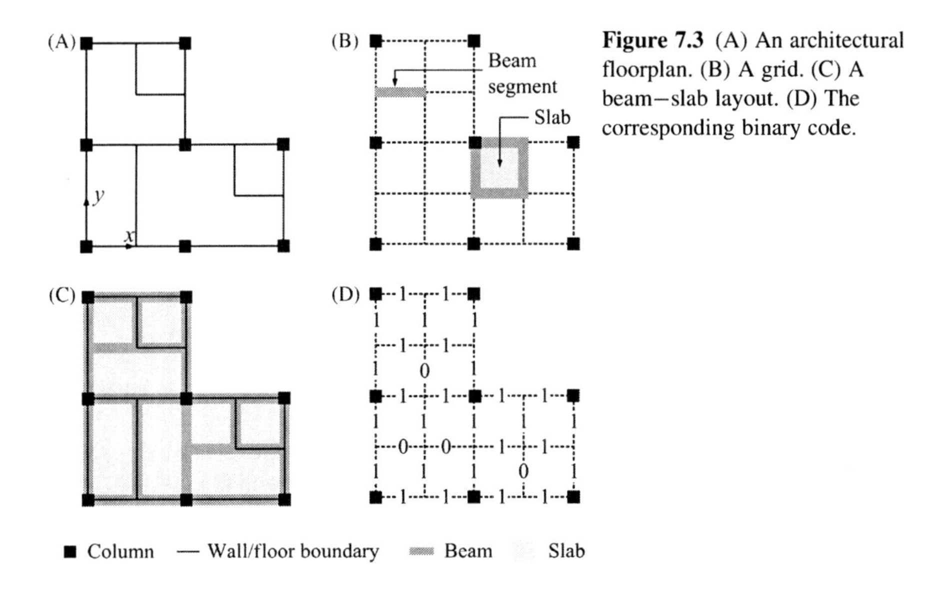

Figure 7.3 (A) An architectural floorplan. (B) A grid. (C) A beam–slab layout. (D) The corresponding binary code.

Each line segment of the grid represents a possible position of a beam segment. A beam segment is defined as one segment of a beam. A beam may consist of one or more segments. Columns are to be used to support beams. Thus, the grid must pass through all columns. In addition, since it is usually desirable to have beams under walls, the grid must also pass through all given wall lines. Here, walls mean permanent walls, which generally require beams to support their weights. The spacing of the grid can be set based on the required degree of precision for beam placement. Figure 7.3 shows an example of an architectural floorplan with a grid. Note that the representations of the floor boundary, columns, walls, beams, and slabs used in this figure will be employed throughout this chapter. Figure 7.3A shows the architectural floorplan, while Figure 7.3B shows the grid whose lines pass through all the columns and wall lines. In addition, Figure 7.3B also shows examples of a slab and a beam segment. Figure 7.3C shows an example of a beam–slab layout built on the grid. Simple binary coding can be used to designate whether there is a beam segment on a grid line segment or not. A one-bit binary code of one and zero can be attached to each grid line segment. In GA terminology, this one-bit binary code of one and zero can be called a one-bit chromosome. If the value of a one-bit chromosome is one, it means that there is a beam segment on that particular grid line segment. On the contrary, if the value is zero, then there is no beam segment at that position. Figure 7.3D shows the corresponding binary code for the beam–slab layout in Figure 7.3C. Obviously, the simple binary coding incorporates all possible patterns, including patterns with T-shaped interior junctions. Unfortunately, it also includes patterns that are not meaningful in the scope of the beam–slab layout design problem. An example of this kind of pattern is shown in Figure 7.4. It can be seen from the figure that the pattern has some beams that are not properly supported. If we simply allow these meaningless or invalid patterns to be directly considered in the design problem without any manipulation or modification, then the simple binary coding yields a one-to-one mapping between the representation code and the corresponding pattern to be directly used in the design problem. Moreover, the time complexity of the translation algorithm for the simple binary coding, with respect to the number of grid line segments, is linear. Certainly, linear-time complexity is the best possible time complexity for this kind of coding. As a result, this simple binary coding fulfills the first, third, and fourth characteristics of good layout representations mentioned earlier. Unfortunately, having coding that includes invalid patterns results in serious difficulty. This is because it is not clear how these invalid patterns from the simple binary coding can be compared with all possible patterns with regard to their quality as beam–slab layouts. Quality evaluation of these invalid patterns is necessary if an optimization method is to be used to solve the problem. During iteration steps in any employed optimization method, intermediate solutions of the optimization process may fall into the region of invalid patterns. If the information of how relatively good or bad these patterns are when compared with all the other patterns and

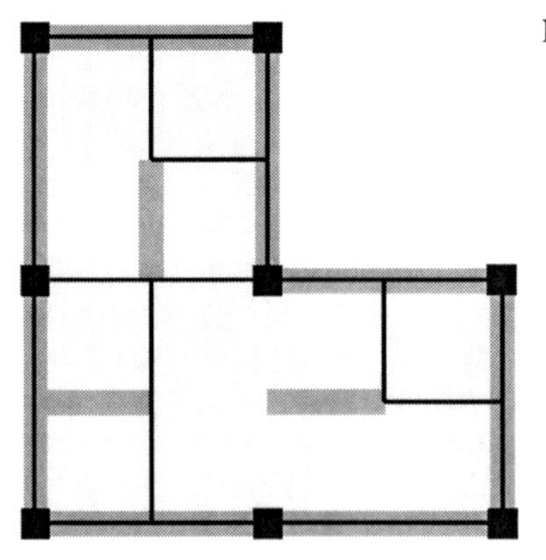

Figure 7.4 An invalid beam pattern.

also among themselves is not available, it will not be possible for the intermediate solutions to informatively move out of the region of invalid patterns. It is therefore necessary to find some ways to evaluate invalid patterns.

7.2.2 Elimination of Invalid Beams

One strategy to solve the problem of invalid patterns is to find a means to interpret all representation codes as valid beam–slab layouts (Nimtawat and Nanakorn, 2009, 2010). Certainly, the simple binary coding with a grid, where each grid line segment represents a possible location of a beam segment, can still be used. However, a translation algorithm from representation codes into corresponding layouts must be created in such a way that every code is interpreted as a valid beam–slab layout. This can be done by first classifying beam segments into two groups, namely valid and invalid beam segments (Nimtawat and Nanakorn, 2010) and defining valid beam–slab layouts as those layouts that have no invalid beam segments. Invalid beam segments include:

1. any isolated beam segments,
2. any beam segments with one free end,
3. any two-beam segments that form an L-shaped interior beam, and
4. any two-beam segments that form a concave L-shaped exterior beam on the outer boundary of the floor.

Figure 7.5 shows examples of invalid beam segments. An isolated beam segment is a beam segment that is not connected to any other beam segments. A beam segment with one free end is a beam segment that has only one end connected to any other beam segment. It is apparent that isolated beam segments cannot be part of any valid beam–slab layout. The rest of the beams in the above list are invalid because they yield nonrectangular slabs, which are not allowed. Classification of beam segments into valid and invalid ones is used in the translation of representation codes into corresponding layouts. In coding, a one-bit chromosome is also attached to each grid line segment. This time, however, the meaning of a one-bit chromosome is not simply defined as having or not having a beam segment. The meaning of a one-bit chromosome is as shown in Table 7.1.

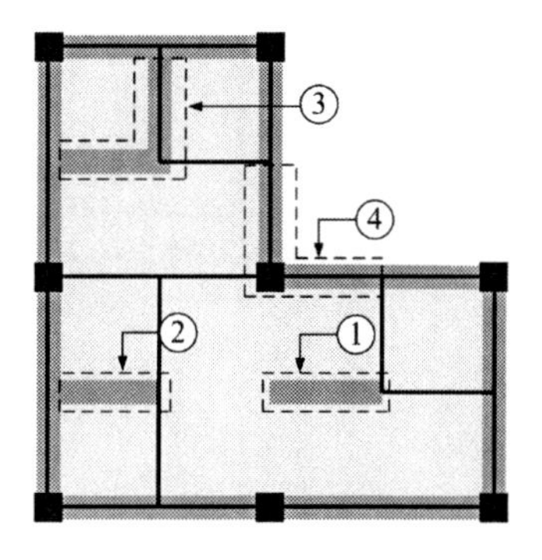

(1) Isolated beam segment

(2) Beam segment with one free end

(3) Two beam segments that form an L-shaped interior beam

(4) Two beam segments that form a concave L-shaped exterior beam

Figure 7.5 Invalid beam segments.

If the value of a one-bit chromosome of a grid line segment is zero, it definitely means that there is no beam segment at that position. However, when the value is equal to 1, the grid line segment will have a beam segment only if that beam segment, if placed there, will become a valid beam segment. To complete the coding of the whole floor, all one-bit chromosomes for all grid line segments will be concatenated to form a binary chromosome string. An algorithm must be created based on Table 7.1 to translate binary chromosome strings into valid beam–slab layouts. In the translation, if the value of a one-bit chromosome is equal to 1, it is necessary to check whether a beam segment, if placed at that position, will be a valid beam segment or not. The problem is that, in order to know whether a beam segment is valid or not, it is necessary to know the existence of beam segments at other locations. This is reciprocally true for every one-bit chromosome whose value is equal to 1. Figure 7.6A shows an example of this problem. If beam segments *C*, *D*, and *E* are considered before *A* and *B*, beam segments *C*, *D*, and *E* can be mistakenly classified as valid beam segments. If beam segments *A* and *B*, which are clearly invalid, are considered before *C*, *D*, and *E*, beam segments *C*, *D*, and *E* will be correctly classified as invalid beam segments. To circumvent this problem, a specific algorithm that determines invalid beam segments and, subsequently, removes them to finally obtain the beam–slab layout must be created. Nimtawat and Nanakorn (2010) used the algorithm shown below.

Valid Beam–Slab Layout Algorithm: Translate a chromosome string into a beam–slab layout

Input: A chromosome string

Output: A beam–slab layout

Table 7.1 Meaning of a One-Bit Chromosome

Value	Type of Beam Segment if Placed at that Location	Meaning
0	—	There is no beam segment
1	Valid	There is a beam segment
	Invalid	There is no beam segment

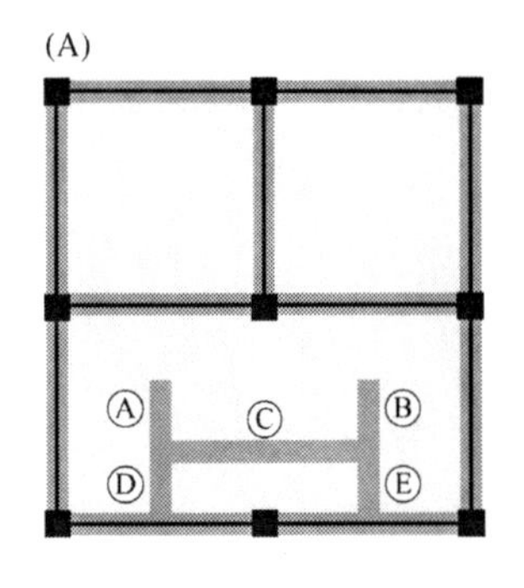

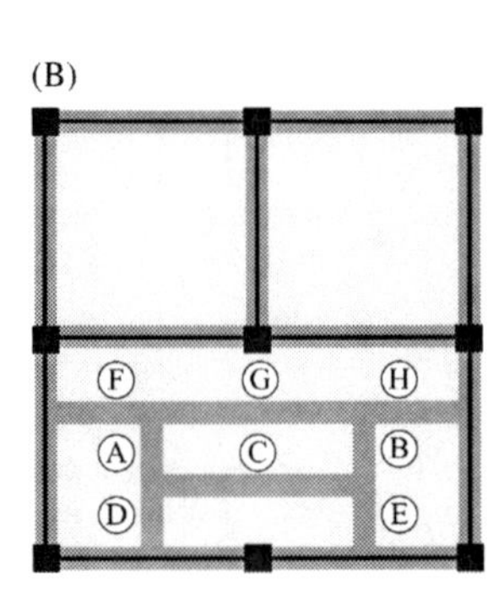

Figure 7.6 (A) Identification of invalid beam segments. (B) Making invalid beam segments valid by adding more beam segments.

- Place beam segments on all grid line segments whose chromosomes are equal to one.
- **While** there is at least one invalid beam **do**
  - Remove all isolated beam segments;
  - Remove all beam segments with one free end;
  - Remove all pairs of beam segments that form an L-shaped interior beam;
  - Remove all pairs of beam segments that form a concave L-shaped exterior beam on the outer boundary of the floor.

Figure 7.7 demonstrates how the aforementioned algorithm works. Figure 7.7A shows the input binary chromosomes and the output beam–slab layout. Figure 7.7B shows the detailed steps of the algorithm. It can be seen that the time complexity of the algorithm, with respect to the number of grid line segments, is quadratic. The quadratic time complexity is still acceptable in most practical cases. In fact, it is always possible to develop an algorithm that adds more beam segments

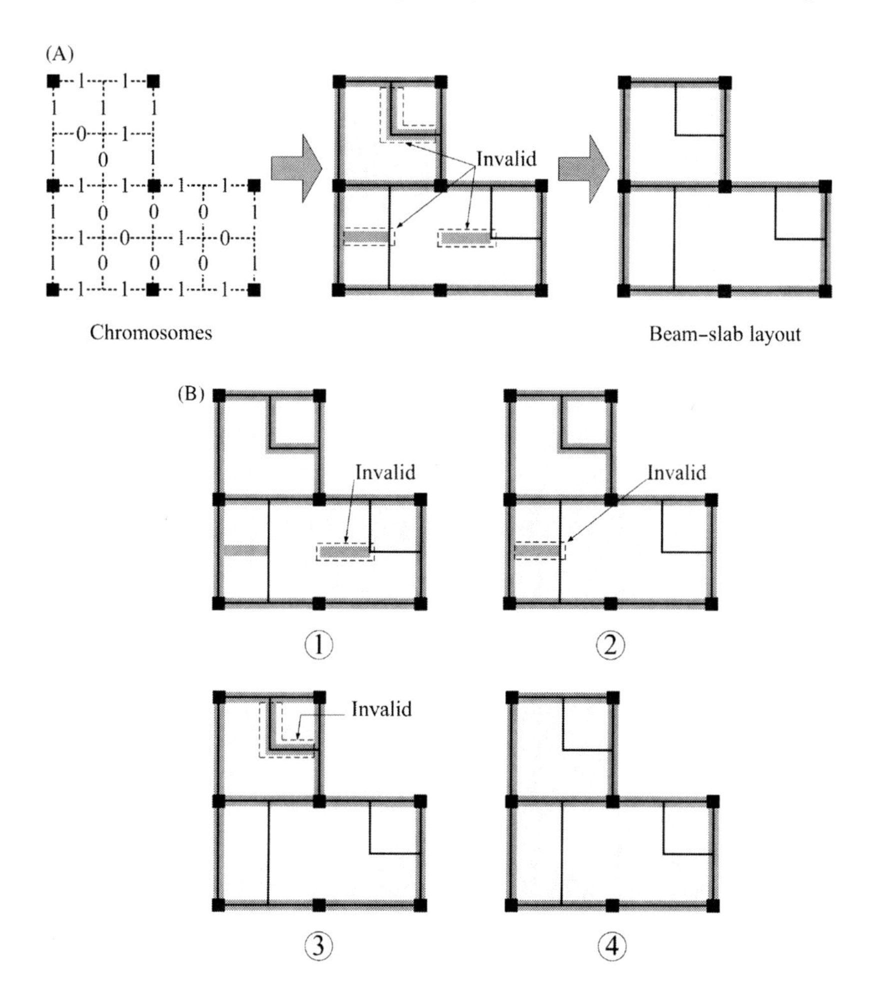

Figure 7.7 (A) Chromosomes and the corresponding beam–slab layout. (B) Decoding steps.

to make invalid segments valid instead of removing them. Figure 7.6B shows an example of this idea. In the figure, if beam segments *F*, *G*, and *H* are added, then *A*, *B*, *C*, *D*, and *E* will become valid beam segments. Unfortunately, there will be generally many different ways to make an invalid beam segment valid and it will be difficult to devise simple steps to efficiently achieve this goal for arbitrary cases. In addition, layouts with larger slabs are generally preferred. Thus, removing invalid beams from a layout seems to be a better choice than adding more beam segments to the layout.

With the translation algorithm shown above, the binary representation of beam–slab layouts now includes only valid beam–slab layouts. As a result, all representation codes can be evaluated and compared. Nevertheless, the above representation scheme incurs two disadvantages. The first disadvantage is that beam–slab layouts obtained from the scheme may not occupy the whole floor area. Figure 7.8 shows an example of this disadvantage. During the decoding process, beam–slab layouts that do not occupy the whole floor area are considered as valid beam–slab layouts. This is reasonable because, although they do not occupy the whole floor area, they can still be evaluated and compared with other layouts. As beam–slab layouts, they are clearly not acceptable and will have to be considered as infeasible layouts in the optimization process. The second disadvantage is that the translation algorithm is not a one-to-one mapping between the representation code and the corresponding layout. Many different chromosome strings may correspond to the same beam–slab layout (Nimtawat and Nanakorn, 2010). In

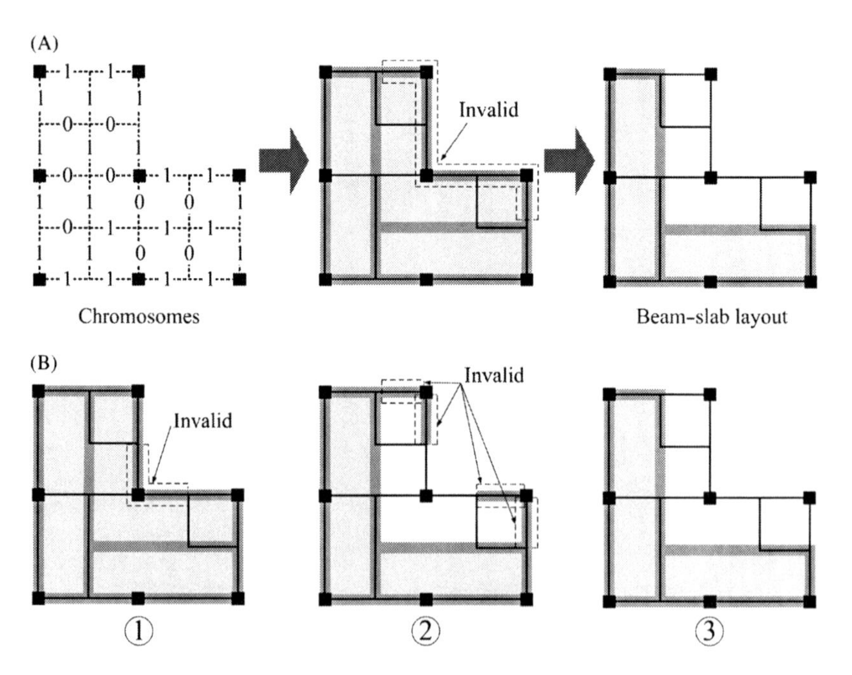

Figure 7.8 (A) A beam–slab layout whose total slab area does not cover the entire floor. (B) Decoding steps.

other words, the translation algorithm is a many-to-one mapping between the representation code and the corresponding layout. Having a many-to-one mapping between the representation code and the corresponding layout simply means that different valid beam–slab layouts may not have the same number of representatives in the search space. This certainly creates a kind of bias in the optimization process. For example, if we simply want to randomly pick up one beam–slab layout from the search space by selecting one representation code from the search space and translating the code into a layout, those beam–slab layouts that have more representatives in the search space will have higher chances to be picked. The bias is toward layouts that have larger slabs and fewer beams as well as layouts that do not occupy the whole floor area. This is because invalid beams are removed from the original pattern of beam segments to obtain the corresponding layout. As a result, the corresponding layout is likely to have fewer beams, and therefore larger slabs. It is also likely that, during the removal of invalid beams, a part of the floor will also be removed. The effects of the two disadvantages can be alleviated by creating, in the optimization process, other biases in the opposite direction.

If we check the above representation of beam–slab layouts against the four characteristics of good layout representations mentioned earlier, the following results can be observed:

1. The representation is able to represent all possible layouts.
2. The representation does not include any invalid patterns.
3. The representation is a many-to-one mapping.
4. The time complexity of the translation algorithm for the representation, with respect to the number of grid line segments, is quadratic.

As mentioned earlier, when the simple binary coding is used without any alteration, it fulfills all the requirements for good layout representations but one. The simple binary coding includes invalid patterns. In order to remove invalid patterns from the simple binary coding, the one-to-one mapping, and the low time complexity are unfortunately sacrificed.

There are always some openings inside a floor. The most obvious openings are stair openings. We can handle openings quite easily by removing grid line segments in openings, so that they will not be considered as possible locations of beams.

7.3 A Representative Optimization Problem

In this section, we discuss how to change the beam–slab layout design task from a heuristic design problem into an algorithmic one by developing its representative optimization problem. To obtain the representative optimization problem, we explicitly create an appropriate objective function as well as appropriate constraints for the problem. However, because the original problem is heuristic, we have to heuristically create these explicit objective function and constraints.

When structural designers design the beam–slab layout for an architectural floorplan whose positions of columns and walls are given, they will certainly use the given positions of columns and walls in their mental reckoning to design the layout. In addition, they will also have to be aware of the boundaries of the floor, which include the external boundary and internal boundaries created by internal openings. The positions of columns and walls tell the designers how beams should be placed while the boundaries of the floor simply tell the designers where the lines of beams must end. The information about the boundaries can be easily included in the representative optimization problem through the construction of the grid as discussed in the previous section. This leaves the positions of columns and walls as the major concerns in the construction of the objective function and constraints. Since the given columns have to be utilized in the most efficient way, we will have to figure out how to evaluate the efficiency of column utilization. It can be argued that the column utilization of a slab is considered better if the slab has more corner columns (Nimtawat and Nanakorn, 2009, 2010). This is because more corner columns imply better load transfer from the slab to the columns via the beams. In addition, the column utilization of the whole floor is considered better if there are fewer slabs in the floor (Nimtawat and Nanakorn, 2009, 2010). Layouts that have fewer slabs also have fewer beams, which generally result in lower construction costs. For these reasons, the objective function F can be written as (Nimtawat and Nanakorn, 2009, 2010):

$$F(\mathbf{x}) = \frac{1}{N_S(\mathbf{x})} \sum_{i=1}^{N_S} S_i(\mathbf{x}) \tag{7.1}$$

Here, the objective function F is a function of the chromosome string denoted by $\mathbf{x}$, which represents a beam–slab layout. In addition, S_i is the score of slab i, and N_S is the total number of slabs in the layout. The slab score S_i is given as 1, 0.75, 0.5, 0.25, or 0 if slab i has 4, 3, 2, 1, or 0 corner columns, respectively. Figure 7.9 shows examples of slabs with different slab scores. The objective function F in Eq. (7.1) exhibits the following behavior:

1. Beam–slab layouts that have fewer slabs yield higher values of the objective function.
2. Beam–slab layouts that have more corner columns yield higher values of the objective function.

Since beam–slab layouts that have fewer slabs and more corner columns are considered better layouts, we will want to maximize the above objective function. The next task is to establish necessary constraints for this maximization problem. Since it is desirable to have beams under walls, we can write a constraint to enforce this condition. It is also practical to allow the maximum dimensions of slabs to be explicitly controlled. The maximum dimensions of slabs to be allowed may depend on, for example, available sizes of precast slabs. In addition, it is not reasonable for a beam–slab floor to have slabs whose dimensions are larger than the maximum column spacing of the floor. Having slabs with dimensions larger than the

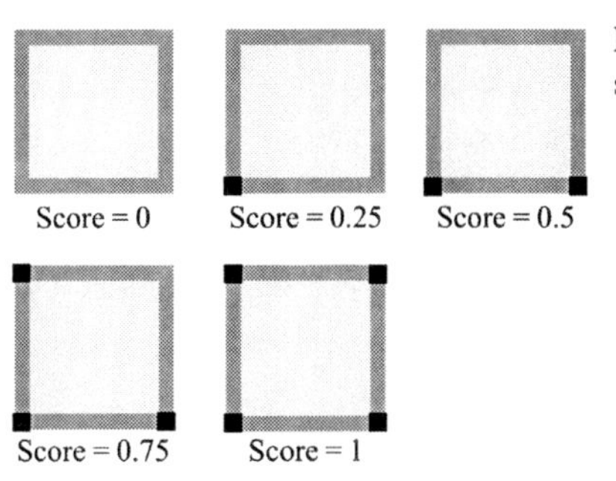

Figure 7.9 Examples of slabs with different scores.

maximum column spacing may simply mean that there can be columns that support no beams. Finally, since the representation of beam–slab layouts includes layouts that do not occupy the whole floor area, these solutions must be considered infeasible and a corresponding constraint must be created. In summary, the following three constraints (Nimtawat and Nanakorn, 2010) can be used:

1. *Wall constraint*: All walls must be directly supported by beams.
2. *Slab constraint*: Each dimension of a slab must not exceed its corresponding prescribed maximum length. Note that it is possible to have more than one prescribed maximum length for slabs. For example, there can be one prescribed maximum length for each direction of the floor.
3. *Floor constraint*: The total slab area must cover the entire floor.

We have to make certain that there will be beams along the boundaries of the floor. This can be achieved by using the wall constraint. Whether or not there are real walls on the external and the internal boundaries of the floor, the boundaries will be treated as walls.

The representative optimization problem for the beam–slab layout design problem can then be written as:

$$\max_{\mathbf{x}} \left[F(\mathbf{x}) = \frac{1}{N_S(\mathbf{x})} \sum_{i=1}^{N_S} S_i(\mathbf{x}) \right] \tag{7.2}$$

subject to

$$L'_W = 0, A'_S = 0, A''_S = 0 \tag{7.3}$$

where L'_W denotes the total length of wall segments that are not directly supported by beams. In addition, A'_S denotes the total area of slabs that have at least one side longer than the corresponding prescribed maximum length and A''_S denotes the total area of the floor that is not covered by slabs.

At this stage, the original heuristic beam–slab layout design problem is converted into an algorithmic design problem, which is described by the above optimization problem. This representative optimization problem can be generally solved by any efficient optimization method. The development of a GA to solve this optimization problem will be discussed in the next section.

7.4 A GA for Beam—Slab Layout Design

7.4.1 Problem Formulation for a GA

To create a GA to solve the representative optimization problem described by Eqs. (7.2) and (7.3), we have to first define how the fitness of a solution is computed. The representative optimization problem is a maximization problem. Thus, the objective function F can be used as the fitness function. We then have to incorporate the three constraints into the fitness function. A penalty function can be employed for this purpose and the augmented fitness function F^a that incorporates the penalty function can be defined as

$$F^a(\mathbf{x}) = F(\mathbf{x}) - P(\mathbf{x}) \tag{7.4}$$

where P denotes the penalty function defined as

$$P(\mathbf{x}) = \lambda E(\mathbf{x}) = \lambda[E_{\text{wall}}(\mathbf{x}) + E_{\text{slab}}(\mathbf{x}) + E_{\text{floor}}(\mathbf{x})] \tag{7.5}$$

In the above definition of the penalty function, λ is a nonnegative factor and E represents the total degree of constraint violation. The total degree of constraint violation E is the summation of the degrees of wall, slab, and floor constraint violation, defined respectively as

$$E_{\text{wall}}(\mathbf{x}) = \frac{L'_{\text{W}}(\mathbf{x})}{L_{\text{W}}} \tag{7.6}$$

$$E_{\text{slab}}(\mathbf{x}) = \frac{A'_{\text{S}}(\mathbf{x})}{A_{\text{S}}} \tag{7.7}$$

$$E_{\text{floor}}(\mathbf{x}) = \frac{A''_{\text{S}}(\mathbf{x})}{A_{\text{S}}} \tag{7.8}$$

In addition to the terms defined earlier, L_{W} denotes the total wall length while A_{S} denotes the total floor area. It can be seen in Eq. (7.4) that each infeasible solution, which violates some of the constraints, will have its fitness reduced by a value of P.

The original optimization problem described by Eqs. (7.2) and (7.3), which is a maximization problem with constraints, is now rewritten as a maximization problem of the augmented fitness function F^a. The new form of the problem has no constraints and, thus, is suitable to be solved by a GA.

7.4.2 Adaptive Penalty and Elitism

In theory, any variant of the GA can be used to solve the above maximization problem. However, special attention has to be given to the bias intrinsic to the representation of beam–slab layouts detailed in Section 7.2. Some measures must be included in the employed algorithm to counter the effect of this bias. Nimtawat and Nanakorn (2009, 2010) modified the simple GA slightly and used it to solve the problem. One key modification is related to the factor λ that is used to provide the significance of the penalty term. They employed an adaptive penalty technique proposed by Nanakorn and Meesomklin (2001) to automatically determine the factor λ in each GA generation. The basic idea of this adaptive penalty technique is to penalize infeasible solutions so that the individual chance of the best infeasible members being selected into the mating pool with respect to that of the average feasible members remains the same in all generations. Note that, in the reproduction process, the augmented fitness values are in general not directly used. Instead, they are scaled into a specified positive range. These scaled fitness values are then used to create the mating pool. The adaptive penalty technique by Nanakorn and Meesomklin (2001) provides the final scaled fitness values to be used in the reproduction process.

The other key modification of the simple GA by Nimtawat and Nanakorn (2009, 2010) is the use of the elitism technique. The representation of beam–slab layouts in Section 7.2 contains the bias toward layouts with larger slabs and fewer beams as well as layouts that do not occupy the whole floor area. To alleviate the effect of the bias, we can create biases in the opposite direction to counter this existing bias. The biases in the opposite direction can be something that encourages solutions that have smaller slabs and more beams. To begin with, we can create a special layout that contains all the possible beam segments (Nimtawat and Nanakorn, 2009, 2010). This layout clearly has the smallest slabs as possible, with respect to the employed grid. In addition, it also has as many beams as possible. With proper grid spacing, this special layout always satisfies all the constraints and is, therefore, a feasible solution. We can then specifically place this layout as an individual in the initial population of the GA. The rest of the individuals in the initial population can be randomly selected as usual. The special layout helps pull the population toward solutions that have smaller slabs and more beams as well as solutions that occupy the entire floor area. However, the effect of the placement of the special layout in the initial population may wear off within a few generations. To prevent this from happening, we can employ the elitism technique in the GA. The main concept of all elitist GAs is that the best solution or solutions of the current generation are placed directly in the population of the subsequent generation regardless of the reproduction, crossover, and mutation operators. However, elitism rules of comparison must be properly created in order for the elitism process to really help counter the coding bias and help find the best feasible solution. It may be argued that the scaled fitness values can be used in the solution comparison to find the elitist solution. However, the scaled fitness values do not provide any information about the constraint violation. It is, in fact, much better to consider various details of every solution in order to find the elitist solution. The following elitism rules of comparison consider many details in order to compare two solutions.

Consider two layouts, Layout_i and Layout_j. Layout_i is better than Layout_j when

1. Layout_i is feasible while Layout_j is not.
2. Both layouts are feasible but Layout_i has higher fitness than Layout_j.
3. Both layouts are feasible and have the same fitness. Nevertheless, Layout_i has a shorter total length of beams than Layout_j.
4. Both layouts are feasible and have the same fitness and total length of beams. Nevertheless, Layout_i has fewer beams than Layout_j. Note that connecting beam segments on the same grid line are counted as one beam.
5. Both layouts are infeasible but Layout_i has a lower total degree of constraint violation than Layout_j.
6. Both layouts are infeasible and have the same total degree of constraint violation. Nevertheless, Layout_i has higher fitness than Layout_j.
7. Both layouts are infeasible and have the same fitness and total degree of constraint violation. Nevertheless, Layout_i has a shorter total length of beams than Layout_j.
8. Both layouts are infeasible and have the same fitness and total degree of constraint violation. In addition, they also have the same total length of beams. Nevertheless, Layout_i has fewer beams than Layout_j.

In the elitism process, the best individual of the current population is found by using the above elitism rules of comparison. If, based on the same rules of comparison, this best individual is better than the existing elitist solution obtained from all past generations, the individual becomes the elitist solution. After that, the worst individual of the generation, obtained also by the same rules, is replaced by the elitist solution. It will always be possible to find the best individual that is feasible by the above elitism rules of comparison because there will always be at least one feasible individual in each generation. This is due to the special layout inserted in the initial population, which is feasible, and the elitism process itself, which simply places the best feasible solution into the next generation.

7.4.3 Algorithm

Besides the adaptive penalty scheme and the elitism process, the rest of the GA for beam–slab layout design can exactly follow the simple GA. This means that three basic GA operators, i.e., reproduction, crossover, and mutation operators, are used. The reproduction operator employs the roulette wheel selection. The crossover operator simply adopts the one-point crossover and the mutation operator utilizes the bitwise mutation.

7.5 Examples

Some examples solved by Nimtawat and Nanakorn (2009, 2010) are shown here to demonstrate that the heuristic beam–slab layout design task can be converted into an algorithmic design task and solved reasonably well by a GA. The geometrical input data of all examples include floor geometry, wall positions, and column positions. In addition, the maximum allowable slab sizes are also required. Since it is practical that there are beams on the external boundary of the floor, beam segments are placed in advance on all grid line segments that represent the external boundary of the floor, and they are subsequently removed from the list of the design variables.

The first example is a simple rectangular floorplan shown in Figure 7.10A (Nimtawat and Nanakorn, 2009). Figure 7.10B shows the grid used in the calculation. The maximum allowable length of a slab is set to 4 m. Figure 7.10C shows the best layout obtained from the algorithm. Figure 7.11 shows a typical evolution of solutions. The second example is a rectilinear floorplan shown in Figure 7.12A

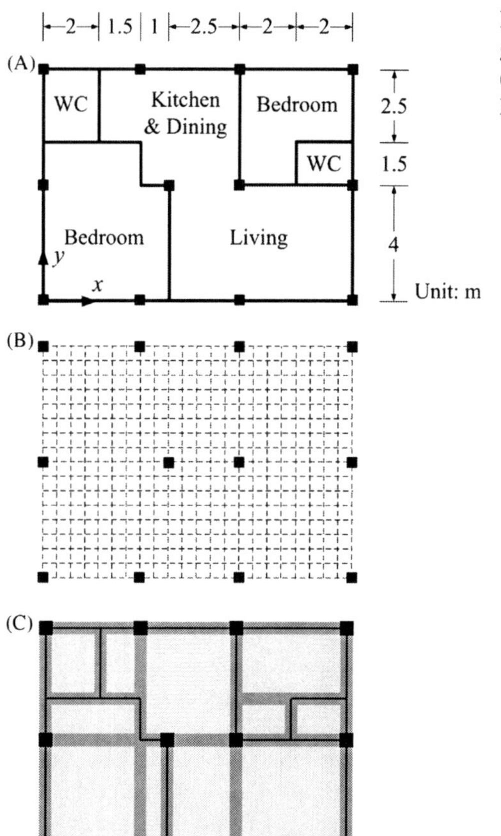

Figure 7.10 The first example: (A) the given floorplan, (B) the grid, and (C) the best solution (Nimtawat and Nanakorn, 2009).

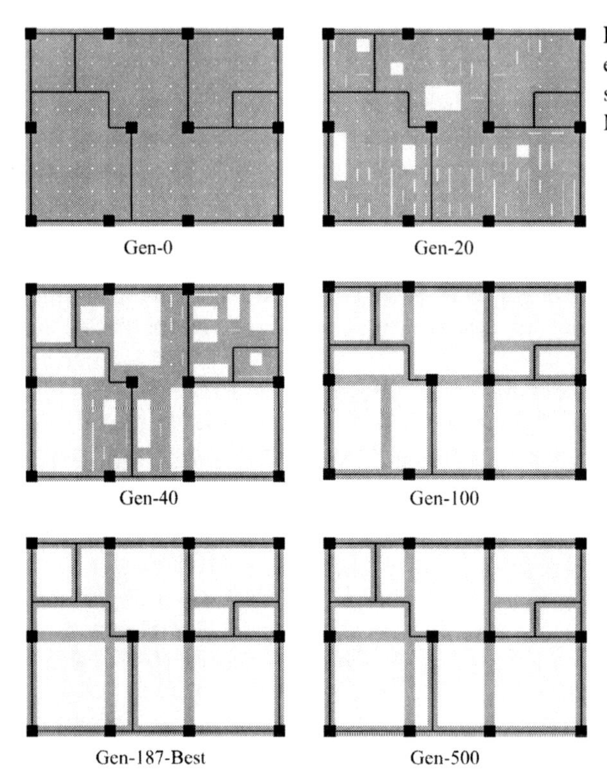

Figure 7.11 The first example: a typical evolution of solutions (Nimtawat and Nanakorn, 2009).

(Nimtawat and Nanakorn, 2010). The floor has a stair opening and, as a result, the employed grid shown in Figure 7.12B does not contain any grid line segments in the opening. The maximum allowable length of a slab is set to 3.5 m, which is the maximum column spacing. Figure 7.12C shows the best layout obtained from the algorithm. The last example is a real floorplan of an existing public building shown in Figure 7.13A (Nimtawat and Nanakorn, 2010). The real structural beam–slab layout of this floor is shown in Figure 7.13B. The chamfered floor corners are changed to right-angle corners to obtain the input of the algorithm shown in Figure 7.14A. There are three stair openings and one lift opening in the floor. Grid line segments are not allowed in these openings. Precast and cast-in-place slabs are used in the real design and the maximum length of the precast slabs is 4 m. The precast slabs in the real design are all aligned with the x direction. To be able to make a comparison with the real design, the maximum allowable slab length in the x direction is set to 4 m in the algorithm. The maximum allowable slab length in the y direction is set to 8 m, which is the maximum column spacing. In this problem, a nonuniform grid is used. The maximum grid spacing is set to 4 m in order that the slab constraint can be satisfied in both directions. The nonuniform grid is constructed by first placing grid lines on all columns and wall lines. After that, the spacing between all grid lines is checked. If any spacing is found to be greater than the maximum grid spacing of 4 m, an additional grid line will be inserted at the

(A)

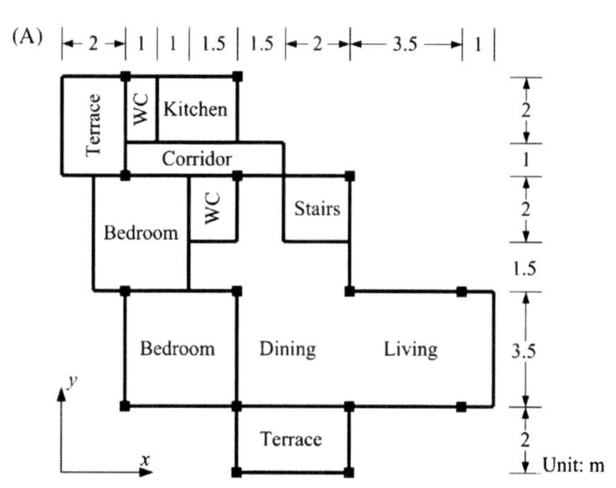

(B)

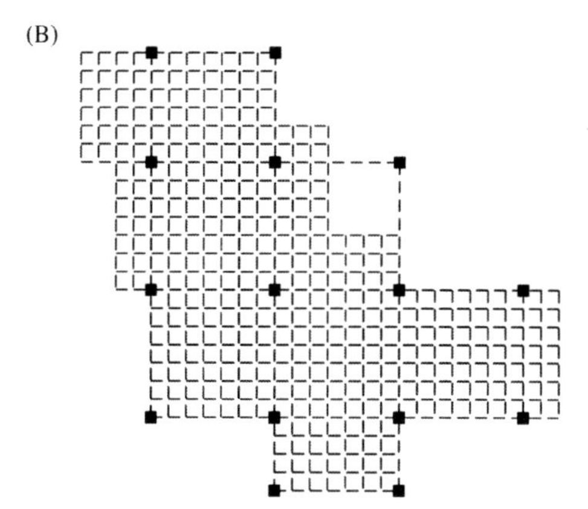

(C)

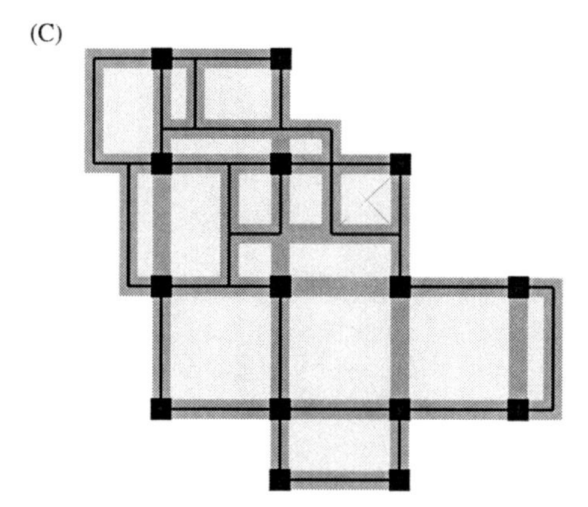

Figure 7.12 The second example: (A) the given floorplan, (B) the grid, and (C) the best solution (Nimtawat and Nanakorn, 2010).

middle of the interval. Figure 7.14B shows the grid employed in this problem. Figure 7.14C shows the best layout obtained from the algorithm.

It can be seen from these results that the algorithm yields acceptably good beam–slab layouts for all the problems. For the last problem, if the differences at

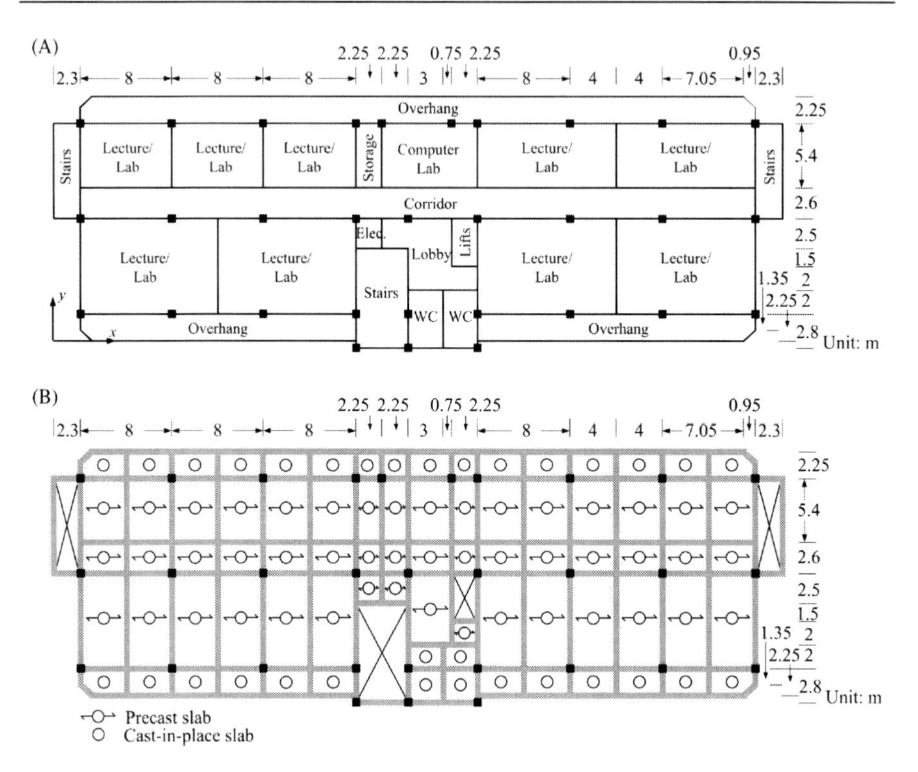

Figure 7.13 The third example: (A) the real architectural floorplan and (B) the real structural floorplan (Nimtawat and Nanakorn, 2010).

the four chamfered corners are disregarded, the obtained layout is exactly the same as the real structural layout shown in Figure 7.13B.

7.6 Future Challenges

We have seen that the entire structural design task of a building is usually subdivided into several intermediate tasks, which can be either heuristic or algorithmic tasks. Since heuristic design tasks do not have explicit relationships between their design goals and design variables, it is not possible to directly automate these tasks by using computers. If an entire structural design task consists of only algorithmic intermediate design tasks, it is possible to change the whole design process into several optimization problems. After that, any efficient and appropriate optimization methods can be used to solve them. In order to remove heuristic intermediate design tasks, we can reintegrate all intermediate design tasks into one big algorithmic task or change all heuristic intermediate design tasks into algorithmic ones. This chapter has discussed how the heuristic beam–slab layout design task can be

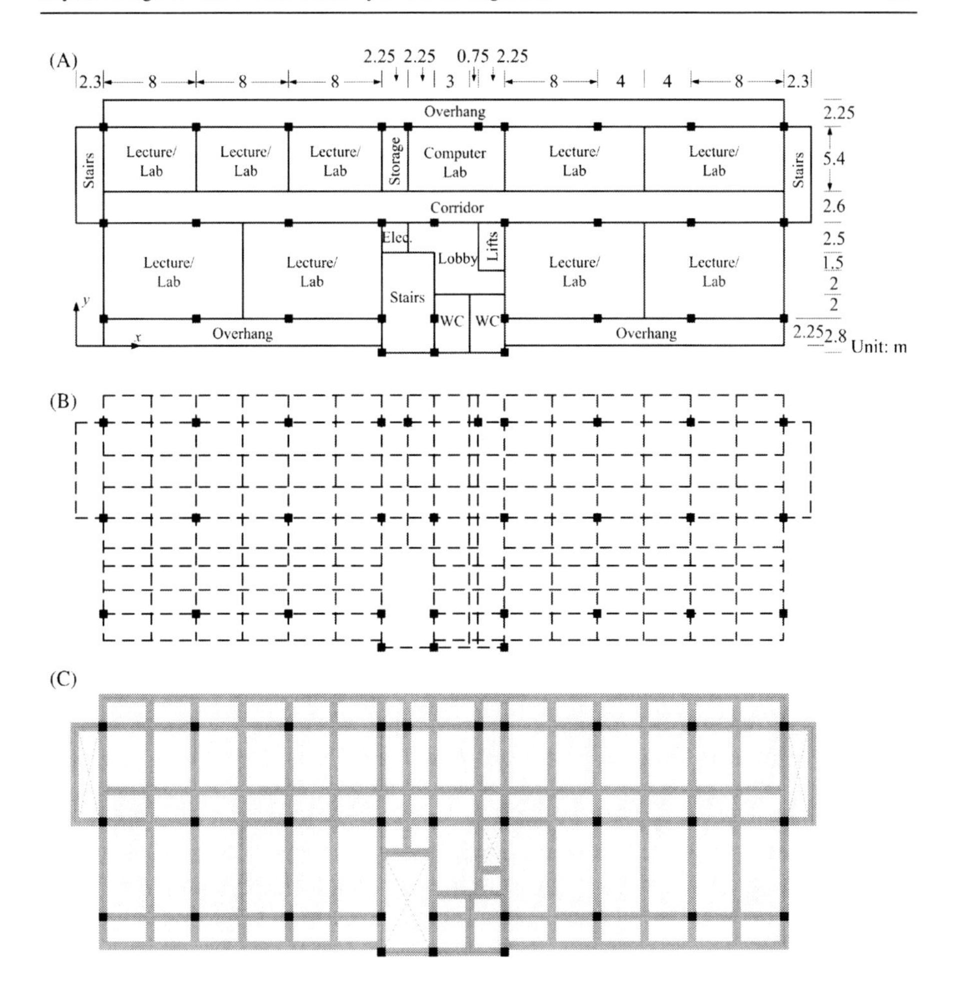

Figure 7.14 The third example: (A) the simplified architectural floorplan, (B) the grid, and (C) the best solution (Nimtawat and Nanakorn, 2010).

converted into an algorithmic design task and subsequently solved by a GA. The main undertakings in this conversion include the development of a representation of beam–slab layouts, the development of a representative optimization problem, and the development of an optimization algorithm, a GA in our case, to be used to solve the representative optimization problem.

In the development of a representation of beam–slab layouts, we want to have a representation that is simple but, at the same time, can represent different beam–slab layouts well. The representation discussed in this chapter is not a one-to-one mapping between the representation code and the corresponding layout. In addition, the time complexity in the translation of representation codes into corresponding layouts is quadratic. The many-to-one mapping results in a bias in the optimization process. Thus, it may be beneficial if an effective representation that is also a one-to-one mapping is developed. In fact, a good representation of beam–slab layouts can be useful for other types of problems, such as the floorplanning problem. As for the time complexity in the translation of representation codes into corresponding layouts, it is obvious that lower time complexity will be advantageous.

In the development of a representative optimization problem, we may want to allow more flexibility in the input or allow the problem scope to be larger. For example, the positions of columns may become part of the design variables. In addition, it is often necessary that the beam–slab layouts of several floors are considered together. This is obviously unavoidable if the positions of columns are to be designed. In this case, the representative optimization problem has to include these floors as one problem. After beam–slab layout design is completed, structural design of all beams and slabs in the layout is the next design task to be performed. Note that, while beam–slab layout design is considered heuristic, structural design of beams and slabs is considered algorithmic. We may want to combine these two successive design tasks into one algorithmic design task and create a representative optimization problem for it. In this way, we can consider the objectives and constraints of layout design and structural member design together and remove some of them that become irrelevant and unnecessary.

Finally, in the development of an optimization algorithm to be used to solve the representative optimization problem, we may want to explore the possibility and benefit of using other efficient global optimization methods besides GAs. We have to keep in mind that the method selected should be appropriate for large problems where many design variables are considered at the same time. This is to make certain that, when more intermediate tasks, which may include both heuristic and algorithmic tasks, are merged to create algorithmic tasks that are more suitable for optimization methods, the selected optimization method must be able to easily handle the resulting larger problems. The ultimate achievement will be when all intermediate design tasks are reintegrated into one big design problem and the resulting problem is solved by an optimization method.

References

Bailey, S.F., Smith, I.F.C., 1994. Case-based preliminary building design. J. Comput. Civ. Eng. 8, 454–467.

Balasa, F., 2000. Modeling non-slicing floorplans with binary trees. IEEE/ACM International Conference on Computer-Aided Design, Digest of Technical Papers, IEEE, San Jose, CA, USA, 13–16.

Dorigo, M., Maniezzo, V., Colorni, A., 1996. Ant system: optimization by a colony of cooperating agents. IEEE Trans. Syst. Man. Cybern. B Cybern. 26, 29–41.

Gandomi, A.H., Yang, X.-S., Alavi, A.H., 2011. Mixed variable structural optimization using firefly algorithm. Comput. Struct. 89, 2325–2336.

Gandomi, A.H., Yang, X.S., Talatahari, S., Alavi, A.H., 2013. Metaheuristic algorithms in modeling and optimization. In: Gandomi, A.H., Yang, X.S., Alavi, A.H., Talatahari, S. (Eds.), Metaheuristic Applications in Structures and Infrastructures. Elsevier, Waltham, MA.

Goldberg, D.E., 1989. Genetic Algorithms in Search, Optimization, and Machine Learning. Addison-Wesley, MA.

Huang, W., Chen, D., Xu, R., 2007. A new heuristic algorithm for rectangle packing. Comput. Oper. Res. 34, 3270–3280.

Kennedy, J., Eberhart, R., 1995. Particle swarm optimization. The 1995 IEEE International Conference on Neural Networks, IEEE, Perth, Australia, 1942–1948.

Kirkpatrick, S., Gelatt Jr., C.D., Vecchi, M.P., 1983. Optimization by simulated annealing. Science. 220, 671–680.

Maher, M.L., 1984. HI-RISE: A Knowledge-Based Expert System for the Preliminary Structural Design of High Rise Buildings. Carnegie-Mellon University, Pittsburg, PA, USA.

Murata, H., Fujiyoshi, K., Nakatake, S., Kajitani, Y., 1996. VLSI module placement based on rectangle-packing by the sequence-pair. IEEE Trans. Comput. Aided Des. Int. 15, 1518–1524.

Nanakorn, P., Meesomklin, K., 2001. An adaptive penalty function in genetic algorithms for structural design optimization. Comput. Struct. 79, 2527–2539.

Nimtawat, A., Nanakorn, P., 2009. Automated layout design of beam-slab floors using a genetic algorithm. Comput. Struct. 87, 1308–1330.

Nimtawat, A., Nanakorn, P., 2010. A genetic algorithm for beam–slab layout design of rectilinear floors. Eng. Struct. 32, 3488–3500.

Park, K.-W., Grierson, D.E., 1999. Pareto-optimal conceptual design of the structural layout of buildings using a multicriteria genetic algorithm. Comput. Aided Civ. Infrastructure Eng. 14, 163–170.

Rafiq, M.Y., Mathews, J.D., Bullock, G.N., 2003. Conceptual building design—evolutionary approach. J. Comput. Civ. Eng. 17, 150–158.

Sahab, M.G., Toropov, V.V., Gandomi, A.H., 2013. A review on traditional and modern structural optimization: problems and techniques. In: Gandomi, A.H., Yang, X.S., Alavi, A.H., Talatahari, S. (Eds.), Metaheuristic Applications in Structures and Infrastructures. Elsevier, Waltham, MA, USA.

Shaw, D., Miles, J., Gray, A., 2008. Determining the structural layout of orthogonal framed buildings. Comput. Struct. 86, 1856–1864.

Sisk, G.M., Miles, J.C., Moore, C.J., 2003. Designer centered development of GA-based DSS for conceptual design of buildings. J. Comput. Civ. Eng. 17, 159–166.

Soibelman, L., Pena-Mora, F., 2000. Distributed multi-reasoning mechanism to support conceptual structural design. J. Struct. Eng. 126, 733–742.

Syrmakezis, C.A., Mikroudis, G.K., 1997. ERDES—an expert system for the aseismic design of buildings. Comput. Struct. 63, 669–684.

Tsakalias, G.E., 1994. KTISMA: a blackboard system for structural model synthesis of asymmetrical skeletal reinforced concrete buildings. In: Topping, B.H.V., Papadrakakis, M. (Eds.), Artificial Intelligence and Object Oriented Approaches for Structural Engineering. Civil-Comp Press, Edinburgh, UK, 15–21.

Yao, B., Chen, H., Cheng, C.K., Graham, R., 2003. Floorplan representations: complexity and connections. ACM Trans. Des. Autom. Electron. Syst. 8, 55–80.

8 Optimum Design of Skeletal Structures via Big Bang–Big Crunch Algorithm

Siamak Talatahari[1] *and Ali Kaveh*[2]

[1]Department of Civil Engineering, University of Tabriz, Tabriz, Iran,
[2]Centre of Excellence for Fundamental Studies in Structural Engineering, Iran University of Science and Technology, Narmak, Tehran, Iran

8.1 Introduction

The optimum design of structures has received considerable attention in the last two decades, despite the existence of major preventive factors. This fact has resulted in the development of many efficient methods for solving structural optimization problems. The optimization methods are divided into two general groups: classical methods and metaheuristic approaches.

Classical optimization methods are often based on mathematical programming and come with some computational drawbacks such as requiring substantial gradient information or continuous objective function, and being dependent on suitable starting points (Kaveh and Talatahari, 2009a). Instead, the metaheuristic search techniques often make use of the ideas inspired from nature and do not suffer the discrepancies of mathematical programming based optimum design methods; thus, global optimum design (or near it) can be archived only with assessment of the objective function.

Genetic algorithms (GAs) (Goldberg, 1998), particle swarm optimization (PSO) (Eberhart and Kennedy, 1995), ant colony optimization (ACO) (Dorigo et al., 1996), imperial competitive algorithm (ICA) (Atashpaz-Gargari and Lucas, 2007), charged system search (CSS) (Kaveh and Talatahari, 2010a) and Big Bang–Big Crunch (BB–BC) (Erol and Eksin, 2006) are some familiar metaheuristic methods. These methods have been applied to many optimization problems because of their high potential for modeling engineering problems in environments that have been resistant to solutions by classic techniques. Application of these metaheuristic algorithms to structural optimization problems includes those of GAs (Camp et al., 1998; Erbatur et al., 2000; Hajela and Lee, 1995; Kameshki and Saka, 2001; Kaveh and Abditehrani, 2004; Kaveh and Kalatjari, 2003; Kaveh and Rahami, 2006; Koumousis and Georgious, 1994; Rajeev and Krishnamoorthy, 1992; Shrestha and

Metaheuristic Applications in Structures and Infrastructures. DOI: http://dx.doi.org/10.1016/B978-0-12-398364-0.00008-5

Ghaboussi, 1998; Toğan and Daloğlu, 2006; Wu and Chow, 1995), ACO (Camp and Bichon, 2004, 2005; Kaveh and Shahrouzi, 2008; Kaveh and Shojaee, 2007; Kaveh and Talatahari 2010b,c; Kaveh et al., 2008a,b, 2010; Serra and Venini, 2006), PSO (Kathiravan and Ganguli, 2007; Li et al., 2007; Perez and Behdinan, 2007; Schutte and Groenwold, 2003; Suresh et al., 2007), ICA (Kaveh and Talatahari, 2010d,e; Mozafari et al., 2010), and CSS (Kaveh and Talatahari, 2010f–h, 2011a–d, 2012). These heuristics often incorporate random variation and selection (Atashpaz-Gargari and Lucas, 2007; Dorigo et al., 1996; Eberhart and Kennedy, 1995; Erol and Eksin, 2006; Goldberg, 1998; Kaveh and Talatahari, 2009a, 2010a). The random selection and the information obtained in each cycle are used to choose new points in the subsequent cycles.

The BB–BC is one of the heuristic algorithms initially suggested by Erol and Eksin (2006) and developed and extended by Kaveh and Talatahari (2009b). This method has a low computational cost and a high convergence speed. Similar to the other evolutionary algorithms, the BB–BC method is a natural evolutionary algorithm. The BB–BC is a heuristic population-based search procedure that incorporates random variation and selection. The random selection and the information obtained in each cycle are used to choose the new points in the subsequent cycles. According to the BB–BC theory, in the BB phase, energy dissipation produces disorder and randomness, whereas, in the BC phase, randomly distributed particles are drawn into an order. The BB–BC method similarly generates random points in the BB phase and shrinks these points to a single representative point via a center of mass in the BC phase. After a number of sequential BBs and BCs, where the distribution of randomness within the search space during the BB phase becomes smaller and smaller about the average point computed during the BC, the algorithm converges to a solution. The BB–BC algorithm has been applied to structural problems including the optimum design of trusses (Kaveh and Talatahari, 2009b) the optimum design of domes (Kaveh and Talatahari, 2010i), the optimum design of frames (Kaveh and Talatahari, 2010j), and parameter estimation in structural systems (Tang et al., 2010).

In this chapter, an efficient hybrid algorithm based on the BB–BC, the PSO, and the suboptimization mechanism (SOM) (Kaveh and Talatahari, 2010b) as a search-space updating technique is developed to find optimum design of truss structures with continuous or discrete domains and frame structures with a discrete search domain. Similar to the simple BB–BC, the present method consists of a BB phase and a BC phase; however, it utilizes the best position of each particle and the best visited position of all particles in addition to the center of mass point.

8.2 Statement of the Optimization Design Problem

Selection of a suitable objective function in optimal design problems is highly significant. In most cases, the objective function not only shows one important feature of a design but it can also contain a combination of different features (Kaveh et al., 2010).

For structural optimization problems, minimization of the weight is often used as the objective function and the optimum design of structures involves a set of design variables that has the minimum weight located in the feasible space which does not violate constraints, as

$$\begin{aligned} &\text{Find:} \quad \mathbf{X} = [x_1, x_2, ..., x_{ng}], \quad x_i \in D_i \\ &\text{to minimize:} \quad W(\mathbf{X}) = \sum_{i=1}^{n} \gamma_i \cdot x_i \cdot L_i \end{aligned} \tag{8.1}$$

$$\text{subject to:} \quad g_i(\mathbf{X}) \le g_i^u \quad i = 1, 2, ..., k \tag{8.2}$$

where $W(\mathbf{X})$ is the weight of the structure, ng is the number of design variables and n is the number of members making up the structure, γ_i is the material density of member i, L_i is the length of member i, D_i is an allowable set of values for the design variable x_i, $g_i(\mathbf{X})$ denotes the constraints considered for the structure containing stress of elements and nodal deflection, and u is utilized to identify the upper bounds.

8.2.1 Constraint Conditions for Truss Structures

Size optimization of truss structures includes finding optimum values for member cross-sectional areas that minimize the structural weight; this also should satisfy inequality constraints that limit design variable sizes and structural responses, as follows:

$$\begin{aligned} \text{subject to:} \quad &\delta_i \le \delta_i^u \quad i = 1, 2, ..., m \\ &\sigma_i \le \sigma_i^u \quad i = 1, 2, ..., n \end{aligned} \tag{8.3}$$

where σ_i and δ_i are the stress of elements and nodal deflection of nodes, respectively and m is the number of nodes.

For truss examples, two types are considered, continuous and discrete problems. In the continuous problems, the design variables can vary continuously in the process of optimization:

$$D_i = \{x_i | x_i \in [x_{i,\min}, \ x_{i,\max}]\} \tag{8.4}$$

where $x_{i,\min}$ and $x_{i,\max}$ are the minimum and maximum allowable values for the design variable i, respectively. If the design variables represent a selection from a predefined countable set, the problem is considered as discrete:

$$D_i = \{d_{i,1}, d_{i,2}, ..., d_{i,r(i)}\} \tag{8.5}$$

where $r(i)$ is the number of available discrete values for the ith design variable.

For continuous problems, the values of the upper bounds for constraints are considered as predefined constant values, while for discrete ones, the stress limitations of the members are imposed according to the provisions of the code ASD-AISC (1991), as follows:

$$\begin{cases} \sigma_i^+ = 0.6F_y & \text{for} \quad \sigma_i \geq 0 \\ \sigma_i^- & \text{for} \quad \sigma_i < 0 \end{cases} \tag{8.6}$$

where σ_i^- is calculated according to the slenderness ratio as:

$$\sigma_i^- = \begin{cases} \left[\left(1 - \dfrac{\lambda_i^2}{2C_c^2}\right)F_y\right] \Big/ \left(\dfrac{5}{3} + \dfrac{3\lambda_i}{8C_c} - \dfrac{\lambda_i^3}{8C_c^3}\right) & \text{for} \quad \lambda_i < C_c \\ \dfrac{12\pi^2 E}{23\lambda_i^2} & \text{for} \quad \lambda_i \geq C_c \end{cases} \tag{8.7}$$

where E is the modulus of elasticity, F_y is the yield stress of steel, C_c is the slenderness ratio dividing the elastic and inelastic buckling regions, and λ_i is the slenderness ratio.

In addition, for these the maximum slenderness ratio is limited to 300 for the tension members, and it is recommended to be 200 for the compression members according to the ASD-AISC (1991) design code provisions.

8.2.2 Constraint Conditions for Steel Frames

Optimal design of frame structures is subjected to the following constraints according to the LFRD-AISC (2001) provisions:

Maximum lateral displacement:

$$\frac{\Delta_T}{H} \leq R \tag{8.8}$$

Inter-story displacements constraints:

$$\frac{d_i}{h_i} \leq R_I, \quad i = 1, 2, \ldots, \text{ns} \tag{8.9}$$

The strength constraints:

$$\begin{aligned} &\frac{P_u}{2\phi_c P_n} + \left(\frac{M_{ux}}{\phi_b M_{nx}} + \frac{M_{uy}}{\phi_b M_{ny}}\right) \leq 1, \quad \text{For} \quad \frac{P_u}{\phi_c P_n} < 0.2 \\ &\frac{P_u}{\phi_c P_n} + \frac{8}{9}\left(\frac{M_{ux}}{\phi_b M_{nx}} + \frac{M_{uy}}{\phi_b M_{ny}}\right) \leq 1, \quad \text{For} \quad \frac{P_u}{\phi_c P_n} \geq 0.2 \end{aligned} \tag{8.10}$$

where Δ_T is the maximum lateral displacement, H is the height of the frame structure, R is the maximum drift index (1/300), d_i is the inter-story drift, h_i is the story height of the ith floor, ns denotes the total number of stories R_I is the inter-story drift index permitted by the code of the practice (1/300), P_u is the required strength (tension or compression), P_n is the nominal axial strength (tension or compression), ϕ_c denotes the resistance factor ($\phi_c = 0.9$ for tension, $\phi_c = 0.85$ for compression), M_{ux} and M_{uy} are the required flexural strengths in the x and y directions, respectively, M_{nx} and M_{ny} are the nominal flexural strengths in the x and y directions, and ϕ_b represents the flexural resistance reduction factor ($\phi_b = 0.90$).

8.2.3 Constraints Handling Approach

For the proposed method, it is essential to transform the constrained optimization problem to an unconstrained one. A detailed review of some constraint-handling approaches is available in Coello (2002) and Michalewicz (1995). In this study, a penalty function method is utilized for handling the design constraints, which is calculated using the following relationships (Kaveh et al., 2008b):

$$\begin{cases} g_i \le g_i^u & \Rightarrow \Phi_g^{(i)} = 0 \\ \text{Otherwise} & \Rightarrow \Phi_g^{(i)} = \dfrac{g_i - g_i^u}{g_i^u} \end{cases} \tag{8.11}$$

The objective function that must be optimized is then defined as:

$$\text{Mer} = \varepsilon_1 \cdot Wd + \varepsilon_2 \cdot \left(\sum \Phi_g^{(i)}\right)^{\varepsilon_3} \tag{8.12}$$

where Mer is the objective function (merit function), ε_1, ε_2, and ε_3 are the coefficients of merit function, and $\Phi_g^{(i)}$ denotes the summation of penalties. In this study, ε_1 and ε_2 are set to 1 and W (weight of structure), respectively, while the value of ε_3 is considered as 0.85 to achieve a feasible solution.

8.3 Review of the Utilized Methods

8.3.1 BB–BC Algorithm

The standard BB–BC method consists of two phases: a BB phase and a BC phase. In the BB phase, candidate solutions are randomly distributed over the search space. Similar to other evolutionary algorithms, initial solutions are spread all over the search space in a uniform manner in the first BB. Erol and Eksin (2006) associated the random nature of the BB to energy dissipation or the transformation from an ordered state (a convergent solution) to a disorder or chaos state (new set of solution candidates).

The BB phase is followed by the BC phase. The BC is a convergence operator that has many inputs but only one output, which is known as the "center of mass" since the only output has been derived by calculating the center of mass. Here, the term mass refers to the inverse of the merit function value. The point representing the center of mass, denoted by $\mathbf{X}_c^k$, is calculated according to:

$$\mathbf{X}_c^{(k,j)} = \frac{\sum_{i=1}^{N}(1/\text{Mer}^i) \cdot \mathbf{X}_i^{(k,j)}}{\sum_{i=1}^{N}(1/\text{Mer}^i)}, \quad j = 1, 2, ..., \text{ng} \tag{8.13}$$

where $\mathbf{X}_i^{(k,j)}$ is the jth component of the ith solution generated in the kth iteration, N is the population size in the BB phase. After the BC phase, the algorithm creates the new solutions to be used as the BB of the next iteration step by using the previous knowledge (center of mass). This is often accomplished by spreading new offsprings around the center of mass using a normal distribution operation in every direction, where the standard deviation of this normal distribution function decreases as the number of iterations of the algorithm increases as:

$$\mathbf{X}_i^{k+1} = \mathbf{X}_c^k + \alpha_1 \frac{r_n \otimes (\mathbf{X}_{\max} - \mathbf{X}_{\min})}{k+1} \tag{8.14}$$

where r_n is a random vector from a standard normal distribution which changes for each candidate and α_1 is a parameter for limiting the size of the search space.

These successive explosion and contraction steps are carried out repeatedly until a stopping criterion has been met. Here, a maximum number of iterations are utilized as a stopping criterion. The pseudocode of the BB–BC algorithm can be summarized as follows:

Step 1. Generating initial candidates in a random manner (considering allowable boundaries).

Step 2. Calculating the value of the objective function for all the candidate solutions.

Step 3. Finding the center of mass.

Step 4. Calculating new candidates around the center of mass.

Step 5. Returning to *Step 2* and repeating the process until the condition for the stopping criterion is fulfilled.

8.3.2 Particle Swarm Optimization

PSO is a stochastic optimization method capable of handling nondifferentiable, nonlinear, and multimodule objective functions. The PSO approach is inspired from the social behavior of bird flocking and fish schooling (Eberhart and Kennedy, 1995). PSO has a population of individuals that move through the search space and each individual has a velocity that acts as an operator to obtain a new set of individuals. Individuals, called particles, adjust their movements depending on both their own experience and the population's experience. Effectively, each particle continuously focuses and refocuses on the effort of its search according to both local and global best. This behavior mimics the cultural adaptation of a biological agent in a swarm: it evaluates its own position based on certain fitness criteria, compares to others, and imitates the best in the entire swarm (Kennedy et al., 2001).

Through an updating process, each particle moves by adding a change velocity $\mathbf{V}_i^{k+1}$ to the current position $\mathbf{X}_i^k$ as follows:

$$\mathbf{X}_i^{k+1} = \mathbf{X}_i^k + \mathbf{V}_i^{k+1} \tag{8.15}$$

The velocity is a combination of three contributing factors:

1. Previous velocity, $\mathbf{V}_i^k$, considering former attempts.
2. Movement in the direction of the local best, $\mathbf{P}_i^k$, using the autobiographical memory.
3. Movement in the direction of the global best, $\mathbf{P}_g^k$, based on the publicized knowledge.

The mathematical relationship can be expressed as:

$$\mathbf{V}_i^{k+1} = \omega \mathbf{V}_i^k + c_1 r_1 (\mathbf{P}_i^k - \mathbf{X}_i^k) + c_2 r_2 (\mathbf{P}_g^k - \mathbf{X}_i^k) \tag{8.16}$$

where ω is an inertia weight to control the influence of the previous velocity, r_1 and r_2 are two random numbers uniformly distributed in the range of (0, 1), and c_1 and c_2 are two acceleration constants. $\mathbf{P}_i^k$ is the best position of the ith particle up to iteration k, and $\mathbf{P}_g^k$ is the best position among all particles in the swarm up to iteration k. $\mathbf{P}_i^k$ and $\mathbf{P}_g^k$ are given by the following equations:

$$\mathbf{P}_i^k = \begin{cases} \mathbf{P}_i^{k-1} & \text{Mer}(\mathbf{X}_i^k) \geq \text{Mer}(\mathbf{P}_i^{k-1}) \\ \mathbf{X}_i^k & \text{Mer}(\mathbf{X}_i^k) < \text{Mer}(\mathbf{P}_i^{k-1}) \end{cases} \tag{8.17}$$

$$\mathbf{P}_g^k = \{\mathbf{P}_i^k | W(\mathbf{P}_i^k) = \min(\text{Mer}(\mathbf{P}_g^{k-1}) \text{ and } \text{Mer}(\mathbf{P}_j^k), \quad j = 1, 2, .., N)\} \tag{8.18}$$

where N is the total number of particles.

The pseudocode of the PSO algorithm can be summarized as follows:

Step 1. Initializing an array of particles with random positions and their associated velocities.

Step 2. Evaluating the fitness function of the particles and updating local best position, $\mathbf{P}_i^k$ according to the best current value of the fitness function.

Step 3. Determining the current global minimum fitness value among the current positions and updating $\mathbf{P}_g^k$, the global best position.

Step 4. Changing the velocities and moving each particle to the new position considering the related velocity.

Step 5. Repeating *Steps 2–4* until a terminating criterion is satisfied.

8.3.3 Sub-Optimization Mechanism

The Sub-Optimization Mechanism (SOM) was first introduced in order to improve the ACO algorithm (Kaveh and Talatahari, 2010b). The SOM is based on the principles of finite element method. The finite element method is one of the major numerical solution techniques that has been developed and applied to solve numerous engineering problems in order to find their approximate solutions. The finite element method requires division of the problem domain into many subdomains and each domain is called a finite element. These finite element patches are considered instead of the main model (Figure 8.1B). As the number of finite elements increases, the approximate solutions obtained with the finite element method will become nearer to the exact solutions; and vice versa, if the number of finite elements is selected as a small number, the number of calculations as well as the accuracy of the solutions will decrease. In the finite element method, to investigate solutions further, some special patches can be divided into smaller sections (Figure 8.1C). Similarly, the SOM divides the search space into subdomains and performs optimization processes with these patches, and then, based on the resulting solutions, the undesirable parts are deleted and the remaining space is divided into smaller parts for more investigation in the next stage.

Therefore, the SOM can be considered as the repetition of the following steps for definite times, nc, (in the stage k of the repetition):

Step 1. Calculating permissible bounds for each variable. If $x_i^{(k-1)}$ is the ith component of the solution obtained from the previous stage $(k-1)$ for the ith variable, then:

$$\begin{cases} \text{If } x_i^{(k-1)} < (1-\beta_1)\cdot x_{i,\min}^{(k-1)} + \beta_1 \cdot x_{i,\max}^{(k-1)} \Rightarrow \begin{cases} x_{i,\min}^{(k)} = x_{i,\min}^{(k-1)} \\ x_{i,\max}^{(k)} = x_{i,\min}^{(k-1)} + 2\cdot\beta_1\cdot(x_{i,\max}^{(k-1)} - x_{i,\min}^{(k-1)}) \end{cases} \\ \text{If } x_i^{(k-1)} > \beta_1 \cdot x_{i,\min}^{(k-1)} + (1-\beta_1)\cdot x_{i,\max}^{(k-1)} \Rightarrow \begin{cases} x_{i,\min}^{(k)} = x_{i,\max}^{(k-1)} - 2\cdot\beta_1\cdot(x_{i,\max}^{(k-1)} - x_{i,\min}^{(k-1)}) \\ x_{i,\max}^{(k)} = x_{i,\max}^{(k-1)} \end{cases} \\ \text{Else} \Rightarrow \begin{cases} x_{i,\min}^{(k)} = x_i^{(k-1)} - \beta_1\cdot(x_{i,\max}^{(k-1)} - x_{i,\min}^{(k-1)}) \\ x_{i,\max}^{(k)} = x_i^{(k-1)} + \beta_1\cdot(x_{i,\max}^{(k-1)} - x_{i,\min}^{(k-1)}) \end{cases} \end{cases} \tag{8.19}$$

(B)

Figure 8.1 The finite element method. (A) the main problem, (B) the finite elements, and (C) the extra finite elements.

where $i = 1,2,\ldots,\text{ng}$, $k = 2,\ldots,\text{nc}$, β_1 is an adjustable factor that determines the amount of the remaining search space for each stage, nc is the maximum number of repetitious stages for SOM, and $x_{i,\min}^{(k)}$ and $x_{i,\max}^{(k)}$ are the minimum and the maximum allowable values for the ith variable at the stage k, respectively. In stage 1, the amounts of $x_{i,\min}^{(1)}$ and $x_{i,\max}^{(1)}$ are set to:

$$x_{i,\min}^{(1)} = x_{i,\min},\ x_{i,\max}^{(1)} = x_{i,\max} \quad i = 1, 2, ..., \text{ng} \tag{8.20}$$

Step 2. Determining the accuracy value for the variables. In each stage, the number of permissible value for each variable is considered as β_2, and therefore the amount of the accuracy rate of each variable equals to:

$$x_i^{*(k)} = \frac{(x_{i,\max}^{(k)} - x_{i,\min}^{(k)})}{(\beta_2 - 1)} \quad i = 1, 2, ..., \text{ng} \tag{8.21}$$

where $x_i^{*(k)}$ is the amount of increase in the ith variable and β_2 is the number of subdomains.

Step 3. Creating a series of allowable values for the variables. The set of allowable values for the variable i can be defined by using Eqs. (8.19) and (8.21) as:

$$x_{i,\min}^{(k)},\ x_{i,\min}^{(k)} + x_i^{*(k)},\ \ldots,\ x_{i,\min}^{(k)} + (\beta_2 - 1) \cdot x_i^{*(k)} = x_{i,\max}^{(k)} \quad i = 1, 2, ..., \text{ng} \tag{8.22}$$

Step 4. Determining the optimum solution of the current stage. The last step is to perform an optimization process using the optimization algorithm when Eq. (8.24) is considered as permissive values for the variables.

The SOM ends when the amount of accuracy rate of the last stage (i.e., $x_i^{*(\text{nc})}$) is less than the amount of accuracy rate of the primary problem (i.e., x_i^*):

$$x_i^{*(\text{nc})} \le x_i^* \quad i = 1, 2, ..., \text{ng} \tag{8.23}$$

SOM improves the search process with updating the search space from one stage to the next stage.

8.4 The Proposed Method

8.4.1 A Continuous Algorithm

The advantages of applying the BB–BC algorithm to structural design are similar to other evolutionary algorithms. The BB–BC is a multiagent and randomized search technique that tests a number of search space points in each cycle. The random selection and the information obtained in each cycle (center of mass) are used to choose new points in subsequent cycles. The BB–BC method has the ability to handle a mixture of discrete and continuous design variables and multiple loading cases.

Although BB–BC performs well in the exploitation (the fine search around a local optimum), it has some problems in the exploration (global investigation of the search space) stage (Kaveh and Talatahari, 2009b). If all of the candidates in the initial BB are collected in a small part of the search space, the BB–BC method may not find the optimum solution and, with a high probability, it may be trapped in that subdomain. One can consider a large number for candidates to avoid this defect, but it causes an increase in the function evaluations as well as the computational costs. This paper uses the PSO capacities to improve the exploration ability of the BB–BC algorithm.

The PSO is motivated from the social behavior of bird flocking and fish schooling, which mimic a population of individuals, called particles, that adjust their movements depending on both their own experience and the population's experience. At each iteration, a particle moves toward a direction computed from the best visited position (local best) and the best visited position of all particles in its neighborhood (global best). The improved BB–BC approach similarly not only uses the center of mass but also utilizes the best position of each candidate ($\mathbf{P}_i^k$) and the best global position ($\mathbf{P}_g^k$) to generate a new solution as:

$$\mathbf{X}_i^{k+1} = \alpha_2 \mathbf{X}_c^k + (1-\alpha_2)(\alpha_3 \mathbf{P}_g^k + (1-\alpha_3)\mathbf{P}_i^k) + \alpha_1 \frac{r_n \otimes (\mathbf{X}_{\max} - \mathbf{X}_{\min})}{k+1} \quad \left\langle \begin{array}{l} i = 1,2,\ldots,\mathrm{ng} \\ j = 1,2,\ldots,N \end{array} \right. \tag{8.24}$$

where $\mathbf{P}_i^k$ is the best position of the ith particle up to the iteration k and $\mathbf{P}_g^k$ is the best position among all candidates up to the iteration k, and α_2 and α_3 are adjustable parameters controlling the influence of the global best and local best on the new position of the candidates, respectively.

Another improvement in the BB–BC method is employing the SOM as an auxiliary tool, which works as a search-space updating mechanism. SOM, based on the principles of the finite element method, divides the search space into subdomains and performs optimization processes with these patches. Then, based on the resulting solutions, the undesirable parts are deleted and the remaining space is divided into smaller parts for additional investigation in the next stage. This process continues until the remaining space becomes smaller than the required size to satisfy accuracy.

This mechanism can be added as the following steps to the BB–BC method:

Step 1. Calculating the bounds for each varible. If $\mathbf{P}_g^{k_{\mathrm{SOM}}}$ is the global best solution obtained from the previous stage ($k_{\mathrm{SOM}} - 1$), then:

$$\begin{cases} \mathbf{X}_{\min}^{(k_{\mathrm{SOM}})} = \mathbf{P}_g^{(k_{\mathrm{SOM}}-1)} - \beta_1 \cdot (\mathbf{X}_{\max}^{(k_{\mathrm{SOM}}-1)} - \mathbf{X}_{\min}^{(k_{\mathrm{SOM}}-1)}) \geq \mathbf{X}_{\min}^{(k_{\mathrm{SOM}}-1)} \\ \mathbf{X}_{\max}^{(k_{\mathrm{SOM}})} = \mathbf{P}_g^{(k_{\mathrm{SOM}}-1)} + \beta_1 \cdot (\mathbf{X}_{\max}^{(k_{\mathrm{SOM}}-1)} - \mathbf{X}_{\min}^{(k_{\mathrm{SOM}}-1)}) \leq \mathbf{X}_{\max,i}^{(k_{\mathrm{SOM}}-1)} \end{cases}, \quad k_{\mathrm{SOM}} = 2,\ldots,\mathrm{nc} \tag{8.25}$$

where α_1 is an adjustable factor that determines the amount of the remaining search space and in this research it is taken as 0.3.

Step 2. Determining the values of the accuracy for the variables as:

$$\mathbf{X}^{*(k_{\mathrm{SOM}})} = \frac{(\mathbf{X}_{\max}^{(k_{\mathrm{SOM}})} - \mathbf{X}_{\min}^{(k_{\mathrm{SOM}})})}{\beta_2 - 1} \tag{8.26}$$

where $\mathbf{X}^{*(k_{\mathrm{SOM}})}$ is the amount of increase in allowable cross-sectional area and α_2 (the number of subdomains) is set to 100.

Step 3. Creating the series of the allowable values for the variables as:

$$\mathbf{X}_{\min}^{(k_{\mathrm{SOM}})},\ \mathbf{X}_{\min}^{(k_{\mathrm{SOM}})} + \mathbf{X}^{*(k_{\mathrm{SOM}})}, \ldots, \mathbf{X}_{\min}^{(k_{\mathrm{SOM}})} + (\beta_2 - 1)\mathbf{X}^{*(k_{\mathrm{SOM}})} \tag{8.27}$$

Step 4. Determining the optimum solution of the stage k_{SOM}. The last step is performing an optimization process using the BB–BC algorithm.

The SOM continues the search process until a solution is obtained with the required accuracy. The SOM performs as a search-space updating rule, which improves the search process by updating the search space from one stage to the next stage. Also, the SOM helps distribute the initial particles in the first BB. Another advantage of the SOM is being able to select a small number of candidates by reducing the search space.

8.4.2 A Discrete Algorithm

One way to solve discrete problems with a continuous algorithm is to use a rounding function, which changes the magnitude of the results by the value of the nearest discrete value. Although this change is simple and efficient, it may reduce the exploration capability of the algorithm (Kaveh and Talatahari, 2009c). Therefore, we utilize the SOM as a search-space updating mechanism considering a continuous search space and a small discrete problem in global and local phases, respectively.

According to this methodology, the optimization process is carried out in two phases, namely the global phase and the local phase. In the global phase, as shown in Figure 8.2, the search space is considered as a continuous domain and the SOM is utilized.

In the local phase, the final optimum design is obtained by considering a determined number of discrete values in the neighborhood of the result gained in the previous phase. The local phase implies that the optimal discrete solution is close to the continuous one, which is usually the case for real engineering structures. Now, a rounding function can be utilized. In this study, eight numbers are selected as the neighborhood discrete values of the global phase result. Figure 8.3 shows the flow chart for the local phase of the discrete variant of the algorithm.

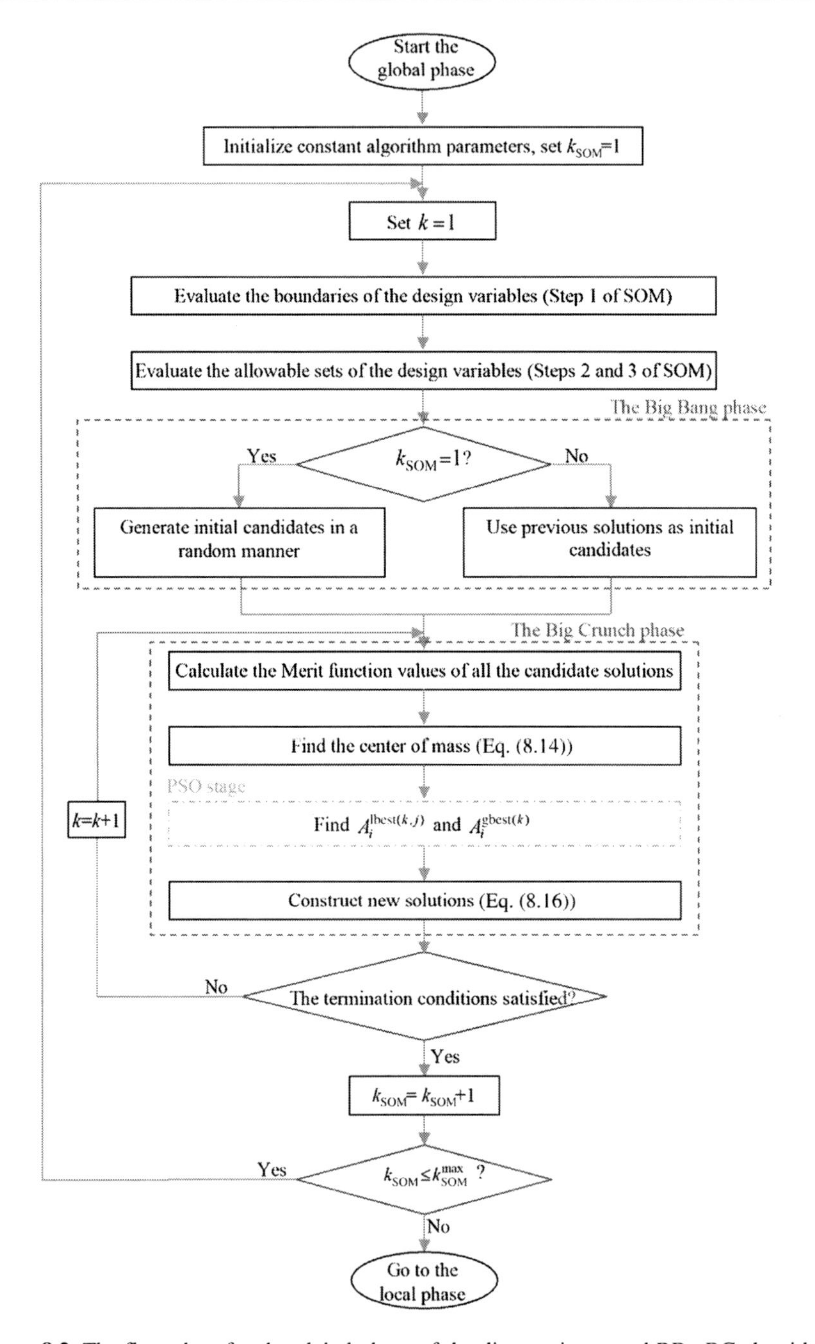

Figure 8.2 The flow chart for the global phase of the discrete improved BB–BC algorithm.

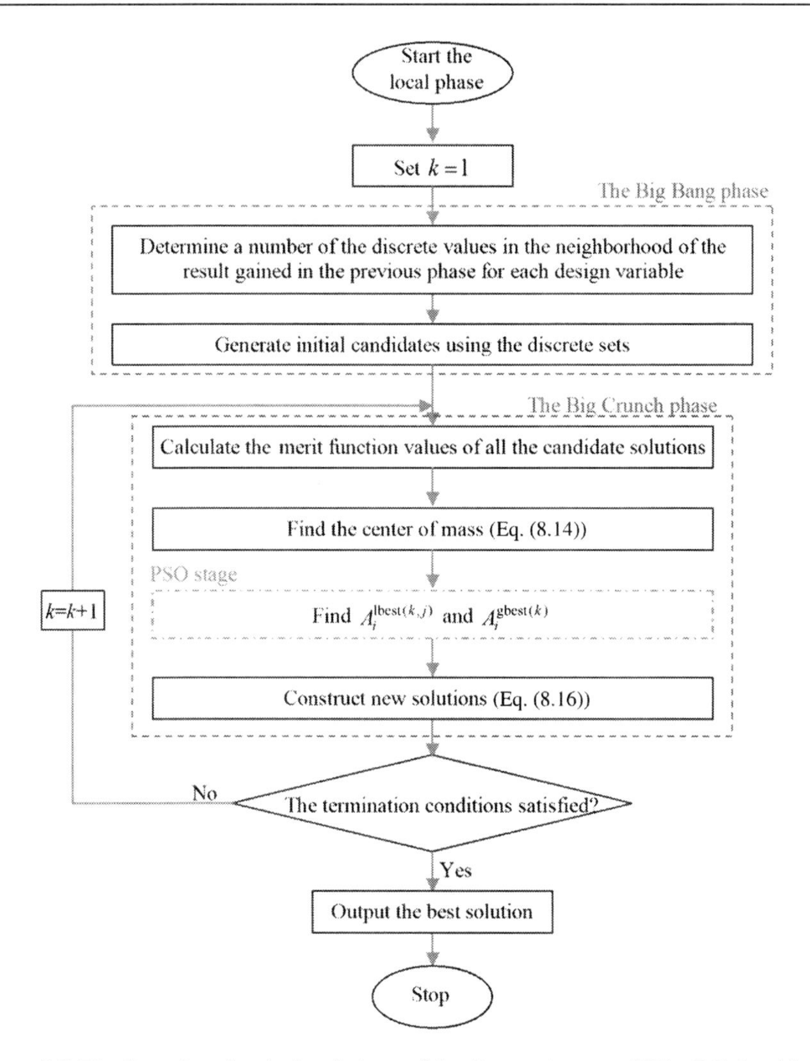

Figure 8.3 The flow chart for the local phase of the discrete improved BB–BC algorithm.

8.5 Design Examples

In this section, six examples with continuous and discrete variables consisting of truss and frame structures are optimized with the proposed method. The final results are compared to the solutions of other methods to demonstrate the efficiency of the present approach. These optimization examples include the following:

With a continuous search domain:

- A square on diagonal double-layer grid.
- A 26-story-tower spatial truss.

With a discrete search domain:

- A 354-bar braced dome truss.
- A 582-bar tower truss.
- A 3-bay, 15-story frame.
- A 3-bay 24-story frame.

For the proposed algorithm, a population of 50 individuals is used for all the examples. The value of the constants α_1, α_2, and α_3 are set to 1.0, 0.40, and 0.80, respectively (Kaveh and Talatahari, 2009b). The algorithms are coded in Matlab and the structures are analyzed using the direct stiffness method.

8.5.1 A Square on Diagonal Double-Layer Grid

A double-layer grid of the type shown in Figure 8.4 with a span of 21 m and the height of 1.5 m is chosen from Salajegheh and Vanderplaats (1986/87). The structure is simply supported at the corner nodes of the bottom layer. The loading is assumed as a uniformly distributed load on the top layer of intensity of 155.5 kg/m^2 and it is transmitted to the joints acting as concentrated vertical loads only. The structure is assumed as pin jointed with an elastic modulus of 210,000 MPa and the

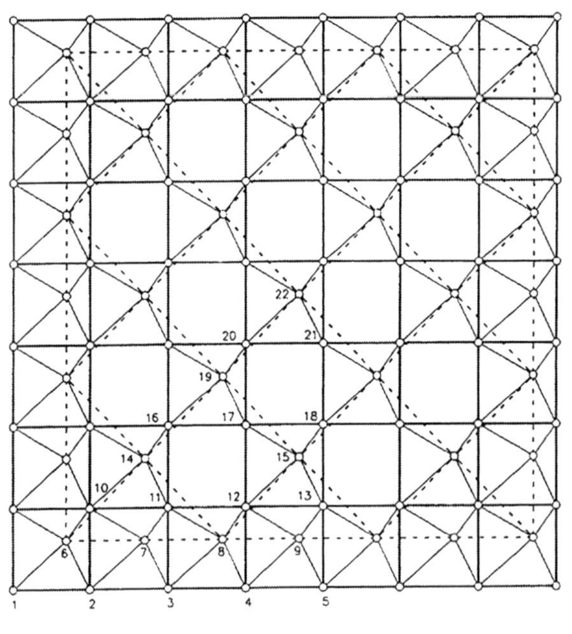

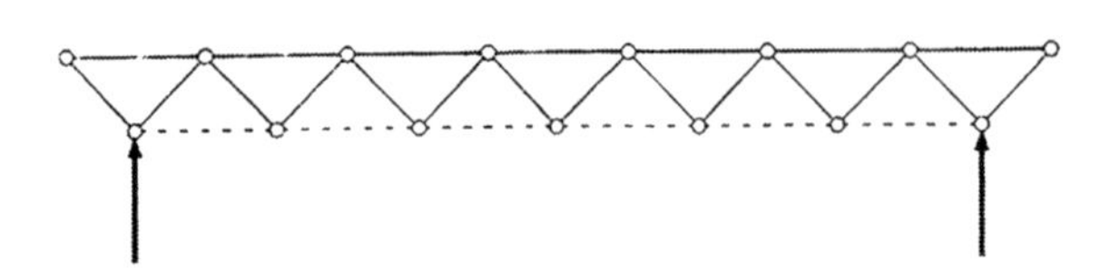

Figure 8.4 A square on diagonal double-layer grid.

material density is assumed as 0.008 kg/cm^3 for all the members. Member areas are linked to maintain symmetry about the four lines of symmetry axes in the plane of the grid. Thus the problem has 47 design variables. The maximum allowable area is considered as 22 cm^2 with a lower limit of 0.1 cm^2.

Stress, Euler buckling, and displacement constraints are considered in this problem. All the elements are subjected to the following stress constraints:

$$-1000 \leq \sigma_i \leq 1400 \text{ kg/cm}^2 \quad i = 1, 2, \ldots, 47 \tag{8.28}$$

where i is the element number. Tubular members are considered with a diameter to thickness ratio of 10. Thus, Euler buckling is considered as:

$$\sigma_i^b = -10.1\ EA_i/8L_i^2 \quad i = 1, 2, ..., 47 \tag{8.29}$$

In addition, some displacement constraints are imposed on the vertical components of the three central joints along the diagonal of the grid (joints 19, 20, and 22) as:

$$-1.5 \leq \delta_i \leq 1.5 \text{ cm} \quad i = 1, 2, 3 \tag{8.30}$$

This example is solved using GA, standard PSO, PSO with passive congregation (PSOPC), BB–BC, and the improved BB–BC algorithm. The number of required iterations for the proposed algorithm is 250 iterations on average, while it is considered as 500 iterations for other methods. The results are presented in Table 8.1. The efficiency of the proposed algorithm in terms of the required optimization time and standard deviation is better than that of other approaches. The optimization time in the improved BB–BC algorithm is 631 s while in the standard BB–BC algorithm, it was 1249 s on a core™ 2 Duo 3.0 GHz CPU, (Kaveh and Talatahari, 2009b). Also, the improved BB–BC algorithm can find the best result in comparison to other algorithms. Figure 8.5 shows the convergence rate of the best and average of 50 runs for the proposed algorithm.

Table 8.1 Performance Comparison for the Square on Diagonal Double-Layer Grid

	Optimal Cross-Sectional Areas (cm^2)				
	GA	**PSO**	**PSOPC**	**BB-BC**	**Present Work**
Best weight (kg)	5236	5814	4951	4636	4413
Average weight (kg)	5614	6917	5162	4762	4508
Standard deviation (kg)	512.6	810.3	352.5	189.5	108.3
Number of analyses	50,000	50,000	50,000	50,000	25,000
Optimization time (s)	1854	1420	1420	1249	631

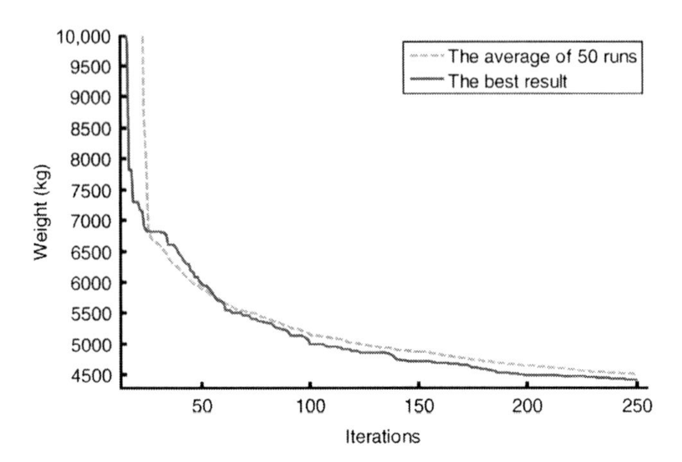

Figure 8.5 The convergence history of the square on diagonal double-layer grid for the improved BB–BC algorithm.

8.5.2 A 26-Story-Tower Spatial Truss

A 26-story-tower space truss containing 942 elements and 244 nodes is considered as the second example. Fifty-nine design variables are used to represent the cross-sectional areas of 59 element groups in this structure, employing the symmetry of the structure. Figure 8.6 shows the geometry and the 59 element groups. The material density is 2767.990 kg/m^3 and the modulus of elasticity is 68,950 MPa. The members are subjected to the stress limits of $\pm$ 172.375 MPa and the four nodes of the top level in the *x*-, *y*-, and *z*-directions are subjected to the displacement limits of $\pm$ 38.10 cm (about 1/250 of the total height of the tower). The allowable cross-sectional areas in this example are selected from 0.6452 cm^2 to 129.032 cm^2. The loading on the structure consists of the following:

1. The vertical load at each node in the first section is equal to -13.344 kN.
2. The vertical load at each node in the second section is equal to -26.688 kN.
3. The vertical load at each node in the third section is equal to -40.032 kN.
4. The horizontal load at each node on the right side in the *x* direction is equal to -4.448 kN.
5. The horizontal load at each node on the left side in the *x* direction is equal to $+6.672$ kN.
6. The horizontal load at each node on the front side in the *y* direction is equal to -4.448 kN.
7. The horizontal load at each node on the back side in the *x* direction is equal to $+4.448$ kN.

The improved BB–BC method achieved a good solution after 30,000 analyses and found an optimum weight of 23,768 kg (Kaveh and Talatahari, 2009b). The best weights for the GA, standard PSO, and BB–BC are 25,556, 27,390, and 24,131 kg,

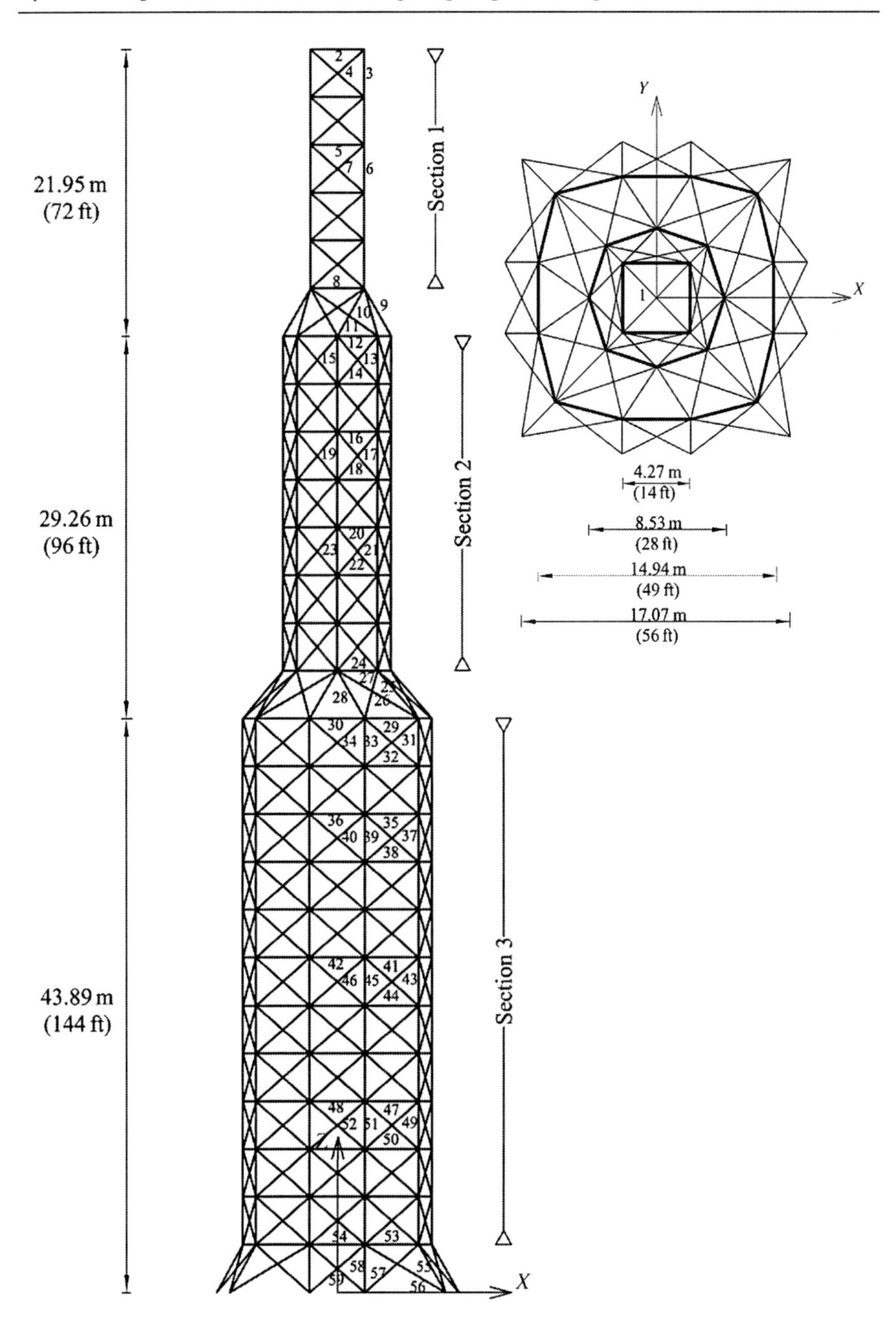

Figure 8.6 A 26-story-tower spatial truss.

Table 8.2 Performance Comparison for the 26-Story-Tower Spatial Truss

	GA	PSO	BB–BC	Present Work
Best weight (kg)	25,556	27,390	24,131	23,768
Average weight (kg)	28,677	34,129	25,041	24,281
Standard deviation (kg)	3012.1	4493.5	1189.0	644.3
Number of analyses	50,000	50,000	50,000	30,000
Optimization time (s)	4450	3640	3162	1926

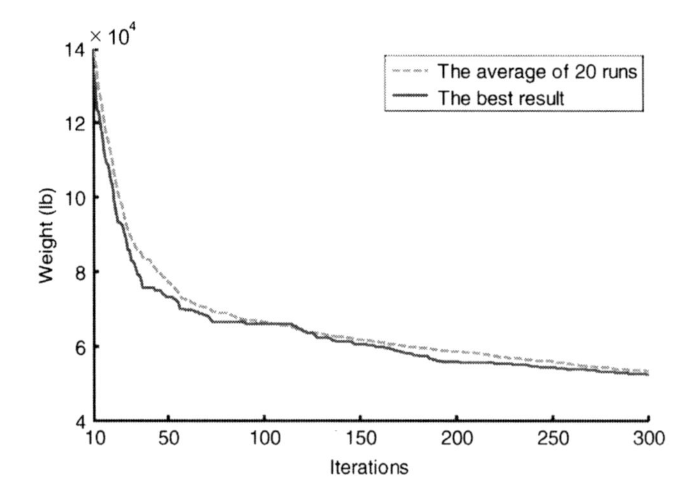

Figure 8.7 The convergence history of the 26-story-tower spatial truss for the improved BB–BC algorithm.

respectively. In addition, the improved BB–BC has better performance in terms of the optimization time, standard deviation, and the average weight. Table 8.2 provides the statistic information for this example. Figure 8.7 shows the best and average of 20 runs of the convergence history for the proposed algorithm.

8.5.3 A 354-Bar Braced Dome Truss

The plan, elevation, and 3D views of a braced dome truss are shown in Figure 8.8. This dome has a 40 m diameter and it is designed for covering the top of an auditorium at an elevation of 10 m. The dome has a height of 8.28 m and consists of 127 joints and 354 members. The 354 members are grouped into 22 independent design variables (Figure 8.8), which are selected from a database of 37 circular hollow sections in the ASD-AISC (1991) steel profile list. For design purposes, the dome is subjected to the following three load cases considering various combinations of dead

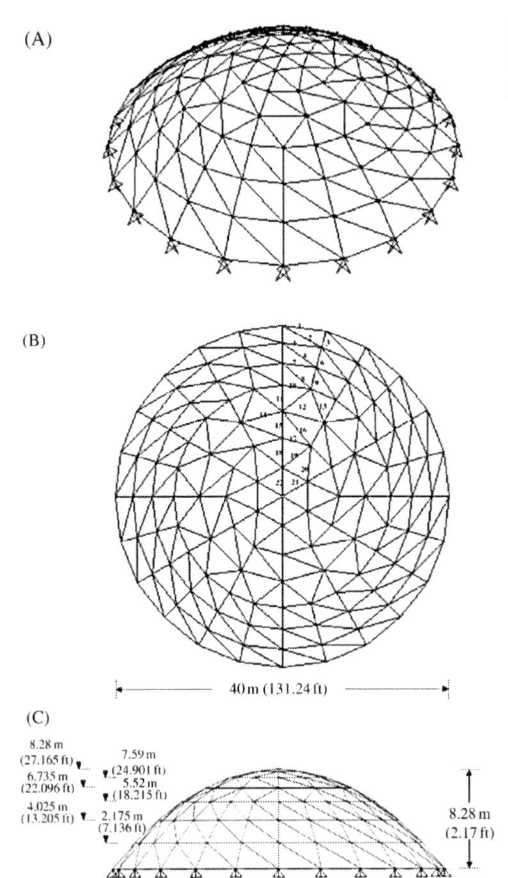

Figure 8.8 A 354-member braced truss dome: (A) 3D view, (B) top view, and (C) side view.

(D), snow (S), and wind (W) loads calculated according to the provisions of the ASCE 7-98 (Minimum Design Loads for Buildings and other Structures, 1998):

1. D + S,
2. D + S + W (with negative internal pressure), and
3. D + S + W (with positive internal pressure).

The load cases resulting from unbalanced snow loads are disregarded in the study. The illustrations of the three load cases are provided in Figure 8.9 (Kaveh and Talatahari, 2010j). It has been assumed that dead and snow loads act on the projected area, while wind load acts on the curved surface area. Sandwich-type aluminum cladding material is used, resulting in an assumed dead load pressure of 200 N/m^2 including the frame elements used for the girts. The equivalent loads for the three loading cases acting on joints are summarized in Table 8.3. The stress and stability limitations of the members are calculated according to the provisions of the ASD-AISC (1991). The displacements of all nodes are limited to 11.1 cm in all directions.

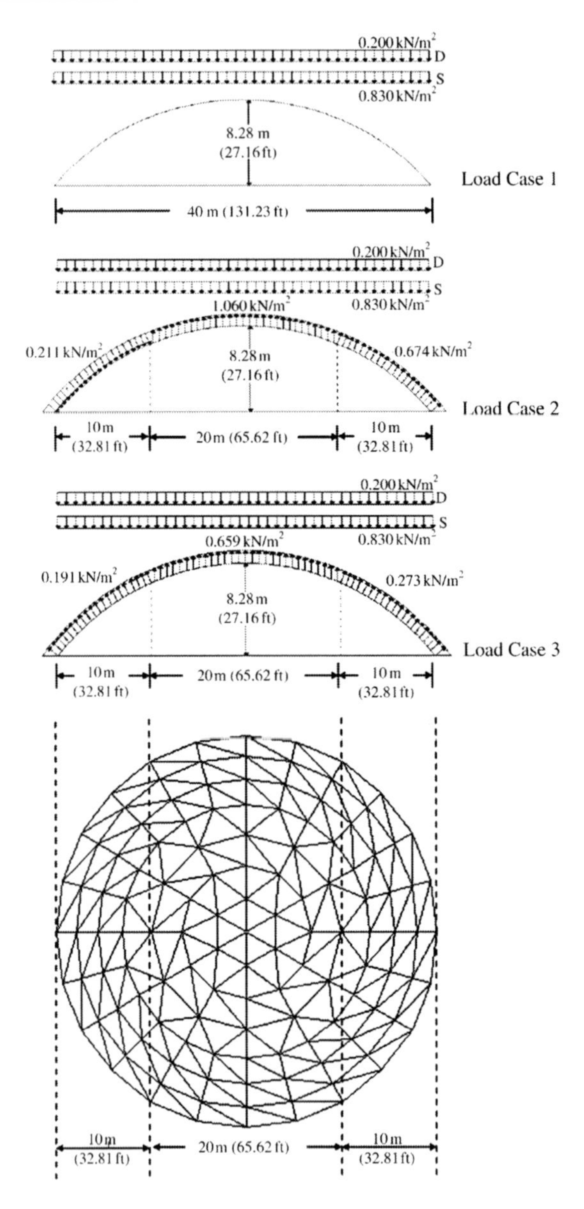

Figure 8.9 The three load cases considered for 354-member braced truss dome.

This example was solved using six metaheuristic algorithms including evolution strategies, simulated annealing, tabu search, ACO, harmony search, and GAs (Hasançebi et al., 2009). The simulated annealing technique resulted in the least weight compared to other heuristic algorithms, which was 14,760.8 kg. The final designs achieved by evolution strategies and PSO methods were both 14,816.3 kg

Table 8.3 Loading Conditions for the 354-Bar Braced Dome Truss

Load Case		P_x (kN)	P_z (kN)
1	All nodes	0.00	10.195
2	Windward quarter	1.257	−11.388
	Center half	0.00	1.708
	Leeward quarter	4.006	−2.960
3	Windward quarter	−1.133	−7.561
	Center half	0.00	−3.023
	Leeward quarter	1.627	−6.786

and are only 0.3% different from the one located by SA. ACO, tabu search, harmony search, and GAs achieved 4%, 8.7%, 8.8%, 12.6% heavier designs, respectively (Hasançebi et al., 2009).

The best result of the improved BB–BC algorithm has the weight as 14,708.9 kg, which is 0.4% lighter than the result of the simulated annealing (Kaveh and Talatahari, 2010j). Table 8.4 gives the comparison of optimal pipe section design results of the 354-bar braced dome truss for the present algorithm and simulated annealing algorithms. For the improved BB–BC result, the stress and stability limitations contrary to node displacements are active constraints. The section of the groups 1, 11, 12, 13, 14, 16, and 19 are determined considering stability limitations which give the weakest section. Therefore, for these element groups, both the simulated annealing method and the improved BB–BC method have found the same sections, while for other elements, the stress constraints determine the sections.

8.5.4 A 582-Bar Tower Truss

A 582-bar tower truss with the height of 80 m is considered as shown in Figure 8.10. The symmetry of the tower around x- and y-axes is considered to group the 582 members into 32 independent size variables. A single load case is considered consisting of the lateral loads of 5.0 kN applied to both x- and y-directions, and a vertical load of −30 kN applied in the z-direction at all the nodes of the tower. A discrete set of 137 economical standard steel sections selected from W-shape profile list based on area and radii of gyration properties is used to size the variables (Hasançebi et al., 2009). The lower and upper bounds on size variables are taken as 39.74 and 1387.09 cm^2, respectively. The stress limitations of the members are imposed according to the relationships of Section 8.2. The limitation of node displacements is 8.0 cm in each direction and the maximum slenderness ratio is limited to 300 and 200 for tension and compression members, respectively.

Table 8.4 Optimal Design Comparison for the 354-Bar Braced Dome Truss

Element Group	Optimal Pipe Sections (cm^2)	
	SA (Hasançebi et al., 2009)	Present Work
1	P2 (6.90)	P2 (6.90 cm^2)
2	P3 (14.39)	P3.5 (17.29)
3	P4 (20.45)	P3 (14.39)
4	P3.5 (17.29)	P3 (14.39)
5	P3 (14.39)	P3 (14.39)
6	P3 (14.39)	P3 (14.39)
7	P3 (14.39)	P3.5 (17.29)
8	P2.5 (10.97)	P3 (14.39)
9	P3 (14.39)	P2.5 (10.97)
10	P3 (14.39)	P3 (14.39)
11	P2.5 (10.97)	P2.5 (10.97)
12	P2.5 (10.97)	P2.5 (10.97)
13	P2.5 (10.97)	P2.5 (10.97)
14	P2.5 (10.97)	P2.5 (10.97)
15	P2.5 (10.97)	P2.5 (10.97)
16	P2.5 (10.97)	P2.5 (10.97)
17	P × 2 (9.55)	P × 2 (9.55)
18	P × 2 (9.55)	P × 2 (9.55)
19	P2 (6.90)	P2 (6.90)
20	P2 (6.90)	P2 (6.90)
21	P2 (6.90)	P2 (6.90)
22	P2 (6.90)	P2 (6.90)
Weight (kg)	14,760.8	14,708.9

As reported by Hasançebi et al. (2009), PSO has obtained the lightest design compared to some other metaheuristic algorithms such as evolution strategies algorithm, simulated annealing, tabu search, ACO, harmony search, and GAs. The evolution strategies technique gives the second best answer, which is only 0.1% heavier than the result of PSO. The other minimum weights obtained by simulated annealing, tabu search, ACO, harmony search, and GA are 0.4%, 1.2%, 1.7%, 3.8%, and 5.7% heavier than the one attained by PSO, respectively (Hasançebi et al., 2009). The authors have solved this problem using a heuristic particle swarm ACO (HPSACO) which is 1.5% lighter than the result of PSO (Kaveh and Talatahari, 2009c).

Table 8.5 gives the best solution vectors of the PSO, HPSACO, and the improved BB–BC algorithms. The optimum result of the new approach is 22.37 m^3, which is close to the result of the PSO (Kaveh and Talatahari, 2010j). The design history graphs are shown in Figure 8.11. The improved BB–BC needs nearly 12,500 analyses to reach a solution, which is less than the 50,000 analyses for PSO (Hasançebi et al., 2009) and more than the 8500 analyses for the HPSACO (Kaveh and Talatahari, 2009c). For the present algorithm, the maximum values of displacements in the x-, y-, and z-directions

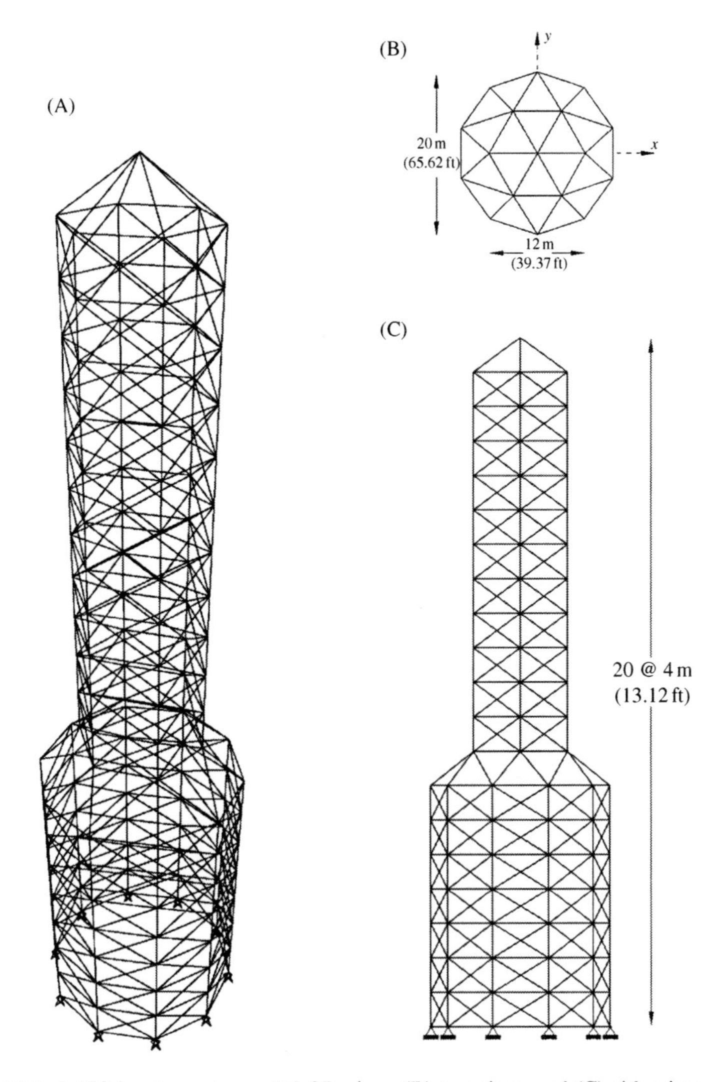

Figure 8.10 A 582-bar tower truss: (A) 3D view, (B) top view, and (C) side view.

are 8.0, 7.61, and 2.35 cm, respectively. The maximum stress and slenderness ratios are 97.67% and 95.36%, respectively.

8.5.5 A 3-Bay 15-Story Frame

The configuration and applied loads of a 3-bay 15-story frame structure (Kaveh and Talatahari, 2009a) is shown in Figure 8.12. The displacement and AISC combined strength constraints are the performance constraint of this frame. The sway

Table 8.5 Optimal Design Comparison for the 582-Bar Tower Truss

Element Group	Optimal W-Shaped Sections		
	PSO (Hasançebi et al., 2009)	**HPSACO (Kaveh and Talatahari, 2009c)**	**Present Work**
1	W8 × 21	W8 × 24	W8 × 24
2	W12 × 79	W12 × 72	W24 × 68
3	W8 × 24	W8 × 28	W8 × 28
4	W10 × 60	W12 × 58	W18 × 60
5	W8 × 24	W8 × 24	W8 × 24
6	W8 × 21	W8 × 24	W8 × 24
7	W8 × 48	W10 × 49	W21 × 48
8	W8 × 24	W8 × 24	W8 × 24
9	W8 × 21	W8 × 24	W10 × 26
10	W10 × 45	W12 × 40	W14 × 38
11	W8 × 24	W12 × 30	W12 × 30
12	W10 × 68	W12 × 72	W12 × 72
13	W14 × 74	W18 × 76	W21 × 73
14	W8 × 48	W10 × 49	W14 × 53
15	W18 × 76	W14 × 82	W18 × 86
16	W8 × 31	W8 × 31	W8 × 31
17	W8 × 21	W14 × 61	W18 × 60
18	W16 × 67	W8 × 24	W8 × 24
19	W8 × 24	W8 × 21	W16 × 36
20	W8 × 21	W12 × 40	W10 × 39
21	W8 × 40	W8 × 24	W8 × 24
22	W8 × 24	W14 × 22	W8 × 24
23	W8 × 21	W8 × 31	W8 × 31
24	W10 × 22	W8 × 28	W8 × 28
25	W8 × 24	W8 × 21	W8 × 21
26	W8 × 21	W8 × 21	W8 × 24
27	W8 × 21	W8 × 24	W8 × 28
28	W8 × 24	W8 × 28	W14 × 22
29	W8 × 21	W16 × 36	W8 × 24
30	W8 × 21	W8 × 24	W8 × 24
31	W8 × 24	W8 × 21	W14 × 22
32	W8 × 24	W8 × 24	W8 × 24
Volume (m^3)	22.3958	22.0607	22.3707

of the top story is limited to 23.5 cm. The material has a modulus of elasticity equal to $E = 200{,}000$ MPa and a yield stress of $F_y = 248.2$ MPa. The effective length factors of the members are calculated as $K_x \geq 0$ for a sway-permitted frame and the out-of-plane effective length factor is specified as $K_y = 1.0$. Each column is considered as nonbraced along its length, and the nonbraced length for each beam member is specified as one-fifth of the span length.

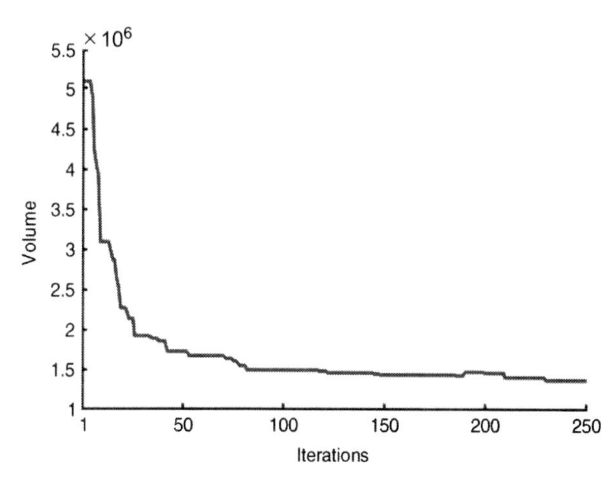

Figure 8.11 Convergence history for the 582-bar tower truss using the improved BB–BC algorithm.

An optimum design of the frame is obtained after 9500 analyses by using the improved BB–BC, having the minimum weight of 434.54 kN (Kaveh and Talatahari, 2010j). The optimum designs for HPSACO, PSOPC, and PSO had the weights of 426.36, 452.34, and 496.68 kN, respectively (Kaveh and Talatahari, 2009a). Table 8.6 summarizes the optimal designs for these algorithms. The global sway at the top story is 11.63 cm, which is less than the maximum sway.

8.5.6 A 3-Bay 24-Story Frame

Figure 8.13 shows the connectivity and the service loading conditions for a 3-bay 24-story frame consisting of 168 members. The frame is designed following the LRFD specification and uses an inter-story drift displacement constraint. The material properties are a modulus of elasticity of $E = 205{,}000$ MPa and a yield stress of $F_y = 230.3$ MPa.

The effective length factors of the members are calculated as $K_x \geq 0$ for a sway-permitted frame and the out-of-plane effective length factor is specified as $K_y = 1.0$. All columns and beams are considered as nonbraced along their lengths. Fabrication conditions are imposed on the construction of the 168-element frame requiring that the same beam section be used in the first and third bay on all floors except the roof beams, resulting in four beam groups. Beginning from the foundation, the exterior columns are combined into one group and the interior columns are combined together in another group over three consecutive stories. This grouping, results in 16 column sections and 4 beam sections for a total of 20 design variables (see Figure 8.13). In this example, each of the four beam element groups is chosen from all 267 W-shapes, while the 16 column element groups are limited to W14 sections (37 W-shapes).

Table 8.7 lists the designs developed by the improved BB–BC algorithm, by the ant colony algorithm (Camp and Bichon, 2005), and by harmony search (Degertekin, 2008). The improved BB–BC algorithm required approximately

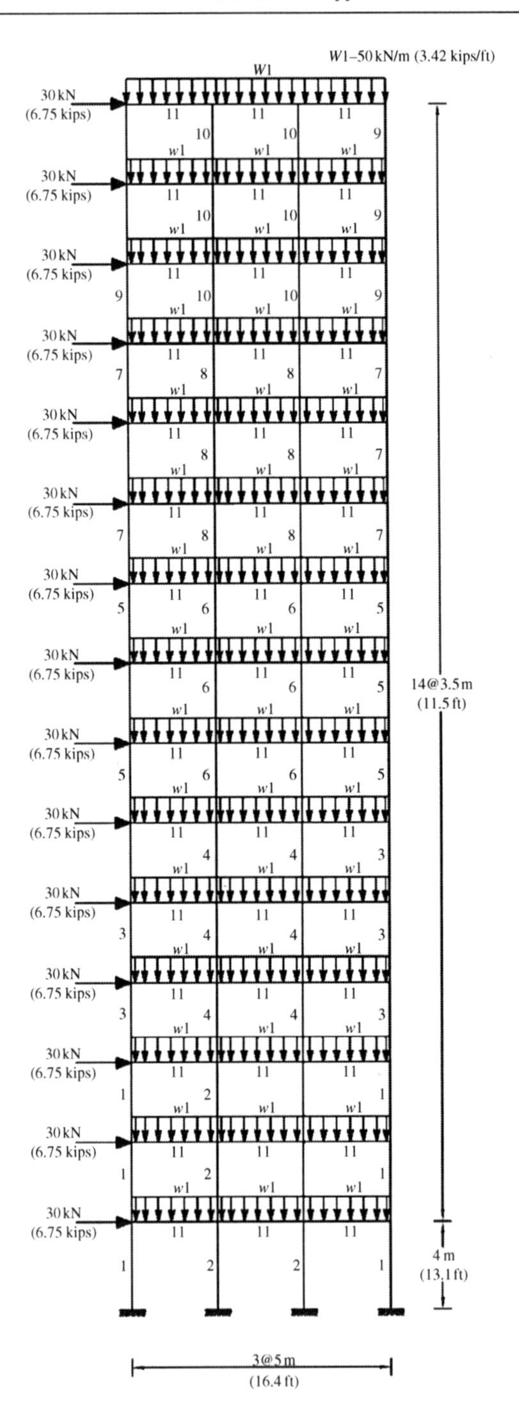

Figure 8.12 A 3-bay 15-story frame.

Table 8.6 Optimal Design Comparison for the 3-Bay 15-Story Frame

Element Group	Optimal W-Shaped Sections			
	PSO (Kaveh and Talatahari, 2009a)	**PSOPC (Kaveh and Talatahari, 2009a)**	**HPSACO (Kaveh and Talatahari, 2009a)**	**Present Work**
1	W33 × 118	W26 × 129	W21 × 111	W24 × 117
2	W33 × 263	W24 × 131	W18 × 158	W21 × 132
3	W24 × 76	W24 × 103	W10 × 88	W12 × 95
4	W36 × 256	W33 × 141	W30 × 116	W18 × 119
5	W21 × 73	W24 × 104	W21 × 83	W21 × 93
6	W18 × 86	W10 × 88	W24 × 103	W18 × 97
7	W18 × 65	W14 × 74	W21 × 55	W18 × 76
8	W21 × 68	W26 × 94	W26 × 114	W18 × 65
9	W18 × 60	W21 × 57	W10 × 33	W18 × 60
10	W18 × 65	W18 × 71	W18 × 46	W10 × 39
11	W21 × 44	W21 × 44	W21 × 44	W21 × 48
Weight (kN)	496.68	452.34	426.36	434.54
The global sway (cm)	10.42	11.36	11.57	11.63

10,500 frame analyses to converge to a solution, which is less than the 15,500 analyses required by ACO (Camp and Bichon, 2005) and the 13,924 analyses required by HS (Degertekin, 2008). Figure 8.14 shows the convergence history for improved BB–BC frame design. The global sway at the top story is 23.52 cm, which is less than the maximum sway.

8.6 Concluding Remarks

This chapter presents a new heuristic population-based search based on the BB–BC theory of the evolution of the universe to optimize skeletal structures. The proposed algorithm hybridizes the BB–BC and PSO algorithm by combining the center of mass, the best position of each candidate, and the best visited position of all the candidates, as an average point in the beginning of each BB phase. Additional improvement is due to the SOM, based on the principles of the finite element method working as a search-space updating technique.

A discrete variant of the optimization approach is also developed by dividing the search process into two phases. Using SOM during the global phase, the problem is first solved by treating all the design variables as continuous; then, in the local phase, a small domain is considered by selecting a few entries from the original discrete domain around the optimal values and the solution of this problem utilizes a rounding function.

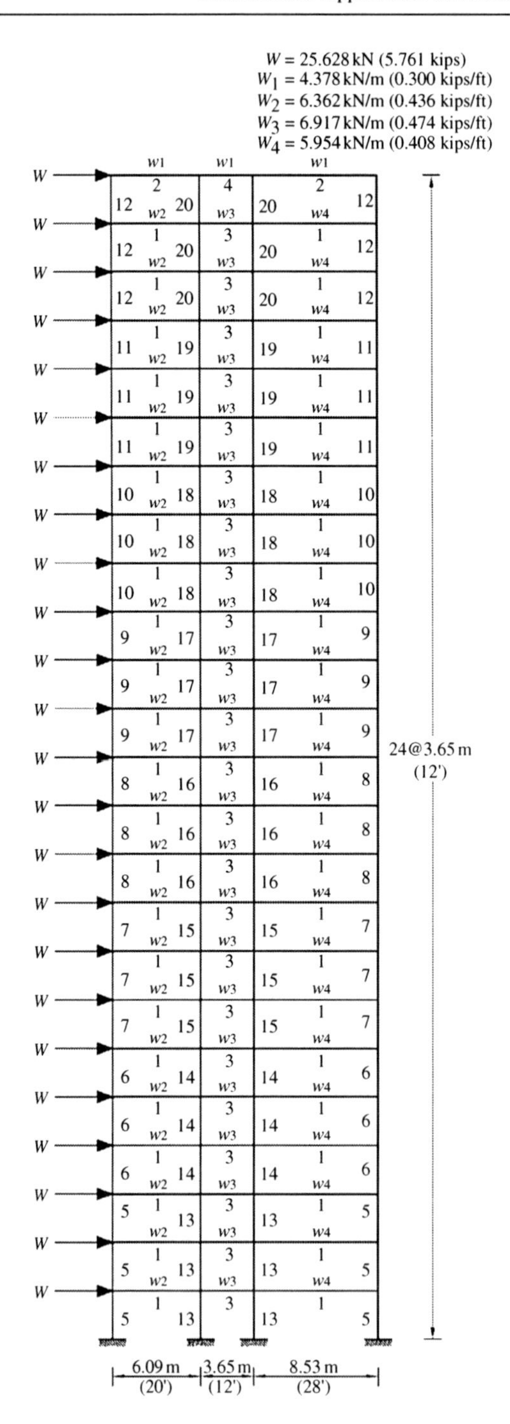

Figure 8.13 A 3-bay 24-story frame.

Table 8.7 Optimal Design Comparison for the 3-Bay 24-Story Frame

Element Group	Optimal W-Shaped Sections		
	ACO (Camp and Bichon, 2005)	HS (Degertekin, 2008)	Present Work
1	W30 × 90	W30 × 90	W30 × 90
2	W8 × 18	W10 × 22	W21 × 48
3	W24 × 55	W18 × 40	W18 × 46
4	W8 × 21	W12 × 16	W8 × 21
5	W14 × 145	W14 × 176	W14 × 176
6	W14 × 132	W14 × 176	W14 × 159
7	W14 × 132	W14 × 132	W14 × 109
8	W14 × 132	W14 × 109	W14 × 90
9	W14 × 68	W14 × 82	W14 × 82
10	W14 × 53	W14 × 74	W14 × 74
11	W14 × 43	W14 × 34	W14 × 38
12	W14 × 43	W14 × 22	W14 × 30
13	W14 × 145	W14 × 145	W14 × 159
14	W14 × 145	W14 × 132	W14 × 132
15	W14 × 120	W14 × 109	W14 × 109
16	W14 × 90	W14 × 82	W14 × 82
17	W14 × 90	W14 × 61	W14 × 68
18	W14 × 61	W14 × 48	W14 × 48
19	W14 × 30	W14 × 30	W14 × 34
20	W14 × 26	W14 × 22	W14 × 26
Weight (kN)	980.63	956.13	960.90

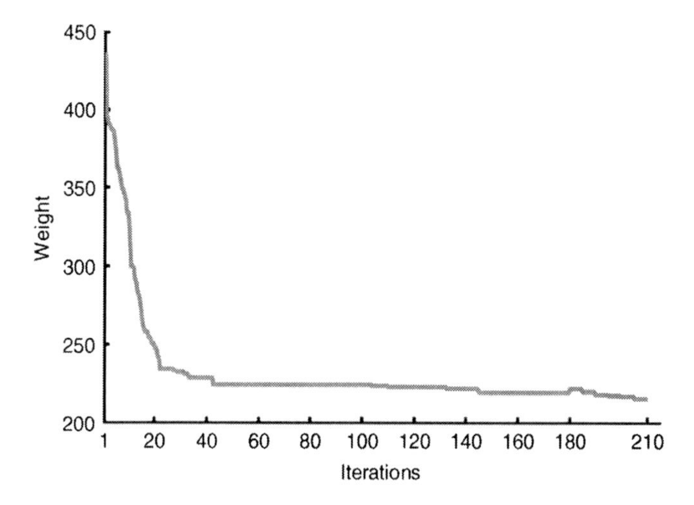

Figure 8.14 The convergence for the 3-bay 24-story frame using the improved BB–BC algorithm.

Six design examples, including four trusses and two frames, are considered to verify the efficiency of the present algorithm. A comparison of the numerical results of the examples using the new method with those obtained by other metaheuristic approaches is performed to demonstrate the robustness of the present algorithm. With respect to the standard BB–BC approach, the improved BB–BC has better solutions and standard deviations. Also, the new method has low computational time and high convergence speed compared to the BB–BC. Specifically, when the number of design variables increases, the improved BB–BC shows better performance. By adding a principle from the PSO to the BB–BC algorithm, we increase the exploration by improving the search ability of the algorithm. As a result, contrary to the other metaheuristic techniques that present convergence difficulty or get trapped at a local optimum in large size structures, the improved BB–BC performs well for the large size structures. On the other hand, increasing the exploration often results in an increase in the number of analyses. This problem is solved by using the SOM, which works as a search-space updating rule and reduces the number analyses for convergence.

References

American Institute of Steel Construction, AISC, 1991. Manual of Steel Construction–Load Resistance Factor Design. AISC, Chicago, IL.

American Institute of Steel Construction (AISC), 2001. Manual of Steel Construction–Load Resistance Factor Design. third ed. AISC, Chicago, IL.

Atashpaz-Gargari E., Lucas C., Imperialist competitive algorithm: an algorithm for optimization inspired by imperialistic competition. In: IEEE Congress on Evolutionary Computation. 2007, Singapore. pp. 4661–4667.

Camp, C.V., Bichon, J., 2004. Design of space trusses using ant colony optimization. J. Struct. Eng. 130 (5), 741–751.

Camp, C.V., Bichon, J., 2005. Design of steel frames using ant colony optimization. J. Struct. Eng. 131, 369–379.

Camp, C.V., Pezeshk, S., Cao, G., 1998. Optimized design of two dimensional structures using a genetic algorithm. J. Struct. Eng. 124 (5), 551–559.

Coello, C.A.C., 2002. Theoretical and numerical constraint-handling techniques used with evolutionary algorithms: a survey of the state of the art. Comput. Methods Appl. Mech. Eng. 191 (11-12), 1245–1287.

Degertekin, S.O., 2008. Optimum design of steel frames using harmony search algorithm. Struct. Multidiscip. Optim. 36, 393–401.

Dorigo, M., Maniezzo, V., Colorni, A., 1996. The ant system: optimization by a colony of cooperating agents. IEEE Trans. Syst. Man Cybern. Part B. 26 (1), 1–13.

Eberhart, R.C., Kennedy, J., 1995. A new optimizer using particle swarm theory. In: Proceedings of the Sixth International Symposium on Micro Machine and Human Science, Nagoya, Japan.

Erbatur, F., Hasancebi, O., Tutuncil, I., Kihc, H., 2000. Optimal design of planar and space structures with genetic algorithms. Comput. Struct. 75, 209–224.

Erol, O.K., Eksin, I., 2006. New optimization method: big bang–big crunch. Adv. Eng. Softw. 37, 106–111.

Goldberg, D.E., 1998. Genetic Algorithm in Search Optimization and Machine Learning. Addison Wesley Publishing Co. Inc., Reading, MA.

Hajela, P., Lee, E., 1995. Genetic algorithms in truss topological optimization. Int. J. Solids Struct. 32 (22), 3341–3357.

Hasançebi, O., Çarbas, S., Dogan, E., Erdal, F., Saka, M.P., 2009. Performance evaluation of metaheuristic search techniques in the optimum design of real size pin jointed structures. Comput. Struct. 87 (5-6), 284–302.

Kameshki, E.S., Saka, M.P., 2001. Optimum design of nonlinear steel frames with semi-rigid connections using a genetic algorithm. Comput. Struct. 79, 1593–1604.

Kathiravan, R., Ganguli, R., 2007. Strength design of composite beam using gradient and particle swarm optimization. Compos. Struct. 81 (4), 471–479.

Kaveh, A., Abditehrani, A., 2004. Design of frames using genetic algorithm, force method and graph theory. Int. J. Numer. Methods Eng. 61, 2555–2565.

Kaveh, A., Kalatjari, V., 2003. Topology optimization of trusses using genetic algorithm, force method, and graph theory. Int. J. Numer. Methods Eng. 58 (5), 771–791.

Kaveh, A., Rahami, H., 2006. Analysis, design and optimization of structures using force method and genetic algorithm. Int. J. Numer. Methods Eng. 65 (10), 1570–1584.

Kaveh, A., Shahrouzi, M., 2008. Dynamic selective pressure using hybrid evolutionary and ant system strategies for structural optimization. Int. J. Numer. Methods Eng. 73 (4), 544–563.

Kaveh, A., Shojaee, S., 2007. Optimal design of skeletal structures using ant colony optimisation. Int. J. Numer. Methods Eng. 70 (5), 563–581.

Kaveh, A., Talatahari, S., 2009a. Hybrid algorithm of harmony search, particle swarm and ant colony for structural design optimization: studies in computational intelligence. In: Geem Z.W. (Ed.), Harmony Search Algorithms for Structural Design Optimization, vol. 239. Springer-Verlag, Berlin, Heidelberg, pp. 159–198.

Kaveh, A., Talatahari, S., 2009b. Size optimization of space trusses using Big Bang–Big Crunch algorithm. Comput. Struct. 87 (17-18), 1129–1140.

Kaveh, A., Talatahari, S., 2009c. A particle swarm ant colony optimization for truss structures with discrete variables. J. Constr. Steel Res. 65 (8-9), 1558–1568.

Kaveh, A., Talatahari, S., 2010a. A novel heuristic optimization method: charged system search. Acta Mech. 213 (3-4), 267–289.

Kaveh, A., Talatahari, S., 2010b. An improved ant colony optimization for constrained engineering design problems. Eng. Comput. 32 (3), 864–873.

Kaveh, A., Talatahari, S., 2010c. An improved ant colony optimization for design of planar steel frames. Eng. Struct. 32 (3), 864–873.

Kaveh, A., Talatahari, S., 2010d. Imperialist competitive algorithm for engineering design problems. Asian J. Civil Eng. 11 (6), 675–697.

Kaveh, A., Talatahari, S., 2010e. Optimum design of skeletal structures using imperialist competitive algorithm. Comput. Struct. 88 (21-22), 1220–1229.

Kaveh, A., Talatahari, S., 2010f. Optimal design of skeletal structures via the charged system search algorithm. Struct. Multidiscip. Optim. 41 (6), 893–911.

Kaveh, A., Talatahari, S., 2010g. Charged system search for optimum grillage systems design using the LRFD-AISC code. J. Constr. Steel Res. 66 (6), 767–771.

Kaveh, A., Talatahari, S., 2010h. A charged system search with a fly to boundary method for discrete optimum design of truss structures. Asian J. Civil Eng. 11 (3), 277–293.

Kaveh, A., Talatahari, S., 2010i. Optimal design of Schwedler and ribbed domes via hybrid Big Bang–Big Crunch algorithm. J. Constr. Steel Res. 66 (3), 412–419.
Kaveh, A., Talatahari, S., 2010j. A discrete Big Bang–Big Crunch algorithm for optimal design of skeletal structures. Asian J. Civil Eng. 11 (1), 103–122.
Kaveh, A., Talatahari, S., 2011a. Geometry and topology optimization of geodesic domes using charged system search. Struct. Multidiscip. Optim. 43 (2), 215–229.
Kaveh, A., Talatahari, S., 2011b. An enhanced charged system search for configuration optimization using the concept of fields of forces. Struct. Multidiscip. Optim. 43 (3), 339–351.
Kaveh, A., Talatahari, S., 2011c. Hybrid charged system search and particle swarm optimization for engineering design problems, engineering computations. Int. J. Comput.-Aided Eng. Softw. 28 (4), 423–440.
Kaveh, A., Talatahari, S., 2011d. Optimization of large-scale truss structures using modified charged system search. Int. J. Optim. Civil Eng. 1 (1), 15–28.
Kaveh, A., Talatahari, S., 2012. Charged system search for optimal design of frame structures. Appl. Soft Comput. 12 (1), 382–393.
Kaveh, A., Hassani, B., Shojaee, S., Tavakkoli, S.M., 2008a. Structural topology optimization using ant colony methodology. Eng. Struct. 30 (9), 2559–2565.
Kaveh, A., Farahmand Azar, B., Talatahari, S., 2008b. Ant colony optimization for design of space trusses. Int. J. Space Struct. 23 (3), 167–181.
Kaveh, A., Farahmand Azar, B., Hadidi, A., Rezazadeh Sorochi, F., Talatahari, S., 2010. Performance-based seismic design of steel frames using ant colony optimization. J. Constr. Steel Res. 66 (4), 566–574.
Kennedy, J., Eberhart, R.C., Shi, Y., 2001. Swarm Intelligence. Morgan Kaufman Publishers, San Francisco, CA.
Koumousis, V.K., Georgious, P.G., 1994. Genetic algorithms in discrete optimization of steel truss roofs. J. Comput. Civil Eng. 8 (3), 309–325.
Li, L.J., Huang, Z.B., Liu, F., Wu, Q.H., 2007. A heuristic particle swarm optimizer for optimization of pin connected structures. Comput. Struct. 85, 340–349.
Michalewicz Z., 1995. A survey of constraint handling techniques in evolutionary computation methods. McDonnell J.R., Reynolds R.G., Fogel D.B. (Eds.), Proceedings of the Fourth Annual Conference on Evolutionary Programming. MIT Press, Cambridge, MA, pp. 135–155.
Minimum Design Loads for Buildings and Other Structures, 1998. American Society of Civil Engineers.
Mozafari, H., Abdi, B., Ayob, A., 2010. Optimization of composite plates based on imperialist competitive algorithm. IJCSE. 2 (9), 2816–2819.
Perez, R.E., Behdinan, K., 2007. Particle swarm approach for structural design optimization. Comput. Struct. 85, 1579–1588.
Rajeev, S., Krishnamoorthy, C.S., 1992. Discrete optimization of structures using genetic algorithms. J. Struct. Eng. 118 (5), 1233–1550.
Salajegheh, E., Vanderplaats, G. N., 1986/87. An efficient approximation method for structural synthesis with reference to space structures. Int. J. Space Struct. 87 (2), 165–175.
Schutte, J.J., Groenwold, A.A., 2003. Sizing design of truss structures using particle swarms. Struct. Multidiscip. Optim. 25, 261–269.
Serra, M., Venini, P., 2006. On some applications of ant colony optimization metaheuristic to plane truss optimization. Struct. Multidiscip. Optim. 32 (6), 499–506.
Shrestha, S.M., Ghaboussi, J., 1998. Evolution of optimization structural shapes using genetic algorithm. J. Struct. Eng. 124 (11), 1331–1338.

Suresh, S., Sujit, P.B., Rao, A.K., 2007. Particle swarm optimization approach for multiobjective composite box-beam design. Compos. Struct. 81 (4), 598–605.

Tang, H., Zhoua, J., Xue, S., Xie, L., Bang-Big., B., 2010. Crunch optimization for parameter estimation in structural systems. Mech. Syst. Signal Proc. 24 (8), 2888–2897.

Toğan, V., Daloğlu, A.T., 2006. Optimization of 3D trusses with adaptive approach in genetic algorithms. Eng. Struct. 28, 1019–1027.

Wu, S.J., Chow, P.T., 1995. Integrated discrete and configuration optimization of trusses using genetic algorithms. Comput. Struct. 55 (4), 695–702.

9 Truss Weight Minimization Using Hybrid Harmony Search and Big Bang–Big Crunch Algorithms

Luciano Lamberti and Carmine Pappalettere

Politecnico di Bari, Dipartimento di Meccanica, Matematica e Management, Viale Japigia, Bari, Italy

9.1 Introduction

Weight minimization of truss structures is an important class of engineering problems often solved with metaheuristic optimization algorithms. Random search allows for the exploration of larger fractions of design space but may entail many structural analyses yielding just marginal improvements in weight or even violating optimization constraints. In order to overcome this problem, a variety of metaheuristic optimization methods inspired by biology, evolution theory, social sciences, music, physics, and astronomy have been developed (see the reviews by Hasancebi et al., 2009, 2010a; Koziel and Yang, 2011; Lamberti and Pappalettere, 2011a; Saka, 2007a): genetic algorithms (GA), simulated annealing (SA), particle swarm optimization (PSO), ant colony optimization (ACO), harmony search (HS), Big Bang–Big Crunch (BB–BC), hunting search (HuS), firefly algorithm (FA), bat algorithm (BA), cuckoo search (CS), teaching–learning-based optimization (TLBO), charged system search (CSS), and so on. Basically, metaheuristic optimization algorithms generate new trial designs by following a random strategy, however, "guided" by the inspiring criterion. Therefore, metaheuristic search is more effective than fully heuristic optimization search.

GA (Erbatur et al., 2000; Galante, 1996; Goldberg, 1989; Hajela and Lee, 1995; Kaveh and Kalatjari, 2002; Rahami et al., 2008; Saka, 2007b) and SA (Balling, 1991; Bennage and Dhingra, 1995; Hasancebi and Erbatur, 2002; Hasancebi et al., 2010b; Kirkpatrick et al., 1983; Lamberti, 2008; Pantelides and Tzan, 2000; Shea et al., 1997; Sonmez, 2007; Van Laarhoven and Aarts, 1987) were the first metaheuristic optimization methods to be applied to structural design problems and are still widely utilized.

The second generation of metaheuristic optimization methods dates back to the last decade and includes population-based algorithms such as PSO (Clerc, 2006; Kennedy and Eberhart, 2001; Li et al., 2007; Perez and Behdinan, 2007; Schutte

Metaheuristic Applications in Structures and Infrastructures. DOI: http://dx.doi.org/10.1016/B978-0-12-398364-0.00009-7

and Groenwold, 2003), ACO (Camp et al., 2005; Dorigo and Stutzle, 2004; Kaveh and Shojaee, 2007; Kaveh and Talatahari, 2009a,b, 2010a; Luh and Lin, 2008; Serra and Venini, 2006), HS (Carbas and Saka, 2012; Degertekin, 2012; Geem et al., 2001; Hasancebi et al., 2010c; Kaveh and Abadi, 2010; Kaveh and Ahangaran, 2012; Lamberti and Pappalettere, 2009; Lee and Geem, 2004, 2005; Saka, 2009), and BB–BC (Camp, 2007; Erol and Eksin, 2006; Kaveh and Talatahari, 2009c, 2010b; Lamberti and Pappalettere, 2011b).

In the past few years, the increasing power of computational devices favored the diffusion of the third generation of metaheuristic algorithms. HuS (Oftadeh et al., 2010), FA (Gandomi et al., 2011; Yang, 2010), BA (Yang and Gandomi, 2012), CS (Gandomi et al., 2012a,b), CSS (Kaveh and Talatahari, 2010c, 2011), and TLBO (Togan, 2012) are the algorithms most recently used in optimization of skeletal structures.

Hybrid optimization algorithms attempt to retain the strength points of each optimization technique included in the formulation (Hwang and He, 2006; Kaveh and Talatahari, 2009a,b; Kaveh and Zolghadr, 2012; Liao, 2010; Luh and Lin, 2008). In general, a hybrid algorithm should preserve the main features of the techniques involved in the formulation and utilize the parts of each technique that better exploit the characteristics of other algorithms. For example, an algorithm may be utilized as a global optimizer while another algorithm may be utilized as a local optimizer, or to escape from local minima, or to store design variables and/or trial designs for improving design in the later stages of the optimization process.

HS reproduces the musical process of searching for a perfect state of harmony while BB–BC mimics the process of evolution of the universe. HS and BB–BC have been proven to be very efficient in weight minimization of skeletal structures. Furthermore, their numerical implementation is very simple. This explains the large number of papers published on this subject. However, the large number of structural analyses required in the optimization process remains an open issue. Furthermore, there is no guarantee that a new trial design can always improve design.

The present authors attempted to overcome the above-mentioned limitations by introducing some gradient information in the metaheuristic search process. In particular, the optimizer is forced to generate new trial designs that must lie on feasible descent directions. This approach, originally introduced in SA (Lamberti, 2008; Lamberti and Pappalettere, 2007), was then extended to HS (Lamberti and Pappalettere, 2009) and BB–BC (Lamberti and Pappalettere, 2011b). The encouraging results obtained in pilot studies on truss weight minimization provided the basis for developing the new optimization framework described in this chapter. The optimization process can be outlined as follows:

1. Generate a random population of feasible trial designs covering the whole design space.
2. Sort population with respect to structural weight; find the best and the worst designs; find the center of mass of the population.
3. Compute cost function gradients at the current best record and perturb design variables always trying to move along feasible descent directions.

4. Evaluate the new trial design and run, eventually, improvement routines.
5. Perform a one-dimensional (1D) probabilistic search (derived from SA) by perturbing variables one at a time.
6. Replace the largest number of candidate designs included in the population as much as possible.
7. Repeat steps 2 through 6 until convergence.

The new optimization framework is hybrid because the metaheuristic search includes some gradient information (i.e., the metaheuristic search operating on candidate designs generated randomly over the entire design space is combined with local search of gradient-based optimization where design variables are perturbed by moving along descent directions), and HS and BB–BC are combined with an SA-type probabilistic search (i.e., trial designs formed by operating on the population generated for HS and BB–BC are eventually improved, accepted, or rejected on the basis of the Metropolis' criterion utilized in SA).

The new optimization algorithms are tested in the large-scale weight minimization problem of a space truss tower including 3586 elements and 280 design variables. Optimization results are compared with the results obtained from the multilevel and multipoint SA algorithm described in Lamberti (2008) and gradient-based commercial optimizers.

Furthermore, a parametric study is carried out to analyze the sensitivity of the new HS and BB–BC algorithms to the population size. The classical sizing optimization problem formulated for the planar 200-bar truss subject to five independent loading conditions (200 design variables) is chosen as the benchmark test. Optimization results are compared with SA and gradient-based commercial optimizers. Comparisons with literature are also presented.

Optimization results clearly demonstrate the efficiency and robustness of the new optimization algorithms developed in this research.

9.2 Statement of the Weight Minimization Problem for a Truss Structure

The weight minimization problem for a truss structure comprised of NOD nodes ($k = 1,\ldots,\text{NOD}$) and NEL elements ($j = 1,\ldots,\text{NEL}$) can be stated as follows:

$$
\begin{aligned}
&\text{Minimize } W(\mathbf{X}) && = \rho g \sum_{j=1}^{\text{NEL}} l_j\, x_j \\
&\text{Subject to} && \begin{cases} u^{\text{L}}_{(x,y,z),k} \le u_{(x,y,z),k,\text{ilc}} \le u^{\text{U}}_{(x,y,z),k} \\ \sigma^{\text{L}}_j \le \sigma_{j,\text{ilc}} \le \sigma^{\text{U}}_j \\ \omega_s \ge \omega^{\text{LIM}}_s \\ x^{\text{L}}_j \le x_j \le x^{\text{U}}_j \end{cases}
\end{aligned}
\tag{9.1}
$$

where:

- x_j is the cross-sectional area of the jth element of the structure included as sizing variable in the optimization process: each sizing variable can range between the corresponding lower bound $x_j^{\rm L}$ and upper bound $x_j^{\rm U}$;
- l_j is the length of the jth element of the structure;
- g is the gravity acceleration value (9.81 m/s^2); ρ is the material density;
- ρ is the material density;
- NLC is the number of independent loading conditions acting on the structure;
- $u_{(x,y,z),k,{\rm ilc}}$ are the displacements of the kth node in the directions x,y,z, varying between $u^{\rm L}{}_{(x,y,z),k}$ and $u^{\rm U}{}_{(x,y,z),k}$, respectively;
- $\sigma_{j,{\rm ilc}}$ is the stress in the jth element, varying between $\sigma_j^{\rm L}$ (compression stress limit may include critical bucking load) and $\sigma_j^{\rm U}$ (allowable tension limit);
- The ilc subscript (ilc = 1, . . .,NLC) indicates that constraints are relative to the "ilc"th loading condition. Constraints are normalized with respect to stress and displacement limits;
- ω_s is the sth natural frequency of the structure that must be greater than limit value $\omega_s^{\rm LIM}$.

If the cross-sectional area of the jth element is comprised of $N_{\rm SEG}^j$ segments with dimensions b_r^j and $h_r^j(r = 1, \ldots N_{\rm SEG}^j)$, it can be expressed as $\sum_{r=1}^{N{\rm SEG}^j} b_r^j \cdot h_r^j$.

For optimization problems including layout variables, Eq. (9.1) can be rewritten as:

$$\text{Minimize } W(\mathbf{X}) = \rho g \sum_{j=1}^{\rm NEL} x_j \sqrt{(x_{j1} - x_{j2})^2 + (y_{j1} - y_{j2})^2 + (z_{j1} - z_{j2})^2} \tag{9.2}$$

where $x_{j1,2}$, $y_{j1,2}$, and $z_{j1,2}$ are the coordinates of the nodes limiting the jth element of the structure.

Variable linking can be adopted by grouping the NEL elements in NGR groups: each group includes elements with identical stiffness properties. This approach allows the number of design variables to be reduced, thus simplifying the optimization process.

9.3 Harmony Search

The HS (Geem et al., 2001; Lee and Geem, 2004, 2005) optimization method reproduces the process of searching for a perfect state of harmony performed by jazz players. The harmony is analogous to the optimum design vector. Musicians' improvisations are analogous to local and global search schemes in the optimization process. The classical HS formulation is now summarized (see also Gandomi et al., 2013; Sahab et al., 2012).

- The HS algorithm operates on N_{POP} candidate designs. The harmony memory [HM] stores the design vectors forming this population. The harmony memory is an [$N_{POP}x$ NDV] matrix defined by Eq. (9.3). Each row represents a candidate design, while columns include values of design variables. Let $\mathbf{X}^r_{HM}$ and W^r_{HM} ($r = 1, \ldots, N_{POP}$), respectively, be the generic design vector and the corresponding cost function value stored in [HM]. Let $\mathbf{X}_{OPT}$ be the current best record with the corresponding structural weight W_{OPT}. The harmony memory should always include feasible designs. Penalty functions can be used to handle constraint violation.

$$[\mathrm{HM}] = \begin{bmatrix} x_1^1 & x_2^1 & \cdots & x_{\mathrm{NDV}-1}^1 & x_{\mathrm{NDV}}^1 \\ x_1^2 & x_2^2 & \cdots & x_{\mathrm{NDV}-1}^2 & x_{\mathrm{NDV}}^2 \\ \cdots & \cdots & \cdots & \cdots & \cdots \\ x_1^{N_{POP-1}} & x_2^{N_{POP-1}} & \cdots & x_{\mathrm{NDV}-1}^{N_{POP-1}} & x_{\mathrm{NDV}}^{N_{POP-1}} \\ x_1^{N_{POP}} & x_2^{N_{POP}} & \cdots & x_{\mathrm{NDV}-1}^{N_{POP}} & x_{\mathrm{NDV}}^{N_{POP}} \end{bmatrix} \tag{9.3}$$

- Each new trial design is generated using three basic rules: (i) random selection, (ii) harmony memory consideration, and (iii) design vector adjustment. In the random selection, each design variable is randomly chosen in the range of values currently stored in [HM]. The harmony memory considering rate (HMCR), ranging between 0 and 1, gives the level of probability of extracting the new variable value right from the set $(x_i^1, x_i^2, \ldots, x_i^{\mathrm{HMS}-1}, x_i^{\mathrm{HMS}})$.
- The trial design $\mathbf{X}' = \{\mathrm{x}'_1, \mathrm{x}'_2, \ldots, \mathrm{x}'_N\}$ must be analyzed to check if optimization variables should be adjusted or not. This process is governed by the pitch adjustment rate (PAR) parameter: there is the probability $(1 - \mathrm{PAR})$ of maintaining the value x'_i previously set.The ith optimization variable can be modified by comparing HMCR and PAR with a random number ρ generated for that design variable. If $\rho > \mathrm{HMCR}$, the new value of design variable is randomly generated. Conversely, if $\rho < \mathrm{HMCR}$, HS extracts one value stored in the harmony memory and checks if that value should be pitch adjusted. If $\rho > \mathrm{PAR}$, the design variable is modified as $(x'_i \pm \mathrm{bw})$ where bw is an arbitrary distance bandwidth.
- If the structural weight/cost of the new harmony $\mathbf{X}'$ is better than that of the worst design stored in the harmony memory, the new trial design replaces the worst design in [HM].
- Each new harmony is generated by following the steps outlined above. The HS process ends when a prespecified number of structural analyses have been executed.

Researchers introduced many ad hoc features to reduce the number of structural analyses. Saka (2009) used an adaptive error strategy to deal with slightly infeasible designs; constraint tolerance is progressively reduced as the search process converges to the optimum. Hasancebi et al. (2010c) developed a scheme to adaptively vary HMCR and PAR; for each new harmony, HS parameters are probabilistically selected about the average values stored in [HM]. Kaveh and Abadi (2010) reprised the concept of linear increase of PAR with the number of generations. A self-adaptive HS algorithm was also utilized in Degertekin (2012). Carbas and Saka (2012) utilized yet another dynamic scheme for parameter updating where HMCR and PAR change more significantly in the current iteration as the ratio of the difference between maximum cost and average cost to the difference between maximum cost and minimum cost increases. Finally, HS was hybridized with other metaheuristic algorithms (Kaveh and Ahangaran, 2012; Kaveh and Talatahari, 2009a,b; Liao, 2010).

However, HS formulations described in the literature have a serious limitation in the fact that there is no guarantee at all that new trial designs can always improve the current best record. To solve this problem, the present authors (Lamberti and Pappalettere, 2009) developed a novel HS formulation where trial designs are generated by including information on cost function gradient. This approach is viable for skeletal structures as gradients often are explicitly available and even constant over design space for sizing variables. The algorithm was modified in this study to make it more general. The new HS algorithm is now described.

9.3.1 Generation, Acceptance/Rejection, and Adjustment of a New Harmony

Let $\mathbf{X}_{\mathrm{OPT}} = \{x_{\mathrm{OPT},1}, x_{\mathrm{OPT},2}, \ldots, x_{\mathrm{OPT},N}\}$ be the current best record extracted from the population of N_{POP} candidate designs. The gradients of cost function with respect to design variables are evaluated at $\mathbf{X}_{\mathrm{OPT}}$. The new value $x_{\mathrm{TR},j}$ assigned to the jth optimization variable is:

$$\begin{array}{l} (x_{\mathrm{OPT},j} - x_j^{\mathrm{L}})\,\partial W/\partial x_j < 0 \Rightarrow x_{\mathrm{TR},j} = x_{\mathrm{OPT},j} + N_{\mathrm{RND},j}\,(x_{\mathrm{OPT},j} - x_j^{\mathrm{L}})\,\mu_j \\ (x_j^{\mathrm{U}} - x_{\mathrm{OPT},j})\,\partial W/\partial x_j < 0 \Rightarrow x_{\mathrm{TR},j} = x_{\mathrm{OPT},j} + N_{\mathrm{RND},j}\,(x_j^{\mathrm{U}} - x_{\mathrm{OPT},j})\,\mu_j \end{array} \quad (j = 1,\ldots,\mathrm{NDV}) \tag{9.4}$$

where $\partial W/\partial x_j$ is the cost function sensitivity with respect to the jth optimization variable currently perturbed; $\mu_j = (\partial W/\partial x_j)/||\overline{\nabla} W(\mathbf{X}_{\mathrm{OPT}})||$; $N_{\mathrm{RND},j}$ is a random number in (0,1). Equation (9.4) is utilized only if $N_{\mathrm{RND},j} > \mathrm{HMCR}$. By setting HMCR as a large value, the classical generation scheme including pitch adjustment is more likely to be used for most variables. In this case, the new value $x_{\mathrm{TR},j}$ is analyzed with the following acceptance/rejection scheme:

$$\begin{array}{l} (\partial W/\partial x_j) \cdot (x_{\mathrm{TR},j} - x_{\mathrm{OPT},j}) < 0 \Rightarrow \mathrm{ACCEPTED} \\ (\partial W/\partial x_j) \cdot (x_{\mathrm{TR},j} - x_{\mathrm{OPT},j}) > 0 \Rightarrow \mathrm{REJECTED} \end{array} \quad (j = 1,\ldots,\mathrm{NDV}) \tag{9.5}$$

Equation (9.4) has been introduced in this research while Eq. (9.5) was the only criterion for accepting/rejecting new harmonies utilized in Lamberti and Pappalettere (2009). Equations (9.4) and (9.5) indicate that trial designs must lie on descent directions. Sensitivities are evaluated at the current best record because any trial design potentially better than $\mathbf{X}_{\mathrm{OPT}}$ is obviously better than the current worst record, which in the classical HS formulation must be replaced.

If the acceptance criterion in Eq. (9.5) is not satisfied, an additional value $x'_{j,\text{add}}$ is generated for the currently analyzed jth variable by using a mirroring strategy about the current best record:

$$x'_{j,\text{add}} = 2 \cdot x_{\text{OPT},\,j} - \eta_{\text{MIRR}} \cdot x_{\text{TR},\,j} \tag{9.6}$$

where η_{MIRR} is a random number, in the interval (0,1), introduced in this research to limit the step size of the jth variable. This reduces the risk that the corrected design may turn infeasible if it tends to reduce cost function too sharply. The new value $x_{j,\text{add}}$ makes the perturbation $\Delta x_j = (x_{\text{TR},\,j} - x_{\text{OPT},\,j})$ change in sign and, hence, the acceptance criterion $\partial W/\partial x_j \cdot \Delta x_j < 0$ gets satisfied.

Classical HS forms a new candidate design by combining selection from harmony memory and pitch-adjusting mechanisms. In Lamberti and Pappalettere (2009), the new harmony resulting from this combination was forced to lie on a descent direction by using Eqs. (9.5) and (9.6). However, both formulations utilize constant values for HMCR and PAR.

In the present study, the new design is generated with a hybrid strategy. If $N_{\text{RND},\,j} > \text{HMCR}$, the new design is generated with Eq. (9.4) directly forcing it to lie on a descent direction. Therefore, it is no longer necessary to use Eqs. (9.5) and (9.6) to check the quality of the new harmony.

Conversely, if $N_{\text{RND},\,j} < \text{HMCR}$, the classical HS generation scheme is utilized but the improvement is that the quality of the new harmony is checked with Eqs. (9.5) and 9.6). If $N_{\text{RND},\,j} > \text{PAR}$, the pitch-adjusting strategy is implemented as follows ($j = 1, \ldots, \text{NDV}$):

$$x_{\text{TR},\,j} = \text{XHS}_{\text{extracted},\,j} + \lambda_{\text{scale}} \cdot N_{\text{RND},\,j} \cdot \frac{(x_j^{\text{U}} - x_j^{\text{L}})}{\text{NG}_{\text{tot}}} \cdot \text{NG}_{\text{pitch,adj}} \tag{9.7}$$

where $\text{XHS}_{\text{extracted},\,j}$ is the value of the jth design variable extracted from [HM], $\text{NG}_{\text{pitch,adj}}$ is the number of trial designs generated via pitch adjustment, and NG_{tot} is the total number of trial designs generated in the optimization search. The scale parameter λ_{scale} is set as:

$$\lambda_{\text{scale}} = \begin{cases} (\text{XHS}_{\text{extracted},\,j} - x_{\text{OPT},\,j}) \cdot \dfrac{\partial W}{\partial x_j} < 0 & \Rightarrow -1/N_{\text{RND},\,j} \\[2ex] (\text{XHS}_{\text{extracted},\,j} - x_{\text{OPT},\,j}) \cdot \dfrac{\partial W}{\partial x_j} > 0 & \Rightarrow -1 \end{cases} \quad (j = 1, \ldots, \text{NDV}) \tag{9.8}$$

In sizing optimization problems of truss structures, gradients of cost functions are constant and positive in sign over the entire design space. Therefore, λ_{scale} must always be equal to -1 regardless that $\text{XHS}_{\text{extracted},\,j}$ is greater or smaller than $\boldsymbol{x}_{\text{OPT},\,j}$.

Equation (9.7) either includes gradient information, as it accounts for the sign of the cost increment $(\text{XHS}_{\text{extracted},\,j} - x_{\text{OPT},\,j}) \cdot \partial W/\partial x_j$ of the jth variable currently perturbed, and information on optimization history, as it includes the ratio between the number of trial designs $NG_{pitch,adj}$ generated via pitch adjustment and the total number of trial designs NG_{tot} generated until that moment. Since $NG_{pitch,adj}/NG_{tot}$ decreases as optimization progresses, the perturbation step defined to pitch adjust each design variable gets finer as the optimum is approached.

A novel scheme for adaptively varying HMCR and PAR is implemented in this study. HMCR and PAR are now randomly generated in each new iteration thus making the present HS algorithm independent of the initial setting of internal parameters. HMCR and PAR are corrected in the different optimization iterations by considering that "global search" involving large perturbations of all design variables should be performed in the early stages of the optimization process where trial designs are far from the optimum and "local search" involving smaller perturbations of a limited set of design variables should be performed as the optimization process approaches the optimum design. HS internal parameters are updated based on the current trend of optimization history. In the qth optimization iteration, we have:

$$\mathrm{HMCR}^{q} = \mathrm{HMCR}^{q}_{\mathrm{extracted}} \cdot \frac{\mathrm{WHS}^{q-1}_{\mathrm{aver,end}}}{\mathrm{WHS}^{q-1}_{\mathrm{aver,init}}} \cdot \frac{\mathrm{NG}_{\mathrm{pitch,adj}}}{\mathrm{NG}_{\mathrm{gradient}}} \tag{9.9}$$

$$\mathrm{PAR}^{q} = \mathrm{PAR}^{q}_{\mathrm{extracted}} \cdot \frac{\mathrm{WHS}^{q-1}_{\mathrm{aver,end}}}{\mathrm{WHS}^{q-1}_{\mathrm{aver,init}}} \cdot \frac{\mathrm{NG}_{\mathrm{gradient}}}{\mathrm{NG}_{\mathrm{pitch,adj}}} \tag{9.10}$$

where $\mathrm{WHS}^{q-1}_{\mathrm{aver,init}}$ and $\mathrm{WHS}^{q-1}_{\mathrm{aver,end}}$ are the average values of cost function for the trial designs included in the harmony memory at the beginning and the end of the previous optimization iteration, respectively (the $\mathrm{WHS}^{q-1}_{\mathrm{aver,end}}/\mathrm{WHS}^{q-1}_{\mathrm{aver,init}}$ ratio is always less than 1) and $\mathrm{NG}_{\mathrm{gradient}}$ is the number of trial designs generated by including gradient information.

Random values $\mathrm{HMCR}^{q}_{\mathrm{extracted}}$ and $\mathrm{PAR}^{q}_{\mathrm{extracted}}$ are defined as:

$$\begin{cases} \mathrm{HMCR}^{q}_{\mathrm{extracted}} = 0.01 + \xi_{\mathrm{HMCR}} \cdot (0.99 - 0.01) \\ \mathrm{PAR}^{q}_{\mathrm{extracted}} = 0.01 + \xi_{\mathrm{PAR}} \cdot (0.99 - 0.01) \end{cases} \tag{9.11}$$

where ξ_{HMCR} and ξ_{PAR} are two random numbers in (0,1). The bounds of 0.01 and 0.99 set in Eq. (9.11) allow including all possible values of internal parameters (Carbas and Saka, 2012).

The rationale behind Eqs. (9.9) and (9.10) is the following. The cost function will decrease more rapidly if large perturbations are assigned to design variables. This can be certainly done by including gradient information in the search process rather than using local refinements based on pitch-adjusting strategy. Since gradient information are directly utilized if $N_{\mathrm{RND},j} > \mathrm{HMCR}$, to increase the probability of including gradient information in the generation of a new harmony, the HMCR value randomly generated is reduced by the $\mathrm{WHS}^{q-1}_{\mathrm{aver,end}}/\mathrm{WHS}^{q-1}_{\mathrm{aver,init}}$ ratio. Hence, the level of probability of using gradient information in the generation of a new harmony becomes strictly dependent on the current trend exhibited by cost function in the optimization process. In fact, HMCR drops down more sharply as cost function decreases more significantly in the previous optimization cycle. In summary, the generation process is forced to be consistent with the current rate of reduction of cost function.

Furthermore, HMCR^q is reduced by $\text{NG}_{\text{pitch,adj}}/\text{NG}_{\text{gradient}}$. If the number of new harmonies generated via pitch-adjusting strategy tends to be smaller with respect to the number of new harmonies generated including gradient information, it is more logical to keep following such a trend.

Similar considerations can be made for the PAR^q parameter. Pitch adjustment is performed if $\text{HMCR} > N_{\text{RND},j} > \text{PAR}$. It is still convenient to include the information on the current rate of reduction of the cost function provided by the $\text{WHS}^{q-1}_{\text{aver,end}}/\text{WHS}^{q-1}_{\text{aver,init}}$ ratio. However, the ratio $\text{NG}_{\text{pitch,adj}}/\text{NG}_{\text{gradient}}$ must be replaced by the ratio $\text{NG}_{\text{gradient}}/\text{NG}_{\text{pitch,adj}}$ as the objective now is to use the pitch-adjusting strategy: the threshold PAR^q will lower more significantly as the number of trial designs $\text{NG}_{\text{pitch,adj}}$ generated via pitch adjustment dominates over (or, at least, tends to) the number of trial designs generated by including gradient information.

The HS strategy must be modified when cost function gradients are not constant over the design space. In such a case, the new harmony is defined as:

$$\mathbf{X}_{\text{TR}} = \mathbf{X}_{\text{OPT}} + \eta_{\text{BEST}} \cdot \mathbf{S}_{\text{BEST}} + \eta_{\text{2ndBEST}} \times \mathbf{S}_{\text{2ndBEST}} + \eta_{\text{FAST}} \cdot \mathbf{S}_{\text{FAST}} \tag{9.12}$$

The new harmony is forced to lie on the descent direction defined by combining the steepest descent direction $\mathbf{S}_{\text{FAST}}$ and the direction $\mathbf{S}_{\text{BEST}}$ for which there is the largest reduction in cost $[W(\mathbf{X}^k_{\text{TR}}) - W(\mathbf{X}_{\text{OPT}})]$ by moving from each candidate design $\mathbf{X}^k_{\mathbf{TR}}$ $(k = 1, \ldots, N_{\text{POP}} - 1)$ stored in the harmony memory toward the current best record $\mathbf{X}_{\text{OPT}}$. The second best direction $\mathbf{S}_{\text{2ndBEST}}$ introduces an elitist search strategy. The steepest descent direction $\mathbf{S}_{\text{FAST}}$ can be found by computing the largest average gradient among "average" gradients $\Delta W^k/\Delta S^k$ for each candidate design $\mathbf{X}^k_{\text{TR}}$ with $\Delta W^k = [W(\mathbf{X}^k_{\text{TR}}) - W(\mathbf{X}_{\text{OPT}})]$; ΔS^k is the distance $\|\mathbf{X}^k_{\text{TR}} - \mathbf{X}_{\text{OPT}}\|$ between the $\mathbf{X}^k_{\text{TR}}$ trial design and the current best record $\mathbf{X}_{\text{OPT}}$.

Equation (9.12) introduces in the HS process a sort of "social behavior" and forces the entire population to move toward the position of the current best record.

9.3.2 Evaluation of the New Trial Design

If the new trial design $\mathbf{X}_{\text{TR}}$ is better than the worst design $\mathbf{X}_{\text{WST}}$ currently stored in the harmony memory, it replaces the worst design in [HM]. The sophisticated generation mechanism developed in this research makes the new trial design have a high probability of improving the current best record. As the optimum design is approached, two situations are likely to occur: (i) $\mathbf{X}_{\text{TR}}$ is worse than the worst design $\mathbf{X}_{\text{WST}}$; (ii) $\mathbf{X}_{\text{TR}}$ is better than $\mathbf{X}_{\text{WST}}$ but violates optimization constraints.

In the former case, another trial design $\mathbf{X}'_{\text{add}}$ is defined by mirroring about the current best record:

$$\mathbf{X}'_{\text{add}} = 2 \cdot \mathbf{X}_{\text{OPT}} - \eta_{\text{MIRR}} \cdot \mathbf{X}_{\text{TR}} \tag{9.13}$$

If $W_{\text{TR}} < W_{\text{WST}}$, but the current trial design $\mathbf{X}_{\text{TR}}$ is infeasible, the HS algorithm checks for all designs stored in [HM] that are worse than $\mathbf{X}_{\text{TR}}$. It may even occur that $\mathbf{X}_{\text{OPT}}$ is worse than the current trial design $\mathbf{X}_{\text{TR}}$. Let us assume that there exist N_{IMPR} designs $\mathbf{X}^k_{\text{IMP}}$ ($k = 1, \ldots, N_{\text{IMPR}}$) worse than the current trial design $\mathbf{X}_{\text{TR}}$. Each direction $\mathbf{S}^k_{\text{IMP}} = [\mathbf{X}_{\text{TR}} - \mathbf{X}^k_{\text{IMP}}]$ is a descent direction with respect to $\mathbf{X}^k_{\text{IMP}}$. Approximate line searches employing fourth order approximations can be performed on each descent direction $\mathbf{S}^k_{\text{IMP}}$ to improve local design.

9.3.3 One-Dimensional SA-Type Probabilistic Search

If all improvement routines failed and/or the cost function did not improve significantly over the last optimization cycles, it is necessary to move away from constraint boundaries and to escape from local minima. For that purpose, a 1D probabilistic search strategy similar to classical SA was implemented.

Let $\mathbf{X}_{\text{TR},j}(x_{\text{OPT},1}, x_{\text{OPT},2}, \ldots, x_{\text{TR},j}, \ldots, x_{\text{OPT},N})$ denote the new trial design generated by perturbing only the jth optimization variable with respect to $\mathbf{X}_{\text{OPT}}$. The difference $\Delta W_{\text{TR},j} = [W(\mathbf{X}_{\text{TR},j}) - W_{\text{OPT}}]$ can be computed. Unlike classical SA, design variables are perturbed taking care to move always along descent directions. If $\partial W/\partial x_j \cdot (x_{\text{TR},j} - x_{\text{OPT},j}) < 0$ is not satisfied, $\text{x}_{\text{TR},j}$ is reset with Eq. (9.6).

If $\Delta W_{\text{TR},j} < 0$ and optimization constraints are satisfied, the new HS algorithm replaces the current best record with the new trial design, updates [HM], and then checks for convergence.

However, if the perturbation step is too large, the trial design $\mathbf{X}_{\text{TR},j}$ may end up infeasible. Let $\Delta G_{\text{MAX}} > 0$ be the largest normalized constraint violation corresponding to the perturbation vector $[\mathbf{X}_{\text{OPT}} - \mathbf{X}_{\text{TR},j}]$. This perturbation vector could be simply rescaled by the ratio $||\mathbf{X}_{\text{OPT}} - \mathbf{X}^*_{TR,j}|| / ||\mathbf{X}_{\text{OPT}} - \mathbf{X}_{\text{TR},j}||$ assuming that constraints vary linearly by moving from the current best record toward the constraint domain boundary on which the trial design $\mathbf{X}^*_{\text{TR},j}$ lies. However, this is not true for a general truss problem and approximate line search may be required to find the actual position of the trial design $\mathbf{X}^*_{\text{TR},j}$.

If it is not possible to generate any feasible design better than $\mathbf{X}_{\text{OPT}}$ but optimization constraints are satisfied, the Metropolis' probability function is used as in classical SA:

$$P(\Delta W_{\text{TR},j}) = \exp\left(\frac{-\Delta W_{\text{TR},j}}{\left(\sum_{r=1}^{\text{NDW}} \Delta W_r/\text{NDW}\right) \cdot T_K}\right) \tag{9.14}$$

where NDW is the number of trial points at which cost function was larger than the current best records found in the optimization process and the ΔW_r terms are the corresponding weight penalties. Each design $\mathbf{X}_{TR,\ j}$ is provisionally accepted or certainly rejected based on the Metropolis' criterion reformulated as:

$$\begin{aligned} &P(\Delta W_{TR,\ j}) > NRD_j \Rightarrow \text{Accept} \\ &P(\Delta W_{TR,\ j}) < NRD_j \Rightarrow \text{Reject} \end{aligned} \tag{9.15}$$

Designs provisionally accepted are included in the database Π. If there are no trial designs for which cost function decreases, the present HS algorithm sets as $\mathbf{X}_{OPT}$ the design $\mathbf{X}_j^{BEST}$ (included in the Π database) minimizing the weight penalty.

9.3.4 Update of the Harmony Memory

Every time a new trial design or a new set of trial designs replace the corresponding elements in [HM], it must be checked whether $\mathbf{X}_{OPT}$ changes or not.

9.3.5 Termination Criterion

The optimization process terminates when the ratio between standard deviation of cost and average cost of designs stored in [HM] is less than 10^{-7}.

The flowchart of the new HS algorithm is shown in Figure 9.1. The hybrid character of the optimization algorithm derived from the combination between metaheuristic search and gradient-based line search, and from the combination between HS search and SA-type probabilistic acceptance/rejection criterion is highlighted in the diagram by using bold lines (i.e., continuous bold lines for tasks involving gradient information and line search, and hatched bold lines for the SA-type search, respectively).

9.4 Big Bang–Big Crunch

The BB–BC algorithm (Erol and Eksin, 2006) reproduces the process of evolution of the universe. Each explosion generates a state of chaos, followed by a state of order lasting until the next explosion. In the optimization process, a set of candidate designs are randomly generated over the design space (i.e., "explosion phase"). The center of mass of these designs is determined as a weighted average (i.e., "contraction phase") where each weighing coefficient depends on the cost function evaluated at a trial design. A new population is generated randomly by perturbing design variables about the center of mass. The explosion/contraction sequence is repeated until no significant change in the cost function occurs.

The inherent simplicity of BB–BC soon attracted structural optimization experts. Camp (2007) utilized BB–BC to optimize space trusses. Sizing optimization of space trusses was carried out in Kaveh and Talatahari (2009c): their BB–BC algorithm not only considered the center of mass as the average point in

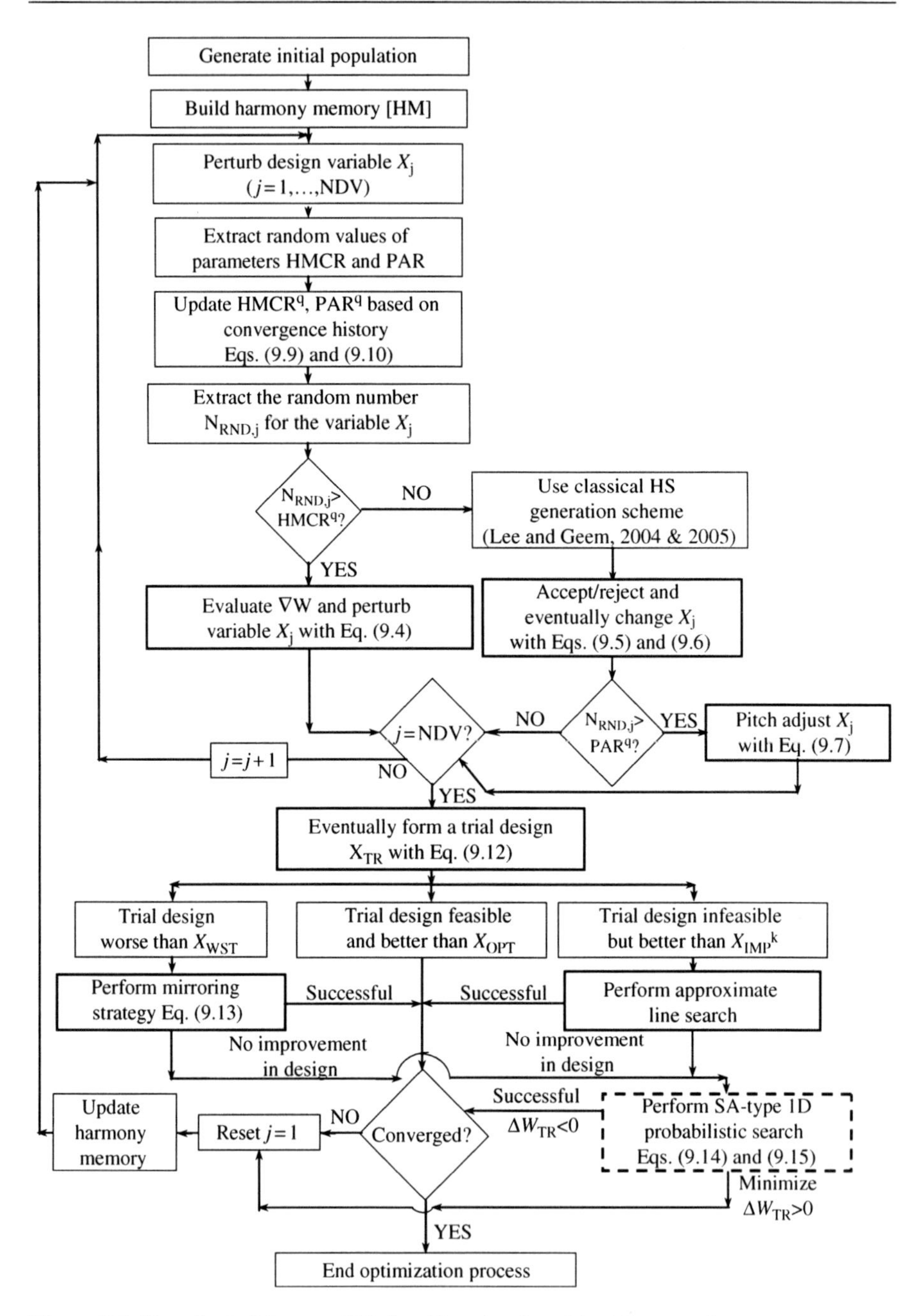

Figure 9.1 Flowchart of the new HS algorithm developed in this research.

the beginning of the BB phase but, similarly to PSO, also utilized the best position of each particle and the best visited position of all particles; furthermore, the sub-optimization mechanism was utilized. Discrete sizing and continuous layout optimization of Schwedler and ribbed domes were then performed (Kaveh and Talatahari, 2010b). Kaveh and Zolghadr (2012) used the explosion phase of BB–BC to escape from traps limiting exploration of the whole design space; this approach was effective for truss weight minimization problems with frequency constraints.

BB–BC implementations presented in literature have a common feature in the fact that new explosions about the center of mass are always performed after that the position of center of mass has been updated (see the review papers by Gandomi et al. (2013) and Sahab et al. (2012)). However, this entails N_{POP} new structural analyses at each new explosion. Furthermore, there is no guarantee that each newly defined center of mass will always lead to improve the current best record. Starting from these considerations, the present authors developed a novel BB–BC formulation (Lamberti and Pappalettere, 2011b), which was further improved in this study. The new optimization algorithm is outlined in the following.

9.4.1 Generation of the Initial Population and Determination of the Center of Mass

There is only one parameter that must be initialized in the BB–BC algorithm: the number of trial designs N_{POP} included in the population. Let $\mathbf{X}^k(x_1^k, x_2^k, \ldots, x_{NDV}^k)$ be the kth design included in the population ($k = 1, \ldots, N_{POP}$). The following generation scheme is utilized to create the initial population for an optimization problem including NDV design variables:

$$x_j^k = x_j^L + \rho_{DG,j}^k \cdot (x_j^U - x_j^L) \quad (j = 1, \ldots, NDV) \tag{9.16}$$

where $\rho_{DG,j}^k$ is a random number chosen in the interval (0,1) for the jth optimization variable considered in the kth generation.

The center of mass $\mathbf{X}_{CM}$ of the population is defined as:

$$x_{CM,j} = \left(\sum_{k=1}^{NPOP} \frac{x_j^k}{W^k}\right) \Bigg/ \left(\sum_{k=1}^{NPOP} \frac{1}{W^k}\right) \quad (j = 1, \ldots, NDV) \tag{9.17}$$

where W^k is the cost function value computed for the kth candidate design.

9.4.2 Evaluation of the Characteristics of the Center of Mass

Equation (9.17) implies that the center of mass always is at least better than the worst design included in the population. Therefore, two cases should be considered: (i) the center of mass is better than the current best record, $W(\mathbf{X}_{CM}) < W(\mathbf{X}_{OPT})$ and (ii) the center of mass is worse than the current best record, $W(\mathbf{X}_{CM}) > W(\mathbf{X}_{OPT})$.

If $W(\mathbf{X}_{CM}) < W(\mathbf{X}_{OPT})$ and $\mathbf{X}_{CM}$ are feasible, the new BB–BC algorithm simply replaces the worst design of the population with the center of mass without performing any new explosion. The position of the center of mass is updated until the condition $W(\mathbf{X}_{CM}) < W(\mathbf{X}_{OPT})$ remains satisfied and $\mathbf{X}_{CM}$ is feasible. The current best record $\mathbf{X}_{OPT}$ changes with respect to the previous iteration as it coincides with the new center of mass. However, the condition $W(\mathbf{X}_{CM}) < W(\mathbf{X}_{OPT})$ is not very likely to occur as the definition of center of mass (Eq. (9.17)) accounts also for the presence of suboptimal designs.

9.4.3 Perturbation of Design Variables

Let us consider the case most likely to occur: the center of mass is worse than the current best record, $W(\mathbf{X}_{CM}) > W(\mathbf{X}_{OPT})$. In general, there may be N_{better} candidate designs included in the population that are actually better than $\mathbf{X}_{CM}$. The descent direction $\mathbf{S}^k = [\mathbf{X}^k_{better} - \mathbf{X}_{CM}]$ $(k = 1, \ldots, N_{better})$ is defined only if it occurs $W(\mathbf{X}_{CM}) > W(\mathbf{X}^k_{better})$. Consequently, the quantity $\Delta W^k = [W(\mathbf{X}_{CM}) - W(\mathbf{X}^k_{better})]$ is the reduction in cost that would be achieved by moving away from the center of mass toward the candidate design $\mathbf{X}^k_{better}$ included in the population.

Figure 9.2 shows the current population, the center of mass $\mathbf{X}_{CM}$, and the three descent directions $\mathbf{S}_{BEST}$, $\mathbf{S}_{2ndBEST}$, and $\mathbf{S}_{FAST}$, respectively, corresponding to the largest, the second largest, and the steepest improvements in design with respect to the center of mass.

The new trial design vector $\mathbf{X}_{TR}$ is generated as follows:

$$\overline{X}_{TR} = \frac{\overline{X}_{CM} + \overline{X}_{OPT}}{2} + \rho_{FAST}\,\overline{S}_{FAST} + \rho_{2ndBEST}\,\overline{S}_{2ndBEST} + \rho_{BEST}\,\overline{S}_{BEST}\frac{\overline{\nabla} W(\overline{X}_{OPT})}{||\overline{\nabla} W(\overline{X}_{OPT})||} \tag{9.18}$$

Equation (9.18) can be written in the scalar form (Eq. (9.19)) for the jth design variable ($j = 1, \ldots,$NDV):

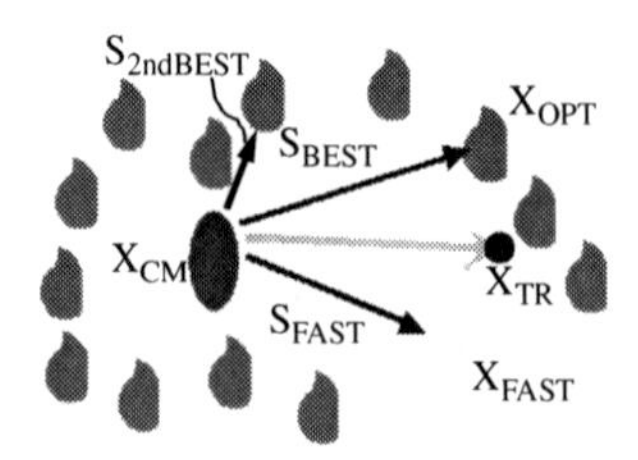

Figure 9.2 Novel mechanism for generating a trial design in BB–BC.

$$x_{\mathrm{TR},j} = \frac{x_{\mathrm{CM},j} + x_{\mathrm{OPT},j}}{2} + \rho_{\mathrm{FAST}}(x_{\mathrm{FAST},j} - x_{\mathrm{CM},j}) + \rho_{\mathrm{2ndBEST}}(x_{\mathrm{2ndBEST},j} - x_{\mathrm{CM},j}) + \rho_{\mathrm{BEST}}(x_{\mathrm{OPT},j} - x_{\mathrm{CM},j})\mu_j \tag{9.19}$$

where ρ_{FAST}, ρ_{2ndBEST}, and ρ_{BEST} are random numbers in the interval (0,1) generated for the descent directions $\mathbf{S}_{\mathrm{BEST}}$, $\mathbf{S}_{\mathrm{2ndBEST}}$, and $\mathbf{S}_{\mathrm{FAST}}$, respectively; each weighting coefficient μ_j is defined as $|\partial W/\partial x_j|/||\overline{\nabla} W(\mathbf{X}_{\mathrm{OPT}})||$; $W_{\mathrm{OPT},l}$ and $W_{\mathrm{OPT},l-1}$ are the last two current best record values taken by the cost function.

Equations (9.18) and (9.19) show that the new trial design $\mathbf{X}_{\mathrm{TR}}$ is generated by perturbing optimization variables with respect to the middle point $\mathbf{X}_{\mathrm{MID}}$ of the segment limited by the center of mass $\mathbf{X}_{\mathrm{CM}}$ and the best design $\mathbf{X}_{\mathrm{OPT}}$. This allows the search process to be maintained close enough to the current best record but, at the same time, far enough from constraint domain boundaries. The latter can reduce the risk of generating infeasible trial points especially when the search process is converging to the optimum design.

The ratio $W_{\mathrm{OPT},l-1}/W_{\mathrm{OPT},l}$ accounts for the current trend exhibited by cost function in the optimization process and forces the optimizer to take movements large enough to maintain at least the current rate of reduction in cost function.

The new BB–BC optimizer developed in this research hence generates new trial designs taking information from several "good" regions of design space. In fact, each new design is formed by combining four descent directions instead of just two or three descent directions as it happened, respectively, in the original BB–BC formulation developed in Lamberti and Pappalettere (2011b) or in the novel HS algorithm described in this chapter. Remarkably, no constraint gradients are required in the definition of the new trial design.

9.4.4 Evaluate the Quality of the New Trial Design, Eventually Use Improvement Routines, and Finally Perform a New Explosion

The quality of the new trial design $\mathbf{X}_{\mathrm{TR}}$ is evaluated by calculating structural weight and constraint margins. If $W(\mathbf{X}_{\mathrm{TR}}) < W(\mathbf{X}_{\mathrm{OPT}})$ and $\mathbf{X}_{\mathrm{TR}}$ is feasible, the new BB–BC algorithm simply replaces the worst design of the population with the new trial design without performing any new explosion. Therefore, unlike classical BB–BC, no explosions are performed until each new trial design keeps improving the current best record. The center of mass is updated with Eq. (9.17) as the new best record replaces the worst design of the population. This strategy allows the number of structural analyses to be considerably reduced with respect to classical BB–BC at least by N_{POP} analyses in each optimization iteration. The optimization algorithm hence becomes a BB–BC scheme with infrequent explosions.

If it happens $W(\mathbf{X}_{\mathrm{TR}}) > W(\mathbf{X}_{\mathrm{OPT}})$ and the trial design $\mathbf{X}_{\mathrm{TR}}$ is feasible, the same mirroring strategy developed for HS in Eq. (9.13) (Section 9.3) including the limitation factor η_{MIRR} is utilized.

The new BB–BC formulation described in this chapter also includes the fourth order approximate line search strategy if $\mathbf{X}_{\mathrm{TR}}$ improves $\mathbf{X}_{\mathrm{OPT}}$ but ends up infeasible and the modified 1D probabilistic search derived from SA if all improvement routines fail.

Each new explosion is performed about the middle point $\mathbf{X}_{\text{MID}}$ between the current best record and the current position of the center of mass, unlike classical BB–BC, where explosions are performed about the current position of center of mass. This elitist strategy drives the search process to better explore the neighborhood of the current best record thus limiting the effect of the presence of less efficient designs intrinsically considered by the definition of center of mass. Intermediate populations are hence generated as follows ($j = 1, \ldots, \text{NDV}$; $k = 1, \ldots, N_{\text{POP}}$):

$$\begin{cases} \rho_{\text{DG},j}^{k} > 0.5 \Rightarrow x_j^k = \dfrac{(x_{\text{CM},j} + x_{\text{OPT},j})}{2} + (\rho_{\text{DG},j}^{k} - 0.5) \cdot \left[x_j^{\text{U}} - \dfrac{(x_{\text{CM},j} + x_{\text{OPT},j})}{2} \right] \\ \rho_{\text{DG},j}^{k} < 0.5 \Rightarrow x_j^k = \dfrac{(x_{\text{CM},j} + x_{\text{OPT},j})}{2} + (\rho_{\text{DG},j}^{k} - 0.5) \cdot \left[\dfrac{(x_{\text{CM},j} + x_{\text{OPT},j})}{2} - x_j^{\text{L}} \right] \end{cases} \tag{9.20}$$

An interesting feature of the present BB–BC algorithm is that, unlike classical BB–BC, perturbation steps assigned to design variables in each new explosion by Eq. (9.20) are not necessarily shrunk as the optimization process progresses. This allows the optimization search not to be confined in regions of design space containing only local minima and hence enhances the metaheuristic algorithm's capability to explore the entire design space.

The flowchart of the new BB–BC algorithm is shown in Figure 9.3. Similarly to the new HS algorithm presented in Section 9.3, the hybrid character of the algorithm results from the combination of metaheuristic search with gradient-based line search and from the combination between BB–BC search and the SA-type probabilistic acceptance/rejection criterion. In the figure, continuous bold lines refer to tasks involving gradient information and line search while hatched bold lines refer to SA-type search.

9.5 Simulated Annealing

SA (Kirkpatrick et al., 1983; Van Laarhoven and Aarts, 1987) simulates the atomic arrangements in liquid or solid materials during cooling. The material reaches the lowest energy level (globally stable condition) as temperature decreases. SA utilizes a rather simple optimization strategy. A trial design is randomly generated and problem functions are evaluated at that point. If the trial point is infeasible, it is rejected and a new trial point is evaluated. If the trial point is feasible and the cost function is smaller than the current best record, then the point is accepted as the new best record. If the trial point is feasible but cost function is larger than current best value, the point is accepted or rejected based on a probabilistic criterion

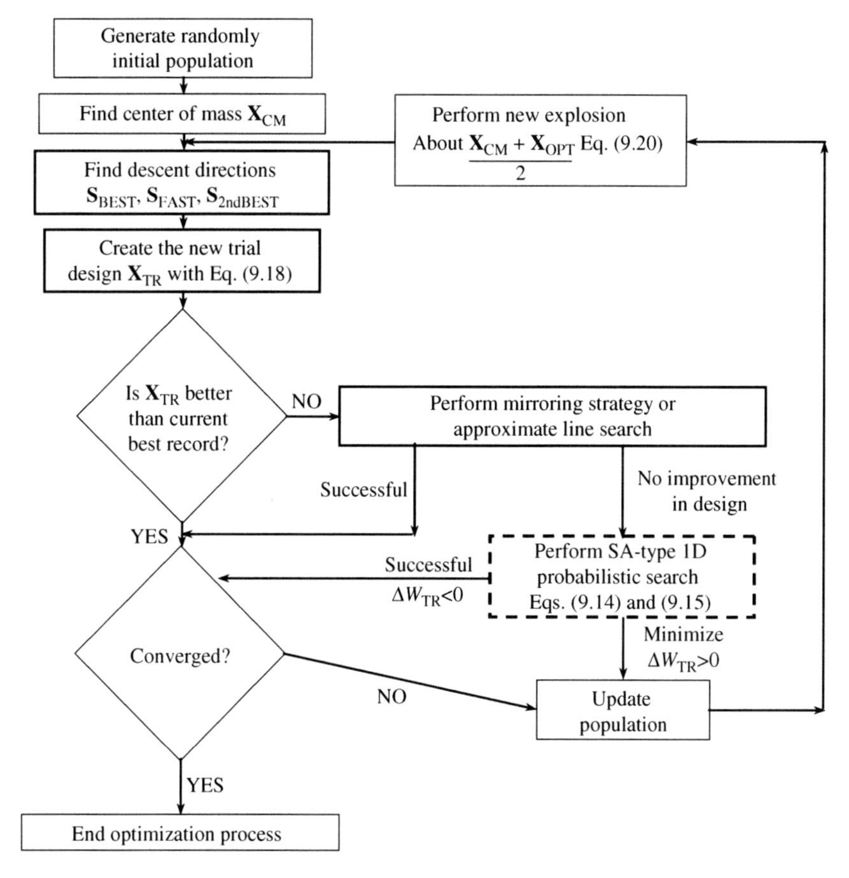

Figure 9.3 Flow chart of the new BB–BC algorithm developed in this research.

that estimates if design may improve in the next function evaluations. The "temperature" parameter is utilized to compute probability. Temperature can be a target value (estimated) for the cost function corresponding to a global minimizer. Initially, a larger target value is selected. As trials progress, the target value is reduced based on a cooling schedule and acceptance probability steadily decreases to zero as temperature is reduced.

SA has been widely utilized in structural optimization problems (Balling, 1991; Bennage and Dhingra, 1995; Chen and Su, 2002; Hasancebi and Erbatur, 2002; Hasancebi et al., 2010b; Lamberti, 2008; Lamberti and Pappalettere, 2007; Pantelides and Tzan, 2000; Shea et al., 1997; Sonmez, 2007) because of its inherent simplicity and ability to find the global optimum even in the presence of many design variables.

The basic formulation of SA was often modified to improve convergence behavior and reduce the number of structural analyses required in the optimization process. There are two main strategies: (i) one-directional ("local") search where variables are perturbed one at a time and (ii) multidirectional ("global") search where all variables are perturbed simultaneously.

Lamberti (2008) and Lamberti and Pappalettere (2007) developed the very sophisticated and powerful multilevel and multipoint SA algorithm taken as the basis of comparison in the present study to evaluate the performance of the new HS and BB–BC formulations. The strength points of the corrected multilevel and multipoint SA (CMLPSA) algorithm in its final form (Lamberti, 2008) are the following: (i) since a population of candidate designs are considered rather than a single trial point, there is no need to restrict the portion of design space currently investigated as the optimum is approached; (ii) each trial point can potentially improve current best record as it lies on a descent direction; (iii) design perturbations are forced to follow the rate of change exhibited by cost function; (iv) search strategy adaptively changes from the global level to the local level based on the convergence history (i.e., feasible, largely or slightly infeasible intermediate designs); (v) CMLPSA checks whether the constraint domain is locally convex or not: this increases further the probability of bypassing local minima yet without using any probabilistic criterion; and (vi) infeasible intermediate designs are handled by performing fourth order approximate line searches in the neighborhood of each feasible design generated when the optimizer tries to move away from infeasible regions.

Similarly to the new HS and BB–BC algorithms presented in Sections 9.3 and 9.4, the hybrid nature of the SA optimization algorithm briefly recalled in this section derives from the combination of metaheuristic search and gradient-based line search strategies.

9.6 Description of Test Problems

By comparing the algorithms described in Sections 9.3 through 9.5, it can be seen that HS and BB–BC operate on a population of N_{POP} candidate designs that must be generated randomly. CMLPSA is similar to BB–BC as it forms a population in the neighborhood of the current best record while BB–BC generates a population about the center of mass.

While the number of candidate designs considered by the SA algorithm basically depends on the number of optimization variables, the population size in HS and BB–BC is set a priori by the user. For this reason, a sensitivity analysis was carried out in this research to investigate the effect of the ratio between population size and number of design variables. For a planar 200-bar truss optimized with 200 design variables, population size was set as $N_{POP} = 20$, 50, 100, 200, 500, and 1000, respectively. The N_{POP}/NDV ratio hence ranged between 0.1 and 5. The large-scale weight minimization problem of a space tower with 3586 elements and 280 sizing variables was then solved with $N_{POP} = 500$ (hence, N_{POP}/NDV = 1.786).

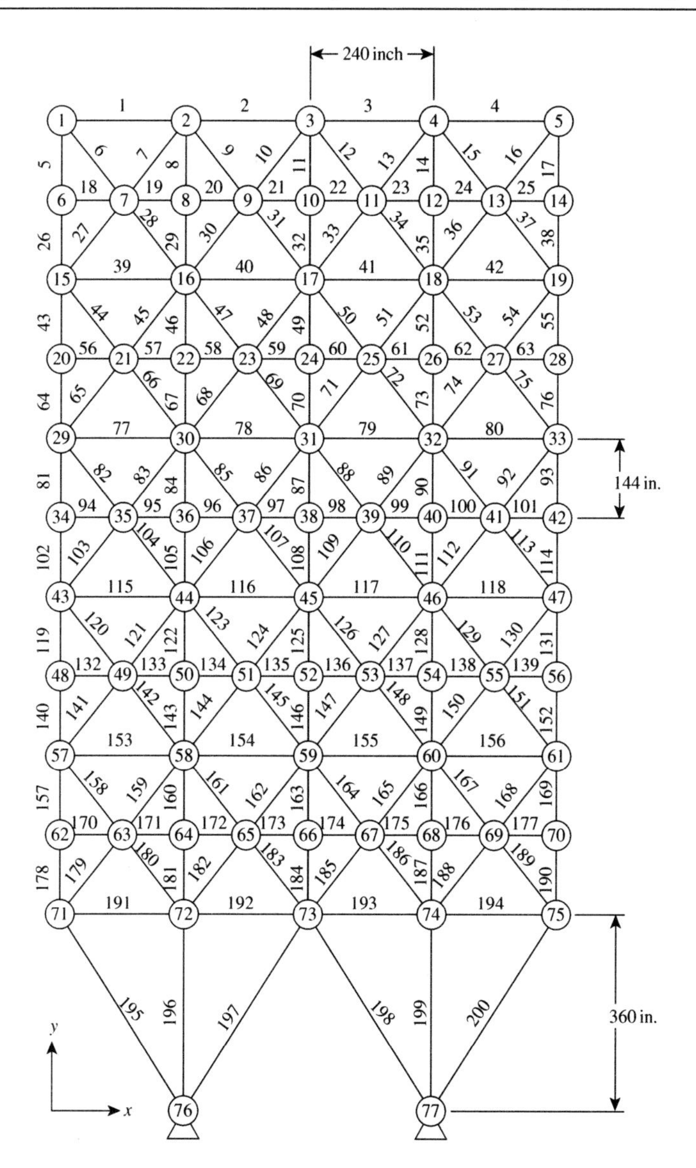

Figure 9.4 Schematic of the planar 200-bar truss structure used in the sensitivity analysis.

9.6.1 Planar 200-Bar Truss Structure Subject to Five Independent Loading Conditions

The planar 200-bar truss structure shown in Figure 9.4 has 77 nodes. The Young's modulus of the material is 206.91 GPa while the mass density is 7833.413 kg/m^3. The structure is optimized with 200 sizing variables corresponding to the cross-sectional area of each element.

The structure is subject to five independent loading conditions:

a. 4.45 kN (i.e., 1000 lbf) acting in the positive *X*-direction at nodes 1, 6, 15, 20, 29, 34, 43, 48, 57, 62, and 71;
b. 44.497 kN (i.e., 10,000 lbf) acting in the negative *Y*-direction at nodes 1, 2, 3, 4, 5, 6, 8, 10, 12, 14, 15, 16, 17, 18, 19, 20, 22, 24, 26, 28, 29, 30, 31, 32, 33, 34, 36, 38, 40, 42, 43, 44, 45, 46, 47, 48, 50, 52, 54, 56, 57, 58, 59, 60, 61, 62, 64, 66, 68, 70, 71, 72, 73, 74, and 75;
c. Loading conditions (a) and (b) acting together;
d. 4.45 kN acting in the negative *X*-direction at nodes 5, 14, 19, 28, 33, 42, 47, 56, 61, 70, and 75;
e. Loading conditions (b) and (d) acting together.

The 200-bar truss problem has been often taken as the benchmark test case in average/large-scale structural optimization problems (Gandomi and Yang, 2011). There are 3500 nonlinear constraints on nodal displacements and member stresses: displacements of all free nodes in both *X*- and *Y*-directions must be less than ± 1.27 cm (i.e., ± 0.5 in). The allowable stress (the same in tension and compression) is 206.91 MPa (i.e., 30,000 psi). Cross-sectional areas can range between 0.64516 (i.e., 0.1 in^2) and 645.16 cm^2 (i.e., 100 in^2).

9.6.2 Spatial 3586-Bar Truss Tower

The spatial 3586-bar truss tower shown in Figure 9.5 has 897 nodes. The Young's modulus E of the material is 68.971 GPa while the mass density is 2767.991 kg/m^3. The tower is 415 m tall and is comprised of five modules and three junction modules. From the top of the spire to the ground level, there are (i) a 15-story square-based pyramid segment of height 60 m, (ii) a 10-story square-based prismatic segment of height 40 m and side length 5 m, (iii) a 15-story octagon-based prismatic segment of height 75 m and radius 5 m, (iv) a 20-story dodecagon-based prismatic segment of height 100 m and radius 8 m, and (v) a 25-story hexadecagon-based prismatic segment of height 125 m and radius 12 m. The three intermediate modules (each 5 m tall) connect adjacent modules having different profiles. Hence, the layout section of the tower is a regular hexadecagon at the ground level and becomes a square in correspondence to the spire located in the top of the tower.

Figure 9.5B presents an assembly view of the tower where modules are represented with different colors. Figure 9.5C–F shows details of the structure and clarifies the story numbering that progresses from the top to the bottom of the tower. Figure 9.5G shows the nodes limiting the cross sections of the tower corresponding to the different modules. Node numbering progresses from the top of the tower to the ground level.

Because of the symmetry of the structure, variable linking can be adopted by grouping cross-sectional areas of truss members into 280 groups; hence, this optimization test case is a large-scale problem including 280 sizing variables.

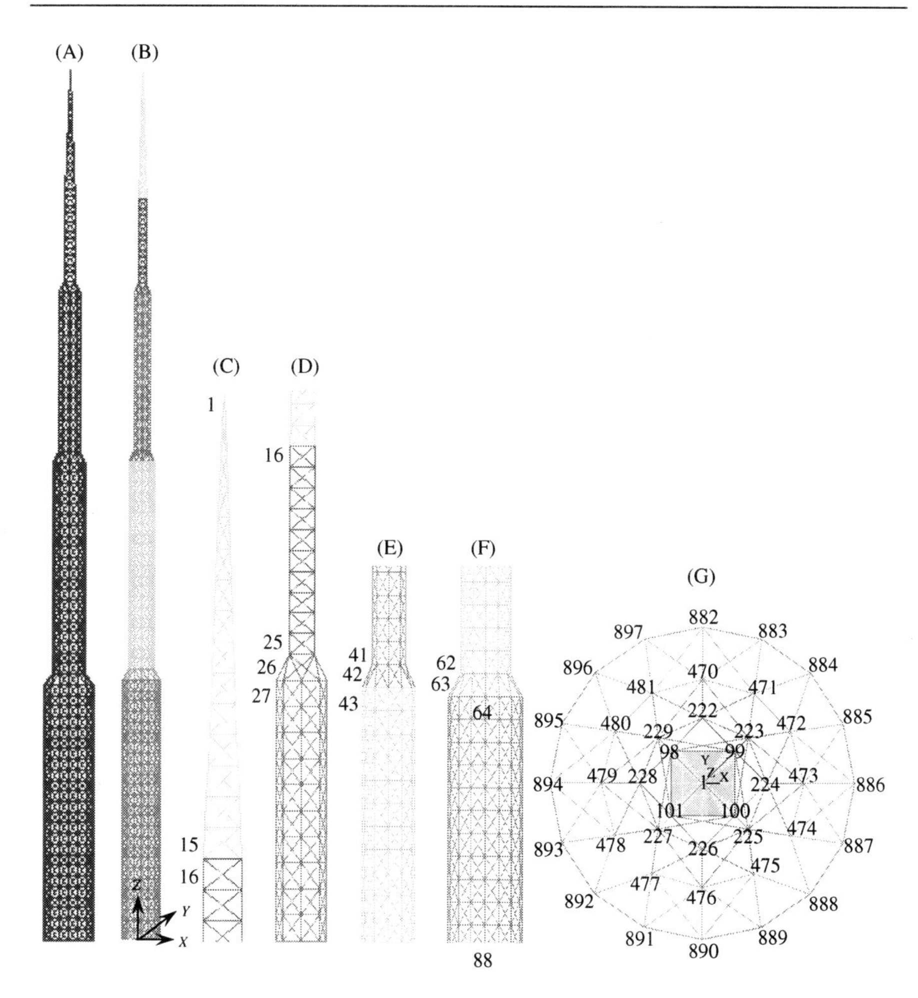

Figure 9.5 Schematic of the spatial 3586-bar truss tower with indication of story numbering and key-nodes: (A) assembly view of the tower; (B) color representation of the modules forming the structure and elements joining modules with different profiles; (C) detail of the square-based pyramid segment lying on the square-based prismatic segment (top of the tower); (D) detail of transition from the square-based prismatic segment to the octagon-based prismatic segment (center of the tower); (E) detail of transition from the octagon-based prismatic segment to the dodecagon-based prismatic segment (center of the tower); (F) detail of transition from the dodecagon-based prismatic segment to the hexadecagon-based prismatic segment (bottom part of the tower); and (G) layout view of the tower.

The tower must carry three independent loading conditions:

1. Concentrated forces of 13.5 kN acting downward on nodes 1 through 61; concentrated forces of 27 kN acting downward on nodes 62 through 101; concentrated forces of 40.5 kN acting downward on nodes 102 through 229; concentrated forces of 54 kN acting

downward on nodes 230 through 481; concentrated forces of 67.5 kN acting downward on nodes 482 through 881.

2. Concentrated forces of 667.2 N acting in the positive X-direction on nodes 2 through 878 of the left side of the tower; concentrated forces of 444.8 N acting in the negative X-direction on nodes 3 through 870 of the right side of the tower.
3. Concentrated forces of 444.8 N acting in the negative Y-direction on nodes 2 through 866 in the rear part of the tower; concentrated forces of 444.8 N acting in the positive Y-direction on nodes 4 through 874 in the front part of the tower.

The optimization problem includes 37,374 nonlinear constraints on nodal displacements, member stresses, and critical buckling loads. The displacements of all free nodes in all directions X, Y, and Z must be less than $\pm$40.64 cm (i.e., $\pm$ 16 in). The allowable tensile stress is 275.9 MPa (i.e., 40,000 psi). Structural members are assumed to be tubular with a nominal diameter to thickness ratio of 100 so that the critical buckling load of the jth member (of cross-sectional area A_j and length l_j) is $-100.01\pi \mathrm{EA}_j/8l_j^2$. The lower bound of cross-sectional areas is 0.64516 cm^2 (i.e., 0.1 in^2) while the upper bound is 1290.32 cm^2 (i.e., 200 in^2).

9.6.3 Implementation Details

The metaheuristic optimization algorithms studied in this research were implemented in Fortran 90. Optimization runs were carried out on a standard Sony VAIO laptop. The initial population of HS and BB–BC algorithms was generated with Eq. (9.16). Intermediate populations generated at each new explosion in BB–BC were instead generated with Eq. (9.20). Initial designs chosen for SA correspond either to the average of the centers of mass determined for the smaller and larger initial populations or to the average of the best designs included in the initial populations. This allowed SA to be run from initial designs that are actually representative of all trial designs initially generated for HS and BB–BC and allowed to compare metaheuristic algorithms on a more homogeneous basis.

Gradient-based optimizations were carried out with the sequential quadratic programming (SQP) routine implemented in the commercial software MATLAB® Version 7.10 (The Mathworks, 2010), and with the sequential linear programming (SLP) and SQP routines implemented in the commercial optimization software DOT™ Version 4.20 (Vanderplaats, 1995). These optimizations were also run on the Sony VAIO laptop. To draw more general conclusions on the relative merits of optimization algorithms/codes, no large-scale problem options eventually available in the commercial codes were activated. Initial points were set similarly to what was done for SA.

9.7 Results of Sensitivity Analysis

Tables 9.1 and 9.2 show the results of the sensitivity analysis obtained, respectively, for the new HS algorithm and the new BB–BC algorithm developed in this research. Optimization results are summarized in Table 9.3, which also

Table 9.1 Planar 200-Bar Truss Subject to Five Independent Loading Conditions: Results of Sensitivity Analysis Carried Out for the New HS Algorithm Proposed in this Research

N_{POP}	**Structural Weight (kg)**	**Optimization Iterations**	**Structural Analyses**	**Constraint Tolerance (%)**	**SA-Type 1D Probabilistic Searches**
20	12,490.377	136	5668	0.00735	4
50	12,490.482	143	6604	0.00750	4
100	12,490.542	139	6360	0.00750	4
200	12,490.332	130	5679	0.00730	3
500	12,490.414	148	5940	0.00643	3
1000	12,490.427	128	5938	0.00560	3

Table 9.2 Planar 200-Bar Truss Subject to Five Independent Loading Conditions: Results of Sensitivity Analysis Carried Out for the New BB–BC Algorithm Proposed in this Research

N_{POP}	**Structural Weight (kg)**	**Optimization Iterations**	**Structural Analyses**	**Constraint Tolerance (%)**	**Number of Explosions**	**SA-Type 1D Probabilistic Searches**
20	12,490.802	145	1924	0.00650	16	4
50	12,490.932	141	2872	0.00625	10	3
100	12,490.536	152	5034	0.00646	10	4
200	12,490.680	164	6874	0.00649	10	4
500	12,490.439	170	7745	0.00559	9	4
1000	12,490.686	166	9460	0.00629	7	4

Table 9.3 Summary of Optimization Results Obtained for the Planar 200-Bar Truss

	HS (Present Study)	**BB–BC (Present Study)**	**Simulated Annealing (Based on Lamberti, 2008)**
Structural weight (kg)	12,490.430 (± 0.0747)	12,490.680 (± 0.1773)	12,492.888
Dispersion of optimized variables (%)	0.3334	0.2397	N/A
Optimization iterations	137 (± 8)	156 (± 12)	149
Structural analyses	6031 (± 377)	5652 (± 2912)	11,726
Constraint tolerance (%)	0.00695 (± 0.000772)	0.00626 (± 0.000347)	None

includes a comparison with SA. The initial design chosen for SA corresponds to the average of the centers of mass of the different initial populations generated by varying N_{POP} from 20 to 1000. The two optimization runs carried out with the SQP routine built into the MATLAB commercial software started from the centers of mass of the initial populations generated for $N_{POP} = 20$ and $N_{POP} = 1000$, respectively.

Although all optimization runs started from regions of design space vary far from the target optimum, it was always possible to find optimized designs much lighter than the initial designs. In particular, Table 9.3 shows that the average optimized weights obtained by HS and BB–BC are 12,490.430 and 12,490.680 kg, respectively. Therefore, the present algorithms converged to slightly better designs than SA, which instead obtained a structural weight of 12,492.888 kg, practically the same as the weight of 12,493.243 kg quoted in Lamberti (2008) starting from a very infeasible point (i.e., cross-sectional areas of all elements were set at their lower bound). Even though the optimized weights quoted in Table 9.3 were penalized by the corresponding constraint tolerances, the updated structural weights obtained by the present HS and BB–BC algorithms (i.e., 12,491.5 and 12,491.7 kg, respectively) would not exceed the structural weight obtained by SA, which corresponds to a fully feasible design.

The SQP-MATLAB optimization routine was stopped after 180 design cycles while trying to reduce constraint violation occurred for some intermediate designs. The corresponding best designs quoted for the commercial optimizer are 14,093.5 and 17,622.8 kg. Optimizations were stopped because all other algorithms had already converged to their optimum design. However, SQP-MATLAB later completed the optimization process, correctly converging to a feasible design very close to the designs optimized by HS, BB–BC, and SA.

Tables 9.1–9.3 also show that the proposed HS and BB–BC algorithms are insensitive to the choice of initial population; in fact, the statistical dispersion observed for the optimized weight is at most 0.177 kg out of about 12,490.5 kg while the largest dispersion on optimized variables is about 0.333%. This happened in spite of the fact that the ratio between the size of the population and the number of design variables varied by 50 times, ranging from 0.1 to 5, and proves the efficiency and robustness of the present algorithms also in optimization problems including hundreds of design variables and thousands of nonlinear constraints.

The optimization process of HS and BB–BC required at most three or four 1D probabilistic searches (Tables 9.1 and 9.2). The number of explosions performed by BB–BC tends to decrease with the population size. This yielded a significant reduction in the computational cost of the proposed algorithm. In fact, the ratio between the number of explosions and the number of optimization iterations ranged between 1/24 and 1/9. For example, in the $N_{POP} = 1000$ case (i.e., the computationally most expensive case), the number of structural analyses that would have been required by classical BB–BC to perform an explosion in each new optimization iteration is 166,000, which is much larger than the 9460 analyses required by the present BB–BC algorithm to complete the entire optimization process.

The average number of optimization cycles required by BB–BC was larger than for HS (i.e., 156 iterations vs. 137 iterations) and also comparable with SA, which found the optimum design in 149 iterations (Table 9.3). No direct relationship

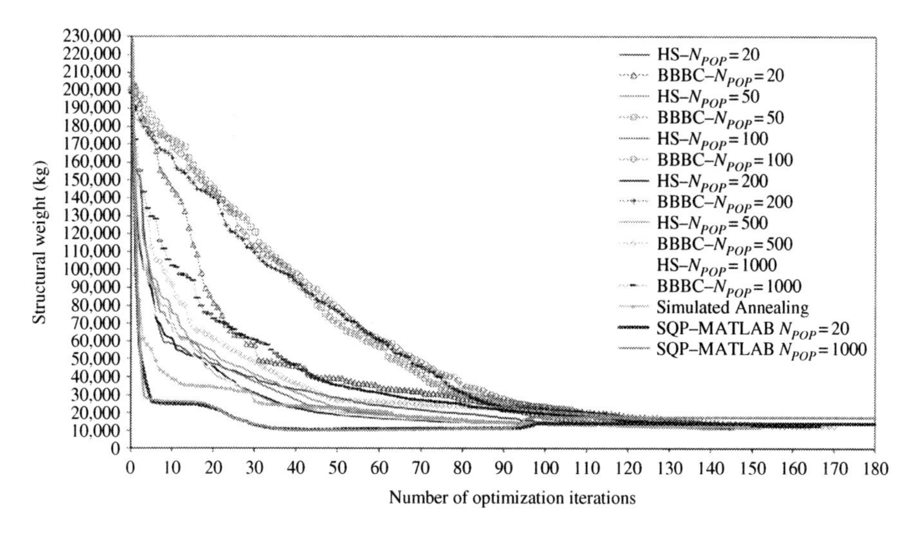

Figure 9.6 Convergence curves recorded in the sensitivity analysis for the planar 200-bar truss.

between the number of optimization iterations and the population size could be established (Tables 9.1 and 9.2).

The number of structural analyses required by BB–BC tends to increase with the size of the population (Table 9.2) and shows a quite large standard deviation (i.e., 51.5%) with respect to the average value of 5652 (Table 9.3). Conversely, the number of structural analyses required by HS is rather insensitive to population size (Tables 9.1 and 9.3).

Table 9.3 shows that HS and BB–BC are computationally much more efficient than SA, as they required practically one-half of the structural analyses required by SA (i.e., between about 5650 and 6030 analyses vs. 11,750 analyses). The average number of structural analyses per optimization cycle is practically the same for HS and BB–BC (i.e., 44 vs. 36), about one-half of the number of structural analyses per optimization cycle of SA (i.e., 79).

The optimization histories recorded for this test problem are plotted in Figure 9.6. Convergence curves recorded for the different populations do not reveal the presence of a direct relationship between population size and convergence speed. However, BB–BC was always slower than HS for all values of N_{POP}. SA was the fastest metaheuristic optimization algorithm in the early stages of the optimization process with a convergence speed comparable with SQP. However, the convergence rate of SA always became similar to the convergence rate of HS after a number of optimization iterations ranging between 35 and 50. The same behavior occurred regardless of population size. Convergence behavior of HS and BB–BC is far more regular than that exhibited by SA: some short steps were seen only for $N_{POP} = 20$.

Table 9.4 Optimization Results Obtained for the Spatial 3586-Bar Truss Tower Problem

	HS (Present Study)	BB–BC (Present Study)	Simulated Annealing (Based on Lamberti, 2008)	SLP-DOT	SQP-MATLAB and DOT
Structural weight (ton)	325.381	325.980	326.185	331.755	326.244
Optimization iterations	98	91	106	72	70
Structural analyses	11,312	14,616	16,240	19,673	18,440
Constraint tolerance (%)	0.0611	0.0618	0.105	None	0.144
Number of explosions	N/A	12	N/A	N/A	N/A
SA-type 1D probabilistic searches	3	2	11	N/A	N/A

9.8 Results of the Large-Scale Optimization Problem

Results of sensitivity analysis presented in Section 9.7 prove the efficiency and robustness of the new HS and BB–BC algorithms, which were rather insensitive to initial population. For this reason, in the large-scale weight minimization problem of the 3586-bar truss tower, the population size was set as $N_{POP} = 500$ (hence, N_{POP}/NDV was only 1.786). Furthermore, setting $N_{POP} = 500$ allowed computational cost of the optimization to be limited.

Table 9.4 presents the optimization results obtained for the 3586-bar truss tower. Although starting points were very far from the target optimum, HS and BB–BC found optimized designs much lighter than the initial design. These optimized designs were lighter and violated optimization constraints less than in the SA case (i.e., about 0.06% vs. 0.105%). All optimized designs were critical with respect to displacement and buckling constraints.

SLP-DOT obtained the heaviest design overall (331.755 tons). However, that was also the only design satisfying optimization constraints. If the optimized designs found by HS and BB–BC were penalized by including the corresponding constraint violations, the structural weight would never exceed 326.2 tons, hence, still 1.67% less than the optimum weight determined by SLP-DOT.

SQP-DOT converged to a slightly heavier design than HS and BB–BC (i.e., 326.244 tons vs. 325.381 tons and 325.980 tons, respectively) but this design violated displacement constraints by 0.144%. By penalizing weight to recover the constraint violation, the optimized structure would weigh 326.7 tons, hence, still 1.52% lighter than in the SLP-DOT case.

The HS and BB–BC algorithms developed in this study were computationally more efficient than SA as they required less structural analyses to converge to the

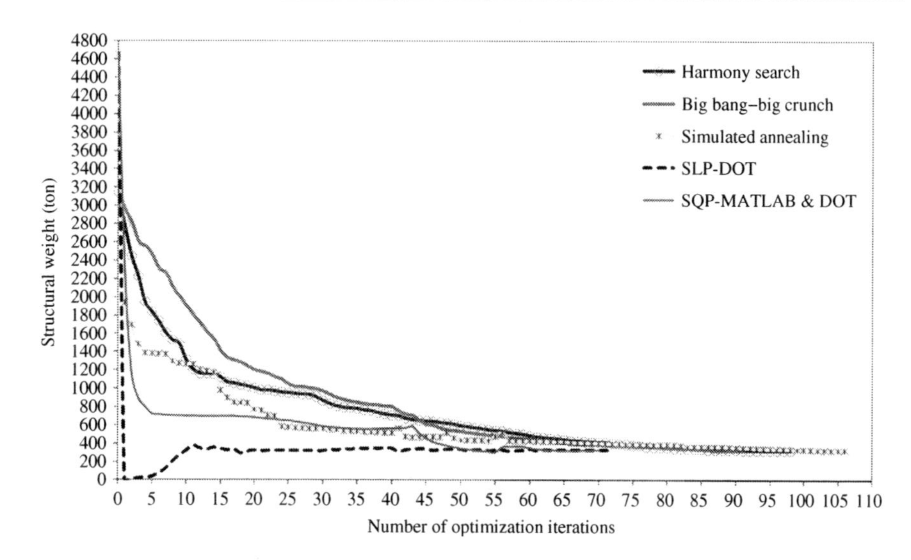

Figure 9.7 Convergence curves recorded for the spatial 3586-bar truss tower problem.

optimum design. HS required overall about 30% less structural analyses than SA and, more specifically, about 25% less structural analyses per design cycle (i.e., about 115 vs. 153).

The number of structural analyses per optimization cycle increased in the case of BB–BC (i.e., about 161) but was still similar to SA. The new BB–BC algorithm required 14,616 structural analyses within 91 iterations. In classical BB–BC, performing a new explosion about the center of mass of population in each optimization iteration would require 45,500 structural analyses, about three times more than the present BB–BC algorithm.

As expected, HS and BB–BC required fewer 1D probabilistic searches than SA: respectively, three or two searches versus 11 searches.

Element cross-sectional areas increased from top to bottom of each segment, thus satisfying the requirement on uniform stiffness put by the presence of the applied loads. Gradient-based optimizers designed more elements close to the lower bound of cross-sectional area.

Convergence curves recorded for the different optimization algorithms are shown in Figure 9.7. HS was much faster than BB–BC in the first 10 optimization iterations. The convergence rates of HS and BB–BC became comparable between the 30th and 40th iteration. BB–BC became faster than HS near the 45th optimization iteration and since that point could always find intermediate designs lighter than HS. Consequently, BB–BC converged to the optimum within 91 design cycles while HS required 98 iterations.

SA was overall the slowest optimizer and required 106 iterations to complete the optimization process. SA is the fastest metaheuristic algorithm in the early stages of the optimization process; its convergence speed is comparable to SLP and SQP (see Figure 9.7). However, the global annealing strategy implemented in the SA optimizer very quickly approached the constraint domain boundaries. Therefore, a 1D search had to be repeatedly performed in order not to generate infeasible designs but to escape from local minima. Steps and oscillatory behavior occurred as the SA optimizer continuously switched from global search to local search, thus penalizing the design when it was necessary to re-enter in a feasible region of design space. Remarkably, between the 10th and 15th iteration, HS found intermediate designs very close to SA and in some cases even better than those found by SA.

SLP reset design variables to their lower bounds in the first iteration, thus generating a very infeasible design. SQP reduced the structural weight by about seven times in the first seven iterations and then showed a large step where cost function decreased slowly until intermediate designs turned infeasible. Constraint violations were recovered by SLP and SQP by increasing structural weight. Such process caused oscillatory behavior until convergence to the optimum design. Both gradient-based optimizers converged to the optimum design in fewer iterations than the metaheuristic algorithms (i.e., 70−72 iterations vs. 91−106 iterations) but required more structural analyses as the number of structural analyses per optimization cycle practically coincides with the number of design variables.

The convergence rate of BB−BC tends to be lower or, at least, never significantly higher than for HS. Both algorithms generate new trial designs by combining descent directions, for which there are the largest and fastest reductions in weight with the descent direction corresponding to the second best design included in the current population. However, HS utilizes a more flexible search strategy that may be adapted according to the current trend of the optimization history (e.g., HS checks whether the trial designs generated by including gradient information were more than the trial designs generated via pitch adjustment, adapts the randomly extracted HMCR, and PAR internal parameters each time a new harmony is generated). Furthermore, the generation mechanism implemented in BB−BC is somehow penalized by the fact that descent directions are defined also with respect to the center of mass, which might retain characteristics of very conservative or inefficient designs.

In the early stages of the optimization process, the presence of the heavier designs included in the population causes the HS/BB−BC search process to be less efficient than SA as the search direction hosting the new trial design may be quite different from the steepest descent direction. Consequently, the new trial design will not lie on a descent direction as steep as in the SA case. Over the long term, multiple line searches and/or combinations of several descent directions made by HS and BB−BC allow for the renewal of the population of candidate designs, always taking information from the largest fraction of the design space as possible and not only from the neighborhood of the current best record. This explains why convergence curves recorded for HS and BB−BC show much shorter steps than SA (or even did not show any step) and practically converged monotonically to the optimum design.

9.9 Summary and Conclusions

This chapter presented two novel metaheuristic optimization formulations for the HS and BB–BC methods. A hybrid search scheme was implemented to improve convergence behavior. The new optimization algorithms are hybrid either because the metaheuristic search includes gradient information (i.e., metaheuristic search strategy operating on candidate designs generated randomly over design space was combined with local search strategy typically used in gradient-based optimization where design variables are perturbed by moving along descent directions) or because the HS and BB–BC include an SA-type probabilistic search (i.e., trial designs formed by operating on the population of candidate designs generated for HS and BB–BC were eventually improved, accepted, or rejected on the basis of the Metropolis' criterion utilized in SA).

The main novelty introduced in this study for the HS/BB–BC algorithms is that any new trial design is always forced to lie on some descent direction defined by combining the descent directions yielding the largest and fastest reductions in structural weight with the descent direction corresponding to the second best design included in the population of candidate designs. As far as it concerns HS specifically, the new formulation does not depend any more on internal parameters like HMCR and PAR that are now randomly generated and then adapted based on the current trend of the optimization process. Another important novelty introduced in BB–BC is that new explosions about the center of mass of the population are no longer performed in each new iteration but are made only when it was not possible to find any descent direction yielding an improvement of the current best record.

A sensitivity study demonstrated that the new HS and BB–BC algorithms developed in this research are superior to other HS (either classical HS formulations or recently published self-adaptive HS formulations that practically are parameter independent), BB–BC (either classical formulations or improved formulations recently developed by the present authors), and SA (multilevel and multipoint formulation also developed by the present authors) algorithms as they always found better designs (with lower structural weight and smaller constraint tolerance) and required fewer structural analyses to converge to the optimum design.

Remarkably, the new HS and BB–BC algorithms described in this chapter are rather insensitive to the choice of the initial population. Since optimized designs did not depend on initial population, it is not necessary to work with very large populations to enhance search capability and/or improve convergence behavior. This allows the number of structural analyses entailed by the optimization process to be substantially reduced with respect to classical metaheuristic optimization algorithms.

The new HS and BB–BC optimization algorithms were then tested in a large-scale weight minimization problem of a space truss tower including 3586 elements and 280 sizing variables, under three independent loading conditions. HS and BB–BC were compared with the powerful multilevel and multipoint SA algorithm developed by the present authors and well-known gradient-based commercial optimizers. Remarkably, the proposed algorithms were the most efficient optimizers in spite of having used a fairly small population with only 500 candidate designs, which is only 1.8 times the number of optimization variables.

The new HS and BB–BC algorithms were superior to SA because they generated new trial designs by combining descent directions that take information from a larger region of design space than in SA. Furthermore, search mechanisms implemented in the new algorithms are very flexible. Besides combining several descent directions and selecting the generation mechanism predominating in the current iteration, the 1D probabilistic search implemented in the new formulations always attempts to generate trial designs lying on descent directions. Furthermore, movements are properly rescaled if any trial design ends up infeasible. The Metropolis' probabilistic acceptance criterion is utilized only if all these tasks failed.

An important result obtained in this study was that the number of structural analyses per optimization iteration is usually smaller than the number of design variables. This is because trial designs are generated by including information on cost function gradients which, in the case of weight minimization problems of truss structures, are available at low computational cost. Hence, the proposed HS and BB–BC algorithms became highly competitive with gradient-based optimizers in terms of computational cost. In fact, even though SLP- and SQP-based routines implemented in commercial software often concluded the optimization process in fewer design cycles, the total number of structural analyses required by the present metaheuristic algorithms was always smaller than for the gradient-based optimizers.

Numerical results indicate that the present BB–BC algorithm was one order of magnitude more efficient than classical BB–BC as the number of explosions is on average 1/10 of the number of design cycles. However, the present BB–BC algorithm still is computationally more expensive than HS as it implements a less flexible mechanism for generating new trial designs. Furthermore, involving the center of mass in the generation of new trial designs also leads to retaining information on the less efficient designs included in the population. A possible strategy to further improve the present BB–BC formulation may be to perform explosions by moving always on descent directions. In this way, all candidate designs would have a higher probability of being better than the current best record. However, this strategy might lead to approach the constraint boundaries too quickly and limit the search to regions of design space hosting local minima. Further investigations will be required to address this question.

In summary, this chapter presented two very efficient and robust hybrid formulations of the HS and BB–BC metaheuristic optimization algorithms. Numerical results obtained in two weight minimization problems of truss structures subject to constraints on nodal displacements, member stresses, and buckling strength proved that the new algorithms presented in this chapter are significantly superior to other advanced metaheuristic algorithms recently published in technical literature and well-known gradient-based optimizers implemented in commercial software packages. However, future investigations should be aimed at improving convergence speed of BB–BC as well as to enhancing the global optimization capability of the present HS and BB–BC formulations.

References

Balling, R.J., 1991. Optimal steel frame design by simulated annealing. J. Struct. Eng. 117 (6), 1780–1795.

Bennage, W.A., Dhingra, A.K., 1995. Single and multi-objective structural optimization in discrete-continuous variables using simulated annealing. Int. J. Numer. Methods Eng. 38 (16), 2553–2573.

Camp, C.V., 2007. Design of space trusses using Big Bang–Big Crunch optimization. J. Struct. Eng. 133 (7), 999–1008.

Camp, C.V., Bichon, J., Stovall, S.P., 2005. Design of steel frames using ant colony optimization. J. Struct. Eng. 131 (3), 369–379.

Carbas, S., Saka, M.P., 2012. Optimum topology design of various geometrically nonlinear latticed domes using improved harmony search method. Struct. Multidiscip. Optim. 45 (3), 377–399.

Chen, T.Y., Su, J.J., 2002. Efficiency improvement of simulated annealing in optimal structural designs. Adv. Eng. Softw. 33 (7 – 10), 675–680.

Clerc, M., 2006. Particle Swarm Optimization. ISTE Publishing Company, London.

Degertekin, S.O., 2012. Improved harmony search algorithms for sizing optimization of truss structures. Comput. Struct. 92–93, 229–241.

Dorigo, M., Stutzle, T., 2004. Ant Colony Optimization. MIT Press, Cambridge, MA.

Erbatur, F., Hasancebi, O., Tutuncu, I., Kilic, H., 2000. Optimal design of planar and space structures with genetic algorithms. Comput. Struct. 75 (2), 209–224.

Erol, O.K., Eksin, I., 2006. New optimization method: Big Bang–Big Crunch. Adv. Eng. Softw. 37 (2), 106–111.

Galante, M., 1996. Genetic algorithms as an approach to optimize real-world trusses. Int. J. Numer. Methods Eng. 39 (3), 361–382.

Gandomi, A.H., Yang, X.S., 2011. Benchmark problems in structural optimization. In: Koziel, S., Yang, X.S. (Eds.), Computational Optimization, Methods and Algorithms, 356. Springer-Verlag, Berlin-Heidelberg, Germany, pp. 259–281. (Studies in Computational Intelligence, Chapter 12).

Gandomi, A.H., Yang, X.S., Alavi, A.H., 2011. Mixed variable structural optimization using Firefly Algorithm. Comput. Struct. 89 (23 – 24), 2325–2336.

Gandomi, A.H., Talatahari, S., Yang, X.S., Deb, S., 2012a. Design optimization of truss structures using cuckoo search algorithm. Struct. Des. Tall Spec. Build. doi: 10.1002/tal.1033.

Gandomi, A.H., Yang, X.S., Alavi, A.H., 2012b. Cuckoo search algorithm: a metaheuristic approach to solve structural optimization problems. Eng. Comput. doi: 10.1007/s00366-011-0241-y.

Gandomi, A.H., Yang, X.S., Talatahari, S., Alavi, A.H., 2013. Metaheuristics in modeling and optimization. In: Gandomi, A.H., Yang, X.S., Talatahari, S., Alavi, A.H. (Eds.), Metaheuristic Applications in Structures and Infrastructures. Elsevier, Waltham, MA (Chapter 1, in press).

Geem, Z.W., Kim, J.H., Loganhatan, G.V., 2001. A new heuristic optimization algorithm: harmony search. Simulation. 76 (2), 60–68.

Goldberg, D.E., 1989. Genetic Algorithms in Search, Operation and Machine Learning. Addison-Wesley, Reading, MA.

Hajela, P., Lee, E., 1995. Genetic algorithm in truss topological optimization. Int. J. Solids Struct. 32 (22), 3341–3357.

Hasancebi, O., Erbatur, F., 2002. Layout optimization of trusses using simulated annealing. Adv. Eng. Softw. 33 (7 − 10), 681−696.

Hasancebi, O., Carbas, S., Dogan, E., Erdal, F., Saka, M.P., 2009. Performance evaluation of metaheuristic search techniques in the optimum design of real size pin jointed structures. Comput. Struct. 87 (5 − 6), 284−302.

Hasancebi, O., Carbas, S., Dogan, E., Erdal, F., Saka, M.P., 2010a. Comparison of non-deterministic search techniques in the optimum design of real size steel frames. Comput. Struct. 88 (17 − 18), 1033−1048.

Hasancebi, O., Carbas, S., Saka, M.P., 2010b. Improving the performance of simulated annealing in structural optimization. Struct. Multidiscip. Optim. 41 (2), 189−203.

Hasancebi, O., Erdal, F., Saka, M.P., 2010c. Adaptive harmony search method for structural optimization. J. Struct. Eng. 136 (4), 419−431.

Hwang, S.F., He, R.S., 2006. Improving real-parameter genetic algorithm with simulated annealing for engineering problems. Adv. Eng. Softw. 37 (6), 406−418.

Kaveh, A., Abadi, A.S.M., 2010. Cost optimization of a composite floor system using an improved harmony search algorithm. J. Constr. Steel Res. 66 (5), 664−669.

Kaveh, A., Ahangaran, M., 2012. Discrete cost optimization of composite floor system using social harmony search model. Appl. Soft Comput. J. 12 (1), 372−381.

Kaveh, A., Kalatjari, V., 2002. Genetic algorithm for discrete sizing optimal design of trusses using the force method. Int. J. Numer. Methods Eng. 55 (1), 55−72.

Kaveh, A., Shojaee, S., 2007. Optimal design of skeletal structures using ant colony optimization. Int. J. Numer. Methods Eng. 70 (5), 563−581.

Kaveh, A., Talatahari, S., 2009a. Particle swarm optimizer, ant colony strategy and harmony search scheme hybridized for optimization of truss structures. Comput. Struct. 87 (5−6), 267−283.

Kaveh, A., Talatahari, S., 2009b. A particle swarm ant colony optimization for truss structures with discrete variables. J. Constr. Steel Res. 65 (8 − 9), 1558−1568.

Kaveh, A., Talatahari, S., 2009c. Size optimization of space trusses using Big Bang−Big Crunch algorithm. Comput. Struct. 87 (17 − 18), 1129−1140.

Kaveh, A., Talatahari, S., 2010a. An improved ant colony optimization for the design of planar steel frames. Eng. Struct. 32 (8), 864−873.

Kaveh, A., Talatahari, S., 2010b. Optimal design of Schwedler and ribbed domes via hybrid Big Bang−Big Crunch algorithm. J. Constr. Steel Res. 66 (3), 412−419.

Kaveh, A., Talatahari, S., 2010c. Optimal design of skeletal structures via the charged system search algorithm. Struct. Multidiscip. Optim. 41 (6), 893−911.

Kaveh, A., Talatahari, S., 2011. Geometry and topology optimization of geodesic domes using charged system search. Struct. Multidiscip. Optim. 43 (2), 215−229.

Kaveh, A., Zolghadr, A., 2012. Truss optimization with natural frequency constraints using a hybridized CSS-BBBC algorithm with trap recognition capability. Comput. Struct. 102−103, 14−27.

Kennedy, J., Eberhart, R.C., 2001. Swarm Intelligence. Morgan Kaufmann, San Francisco, CA.

Kirkpatrick, S., Gelatt, C.D., Vecchi, M.P., 1983. Optimization by simulated annealing. Science. 220 (4598), 671−680.

Koziel, S., Yang, X.S. (Eds.), 2011. Computational Optimization, Methods and Algorithms, vol. 356. Springer-Verlag, Germany (Studies in Computational Intelligence).

Lamberti, L., 2008. An efficient simulated annealing algorithm for design optimization of truss structures. Comput. Struct. 86 (19 − 20), 1936−1953.

Lamberti, L., Pappalettere, C., 2007. Weight optimization of skeletal structures with multi-point simulated annealing. Comput. Modeling Eng. Sci. 18 (3), 183–221.
Lamberti, L., Pappalettere, C., 2009. An improved harmony-search algorithm for truss structure optimization. In: Topping, B.V.H., Costa Neves, L.F., Barros, R.C. (Eds.), Proceedings of the Twelfth International Conference on Civil, Structural and Environmental Engineering Computing. Civil Comp Press, Dun Eaglais.
Lamberti, L., Pappalettere, C., 2011a. Metaheuristic design optimization of skeletal structures: a review. Comput. Technol. Rev. 4, 1–32.
Lamberti, L., Pappalettere, C., 2011b. A fast Big Bang – Big Crunch optimization algorithm for weight minimization of truss structures. In: Tsompanakis, Y., Topping, B.H.V. (Eds.), Proceedings of the Second International Conference on Soft Computing Technology in Civil, Structural and Environmental Engineering. Civil Comp Press, Dun Eaglais.
Lee, K.S., Geem, Z.W., 2004. A new structural optimization method based on the harmony search algorithm. Comput. Struct. 82 (9–10), 781–798.
Lee, K.S., Geem, Z.W., 2005. A new meta-heuristic algorithm for continuous engineering optimization: harmony search theory and practice. Comput. Methods Appl. Mech. Eng. 194 (36 – 38), 3902–3933.
Li, J.L., Huang, Z.B., Liu, F., Wu, Q.H., 2007. A heuristic particle swarm optimizer for optimization of pin connected structures. Comput. Struct. 85 (7–8), 340–349.
Liao, T.W., 2010. Two hybrid differential evolution algorithms for engineering design optimization. Appl. Soft Comput. 10 (4), 1188–1199.
Luh, G.C., Lin, C.Y., 2008. Optimal design of truss structures using ant algorithm. Struct. Multidiscip. Optim. 36 (4), 365–379.
Oftadeh, R., Mahjoob, M.J., Shariatpanahi, M., 2010. A novel meta-heuristic optimization algorithm inspired by group hunting of animals: hunting search. Comput. Math. Appl. 60 (7), 2087–2098.
Pantelides, C.P., Tzan, S.R., 2000. Modified iterated simulated annealing algorithm for structural synthesis. Adv. Eng. Softw. 31 (6), 391–400.
Perez, R.E, Behdinan, K., 2007. Particle swarm approach for structural design optimization. Comput. Struct. 85 (19 – 20), 1579–1588.
Rahami, H., Kaveh, A., Gholipour, Y., 2008. Sizing, geometry and topology optimization of trusses via force method and genetic algorithm. Eng. Struct. 30 (9), 2360–2369.
Sahab, M.G., Toropov, V.V., Gandomi, A.H., 2012. A review on traditional and modern structural optimization: problems and techniques. In: Gandomi, A.H., Yang, X.S., Talatahari, S., Alavi, A.H. (Eds.), *Metaheuristic Applicationsin Structures and Infrastructures*. Elsevier, The Netherlands (Chapter 2, in press).
Saka, M.P., 2007a. Optimum design of steel frames using stochastic search techniques based on natural phenomena: a review. In: Topping, B.H.V. (Ed.), Civil Engineering Computations: Tools and Techniques. Saxe-Coburg Publications, Stirlingshire, UK, pp. 105–147. (Chapter 6).
Saka, M.P., 2007b. Optimum topological design of geometrically nonlinear single layer latticed domes using coupled genetic algorithm. Comput. Struct. 85 (21–22), 1635–1646.
Saka, M.P., 2009. Optimum design of steel sway frames to BS5950 using harmony search algorithm. J. Constr. Steel Res. 65 (1), 36–43.
Schutte, J.F., Groenwold, A.A., 2003. Sizing design of truss structures using particle swarms. Struct. Multidiscip. Optim. 25 (4), 261–269.

Serra, M., Venini, P., 2006. On some applications of ant colony to plane truss optimization. Struct. Multidiscip. Optim. 32 (2), 499–506.

Shea, K., Cagan, J., Fenves, S.J., 1997. A shape annealing approach to optimal truss design with dynamic grouping of members. J. Mech. Des. 119 (3), 388–394.

Sonmez, F.O., 2007. Shape optimization of 2D structures using simulated annealing. Comput. Methods Appl. Mech. Eng. 196 (35 – 36), 3279–3299.

The MathWorks, 2010. *MATLAB® Version 7.10*. Austin, TX.

Togan, V., 2012. Design of planar steel frames using teaching–learning based optimization. Eng. Struct. 34, 225–232.

Vanderplaats, G.N., 1995. *DOT® Users Manual, Version 4.20*. VR&D Inc., Colorado Springs, Co, USA.

Van Laarhoven, P.J.M., Aarts, E.H.L., 1987. Simulated Annealing: Theory and Applications. Kluwer Academic Publishers, Dordrecht, The Netherlands.

Yang, X.S., 2010. Firefly algorithm, Levy flights and global optimization. In: Bramer, M., Ellis, R., Petridis, M. (Eds.), Research and Development in Intelligent Systems XXVI. Springer, London, pp. 209–218.

Yang, X.S., Gandomi, A.H., 2012. Bat algorithm: a novel approach for global engineering optimization. Eng. Comput. 29 (5), 464–483.

10 Graph Theory in Evolutionary Truss Design Optimization

Benoît Descamps and Rajan Filomeno Coelho

Building, Architecture & Town planning (BATir) Department, Université libre de Bruxelles, Brussels, Belgium

10.1 Introduction

The structural design process is traditionally based on the engineer's experience. A model is created, tested, and updated in a series of structural analyses in order to achieve an optimal design. This time-consuming, empirical process might become cumbersome and must be ultimately discarded for large-scale structures. As a remedy, researchers have continuously developed original methods over decades to solve these issues according to the designer's criteria. These methods are commonly used in aeronautics and automotive industries (Breitkopf and Filomeno Coelho, 2010), but they also meet a growing interest in civil and architectural engineering.

The design problem consists of finding the best material distribution over a given design domain approximated by finite elements. The structural optimization problem involves three aspects of structural design—sizing, shape, topology—which are classified according to the types of variables involved. This chapter focuses on sizing and topology optimization of truss structures using the benefits of graph theory. Based on this parameterization, the stress-based mass minimization is a classical problem in the literature. To solve it, the ground structure approach enables the determination of a certain set of connections between a fixed grid of nodal points (Dorn et al., 1964). If cross-sectional areas are allowed to reach zero in a continuous way, the standard theory of linear programming holds and the problem is well suited for mathematical programming. Also, one can derive from the necessary conditions of optimality that the globally optimal solution is fully stressed (Bendsøe and Sigmund, 2003). Nonetheless, such solutions are, in general, unstable and collapse if any perturbation in the external loading occurs. Moreover, due to manufacturing standards, cross sections are often selected from a finite set of discrete values, and topology is preferably defined using a binary variable associated with each truss member that specifies the connection status ($=1$) indicating its presence or ($=0$) its absence.

Metaheuristic Applications in Structures and Infrastructures. DOI: http://dx.doi.org/10.1016/B978-0-12-398364-0.00010-3

Due to the presence of mixed variables, discontinuity and nondifferentiability of response functions, and of nonconvexity of the design space (hence, the existence of several local minima), metaheuristic algorithms are adopted to solve the optimization problem. Among many others, evolutionary algorithms (EAs) have been the subject of intense research. These algorithms attempt to drill the secrets of nature by abstracting the adaptive processes of biological systems. Besides the appealing beauty-of-nature argument, these techniques are based on solid theoretical and empirical evidence (Holland, 1974), demonstrating their effectiveness in solving versatile optimization problems that are intractable by other methods.

Truss optimization by EAs raises two key issues: the way of coding structural configurations has a major impact on the convergence process; and the random generation of truss topologies inevitably involves a certain proportion of invalid configurations, therefore, requiring an in-depth study. Goldberg (1989) claims that the best representation is *the smallest alphabet that permits a natural expression of the problem*. By definition, truss topology deals with the study of pairwise relations among nodes. In that regard, graph theory provides a valuable insight. This mathematical abstraction offers an excellent representation of truss assembly. Giger and Ermanni (2006) proposed the first graph-based parameterization for EAs expanded as a matrix encoding. The method succeeds not only in expressing the ground structure in a consistent way (thus easy-to-implement) but also in bringing out topological properties to avoid unstable configurations. It results into an integrated design process that combines advantages of different disciplines.

The content of the chapter is as follows: Section 10.2 introduces the truss design problem and discusses several optimization methods to solve it. Then, Section 10.3 introduces a systematic terminology and characterization of graph theory so that it can be extended to other applications. The corresponding matrix encoding for truss structures is derived. Section 10.4 describes the main steps of the EA developed for numerical tests. In particular, special care is devoted to several crossover and mutation operators based on the matrix encoding. Also, the strategy to resolve unstable configurations is explained. Three applications illustrate the implementation in Section 10.5. Finally, the conclusions are drawn in Section 10.6.

10.2 Truss Design

10.2.1 Equilibrium Equations

Although several aspects of the method developed in this chapter are quite general and easily applied to most types of structures, trusses are chosen for several reasons (Kirsch and Rozvany, 1994):

- a wide range of structures can be approximated by trusses, especially at the conceptual stage;
- the analysis code is easily written, allowing the designer to focus on the optimization part;

- they provide excellent test cases for optimization by highlighting complex structural issues in a simple manner.

Let us focus first on the equilibrium formulation. Using the standard finite element concepts, we consider a pin-joint assembly of truss members subject to axial forces and devoid of imperfections. By denoting d the spatial dimension (2D or 3D) and N_s the number of support reactions, the number of degrees of freedom considering the boundary conditions is given by $N_d = d \cdot N_n - N_s$. If N_b is the number of members, the static equilibrium equations are written in matrix form as (Petersson, 2001):

$$\mathbf{Rt} = \mathbf{p} \Leftrightarrow \sum_{e=1}^{N_b} t_e \mathbf{r}_e = \mathbf{p}, \quad e = 1,\ldots,N_b \tag{10.1}$$

where $\mathbf{t} \in \mathbb{R}^{N_b}$ is the vector collecting each member force $t_e \in \mathbb{R}$, $\mathbf{p} \in \mathbb{R}^{N_d}$ is the external force vector, and $\mathbf{r}_e \in \mathbb{R}^{N_d}$ are the bar direction cosine vectors; $\mathbf{R} \in \mathbb{R}^{N_d \times N_b}$ is called the equilibrium matrix. Three different cases are identified:

1. $N_b < N_d$, the system is underdetermined and, in general, there is no solution for $\mathbf{t}$ in Eq. (10.1); unless an appropriate configuration under prestress (for instance, the cable-net structure) is included, a stable equilibrium state cannot be reached.
2. $N_b = N_d$, the system is determined and there exists exactly one solution for $\mathbf{t}$ in Eq. (10.1); the truss is statically determinate and the internal force distribution is independent of material considerations.
3. $N_b > N_d$, the system is overdetermined and, in general, there are infinite solutions for $\mathbf{t}$ in Eq. (10.1); the structure is statically indeterminate and compatibility conditions are required to find one solution to the equilibrium problem.

Equation (10.1) can be solely used to state the static equilibrium. For more practical designs and redundant structures, information on kinematics is needed. For this purpose, we introduce the compatibility conditions between the nodal displacement vector $\mathbf{u} \in \mathbb{R}^{N_d}$ and the member elongation vector $\mathbf{e} \in \mathbb{R}^{N_b}$:

$$\mathbf{R}^{\mathrm{T}}\mathbf{u} = \mathbf{e} \Leftrightarrow \mathbf{r}_e^{\mathrm{T}}\mathbf{u} = \varepsilon_e l_e, \quad e = 1,\ldots,N_b \tag{10.2}$$

with the Cauchy strain denoted $\varepsilon_e \in \mathbb{R}$ and $l_e \in \mathbb{R}$ the member length. A similar discussion to Eq. (10.1) can be done for Eq. (10.2) (see Pellegrino (1993) for more details). Under assumption of a linear elastic material, the constitutive equations are given by the Hook's law:

$$\sigma_e = E_e \varepsilon_e, \quad e = 1,\ldots,N_b \tag{10.3}$$

For overdetermined systems, a unique solution can be found by nesting Eqs. (10.1)–(10.3). With this aim, we start from the definition of the axial stress $\sigma_e \in \mathbb{R}$ and the cross-section area $a_e \in \mathbb{R}$:

$$t_e = \sigma_e a_e, \quad e = 1,\ldots,N_b \tag{10.4}$$

Then, by introducing successively the constitutive law, Eq. (10.3), and the compatibility conditions, Eq. (10.2), one gets

$$t_e = E_e a_e \varepsilon_e = \frac{E_e a_e}{l_e} \mathbf{r}_e^{\mathrm{T}} \mathbf{u}, \quad e = 1, \ldots, N_{\mathrm{b}} \tag{10.5}$$

This equation can be later introduced in the static equilibrium Eq. (10.1) as

$$\mathbf{RDR}^{\mathrm{T}}\mathbf{u} = \mathbf{p} \Leftrightarrow \sum_{e=1}^{N_{\mathrm{b}}} \frac{E_e a_e}{l_e} (\mathbf{r}_e^{\mathrm{T}} \mathbf{u}) \mathbf{r}_e = \mathbf{p}, \quad e = 1, \ldots, N_{\mathrm{b}} \tag{10.6}$$

with

$$\mathbf{D} = \mathrm{diag}\left(\frac{E_e a_e}{l_e}\right) \in \mathbb{R}^{N_{\mathrm{b}} \times N_{\mathrm{b}}}, \quad e = 1, \ldots, N_{\mathrm{b}} \tag{10.7}$$

By setting $\mathbf{K} = \mathbf{RDR}^{\mathrm{T}}$, we obtain the well-known displacement-based formulation:

$$\mathbf{Ku} = \mathbf{p} \tag{10.8}$$

The stiffness matrix $\mathbf{k} \in \mathbb{R}^{N_{\mathrm{d}} \times N_{\mathrm{d}}}$ is positive semidefinite and thus invertible, ensuring the uniqueness of the solution $\mathbf{u}$.

10.2.2 Formulation of the Optimization Problem

A general structural design problem can be formulated as a mathematical optimization problem: it consists of finding the set of bounded design variables $\mathbf{x} \in X$ that would minimize an objective function $f \in \mathbb{R}$ subject to nonlinear inequality constraints $\mathbf{g} \in \mathbb{R}^{N_{\mathrm{g}}}$, and nonlinear equality constraints $\mathbf{h} \in \mathbb{R}^{N_{\mathrm{h}}}$. Note that each evaluation requires the solution of an entire system of equilibrium equations. In summary, the optimization problem is expressed in the following mathematical formulation:

$$\min_{\mathbf{x}} \{ f(\mathbf{x}) | \mathbf{g}(\mathbf{x}) \leq \mathbf{0}, \quad \mathbf{h}(\mathbf{x}) = \mathbf{0} \} \tag{10.9}$$

Typical objective functions include the mass, the compliance, element stresses (all to be minimized), or the lowest eigenfrequency (to be maximized). Additionally, basic structural constraints are stress, displacement, and instability. Regarding the design variables, we distinguish three classes of structural optimization problems applied to truss structures (Figure 10.1):

1. *Sizing optimization* of truss structures is concerned with cross-section variables such as area and area moment of inertia. The variables can be continuous, discrete, or integer.

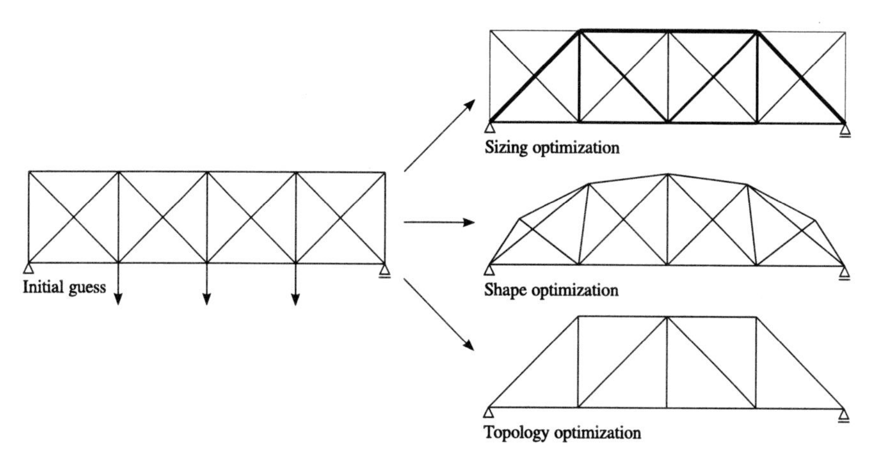

Figure 10.1 Starting from an initial guess, different outcomes for sizing, shape, and topology optimization.

2. *Shape optimization* acts on geometrical variables, either directly on finite element nodes or by means of computer-aided geometrical design functions, e.g., nonuniform rational basis spline.
3. *Topology optimization* of discrete systems fundamentally differs from continuum structures. For truss structures, topology optimization specifies the presence or removal of specific members. The type of variable associated to this parameterization is inherently binary. As mentioned in the introduction, several works merge sizing and topology concepts by allowing cross sections to reach zero.

The aforementioned classification is subject to exceptions. For instance, by reformulating the equilibrium equations and solving them as a structural form-finding problem, the axial forces may become shape variables (Descamps et al., 2011).

At first glance, any combinations of objectives, constraints, and variables are possible. Yet, only a few of them are meaningful. For instance, topology optimization of compliance requires volume constraints; sizing optimization for the mass minimization problem must be performed in conjunction with stress constraints. Additionally, several constraints can lead to a more realistic design but at the cost of complicating the optimization problem. The role of designer is to draw upon the trade-off between complexity of the problem and quality of the solution.

For truss structures, the mass minimization problem subject to stress, displacement, and local buckling constraints is explicitly written as

$$\min\left\{\sum_{e=1}^{N_\mathrm{b}}\rho_e l_e a_e \middle| \frac{|\sigma_e|}{\sigma_{\max}}-1\leq 0, \frac{|u_i|}{u_{\max}}-1\leq 0, \frac{\sigma_e^-}{\sigma_{\mathrm{crit}}}-1\leq 0, 0\leq a_e\right\},$$

$$e=1,\ldots,N_\mathrm{b},\quad i=1,\ldots,N_\mathrm{d} \tag{10.10}$$

In the mass function, ρ_e is the mass density, l_e is the member length, and a_e is the cross-sectional area. The normalized, nonlinear inequality constraints are related to element stress σ_e, nodal displacement u_i, and the local stability of each compression member σ_e^-. If I_e is the smallest cross-sectional moment of inertia of a truss member and E_e its elastic modulus, the critical stress for the case of elastic buckling is given by the Euler formula:

$$\sigma_{\text{crit}} = \frac{\pi^2 E_e I_e}{a_e l_e^2}, \quad e = 1, \ldots, N_b \tag{10.11}$$

10.2.3 Optimization Methods

In the realm of optimization methods, part of them—the mathematical programming algorithms—are deterministic and take into consideration properties of the mathematical problem, generally by means of derivatives. As such, they are local in scope and optimality conditions ensure the finding of a local minimum. These smooth techniques offer accurate and efficient solutions for optimization problems where assumptions of continuity and existence of derivatives are met. Otherwise, nonsmooth algorithms are required, leading to a rapid degradation of computational performance.

In contrast, metaheuristics are general-purpose optimization algorithms. With very limited *a priori* knowledge, they apply stochastic rules to iteratively converge toward an optimal solution. In theory, they would be able to reach the global optimum; yet, there is no guarantee of finding it. But their robustness on a broad range of problems substitutes for the high performance of gradient-based algorithms. Albeit computationally demanding, they have shown their effectiveness in countless structural design problems. Their two major components are intensification and diversification: while the former means concentrating efforts in the neighborhood of the current solution, the latter tends to widely explore the design space. In the sequel, we briefly describe some of their applications in truss design. Due to their widespread availability, the enumeration is by no means comprehensive and mainly focuses on recent investigations.

The class of trajectory-based metaheuristics assigns a single agent to run through the design space. Starting from an arbitrary initial location, these intensification techniques evaluate several directions at each iteration; a downhill solution is always accepted, while uphill solutions are considered with a certain probability in order to avoid getting stuck in local optima. Independently proposed by Kirkpatrick et al. (1983) and Černý (1985), the most popular and powerful is probably the simulated annealing (SA): it relies on the analogy between the cooling process toward a minimal state energy and the convergence toward the global optimum. The method is originally developed for combinatorial problems, but recent investigations in truss design have demonstrated its applicability to sizing optimization problems with discrete (Xiang et al., 2009) and continuous variables (Lamberti, 2008), as well as in mixed sizing, shape, and topology optimization (Hasançebi and Erbatur, 2002). Nevertheless, driving parameters must be set carefully. The tabu search (TS) (Glover, 1989, 1990) is another interesting algorithm that looks at a better solution around the current point while keeping track (using a taboo list) of visited regions in the design space. The algorithm is best suited for local search in combinatorial problems, thus preferably hybridized with more exploratory algorithms in structural design problems (Degertekin et al., 2008).

Using several agents rather than a single one, population-based metaheuristics serve for exploration purposes. By observation of natural systems, the process suggests improving a group by promoting interaction between agents. According to the famous holistic expression "the whole is better than the sum of its parts," natural systems are viewed as a unit. Hence, treating them solely in terms of individuals is meaningless.

Some nature-inspired methods mimic the social behavior of living species. They postulate that better solutions can be found using swarm intelligence (SI) rather than those from a single agent. The particle swarm optimization algorithm imitates the social behavior of birds in a flock. The algorithm shows good performance for large truss optimization problems (Gomes, 2011; Luh and Lin, 2008a). Similarly, (Dorigo and Stützle 2004) first proposed the ant colony optimization that simulates the search for the shortest path to food by ant colonies. Due to the discrete nature of the algorithm, sizing optimization with discrete sections has been investigated (Kaveh and Shojaee, 2007; Luh and Lin, 2008b). Similarly, the bee algorithm is a metaphor of the foraging of bees and has been applied to sizing optimization with continuous design variables (Sonmez, 2011). Also, more recent developments in the field are concerned with bat algorithms (BA) (Yang and Gandomi, 2012), firefly algorithms (FA) (Gandomi et al., 2011a; Yang, 2008), and cuckoo search (CS) algorithms (Gandomi et al., 2011b; Yang and Deb, 2010), all of which have successfully been applied to numerous analytical structural optimization cases, Further investigations would be concerned with large-scale truss design.

The family of EAs—which encompasses genetic algorithms (GA), genetic programming (GP), evolution strategy (ES), and evolutionary programming (EP)—is inspired by the natural evolution of biological species. For a good overview, the interested reader is referred to Bäck et al. (1997). These algorithms are the most popular class of metaheuristics up to now and have been used in many research fields. In particular, GA have been widely applied to truss optimization with respect to sizing, shape, and topology, including mixed variables (Coello Coello et al., 1994). A few publications have been reported regarding ES for truss design because of the continuous nature of the algorithm (Hasançebi, 2007).

Finally, a worthwhile reading would be Hasançebi et al. (2009). It compares several well-known algorithms in the case of continuous sizing optimization of truss structures. It follows from this study that SA and evolutionary strategy work best. However, it is well known that the efficiency of metaheuristics is highly related to the implementation issues. In the following, EAs will be investigated, thanks to the excellent consistency with graph theory, as described in the next section.

10.3 Graph Theory

10.3.1 Basic Terminology

Graph theory is present in a myriad of applications such as biochemistry, electrical engineering, data organization, and operation research (Pirzada and Dharwadker, 2007; Shirinivas et al., 2009). The introduction of graph theory in computational mechanics is mainly due to the consistency with the finite element discretization. Kaveh and coauthors (Kaveh, 2004; Kaveh and Rahami Bondarabady, 2002) investigated the node ordering problem of finite element mesh. They also take advantage of graph theory with the force method in truss topology optimization (Kaveh and Kalatjari, 1999, 2003). In topology optimization of continuous structures, Wang and Tai (2004, 2005) proposed a methodology to generate valid paths through the design domain. Using EAs, Giger and Ermanni (2006) first proposed the graph-based parameterization concept for sizing and topology optimization of truss structures. This groundbreaking approach significantly modified the way of coding an individual: matrix representation is used instead of the conventional chromosome expanded as a vector encoding.

Prior to deeper explanations, we take the opportunity to briefly introduce some basic concepts related to graph theory necessary for the remainder of Section 10.3. A graph is a mathematical abstraction to model pairwise relations between similar objects. Rigorously (Foulds, 1994), a graph $G(V,E)$ is an ordered pair, where $V = \{v_1, \ldots, v_i, \ldots\}$, $i = 1, \ldots, N_n$ is the set of *vertices*, and $E = \{e_1, \ldots, e_e, \ldots\}$, $e = 1, \ldots, N_b$ is the set of unordered pairs of distinct vertices of V. Each element $e_{ij} = (v_i, v_j)$, $i,j = 1, \ldots, N_n$ is called an *edge*. The *size* of $G(V,E)$ is defined by the number of edges while the *order* is defined by the number of vertices. Figure 10.2 gives an example of the graph defined by Eq. (10.12).

$$\begin{aligned} V &= \{v_1, v_2, v_3, v_4, v_5\} \\ E &= \{e_1, e_2, e_3, e_4, e_5, e_6\} \\ &= \{(v_3, v_3), (v_3, v_4), (v_3, v_4), (v_2, v_4), (v_2, v_3), (v_1, v_2)\} \end{aligned} \tag{10.12}$$

This example deserves explanatory comments. Two vertices are said to be *adjacent* if there exists an edge connecting these vertices, e.g., e_2, e_3, e_4, e_5, e_6. Inversely, a vertex is called *isolated* vertex if it is not connected to any edge, e.g., v_5.

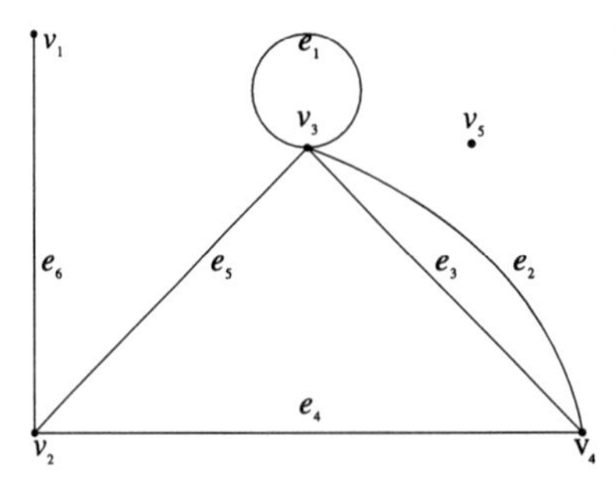

Figure 10.2 Example of graph $G(V,E)$.

A *loop* is an edge that connects a vertex to itself, e.g., e_1. Also two edges are said to be *parallel* when they are connecting the same nodes, e.g., e_2 is parallel to e_3. Note that the example is a *labeled* graph because edges and vertices are distinctly identified by integer numbers. Finally, analogously to colored graph, one can associate properties to each edge through nonnegative *weights*; their sum provides the total weight of the graph. The aforementioned concepts lay the basis for the application to finite elements.

10.3.2 Finite Element Representation

Nowadays, the finite element method for structural analysis is commonly used by scientists and engineers. The mesh discretization of a continuous domain relies on elements interconnected by nodes and characterized by material properties. From this description, they appear well suited for graph representation. Figure 10.3 shows some examples of how graph theory can be applied to represent finite element models. A one-dimensional element with two end-nodes (e.g., a truss) was selected for illustration purposes, but the principle can be extended to higher dimensional finite elements (Giger et al., 2008; Sauter et al., 2008). The representation of the model through graph theory depends on the information considered.

Let us consider Figure 10.3 describing a truss assembly (dash lines with black points) and its associated graphs (solid lines with open points). According to the terminology of Giger et al. (2008), the one-dimensional finite element assembly (Figure 10.3A) can first be abstracted by a *node–node graph* representation, which contains the contiguity information between nodes: the *node vertices* are connected by the so-called *node–node edges* (Figure 10.3B). Alternatively, the structure can be described using *element–element graph* representation. In that case, the connectivity between two *element vertices* is given by the *element–element edge* (Figure 10.3C). Finally, the relation between *node vertices* and *element vertices* is established using *node–element graph* representation (Figure 10.3D). Due to the similarity between truss model and *node–node graph* representation, a *node vertex*

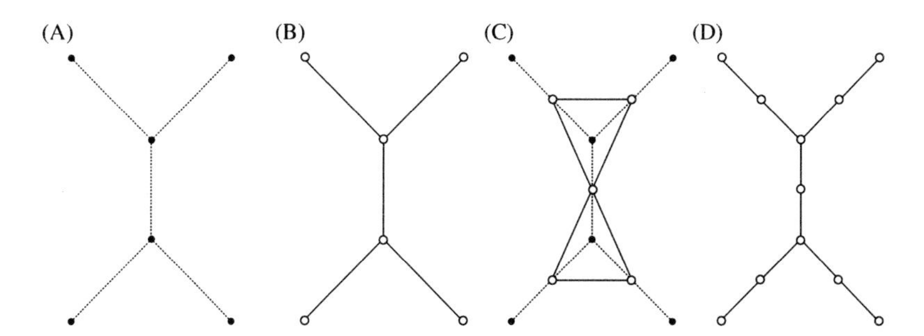

Figure 10.3 Graph abstraction of one-dimensional finite element assembly: (A) truss model, (B) node–node graph, (C) element–element graph, and (D) node–element graph.

and a *node–node edge* of a truss structure will respectively be referred as *vertex* and *edge* in the remainder of this chapter.

10.3.3 Weighted Adjacency Matrix

The information necessary to describe a truss assembly is a unique identifier for each node of the ground structure, the set of connected truss members, as well as the associated structural component sizes, although other member properties can be introduced (Descamps et al., 2010). To synthesize these data within a unified encoding system, let us first define the adjacency notion (Godsil and Royle, 2001).

In graph theory, the *adjacency matrix* **B** of a finite graph $G(V,E)$ is the $N_b \times N_b$ Boolean matrix in which the status $b_{ij} \in \{0,1\}$ is equal to 1 if the vertices v_i and v_j are adjacent, and 0 otherwise. Using real numbers instead of Boolean ones, a weight w_{ij} (for instance, the cross-section properties) can be associated to each edge. Hence, the *weighted adjacency matrix* **W** of a finite graph $G(V,E)$ with $w_{ij} \neq 0$ indicates that the vertices v_i and v_j are adjacent with an edge weight w_{ij}. The case where both vertices are not adjacent is still the same, i.e., $w_{ij} = 0$. Figure 10.4 illustrates the difference between a real-coded vector representation and a matrix encoding for a given design.

In the literature, the weighted adjacency matrix is also called *node matrix encoding* (Su et al., 2009). A close inspection of this matrix reveals symmetry (i.e., $b_{ij} = b_{ji}$ and $w_{ij} = w_{ji}$). The sparsity can be advantageously exploited to minimize memory requirements. Additionally, the size of the weighted adjacency matrix is constant and does not depend on the number of edges, enabling common operations on different individuals, as described in the next section.

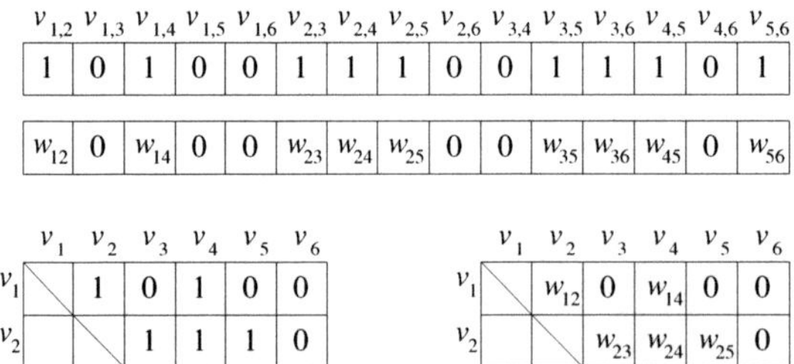

$v_{1,2}$	$v_{1,3}$	$v_{1,4}$	$v_{1,5}$	$v_{1,6}$	$v_{2,3}$	$v_{2,4}$	$v_{2,5}$	$v_{2,6}$	$v_{3,4}$	$v_{3,5}$	$v_{3,6}$	$v_{4,5}$	$v_{4,6}$	$v_{5,6}$
1	0	1	0	0	1	1	1	0	0	1	1	1	0	1
w_{12}	0	w_{14}	0	0	w_{23}	w_{24}	w_{25}	0	0	w_{35}	w_{36}	w_{45}	0	w_{56}

	v_1	v_2	v_3	v_4	v_5	v_6
v_1		1	0	1	0	0
v_2			1	1	1	0
v_3				0	1	1
v_4					1	0
v_5						1
v_6						

	v_1	v_2	v_3	v_4	v_5	v_6
v_1		w_{12}	0	w_{14}	0	0
v_2			w_{23}	w_{24}	w_{25}	0
v_3				0	w_{35}	w_{36}
v_4					w_{45}	0
v_5						w_{56}
v_6						

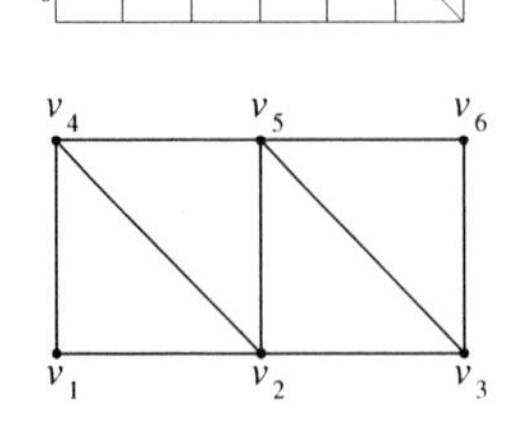

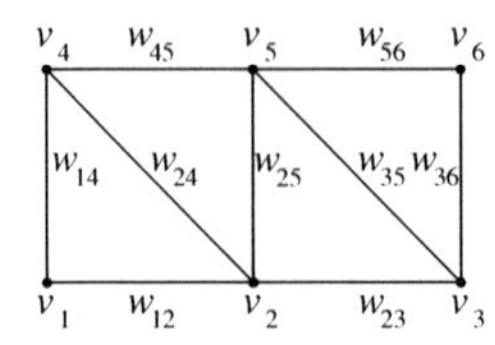

Figure 10.4 Example of matrix encoding for a truss assembly—on the top, the conventional real-coded vector encoding; on the left, the adjacency matrix representation; on the right, the weighted adjacency matrix.

10.4 Evolutionary Algorithm

10.4.1 Outline

EAs are based on Darwin's postulate about evolution of biological species. The underlying principle is an iterative process, based on nature-inspired rules, that converges toward good solutions. The principle of the survival of the fittest introduces a notion of competition among individuals. Common operations are reproduction, mutation, crossover, and selection. The difference in the different classes of EAs lies in their implementation procedures, with a more or less emphasis to the aforementioned operators. Actual trends break down the barriers among the different subclasses. Although the algorithm described here is close to GA, it also borrows some features from the evolution strategies. For this reason, it is wiser to refer to them as EAs without further distinction. These algorithms are also applicable to multimodal, multiobjective, and multiconstraint optimization problems.

The procedure can be summarized as follows (Talbi, 2009; Yang, 2010). An initial population of individuals is randomly selected. Each individual is an encoded version (the *genotype*) of a trial design. The fitness function associates a score to each individual in order to describe its adaptability to survive in its environment, which corresponds to a low-objective function value and admissible constraints (in a single-objective optimization context). In each generation, the fitness of individuals is evaluated, and the most adapted ones are endowed with a higher probability to be selected as parents for the next generation. A reproduction scheme—crossover and mutation—generates new offsprings. Then, the replacement determines which individuals will survive in the next generation. This process is repeated again until the stopping criterion is satisfied. The main steps of an EA are depicted in Figure 10.5. Hereafter, every operator is explained with an emphasis on graph specificities.

10.4.2 Representation

Based traditionally on chromosome representation (i.e., a concatenated vector), the method for storing genetic information of individuals is revisited hereafter. Using the weighted adjacency matrix, we have one of the most compact formalism to synthesize both sizing and topology information, in line with the truss assembly model. The selection of the numeral system is subject to a long and tense debate among the proponents of binary versus real-coding. Although the binary alphabet is the simplest one, real-coding provides a simple way of dealing with the actual numbers.

Goldberg (1991) lays the theoretical background that justifies the empirical success of real-coded genetic representation (Dede et al., 2011; Tang et al., 2005). Nonetheless, it also predicts possible blocking in finding the global optimum and encourages the development of operators to circumvent this effect. Indeed, Wright (1991) pointed out that binary crossover can be viewed as a real-coded crossover

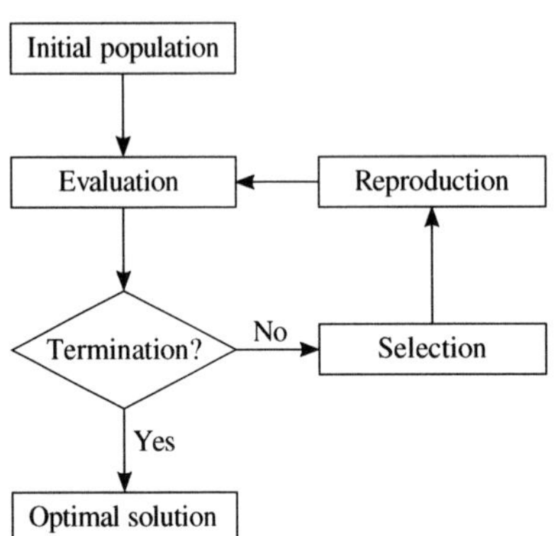

Figure 10.5 General flowchart of the proposed EA.

plus a mutation (see Sections 10.4.7, 10.4.8). Thus, the probability of occurrence of the mutation operator should be increased as well to maintain diversity.

Starting from real-coding, if cross sections should be selected among a discrete set of options $S = \{s_1, \ldots, s_k, \ldots\}$, $k = 1, \ldots, N_s$, the mapping between the continuous edge weights w_{ij} and the discrete cross sections s_k is simply created by a rounding up operation as follows:

$$s_k = \lceil w_{ij} \rceil = \min\{s_k \in \mathbb{Z} | w_{ij} \leq s_k, 0 \leq w_{ij} \leq N_s\}, \quad k = 1, \ldots, N_s \tag{10.13}$$

10.4.3 Initial Population

The evolutionary process starts from a population of individuals generated randomly. This first step is designed on purpose to widely explore the search space. Several possible strategies of population initialization are available in the literature, namely random generation, parallel diversification, sequential diversification, or heuristic initialization. The random generation is generally sufficient to maintain a certain level of diversity. To create a new individual, i.e., a genotype representing the whole truss structure, the adjacency matrix b_{ij} is randomly generated. Then, a weight w_{ij} is assigned to the edge e_{ij} by:

$$w_{ij} = w^l + \text{rand}[0, 1]_{ij}(w^u - w^l), \quad i, j = 1, \ldots, N_n \tag{10.14}$$

10.4.4 Kinematic Stability

Truss topology optimization is a tricky problem. A typical bottleneck of the random generation is the presence of mechanisms, i.e., the structure is subject to rigid body motions. Therefore, the direct transfer of unstable structures from the EA to the finite element analysis leads to serious numerical difficulties if no preventive measure is taken.

In order to detect unstable configurations, some works (Deb and Gulati, 2001; Hajela and Lee, 1994) checked the condition number of the stiffness matrix by computing its eigenvalues. If the condition number is close to one, the matrix is well conditioned and the structure is stable. In contrast, a value close to zero indicates the presence of singularities. In the latter case, the fitness of the unstable configuration is highly penalized, making the individual unfit to its environment, and thus ineligible in the selection procedure. However, the probability of occurrence of this phenomenon is dramatically increased in large-scale problems, leading to unaffordable computational time.

Kawamura et al. (2002) proposed an interesting strategy to generate stable configurations. The process starts with a stable triangle (2D) or tetrahedron (3D). Progressively, the structure grows up around this kernel. It results in stable triangulated structures. Nonetheless, the selection of predefined modules can hinder the power of EAs to widely explore the design space for innovative solutions. Hence, Richardson et al. (2012) advocates the application of this strategy for a certain proportion of the population only.

Alternatively, one can make use of the well-known necessary, yet not sufficient, Maxwell's criterion to ensure that, at least, there exists one solution to the static equilibrium problem:

$$N_b + N_s - d \cdot N_n \geq 0 \tag{10.15}$$

where, once again, N_b is the number of members, N_s the number of support reactions, d the spatial dimension (2D/3D), and N_n the number of nodes. Consequently, each individual is regenerated until this criterion is enforced. Obviously, this approach is not sufficient to ensure the stability of any configuration, but it provides a fast tool to tend to stable assembly.

Using graph theory, we are able to detect numerous invalid configurations. The causes of singularities are of different natures. First, some of them are related to invalid adjacency graphs (i.e., invalid topologies), thus independent from any geometrical consideration. The reason is the disconnection or insufficient connections of graph objects such as:

- subgraphs (Figure 10.6A),
- loaded vertex (Figure 10.6B),
- supported vertex (Figure 10.6C),
- needless edge (Figure 10.6D).

Since the presence of supported and loaded vertices is mandatory, they are defined as *basic* vertices. In theory, the graph object is said to be *connected* if there is a path such that any vertex is accessible from another vertex. A simple way to detect disconnected graph is to count the number of incident edges on a vertex. If $e_{i\bullet}$ denotes all edges associated to the vertex i, their number $N_{i\bullet}$ is simply given by:

$$N_{i\bullet} = \sum_j (b_{ij} + b_{ji}) \tag{10.16}$$

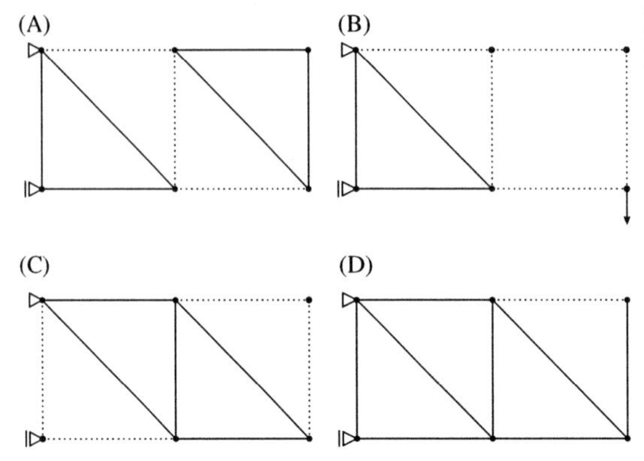

Figure 10.6 Example of the two-point crossover for matrix encoding: (A) subgraph, (B) loaded vertex, (C) supported vertex, and (D) needless edge.

Additionally, geometrical considerations, such as collinearity and coplanarity, bring out other types of singularities which cannot be detected only by graph theory. In 2D, two truss elements are collinear if condition $c_{2D} = 0$ holds, with c_{2D} being the area formed by their three nodes (denoted 1, 2, 3) and computed by:

$$c_{2D} = \begin{vmatrix} x_1 & y_1 & 1 \\ x_2 & y_2 & 1 \\ x_3 & y_3 & 1 \end{vmatrix} \tag{10.17}$$

In 3D, three truss elements are coplanar if condition $c_{3D} = 0$ holds, with c_{3D} being the volume formed by their four nodes (denoted 1, 2, 3, 4) and computed by:

$$c_{3D} = \begin{vmatrix} x_1 & y_1 & z_1 & 1 \\ x_2 & y_2 & z_2 & 1 \\ x_3 & y_3 & z_3 & 1 \\ x_4 & y_4 & z_4 & 1 \end{vmatrix} \tag{10.18}$$

Therefore, the graph theory combined with collinearity (or coplanarity) checks provides a clear view of stability. The subsequent repair procedure consists of either adding or removing relevant edges according to a given probability. If nearly singular configurations arise, the matrix is still invertible but leads to poor accuracy. The results of linear finite element analyses are inaccurate; large values of displacements and stresses are obtained that lead to infeasible individuals. In summary, the sequential procedure is as follows:

a. generate truss assembly randomly until Maxwell's rule is enforced,
b. check if all basic vertices are included in the structure; add edge if needed,
c. check for unstable configuration using graph-based heuristic rules and repair if needed,
d. a nearly singular configuration simply makes an individual design infeasible.

10.4.5 Evaluation

After setting the generation problem for any trial truss configuration, the next step lies in the assessment of the objective function and constraints. Each individual is evaluated on the basis of the following optimization problem:

$$\min_{\substack{b_{ij}\in\{0,1\}\\ w_{ij}\in\mathbb{R}}} \{f(b_{ij},w_{ij})\,|g_m(b_{ij},w_{ij})\le 0, h_n(b_{ij},w_{ij})=0, 0\le w_{ij}\le N_s\} \tag{10.19}$$

which for mass minimization takes the form

$$\min_{\substack{b_{ij}\in\{0,1\}\\ w_{ij}\in\mathbb{R}}} \left\{\sum_{i=1}^{N_v}\sum_{j=1}^{N_v} b_{ij}\rho_{ij}l_{ij}a(w_{ij})\left|\frac{|\sigma_{ij}|}{\sigma_{\max}}-1\le 0, \frac{|u_i|}{u_{\max}}\right.-1\le 0, \frac{\sigma_{ij}^{-}}{\sigma_{\text{crit}}}-1\le 0, 0\le w_{ij}\le N_s\right\} \tag{10.20}$$

To implement this problem, a score is assigned to every individual according to the fitness function. For unconstrained optimization, the fitness function can simply be equivalent to the objective function. However, in the presence of constraints, more sophisticated techniques are mandatory. Coello Coello (1999) classified constraint-handling methods as follows: direct approach (or death penalty), penalization methods, methods based on special individual representation, repair methods, methods based on separation of objective and constraints, and hybrid methods.

In the direct approach, individuals that violate constraints are directly discarded and replaced by new individuals. However, this straightforward process might reject potentially high-performance individuals which slightly violate constraints, or which could bring useful genetic information for the future generations. That results in nonoptimal solutions, especially because those optimal points are very often located at the boundary of one or intersection of two or more constraints in the design space.

Among the remaining constraint-handling methods, the penalty methods are fairly general, easy-to-implement, and sufficient for application to a wide range of problems. Penalization consists of degrading the fitness of an individual design according to the level of constraint violation, thus making infeasible individuals able to compete with feasible ones. Practically, the fitness Φ is obtained by penalizing the objective function f with the sum of constraint violations (in absolute value) multiplied by large penalty vectors μ_m, ν_n such that

$$\Phi(b_{ij},w_{ij}) = f(b_{ij},w_{ij}) + \sum_{m=1}^{N_g}\mu_m \max[0, g_m(b_{ij},w_{ij})] + \sum_{n=1}^{N_h}\nu_n \max[0, |h_n(b_{ij},w_{ij})| - \xi] \tag{10.21}$$

In Eq. (10.21), the relaxation term ξ is a very small positive value that makes slightly violated equality constraints feasible by transforming the equality constraints into inequalities. Note that the penalty term can be constant, dynamic, or adaptive (Toğan and Daloğlu, 2006).

10.4.6 Selection

According to the principle of survival of the fittest, a proportionate random selection over individuals is performed. High-scoring individuals have a higher probability to be selected as parent for the next generation with the aim of moving toward better solutions.

Different selection strategies have been proposed in the literature, such as the roulette wheel selection, the Boltzmann selection, the tournament selection, the rank-based selection, or the steady-state selection. The tournament selection is adopted here because of its insensitivity to scaling effect, its fastness, and its ease of implementation.

In tournament selection, a prescribed number of individuals are randomly selected from the population and the best among the subset are deterministically chosen as candidates for the next generation. Therefore, the selection pressure is driven by the number of selected individuals, which draws the trade-off between intensification and diversification of search. Empirical evidence suggests speeding up the optimization process by prescribing large values, but at the risk of premature convergence. In contrast, the lower the number of selected individuals, the greater the diversity, at the expense of slower convergence rate.

10.4.7 Crossover

While the previous genetic operators are common to most EA implementations, the crossover techniques described in this chapter are directly derived from the truss coding based on a graph theory approach. In the reproduction phase, the crossover is performed according to a high probability: two parents are paired up, their genotypes are crossed over, and a new offspring is generated with inherited characteristics. The crossover operations mainly depend on the genotype. A distinction has to be made between the adjacency crossover operators and the weight crossover operators.

First, let us consider the adjacency crossover operators. Four operators are considered:

1. *One-point crossover*: This operator acts like the single-point crossover in binary chromosome representation. A split point is randomly chosen on the diagonal. Then the matrix is divided to four submatrices and the rightmost lower submatrices of both parents are exchanged.
2. *Two-point crossover*: Close to the one-point crossover, a second split point is chosen on the diagonal so that the block is swapped with that of the other parent.
3. *Uniform crossover*: This third crossover operator is an extreme case. For each edge, the crossover is performed with a given mixing probability of 0.5. Unlike previous methods,

this crossover shows poor performance by constant remixing of the individuals. Nevertheless, this crossover type is sometimes interesting for exploratory purposes.

4. *Vertex crossover*: This crossover operator switches the incident edges of a random vertex with that of the second parent. In the adjacency matrices, this means that the row and column of the node are exchanged.

In our numerical experiments, the two-point crossover leads to the best results. In Figure 10.7, one can observe the excellent consistency between the matrix encoding and the structural model.

Regarding the weight, the crossover of two selected parents is performed by combining their edge weight in order to generate two children with inherent characteristics. Let $w_{ij,1}$ and $w_{ij,2}$ be the weight edges of the first and the second parent,

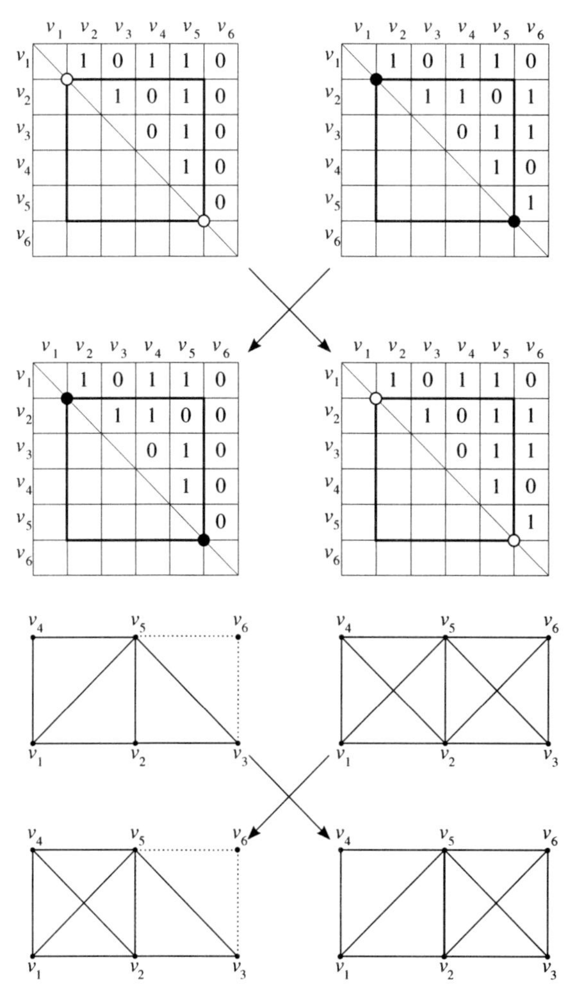

Figure 10.7 Example of the flipping edge mutation for matrix encoding.

respectively. Similarly, $w'_{ij,1}$ and $w'_{ij,2}$ are the weight edges of each child and given by:

$$\begin{aligned} w'_{ij,1} &= rw_{ij,1} + (1-r)w_{ij,2} \\ w'_{ij,2} &= rw_{ij,2} + (1-r)w_{ij,1} \end{aligned} \tag{10.22}$$

where $r = \text{rand}[0,1]_{ij}$. In the segment-type operator, r is chosen once and for all, while the hypercube-type operator generates a new operator by edge.

10.4.8 Mutation

The other reproduction mechanism is the mutation operator performed with a low probability. Among living species, individuals may accidentally mutate over time, which can provide new genetic material potentially better suited to the environment. The aim of the mutation operator is thus to explore the neighborhood of current solutions, or to roam the undiscovered regions of the search space. The different adjacency mutation operators are:

- *Flipping edge*: This operator switches the adjacency of an edge by changing vertex with another randomly determined vertex. The operation slightly modifies the individuals. Hence, it is well suited when applied to high-scoring individuals.
- *Connection/disconnection of vertex*: An interesting operator against oversized individuals is the disconnection of a vertex and its incidents edges. Inversely, weakly connected individuals are enhanced by the inverse operation.
- *Switching edges*: Similar to the random edge crossover, a Boolean matrix similar to the adjacent matrix is randomly generated. For each edge, if the corresponding value is one, the edge is removed or added according to its adjacency status. Again, this operator must be used with caution since it might break up the evolutionary process.

In our numerical experiments, the flipping edge operator leads to the best results (Figure 10.8).

The weight mutation operator consists of applying a small perturbation ζ on the edge weight w_{ij} of the parent to obtain the edge weight of the child w'_{ij}:

$$w'_{ij} = w_{ij} + \zeta \tag{10.23}$$

where ζ is a random number that follows a Gaussian distribution of mean 0 and standard deviation σ_G.

10.4.9 Replacement

The replacement mechanism is related to the way of selecting individuals for the next generation. EAs based on genetic principles have recourse to a deterministic approach that automatically replaces all parents by children.

In order to keep track of the best solutions found so far, the highest scoring individual (or a very few top scorers) is (are) automatically stored in separate slot(s)

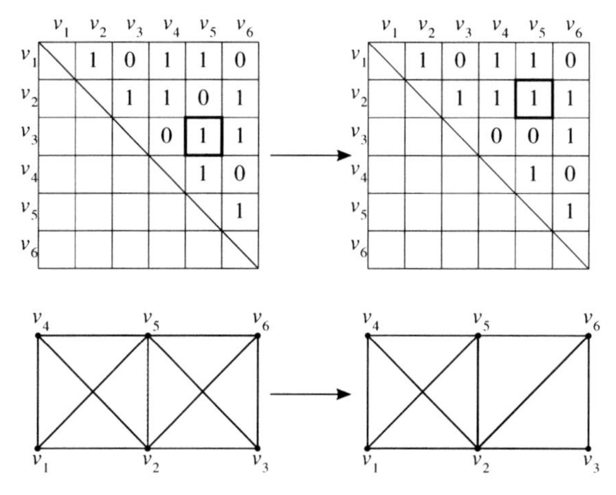

Figure 10.8 Example of topological singularities due to insufficient.

for the next generation, without allowing any reproduction operators on them. This procedure, called elitism, avoids the loss of best individuals over generations.

10.5 Application

The previous sections lay the foundations of graph-based EAs for truss design. Three numerical applications of simultaneous sizing and topology optimization are now presented to illustrate the benefits of this truss-oriented metaheuristic approach.

To select cross sections, Table 10.1 extracts data from a catalog of circular tubes.

In the examples, the same structural properties are assigned (Table 10.2).

10.5.1 Free-Form Tower

The first example is on the design of a small free-form tower under compressive loads. Albeit of small scale, the design by hand is made difficult because of the complex 3D shape and asymmetry. The 12 m high ground structure consisting of 36 members carries a vertical loading of 10 kN on the three top nodes. Due to the slenderness of truss members coupled with a predominant compressive state, the buckling is of prominent issue. Accordingly, results for two different cases are considered and compared: either with buckling constraints or without. Contrariwise, displacement restrictions are inconsequential. In Figure 10.9A, the ground structure contains three triangles of topologically fixed members (colored in black) that preserve the enclosure while the other members (colored in gray) are allowed to vary.

Regarding the evolutionary parameters, the initial pool is populated by 250 individuals, reproduced over a maximum of 200 generations. The crossover and mutation rates are set to 0.8 and 0.2, respectively. By tournament selection, four

Table 10.1 Discrete Set of Cross Sections

Identifier	Diameter (mm)	Area (cm^2)	Moment of Inertia (cm^4)
1	3.37	2.81	3.36
2	4.24	3.25	6.46
3	4.83	3.60	9.50
4	6.03	5.23	21.60
5	7.61	6.67	44.70
6	8.89	8.09	74.73
7	10.16	9.29	112.98
8	11.43	10.48	162.47
9	13.30	16.20	338.00
10	13.97	17.04	392.66
11	15.24	18.60	514.00
12	16.83	20.04	696.74
13	19.37	23.83	1,072.25
14	21.91	27.02	1,563.04
15	27.30	42.08	3,778.90
16	32.39	59.89	7,568.63
17	35.56	65.86	10,065.45
18	40.64	75.44	15,120.66
19	45.72	89.20	22,672.83
20	50.80	99.25	31,230.65

Table 10.2 Material Properties for Both Examples

Material Property	Units	Value
Elastic modulus, E	N/m^2	210×10^9
Mass density, ρ	kg/m^3	785×10^1
Allowable stress, σ_{max}	N/m^2	235×10^6
Maximum deflection, u_{max}	m	3×10^{-2}

individuals are picked. Knowing that the average mass is about half a ton and the normalized constraints are around unity, the constant penalty term for nonlinear inequality constraints is set to 10^3 in order to increase the effect of the penalty term.

As expected, both optimal solutions are outwardly different: the case without buckling constraints (Figure 10.9B) is quite sparse with 29 members while the configuration subject to buckling constraints (Figure 10.9C) is more densely connected with 32 members. By taking instabilities into account, the optimal mass rises from 246 to 338 kg, hence 37% heavier (Figure 10.10).

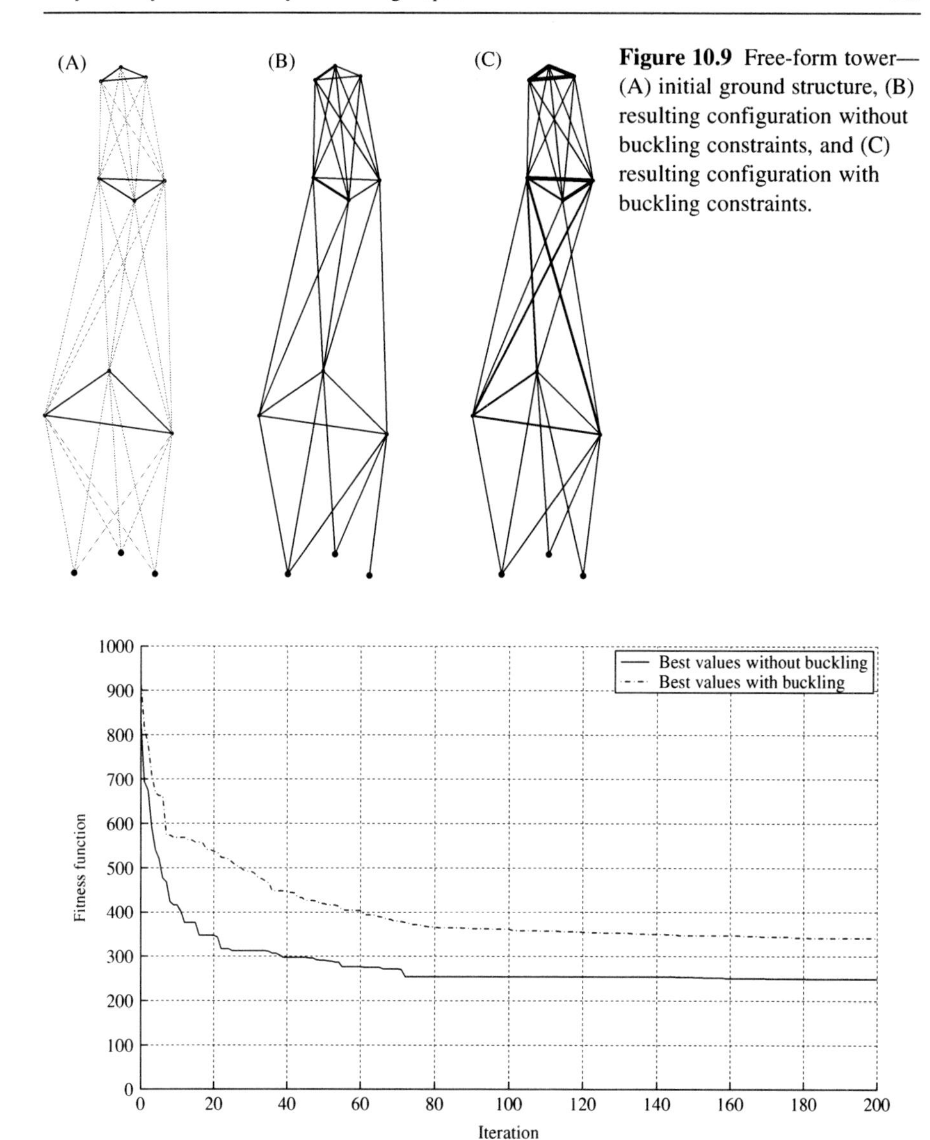

Figure 10.9 Free-form tower—(A) initial ground structure, (B) resulting configuration without buckling constraints, and (C) resulting configuration with buckling constraints.

Figure 10.10 Convergence curves for the free-form tower optimization according to buckling considerations.

10.5.2 Bridge Structure

The second application consists of designing a bridge truss structure. The span of 15 m is regularly spaced with one-by-one meter modules, and supported at both extremities. The 47-node structure is subject to a downward vertical load of 5 kN applied to each node on top. The ground structure consists of 171 members.

The equivalent bisupported truss indicates that displacement constraints are almost active at the optimum (not exactly because of discrete variables). This suggests comparing both cases: with and without vertical displacement limits of 0.05 m at each node.

To parameterize this larger problem, it is acceptable to assume that the three main truss segments are automatically present in the final solution (black members in Figure 10.11A), for which the geometry is defined according to static principles. This assumption avoids many unnecessary repairs of infeasible individuals due to disconnection of loaded nodes. Only bracing members are topologically varied (gray members in Figure 10.11A). As long members lead to oversized and inefficient sizing, it is reasonable to restrict the adjacency scheme to a star graph: potential connections are restricted to the first order of vicinity, i.e., the closest nodes. Consequently, buckling constraints are not critical. Note that the 124 potential members involve $2^{124} \approx 2.10^{37}$ possible adjacency matrices.

To parameterize this larger problem, the initial pool is populated by 800 individuals, reproduced over 500 generations. The crossover and mutation rates are set to 0.9 and 0.1, respectively. By tournament selection, only 5 individuals are picked in order to avoid premature convergence. With an average mass of about one ton, the penalty term is set to 10^4.

Displacement-based design leads to stiffer configurations, which—in a sizing and topology optimization context—means more material. These ascertainments must be moderated because self-weight is not considered. Nevertheless, the fact remains that the solution process gives a mass of 957 kg without displacement

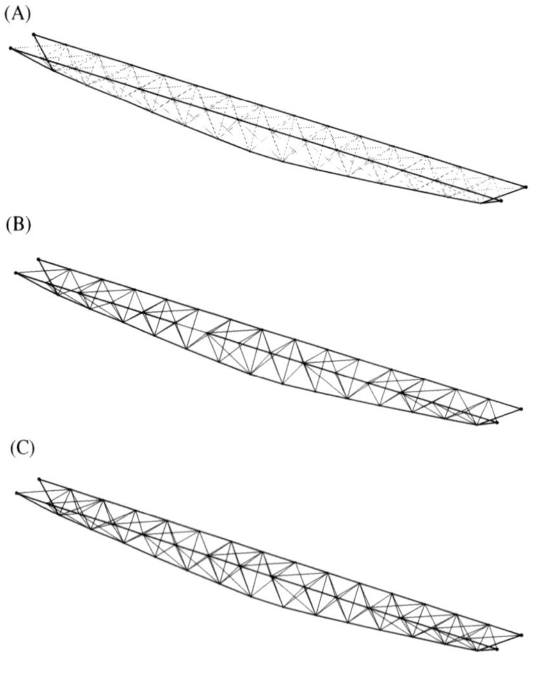

Figure 10.11 Bridge structure—(A) initial ground structure, (B) resulting configuration without displacement constraints, and (C) resulting configuration with displacement constraints.

constraints (Figure 10.11B), and 1302 kg with displacement constraints (Figure 10.11C). The former contains 134 members, whereas the latter includes 156 members.

The convergence curves of the fitness function for the best and average values are depicted in Figure 10.12. The algorithm shows a fast improvement during the first 100 iterations, mainly due to the search for an efficient topology. The fine-tuning is mainly concerned with small sizing and topology enhancements during the remainders. To speed up the optimization process, a local search could be considered, as in memetic algorithms (this is beyond the scope of this chapter).

10.5.3 Double-Layer Truss Grid

The last example is concerned with the design of a double-layer truss grid (Figure 10.13). The plan is based on a 10 m square grid, regularly spaced with two-by-two meter modules, and attached to the perimeter where displacements are blocked in the three spatial dimensions. The 61-node structure is subject to a downward vertical load of 5 kN applied to each node on the lower layer. While we assume that the topology of the lower layer members is kept fixed (colored in black), we compare now the reduced problem where the topology on the upper layer is fixed (Figure 10.13A), and the larger problem that may change them as well (Figure 10.13B). Again, the intermediate members (color in gray) are allowed to vary. This implies 98 or 138 potential members accordingly.

The parameters of the EA are similar to the previous example, except for the number of individuals; 800 for the reduced problem, 1200 for the larger problem. Although computationally more expensive, the solution for the larger problem

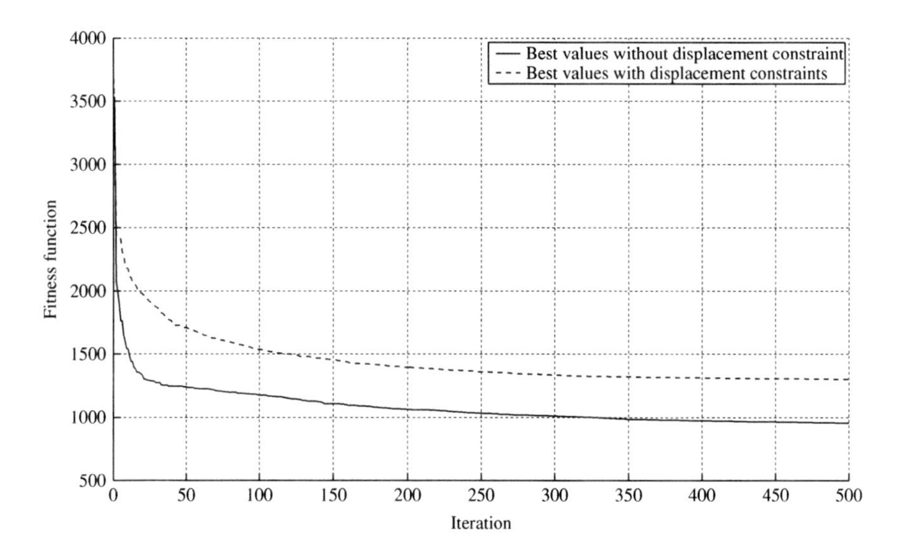

Figure 10.12 Convergence curves for the bridge structure according to displacement considerations.

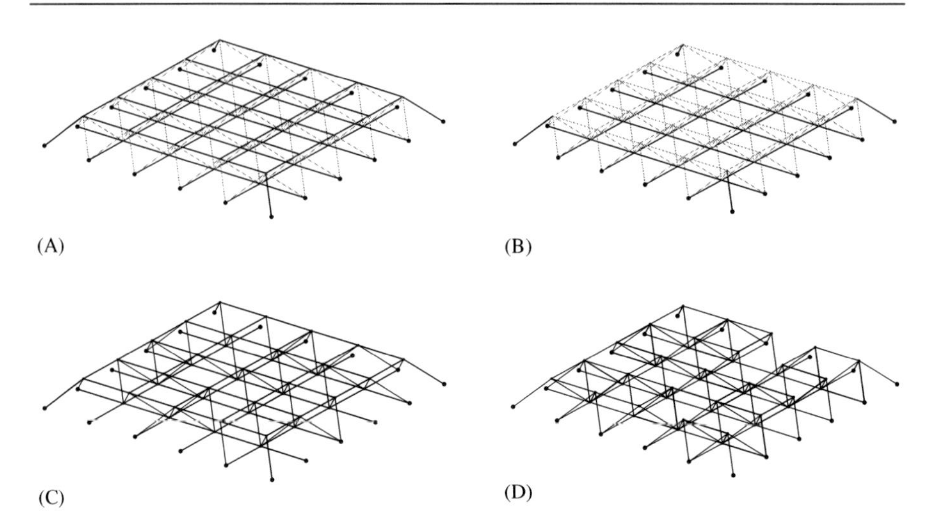

Figure 10.13 Double-layer truss grid—(A) initial ground structure for the reduced problem, (B) initial ground structure for the larger problem, (C) resulting configuration for the reduced problem, and (D) resulting configuration for the larger problem.

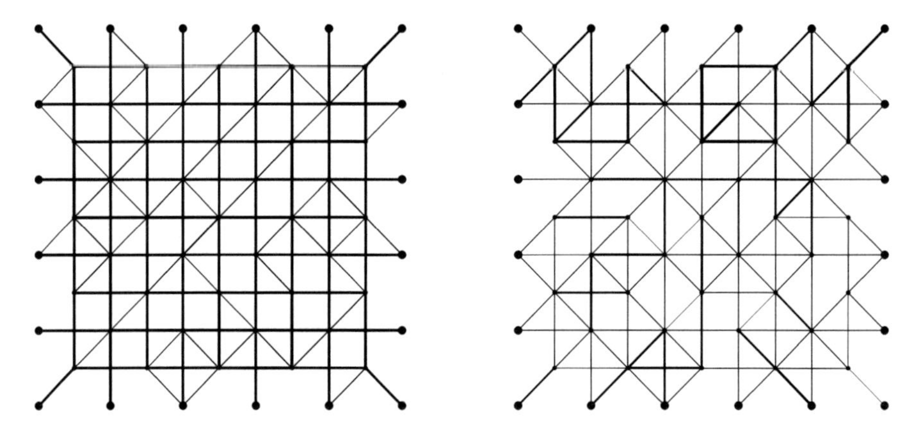

Figure 10.14 Comparison of solutions for the reduced (left) and larger (right) problem in plan.

(Figure 10.13D) results in a lighter solution; 1230 kg versus 1561 kg for the reduced problem (Figure 10.13C). Both are also compared in plan (Figure 10.14). Note that the stability for the larger problem is ensured because horizontal displacements on the perimeter are blocked.

The convergence curves of the fitness function for the best and average values are depicted in Figure 10.15. It is important to keep in mind that although quite

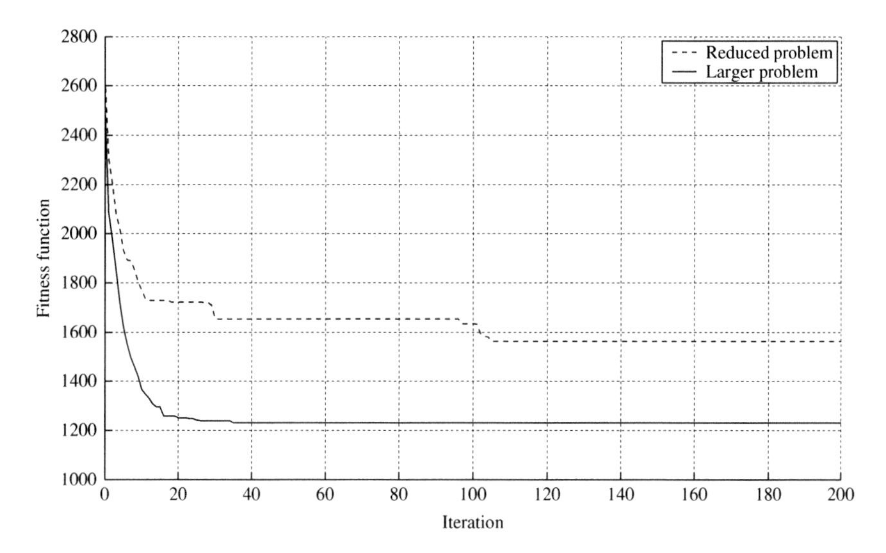

Figure 10.15 Convergence curves for the double-layer truss grid optimization for different topology parameterizations.

fast, the process converges to a point which is not the global optimum. Hence, there are still some rooms for improvement.

10.6 Conclusions

The aim of this chapter is to provide a unified framework to bridge the gap between graph theory, EAs, and truss optimization. The selection of EAs for sizing and topology optimization of truss structures is motivated by their ability to concurrently address operations on discrete and binary variablcs in possible nonconvex design space. In this context, graph abstraction aids in stating both design encoding and structure representation in a consistent way, so that the subsequent evolutionary operators are naturally carried out. By providing a clear picture of truss topological properties, it also highlights structural deficiencies and serves as a sound basis for the repair method. Several applications illustrate their versatility in formulating various structural design problems. It follows from this study that the transfer of methodological ideas can be further extended to higher-order finite elements, as indicated by recent investigations.

References

Bäck, T., Fogel, D.B., Michalewicz, Z. (Eds.), 1997. Handbook of Evolutionary Computation. Oxford University Press and IOP Publishing Ltd., Bristol, UK.

Bendsøe, M., Sigmund, O. (Eds.), 2003. Topology optimization: Theory, Methods and Applications. second ed. Springer, Berlin, Heidelberg, New York.
Breitkopf, P., Filomeno Coelho, R. (Eds.), 2010. Multidisciplinary Design Optimization in Computational Mechanics. ISTE/John Wiley & Sons, Chippenham.
Černý, V., 1985. Thermodynamical approach to the traveling salesman problem: an efficient simulation algorithm. J. Optim. Theory Appl. 45, 41–51.
Coello Coello, C., 1999. A Survey of Constraint Handling Techniques Used with Evolutionary Algorithms. Laboratorio Nacional de Informática Avanzada (Technical Report Lania-RI-99-04), Mexico City, Mexico.
Coello Coello, C.A., Rudnick, M., Christiansen, A.D., 1994. Using genetic algorithms for optimal design of trusses. In: Sixth International Conference on Tools with Artificial Intelligence, Los Angeles, USA.
Deb, K., Gulati, S., 2001. Design of truss-structures for minimum weight using genetic algorithms. Finite Elem. Anal. Des. 37, 447–465.
Dede, T., Bekiroğlu, S., Ayvaz, Y., 2011. Weight minimization of trusses with genetic algorithm. Appl. Soft Comput. 11, 2565–2575.
Degertekin, S.O., Saka, M.P., Hayalioglu, M.S., 2008. Optimal load and resistance factor design of geometrically nonlinear steel space frames via tabu search and genetic algorithm. Eng. Struct. 30, 197–205.
Descamps, B., Bouillard, Ph., Ney, L., Filomeno Coelho, R., 2010. Shape and topology optimization by the force density method and hybrid genetic algorithms. In: Proceedings of the International Association for Shell and Spatial Structures (IASS) Symposium 2010, Shanghai, China.
Descamps, B., Filomeno Coelho, R., Ney, L., Bouillard, Ph., 2011. Multicriteria optimization of lightweight bridge structures with a constrained force density method. Comput. Struct. 89, 277–284.
Dorn, W.S., Gommory, R.E., Greenberg, H.J., 1964. Automatic design of optimal structures. J. Mécanique. 3, 25–52.
Dorigo, M., Stützle, T. (Eds.), 2004. Ant Colony Optimization. MIT Press, Cambridge, USA.
Foulds, L.R. (Ed.), 1994. Graph Theory Applications. Springer, Berlin, Heidelberg, New York.
Gandomi, A.H., Yang, X.-S., Alavi, A.H., 2011a. Mixed variable structural optimization using firefly algorithm. Comput. Struct. 89, 2325–2336.
Gandomi, A.H., Yang, X.-S., Alavi, A.H., 2011b. Cuckoo Search Algorithm: A Metaheuristic Approach to Solve Structural Optimization Problems. Eng. Comput. doi: 10.1007/s00366-011-0241-y.
Giger, M., Ermanni, P., 2006. Evolutionary truss topology optimization using a graph-based parameterization concept. Struct. Multidiscip. Optim. 32, 313–326.
Giger, M., Keller, D., Ermanni, P., 2008. A graph-based parameterization concept for global laminate optimization. Struct. Multidiscip. Optim. 36, 289–305.
Glover, F., 1989. Tabu search—Part 1. ORSA J. Comput. 1, 190–206.
Glover, F., 1990. Tabu search—Part 2. ORSA J. Comput. 2, 4–32.
Godsil, C., Royle, G. (Eds.), 2001. Algebraic Graph Theory. Springer, Berlin, Heidelberg, New York.
Goldberg, D., (Ed.), 1989. Genetic Algorithms in Search, Optimization and Machine Learning. Addison-Wesley Longman, Boston, MA.
Goldberg, D., 1991. Real-coded genetic algorithms, virtual alphabet, and blocking. Complex Syst. 5, 139–167.

Gomes, H.M., 2011. Truss optimization with dynamic constraints using particle swarm algorithm. Expert Syst. Appl. 38, 957–968.

Hajela, P., Lee, E., 1994. Genetic algorithm in truss topological optimization. Int. J. Solids Struct. 32, 3341–3357.

Hasançebi, O., 2007. Optimization of truss bridges within a specified design domain using evolution strategies. Eng. Optim. 39, 737–756.

Hasançebi, O., Erbatur, F., 2002. Layout optimisation of trusses using simulated annealing. Adv. Eng. Softw. 33, 681–696.

Hasançebi, O., Çarbaş, S., Doğan, E., Erdal, F., Saka, M.P., 2009. Performance evaluation of metaheuristic search techniques in the optimum design of real size pin jointed structures. Comput. Struct. 87, 284–302.

Holland, J. (Ed.), 1975. Adaptation in Natural and Artificial Systems. University of Michigan Press, Ann Arbor, MI.

Kaveh, A. (Ed.), 2004. Structural Mechanics: Graph and Matrix Methods. 3ed. Research Studies Press, Baldock.

Kaveh, A., Kalatjari, V., 1999. Topology optimization of trusses using the graph theory. In: First Conference of the Iranian Society of Civil Engineers, Tehran, Iran.

Kaveh, A., Kalatjari, V., 2003. Topology optimization of trusses using genetic algorithm, force method and graph theory. Int. J. Numer. Methods Eng. 58, 771–791.

Kaveh, A., Rahami Bondarabady, H.A., 2002. A multi-level finite element nodal ordering using algebraic theory. Finite Elem. Anal. Des. 38, 245–261.

Kaveh, A., Shojaee, S., 2007. Optimal design of skeletal structures using ant colony optimization. Int. J. Numer. Methods Eng. 70, 563–581.

Kawamura, H., Ohmori, H., Kito, N., 2002. Truss topology optimization by a modified genetic algorithm. Struct. Multidiscip. Optim. 23, 467–472.

Kirkpatrick, S., Gelatt, C.D., Vecchi, M.P., 1983. Optimization by simulated annealing. Science. 220, 671–680.

Kirsch, U., Rozvany, G.I.N., 1994. Alternative formulations of structural optimization. Struct. Optim. 7, 32–41.

Lamberti, L., 2008. An efficient simulated annealing for design optimization of truss structures. Comput. Struct. 86, 1936–1953.

Luh, G.-C., Lin, C.-Y., 2008a. Optimal design of truss-structures using particle swarm optimization. Comput. Struct. 89, 2221–2232.

Luh, G.-C., Lin, C.-Y., 2008b. Optimal design of truss-structures using ant colony optimization. Struct. Multidiscip. Optim. 36, 365–379.

Pellegrino, S., 1993. Structural analysis by the singular value decomposition of the equilibrium matrix. Int. J. Solids Struct. 30, 3025–3035.

Petersson, J., 2001. On the continuity of the design-to-state mappings for trusses with variable topology. Int. J. Eng. Sci. 39, 1119–1141.

Pirzada, S., Dharwadker, A., 2007. Applications of graph theory. J. Korean Soc. Ind. Appl. Math. 11, 19–38.

Richardson, J.N., Adriaenssens, S., Bouillard, Ph., Filomeno Coelho, R., 2012. Multiobjective topology optimization of truss structures with kinematic stability repair. Struct. Multidiscip. Optim. 46, 513–532.

Sauter, M., Kress, G., Giger, M., Ermanni, P., 2008. Complex-shaped beam element and graph-based optimization of compliant mechanisms. Struct. Multidiscip. Optim. 36, 429–442.

Shirinivas, S.G., Vetrivel, S., Elango, N.M., 2009. Application of graph theory in computer science: an overview. Int. J. Eng. Sci. Technol. 2, 4610–4621.

Sonmez, M., 2011. Artificial bee colony algorithm for optimization of truss structures. Appl. Soft Comput. 11, 2406–2418.

Su, R.Y., Gui, L.J., Fan, Z., 2009. Topology and sizing optimization of truss structures using adaptive genetic algorithm with node matrix encoding. In: The Fifth International Conference on Natural Computation, Tianjin, China.

Talbi, E. (Ed.), 2009. Metaheuristics: From Design to Implementation. John Wiley & Sons, New Jersey.

Tang, W., Tong, L., Gu, Y., 2005. Improved genetic algorithm for design optimization of truss structures with sizing, shape, and topology variables. Int. J. Numer. Methods Eng. 62, 1737–1762.

Toğan, V., Daloğlu, A.T., 2006. Optimization of 3D trusses with adaptive approach in genetic algorithms. Eng. Struct. 28, 1019–1027.

Wang, S.Y., Tai, K., 2004. Graph representation for structural topology optimization using genetic algorithms. Comput. Struct. 82, 1609–1622.

Wang, S.Y., Tai, K., 2005. Structural topology design optimization using genetic algorithms with a bit-array representation. Comput. Methods Appl. Mech. Eng. 194, 3749–3770.

Wright, A., 1991. Genetic algorithms for real parameter optimization. In: Foundation of Genetic Algorithms 1. Morgan Kaufmann, San Mateo, CA.

Xiang, B.-W., Chen, R.-Q., Zhang, T., 2009. Optimization of trusses using simulated annealing for discrete variables. In: International Conference on Image Analysis and Signal Processing (IASP), Taizhou, China.

Yang, X.S., 2008. Nature-Inspired Metaheuristic Algorithms. Luniver Press, Frome.

Yang, X.S., Gandomi, A.H., 2012. Bat Algorithm: A Novel Approach for Global Engineering Optimization. Eng. Computation. 29, 464–483.

Yang, X.S., Deb, S., 2010. Engineering opitmisation by cuckoo search. Int. J. Math. Model. Numer. Optim. 1 (4), 330–343.

11 Element Exchange Method for Stochastic Topology Optimization

Mohammad Rouhi and Masoud Rais-Rohani

Mississippi State University, Starkville, MS, USA

11.1 Introduction

Topology optimization seeks the optimum distribution of a fixed volume of material over a selected design domain that satisfies the specified boundary conditions and design constraints. An illustrative example is shown in Figure 11.1. An optimum topology could be defined by such criteria as displacement or stress; however, it is commonly based on the minimization of structural compliance or strain energy, which results in an optimal load path from the loading points to the structural supports. It should be noted that in topology optimization, both the shape of the exterior boundary and configuration of the interior boundaries (i.e., holes, cutouts) can be optimized together.

Following the pioneering work by Bendsøe and Kikuchi (1988) and Rozvany et al. (1992), topology optimization has gained considerable attention in the past two decades with applications in structural and material design (Bendsøe and Sigmund, 2003), mechanism (Sigmund, 1997), and micro-electro-mechanical systems design (Sigmund, 2000, 2001a). Bendsøe et al. (2005) provided a review of the recent developments in topology optimization for application in various design problems related to laminated composite structures, heat transfer, fluids, acoustics, electromagnetism, and photonics.

In the so-called "material distribution method" (Bendsøe and Kikuchi, 1988), which is the basis for the design parameterization in topology optimization, the goal is to create regions of uniform material distribution to minimize a specific structural property (e.g., compliance). In this method, a discretized or finite element (FE) model of the structural domain (Figure 11.2) is used to perform the structural analysis and optimization.

By treating the nondimensional density of each element as an independent design variable and relating the other physical and engineering properties to the element density, a parameterized model is developed that can be used to find such properties as stiffness, thermal conductivity, magnetic permeability, porosity,

Metaheuristic Applications in Structures and Infrastructures. DOI: http://dx.doi.org/10.1016/B978-0-12-398364-0.00011-5

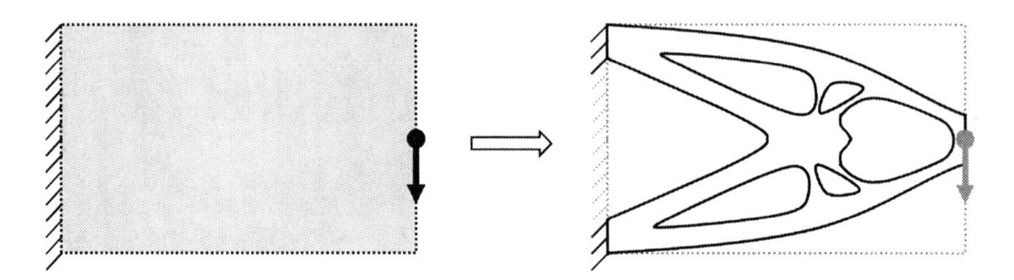

Figure 11.1 Illustrative example of material distribution through topology optimization.

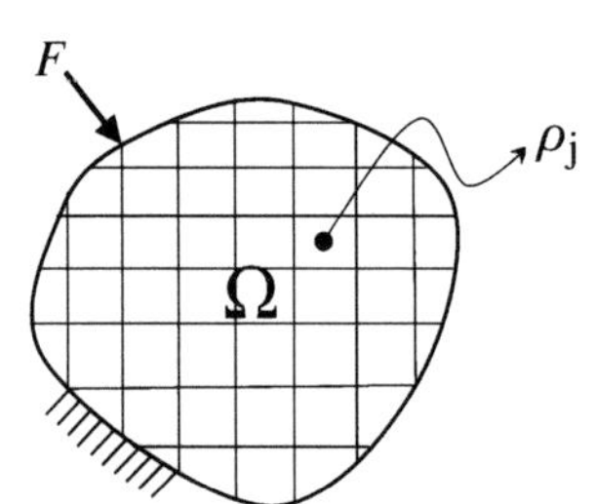

Figure 11.2 Discretized model of a structural domain with specified boundary conditions.

and so on. Theoretically, the nondimensional density takes a value of one or zero for a solid or void element, respectively. Given an initial distribution of a specified amount of mass, a structural analysis (e.g., finite element analysis (FEA)) is performed to evaluate the response characteristics of the structure. Depending on the topology optimization methodology used, the distribution of solid and void elements is updated and another structural analysis is performed. The sequence of analysis and optimization is continued until the solution convergences with the emergence of the optimal topology.

For a continuum structure represented by a domain of finite elements and associated boundary conditions, the topology optimization problem can be formulated as:

$$\begin{aligned} \min \quad & f(\rho) = \mathbf{u}^{\mathrm{T}} K \mathbf{u} = \sum_{j=1}^{N_e} u_j^{\mathrm{T}} K_j u_j = \sum_{j=1}^{N_e} 2\mathcal{E}_j \\ \text{s.t.} \quad & \sum_{j=1}^{N_e} \rho_j V_j \leq V_0 \\ & 0 < \rho_{\min} \leq \rho \leq 1.0 \end{aligned} \tag{11.1}$$

where $f(\boldsymbol{\rho})$ is the total strain energy, $\boldsymbol{\rho} = (\rho_1, \rho_2, \ldots, \rho_{N_e})^{\mathrm{T}}$ is the vector of design variables representing nondimensional element densities, $\mathbf{u} = (u_1, u_2, \ldots, u_{N_e})^{\mathrm{T}}$ is the vector of global generalized nodal displacements, K is the global stiffness matrix, with u_j, K_j, ρ_j, and V_j as the displacement vector, stiffness matrix, nondimensional density, and volume of the jth element, respectively. The constraint in

Eq. (11.1) imposes an upper bound on the acceptable volume fraction in the design domain. The element stiffness and density follow the relationship:

$$K_j = \rho_j^p K_e \tag{11.2}$$

where K_e is the stiffness matrix of the jth element if it is solid and p is the penalization power, which can take any value greater than one. Therefore, depending on the value of ρ_j, the element stiffness matrix can vary in magnitude from a minimum (void element) to a maximum (solid element) as proposed in solid isotropic microstructure with penalization (SIMP) method (Rozvany et al., 1992). However, to avoid having an ill-conditioned stiffness matrix, the void elements are given a small density, $\rho_{\min} > 0$. In this way, the intermediate densities are penalized by power p; typically, using $p \geq 3$ results in a black-and-white (solid-and-void) topology that is very desirable in structural topology optimization (Sigmund, 1994).

As seen in Eq. (11.1), the number of design variables in the optimization problem is the same as the number of finite elements in the model. Therefore, the choice of optimization algorithm and the number of FEA calculations are crucial in the search for optimum topology.

11.2 Overview of Topology Optimization Methods

Topology optimization methods can be divided into two main categories: derivative-based and derivative-free methods. In derivative-based optimization, the design variables must be continuous to allow the calculation of the first-order or possibly second-order derivatives of the differentiable response functions and the use of mathematical programming techniques for solution of the optimization problem. In derivative-free approaches, the design variables can take discrete values and the methods rely on repeated function evaluations using a stochastic or population-based algorithm. A brief overview of the methods that are used as reference for comparison of results in this chapter is presented below.

A commonly used derivative-based topology optimization method is SIMP (Bendsøe and Sigmund, 2003; Rozvany et al., 1992; Sigmund 1994; Zhou and Rozvany, 1991) with the corresponding algorithm summarized in Figure 11.3. In the initialization step, the geometry, FE mesh, boundary conditions, and the density distribution are specified. A uniform density distribution that is consistent with the specified volume fraction is commonly used. The subsequent steps in the iterative optimization process involve FEA, sensitivity analysis, filtering, and updating of the design variables using either the optimality criteria (Bendsøe and Sigmund, 2003) or the method of moving asymptotes (Svanberg, 1987). The updated design variables and the resulting topology will be analyzed again and the process of analysis and optimization is repeated until convergence is reached.

The SIMP method has become a very popular approach as it is simple to implement, computationally efficient, and easy to integrate with general-purpose FEA codes (Rozvany, 2009). Given its reliance on a derivative-based optimization algorithm, SIMP can provide a locally optimum solution that depends on the choice of initial design.

Recent improvements to the SIMP method include the use of mesh-independency filtering (Bendsøe and Sigmund, 2003; Sigmund 2001b), higher-order finite elements (Diaz and Sigmund, 1995; Jog and Haber, 1996), perimeter constraint on the density function (Haber et al., 1996), and alternative density-stiffness interpolation schemes (Guo and Gu, 2004). Commonly used filtering techniques adjust either the sensitivity derivatives of the objective function with respect to the design variables or the design variables themselves in order to eliminate the checkerboard effect. Bruns (2005) introduced a method based on hyperbolic sine functions to remedy the drawbacks of both of these filtering approaches while capitalizing on the advantages of each approach. Using hyperbolic sine functions, the intermediate density material is made less volumetrically effective than solid or void elements and consequently results in unambiguous and predominantly solid−void designs. By adding a new constraint to the topology optimization problem, labeled the sum of the reciprocal variables, Fuchs et al. (2005) produced sharper 0-1 solutions than the SIMP with greater stiffness for the same amount of material. Efforts to produce better design topologies include relaxation or restriction of the design problem and discretization of the original topology optimization problem combined with heuristic rules to avoid unwanted effects such as checkerboards (Sigmund and Petersson 1998).

Recent advances in the application of derivative-free (metaheuristic, direct search) approaches to topology optimization problems include the simulated biological growth (Mattheck and Burkhardt, 1990), particle swarm optimization (PSO)

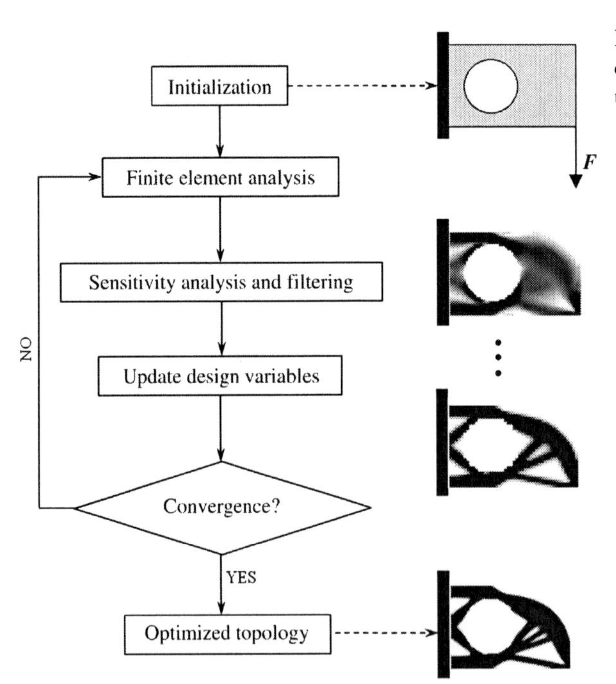

Figure 11.3 General scheme of topology optimization using SIMP.

(Fourie and Groenwold, 2001), evolutionary structural optimization (ESO) (Xie and Steven, 1993), bidirectional ESO (BESO) (Querin et al., 1998), and metamorphic development (MD) (Liu et al., 2000). Some of these methods, together with other algorithms that mimic natural phenomena such as genetic algorithm (GA) (Goldberg, 1989) and cellular automata (Kita and Toyoda, 1999), have been used in sizing and shape optimization problems as well. The use of binary design variables enables these methods to produce a black–white (solid–void) optimal topology that excludes any gray (i.e., fuzzy or intermediate density) regions without using any filtering technique. Another advantage of the direct search methods is their nonlocal search algorithms that can lead to a better solution than the local optimum found by derivative-based optimizers in the vicinity of the initial design point. However, due to the need for a large number of function evaluations for the multitude of candidate designs, the direct search methods tend to be computationally inefficient (Bruns, 2005; Fuchs et al., 2005; Mattheck and Burkhardt, 1990; Rozvany, 2009). To remedy the checkerboard problem, derivative-free methods also resort to using mostly heuristic schemes that, while increasing the possibility of finding a global optimum topology, suffer from some shortcomings including relatively high computational cost (Fourie and Groenwold, 2001; Jakiela et al., 2000; Rozvany, 2009).

The recently developed element exchange method (EEM) (Rouhi et al., 2010) has many of the same advantages of the other derivative-free methods but with noticeably better computational efficiency. Named after the principal operation in its topology optimization strategy, EEM falls under the same category as BESO and GA due to the use of heuristic relationships, but it has certain features that are quite distinct from the other two methods. In the remaining portion of this chapter, various aspects of EEM, including the element exchange strategy, checkerboard control procedure, convergence criteria, and the algorithmic parameters used in conjunction with different operations in EEM, are presented and discussed. Several example problems are used to compare the EEM-based solutions with those of the other topology optimization methods.

11.3 Element Exchange Method

To describe the basic principles of EEM, a simple structural system is considered. As shown in Figure 11.4, the system is idealized by a combination of four linearly elastic springs and associated boundary conditions for compliance minimization.

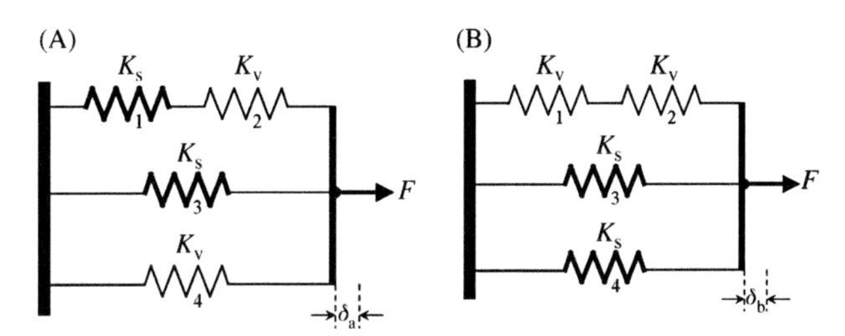

Figure 11.4 Spring system (A) before and (B) after element exchange operation.

The total strain energy, $\mathcal{E}_t$ stored in the system due to the applied force F, is simply the sum of energy stored in individual springs found as:

$$\mathcal{E}_t = \sum_{j=1}^{4} \mathcal{E}_j = \frac{1}{2}\sum_{j=1}^{4} K_j \delta_j^2 \tag{11.3}$$

where $\mathcal{E}_t$ is the energy in the jth spring defined in terms of the corresponding stiffness, K_j, and elongation, δ_j.

Assuming that only two springs can be used for minimizing the strain energy of the system in this example, the problem becomes the placement optimization of the two springs in the four candidate locations. The two springs that are kept would create an optimal load path between the loaded and supported ends of the system. For simplicity, a "solid" spring is assumed to have a stiffness of K_s while a "void" spring has a stiffness of $K_v = 0.001\ K_s$. Using the initial distribution of springs shown in Figure 11.4A and knowing that springs 1 and 2 are under equal axial force, elongations are found to be:

$$\delta_1 = \frac{\delta_a}{1001} \approx \frac{\delta_a}{1000};\ \delta_2 = \frac{1000}{1001}\delta_a \approx \delta_a\ ;\ \delta_3 = \delta_4 = \delta_a$$

Substituting these values into Eq. (11.3) gives:

$$\mathcal{E}_t = \frac{K_s}{2}\left[1\times 10^{-6} + 1\times 10^{-3} + 1\times 10^{0} + 1\times 10^{-3}\right]\delta_a^2 \approx \frac{1}{2}K_s\delta_a^2 = \frac{1}{2}\frac{F^2}{K_s}$$

Since spring 1 is a solid spring with the lowest strain energy between the two solid springs, it will be converted into a void spring in the next iteration while spring 4—representing a void spring having the highest strain energy between the two void springs—will be converted into a solid spring. The updated layout is shown in Figure 11.4B with the total strain energy being:

$$\mathcal{E}_t = \frac{K_s}{2}\left[2.5\times 10^{-4} + 2.5\times 10^{-4} + 1\times 10^{0} + 1\times 10^{0}\right]\delta_b^2 \approx K_s\delta_b^2 = \frac{1}{4}\frac{F^2}{K_s}$$

While the number of solid springs is kept constant, the element exchange reduced the total strain energy of the system by 50% resulting in greater stiffness and smaller compliance.

Extending this concept to a continuum domain represented by a FE mesh would make it possible to use EEM as part of a more general algorithm and solution procedure for finding the optimal topology.

11.3.1 EEM Algorithm

In the topology optimization problem described by Eq. (11.1), ρ_j is treated as a discrete design variable with $E_j = \rho_j E$, where E is the Young's modulus of the solid material. For a design domain defined by a uniform mesh, the volume fraction constraint in Eq. (11.1) becomes an equality constraint that has to be strictly satisfied in every iteration; whereas, in the case of a nonuniform mesh, a small fluctuation of volume is tolerated around the prescribed limit, V_0.

The EEM algorithm is shown in Figure 11.5. Once the design domain and boundary conditions are defined and an FE mesh is generated, the number of solid elements $N_s = MV_0$ is randomly distributed throughout the design domain. This operation divides the domain into two sets of distributed elements, one solid (i.e., solid set) and one void (i.e., void set). All elements in the void set are given a nondimensional minimum material density of $\rho_{min} = 0.001$ while those in the solid set have $\rho = 1$. The EEM parameters and convergence criteria are also defined before the EEM loop begins.

A static FEA is performed to find the strain energy of the individual elements as well as the total strain energy of the structure as a whole. The convergence check is skipped in the first iteration. While keeping the volume fraction fixed, a subset of less-effective elements among those in the solid set is converted into void while a subset of more-effective elements among those in the void set is turned into solid according to the operation procedure described later in this section. In the case of a uniform mesh, where all elements are geometrically identical, volume fraction remains constant by simply setting the number of elements in the two subsets equal

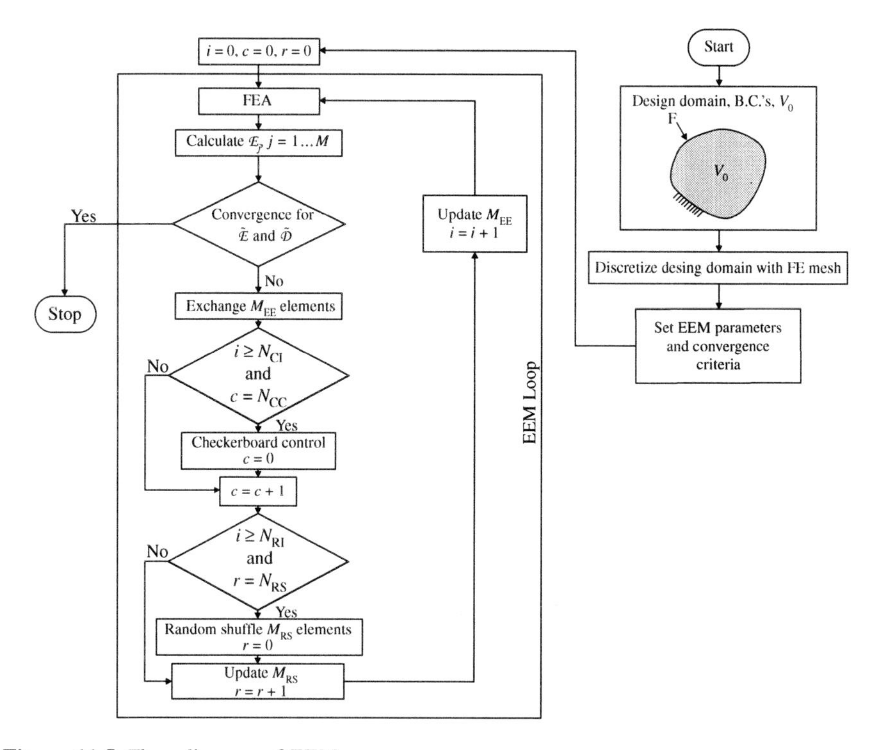

Figure 11.5 Flow diagram of EEM.

to each other. In the case of nonuniform mesh, the elements in the two subsets are converted one by one until the exchange volume is balanced. A small difference in volume is possible because of using discrete density and variation in elements' geometry, which may require the relaxation of the volume fraction constraint. Since the type of mesh used does not change the overall framework of EEM, henceforth, the mesh is assumed to be uniform.

After completion of the predefined number of FEA and element exchange operations, checkerboard control and random shuffle steps are introduced before proceeding to the next round of element exchange operations and FEA. Random shuffle is similar to the mutation operation in GA and it is used to diversify the search and increase the chances of finding the global optimum design. The principal operations of EEM (i.e., FEA, element exchange, checkerboard control, and random shuffle) are repeated at different intervals until the convergence criteria are satisfied. Similar to other stochastic methods, a limit is imposed on the number of iterations to stop the program if the convergence criteria cannot be achieved.

The evolution of topology in EEM from the initialization of the discretized domain to optimum (minimum compliance) design is shown in Figure 11.6 for a tip-loaded cantilevered beam with a circular hole. The normalized strain energy variation history plotted in Figure 11.6 illustrates a typical convergence pattern in EEM. The spikes that appear at different intervals (abrupt change in strain energy) are mostly due to the random shuffle operation, although it is also possible to be the result of a routine element exchange operation.

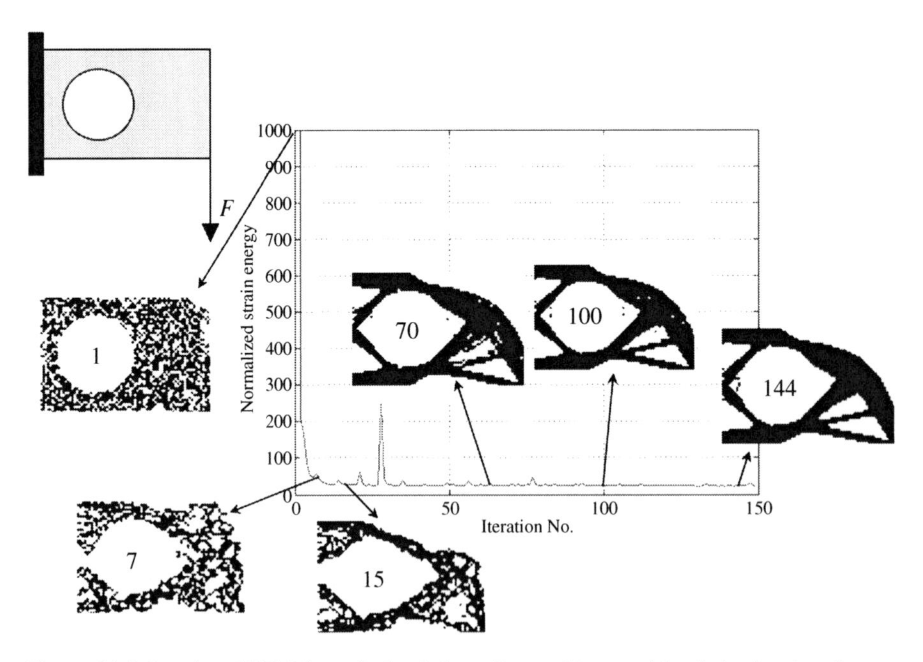

Figure 11.6 Results of EEM for a tip-loaded cantilevered beam with a hole showing the evolution in topology and the normalized strain energy convergence history.

Since the initial distribution of solid elements is selected at random and some of the operations performed at different stages are stochastic, different solution paths may be taken by EEM for the same topology optimization problem. As a consequence, if a problem has multiple local optima, especially with nearly equal objective function values (Kutylowski, 2002), it is possible for the EEM solution to converge to any one of the local optima with potentially different solid–void material distributions.

Unlike in ESO (Querin et al., 1998; Xie and Steven, 1993), the void elements can also be converted into solid and vice versa. Due to the fact that the void elements have small but nonzero stiffness and density, even if the initial random distribution of solid elements does not provide a continuous load path (infeasible topology) between the loading points and supports, EEM gradually connects all the solid elements in its search for an optimum load path (topology). Besides, in EEM the solid and void elements involved in the exchange operation are not limited to any particular region of the design domain as is the case with BESO (Querin et al., 1998). These features, along with the random shuffle and overall topology optimization scheme, help distinguish EEM from both ESO and BESO. The EEM algorithm is readily applicable to any two- or three-dimensional domain and boundary conditions, irrespective of its geometric or loading complexity.

For compliance minimization problems, EEM is very easy to implement since the total strain energy of the system is a simple summation of the strain energy of individual elements distributed in the design domain. As such, it is possible to use the element strain energy to measure the relative influence of individual solid or void elements with respect to other elements in the same set as well as on the objective function. In the case when a different objective function is chosen, or a new design constraint is added to the topology optimization problem, it is necessary to use a different metric to determine the exchange criterion.

11.3.2 Element Exchange

If in every round of the element exchange operation the number of elements in the two subsets that are converted from solid to void and vice versa is kept constant, the solution may not converge. This sort of behavior has also been observed in other metaheuristic approaches (Huang and Xie, 2007). To alleviate this problem, the parameter M_{EE} representing the number of elements to be exchanged in each iteration is introduced such that for a linear reduction scheme, it takes the form:

$$M_{\mathrm{EE}}^{k} = \mathrm{int}\left[M_{\mathrm{EE-max}} - i\left(\frac{M_{\mathrm{EE-max}} - M_{\mathrm{EE-min}}}{N_{\max}}\right)\right] \tag{11.4}$$

where M_{EE} is gradually reduced from its maximum value $M_{\mathrm{EE\text{-}max}}$ at the beginning ($i = 0$) to its minimum value $M_{\mathrm{EE\text{-}min}}$ at $i = N_{\max}$ with i and $N_{\max}$ denoting the iteration counter and the prescribed maximum number of EEM iterations, respectively. The parameters $N_{\max}$, $M_{\mathrm{EE\text{-}max}}$, and $M_{\mathrm{EE\text{-}min}}$ can have a wide range of acceptable values as will be noted later in the discussion of the example problems. Their individual influence on solution accuracy and rate of convergence are also addressed later in this chapter.

11.3.3 Checkerboard Control

Depending on the specified volume fraction and the proximity of strain energy levels for different distributions of the same volume of material, it is possible to encounter a checkerboard pattern, which is generally undesirable in topology optimization. In Figure 11.7, the checkerboard regions are clearly visible, which can be verbally described as void/solid elements that do not share edges with similar elements.

Diaz and Sigmund (1995) showed that a checkerboard pattern occurs because it has a numerically induced (artificially) high stiffness compared with a material having a uniform material distribution. The easiest way to prevent a checkerboard is to use higher-order elements (8- or 9-node elements for two-dimensional cases) (Jog and Haber, 1996). This, however, drastically increases the computational time (Sigmund, 1994). There are several checkerboard-prevention schemes that are almost all based on heuristics (Bendsøe et al., 2005; Sigmund and Petersson, 1998). Smoothing the optimal topology through image-processing techniques is one of these methods (Sigmund, 1994; Sigmund and Petersson, 1998). Another more popular checkerboard control technique is filtering, where the modified design sensitivities are used in each iteration of the algorithm for solving the discretized problem (Sigmund, 1994, 2001b). The filter makes design sensitivity of each element dependent on a weighted average of that specific element and its eight neighboring (contacting) quadrilateral elements.

Such filtering techniques that rely on design sensitivities cannot be used in EEM since it is a derivative-free approach. Instead, solid checkerboard elements are identified as solid elements that share edges with void elements as shown in Figure 11.8A, whereas void checkerboard elements are the exact opposite as illustrated in Figure 11.8B. Whether the dashed elements in Figure 11.8A and B are solid or void will not alter the checkerboard condition.

Since in EEM the initial solid set of elements is distributed in a random manner, the initial topology may have multiple checkerboard regions such as those shown in Figure 11.8C. Given the iterative and stochastic nature of EEM, it is not necessary to eliminate the checkerboard regions in every iteration as that would be unnecessary and rather inefficient. Instead, EEM allows the exchange operation to proceed for several iterations before the active search for checkerboard patterns is conducted using an interval N_{CC} that can be set 1–5% of N_{max}. At every N_{CC}

Figure 11.7 Illustration of checkerboard pattern.

iteration, the solid checkerboard elements are converted into voids and vice versa as shown in Figure 11.8D and E. In the case of an unequal number of solid and void checkerboard elements, the difference is randomly redistributed in the design domain to maintain the specified volume fraction. This random redistribution may result in formation of small, new checkerboard regions, which tend to gradually diminish as the EEM procedure is continued. The checkerboard elimination procedure in EEM is heuristic and checkerboard elements are removed regardless of their impact on the overall compliance of the structure.

The checkerboard regions are affected by the mesh size of the discretized design domain. Exact analytical solutions for Michell truss structures (Lewinski and Rozvany, 2008; Lewinski et al., 1994; Rozvany et al., 2006) show many narrow connecting branches in the optimal layout. When the mesh is relatively coarse, the elements in some of those branches would have a pixilated appearance and are identified as checkerboard elements to be eliminated. Later in this chapter, it will be shown that through mesh refinement EEM is able to produce a topology that asymptotically approaches the analytical solution.

11.3.4 Random Shuffle

Random shuffle is another essential operation in EEM. In this operation, a subset of solid elements is randomly redistributed in the void regions of the design domain. Analogous to the mutation operation in GA (Goldberg, 1989; Jakiela et al., 2000) or craziness in PSO (Fourie and Groenwold, 2001), it helps prevent premature convergence due to insufficient exploration of the design domain in search of the optimum design. It also helps EEM converge faster by alleviating the occasional back and forth alternation (oscillatory exchange) in a subset of elements from solid to void and back to solid in successive element exchange operations. The qualitative effect of random shuffle on the solution is shown later in this chapter. Random shuffle may generally result in an abrupt change in stiffness (Rouhi et al., 2010) as well as the total strain energy (see the spikes in Figure 11.6)

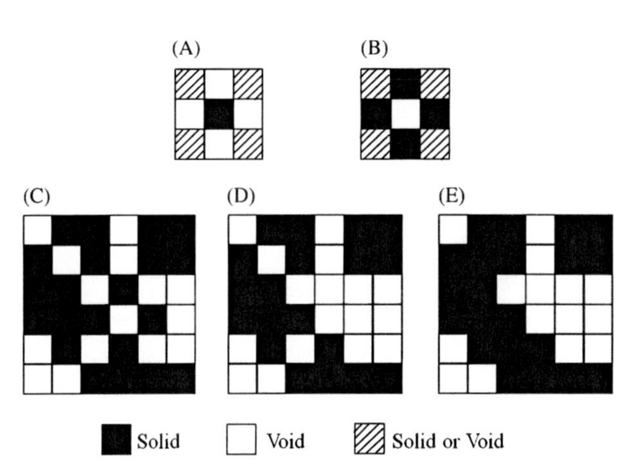

Figure 11.8 (A) solid checkerboard, (B) void checkerboard, (C) topology before checkerboard control, (D) after solid checkerboard elimination, and (E) after void checkerboard elimination.

and creation of small checkerboard regions. Nevertheless, none of them are detrimental since they are rectified by element exchange operations and checkerboard control in the subsequent iterations of EEM.

Random shuffle occurs after every N_{RS} iterations that can be equal to 2−10% of N_{max}. Following a similar rationale in conjunction with Eq. (11.4), the number of elements participating in the random shuffle, M_{RS}, varies in the optimization process using the expression:

$$M_{RS}^{k} = \text{int}\left[M_{RS-max} - i\left(\frac{M_{RS-max} - M_{RS-min}}{N_{max}}\right)\right] \tag{11.5}$$

where $M_{RS\text{-}max}$ is the maximum number of participating elements at the beginning ($i = 0$) and $M_{RS\text{-}min}$ is the minimum number. Although it is possible to set $M_{RS\text{-}max} = M_{EE\text{-}max}$ and $M_{RS\text{-}min} = M_{EE\text{-}min}$, there is no limitation otherwise.

11.3.5 Passive Elements

As a design constraint, there may be some subregions in the structure whose geometry, properties, and locations are not allowed to change during topology optimization. In that case, it would be convenient to use an overall uniform mesh and assign the elements in the fixed subregions $\rho_{pi} = \rho_{min}$, if forming a hole or cutout, or $\rho_{pi} = 1.0$, if forming a solid fixed section. These elements are referred to as passive elements and are not allowed to participate in any of the operations during the EEM process. For example, in the model shown in Figure 11.6, the mesh consisted of a rectangular domain with the elements inside the circular hole treated as passive.

11.3.6 Convergence Criteria

EEM generally benefits from increasing the number of iterations in order to find a more refined optimal topology but at a higher computational cost due to increased number of FEA. However, besides N_{max}, two convergence criteria are also imposed.

The first criterion examines the relative difference in the element strain energy distributions in two consecutive elite topologies. Elite topology refers to the topology with the lowest strain energy up to the current iteration. Mathematically, this criterion is expressed as:

$$\frac{||\tilde{\mathcal{E}}_{ce} - \tilde{\mathcal{E}}_{pe}||}{||\tilde{\mathcal{E}}_{pe}||} \leq \varepsilon_E \tag{11.6}$$

where $\tilde{\mathcal{E}}$ represents the vector of strain energy distribution with subscripts "ce" and "pe" referring to the current and previous elite topologies, respectively, and ε_E is an acceptable tolerance on the relative strain energy difference.

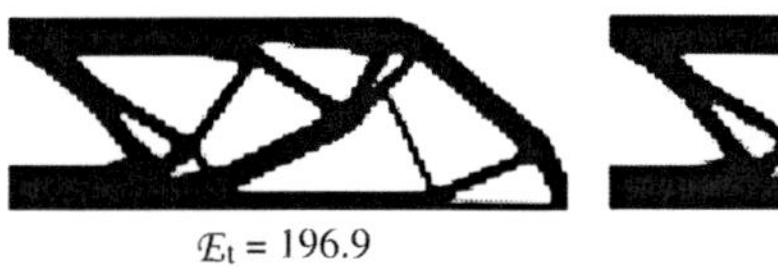

Figure 11.9 Two different topologies with nearly identical strain energy values.

Figure 11.9 shows an example of two topologies found using SIMP with almost equal total strain energy values but with somewhat different topologies. The same condition can also occur when using heuristic topology optimization methods. Therefore, an additional convergence criterion is introduced to compare the material distribution in two consecutive elite topologies. Vector $\tilde{\mathbf{D}}$ defines the domain topology in which the individual components have binary values depending on the solid (1) or void (0) property of the corresponding elements. Based on this definition, the topology convergence criterion for two consecutive elite topologies is defined as:

$$\frac{||\tilde{\mathbf{D}}_{ce}-\tilde{\mathbf{D}}_{pe}||}{||\tilde{\mathbf{D}}_{pe}||} \leq \varepsilon_t \qquad (11.7)$$

where ε_t is an acceptable tolerance on the relative topology difference.

11.4 EEM Application

A number of example problems are solved using EEM with the results compared with those obtained using other optimization approaches. Each two-dimensional domain is defined by n_x, n_y, and V_0 representing the number of finite elements in the x and y directions and the prescribed volume fraction, respectively. In all examples, the nodal displacements and element stiffness are normalized with respect to element size and Young's modulus of the material to normalize the strain energy, and all design problems are optimized for minimum normalized strain energy denoted by $\mathcal{E}_t$.

The number of function calls (FEA) during the solution process is equal to the number of iterations (N) when using EEM; however, this may not be the case for the other methods considered. Regardless of the optimization method used, the majority of the computational time is spent on the compliance calculation via FEA. Hence, the total number of function calls can be used as a fairly accurate measure of computational efficiency. Unlike the EEM or SIMP methods, in which the FE solution is called once in each iteration and consequently the required computational time is proportional to the number of iterations, in population-based methods, such as GA or PSO, computational time is proportional to the number of iterations times the population size, which makes them computationally orders of magnitude less efficient.

Example 11.1

Topology optimization of a simply supported beam is considered. The beam is under a point force in the middle at the top as shown in Table 11.1. Only one half of the physical domain is modeled because of the overall symmetry. The final topologies along with the N and $\mathcal{E}_t$ values are also given in Table 11.1 for $(n_x, n_y, V_0) = (90, 30, 0.5)$. The r value in Table 11.1

Table 11.1 Comparison of EEM and SIMP Results for Example 11.1

Design Domain and Boundary Conditions

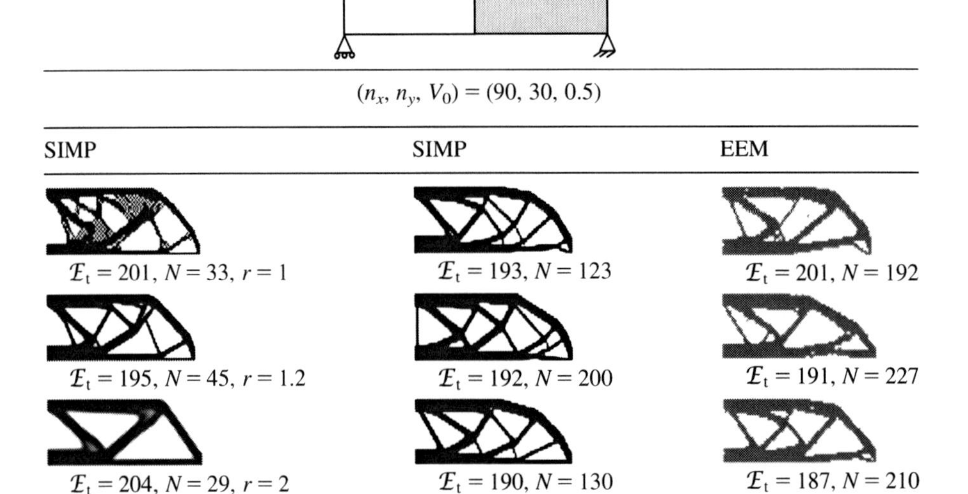

$(n_x, n_y, V_0) = (90, 30, 0.5)$

SIMP	SIMP	EEM
$\mathcal{E}_t = 201, N = 33, r = 1$	$\mathcal{E}_t = 193, N = 123$	$\mathcal{E}_t = 201, N = 192$
$\mathcal{E}_t = 195, N = 45, r = 1.2$	$\mathcal{E}_t = 192, N = 200$	$\mathcal{E}_t = 191, N = 227$
$\mathcal{E}_t = 204, N = 29, r = 2$	$\mathcal{E}_t = 190, N = 130$	$\mathcal{E}_t = 187, N = 210$

represents the filtering radius in the SIMP-based solutions. Results in the first column are based on different filtering radii and identical initial designs, whereas those in the second column are based on different initial designs but equal filtering radius ($r = 1.2$).

The SIMP-based solutions are obtained using the algorithm provided by Sigmund (2001b). The effect of initial design and filtering radius on the final topology is noticeable. The EEM results, on the other hand, represent three converged solutions using the same values for all EEM algorithmic parameters. The diversity in results is caused by the existence of multiple local optima in this problem.

Generally, the closer the topology is to the analytical solution (ideal Michell truss structure in this case), the lower the strain energy. This resemblance (Lewinski et al., 1994) is more visible in the models with the lowest strain energy than the rest in Table 11.1.

It is worth noting that in its current implementation, EEM does not check for the presence and removal of a few solid or void (floating) elements that appear as specks in the final topology. These elements remain as a result of random shuffle for volume fraction balance after checkerboard control. Since these elements have no considerable impact on the final topology or in the total strain energy, no attempt is made to remove them in any of the presented solutions. Due to the stochastic nature of EEM, another solution to the same optimization problem may show floating elements in different regions or none at all.

Example 11.2

This example is a variant of Example 1 with the load location changed to the bottom and the horizontal displacement at the left support constrained similar to the one on the right side. Table 11.2 shows the optimum topology and the related data for the EEM solution as well as those reported in the literature using BESO (Querin et al., 1998) and level set method (Wang and Wang, 2003). The results of EEM and level set are for $(n_x, n_y, V_0) = (61, 62, 0.31)$ whereas that for BESO is for $(n_x, n_y, V_0) = (31, 32, 0.25)$. The value of $\mathcal{E}_t$ was not reported for the other two solutions and for the BESO result, the number of iterations reported is the "steady state" condition that can be considerably smaller than the number of FE solutions. The final topologies obtained by the three methods are in general agreement with the difference being the location of the member closest to the axis of symmetry. For the level set solution, this member appears to fall on the axis of symmetry, whereas for the other two, it is slightly shifted to one side resulting in the formation of one extra member in the BESO and EEM topologies.

Table 11.2 Comparison of EEM Results with Level Set and BESO Solutions for Example 11.2

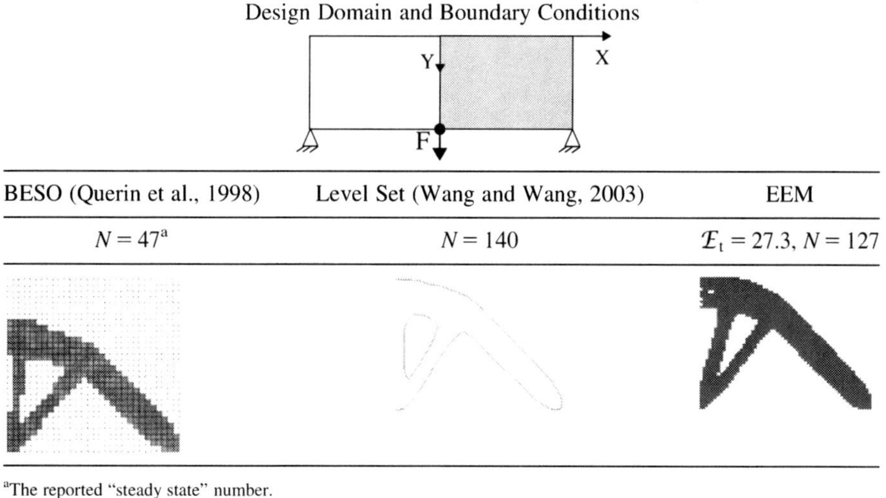

Design Domain and Boundary Conditions		
BESO (Querin et al., 1998)	Level Set (Wang and Wang, 2003)	EEM
$N = 47$[a]	$N = 140$	$\mathcal{E}_t = 27.3$, $N = 127$

[a]The reported "steady state" number.

Example 11.3

This example considers topology optimization of a rectangular cantilevered beam model with a tip load acting at the bottom. The optimization results obtained by EEM are compared with those of SIMP in Table 11.3 for two different mesh sizes. The filtering radius in the case of SIMP is set to $r = 1.2$. The results show fairly similar values for $\mathcal{E}_t$, but the

Table 11.3 Comparison of EEM and SIMP Results for Example 11.3

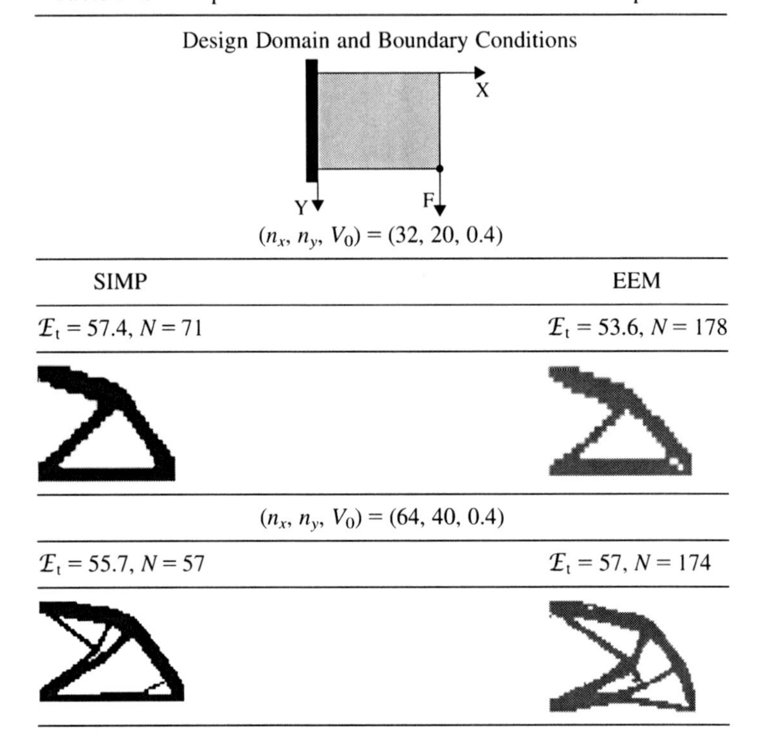

Design Domain and Boundary Conditions	
$(n_x, n_y, V_0) = (32, 20, 0.4)$	
SIMP	EEM
$\mathcal{E}_t = 57.4, N = 71$	$\mathcal{E}_t = 53.6, N = 178$
$(n_x, n_y, V_0) = (64, 40, 0.4)$	
$\mathcal{E}_t = 55.7, N = 57$	$\mathcal{E}_t = 57, N = 174$

iteration numbers are different. Refining the mesh pushes the final topology toward the analytical solution in both EEM and SIMP methods, however, with minimal change in $\mathcal{E}_t$.

In the case of EEM, the higher $\mathcal{E}_t$ value for the refined mesh, which is a more Michell-like structure (Lewinski et al., 1994), does not imply inferiority of the solution compared with the one for the coarse mesh. Rather, it is due to the fact that the FEA solution and $\mathcal{E}_t$ value for the case of fine mesh is more accurate than that for the coarse mesh. Discretizing the optimal layout (ground elements) for both cases at the same element size to increase the accuracy of the FEA solution would reveal the smaller strain energy of the more Michell-like topology at the same volume fraction.

Example 11.4

This example considers the same problem as in Example 3 with the tip load placed half way between the top and bottom sides of the beam. The optimum topology and other results are shown in Table 11.4 and compared with those found using the enhanced GA approach (Wang et al., 2006). EEM is used to optimize the topology for two different mesh densities but at the same volume fraction. The results show a similar development of the final topology but a more accurate solution for the refined mesh. In spite of the fact

Table 11.4 Comparison of EEM Results with Enhanced GA Solution for Example 11.4

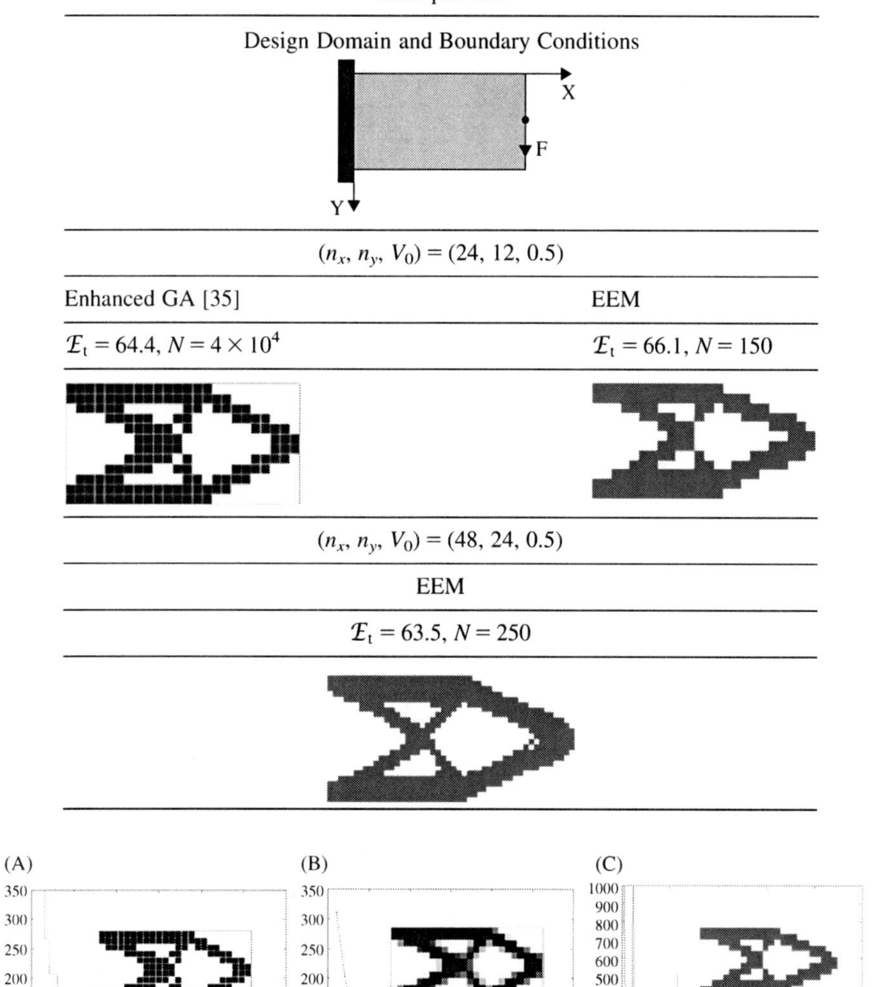

Design Domain and Boundary Conditions	
$(n_x, n_y, V_0) = (24, 12, 0.5)$	
Enhanced GA [35]	EEM
$\mathcal{E}_t = 64.4$, $N = 4 \times 10^4$	$\mathcal{E}_t = 66.1$, $N = 150$
$(n_x, n_y, V_0) = (48, 24, 0.5)$	
EEM	
$\mathcal{E}_t = 63.5$, $N = 250$	

Figure 11.10 Compliance convergence history and final topology for (A) enhanced GA (Wang et al., 2006), (B) SIMP, and (C) EEM.

that the final topology and $\mathcal{E}_t$ values are nearly the same, EEM converges at least 160 times faster. Generally, GA-based solutions may require 10–100 times more function calls compared with homogenization-based solutions (Jakiela et al., 2000).

Figure 11.10 compares the computational efficiency of EEM, enhanced GA, and SIMP in their search for the optimum topology. EEM is significantly more efficient than GA but less efficient than SIMP. It is also worth noting that for GA, the number of iterations times the population size gives the total number of function calls (Wang et al., 2006).

Example 11.5

This example considers topology optimization of a cantilevered beam that is shorter in length than height using the same loading condition as in Example 4. Table 11.5 shows the results of EEM compared with PSO-based solutions (Fourie and Groenwold, 2001) for two different mesh densities. The $\mathcal{E}_t$ values for the EEM results are shown in Table 11.5, but those for the PSO solutions were not reported (Fourie and Groenwold, 2001).

The results of EEM for the coarse mesh appear superior to those of PSO in terms of both solution accuracy and computational efficiency. For the fine mesh, the optimum topologies are almost identical; however, EEM converges 100 times faster than PSO with no loss of accuracy. The actual efficiency would be much higher since, unlike PSO, EEM is not population based.

Table 11.5 Comparison of EEM Results with PSO Solutions for Example 11.5

Design Domain and Boundary Conditions			
$(n_x, n_y, V_0) = (20, 47, 0.5)$		$(n_x, n_y, V_0) = (40, 94, 0.5)$	
EEM	PSO (Fourie and Groenwold, 2001)	EEM	PSO (Fourie and Groenwold, 2001)
$\mathcal{E}_t = 3.0$, $N = 100$	$N = 10^5$ 	$\mathcal{E}_t = 5.1$, $N = 103$ 	$N = 10^3$ 
	Continuous Density		Binary Density

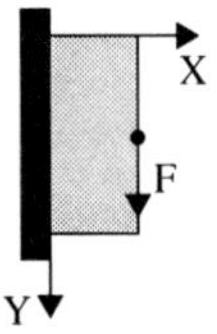

Example 11.6

In this example, a tip-loaded cantilevered beam is optimized for two different load cases. In one case, the beam is under an upward tip load at the top whereas in the other case, it is under a downward tip load at the bottom. Table 11.6 shows the model and the results found using EEM and SIMP with $r = 1.2$.

The E_t values shown in Table 11.6 represent the sum for the two load cases. As a result, the additive form of the objective function is retained and the relationship between element strain energy and element exchange operation remains unchanged. Thus, the exchange operation is performed based on the elements' strain energy from the two load cases combined. The symmetric layout of the final topology is due to F_1 and F_2 being equal in magnitude. Although the $\mathcal{E}_t$ values are nearly equal, the EEM and SIMP topologies have some distinct differences. As discussed earlier in this section, the specks seen in the EEM topology are the residue or the floating elements from the last random shuffle operation.

Table 11.6 Comparison of EEM Results with SIMP Solution for Example 6

Design Domain and Boundary Conditions	
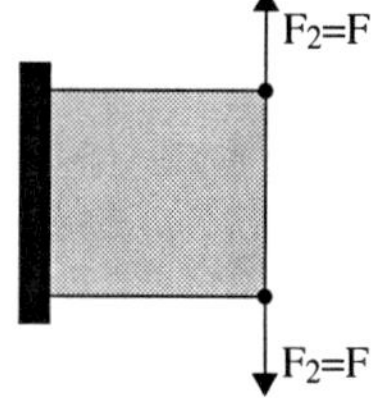 (F_1 and F_2 represent separate load cases)	
$(n_x, n_y, V_0) = (50, 50, 0.4)$	
EEM	SIMP
$\mathcal{E}_t = 60.9$, $N = 104$	$\mathcal{E}_t = 61.3$, $N = 60$
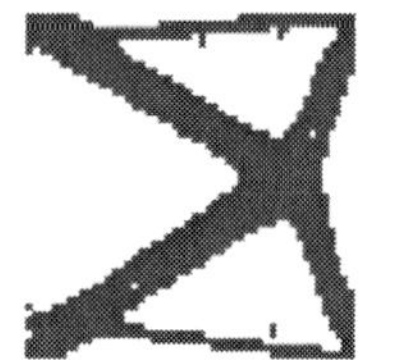	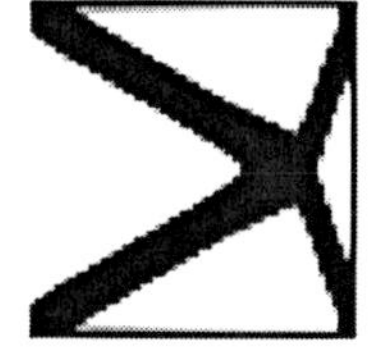

Example 11.7

In this example, an L-shaped cantilevered beam is loaded by a distributed force at the tip over the middle one-third portion of the free end as shown in Figure 11.11A. The optimum topology found by EEM for $(n_x, n_y, V_0) = (90, 90, 0.35)$ is shown in Figure 11.11C and is compared to that found using the neighborhood search method (Svanberg and Werme, 2005) in Figure 11.11B requiring 378 iterations. For the same mesh size, EEM found the optimum topology with $\mathcal{E}_t = 85.1$ after 167 iterations. In both cases, the larger members show the average orientation for the more finely distributed members over the same domain (Lewinski and Rozvany, 2008).

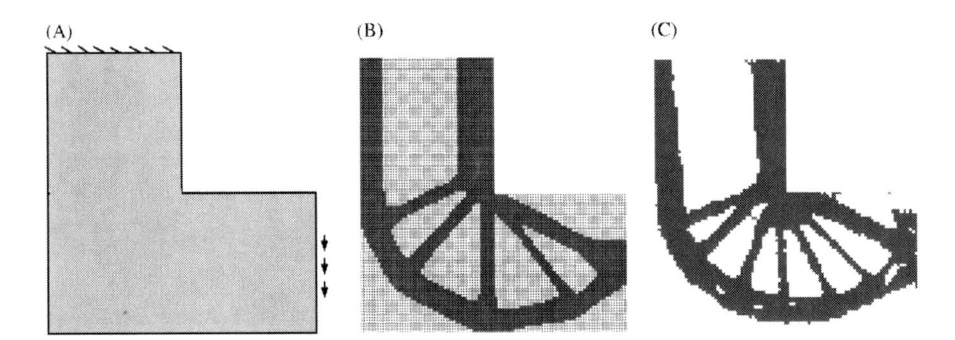

Figure 11.11 (A) L-shaped domain with distributed tip load and optimal layouts based on (B) neighborhood search method (Svanberg and Werme, 2005) and (C) EEM.

Example 11.8

In this example, the loading condition on the L-shaped beam in Example 7 is changed into a downward-concentrated load acting at the tip on the bottom side as shown in Figure 11.12A. The optimum topology using the exact analytical solution (Lewinski and Rozvany, 2008; Rozvany et al., 2006) is shown in Figure 11.12B and compared with the results found by EEM for $(n_x, n_y, V_0) = (90, 90, 0.35)$ and $(n_x, n_y, V_0) = (250, 250, 0.35)$ as shown in Figure 11.12C and D, respectively. Recognizing that it would be very difficult to match the analytical solution with unlimited number of bars by a discretized FE model with a limited number of elements and that the boundary conditions along the supported ends are different, the EEM solutions show trends that appear to be in reasonable agreement with the exact solution. The level of agreement improves through mesh refinement.

Besides the examples presented in this chapter, EEM has also been applied to topology optimization of three-dimensional structures (Rouhi et al., 2010) and more recently to blank optimization in a sheet forming process (DorMohammadi et al., 2012).

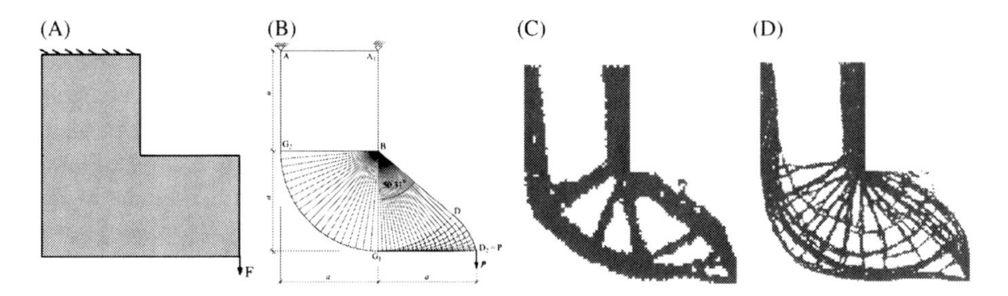

Figure 11.12 (A) L-shaped domain with single tip load and optimal layouts based on (B) exact analytical solution (Wang et al., 2006), (C) EEM with coarse mesh, and (D) EEM with fine mesh.

11.5 Influence of EEM Operations and Parameters on Optimization Results

The effects of operations such as checkerboard control and random shuffle are explored by considering the cantilevered beam problem in Example 3. The model in Figure 11.13A is to be optimized using the initial distribution of solid elements and associated volume fraction (0.4) as shown in Figure 11.13B. With both the checkerboard control and random shuffle operations included, the optimum topology is that shown in Figure 11.13C with $\mathcal{E}_t = 54$. However, without the checkerboard control, the optimum topology changes to that shown in Figure 11.13D having $\mathcal{E}_t = 55.5$, and the result without the random shuffle operation is that in Figure 11.13E indicating a local optimum when its final layout and strain energy ($\mathcal{E}_t = 58.5$) are compared with the ones in Figure 11.13C.

As mentioned earlier in this chapter, the appropriate values for the EEM input parameters must be specified before starting the solution process. These parameters include V_0 (the volume fraction), M (the number of finite elements in the model),

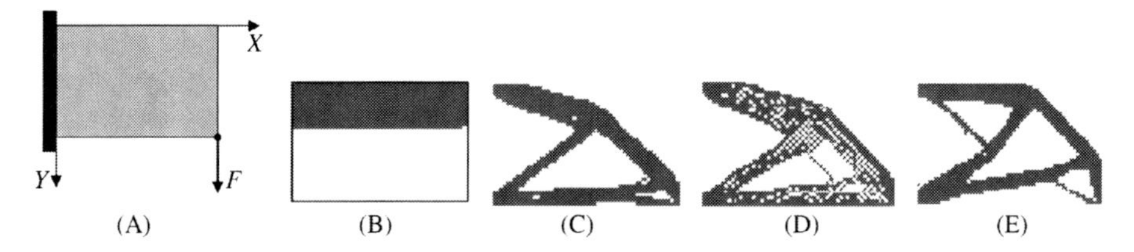

Figure 11.13 (A) The design domain and BCs, (B) initial distribution of solid elements, (C) EEM result with and (D) EEM result without checkerboard control operation, and (E) EEM result without random shuffle operation.

N_{max} (the maximum allowable number of iterations in the EEM solution), ε_E (total strain energy convergence parameter), ε_t (element topology or density distribution convergence parameter), $M_{EE\text{-}max}$ (the maximum number of exchanged elements), $M_{EE\text{-}min}$ (the minimum number of exchanged elements), $M_{RS\text{-}max}$ (the maximum number of solid elements participating in the random shuffle), $M_{RS\text{-}min}$ (the minimum number of solid elements participating in the random shuffle), N_{CI} (the number of completed iterations before starting checkerboard control), N_{CC} (the number of iterations or interval between checkerboard control operations), and N_{RS} (the number of iterations or interval between random shuffle operations). All other parameters not listed above are calculated from these input parameters.

Generally speaking, the EEM search for optimum topology becomes more rigorous by increasing the value of N_{max} and decreasing the values of ε_E and ε_t at the expense of increased computational time. As with the other stochastic methods, if N_{max} is too small, EEM may not be able to find an optimum solution. On the other hand, if ε_E and ε_t are too small, convergence may be hard to achieve. As shown in Figure 11.6, EEM finds the basic layout of the optimum topology in about 20 iterations with the remaining iterations devoted to refinement of the topology. However, when the design domain includes multiple local optima, then it is possible for the topology to vary widely during the solution sequence.

Since element exchange is the main operation in EEM, the value selected for M_{EE} is important. As noted previously, M_{EE} cannot be treated as a constant and must be gradually reduced from its prescribed maximum value ($M_{EE\text{-}max}$) at the very beginning to its imposed minimum ($M_{EE\text{-}min}$) at the end. Based on the multitude of problems examined, choosing $M_{EE\text{-}max} \sim 5-10\%$ and $M_{EE\text{-}min} \sim 0.2-0.4\%$ of the solid elements (MV_0) would be appropriate. Generally speaking, selecting a relatively large value for $M_{EE\text{-}max}$ will lead to the formation of the main load path in the early stages of the optimization process. However, it may also reduce the computational efficiency due to participation of a large number of elements in oscillatory exchange phenomenon discussed earlier. Likewise, choosing a relatively small value for $M_{EE\text{-}min}$ makes the solution easier to converge with a more refined final topology.

Similar to element exchange, the number of elements (M_{RS}) and the interval (N_{RS}) selected for random shuffle are crucial to the success of EEM. Equation (11.5) provides an acceptable reduction scheme for M_{RS} in the range $M_{RS\text{-}min} \leq M_{RS} \leq M_{RS\text{-}max}$. While starting with $M_{RS} = M_{RS\text{-}max}$ widens the domain of exploration for optimum design (similar to a larger coefficient for the particle's velocity in PSO (Fourie and Groenwold, 2001)), it may reduce the computational efficiency by exchanging a large number of elements in a random fashion. On the other hand, a small $M_{RS\text{-}min}$ value helps with convergence while enhancing the final topology. For problems with a large number of solid elements (i.e., large volume fraction) or large MV_0, EEM is less likely to get trapped in a loop (see Section 11.3.1) and the step size for random distribution can be increased to improve the computational efficiency. Choosing $N_{RS} \approx 2-5\%$ N_{max} is found to be sufficient to help EEM not to get trapped in a local optimum while reducing the number of elements involved in oscillatory exchange.

Table 11.7 List of Values Selected for EEM Parameters in the Example Problems Presented

Case	V_0	$n_x \times n_y$	N_{max}	N_{RS}	N_{CC}	N_{CI}	$M_{EE\text{-}max}$ [a]	$M_{EE\text{-}min}$ [a]	ε_E	ε_t
Table 11.1	0.5	90×30	250	7	3	15	60	5	1e-6	1e-3
Table 11.2	0.31	61×62	200	7	3	20	60	5	1e-6	1e-3
Table 11.3—Coarse	0.4	32×20	200	7	3	10	20	3	1e-6	1e-3
Table 11.3—Fine	0.4	64×40	200	7	3	20	100	5	1e-6	1e-3
Table 11.4—Coarse	0.5	24×12	200	7	3	10	20	2	1e-6	1e-3
Table 11.4—Fine	0.5	48×24	300	7	3	20	50	5	1e-6	1e-3
Table 11.5—Coarse	0.5	20×47	150	5	3	10	25	3	1e-6	1e-3
Table 11.5—Fine	0.5	40×94	150	7	3	20	100	5	1e-6	1e-3
Table 11.6	0.4	50×50	150	7	3	20	75	5	1e-6	1e-3
Figure 11.11	0.35	90×90	200	13	4	20	500	15	1e-9	1e-6
Figure 11.12C	0.35	90×90	250	13	4	20	250	10	1e-9	1e-6
Figure 11.12D	0.35	250×250	250	13	4	20	1000	15	1e-9	1e-6

[a] $M_{RS\text{-}max} = M_{EE\text{-}max}$, $M_{RS\text{-}min} = M_{EE\text{-}min}$ in all cases.

As shown in Eq. (11.4), a large value for N_{max} decreases the rate of reduction in M_{EE}, which increases the number of EEM iterations. Although it delays the finding of the final topology, it makes entrapment in a local minimum less likely.

The value selected for N_{CI} should provide sufficient opportunity for the element exchange operation to improve upon the initial topology (random distribution of solid elements), which may include many checkerboard regions. Similarly, N_{CC} should not coincide directly with N_{RS}, as random shuffle is likely to cause the creation of checkerboard elements. Choosing $N_{CI} \approx 5-10\%$ N_{max} and $N_{CC} \approx 1\%$ N_{max} are found to be sufficient to allow both the element exchange and random shuffle operations to improve the design before the accumulated checkerboard elements are identified and eliminated by the checkerboard control procedure.

All of these parameters are flexible, and reasonable deviations from the suggested values may not dramatically affect the final results. The values selected for the EEM parameters in all of the example problems presented earlier are shown in Table 11.7.

11.6 Conclusion

A new stochastic direct search topology optimization method has been developed and used for compliance minimization problems subject to a volume fraction constraint. The nondimensional density of each finite element is treated as a binary design variable with a linear element density-stiffness relationship. The basic principle behind the proposed method is that by exchanging the low-strain-energy solid elements with the high-strain-energy void elements from one iteration to the next, an optimum topology will emerge. The EEM provides converged solutions resulting in minimum strain energy. However, depending on the selected mesh density and the desired level of clarity in the final topology, the number of iterations required for convergence may vary. Through the solution of several example problems, the accuracy and efficiency of EEM were examined and compared with different derivative-based and derivative-free methods reported in the literature.

In general, the EEM method is easy to implement and can be directly coupled with any FE code. Unlike the derivative-based methods, it requires no filtering and the resulting solid–void solution satisfies the imposed volume fraction. The checkerboard control and the random shuffle algorithms help increase the solution fidelity and accuracy. EEM is also found to be significantly more efficient than many other derivative-free methods, such as GA and PSO, reported in the literature.

References

Bendsøe, M.P., Kikuchi, N., 1988. Generating optimal topologies in structural design using a homogenization method. Comput. Method Appl. Mech. Eng. 71 (2), 197–224.

Bendsøe, M.P., Lund, E., Olhoff, N., Sigmund, O., 2005. Topology optimization—broadening the areas of application. Control Cybern. 34 (1), 7–35.

Bendsøe, M.P., Sigmund, O., 2003. Topology Optimization, Theory, Methods and Applications. Springer-Verlag Berlin Heidelberg.

Bruns, T.E., 2005. A reevaluation of the SIMP method with filtering and an alternative formulation for solid-void topology optimization. Struct. Multi. Optim. 30 (6), 428–436.

Diaz, A.R., Sigmund, O., 1995. Checkerboard patterns in layout optimization. Struct. Optim. 10, 40–45.

DorMohammadi, S., Rouhi, M., Rais-Rohani, M., 2012. Topology optimization of blank geometry for the sheet forming process. Proceedings of the ASME 2012 International Design Engineering Technical Conferences. 12–15 August, Chicago, IL.

Fourie, P.C., Groenwold, A. A., 2001. The particle swarm algorithm in topology optimization. In: Proceedings of the Fourth World Congress of Structural and Multidisciplinary Optimization. Dalian, China.

Fuchs, M.B, Jiny, S., Peleg, N., 2005. The SRV constraint for 0/1 topological design. Struct. Multi. Optim. 30 (4), 320–326.

Goldberg, D.E., 1989. Genetic Algorithms in Search, Optimization, and Machine Learning. Addison-Wesley, New York, NY.

Guo, X., Gu, Y.X., 2004. A new density-stiffness interpolation scheme for topology optimization of continuum structures. Eng. Comput. 21 (1), 9–22.

Haber, R.B., Jog, C.S., Bendsoe, M.P., 1996. A new approach to variable-topology design using a constraint on the perimeter. Struct. Optim. 11 (1-2), 11–12.

Huang, X., Xie, Y.M., 2007. Convergent and mesh-independent solutions for the bi-directional evolutionary structural optimization method. Finite Elem. Anal. Des. 43, 1039–1049.

Jakiela, M.J., Chapman, C., Duda, J., Adewuya, A., Saitou, K., 2000. Continuum structural topology design with genetic algorithms. Comput. Methods Appl. Mech. Eng. 186 (2-4), 339–356.

Jog, C.S., Haber, R.B., 1996. Stability of finite element models for distributed parameter optimization and topology design. Comput. Methods Appl. Mech. Eng. 130 (3), 203–226.

Kita, E., Toyoda, T., 1999. Structural optimization using local rules. In: Proceedings of Third World Congress of Structural and Multidisciplinary Optimization. Paper no. 30-SMD-3, May, Niagara Falls, NY.

Kutylowski, R., 2002. On nonuniqueness solutions in topology optimization. Struct. Multi. Optim. 23 (5), 398–403.

Lewinski, T., Rozvany, G.I.N., 2008. Exact analytical solutions for some popular benchmark problems in topology optimization III: L-shaped domains. Struct. Multi. Optim. 35 (2), 165–174.

Lewinski, T., Zhou, M., Rozvany, G.I.N., 1994. Extended exact solutions for least-weight truss layouts-part I: cantilever with a horizontal axis of symmetry. Int. J. Mech. Sci. 36 (5), 375–398.

Liu, J.S., Parks, G.T., Clarkson, P.J., 2000. Metamorphic development: a new topology optimization method for continuum structures. Struct. Multi. Optim. 20 (4), 288–300.

Mattheck, C., Burkhardt, S., 1990. A new method of structural shape optimization based on biological growth. Int. J. Fatigue. 12 (3), 185–190.

Querin, O.M., Steven, G.P., Xie, Y.M., 1998. Evolutionary structural optimization (ESO) using a bidirectional algorithm. Comput. Struct. 15 (8), 1031–1048.

Rouhi, M., Rais-Rohani, M., Williams, T., 2010. Element exchange method for topology optimization. Struct. Multi. Optim. 42 (2), 215–231.

Rozvany, G.I.N., 2009. A critical review of established methods of structural topology optimization. Struct. Multi. Optim. 37 (3), 217–237.

Rozvany, G.I.N., Zhou, M., Birker, T., 1992. Generalized shape optimization without homogenization. Struct. Optim. 6, 200–204.

Rozvany, G.I.N., Lewinski, T., Querin, O. M., Logo, J., 2006. Quality control in topology optimization using analytically derived benchmarks. Eleventh AIAA/ISSMO Multidisciplinary Analysis and Optimization Conference. 6–8 September, Portsmouth, VA.

Sigmund, O., 1994. Design of Material Structures Using Topology Optimization. Ph.D. Thesis. Department of Solid Mechanics, DTU (DCAMM Report S.69).

Sigmund, O., 1997. On the design of compliant mechanisms using topology optimization. Mech. Struct. Mach. 25 (4), 493–524.

Sigmund, O., 2000. Optimum design of microelectromechanical systems: mechanics for a new millennium. Proceedings of the 20th International Congress of Theoretical and Applied Mechanics. 27 August–2 September, Chicago, IL, pp. 505–520.

Sigmund, O., 2001b. A 99 line topology optimization code written in MATLAB. Struct. Multi. Optim. 21 (2), 120–127.

Sigmund, O., 2001a. Design of multiphysics actuators using topology optimization—Part I: one-material structures. Comput. Method Appl. Mech. Eng. 190 (49-50), 6577–6604.

Sigmund, O., Petersson, J., 1998. Numerical instabilities in topology optimization: a survey on procedures dealing with checkerboards, mesh-dependencies and local minima. Struct. Optim. 16 (1), 68–75.

Svanberg, K., 1987. Method of moving asymptotes—a new method for structural optimization. Int. J. Num. Method Eng. 24, 359–373.

Svanberg, K., Werme, M., 2005. A hierarchical neighborhood search method for topology optimization. Struct. Multi. Optim. 29, 325–340.

Wang, M.Y. and Wang, X., 2003. Level set models for structural topology optimization. Proceedings of DETC'03, ASME 2003 Design Engineering Technical Conferences and ASME 2003 Design Engineering Technical Conferences and Computers and Information in Engineering Conference. 2–6 September, Chicago, IL.

Wang, S.Y., Tai, K., Wang, M.Y., 2006. An enhanced genetic algorithm for structural topology optimization. Int. J. Numer. Meth. Eng. 65 (1), 18–44.

Xie, Y.M., Steven, G.P., 1993. A Simple evolutionary procedure for structural optimization. Eng. Comput. 49 (5), 885–896.

Zhou, M., Rozvany, G.I.N., 1991. The COC algorithm, part II: topological, geometry and generalized shape optimization. Comput. Method Appl. Mech. Eng. 89 (1-3), 197–224.

Part Two

Structural Control and Identification

12 Evolutionary Path-Dependent Damper Optimization for Variable Building Stiffness Distributions

Izuru Takewaki and Kohei Fujita

Department of Architecture and Architectural Engineering
Kyoto University, Kyoto, Japan

12.1 Introduction

Supplemental passive dampers are widely accepted as effective vibration control devices for high-rise buildings. In the early stages of development of passive structural control, the installation itself of supplemental dampers was the major objective. It appears natural that, after extensive developments of various damper systems, another target was directed to the development of smart and effective installation of such dampers.

However, research on optimal damper placement has been limited. Several studies have dealt with this subject in the early stage. De Silva (1981) presented a gradient algorithm for the optimal design of discrete dampers in the vibration control of a class of flexible systems. Constantinou and Tadjbakhsh (1983) derived the optimum damping coefficient for a damper placed on the first story of a shear building subjected to horizontal ground motions. Gurgoze and Muller (1992) presented a numerical optimal design method for a single viscous damper (VD) in a prescribed linear multi-degree-of-freedom system. Hahn and Sathiavageeswaran (1992) performed parametric studies on the effects of damper distribution on the earthquake response of buildings, and showed that, for a building with uniform story stiffnesses, dampers should be added to floors of the lower half of the building. Tsuji and Nakamura (1996) proposed an algorithm to find both the optimal story stiffness and damper distributions for a shear building subjected to the spectrum-compatible ground motions.

Rather recently, Takewaki (1997) developed another optimal method for the smart damper placement with the help of the concepts of inverse problem approaches and optimal criteria-based design approaches. He solved the problem of optimal damper placement by deriving the optimality criteria and then by developing an incremental inverse problem approach. Subsequently, Takewaki and

Metaheuristic Applications in Structures and Infrastructures. DOI: http://dx.doi.org/10.1016/B978-0-12-398364-0.00012-7

Yoshitomi (1998) and Takewaki (2000) introduced a different approach based on the concept of optimal sensitivity. The optimal quantity of passive dampers is obtained automatically together with the optimal placement through this new method. The essence of these approaches is summarized in Takewaki (2009).

In the meanwhile, significant works have been developed by many researchers (Attard, 2007; Cimellaro, 2007; Cimellaro and Retamales, 2007; Garcia, 2001; Kiu et al., 2004; Lavan and Levy, 2005, 2006; Levy and Lavan, 2006; Liu et al., 2003; Marano et al., 2007; Trombetti and Silvestri, 2004, 2007; Uetani et al., 2003; Viola and Guidi, 2008). Most of these studies have developed new optimal design methods of supplemental dampers and proposed effective and useful approaches.

In this paper, an evolutionary method is proposed for finding the optimal design of both dampers and their supporting members to minimize an objective function of a linear multistory structure subjected to critical resonant ground input. While a great deal of research has been done on the design of passive dampers themselves, research on damper design including supporting members is very limited. The objective function is taken as the sum of the mean squares of the inter-story drifts. From the practical point of view, two different problems of optimal damper placement considering the effect of supporting members are investigated. In the first part, the stiffness of each supporting member is treated as another design variable and the axial force of each supporting member is constrained within a variable upper limit (i.e., the yield force). When the condition on the axial force is active, both damper capacities and stiffnesses of supporting members should be updated to satisfy the stationarity conditions. In the second part, the stiffness of supporting members is given as a fixed design value and the axial force of each supporting member is constrained within a fixed upper limit. In the second problem, the damper placement should be determined according to not only the stationary conditions but also the conditions on the axial force of supporting members. Since the optimal damper placement is dependent on the initial design and the treatment of supporting members, the obtained damper placement should be said to be path-dependent. This characteristic will be explored through numerical examples.

12.2 Concept of Adaptive Sensitivity

In this paper, the new concept of adaptive sensitivity of optimal damper placement is introduced. The optimal solution such as optimal damper placement is effective only for preassigned design parameters, e.g., stiffness of framed structures. However, uncertainties with respect to structural properties caused by construction error or performance deterioration due to aging cannot be neglected. For this reason, more reliable optimal design methodologies are desired. The adaptive sensitivity of optimal damper placement is aimed at investigating the sensitivity of optimal damper placement considering multiple

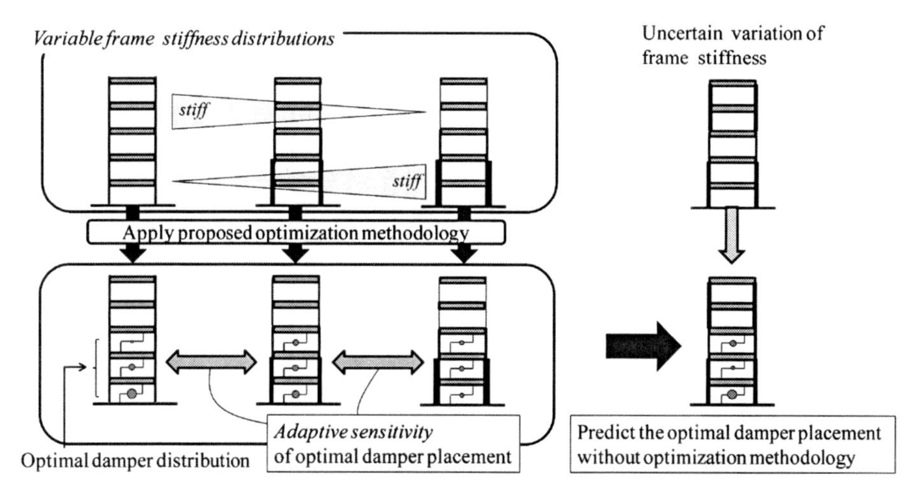

Figure 12.1 Concept of adaptive sensitivity of optimal damper placement.

variations of other design parameters, e.g., variable frame stiffness variations. Once the adaptive sensitivity of the optimal damper placement is obtained, a wider design decision can be performed in the next phase. Figure 12.1 shows the concept of adaptive sensitivity.

In this paper, the method of analysis of the adaptive sensitivity is proposed. The effectiveness of the adaptive sensitivity and the relationship between the optimal damper placement and frame stiffness distributions are discussed later.

12.3 Structural Model with Passive Dampers

Consider an N-story planar shear building model with both VD and supporting members as shown in Figure 12.2. Let M_i, k_{Fi}, k_{bi}, c_{Fi}, and c_{Di} ($i = 1,\ldots, N$) denote the floor mass, the story stiffness of the frame, the stiffness of the supporting member, the structural damping coefficient, and the additional damping coefficient by the passive damper in the ith story, respectively. k_{Fi} and c_{Fi} are placed in parallel and k_{bi} and c_{Di} are in series as shown in Figure 12.2.

The supplemental damper with a supporting member is treated as a detailed model in which a small lumped mass m_i is allocated between the damper and the supporting member as shown in Figure 12.3. Let u_{Fi} and u_{bi} denote the ith story floor displacement and the displacement of the lumped mass relative to the ground, respectively. This detailed damper model has two degrees of freedom in each story. $\mathbf{U}(\omega)$ and $\ddot{U}_g(\omega)$ denote the Fourier transforms of the horizontal displacement vector $\mathbf{u}(t) = \{u_{F1}, u_{b1}, \ldots, u_{FN}, u_{bN}\}$ and the horizontal acceleration $\ddot{u}_g(t)$ of the ground motion. The equations of

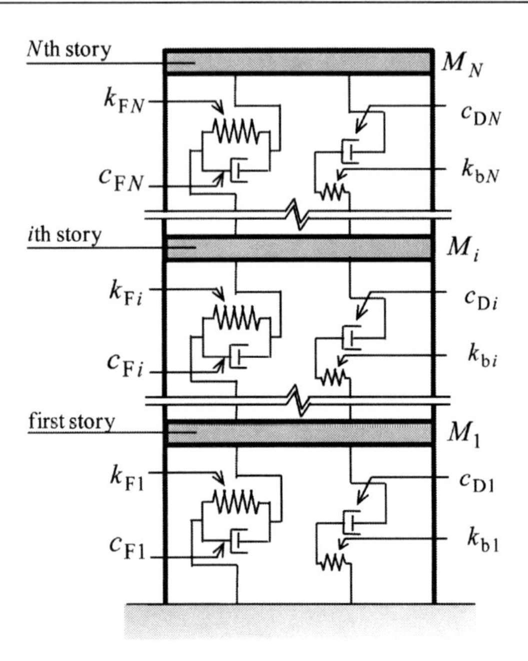

Figure 12.2 *N*-story shear building model with passive dampers and supporting members.

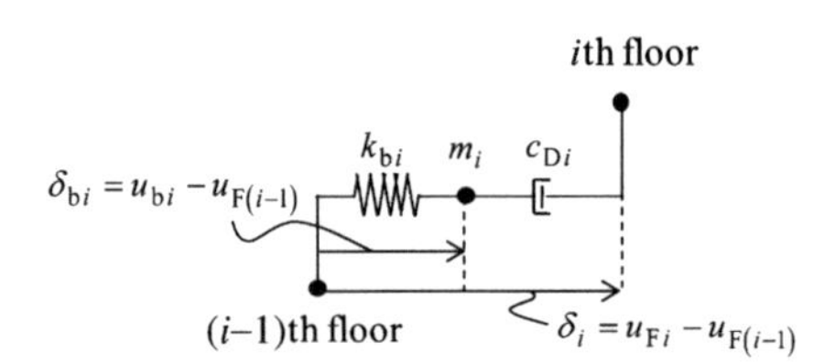

Figure 12.3 Damper unit with supporting member.

motion for the building without VD subjected to $\ddot{u}_g(t)$ can be expressed in frequency domain by:

$$(-\omega^2\mathbf{M} + \mathrm{i}\omega\mathbf{C} + \mathbf{K})\,\mathbf{U}(\omega) = -\mathbf{Mr}\ddot{U}_g(\omega) \tag{12.1}$$

where **M**, **C**, and **K** are the system mass, structural damping, and stiffness matrices, respectively. The components of these matrices are linear combination of M_i, m_i, k_{Fi}, k_{bi} and c_{Fi}. Furthermore, $\mathbf{r} = \{1,\ldots,1\}^{\mathrm{T}}$ is the influence coefficient vector.

When passive VD are added to the building, Eq. (12.1) can be modified to:

$$\{-\omega^2\mathbf{M} + \mathrm{i}\omega(\mathbf{C} + \mathbf{C}_\mathrm{D}) + \mathbf{K}\}\,\mathbf{U}(\omega) = -\mathbf{Mr}\ddot{U}_g(\omega) \tag{12.2}$$

where $\mathbf{C}_\mathrm{D}$ denotes the damping matrix by the added VD. The detailed expressions of the system matrices are shown in Fujita et al. (2010). Equation (12.2) can be described simply as:

$$\mathbf{AU}(\omega) = \mathbf{B}\ddot{U}_\mathrm{g}(\omega) \tag{12.3}$$

where

$$\mathbf{A} = -\omega^2\mathbf{M} + \mathrm{i}\omega(\mathbf{C} + \mathbf{C}_\mathrm{D}) + \mathbf{K},\ \mathbf{B} = -\mathbf{Mr} \tag{12.4}$$

The Fourier transforms $\mathbf{D}(\omega) = \{D_1, \ldots, D_N\}^\mathrm{T}$ of the inter-story drifts can then be derived as:

$$\mathbf{D}(\omega) = \mathbf{TU}(\omega) \tag{12.5}$$

where $\mathbf{T}$ is a constant transformation matrix consisting of 1, -1, and 0.

The transfer functions $\mathbf{H}_\mathrm{D}(\omega) = \{H_{\mathrm{D}_1}(\omega), \ldots, H_{\mathrm{D}_N}(\omega)\}^\mathrm{T}$ of the inter-story drifts can be derived by:

$$\mathbf{H}_\mathrm{D}(\omega) = \mathbf{TA}^{-1}\mathbf{B} \tag{12.6}$$

It is remarkable that, since the mean squares of the response can be evaluated by multiplying the power spectral density (PSD) function $S_\mathrm{g}(\omega)$ of the horizontal acceleration $\ddot{u}_\mathrm{g}(t)$ on the squared transfer function and integrating that in the frequency domain, the amplitude of the transfer function is meaningful for designing the building. The mean squares of the inter-story drift $\delta_i(t)$ is derived by:

$$E[\delta_i(t)^2] \cong \int_{-\infty}^{\infty} |\mathbf{H}_\mathrm{D}(\omega)|^2 S_\mathrm{g}(\omega) d\omega \tag{12.7}$$

The maximum response in time domain can then be evaluated by multiplying a peak factor ρ.

$$\max_t |\delta_i(t)| = \rho \left\{ \int_{-\infty}^{\infty} |H_{\mathrm{D}i}(\omega)|^2 S_\mathrm{g}(\omega) d\omega \right\}^{1/2} \tag{12.8}$$

It is well understood that the stiffness of supporting members should be large enough to ensure the damper effectiveness. When the force $N_{\mathrm{b}i}$ in the supporting member exceeds a limit, an unpreferable influence may be induced to the main frame. To avoid this, the maximum axial force of the supporting member should be constrained within a limit, e.g., a criterion on the yield force of the supporting member. The maximum value of the axial force of the supporting member can be evaluated by:

$$\max_t |N_{\mathrm{b}i}(t)| = \rho k_{\mathrm{b}i} \left\{ \int_{-\infty}^{\infty} |H_{\mathrm{B}i}(\omega)|^2 S_\mathrm{g}(\omega) d\omega \right\}^{1/2} \tag{12.9}$$

where $H_{\mathrm{B}i}(\omega)$ is the ith component of the transfer function $\mathbf{H}_{\mathrm{B}}(\omega)$ of the supporting member displacement relative to the floor. $\mathbf{H}_{\mathrm{B}}(\omega)$ can be derived by replacing $\mathbf{T}$ with a transformation matrix $\mathbf{T}_{\mathrm{B}}$ from the nodal displacements to the relative displacements between both ends of the supporting members.

12.4 Critical Excitation for Variable Design

In the seismic-resistant design of important structures, time-history response analyses are often used for a set of recorded ground motions. However, it is well recognized that the ground motions include various uncertainties of different levels. In order to overcome these uncertainties, more reliable and robust structural design methods are needed. The critical excitation method (Drenick, 1970; Takewaki, 2006) is adopted in this paper. Takewaki (2002) introduced the concept of variable critical excitation, which has a variable resonant frequency close to the fundamental natural frequency of the structure with varied stiffnesses of the supplemental dampers. Based on this concept, a problem is posed here such that the optimal damper placement is identified together with the selection of the optimal stiffness of the supporting members.

Let $S_{\mathrm{g}}(\omega)$ denotes the PSD function of the input ground acceleration $\ddot{u}_{\mathrm{g}}(t)$. The constraints on $S_{\mathrm{g}}(\omega)$ are the power of the PSD function (i.e., the area under the PSD function representing the input variance) and the intensity of the PSD function (i.e., the maximum or the peak value of the PSD function), described by:

$$\int_{-\infty}^{\infty} S_{\mathrm{g}}(\omega)d\omega \le \overline{S}, \quad \sup S_{\mathrm{g}}(\omega) \le \overline{s} \tag{12.10}$$

where $\overline{S}$ and $\overline{s}$ are the limits on the power and intensity, respectively. These parameters of the critical excitation are determined from the analyses of recorded ground motions. The shape of the PSD function as a solution to this problem is

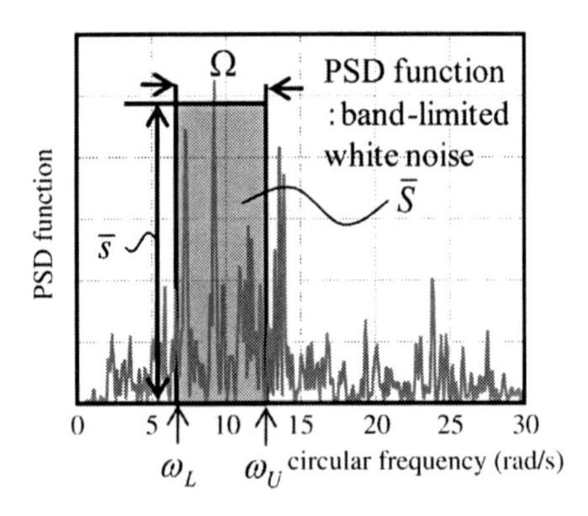

Figure 12.4 Power spectrum density function.

assumed to be a Dirac delta function or a band-limited white noise. A band-limited white noise is shown in Figure 12.4, where a frequency bandwidth Ω and upper and lower bounds ω_U, ω_L of frequency are obtained from the given parameters $\overline{S}$ and $\overline{s}$.

12.5 Optimal Design Problem

In this section, two different optimal design problems are introduced where the optimal design can be determined to minimize the particular response quantity for the N-story shear building model subjected to the critical excitation. The primary objective function f to be minimized in both optimal design problems is given by the sum of the mean squares of inter-story drifts.

12.5.1 Performance-Based Optimal Design with Multiple Design Parameters

The first problem of the optimal design for the N-story shear building model is to find both the optimal distribution of VD capacities $\mathbf{C}_\mathrm{d} = \{c_{\mathrm{D}1}, \ldots, c_{\mathrm{D}N}\}$ and the optimal stiffness of supporting members $\mathbf{k}_\mathrm{b} = \{k_{\mathrm{b}1}, \ldots, k_{\mathrm{b}N}\}$ so as to minimize the objective function f. The problem can be stated as:

Find: $\mathbf{C}_\mathrm{d}, \mathbf{k}_\mathrm{b}$

$$\text{so as to minimize:} \; f = \sum_{i=1}^{N} \delta_i^2(\mathbf{C_d}, \mathbf{k_b}) \tag{12.11}$$

$$\text{subject to:} \; \sum_{i=1}^{N} c_{\mathrm{D}i} = \overline{W}, \quad N_{\mathrm{b}i}(\mathbf{C_d}, \mathbf{k_b}) \leq \overline{P}_{\mathrm{y}i}(k_{\mathrm{b}i}), \quad 0 \leq c_{\mathrm{D}i} \; (i = 1, \ldots, N)$$

The first constraint is on the total damper capacity $\overline{W}$, which can be determined by the condition on the maximum inter-story deformation angle described by:

$$\max_i \left(\rho \sqrt{\delta_i^2(\mathbf{C}_\mathrm{d})} / h_i \right) \leq R_\mathrm{D} \tag{12.12}$$

In Eq. (12.12), $\rho\sqrt{\delta_i^2}/h_i$ and R_D denote the maximum inter-story deformation angle and its limit, respectively. The supplemental dampers have to be added until Eq. (12.12) is satisfied. The second constraint is on the axial force of the supporting members. In Eq. (12.11), $\overline{P}_{\mathrm{y}i}$ is the yield force of the ith story supporting member. It should be noted that the upper limit with respect to the yield force is to be updated according to the variation of stiffness $\mathbf{k}_\mathrm{b}$ of the supporting members.

12.5.2 Performance-Based Optimal Damper Placement for Given Supporting Members

From the structural design point of view, it may not be preferable that the supporting members are designed independently in each story. The second problem of the optimal design for the N-story shear building model is to find the optimal distribution of VD capacities $\mathbf{C}_\mathrm{d} = \{c_{\mathrm{D}1}, \ldots, c_{\mathrm{D}N}\}$ so as to minimize the objective function f. The stiffness of supporting members is assumed to be given and fixed at an initial value.

Find : $\mathbf{C}_\mathrm{d}$

$$\text{so as to minimize:} \; f = \sum_{i=1}^{N} \delta_i^2(\mathbf{C_d}) \tag{12.13}$$

$$\text{subject to:} \; \sum_{i=1}^{N} c_{\mathrm{D}i} = \overline{W}, \quad N_{\mathrm{b}i}(\mathbf{C}_\mathrm{d}) \leq \overline{P}_{\mathrm{y}i}, \quad 0 \leq c_{\mathrm{D}i} \quad (i = 1, \ldots, N)$$

The upper limit of the axial force N_b of supporting members, that is, yield force $\overline{\mathbf{P}}_\mathrm{y} = \{\overline{P}_{\mathrm{y}1}, \ldots, \overline{P}_{\mathrm{y}N}\}$ of the supporting member, is fixed and does not change.

12.6 Optimality Conditions

The generalized Lagrangian L for the first optimal design problem can be defined as:

$$L(\mathbf{C}_\mathrm{d}, \mathbf{k}_\mathrm{b}, \lambda, \boldsymbol{\mu}, \boldsymbol{\kappa}) = f + \lambda\left(\sum_{i=1}^{N} c_{\mathrm{D}i} - \overline{W}\right) + \sum_{i=1}^{N} \mu_i(0 - c_{\mathrm{D}i}) + \sum_{i=1}^{N} \kappa_i(N_{\mathrm{b}i} - \overline{P}_{\mathrm{y}i}) \tag{12.14}$$

where λ, $\boldsymbol{\mu} = \{\mu_i\}$, and $\boldsymbol{\kappa} = \{\kappa_i\}$ are the Lagrange multipliers. The principal optimality conditions for this problem without active upper and lower bound conditions on $\mathbf{C}_\mathrm{d}$ and N_b may be derived from the stationarity conditions of $L(\boldsymbol{\mu} = \mathbf{0}, \boldsymbol{\kappa} = \mathbf{0})$ with respect to $\mathbf{C}_\mathrm{d}$ and $\mathbf{k}_\mathrm{b}$.

$$f_{,j} + \lambda = 0 \quad \text{for} \quad \sum_{i=1}^{N}(c_{\mathrm{D}i}) - \overline{W} = 0, \quad 0 < c_{\mathrm{D}j}, \quad N_{\mathrm{b}j} < \overline{P}_{\mathrm{y}j} \tag{12.15}$$

$$f^{,j} + \lambda = 0 \quad \text{for} \quad \sum_{i=1}^{N}(c_{\mathrm{D}i}) - \overline{W} = 0, \quad 0 < c_{\mathrm{D}j}, \quad N_{\mathrm{b}j} < \overline{P}_{\mathrm{y}j} \tag{12.16}$$

The symbols $(f)_{,j}$ and $(f)^{,j}$ denote the partial differentiation with respect to $c_{\mathrm{D}j}$ and $k_{\mathrm{b}j}$. Equation (12.15) indicates that the derivative of the objective function with respect to c_{Dj} is constant if the conditions $\sum_{i=1}^{N}(c_{Di}) - \overline{W} = 0$ and $0 < c_{Dj}$ are satisfied. On the other hand, Eq. (12.16) implies that the derivative of the objective function with respect to k_{bj} is constant if the condition $N_{bj} < \overline{P}_{yj}$ is satisfied.

In the process of increasing the quantity of VD in each story, the axial force of the supporting member usually increases. When the constraint on axial force of the supporting member is active, the optimality conditions should be modified by the stationarity conditions $L(\mu = 0)$ as follows:

$$f_{,j} + \lambda + \sum_{i=q_1}^{q_n}(\kappa_i N_{bi,j}) = 0 \quad \text{for} \quad \sum_{i=1}^{N}(c_{Di}) - \overline{W} = 0,\ 0 < c_{Dj},\ N_{bi} = \overline{P}_{yi} \quad (i = q_1, \ldots, q_n) \tag{12.17}$$

$$f^{,j} + \sum_{i=q_1}^{q_n} \kappa_i (N_{bi}{}^{,j} - \overline{P}_{yi}{}^{,j}) = 0 \quad \text{for} \quad N_{bi} = \overline{P}_{yi} \quad (i = q_1, \ldots, q_n) \tag{12.18}$$

where n and q_i denote the number of stories and their locations having active constraints on axial force of the supporting member. In Eq. (12.18), it is assumed that the partial differentiation of ith story axial force N_{bi} with respect to other story's supporting member stiffness k_{bj} can be neglected. In addition, the yield force $\overline{P}_{yj}$ of each supporting member is assumed to be a function of only the stiffness k_{bj} of that supporting member. As a result, Eq. (12.18) can be reduced to:

$$f^{,j} + \kappa_j (N_{bj}^{,j} - \overline{P}_{yj}^{,j}) = 0 \tag{12.19}$$

It is noted that the Lagrange multiplier κ_j can be evaluated directly from Eq. (12.19). When the other constraint on upper bound of damper's area is active, the optimality conditions should be modified by:

$$f_{,j} + \sum_{i=q_1}^{q_n}(\kappa_i N_{bi,j}) \leq 0 \quad \text{for} \quad \sum_{i=1}^{N}(c_{Di}) - \overline{W} = 0, \quad N_{bi} = \overline{P}_{yi} \quad (i = q_1, \ldots, q_n) \tag{12.20}$$

On the other hand, in the second optimal design problem, we need not include the optimality conditions with respect to the stiffness $\mathbf{k}_b$ of the supporting members.

12.7 Solution Procedure of Optimal Design Problem

A gradient-based evolutionary solution algorithm has been presented for the problem of optimal damper placement (Takewaki et al., 1999). Since it is quite beneficial to obtain the optimal damper placement for various capacities of dampers, the total damper quantity $\overline{W}$ is increased gradually. In this section, both the solution procedures of the optimal design problems are presented.

12.7.1 Solution Procedure for the Performance-Based Optimal Design with Multiple Design Parameters

The initial model is assumed to have no supplemental dampers, i.e., $c_{Di} = 0 \quad (i = 1, \ldots, N)$. Additional VD is distributed via the steepest direction search algorithm (Takewaki, 2009). Let $\Delta c_D = \{\Delta c_{Di}\}$ and ΔW denote the increment of VD damping capacity and the increment of the sum of VD damping capacity, respectively. When ΔW is given, it is needed to find the optimal placement to decrease the objective function most effectively. For this purpose, the first- and the second-order sensitivities of the objective function with respect to the design variables $\mathbf{C}_d$ and $\mathbf{k}_b$ are necessary. In the concerned optimal design problem, those sensitivities $f_{,j}$, $f^{,j}$, $f_{,jk}$, $f_{,j}^{,k}$, and $f^{,jk}$ can be derived explicitly. The solution procedure may be summarized as follows:

Step 0. Design the main frame without VD under the system dependent critical excitation.

Step 1. Calculate the fundamental natural circular frequency ω_0 of the current structural system.

Step 2. Create the critical PSD function $S_g(\omega)$ as a band-limited white noise.

Step 3. Evaluate the axial force N_{bi} of the supporting member and count the number n of stories in which N_{bi} reaches its yield axial force $\overline{P}_{yi}$.

Step 4. Identify the location of the story where the absolute value of the first-order sensitivity $f_{,j}$ of the objective function is maximized.

Step 5. Count the number m of stories where $\max_j |f_{,j}|$ coincides in the multiple stories.

The above global procedures can further be subdivided into four different domains depending on the values m and n. An appropriate set of optimality conditions should be selected from Eqs. (12.15) to (12.18).

Step 6A. When $m = 1, \quad n = 0$: The increment ΔW of VD is added only to the specific story attaining $\max_j |f_{,j}|$.

Step 6B. When $m \geq 2, \quad n = 0$: The optimal damper distribution $c_{Di} \quad (i = p_1, \ldots, p_m)$ has to be computed and updated to keep the coincidence of the multiple first-order sensitivities.

Step 6C. When $m = 1, \quad n = 1$: The stiffness k_{bi} of the supporting member is increased to prevent N_{bi} from exceeding the yield axial force $\overline{P}_{yi}$.

Step 6D. When $m \geq 2$, $n \geq 1$: All the optimality conditions have to be satisfied. The multiple first-order sensitivities coincide due to Eq. (12.17) and the corresponding k_{bi} is increased to satisfy $N_{bi} = \overline{P}_{yi}$.

Step 7. Update design variables $\mathbf{C}_d$ and $\mathbf{k}_b$ according to the optimality conditions.

Step 8. Repeat step 1 through step 7 until the performance criterion, that is, Eq. (12.12), is satisfied.

For clarification, the procedure of the steps 6C and 6D are explained in more detail below.

Step 6C. When N_{bj} attains its upper bound, $\overline{P}_{yj}$, k_{bj} has to be increased so as to keep the increment ΔN_{bj} coinciding with $\Delta \overline{P}_{yj}$. This requires the following equation:

$$dN_{bj} = d\overline{P}_{yj} \Rightarrow \sum_{i=q_1}^{q_n} (N_{bj,i} \Delta c_{Di}) + N_{bj}^{,j} \Delta k_{bj} = \overline{P}_{yj}^{,j} \Delta k_{bj} \quad (j = q_1, ..., q_n) \tag{12.21}$$

Here in the assumption discussed below in Eq. (12.18) is employed again. When $n = 1$, the increment Δk_{bj} of the stiffness of the supporting member can be derived by:

$$\Delta k_{bq_1} = -N_{bq_1,q_1} \Delta c_{Dq_1} / (N_{bq_1}^{,q_1} - \overline{P}_{yq_1}^{,q_1}) \tag{12.22}$$

Step 6D. The distribution of VD to the multiple stories should be conducted for ΔW. In this case, the number of unknown variables is $m + n$. From Eq. (12.21), n equations with respect to $\mathbf{k}_b$ can be derived. More m equations are needed. Successive satisfaction of Eq. (12.17) requires that:

$$\begin{aligned} &\sum_{i=p_1}^{p_m} f_{ji} \Delta c_{Di} + \sum_{i=q_1}^{q_n} f_{j}^{,i} \Delta k_{bi} + \sum_{i=p_1}^{p_m} \sum_{k=q_1}^{q_n} (\kappa_k N_{bk,j})_{,i} \Delta c_{Di} \\ &\quad + \sum_{i=q_1}^{q_n} \sum_{k=q_1}^{q_n} (\kappa_k N_{bk,j})^{,i} \Delta k_{bi} = 0 \quad (j = q_1, \ldots, q_n) \end{aligned} \tag{12.23}$$

where κ_k in Eq. (12.23) can be derived from Eq. (12.19) as:

$$\kappa_k = -f^{,k} / (N_{bk}^{,k} - \overline{P}_{yk}^{,k}) \quad (k = q_1, \ldots, q_n) \tag{12.24}$$

After the multiple optimality conditions are updated, the first-order sensitivities should continue to be satisfied. To achieve this, the following equation can be derived by substituting Eq. (12.24) into Eq. (12.25).

$$\begin{aligned} &\sum_{i=p_1}^{p_m} \left[f_{,ji} - \sum_{k=q_1}^{q_n} \{ f^{,k} N_{bk,j} / (N_{bk}^{,k} - \overline{P}_{yk}^{,k}) \}_{,i} \right] \Delta c_{Di} \\ &+ \sum_{i=q_1}^{q_n} \left[f_{,j}^{,i} - \sum_{k=q_1}^{q_n} \{ f^{,k} N_{bk,j} / (N_{bk}^{,k} - \overline{P}_{yk}^{,k}) \}^{,i} \right] \Delta k_{bi} = \text{const.} \quad (j = q_1, ..., q_n) \end{aligned} \tag{12.25}$$

In case of using Eq. (12.17), the following equations should be employed in place of Eq. (12.15).

$$\sum_{i=p_1}^{p_m} (f_{,ji} \Delta c_{Di}) + \sum_{i=q_1}^{q_n} (f_{,j}^{,i} \Delta k_{bi}) = \text{const.} \quad (j = p_{n+1}, ..., p_m) \tag{12.26}$$

After some manipulation in Eq. (12.26), we can derive $m-1$ equations with respect to $\mathbf{C}_\mathrm{d}$ and $\mathbf{k}_\mathrm{b}$. In addition, the sum of the increment Δc_D have to be the same as ΔW, i.e., $\sum_{i=p_1}^{p_m} \Delta c_{\mathrm{D}i} = \Delta W$.

Finally, we can derive the following simultaneous linear equations for the increments of $\mathbf{C}_\mathrm{d}$ and $\mathbf{k}_\mathrm{b}$.

$$\begin{bmatrix} \alpha_{11} & \cdots & \alpha_{1m} & \beta_{11} & \cdots & \beta_{1n} \\ \vdots & \ddots & \vdots & \vdots & \ddots & \vdots \\ \alpha_{m-1,1} & \cdots & \alpha_{m-1,m} & \beta_{m-1,1} & \cdots & \beta_{m-1,n} \\ N_{\mathrm{b}q_1,p_1} & \cdots & N_{\mathrm{b}q_1,p_m} & N_{\mathrm{b}_{q_1}}^{,q_1} - \overline{P}_{\mathrm{y}q_1}^{,q_1} & \cdots & N_{\mathrm{b}_{q_1}}^{,q_n} - \overline{P}_{\mathrm{y}q_1}^{,q_n} \\ \vdots & \ddots & \vdots & \vdots & \ddots & \vdots \\ N_{\mathrm{b}q_n,p_1} & \cdots & N_{\mathrm{b}q_n,p_m} & N_{\mathrm{b}q_n}^{,q_1} - \overline{P}_{\mathrm{y}q_n}^{,q_1} & \cdots & N_{\mathrm{b}q_n}^{,q_n} - \overline{P}_{\mathrm{y}q_n}^{,q_n} \\ 1 & \cdots & 1 & 0 & \cdots & 0 \end{bmatrix} \begin{Bmatrix} \Delta c_{\mathrm{D}p_1} \\ \vdots \\ \Delta c_{\mathrm{D}p_m} \\ \Delta k_{\mathrm{b}q_1} \\ \vdots \\ \Delta k_{\mathrm{b}q_n} \end{Bmatrix} = \begin{Bmatrix} 0 \\ \vdots \\ 0 \\ \Delta W \end{Bmatrix} \tag{12.27}$$

where α_{ij} $(i = 1, 2, \ldots, m-1;\ j = 1, 2, \ldots, m)$ and β_{ij} $(i = 1, 2, \ldots, m-1;$ $j = 1, 2, \ldots, n)$ are described by:

$$\begin{aligned} \alpha_{ij} =& f_{,q_1p_j} - f_{,q_{i+1}p_j} \\ &+ \sum_{k=q_1}^{q_n} \frac{1}{N_{\mathrm{b}k}^{,k} - \overline{P}_{\mathrm{y}k}^{,k}} \begin{Bmatrix} -f_{,p_j}^{,k}(N_{\mathrm{b}k,q_1} - N_{\mathrm{b}k,q_i}) - f^{,k}(N_{\mathrm{b}k,q_1p_j} - N_{\mathrm{b}k,q_ip_j}) \\ + f^{,k} N_{\mathrm{b}k,p_j}^{,k}(N_{\mathrm{b}k,q_1} - f^{,k} N_{\mathrm{b}k,q_i})/(N_{\mathrm{b}k}^{,k} - \overline{P}_{\mathrm{y}k}^{,k}) \end{Bmatrix} \\ & (i = 1, \ldots, n-1 \quad j = 1, \ldots, m) \end{aligned} \tag{12.28a}$$

$$\alpha_{ij} = f_{,p_np_j} - f_{,p_{i+1}p_j} \quad (i = n, \ldots, m-1 \quad j = 1, \ldots, m) \tag{12.28b}$$

$$\begin{aligned} \beta_{ij} =& f_{,p_1}^{,q_j} - f_{,p_{i+1}}^{,q_j} \\ &+ \sum_{k=q_1}^{q_n} \frac{1}{N_{\mathrm{b}k}^{,k} - \overline{P}_{\mathrm{y}k}^{,k}} \begin{Bmatrix} -f^{,kq_j}(N_{\mathrm{b}k,q_1} - N_{\mathrm{b}k,q_{i+1}}) - f^{,k}(N_{\mathrm{b}k,q_1}^{,q_j} - N_{\mathrm{b}k,q_{i+1}}^{,q_j}) \\ + N_{\mathrm{b}k}^{,kq_j}(f^{,k} N_{\mathrm{b}k,q_1} - f^{,k} N_{\mathrm{b}k,q_{i+1}})/(N_{\mathrm{b}k}^{,k} - \overline{P}_{\mathrm{y}k}^{,k}) \end{Bmatrix} \\ & (i = 1, \ldots, n-1 \quad j = 1, \ldots, n) \end{aligned} \tag{12.28c}$$

$$\beta_{ij} = f_{,p_n}^{,q_j} - f_{,p_{i+1}}^{,q_j} \quad (i = n, \ldots, m-1 \quad j = 1, \ldots, n) \tag{12.28d}$$

12.7.2 Solution Procedure for the Optimal Design Problem with Given Supporting Members

Since the design parameter for the optimal design is only the distribution of the damper capacity $\mathbf{C}_\mathrm{d}$ in the second optimal design problem, the optimality conditions with respect to the stiffness of supporting member $\mathbf{k}_\mathrm{b}$ such as Eqs. (12.16) and (12.18) are unnecessary. For this reason, the optimization procedure after the step 6B in the previous flowchart is different from the present optimal design problem. When the condition with respect to the axial force N_{bi} of the supporting member is active, the distribution of the damper capacity $\mathbf{C}_d$ is required to be updated to keep N_{bi} within a preassigned limit. The procedure after the step 6B has to be replaced by the following:

Step 6C. When $m = 1$, $n \geq 1$: The additional VD should be distributed to multiple stories to prevent N_{bi} from exceeding the yield axial force $\overline{P}_{yi}$.

Step 6D. When $m \geq 2$, $n \geq 1$: Both the optimality condition and the condition on the axial force N_b have to be satisfied. This corresponds to the conditions that the multiple first-order sensitivities coincide and the corresponding k_{bi} is increased to satisfy $N_{bi} = \overline{P}_{yi}$.

Step 7. Update design variables $\mathbf{C}_d$ according to the optimality conditions or the condition on upper limit of the axial force N_b.

Step 8. Repeat step 1 through step 7 until the performance criterion is satisfied.

The detailed procedures in steps 6C and 6D of the present problem are explained in the following:

Step 6C. When N_{bj} attains its upper bound $\overline{P}_{yj}$, adding VD to the ith story should be restrained regardless of the gradient vector $f_{,i}$ of the objective function. The increment ΔN_{bj} of the axial force in the ith story should be zero to prevent N_{bi} from exceeding the yield axial force $\overline{P}_{yi}$. To accomplish this, it is needed to add VD to another story including a current ith story identified in step 4. This requires the following equation of the increment Δc_D of VD:

$$\sum_{i=q_1}^{q_{n+1}} (\Delta c_{Di} N_{bj,i}) = 0 \quad (j = 1, \ldots, n) \tag{12.29}$$

where q_{n+1} corresponds to the story location having the $(n+1)$th largest value in the first-order sensitivity $f_{,i}$ of the objective function. In addition, the sum of the increment Δc_D should satisfy Eq. (12.27). In this procedure, VD may be distributed to multiple stories contrary to the optimality conditions to satisfy the criterion requirement. For this reason, the number n may increase even if $m = 1$. When $n \geq 2$, the number of unknown values of Δc_D also becomes n. Finally, the following simultaneous linear equation for the increment Δc_D can be derived from the Eqs. (12.26) and (12.29).

$$\begin{bmatrix} N_{q_1,q_1} & \cdots & N_{q_1,q_{n+1}} \\ \vdots & & \vdots \\ N_{q_n,q_1} & \cdots & N_{q_n,q_{n+1}} \\ 1 & \cdots & 1 \end{bmatrix} \begin{Bmatrix} \Delta c_{Dq_1} \\ \vdots \\ \Delta c_{Dq_{n+1}} \end{Bmatrix} = \begin{Bmatrix} 0 \\ \vdots \\ 0 \\ \Delta W \end{Bmatrix} \tag{12.30}$$

Step 6D. When $m \geq 2$ and $n \geq 1$, both the optimality condition with respect to c_D and the condition on the axial force N_b of the supporting member should be satisfied. In this case, the number of unknown variables is the larger value of either m

or $n+1$. For instance, in the case of $m=2$, $n=1$, we need two equations with respect to Δc_{D}. As shown in the procedure of step 6C in the previous optimal design problem, successive satisfaction of Eq. (12.17) can be performed more simply by:

$$\sum_{i=p_1}^{p_m}(f_{,ji}\Delta c_{\mathrm{D}i}) + \sum_{i=p_1}^{p_m}\sum_{k=q_1}^{q_n}\{(\kappa_k N_{\mathrm{b}k,j})_{,i}\Delta c_{\mathrm{D}i}\} = 0 \quad (j=p_1,\ldots,p_m) \tag{12.31}$$

To achieve the coincidence of the first-order sensitivities among the multiple stories, we can derive $m-1$ equations from Eq. (12.31). As for the condition on N_{b}, the increment Δc_{D} should be satisfied by Eq. (12.29). Finally, we can derive the following simultaneous linear equation:

$$\begin{bmatrix} \alpha_{11} & \cdots & \alpha_{1n+1} \\ N_{q_1,q_1} & \cdots & N_{q_1,q_{n+1}} \\ \vdots & \ddots & \vdots \\ N_{q_n,q_1} & \cdots & N_{q_n,q_{n+1}} \end{bmatrix} \begin{Bmatrix} \Delta c_{\mathrm{D}q_1} \\ \vdots \\ \Delta c_{\mathrm{D}q_{n+1}} \end{Bmatrix} = 0 \tag{12.32}$$

where α_{1i} ($i=1,\ldots,n+1$) is described by:

$$\alpha_{1j} = f_{,q_1p_j} - f_{,q_2p_j} + \sum_{k=q_1}^{q_n}(\kappa_{k,i}N_{\mathrm{b}k,j} + \kappa_k N_{\mathrm{b}k,ji}) \quad (j=p_1,\ldots,p_m) \tag{12.33}$$

It can be mentioned that the condition on the sum of the increments Δc_{D} described by Eq. (12.27) is not included in Eq. (12.33). This means that the solution of the optimal distribution obtained by Eq. (12.33) does not guarantee the increase of the sum of the additional VD. If this phenomenon occurs, we cannot decrease the objective function any more via the optimality conditions. For solving this problem, some numerical manipulation may be needed. A temporal procedure presented in this paper is that the additional VD is assumed to be temporally added to the story identified by the first-order sensitivity $f_{,i}$ of the objective function without considering the concerned story location where the condition on N_{b} has been active. After this temporal procedure is executed, the coincidence of the first-order sensitivity $f_{,i}$ of the objective function may be disturbed, i.e., the number m decreases, and the procedure of optimal damper placement is directed to step 6B or step 6C.

12.8 Numerical Examples

In this section, numerical examples of the adaptive sensitivity of optimal damper placement are presented for variable distributions of frame stiffness. Five-story building models are used to demonstrate the usefulness and validity of the proposed method.

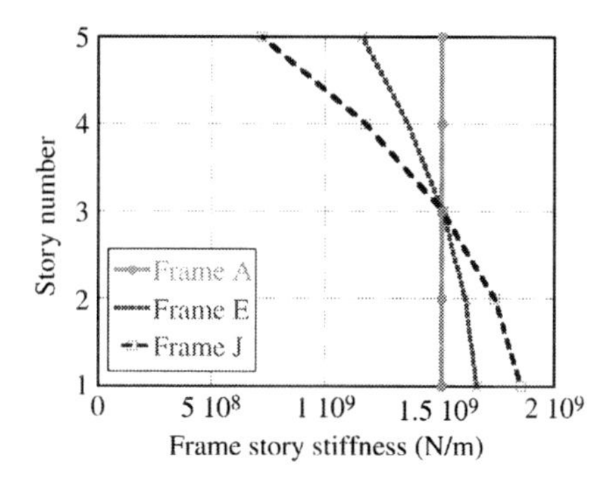

Figure 12.5 Various distributions of frame stiffness.

The floor masses are identical in all stories, i.e., $m_i = 5.12 \times 10^5$ kg $\quad (i = 1, \ldots, 5)$. The initial uniform frame story stiffness $k_{Fi} = 1.51 \times 10^9$ N/m is given. The fundamental natural period approximately attains 0.4 s. For investigating the adaptive sensitivities with respect to the variable distributions of frame stiffness, 10 different structural models are compared, called "Frame A" to "Frame J," which have different distributions of frame stiffness $\boldsymbol{k}_F$. Figure 12.5 shows the comparison of the distributions of $\boldsymbol{k}_F$ of "Frame A," "Frame E," and "Frame J." The stiffness distributions of the other models are given by interpolating those frames in Figure 12.5. Table 12.1 shows the story stiffness distribution from "Frame A" to "Frame J." The structural damping ratio of the main frame is assumed to be 0.02 (stiffness-proportional damping). The ratio $r_s = k_{bi}/k_{Fi}$ is given by 0.5 or 3.0. The initial stiffness of the supporting member is given by selecting an appropriate ratio r_s. The total damper area $\overline{W}$ is gradually increased until the performance criterion with respect to the maximum inter-story deformation angle is satisfied. The peak factor for the axial force of supporting members is given as 2.5 in all examples. The parameters $\overline{s}$ and $\overline{S}$ of the input critical acceleration are estimated from the NS component of the 1940 El Centro earthquake with PGA (Peak Ground Acceleration) $= 0.32$ g (namely, $\overline{s} = 0.02$ m^2/s^3 and $\overline{S} = 0.04$ m^2/s^4). The performance criterion R_D is 1/150 and the corresponding maximum value of inter-story drift is 0.0267 m where the story height is 4.0 m. The optimization procedure is continued to satisfy the performance criterion on the maximum inter-story drift.

12.8.1 Adaptive Sensitivity of the Optimal Damper Placement with Multiple Design Parameters

The sensitivity is investigated of the optimal damper placement for variable frame stiffness distribution.

Figure 12.6 shows the absolute value of the first-order sensitivity of the objective function f with respect to the damper capacity $\mathbf{C}_d$ for "Frame E" and $r_s = 3.0$. In the initial phase of the optimization procedure, $\max_j |f_{,j}|$ is attained only in the

Table 12.1 Story Stiffness Distribution ($\times 10^9$ N/m) (Frames A–J)

Frame Models	First Story	Second Story	Third Story	Fourth Story	Fifth Story
Frame A	1.51	1.51	1.51	1.51	1.51
Frame B	1.55	1.54	1.51	1.47	1.42
Frame C	1.59	1.56	1.51	1.44	1.33
Frame D	1.63	1.59	1.51	1.40	1.25
Frame E	1.66	1.61	1.51	1.36	1.16
Frame F	1.70	1.64	1.51	1.32	1.07
Frame G	1.74	1.67	1.51	1.29	0.98
Frame H	1.78	1.69	1.51	1.25	0.90
Frame I	1.82	1.72	1.51	1.21	0.81
Frame J	1.86	1.74	1.52	1.17	0.72

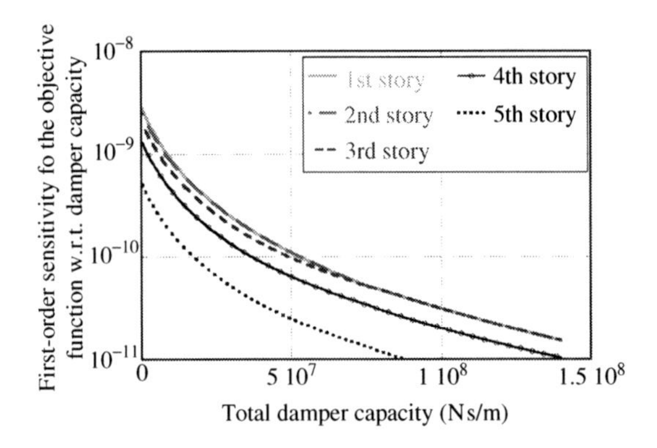

Figure 12.6 First-order sensitivity of objective function with respect to damper capacity (Frame E).

first story. On the other hand, after the occurrence of multiple coincidence of $\max_j |f_{,j}|$, they should continue to be satisfied as the optimality condition.

When $m = 3$, that is, the multiple coincidence in the sensitivity f with respect to the damper capacity C_d from first to third story, a nonmonotonic optimal damper placement path appears, where the main parameter is switched to the damping capacity of a particular story.

Figures 12.7–12.9 indicate the variations of the optimal distribution of damper capacity, the optimal distribution of supporting member stiffness, and the variation of the axial force of supporting members, respectively, with respect to the varied total damper capacity for "Frame E" and $r_s = 3.0$. It can be observed that passive dampers are distributed optimally in the structural model according to the variation of the first-order sensitivity $f_{,i}$ and the supporting member stiffness is increased to prevent the axial force N_b of the supporting members from exceeding the yield axial force $\overline{P}_{yi}$. Figure 12.10 shows the comparison of the variation of the objective function, that is, the sum of the mean squares of the inter-story drifts, for various

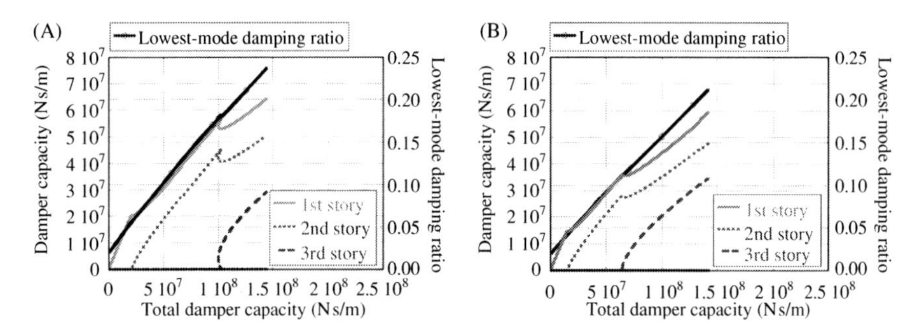

Figure 12.7 Variations of optimal damper placement and lowest-mode damping ratio with respect to varied total damper capacity: (A) Frame A and (B) Frame E.

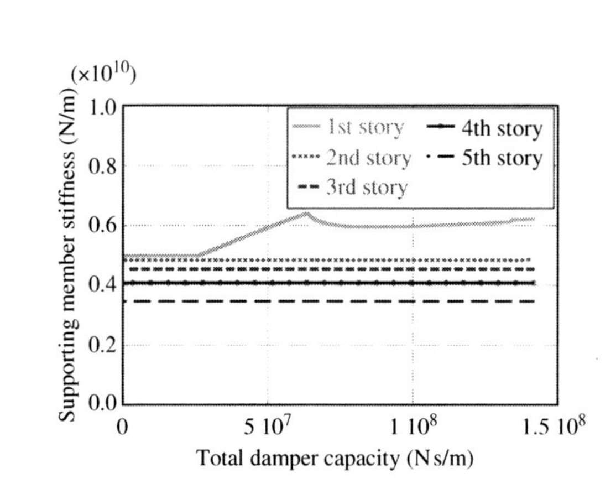

Figure 12.8 Variation of supporting member stiffness.

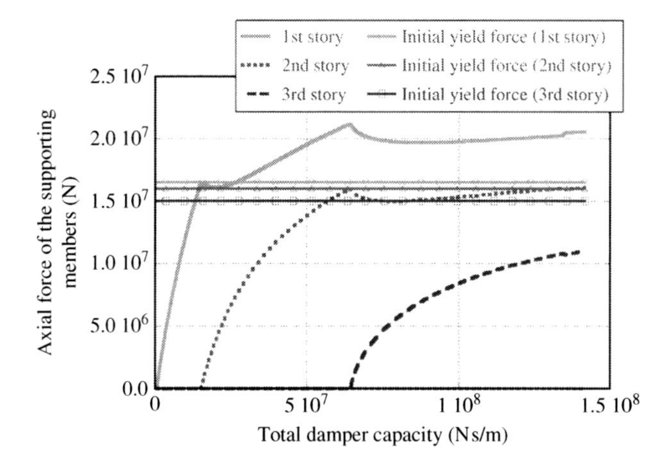

Figure 12.9 Variation of axial force of supporting member.

structural models (frame stiffness is fixed as "Frame E") with randomly placed VD. The supporting member stiffnesses are treated here so as to be the same as the optimal one. From Figure 12.10, the optimal damper placement derived in the proposed methodology decreases the objective function most effectively and gives the global optimal solution. These figures verify the validity of the proposed optimization methodology.

Figure 12.11 shows the variation of the optimal damper placement, that is, adaptive sensitivity of the optimal solutions, with respect to different structural models with variable frame stiffness. It can be observed that, when the number of stories with the supplemental dampers is the same, the variation of the optimal damper placement for variable frame stiffness from "Frame A" to "Frame H" can be regarded to be varied almost linearly. On the other hand, from "Frame H" to "Frame J," the additional damper is also distributed to the fourth story. In this case, the added dampers to the first, second, and third stories are also decreased compared with "Frame H," but the deviation of the optimal damper placement with respect to the variation of frame stiffness seemed to be different from that of the previous phase, i.e., frame stiffness from "Frame A" to "Frame H."

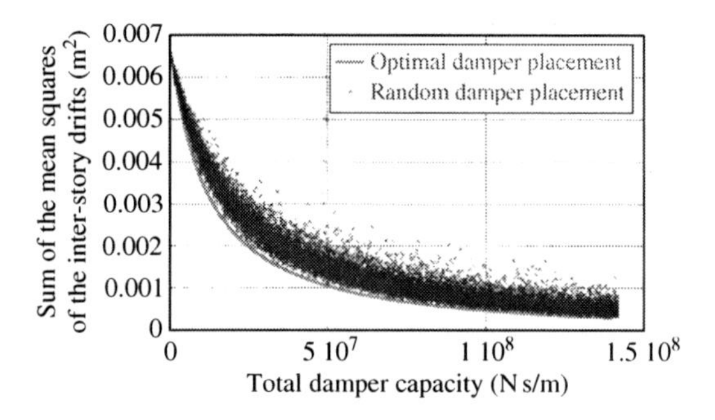

Figure 12.10 Comparison of objective functions for random damper placement.

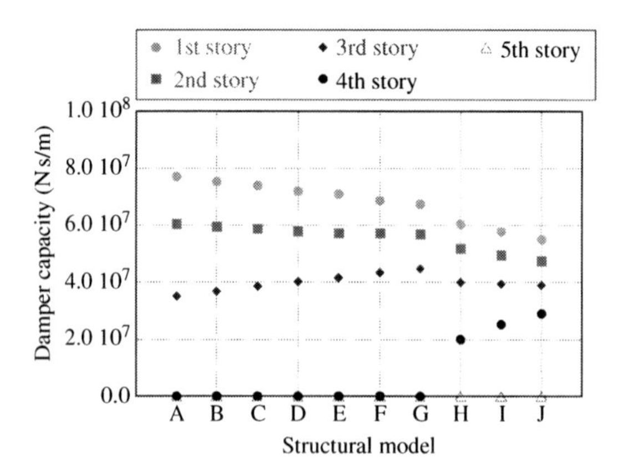

Figure 12.11 Adaptive sensitivity of optimal damper placement for variable frame stiffness.

12.8.2 Optimal Damper Placement for Given Supporting Members

Examples for the second problem are presented. The properties of structural models are taken as "Frame A" and "Frame E," $r_s = 3.0$, respectively. In case of $r_s = 0.5$, the optimal damper solution has not been derived numerically since the constraint on N_b may restrict the placement of supplemental VD.

Figure 12.12A and B shows the variation of the optimal damper distribution. Comparing Figure 12.12 with the results of the previous problem in Figure 12.7, the additional VD is distributed to multiple stories in an early phase. These facts indicate that the constraint on N_b affects the optimization procedure.

Figures 12.13 and 12.14 show the variation of N_b of the supporting member with respect to varied total damper capacity and the variation of the first-order sensitivity of the objective function, respectively. In Figure 12.13, constant limits of the axial forces are also indicated. Since the limit of N_b for the ith story is a function of only the supporting member stiffness of the ith story, the upper limits may vary depending on the distribution of frame stiffness as shown in Figure 12.13A and B. By comparing Figures 12.12 and 12.13, the dampers are distributed to

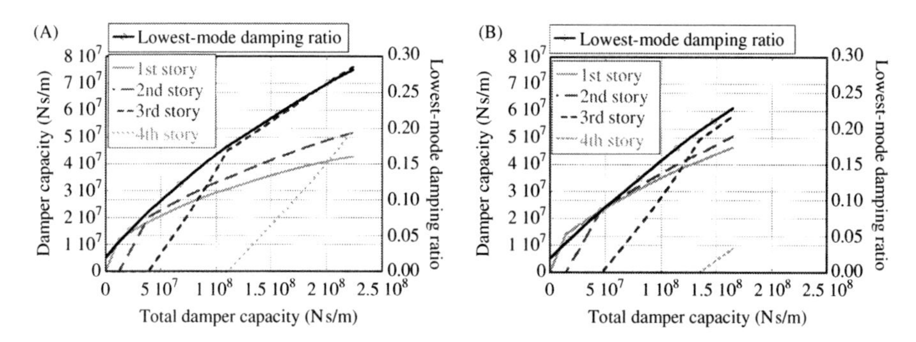

Figure 12.12 Variation of optimal damper placement and damping ratio: (A) Frame A (uniform stiffness distribution) and (B) Frame E.

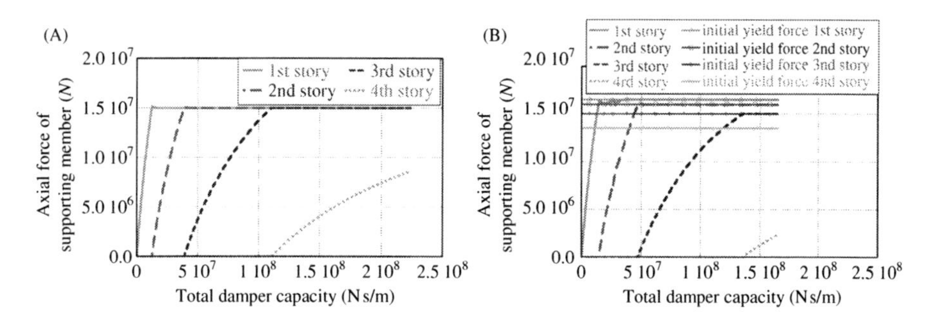

Figure 12.13 Variation of axial forces of supporting members: (A) Frame A (uniform stiffness distribution) and (B) Frame E.

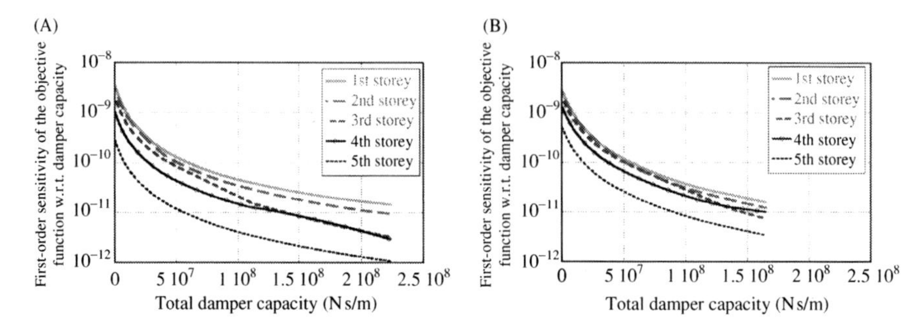

Figure 12.14 Absolute value of first-order sensitivity of sum of mean squares of inter-story drift with respect to varied total damper capacity: (A) Frame A (uniform stiffness distribution) and (B) Frame E.

multiple stories to avoid the violation on N_b. Furthermore, in Figure 12.14B, although coincidence of the first-order sensitivities on the both first and second stories occurs in an early phase, the optimality conditions in terms of the gradient vector of the objective function with respect to $\mathbf{c}_D$ seems to be violated after the condition with respect to N_b is active. For these reasons, it can be concluded that more quantity of damper is needed in the concerned optimal design problem compared with the previous one. Particularly, the increase in the damper quantity is remarkable in Frame A by comparing Figure 12.14A with Figure 12.7A.

12.9 Conclusions

The conclusions may be stated as follows:

1. An evolutionary optimization methodology of damper placement and supporting members has been proposed to minimize the sum of the mean squares of inter-story drifts under random input. Based on the proposed optimization methodology, two different treatments with respect to the supporting member have been introduced. Simultaneous design consideration of dampers and supporting members is a new aspect of the theoretical development and practicality.
2. A new concept of adaptive sensitivities of optimal damper placement for variable frame stiffness variation has been introduced by using the proposed optimization methodology under uncertainty. Information of the variation of optimal damper placement according to the variation of variable design parameters is very useful in the damper design under uncertainty.
3. A gradient-based evolutionary optimization technique is developed by using the Lagrange multiplier optimization technique. Simultaneous satisfaction of the optimality criterion on placement of passive dampers and the constraint on forces of the supporting members has been guaranteed theoretically and demonstrated through numerical examples.
4. While the global optimality of the damper placement has been demonstrated numerically for a fixed set of supporting member stiffnesses, the global optimality for a broader design domain including the supporting member design domain may be arguable. In this

sense, the proposed optimization methodology of damper placement and supporting members should be said to be path-dependent. A further investigation may be necessary for a deeper understanding of the path-dependency.

Acknowledgments

Part of the present work is supported by the Grant-in-Aid for Scientific Research of Japan Society for the Promotion of Science (Nos.18360264 and 21360267) and the Grant-in-Aid for JSPS Fellows (No. 21•364). This support is greatly appreciated.

References

Attard, T.L., 2007. Controlling all interstory displacements in highly nonlinear steel buildings using optimal viscous damping. J. Struct. Eng. 133 (9), 1331–1340.

Cimellaro, G.P., 2007. Simultaneous stiffness-damping optimization of structures with respect to acceleration, displacement and base shear. Eng. Struct. 29, 2853–2870.

Cimellaro, G.P., Retamales, R., 2007. Optimal softening and damping design for buildings. Structural Control and Health Monitoring. 14 (6), 831–857.

Constantinou, M.C., Tadjbakhsh, I.G., 1983. Optimum design of a first story damping system. Comput. Struct. 17 (**2**), 305–310.

De Silva, C.W., 1981. An algorithm for the optimal design of passive vibration controllers for flexible systems. J. Sound Vib. 74 (4), 495–502.

Drenick, R.F., 1970. Model-free design of aseismic structures. J. Eng. Mech. Div. 96 (EM4), 483–493.

Fujita, K., Moustafa, A., Takewaki, I., 2010. Optimal placement of viscoelastic dampers and supporting members under variable critical excitations. Earthq. Struct. 1 (1), 43–67.

Garcia, D.L., 2001. A simple method for the design of optimal damper configurations in MDOF structures. Earthq. Spectra. 17 (3), 387–398.

Gurgoze, M., Muller, P.C., 1992. Optimal positioning of dampers in multi-body systems. J. Sound Vib. 158 (**3**), 517–530.

Hahn, G.D., Sathiavageeswaran, K.R., 1992. Effects of added-damper distribution on the seismic response of buildings. Comput. Struct. 43 (**5**), 941–950.

Kiu, W., Tong, M., Wu, Y., Lee, G., 2004. Optimized damping device configuration design of a steel frame structure based on building performance indices. Earthq. Spectra. 20 (1), 67–89.

Lavan, O., Levy, R., 2005. Optimal design of supplemental viscous dampers for irregular shear-frames in the presence of yielding. Earthq. Eng. Struct. Dyn. 34 (8), 889–907.

Lavan, O., Levy, R., 2006. Optimal design of supplemental viscous dampers for linear framed structures. Earthq. Eng. Struct. Dyn. 35 (3), 337–356.

Levy, R., Lavan, O., 2006. Fully stressed design of passive controllers in framed structures for seismic loadings. Struct. Multidisc. Optim. 32 (6), 485–498.

Liu, W., Tong, M., Wu, X., Lee, G., 2003. Object-oriented modeling of structural analysis and design with application to damping device configuration. J. Comput. Civil Eng. 17 (2), 113–122.

Marano, G.C., Trentadue, F., Greco, R., 2007. Stochastic optimum design criterion for linear damper devices for seismic protection of building. Struct. Multidisc. Optim. 33, 441–455.

Takewaki, I., 1997. Optimal damper placement for minimum transfer functions. Earthq. Eng. Struct. Dyn. 26 (11), 1113–1124.

Takewaki, I., 2000. Optimal damper placement for critical excitation. Prob. Eng. Mech. 15 (4), 317–325.

Takewaki, I., 2002. Robust building stiffness design for variables critical excitations. J. Struct. Eng. 128 (12), 1565–1574.

Takewaki, I., 2006. Critical Excitation Methods in Earthquake Engineering. Elsevier Science, Amsterdam.

Takewaki, I., 2009. Building Control with Passive Dampers: Optimal Performance-Based Design for Earthquakes. John Wiley & Sons, Singapore.

Takewaki, I., Yoshitomi, S., 1998. Effects of support stiffnesses on optimal damper placement for a planar building frame. J. Struct. Des. Tall Build. 7 (4), 323–336.

Takewaki, I., Yoshitomi, S., Uetani, K., Tsuji, M., 1999. Non-monotonic optimal damper placement via steepest direction search. Earthq. Eng. Struct. Dyn. 28 (6), 655–670.

Trombetti, T., Silvestri, S., 2004. Added viscous dampers in shear-type structures: the effectiveness of mass proportional damping. J. Earthq. Eng. 8 (2), 275–313.

Trombetti, T., Silvestri, S., 2007. Novel schemes for inserting seismic dampers in shear-type systems based upon the mass proportional component of the Rayleigh damping matrix. J. Sound Vib. 302 (3), 486–526.

Tsuji, M., Nakamura, T., 1996. Optimum viscous dampers for stiffness design of shear buildings. J. Struct. Des. Tall Build. 5, 217–234.

Uetani, K., Tsuji, M., Takewaki, I., 2003. Application of optimum design method to practical building frames with viscous dampers and hysteretic dampers. Eng. Struct. 25 (5), 579–592.

Viola, E., Guidi, F., 2008. Influence of the supporting braces on the dynamic control of buildings with added viscous dampers. Struct. Control Health Monit. 16 (3), 267–286.

13 Application of Genetic Algorithms in Ground Motion Selection for Seismic Analysis

Alireza Azarbakht and Mehdi Mousavi

Department of Civil Engineering, Faculty of Engineering, Arak University, Arak, Iran

13.1 An Introduction to Structural Nonlinear Response-History Analysis

13.1.1 The Role of Dynamic Analysis in Performance-Based Earthquake Engineering

For several years, researchers and practitioners have been trying to develop a set of guidelines for adapting the design procedure to the real behavior of structures. Performance-based earthquake engineering (PBEE) is one of the major improvements in that effort (Cornell and Krawinkler, 2000). PBEE is a relatively new approach that has recently evolved, and the traditional methods are going to be replaced by this new method. The main goal of this approach is to ensure that the design meets a set of performance objectives expected from a certain structure during future scenario earthquakes (Cornell et al., 2002). The various components of PBEE can be listed as: (i) definition of performance objectives, (ii) general design methodology, (iii) selection of ground motion records, and (iv) demand and capacity evaluations (Ghobarah, 2001). A set of performance objectives must be satisfied when the structure is subjected to various hazard levels. The stated performance objectives are usually immediate occupancy, life safety, and collapse prevention (FEMA356, 2000). The level of expected performance for a particular building is determined based on its function. The various limit states that may be used to evaluate the performance objectives can be interstory drift ratio, inelastic rotation, number of failed connections, displacement ductility, or energy dissipation.

A common method for estimating and/or assessing structural performance, which has been proposed by the Pacific Earthquake Engineering Research Center (PEER), consists of four major steps: hazard analysis, structural analysis, damage analysis, and loss analysis. This procedure, which is called the PEER PBEE

Metaheuristic Applications in Structures and Infrastructures. DOI: http://dx.doi.org/10.1016/B978-0-12-398364-0.00013-9

framework, is a general form of the total probability theorem (Benjamin and Cornell, 1970). One of the key parts of this framework is the engineering demand parameter (EDP), which is formulated by statistically analyzing the results of a set of nonlinear response-history analyses of a given structure under the effect of expected earthquake ground motion records in a particular region. For this purpose, incremental dynamic analysis (IDA) has recently emerged, which offers seismic demand and capacity prediction capability by using a series of nonlinear dynamic analyses under properly scaled ground motion records (Vamvatsikos and Cornell, 2002).

The main aim of the PBEE is to assess the annual probability of exceeding a predefined limit state, which is termed λ_{EDP}. To evaluate λ_{EDP} at a designated site, the uncertainties in the ground motions and the nonlinear structural response need to be taken into account. A Monte Carlo simulation can be utilized, but this approach requires computationally extensive analyses in order to evaluate λ_{EDP} (Collins et al., 1996; Han and Wen, 1997; Jalayer et al., 2004; Wen, 2000). For a given fault i, this method requires the computation of $\nu_i P[\text{EDP} > x|\text{fault } i]$, where ν_i is the mean annual rate of occurrence of an earthquake above a threshold magnitude of fault i, and $P[\text{EDP} > x|\text{fault } i]$ is the probability of exceeding the response or EDP level X given an event for fault i. This term can be calculated from the nonlinear dynamic analysis results, i.e., from, for example, synthetic records of random magnitude and the location on the fault. The summation from all the sources in the region (assuming the earthquake faults occur independently) is:

$$\lambda_{\text{EDP}}(x) = \sum_i \nu_i P[\text{EDP} > x|\text{fault } i] \tag{13.1}$$

Given that the dispersion of response for a fault (or simply a given magnitude, M_w, and a source-to-site distance, R) is about 0.8 or more, the required sample size is $(0.8/0.1)^2 = 64$ in order to estimate the median EDP within a standard error of 10%. The computation also requires at least 20 or more M_w and R pairs to adequately cover the seismic source contributions in the region. This procedure ultimately requires thousands of records to be simulated and analyzed through the structure in order to obtain accurate estimates of the extreme responses and ground motions. To improve efficiency in the calculation, the "deaggregation" method, which selects only the magnitudes, M_w, and the source-to-site distances, R, that contribute the most to λ_{EDP} can be utilized (McGuire, 1995).

Another approach that can be used to calculate the seismic performance of structures is to use a structure and the response-specific attenuation relationship as a function of M_w and R. This concept is similar to the conventional probabilistic seismic hazard analysis (PSHA) (McGuire, 1995); merely the ground motion attenuation relationship is replaced by a more structure-specific one:

$$\lambda_{\text{EDP}}(x) = \sum_i \nu_i \iint P[\text{EDP} > x|m_w, r] \cdot f_{M_w,R}(m, r) \cdot \text{d}r \cdot \text{d}m \tag{13.2}$$

where $f_{M_w,R}(m,r)$ is the joint probability density function of the M_w and R of a given fault. Large suites of records (from various M_w, R, fault mechanisms, and so on) are needed to obtain an accurate structure-specific attenuation model (Mousavi et al., 2012), which is found by a regression analysis of EDP performed upon M_w and R. The implicit assumptions in this method are (i) the functional form of the regression equation, Eq. (13.2), (ii) the lack of dependence of EDP on the source characteristics not contained in the vector of independent variables (e.g., rupture duration), and (iii) the geometry of the fault relative to the site (Cornell, 1996; McGuire and Cornell, 1974; Sewell, 1989). The drawback is that this method still needs hundreds of analyses in order to be able to obtain a reliable estimate for the structure- and response-specific attenuation model. On the other hand, the number of records is relatively small compared to that of the simulation-based method, and there are sufficient real records in the catalogs to avoid the use of synthetic records. Cornell and co-workers (Bazzurro, 1998; Luco, 2002; Shome and Cornell, 1999) simplified the problem further by decoupling the ground motion hazard and nonlinear dynamic analyses via an intermediate variable known as the ground motion intensity measure (IM). A conventional IM is the peak ground acceleration (PGA) or, a little more structure-specific, the pseudospectral acceleration corresponding to the first-mode period (denoted as $Sa(T_1)$ or simply Sa). The benefit of this method is that the number of records needed can be substantially reduced because most of the uncertainties in EDP are concentrated in λ_{IM}, which is found by conventional PSHA, leaving a small variation of EDP given that IM is estimated from the dynamic analyses (while still obtaining accurate estimates for the marginal EDP distribution of the structure at a site). Typical values for the dispersion of EDP conditioned on IM associated with large ductility levels are about 0.3–0.4, implying a necessary sample size in the order of 10 (i.e., $(0.35/0.1)^2$) records to estimate the median EDP within a standard error of 10%. Assuming that five to six IM levels need to be analyzed, the total number of required analyses is about 50–60. By using the total probability theorem (Benjamin and Cornell, 1970), $\lambda_{EDP}(x)$ representing all M_w and R scenarios from the causative faults can be expressed as:

$$\lambda_{EDP}(x) = \int P[EDP > x|im] \cdot d\lambda_{IM}(im) \tag{13.3}$$

where $d\lambda_{IM}(im) = \lambda_{IM}(im) - d\lambda_{IM}(im + dim)$ is the differential of the ground motion hazard curve in terms of IM. $\lambda_{IM}(im)$ is typically determined by seismologists. It is approximately the mean annual frequency of IM = im, where dim is a small increment in the ground motion intensity. Note that this site-specific seismic hazard reflects all M_w and R scenarios. A major portion of the specific nature can thus be captured at the site. The question still remains, which IM should be selected. This seismic performance assessment approach is known as the IM-based probabilistic seismic demand analysis (PSDA) (Shome and Cornell, 1999). Note that it has been applied in various fields to diverse degrees. For example, the US nuclear power industry has used the seismic probabilistic risk assessments for more than two decades, applying it to all plants in the United States (Hickman, 1983).

This framework was also implemented in the probabilistic assessment of steel moment-resisting frames (FEMA350, 2000) and other guidelines (ATC40, 1996). The principal assumption (for IM-based PSDA) is the "sufficiency" property of the IM (Luco and Cornell, 2007), which requires that the probability distribution of the structural response of interest given an IM is conditionally independent of the other ground motion parameters (i.e., M_w, distance, epsilon, the fault mechanism, and so on). This assumption ultimately implies that a detailed record selection is not necessary (i.e., any ground motion records from any M_w, distance, epsilon, fault mechanism, and so on can be selected). If the selected IM is not sufficient, a full conditional probability distribution of EDP needs to be used to ensure the accuracy of the PSDA.

$$\lambda_{\mathrm{EDP}}(x)=\iint P[\mathrm{EDP}>x|\mathrm{im},m_w,r,\varepsilon,\ldots]\cdot f_{M_W,R,\varepsilon,\ldots|\mathrm{IM}}(m_w,r,\varepsilon,\ldots|\mathrm{im})\cdot \mathrm{d}(m_w,r,\varepsilon,\ldots)\cdot \mathrm{d}\lambda_{\mathrm{IM}}(\mathrm{im}) \tag{13.4}$$

where the ground motion epsilon (ε) is a proxy for the deviation of an IM (e.g., Sa) of as-recorded ground motion relative to the predicted (median) value calculated from the attenuation model. $f_{M_W,R,\varepsilon,\ldots|\mathrm{IM}}(m_w,r,\varepsilon,\ldots|\mathrm{im})$ is the joint probability distribution function of M_w, R, ε, and other ground motion parameters at a given IM level, which can be obtained from the results of PSHA disaggregation. If an insufficient IM is used, and the selected records do not represent the hazard at the site, the seismic performance estimation will be biased. The choice of an appropriate IM is essential in obtaining an accurate estimate for the seismic performance of structures. The efficiency, sufficiency, and bias in the response, when scaling records, depend on the chosen IM. With a sufficient IM, however, evaluation of the seismic performance of a structure can be estimated using Eq. (13.4), simply because $P[\mathrm{EDP}>x|\mathrm{im},m_w,r,\varepsilon$, etc.] is functionally independent of the above-mentioned ground motion parameters. Two other important properties of an IM are the "efficiency" and the "scaling robustness." The former implies that a more efficient IM can reduce the number of nonlinear dynamic analyses but still achieve the same accuracy in seismic performance estimation, whereas, the latter implies that there is no statistically strong relationship between the structural response and the scale factors used in scaling the amplitude of the records. By definition, rare or extreme ground motions are scarce. Real (as-recorded) ground motions will, instead, be used and scaled. An IM is introduced to quantify the scaling of ground motion records. Scaling records to a common elastic value (i.e., using elastic-based IMs) can result in the spectral shape being largely influenced by ε, M_w, soil type, and so on, which impacts the nonlinear response of structures (Tothong, 2007). If scaling factors are, however, determined using inelastic-based IMs (i.e., inelastic spectral displacement), the influence from the spectral shape (vis-à vis ε and M_w) will be less prominent. This is mainly because the positive slope (relative to the median spectra) of more aggressive records will be scaled less than the negative slope records (Tothong, 2007) in order to achieve a common structural response level.

13.1.2 Selection of Ground Motion Records as an Important Challenge in PBEE

The increasing availability of strong ground motion records, and the relative ease with which they can be obtained compared to synthetic or artificial records, makes the use of real records an even more attractive option for defining the input to dynamic analyses in structural engineering (Bommer, 2004). The current state of the best practice (Commission, 2001) in selecting ground motion records for assessing the nonlinear demand of structures is based on magnitude and distance (M_w and R). There is, however, a general lack of guidelines on procedures for the selection of appropriate suites of records for this purpose, and seismic design regulations are particularly poor in this respect (i.e., they do not recommend any explicit precedence criteria for record selection). Usually, to use the corresponding critical or the average actions, the selection of three or seven spectrum-compatible records is recommended by the regulations, respectively.

For earthquake-resistant design and also for the seismic assessment of existing structures, earthquake-induced ground shaking is generally represented in the form of a response spectrum of acceleration, displacement, and velocity. The equivalent lateral force is then obtained by scaling the design elastic spectrum by some factors that take into account the influence of the inelastic structural response. However, there are situations in which the simulation of structural response using a scaled elastic response spectrum is not considered appropriate, and a fully dynamic analysis is required. These situations may include the following: buildings designed for a high degree of ductility; structures with a configuration in plan or elevation that is highly irregular; structures for which higher modes are likely to be excited; critical structures, the failure of which would cause unacceptable harm or disruption; and structures with special features such as base isolation (Bommer, 2004). Faced with these special situations, employment of a response-history analysis, for which the requirements are (i) an appropriate nonlinear model for the structure and (ii) a suitable suite of ground motion records to represent the seismic excitation, is necessary. In general, three basic options are available to the engineer in terms of obtaining ground motion records:

1. Using of artificial spectrum-compatible records generated using programs such as Simulation of earthquake (SIMQKE) (Gasparini and Vanmarcke, 1979). The approach employed in SIMQKE is to generate a power spectral density function from the smoothed response spectrum and then to derive sinusoidal signals having random phase angles and amplitudes. The sinusoidal motions are then summed and an iterative procedure can be invoked to improve the match with the target response spectrum by calculating the ratio between the target and the actual response ordinates at selected frequencies. The power spectral density function is then adjusted by the square of this ratio, and a new motion is generated. The attractiveness of such an approach is obvious since it is possible to obtain acceleration time series that are almost completely compatible with the elastic design spectrum, which in some cases will be the only information available to the design engineer regarding the nature of the ground motions to be considered. However, it is now widely accepted that the use of such artificial records, particularly for nonlinear analyses, is problematic (Naeim and Lew, 1995). The basic problem with spectrum-compatible

artificial records is that they generally have an excessive number of cycles of strong motion and thus possess an unreasonably high energy content. These types of records are not considered to be suitable for use in nonlinear analyses. In addition to the problems associated with how these artificial records are generated, there can also be difficulties that arise from matching the acceleration time series to the entire elastic design spectrum. The latter will generally be a uniform hazard spectrum (UHS), included in seismic design codes, obtained from PSHA, and therefore enveloping ground motions from several seismic sources (Bommer et al., 2000; Reiter, 1990). Naeim and Lew (1995) asserted that: "To generate an acceleration response-history to be compatible to a PSHA-generated design spectrum is neither reasonable nor realistic."

2. The second category of ground motion records available to the engineer consists of synthetic accelerograms generated from seismological source models and accounts for path and site effects. These models range from point source stochastic simulations through their extension to finite sources to fully dynamic models of stress release, although the latter are still under development. Programs for some of the many methods of ground motion generation that have been developed (Beresnev and Atkinson, 1998; Boore, 2003; Zeng et al., 1994) are freely available, but their application, in terms of defining the many parameters required to characterize the earthquake source, will generally require that the engineer engages the services of a specialist consultant in engineering seismology.
3. The third category of records consists of real ground motion records, which have been recorded during the past real earthquakes, which are, by definition, free from the problems associated with artificial and synthetic records. Real strong-motion records are now easily accessible in large numbers, and their retrieval and manipulation are relatively straightforward, so that the design engineer will often be able to prepare a suite of records without the services of an engineering seismologist.

In general, the use of real ground motion records in regions where they are accessible is superior. On the other hand, artificial and synthetic records are appropriate for use in regions where the real record catalog is lacking or limited (e.g., the eastern part of the United States).

In the following sections, first the genetic algorithm (GA), as one of the popular metaheuristics, is reviewed, and the application of GA in record selection practice is presented with some examples.

13.2 A Snapshot of the Genetic Algorithm as One of the Popular Metaheuristics

Traditionally, there are two major classes of optimization algorithms that have been classified into calculus-based techniques and enumerative techniques (Goldberg, 1989). Calculus-based optimization techniques employ a gradient-directed searching mechanism, starting from an initial guessed solution, and are therefore local in scope. Although these techniques are well developed, they have significant drawbacks. Indeed, for ill-defined or multimodal objective functions, instability and/or local optima are usually obtained. Additionally, since the objective function is often problem-oriented, the implementation of these techniques can be very complex. Many enumerative schemes have been suggested to handle the local optima problem but at the expense of computational inefficiency.

The second class of optimization technique, which has achieved increasing popularity, is the random/probabilistic search algorithm. More particularly, the 1970s saw the emergence of evolutionary algorithms that employ mechanisms of natural selection to solve optimization problems (Mitchell, 1996). These algorithms were thought to be the answer to the question of how the search should be organized in order to achieve a high likelihood of locating a near-optimal solution.

A GA is a simulation of the natural evolutionary processes in order to solve search and optimization for both constrained and unconstrained problems. It has taken a long time, however, for this subject to become mature enough to be used as a practical tool. The pioneering work performed by Goldberg (1989), together with the availability of high-speed computers, paved the way for the application of GA to the solving of engineering problems. GA families have proven themselves to be reliable computational search and optimization procedures for complex objectives involving a large number of variables. In structural and earthquake engineering, GAs have been used over the last decade in design optimization of nonlinear structures (Pezeshk et al., 2000), in the selection and scaling of the earthquake ground motion records (Azarbakht and Dolšek, 2007; Azarbakht and Dolsek, 2011; Ghafory-Ashtiany et al., 2011; Kayhani et al., 2012; Mousavi et al., 2011; Naeim et al., 2004), and in many other applications (Gandomi and Alavi, 2012; Gandomi et al., 2011; Gandomi et al., 2012; Yang and Gandomi, 2012).

Similar to all evolutionary algorithms, GA is a search procedure modeled on the mechanics of natural selection rather than on the simulated reasoning process developed by Holland (1975). These algorithms were originally used for the study of artificial systems. Today, there are many different GAs (Mitchell, 1996). There are no rigorous definitions of GAs accepted by all in the evolutionary computation community, which might differentiate GAs from other evolutionary computation methods. Indeed, some currently used GAs can be very far from Holland's original concept. However, it can be said that most methods called "GAs" have at least the following elements in common: populations of individuals, selection according to the individuals' fitness, crossover to produce new individuals, random mutation of new individuals, and the replacement of populations. For more details, refer to Chapters 1 and 2.

13.3 Code-Conforming Ground Motion Selection

In this section, the application of GA in a code-conforming approach of ground motion selection is explained. "Spectral matching to a target spectrum" is a common strategy of structural design codes for ground motion selection. Following a brief review on the target spectrum definition, the code instructions for ground motion selection are described, and, finally, the application of a GA is studied as an efficient tool for the handling of code instructions.

13.3.1 Code-Based Target Spectrum

Different seismic codes prescribe general guidelines for ground motion selection for nonlinear dynamic analysis purposes. The main idea of seismic codes in ground motion selection is to reconstruct the conditions that are consistent with the seismic hazard, as well as the soil conditions of the site of a given structure. In areas with a high potential of seismic activities, the objective site is influenced by different seismic sources, which may have enough potential to generate great earthquakes in the future (Ebrahimian et al., 2012). There is a huge amount of uncertainty about the location, size, and resulting ground motions of future earthquakes. PSHA aims to quantify these uncertainties, and combine them to produce an explicit description of the distribution of future ground motion that may occur at a site (McGuire, 1995).

The major selection criterion in most contemporary codes such as ASCE7-5 (2005) and FEMA356 (2000) is that acceptable ground motions should be compatible with a code-prescribed smooth target spectrum. This spectrum represents the seismological aspects as well as the geotechnical conditions of the site. Most codes suggest a simple instruction for the construction of a design spectrum, which can also be used as a target spectrum for ground motion selection. As an alternative, UHS has been proposed by codes as the target spectrum. UHS is one of the most popular outputs of a PSHA procedure, and it is defined as the locus of points such that the spectral acceleration value at each period has an exceedance probability equal to the specified target probability. Figure 13.1 shows the UHS for an arbitrarily selected site for an exceedance probability of 2% in 50 years, which is equivalent to a return period of 2475 years.

The spectral acceleration of an arbitrarily selected ground motion record is also shown in Figure 13.1. Either the design spectrum or UHS can be used as the target spectrum in ground motion selection, according to the instructions described in the following section.

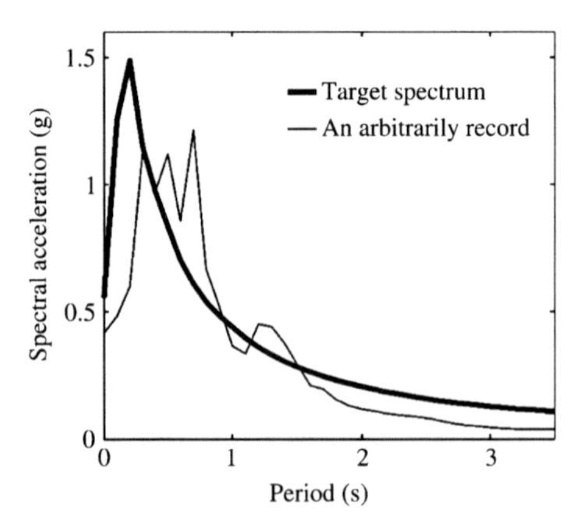

Figure 13.1 The UHS for an arbitrarily selected site, the record of the Superstition Hills event (1987) at the El Centro station.

13.3.2 Instructions for Ground Motion Selection

As shown in Figure 13.1, the target spectrum is significantly different from the arbitrarily selected ground motion. Due to the random nature of ground motions, it is seldom that a single ground motion can be found that exactly matches the target spectrum. As a result, all seismic design codes and guidelines require scaling of a number of selected ground motions in which they match or exceed the controlling design spectrum within the period range of interest. The most practical procedure is to select seven ground motions whose mean spectrum matches the target spectrum in the period range of $0.2T^*$ to $1.5T^*$ s. T^* is the dominant period of the given structure. Supposing $T^* = 1.0$ s, the matching range is considered to be 0.2–1.5 s in all of this section.

13.3.3 Definition of the Problem

Let us consider the ground motion records set defined in the Applied Technology Council Project as a procedure to validate the provisions for seismic structural design (ATC63, 2008). The detailed information about the records, as well as the complete list of selection criteria, are provided in ATC63 (2008). This set of ground motions is listed in Table 13.1.

Figure 13.2 shows the magnitude-distance distribution as well as the 5% damped acceleration response spectra of the reference set.

Now, the problem is how to select a subset with seven records from the reference set with the aim of matching the resulting mean spectrum with the target spectrum that is shown in Figure 13.1. According to the size of the reference set, there are:

$$\binom{44}{7} = \frac{44!}{37!\ 7!} = 38320568$$

feasible solutions for this objective. As an example, an arbitrarily selected subset has been defined and the corresponding records are plotted in Figure 13.3. The vector [35 3 40 11 6 22] represents the indices of the arbitrarily selected records.

As shown in Figure 13.3, the mean spectrum of the arbitrarily selected subset differs significantly from the target spectrum. Clearly, the use of conventional optimization techniques for the selection of the best feasible subset would take an enormous number of computations and would thus be infeasible. Conversely, a GA can converge with a reasonable computing effort and a fairly short computing time.

13.3.4 GA Solution

As the first step, a fitness function needs to be defined. The area located between the target spectrum and the mean spectrum of the random subset is defined as the fitness function that needs to be minimized. As also shown in Figure 13.4, this error function has been calculated for a specific matching range, i.e., 0.2–1.5 s.

Table 13.1 List of Reference Set of Ground Motion Records

Index	Magnitude	Year	Event	Station Name	Component
1	6.7	1994	Northridge	Beverly Hills−14145	H1
2	6.7	1994	Northridge	Beverly Hills−14145	H2
3	6.7	1994	Northridge	Canyon Count-W Lost	H1
4	6.7	1994	Northridge	Canyon Count-W Lost	H2
5	7.1	1999	Duzce, Turkey	Bolu	H1
6	7.1	1999	Duzce, Turkey	Bolu	H2
7	7.1	1999	Hector Mine	Hector	H1
8	7.1	1999	Hector Mine	Hector	H2
9	6.5	1979	Imperial Valley	Delta	H1
10	6.5	1979	Imperial Valley	Delta	H2
11	6.5	1979	Imperial Valley	El Centro Arr #11	H1
12	6.5	1979	Imperial Valley	El Centro Arr #11	H2
13	6.9	1995	Kobe, Japan	Nishi-Akashi	H1
14	6.9	1995	Kobe, Japan	Nishi-Akashi	H2
15	6.9	1995	Kobe, Japan	Shin-Osaka	H1
16	6.9	1995	Kobe, Japan	Shin-Osaka	H2
17	7.5	1999	Kocaeli, Turkey	Duzce	H1
18	7.5	1999	Kocaeli, Turkey	Duzce	H2
19	7.5	1999	Kocaeli, Turkey	Arcelik	H1
20	7.5	1999	Kocaeli, Turkey	Arcelik	H2
21	7.3	1992	Landers	Yermo Fire Stat	H1
22	7.3	1992	Landers	Yermo Fire Stat	H2
23	7.3	1992	Landers	Coolwater	H1
24	7.3	1992	Landers	Coolwater	H2
25	6.9	1989	Loma Prieta	Capitola	H1
26	6.9	1989	Loma Prieta	Capitola	H2
27	6.9	1989	Loma Prieta	Gilroy Array #3	H1
28	6.9	1989	Loma Prieta	Gilroy Array #3	H2
29	7.4	1990	Manjil, Iran	Abbar	H1
30	7.4	1990	Manjil, Iran	Abbar	H2
31	6.5	1987	Superstition Hills	El Centro Imp. Cent	H1
32	6.5	1987	Superstition Hills	El Centro Imp. Cent	H2
33	6.5	1987	Superstition Hills	Poe Road (temp)	H1
34	6.5	1987	Superstition Hills	Poe Road (temp)	H2
35	7	1992	Cape Mendocino	Rio Dell Over−FF	H1
36	7	1992	Cape Mendocino	Rio Dell Over−FF	H2
37	7.6	1999	Chi-Chi, Taiwan	CHY101	H1
38	7.6	1999	Chi-Chi, Taiwan	CHY101	H2
39	7.6	1999	Chi-Chi, Taiwan	TCU045	H1
40	7.6	1999	Chi-Chi, Taiwan	TCU045	H2
41	6.6	1971	San Fernando	LA, Hollywood FF	H1
42	6.6	1971	San Fernando	LA, Hollywood FF	H2
43	6.5	1976	Friuli, Italy	Tolmezzo	H1
44	6.5	1976	Friuli, Italy	Tolmezzo	H2

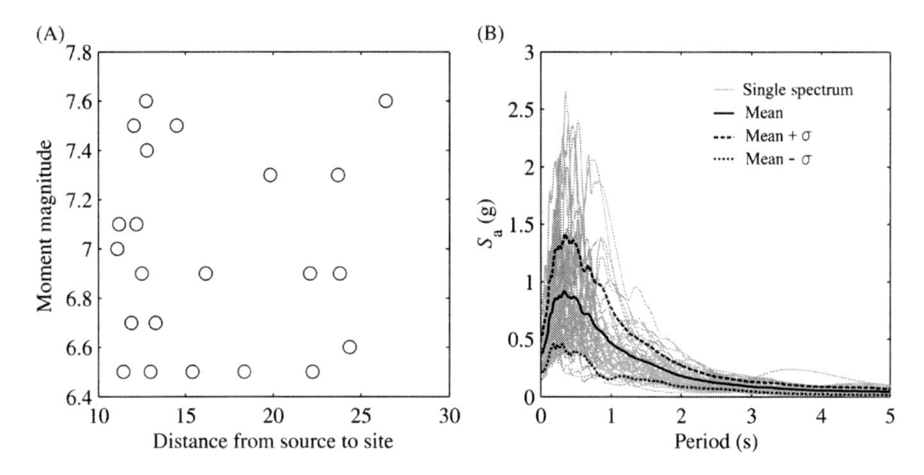

Figure 13.2 (A) The magnitude-distance distribution and (B) the 5% damped acceleration response spectra of the reference set of ground motion records.

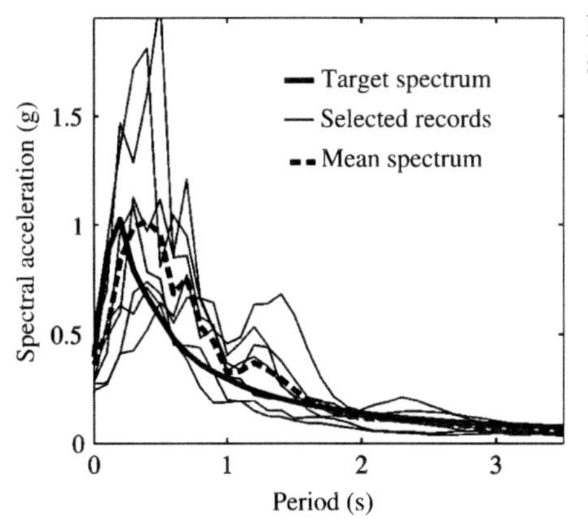

Figure 13.3 Selection of an arbitrarily subset with seven records.

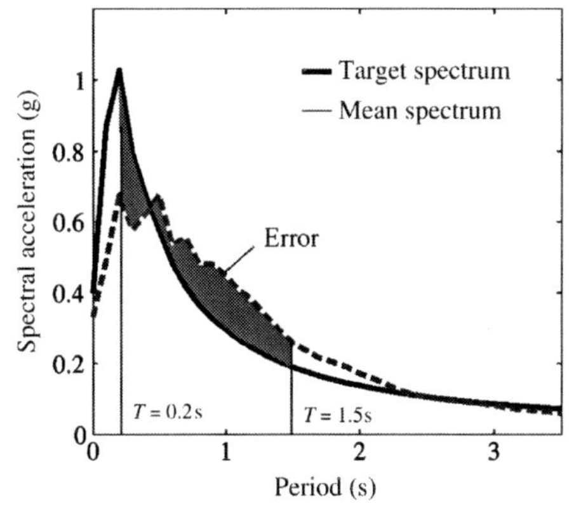

Figure 13.4 Definition of the error function.

According to the coordinates of the problem, seven integer variables, which indicate the indices of the selected records, create the decision space. The integer variables shall take a value between 1 and 44, in which duplication is not permissible.

The GA-based selection and scaling of records were first proposed by Naeim et al. (2004). Any arbitrary combination of seven records and seven scaling factors was defined as a single individual or chromosome (as shown in Figure 13.5). The objective is to create the best individual using the pool of earthquake records in the database and scaling factors. Therefore, each individual has 14 subdivisions to represent each variable (seven for the identification of records in the database and seven for the identification of the corresponding scale factors). A length of 10 binary digits is assigned to each subdivision, making the total length of each individual equal to 140 binary digits. This can be changed and longer binary strings can be used to accommodate larger earthquake record databases. The first seven binary substrings provide the positions of the seven records in the database. The remaining seven substrings represent the corresponding scaling factors (Figure 13.5). Due to certain complexities of this procedure, at this point, further details will be skipped.

Here, another simple procedure is proposed for the selection of ground motions by using a GA tool. It may be recalled that the decision space for record selection consists of integer values within the range between 1 and 44 and also that the duplication of values is not permissible (one record cannot be selected twice or more within a subset). These two issues create great concerns in the programming of the GA for the record selection. As an efficient solution for this problem, suppose that a random vector with a size 44 has been created. After sorting this

The binary strings: variables 1 through 14, Individual No. 100

Generation 1:

1001101000 | 0111100110 | 1110011100 | 0100110101 | 0111101011 | 0000010101 | 0000000101 |
0001010111 | 1010110011 | 1011010111 | 1001101001 | 1000001101 | 1000000111 | 0101011000 |

Generation 150:

0010101011 | 1111001010 | 1001001000 | 1100100101 | 0000111110 | 1101110000 | 1111111000 |
0101010100 | 1011001001 | 0010111111 | 0101011111 | 1100000111 | 1100011011 | 1111111000 |

Generation 300:

1000011100 | 0110110101 | 0010100010 | 1011010011 | 0001011010 | 0000111101 | 1001001011 |
0010001100 | 1001111110 | 1010100111 | 1100001100 | 0011111110 | 1001001000 | 1011111101 |

Set of decoded variables or solution vector: variables 1 through 14, Individual No. 100

Generation 1:

1,090 | 959 | 1,400 | 782 | 964 | 494 | 478 |
0.59 | 1.18 | 1.21 | 1.10 | 1.01 | 1.01 | 0.84 |

Generation 150:

644 | 1,440 | 1,060 | 1,280 | 535 | 1,350 | 1,490 |
0.83 | 1.20 | 0.69 | 0.84 | 1.26 | 1.28 | 1.49 |

Generation 300:

1,010 | 910 | 635 | 1,200 | 563 | 534 | 1,060 |
0.64 | 1.12 | 1.16 | 1.26 | 0.75 | 1.07 | 1.25 |

Figure 13.5 Binary chromosomes and decoded representation of an individual in the procedure of Naeim et al. (2004).

random vector, the rank of the first seven records can be accounted as a feasible solution for the selection of the subset. It is clear that both of the aforementioned criteria have been met by this trick. For more clarification, suppose that a random real vector A has been produced as below:

A = [− 0.79, 1.01, −0.13, −0.71, 1.35, −0.22, −0.58, −0.29, −0.84, − 1.12, 2.52, 1.65, 0.30, − 1.25, −0.86, −0.17, 0.79, − 1.33, −2.32, − 1.44, 0.33, 0.39, 0.45, −0.13, 0.18, −0.47, 0.86, − 1.36, 0.45, −0.84, −0.33, 0.55, 1.03, − 1.11, 1.26, 0.66, −0.06, −0.19, −0.21, −0.30, 0.02, 0.05, 0.82, 1.52]

Vector B shows the rank of each component of vector A in an ascending sort:

B = [19, 20, 28, 18, 14, 10, 34, 15, 9, 30, 1, 4, 7, 26, 31, 40, 8, 6, 39, 38, 16, 3, 24, 37, 41, 42, 25, 13, 21, 22, 23, 29, 32, 36, 17, 43, 27, 2, 33, 35, 5, 44, 12, 11]

The first seven components of this vector, [19, 20, 28, 18, 14, 10, 34], show the indices of the selected records. Also, another real vector with the size of seven can be generated to indicate the corresponding scale factor of each record. Thus, a single chromosome with 51 genes can be defined to represent each variable. Forty-four genes are assigned to find the best subset and seven genes for the identification of the corresponding scale factor. Associated with each random chromosome, the fitness function can be evaluated. The GA parameters consist of the population size, the number of generations, the crossover ratio, and the mutation ratio. The default values have been successfully defined in the MATLAB software package, although other values may also provide promising results.

- Acceptable scale factor range: 0.5−1.5
- Population of individuals: 200
- Number of generations: 300
- Crossover range: 0.65
- Mutation ratio: 0.025

The proposed procedure was applied to find the best subset of ground motions from Table 13.1. Minimization of the fitness value as a function of the number of generations is shown in Figure 13.6.

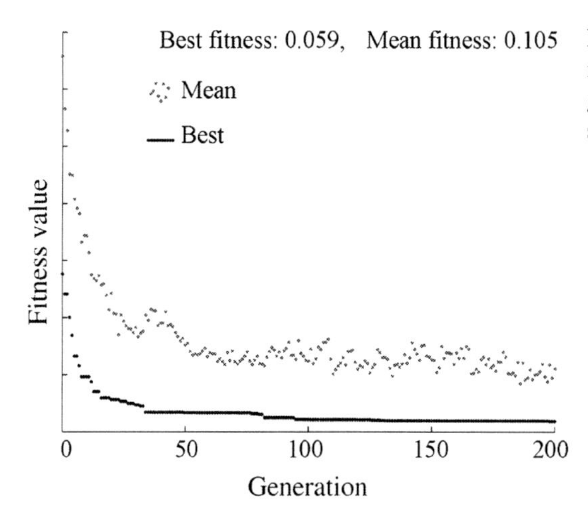

Figure 13.6 The mean and minimum fitness values in each generation for the code-conforming selection of records.

Table 13.2 Final Result of the Code-Conforming Selection

Record Index	Scaling Factor
2	1.04
9	0.59
10	0.69
21	0.83
28	0.92
38	1.08
40	1.26

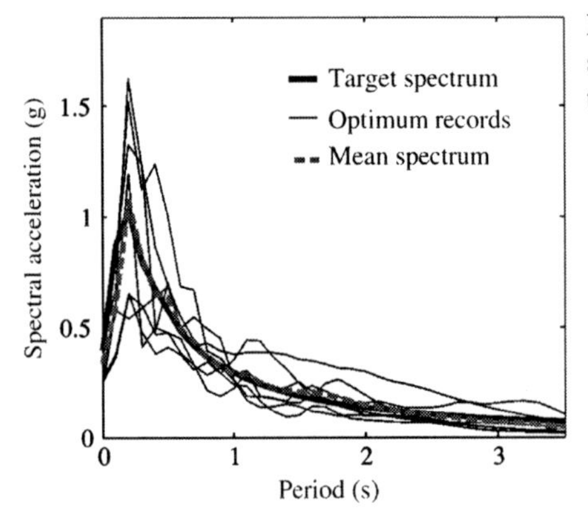

Figure 13.7 The spectrum of the optimum subset of records, the mean spectrum, and the target spectrum.

Table 13.2 indicates the selected records with the corresponding scaling factors resulting from the GA application. The selected records, the mean spectrum, and the target spectrum are plotted in Figure 13.7. It is clear that the mean and target spectrum have been matched together with the desired range of periods.

Here, the application of a GA in code-conforming record selection is covered. Ground motion selection in PBEE, as a modern approach, is discussed in the next section and a GA solution is presented.

13.4 Ground Motion Record Selection in PBEE

13.4.1 Definition of the Subject

The problem of the selection of earthquake ground motion records is still an interesting area for research. The median nonlinear response and its dispersion can vary significantly based on the record selection. Hence, the performance evaluation will be sensitive to the records that are selected. The common codes recommend using three (with critical actions) or seven (with average actions) design-spectrum-compatible records for the dynamic nonlinear analysis (ICBO2000, 2000, Chapter 16; UBC97, 1997, Chapter 16) as described in the previous section. However, the structural response based on different selection sets might still be different by using this approach. This lack of consistency motivated some researchers to propose new methods for ground motion selection in order to create a link between the record selection and the structural response. On the other hand, the use of the real records (beyond using compatible design-spectrum approaches) is not explicitly clear in the codes. Some bias in the results may arise if spectrum-compatible ground motion records are used. In this section, an attempt has been made to introduce a new methodology for the use of real records instead of compatible records. The methodology is described here in order to predict the three common fractile IDA curves (i.e., the 16%, 50%, and 84% fractiles). It is shown that the proposed methodology can be performed for a certain selection of IM and damage measure (DM) for a particular structure.

Progressive incremental dynamic analysis (PIDA) is a type of IDA analysis that is based on a limited preselected number of ground motion records, whose aim is to achieve the same summarized IDA curves that can be obtained from a larger number of ground motion records within an acceptable tolerance. PIDA has been proposed in this section in order to reduce the number of selected ground motion records for the sufficiently accurate prediction of summarized IDA curves. The procedure has been tested for a three-story concrete moment-resisting frame building (Fajfar et al., 2006), and the results obtained are presented. The set of ground motion records for IDA analysis consists of 30 free-field ground motion records. It is shown that the summarized IDA curves can be predicted with acceptable accuracy by employing only six ground motion records instead of 30, which is the total number of ground motion records for the predefined set.

13.4.2 PIDA Methodology

The aim of the proposed methodology is to decrease the number of ground motion records needed for the prediction of the summarized IDA curves (16%, 50%, and 84% fractiles). The main steps of the modified methodology are presented in Figure 13.1 and can be described as follows:

1. Select a set of ground motion records based on the earthquake scenario. This is the same step as in an IDA analysis (Vamvatsikos and Cornell, 2002). The number of records within the given set can be high since there is no need to compute the seismic response of the Multi-Degree-Of-Freedom (MDOF) model for all records in order to obtain a good prediction of the summarized IDA curves. The MDOF model will be analyzed only for a few preselected ground motion records.
2. Create an MDOF mathematical model that can be used for the simulation of the realistic seismic response of the structure under investigation.
3. Define a simple mathematical model, e.g., an Single-Degree-Of-Freedom (SDOF) model. This model should be a good representative of the linear and nonlinear characteristics of the MDOF mathematical model, yet simple enough for it to be possible to perform a large

number of nonlinear response-history analyses, without the need to perform time-consuming calculations.

4. Compute single-record IDA curves for the simple model, for all the ground motion records within the given set. Because of the simplicity of the chosen simple model, this should not be a time-consuming task. These IDA curves can be calculated for any preselected IM (e.g., PGA or the spectral acceleration corresponding to the first period of the simple model).
5. Arrange the ground motion records, based on the results obtained in step 4, within the given set in order to obtain a good precedence list. This is an optimization problem, which is explained in Section 13.4.3. The objective of the optimization is to minimize the differences between the "original" and the "selected" summarized IDA curves. The "original" summarized IDA curves are obtained from all the single-record IDA curves (step 4), whereas the "selected" summarized IDA curves are obtained only for the first s ground motion records from the precedence list, where s is the number of "selected" ground motion records.
6. Compute three single-record IDA curves for the MDOF model, starting with the first three records from the precedence list. After computation of the single-record IDA curves for the sth record from the precedence list (where s is a number greater than or equal to three), compute the "selected" summarized IDA curves and compare them with the "selected" summarized IDA curves obtained from the (s-3)th records.
7. Repeat step 6 until the difference between the "selected" summarized IDA curves, determined for the sth and (s-3)th records, is less than the acceptable tolerance; then stop performing the IDA analysis on the MDOF model.
8. The "selected" summarized IDA curves, calculated from the s single-record IDA curves, can be used for further seismic performance assessments.

The described procedure can significantly reduce the number of nonlinear response-history analyses needed to predict the summarized IDA curves with sufficient accuracy.

13.4.3 Precedence List of Ground Motion Records

The precedence list of ground motion records was determined for the selected set of ground motion records by employing a GA technique (Goldberg, 1989). The input data for determining the precedence list are the "original" summarized IDA curves (corresponding to the 16%, 50%, and 84% fractiles), single-record IDA curves, both determined on the basis of IDA analysis for the simple model (e.g., the SDOF model), and the corresponding ID numbers of the ground motion records (Table 13.3). The precedence list of the ground motion records is obtained by rearranging the ID numbers of the ground motion records (Table 13.3) in order to minimize the fitness function Z:

$$Z = \frac{1}{n/3}\sum_{s=3}^{n} V(s) = \frac{1}{n/3}\sum_{s=3i}^{n}\left[\sum_{f=1}^{3}\mathrm{Error}(s,f)\right], \quad i = 1,2,3,\ldots \tag{13.5}$$

The fitness function is defined as the summation of the so-called "partial" fitness function $V(s)$ normalized by $n/3$, where n is the number of ground motion

Table 13.3 The Free-Field Set of Ground Motion Records

Event, Year, M_W[a]	ID	Station	φ°[b]	Soil[c]	R[d]	PGA
Loma Prieta, 1989, 6.9	1	Agnews State Hospital	090	C,D	28.2	0.159
	2	Hollister Diff. Array	255	-,D	25.8	0.279
	3	Anderson Dam Downstrm	270	B,D	21.4	0.244
	4	Coyote Lake Dam Downstrm	289	B,D	22.3	0.179
	5	Sunnyvale Colton Ave	270	C,D	28.8	0.207
	6	Anderson Dam Downstrm	360	B,D	21.4	0.24
	7	Hollister South & Pine	000	-,D	28.8	0.371
	8	Sunnyvale Colton Ave	360	C,D	28.8	0.209
	9	Halls Valley	090	C,C	31.6	0.103
	10	WAHO	000	-,D	16.9	0.37
	11	Hollister Diff. Array	165	-,D	25.8	0.269
	12	WAHO	090	-,D	16.9	0.638
Northridge, 1994, 6.7	13	LA, Baldwin Hills	090	B,B	31.3	0.239
	14	LA, Hollywood Storage FF	360	C,D	25.5	0.358
Imperial Valley, 1979, 6.5	15	Computertas	285	C,D	32.6	0.147
	16	Plaster City	135	C,D	31.7	0.057
	17	El Centro Array # 12	140	C,D	18.2	0.143
	18	Cucapah	085	C,D	23.6	0.309
	19	Chihuahua	012	C,D	28.7	0.27
	20	El Centro Array # 13	140	C,D	21.9	0.117
	21	Westmoreland Fire Station	090	C,D	15.1	0.074
	22	Chihuahua	282	C,D	28.7	0.254
	23	El Centro Array # 13	230	C,D	21.9	0.139
	24	Westmoreland Fire Station	180	C,D	15.1	0.11
	25	Computertas	015	C,D	32.6	0.186
	26	Plaster City	045	C,D	31.7	0.042
San Fernando, 1971, 6.6	27	LA, Hollywood Stor. Lot	180	C,D	21.2	0.174
	28	LA, Hollywood Stor. Lot	090	C,D	21.2	0.21
Superstition Hills, 1987, 6.7	29	Wildlife Liquefaction Array	090	C,D	24.4	0.18
	30	Wildlife Liquefaction Array	360	C,D	24.4	0.2

[a]Moment magnitude.
[b]Component.
[c]USGS, Geomatrix soil class.
[d]Closest distance to fault rupture expressed in kilometers.

records in the set. Z therefore can be interpreted as the average "partial" fitness function $V(s)$. The "partial" fitness function, $V(s)$, is defined as the cumulative error for the three fractile curves (f = 16%, 50%, 84%), which are the subject of the optimization. However, minimization of the "partial" fitness function means the selection of those s ground motion records that are the best representatives of the "original" summarized IDA curves (the 16%, 50%, and 84% fractiles) determined on the basis of IDA analysis for the simple model (e.g., SDOF model). The Error(s,f), which is called the error function, is defined as the normalized area, which is determined based on the difference between the "original" and the "selected" summarized IDA curves, which can be 16%, 50%, or 84% fractiles. The error function is a function of a particular fractile curve f and of the s selected ground motion records for which the "selected" summarized IDA curve is determined. Note, as explained in the methodology, that the "original" summarized IDA curves (16%, 50%, and 84% fractiles) are obtained from all the single-record IDA curves, whereas the "selected" summarized IDA curves are obtained for just the first s ground motion records from the precedence list, where s is equal to or greater than three since three fractile curves (16%, 50%, and 84%) can be predicted at least with three ground motion records.

The normalized area between the "original" and the "selected" summarized IDA curves, expressed as a percentage, can be calculated as:

$$\text{Error}(s,f) = 100 \times \frac{\int_0^{\text{DM}_{\max}(s,f)} |\Delta\text{IM}(s,f)| \, \text{dDM}}{\int_0^{\text{DM}_{\max,\text{or}}(f)} \text{IM}_{\text{or}}(f) \, \text{dDM}} \tag{13.6}$$

where DM is the damage measure, IM is an intensity measure for the IDA analysis, $\Delta\text{IM}(s,f)$ is the difference in the IM corresponding to the "original" and "selected" f summarized IDA curves, and $\text{DM}_{\max}(s,f)$ is the maximum DM, as presented in Figure 13.8. The parameter $\Delta\text{IM}(s,f)$ depends on the s ground motion records, which are used to determine the "selected" f summarized IDA curve, and also on the DM, as shown in Figure 13.8. The maximum damage measure $\text{DM}_{\max}(s,f)$ is usually defined by the capacity point on the "original" or "selected" summarized IDA curves. This measure also depends on the number of selected ground motion records s. The original maximum damage measure $\text{DM}_{\max,\text{or}}(f)$ is usually defined by the capacity point on the "original" summarized IDA curves and $\text{IM}_{\text{or}}(f)$ is the IM of the "original" f summarized IDA curves. Different possibilities of the relationship between the "original" and "selected" summarized IDA curves and the explained parameters of Eq. (13.6) are presented in Figure 13.8.

The GA optimization technique that is used in this section cannot determine the exact global optima, but it can usually find a solution near the global optimum. The precedence list of ground motion records might be different in a new run, but the predicted summarized IDA curves do not change significantly since the solution is always near enough to the global optima.

13.4.4 Example

In order to demonstrate the applicability of the proposed methodology, a precedence list of ground motion records was determined to predict the summarized IDA curves (16%, 50%, and 84% fractiles) for a three-story reinforced concrete frame building. Only a limited number of ground motion records were used. The precedence list was determined for a set that included 30 free-field ground motion records. The spectral acceleration corresponding to the first mode period of the considered structure was selected in the example as an IM. The maximum interstory drift ratio of the building was chosen as the damage measure. The results are

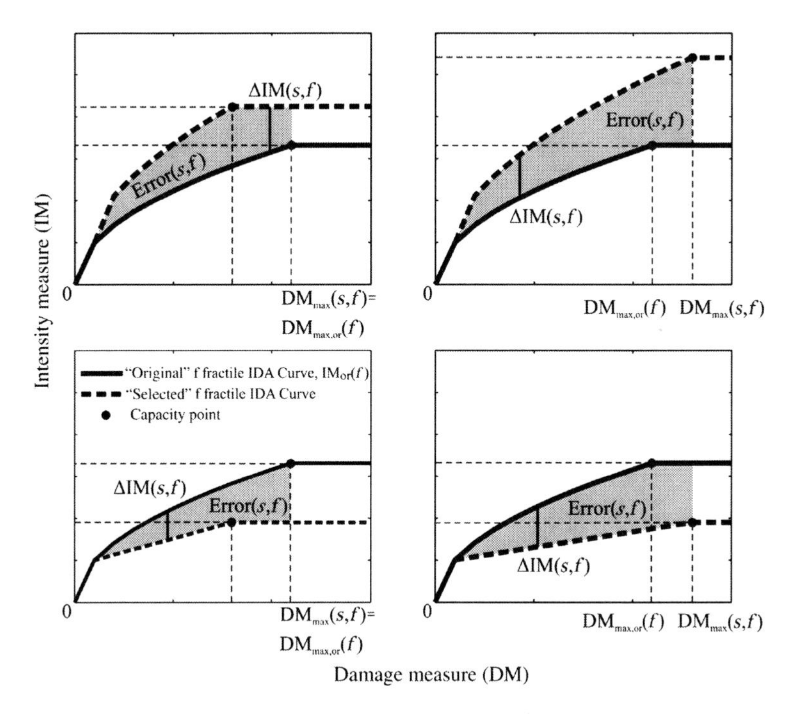

Figure 13.8 Schematic definition of $DM_{max}(s,f)$ and Error(s,f), shown hatched, based on six possible conditions of the "original" summarized IDA curves and the "selected" summarized IDA curve, which is determined based on s selected ground motion records (Azarbakht and Dolšek, 2007).

presented in terms of the "selected" summarized IDA curves and compared with the "original" summarized IDA curves.

13.4.4.1 The Test Structure and Selected Ground Motion Records

This structure (referred to in the following text as the SPEAR building) was a three-story asymmetric reinforced concrete frame building, for which a pseudodynamic experiment was performed at full scale at the ELSA Laboratory, within the European research project SPEAR (Seismic performance assessment and rehabilitation of existing buildings) (Negro et al., 2004). This building was designed for gravity loads only. The elevation and plan view of the SPEAR building are shown in Figure 13.9, together with details about the typical reinforcement in its beams and columns. The so-called posttest mathematical model (Fajfar et al., 2006), created within the OpenSees program (PEER), was employed for the analyses performed in this study. This mathematical model consists of beam and column elements whose flexural behavior was modeled by one-component lumped plasticity elements, consisting of an elastic beam and two inelastic rotational hinges (defined by the moment-rotation relationship). In the case of the beams, the plastic

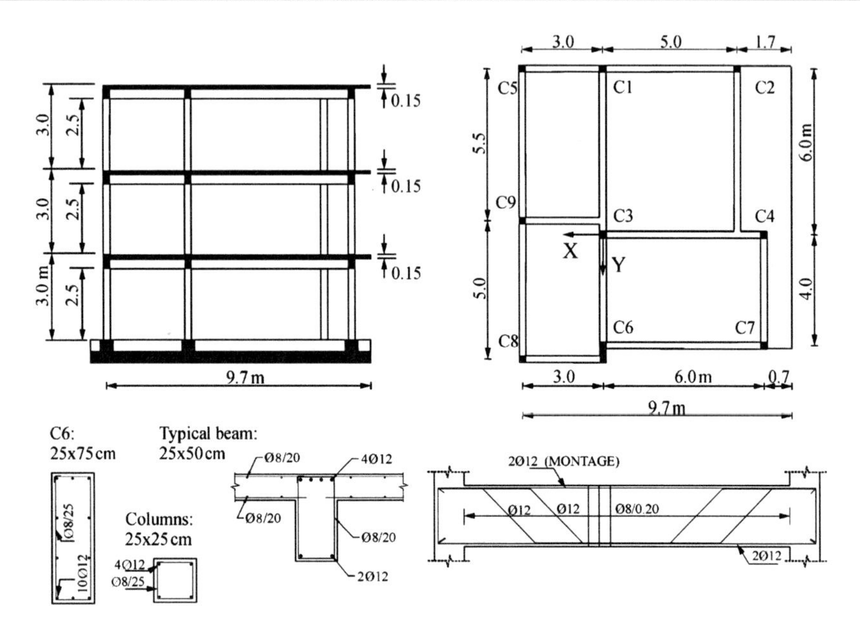

Figure 13.9 Elevation and plan view of the SPEAR building, showing typical reinforcement details (Fajfar et al., 2006).

hinge was used for major axis bending only. In the case of the columns, two independent plastic hinges for bending about the two principal axes were used. The moment-rotation envelope for inelastic rotational hinges was determined based on the axial force from the vertical load and on zero axial force for the hinges in the columns and beams, respectively. The maximum story drift time histories observed in the experiment are presented in Figure 13.10 and compared with the calculated results. A more detailed explanation of the model and a comparison with experimental results can be found in (Dolšek and Fajfar, 2005). The input files of the mathematical model of the SPEAR building are available at www.ikpir.com\projects\spear. A more detailed explanation of the model and a comparison with the experimental results can be found in Fajfar et al. (2006).

13.4.4.2 IDA Analysis for a Simple Mathematical Model

The simple mathematical model is introduced by an SDOF model, which is based on the results of pushover analysis of the MDOF model. Pushover analysis of the MDOF model was performed for the weak direction only since the mathematical model of the test structure, too, was subjected only to ground motion records in the weak direction. The load pattern employed in the pushover analysis corresponded to the dominant mode shape in the weak direction. The pushover curve and the idealized base shear-top displacement relationship is presented in Figure 13.10A.

The SDOF model was then defined based on the approach presented in Fajfar (1999). The force-displacement envelope of the SDOF model was obtained by

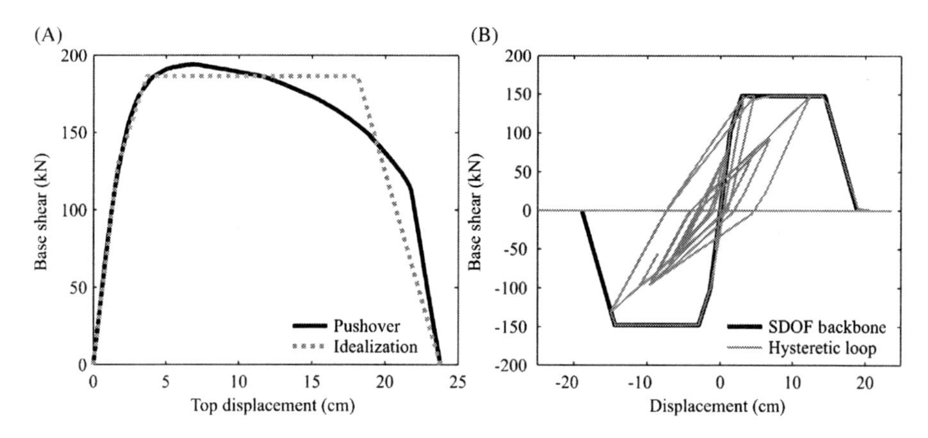

Figure 13.10 (A) The pushover curve, the idealized base shear-top displacement relationship and (B) the force-displacement relationship for the SDOF model, with typical hysteretic behavior (Azarbakht and Dolšek, 2007).

dividing the forces and displacements of the idealized pushover curve (Figure 13.10A) by a transformation factor, which in this example, is equal to 1.26. The period of the SDOF model is 0.85 s. The hysteretic behavior of the SDOF model, as presented in Figure 13.10B, was selected in order to properly simulate the hysteretic behavior of the MDOF model (Takeda et al., 1970).

The IDA analysis for the selected set of ground motion records (Table 13.3) was then performed on the SDOF model. The results are presented in terms of single-record IDA curves. In addition to the single-record IDA curves, based on the response of the SDOF model, the corresponding ID numbers of the ground motion records and the summarized IDA curves are needed in order to determine the precedence list of ground motion records.

The records listed in Table 13.3 were selected for the nonlinear dynamic analysis (PEER, 2008).

13.4.4.3 The Precedence List of Ground Motion Records

The precedence list of ground motion records were determined by employing the GA-based optimization technique, which has been described in Goldberg (1989). The input data for determining the precedence list consisted of the single-record IDA curves, the corresponding ID numbers of the ground motion records (Table 13.3), and the fractile IDA curves. The time for determining the precedence list of ground motion records, together with the IDA analysis performed for the SDOF model, was less than 60 min in the case of GA. This is even much less than the time needed to determine one single-record IDA curve of the MDOF model.

The precedence list of the ground motion records is basically obtained by rearranging the ID numbers of the ground motion records (Table 13.3) in order to minimize the fitness function defined by Eq. (13.5). The ID numbers and the

Table 13.4 The Precedence List of Ground Motion Records

Event, Year, M_W[a]	ID	Priority List (GA Technique)	Priority List (Simple Method)
Loma Prieta, 1989, 6.9	1	18	13
	2	2	1
	3	3	5
	4	5	9
	5	8	21
	6	17	19
	7	6	16
	8	14	12
	9	12	11
	10	29	17
	11	11	18
	12	13	4
Northridge, 1994, 6.7	13	7	7
	14	19	28
Imperial Valley, 1979, 6.5	15	21	26
	16	24	24
	17	10	10
	18	22	2
	19	27	14
	20	23	25
	21	28	29
	22	30	30
	23	1	3
	24	20	27
	25	9	22
	26	4	23
San Fernando, 1971, 6.6	27	16	20
	28	15	6
Superstition Hills, 1987, 6.7	29	26	8
	30	25	15

[a]Moment magnitude.

precedence list of ground motion records are presented in Table 13.4 for the GA techniques.

13.4.4.4 IDA Analysis for the MDOF Model

Analysis of the MDOF model, based on steps 6 and 7 of the proposed methodology, is illustrated in this section, whereas, step 8, the seismic performance assessment of the test structure, is outside the scope of this chapter. The results are presented in terms of the maximum inter-story drift ratio, which is usually a sufficiently good EDP for defining different limit states (performance levels).

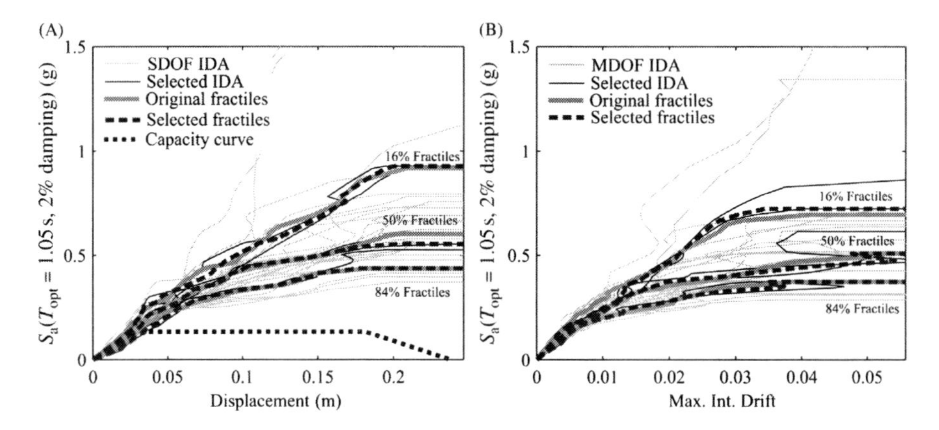

Figure 13.11 The comparison of "selected" summarized IDA curves using first six ground motion records from the precedence list (Table 13.4) with the "original" summarized IDA curves (A) for the SDOF model and (B) for the MDOF model.

The "selected" summarized IDA curves determined for the first six ground motion records from the precedence list are presented in Figure 13.11 and compared with the original summarized IDA curves. In Figure 13.11A, a comparison has been made on the basis of the IDA analysis for the SDOF model, whereas, in Figure 13.11B, the results are shown for the MDOF model. Good agreement between the summarized IDA curves can be observed, although a small number of ground motion records, six in this case, were used.

13.5 Conclusions

The application of GA has been introduced for the purpose of ground motion selection. The code-conforming procedures have been reviewed and a simple example has been presented. Also, a methodology has been proposed in order to predict the summarized IDA curves with only a limited number of ground motion records from a given set of records. For this purpose, the concept of a precedence list of ground motion records has been introduced. Determination of the precedence list of ground motion records is an optimization problem, which has been solved by a GA technique. In the proposed methodology, as in other simplified methods, the response of a simple (e.g., SDOF) model is taken into account. Such an approach is not computationally demanding and can substantially decrease the number of nonlinear dynamic analyses needed for sufficiently accurate prediction of the summarized IDA curves.

The methodology was applied to the analysis of a three-story reinforced concrete frame building, using a set of 30 ground motion records. It was shown that, for this particular example, the 16%, 50%, and 84% summarized IDA curves can be predicted with acceptable accuracy by employing only six ground motion records instead of 30, which is the number of all the ground motion records in the set of records.

References

ASCE7-5, 2005. Minimum Design Loads for Buildings and Other Structures. American Society of Civil Engineers/Structural Engineering Institute, Reston, VA.

ATC40, 1996. Methodology for Seismic Evaluation and Retrofit of Existing Concrete Buildings. Applied Technology Council, Redwood City, CA.

ATC63, 2008. Recommended Methodology for Quantification of Building System Performance and Response Parameters. Applied Technology Council, Redwood City, CA.

Azarbakht, A., Dolšek, M., 2007. Prediction of the median IDA curve by employing a limited number of ground motion records. Earthq. Eng. Struct. Dyn. 36, 2401–2421.

Azarbakht, A., Dolsek, M., 2011. Progressive incremental dynamic analysis for first-mode dominated structures. J. Struct. Eng. 137, 445–455.

Bazzurro, P., 1998. *Probabilistic SeismicDemand Analysis*. Department of Civil and Environmental Engineering, Stanford University, USA.

Benjamin, J.R., Cornell, C.A., 1970. Probability, Statistics, and Decision for Civil Engineers. McGraw Hill, Inc., New York, NY.

Beresnev, I.A., Atkinson, G.M., 1998. FINSIM: A FORTRAN program for simulating stochastic acceleration time histories from finite faults. Seismol. Res. Lett. 69, 27–32.

Bommer, J., 2004. The Use of Real Earthquake Accelerograms as Input to Dynamic Analysis. Imperial College Press, London, UK.

Bommer, J.J., Scott, S.G., Sarma, S.K., 2000. Hazard-consistent earthquake scenarios. Soil Dyn. Earthq. Eng. 19, 219–231.

Boore, D.M., 2003. Simulation of ground motion using the stochastic method. Pure Appl. Geophys. 160, 635–676.

Collins, K.R., Wen, Y., Foutch, D.A., 1996. Dual-level seismic design: a reliability based methodology. Earthq. Eng. Struc. Dyn. 25, 1433–1467.

Commission, U.N.R., 2001. Technical Basis for Revision of Regulatory Guidance on Design Ground Motions: Hazard- and Risk-Consistent Ground Motion Spectra Guidelines. Government Printing Office, Washington, DC.

Cornell, C.A., 1996. Reliability-based earthquake-resistant design: the future. In: Proceedings of the Eleventh World Conference on Earthquake Engineering. Acapulco, Mexico.

Cornell, CA., Krawinkler H., 2000. Progress and challenges in seismic performance assessment. PEER Center News 3 (2).

Cornell, C.A., Jalayer, F., Hamburger, R.O., Foutch, D.A., 2002. The probabilistic basis for the 2000 SAC/FEMA steel moment frame guidelines. J. Struct. Eng. 128, 526–533.

Dolšek, M., Fajfar, P., 2005. Simplified non-linear seismic analysis of infilled reinforced concrete frames. Earthq. Eng. Struct. Dyn. 34, 49–66.

Ebrahimian, H., Azarbakht, A., Tabandeh, A., Golafshani, A., 2012. The exact and approximate conditional spectra in the multi-seismic sources regions. Soil Dyn. Earthq. Eng. 39, 61–77.

Fajfar, P., 1999. Capacity spectrum method based on inelastic demand spectra. Earthq. Eng. Struct. Dyn. 28, 979–993.

Fajfar, P., Dolsek, M., Marusic, D., Stratan, A., 2006. Pre- and post-test mathematical modeling of a plan-asymmetric reinforced concrete frame building. Earthq. Eng. Struct. Dyn. 35, 1359–1379.

FEMA350, 2000. Recommended Seismic Design Criteria for New Steel Moment-Frame Buildings. SAC Joint Venture. Prepared for the Federal Emergency Management Agency, Washington, DC.

FEMA356, 2000. Recommended Seismic Design Criteria for New Steel Moment-Frame Buildings. SAC Joint Venture. Prepared for the Federal Emergency Management Agency, Washington, DC.

Gandomi, A., Alavi, A., 2012. Krill Herd: A New Bio-Inspired Optimization Algorithm. Communications in Nonlinear Science and Numerical Simulation. 17 (12), 4831–4845.

Gandomi, A., Yang, X.S., Alavi, A., 2011. Mixed variable structural optimization using firefly algorithm. Comput. Struct. 89 (23–24), 2325–2336.

Gandomi, A., Yang, X.S., Alavi, A., 2012. Cuckoo search algorithm: a metaheuristic approach to solve structural optimization problems. Eng. Comput. doi: 10.1007/s00366-011-0241-y.

Gasparini, D.A., Vanmarcke, E.H., 1979. Simulated Earthquake Motions Compatible with Prescribed Response Spectra. Department of Civil Engineering, MIT, Cambridge, MA.

Ghafory-Ashtiany, M., Mousavi, M., Azarbakht, A., 2011. Strong ground motion record selection for the reliable prediction of the mean seismic collapse capacity of a structure group. Earthq. Eng. Struct. Dyn. 40, 691–798.

Ghobarah, A., 2001. Performance-based design in earthquake engineering: state of development. Eng. Struct. 23, 878–884.

Goldberg, D., 1989. Genetic Algorithms in Search, Optimization, and Machine Learning. Addison-Wesley, Reading, MA.

Han, S.W., Wen, Y.K., 1997. Method of reliability-based seismic design. I: equivalent nonlinear systems. J. Struct. Eng. 123, 256–263.

Hickman, J.W., 1983. PRA Procedures Guide: A Guide to the Performance of Probabilistic Risk Assessments for Nuclear Power Plants. Office of Nuclear Regulatory Research, U.S. Nuclear Regulatory Commission, Washington, DC.

Holland, H.J., 1975. Adaptation in Natural and Artificial Systems: An Introductory Analysis with Applications to Biology, Control and Artificial Intelligence. University of Michigan Press, Ann Arbor, MI.

ICBO2000, 2000. International Building Code (IBC). Publisher, Whittier, CA.

Jalayer, F., Beck, J.L., Porter, K.A., Hall, J.F., 2004. Effects of ground motion uncertainty on predicting the response of an existing RC frame structure. In: Proceedings of the 13th World Conference on Earthquake Engineering. Vancouver, Canada.

Kayhani, H., Azarbakht, A., Ghafory-Ashtiany, M., 2012. Estimating the annual probability of failure using improved progressive incremental dynamic analysis of structural systems. The structural design of tall and special buildings Doi: 10.1002/tal.1006.

Luco, N., 2002. Probabilistic Seismic Demand Analysis, SMRF Connection Fractures, and Nearsource Effects. Ph.D. Dissertation, Stanford University, USA.

Luco, N., Cornell, C.A., 2007. Structure-specific scalar intensity measures for near-source and ordinary earthquake ground motions. Earthq. Spectra. 23, 357–392.

McGuire, R., 1995. Probabilistic seismic hazard analysis and design earthquakes: closing the loop. Bull. Seismol. Soc. Am. 85, 1275–1284.

McGuire, R.K., Cornell, C.A., 1974. Seismic Structural Response Risk Analysis, Incorporating Peak Response Regressions on Earthquake Magnitude and Distance. Department of Civil Engineering, Massachusetts Institute of Technology, Cambridge, MA.

Mitchell, M., 1996. An Introduction to Genetic Algorithms. Bradford Publishing, Complex Adaptative System Series. The MIT Press, Massachusetts Institute of Technology, Cambridge, Massachusetts.

Mousavi, M., Ghafory-Ashtiany, M., Azarbakht, A., 2011. A new indicator of elastic spectral shape for the reliable selection of ground motion records. Earthq. Eng. Struct. Dyn. 40, 1403–1416.

Mousavi, M., Ansari, A., Zafarani, H., Azarbakht, A., 2012. Selection of Ground Motion Prediction Models for Seismic Hazard Analyzes in Zagros Region, Iran. J. Earthquake Eng. 16, 1184–1207.

Naeim, F., Lew, M., 1995. On the use of design spectrum compatible time histories. Earthq. Spectra. 11, 111–127.

Naeim, F., Alimoradi, A., Pezeshk, S., 2004. Selection and scaling of ground motion earthquakes for structural design using genetic algorithms. Earthq. Spectra. 20, 413–426.

Negro, P., Mola, E., Molina, F.J., Magonette, G.E., 2004. Full-scale testing of a torsionally unbalanced three-storey nonseismic RC frame. Proceedings of the 13th World Conference on Earthquake Engineering. Vancouver, Canada.

PEER Open System For Earthquake Engineering Simulation (OpenSees). 2006 Pacific Earthquake Engineering Research Center (PEER). Available from: <http://opensees.berkeley.edu/>. (last accessed 27.08.2012).

PEER. 2008. Next generation attenuation (NGA) models. Available from: <http://peer.berkeley.edu/ngawest/nga_models.html/>.

Pezeshk, S., Camp, C.V., Chen, D., 2000. Design of framed structures by genetic optimization. J. Struct. Eng. 126, 382–388.

Reiter, L., 1990. Earthquake Hazard Analysis: Issues and Insights. Columbia University Press, New York, NY, pp. 254.

Sewell, R.T., 1989. Damage Effectiveness of Earthquake Ground Motion: Characterizations Based on the Performance of Structures and Equipment. Ph.D., Stanford University.

Shome, N., Cornell, C., 1999. Probabilistic seismic demand analysis of nonlinear structures. RMS-35. RMS Program, Stanford, CA. pp. 320. Available from: <http://www.stanford.edu\group\rms\> (accessed 14.03.05).

Takeda, T., Sozen, M., Nielsen, N., 1970. Reinforced concrete response to simulated earthquakes. J. Struct. Div. 96, 2557–2573.

Tothong, P., Luco, N. 2007. Probabilistic Seismic Demand Analysis Using Advanced Ground Motion Intensity Measures. J. Earthquake Eng. & Structural Dynamics, 36 (13), 1837–1860.

UBC97, 1997. Structural Engineering Design and Provisions. International Conference of Building Officials (ICBO), Whittier, CA.

Vamvatsikos, D., Cornell, C., 2002. Incremental dynamic analysis. Earthq. Eng. Struct. Dyn. 31, 491–514.

Wen, Y.K., 2000. Reliability and performance based design. The Eight ASCE Specialty Conference on Probabilistic Mechanics and Structural Reliability. University of Notre Dame, South Bend, IN.

Yang, X.S., Gandomi, A., 2012. Bat algorithm: a novel approach for global engineering optimization. Eng. Comput. 29 (5), 464–483.

Zeng, Y., Anderson, J.G., Yu, G., 1994. A composite source model for computing realistic synthetic strong ground motions. Geophys. Res. Lett. 21, 725–728.

14 Optimization of Tuned Mass Damper with Harmony Search

Gebrail Bekdaş and Sinan Melih Nigdeli

Department of Civil Engineering, Istanbul University, Avcılar, Istanbul, Turkey

14.1 Introduction

Metaheuristic algorithms simulate natural phenomena in order to find the maximum or minimum of an objective function under some specific restrictions. These optimization algorithms are usually inspired by observations of natural processes for solving problems, including engineering problems. For example, the inspiration and name of the simulated annealing (SA) method comes from the process of annealing that involves heating and controlled cooling of a material in order to increase the size of its crystals and reduce their defects (Kirkpatrick et al., 1983). The genetic algorithms (GA) are inspired by natural evolutionary processes, such as inheritance, mutation, selection, and crossover (Goldberg, 1989; Holland, 1975). The behavior of bees was imitated for several algorithms such as the honey bee algorithm (HBA) (Nakrani and Tovey, 2004), virtual bee algorithm (VBA) (Yang, 2005), honey bee mating optimization (HBMO) (Afshar et al., 2007), and artificial bee colony (ABC) optimization algorithm (Karaboga, 2005). Other natural phenomena that have inspired optimization algorithms include the human memory for tabu search (TS) (Glover, 1977), the movement of members in a bird flock or fish school for particle swarm optimization (PSO) (Kennedy and Eberhart,1995), the behavior of ants seeking a path between colony and food for ant colony optimization (ACO) (Dorigo et al., 1996), the common theories of the evolution of the universe for the Big Bang–Big Crunch (BB–BC) method (Erol and Eksin, 2006), the flashing characteristic of fireflies for the firefly algorithm (FA) (Yang, 2008), the combination of brood parasitic behavior of cuckoo species and Levy flight behavior for cuckoo search (CS) (Gandomi et al., 2012a), and the echolocation characteristic of microbats for bat algorithm (BA) (Yang and Gandomi, 2012). Recently, a new bioinspired optimization algorithm called the krill herd (KH) algorithm was developed from the observations of herding behavior of krill individuals (Gandomi and Alavi, 2012).

The harmony search (HS) algorithm, which was developed by Geem et al. (2001), is a memory-based random search method, which imitates the music

Metaheuristic Applications in Structures and Infrastructures. DOI: http://dx.doi.org/10.1016/B978-0-12-398364-0.00014-0

performance process. In this process, a musician tries to find a pleasing harmony, which is a perfect state for appreciation of the audience. Like a musician, a researcher tries to find a global solution (a perfect state) for maximum performance with a low cost. Compared to other metaheuristic methods, HS is not complex because a stochastic random search is used instead of a gradient search. HS is not a hill-climbing algorithm so the local optimal solution does not occur in solving problems. The HS algorithm is capable of solving problems with discrete and continuous variables (Lee and Geem, 2005; Lee et al., 2005). In the HS algorithm, the number of iterations can be reduced by using stochastic derivatives. In scientific and engineering problems, when the function's mathematical derivative cannot be analytically obtained or the function's type is stepwise or conditionwise, the usage of stochastic derivatives is important (Geem, 2008).

The parameters of HS offer an effective solution and process time. One of the parameters is a range in which the optimum solutions are searched for maximizing or minimizing the objective function. Selecting a wide range can increase the optimization time, while a tight range may delay the finding of an optimum solution. The basic form of HS method contains specific parameters like other optimization algorithms. These parameters are harmony memory size (HMS), harmony memory considering rate (HMCR), and pitch adjusting rate (PAR). If these parameters are not suitable to the problem, the duration of the optimization process may increase. For that reason, the HS algorithm was also improved with parameter setting free techniques (Geem and Sim, 2010; Hasancebi et al., 2010). Except for the parameter setting procedure, the HS algorithm can be defined in four main steps. These steps are generation of initial harmony memory (HM), generation of a new harmony, comparison of the new harmony with the HM, and checking the stopping criteria, respectively. Also, a new type of HS algorithm that automatically selecting the proper pitch adjustment strategy based on its HM was developed in order to maintain a proper balance between diversification and intensification throughout the search process (Yadav et al., 2012). Detailed descriptions of metaheuristic methods and the HS algorithm can be found in Chapters 1 (Gandomi et al., 2012b) and 2 (Sahab et al., 2012).

14.2 A Passive Structural Control Device: Tuned Mass Damper

Passive control systems are effective for damping vibration with their mechanical components. Their performance is limited by their materials and properties. By using the optimum properties, it is possible to reduce structural responses economically when structures are subjected to unstable excitations that result from earthquakes and strong winds. Passive control of structures is a popular subject, especially in the area of retrofit. The application of passive control includes yielding steel energy absorbing devices, base isolation systems, tuned mass dampers (TMDs), friction and viscoelastic dampers, fiber-reinforced plastics (FRP), and even reinforced concrete walls.

Figure 14.1 The TV Tower in Berlin.

The TMD is a passive-control system consisting of mass, dampers, and stiffness members sustained by springs, cables, and isolators. The duty of this device at a structure is to reduce earthquake and wind-induced vibrations. TMDs have been applied to tower structures under the risk of strong winds. These include the Citigroup Center in New York City, Yokohama Landmark Tower in Yokohama, Burj Al Arab in Dubai, Trump World Tower in New York City, Taipei 101 in Taipei, The TV Tower in Berlin, and the Theme Building in Los Angeles. TMDs were applied to some of these structures after their construction. The TV Tower in Berlin contains a TMD consisting of a 1.5 t metal rod suspended by three cables and four hydraulic telescopic shock absorbers (Figure 14.1). TMDs can be used for seismic retrofit of landmark structures without changing their appearance. For example, the Theme Building at the Los Angeles International Airport (LAX Theme Building) was retrofitted by adding a TMD at the top of the concrete core (Figure 14.2). In order to get approximately 30–40% reduction in the responses, a TMD with 20% mass ratio, eight lead rubber bearings, and eight fluid viscous dampers were used (Miyamoto et al., 2011).

The basic form of TMDs was invented by Frahm in 1909 for damping resonant vibrations. These devices had only a mass with stiffness members without any

Figure 14.2 The LAX Theme Building (left), implemented damper and isolator (right).

damping device. For that reason, this device was effective only when the absorber's natural frequency was very close to the excitation frequency (Frahm, 1911). With a certain amount of damping, it is possible to obtain beneficial results under changing excitation frequency. Ormondroyd and Den Hartog (1928) attached damper devices to the initial form of TMD and the form of a regular TMD was created. For the best performance of the device, the properties of the mechanical components of TMD must be tuned.

14.2.1 A Brief Review of Studies on Parameter Estimation of TMDs

The basic tuning method for TMDs is to equalize the natural frequencies of the main structure and TMD. However, the optimum ratio between the frequencies may vary under random vibrations. In addition to that, the optimum damping and mass may be investigated for the best performance. Researchers developed several methods around that subject and the studies have been continuing in order to obtain the most ideal results.

Den Hartog (1947) developed closed form expressions of optimum damper parameters including frequency ratio and damping ratio of TMD. These expressions of optimum frequency ratio (f_{opt}) and damping ratio ($\xi_{d_{\text{opt}}}$) of a TMD that minimize the steady-state response of an undamped single degree of freedom (SDOF) main mass subjected to a harmonic excitation can be seen in Eqs. (14.1) and (14.2), respectively. The frequency ratio is the ratio between the frequencies of TMD ($w_{d,\text{opt}}$) and SDOF system (w_s). The expressions of Den Hartog depend on a mass ratio (μ) (Eq. (14.3)).

$$f_{\text{opt}} = \frac{w_{d,\text{opt}}}{w_s} = \frac{1}{1+\mu} \tag{14.1}$$

$$\xi_{d_{\text{opt}}} = \frac{c_d}{2m_d w_{d,\text{opt}}} = \sqrt{\frac{3\mu}{8(1+\mu)}} \tag{14.2}$$

The mass ratio is the ratio between the masses of TMD (m_d) and SDOF system (m_s). The stiffness coefficient of TMD (k_d) can be obtained by using Eq. (14.4) if the unit of natural frequency of TMD is radian per second.

$$\mu = \frac{m_d}{m_s} \tag{14.3}$$

$$k_d = w_{d,\text{opt}}^2 m_d \tag{14.4}$$

After the study by Den Hartog, damping in the main system was taken into account in several studies (Bishop and Welbourn, 1952; Falcon et al., 1967; Ioi and Ikeda, 1978; Snowdon, 1959). Warburton and Ayorinde (1980) proved that complex systems can be thought of as an equivalent SDOF system in optimization processes if its natural frequencies are well separated. A frequency locus method was used to obtain optimum damper parameters by Thompson (1981). Warburton (1982) advised simple expressions for optimum TMD parameters for an undamped SDOF main system under harmonic and white noise random excitations. The optimum parameters of TMD under random acceleration excitation with white noise spectral density are given in Eqs. (14.5) and (14.6).

$$f_{\text{opt}} = \frac{\sqrt{1-(\mu/2)}}{1+\mu} \tag{14.5}$$

$$\xi_{d_{\text{opt}}} = \sqrt{\frac{\mu(1-(\mu/4)}{4(1+\mu)(1-\mu/2)}} \tag{14.6}$$

Sadek et al. (1997) searched numerically f_{opt} and $\xi_{d_{\text{opt}}}$ values for different mass ratios and main system damping ratio. The expressions of f_{opt} and $\xi_{d_{\text{opt}}}$, which were found by using a curve fitting method, are given in Eqs. (14.7) and (14.8). In these expressions, the damping ratio of the main system (ξ) was also taken into account.

$$f_{\text{opt}} = \frac{1}{1+\mu}\left[1 - \xi\sqrt{\frac{\mu}{1+\mu}}\right] \tag{14.7}$$

$$\xi_{d_{\text{opt}}} = \frac{\xi}{1+\mu} + \sqrt{\frac{\mu}{1+\mu}} \tag{14.8}$$

Also, multiple degree of freedom (MDOF) systems with three different structural models were investigated by Sadek et al. (1997). If the amplitude of the first mode shape for a unit modal participation factor at the location of the TMD is represented with Φ, the tuning ratio for the MDOF system is nearly equal to the tuning ratio for a SDOF system for a mass ratio of $\mu\Phi$. Also, the optimum damping ratio for a MDOF system is approximately equal to the optimum value computed for a SDOF system multiplied by Φ. In Eqs. (14.9) and (14.10), the closed form formulas are given for the optimization of TMD parameters for a damped MDOF structure.

$$f_{\text{opt}} = \frac{1}{1+\mu\Phi}\left[1 - \xi\sqrt{\frac{\mu\Phi}{1+\mu\Phi}}\right] \tag{14.9}$$

$$\xi_{d_{\text{opt}}} = \Phi\left[\frac{\xi}{1+\mu} + \sqrt{\frac{\mu}{1+\mu}}\right] \tag{14.10}$$

Kareem (1997) investigated the use of TMDs for base isolated buildings, and compared with different layouts of dampers. Rana and Soong (1998) suggested a numerical optimization because an inherent damping of structure is a restriction to derive close form expressions for the optimum design of TMD. In their study, numerical trials were used to control a single structural mode. But also, controlling multiple structural modes using multiple tuned mass dampers (MTMDs) was investigated.

Carotti and Turci (1999) introduced a geometric formalism based on the use of phasors in the Argand–Gauss plane for the optimization of inertial tuned dampers. Chang (1999) proposed optimum TMD formulas in closed forms for SDOF systems under wind and earthquake loadings. Lin et al. (2001) employed an extended random decrement method for buildings with TMD by investigating displacement and acceleration response spectra for structures with and without TMD. Aldemir (2006) optimally designed semiactive TMDs using a Magnetorheological (MR) damper in order to reduce peak responses of a SDOF structure subjected to various seismic inputs. In order to decrease the performance index value, a numerical optimization algorithm was developed for the optimization of TMD by Lee et al. (2006). By using a base acceleration modeled as Gaussian white-noise random process as an external excitation, explicit mathematic expressions for optimum parameters of TMD were obtained in the study of Bakre and Jangid (2007). Rüdinger (2007) studied the effect of nonlinear viscous damping for TMDs. Hoang et al. (2008) obtained simple formulas of the optimal frequency and damping ratios and investigated the seismic retrofit of long-span truss bridges with TMD. As a result of their study, the optimum TMD has a lower frequency and higher damping ratio when the mass ratio increases. In the study of Weber and Feltrin (2010), experiments for different types of bridges with TMD were conducted to realize the long-term behavior of TMDs. A hybrid tracking controller with a TMD was proposed by Bozer and Altay (2011) in order to set the frequency of TMD by tracking the response of a vibration.

Metaheuristic algorithms have widely been used for the optimization of TMDs; these include evolutionary algorithms such as GA (Desu et al., 2006; Hadi and

Arfiadi, 1998; Marano et al., 2010; Pourzeynali et al., 2007; Singh et al., 2002) and bionic algorithms (Steinbuch, 2011); PSO (Leung and Zhang, 2009; Leung et al., 2008); and HS (Bekdaş and Nigdeli, 2011a,b; Nigdeli and Bekdaş, 2011).

Hadi and Arfiadi (1998) used GA in order to find the optimum stiffness and damping of the TMD for a MDOF structure without specifying the modes to be controlled. The mass ratio is a preselected value in most of the studies. Marano et al. (2010) used a SDOF main structure to develop two different optimization criteria for the aim of minimizing the main system displacement or the inertial acceleration to find optimum TMD parameters including mass ratio by using GA.

Leung and Zhang (2009) obtained explicit expressions of the optimum TMD parameters for structures under external force, harmonic base excitation, and Gaussian white noise. For example, under white-noise base excitation, the optimum frequency ratio and damping ratio of TMD for a damped SDOF system obtained by using PSO algorithm are Eqs. (14.11) and (14.12), respectively.

$$\begin{aligned} f_{\mathrm{opt}} = {} & \frac{\sqrt{1-(\mu/2)}}{1+\mu} + (-4.9453 + 20.2319\sqrt{\mu} - 37.9419\mu)\sqrt{\mu}\xi \\ & + (-4.8287 + 25.0000\sqrt{\mu})\sqrt{\mu}\xi^2 \end{aligned} \tag{14.11}$$

$$\xi_{d_{\mathrm{opt}}} = \sqrt{\frac{\mu(1-\mu/4)}{4(1+\mu)(1-\mu/2)}} - 5.3024\xi^2\mu \tag{14.12}$$

14.2.2 Equations of Motion for Structure with TMD

The classical equations of motion of a MDOF shear building are summarized in this section. A shear building model can be represented by assuming rigid floor slabs. Thus, no rotation takes places at the joint and only lateral displacements occur at the floor levels resulting from the flexibility of the columns. In Figure 14.3, a shear building model with a single TMD on the top of the structure can be seen. The mass, stiffness, damping coefficient, and horizontal displacement of the ith story of the building are represented by the notations m_i, k_i, c_i, and x_i, respectively if $i = 1, 2, \ldots, N$. For a TMD installed on top of a building, the notations m_d, k_d, and c_d are mass, stiffness, and damping coefficient, respectively, of TMD. The displacement of TMD with respect to ground is x_d. The equations of motion of a MDOF linear system subjected to external loading $\mathbf{P}(t)$ can be written as:

$$\mathbf{M}\ddot{\mathbf{x}}(t) + \mathbf{C}\dot{\mathbf{x}}(t) + \mathbf{K}\mathbf{x}(t) = \mathbf{P}(t) \tag{14.13}$$

where $\mathbf{M}$, $\mathbf{C}$, $\mathbf{K}$, and $\mathbf{x}(t)$ represent mass, damping, stiffness matrices, and the vector of the corresponding displacements with respect to ground, respectively. These matrices and the vector are given in Eqs. (14.14)–(14.17) for an N degrees

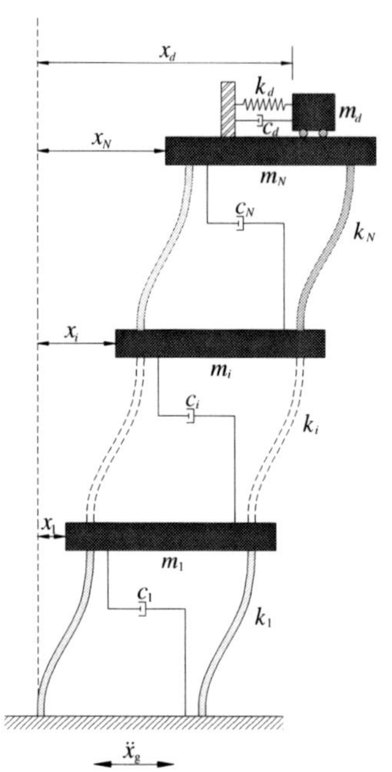

Figure 14.3 System model of multistory building structure with single TMD.

of freedom shear building. These matrices are for the TMD controlled structure. In these matrices, the terms related with TMD are taken zero and the last row and columns are omitted to find the matrices of uncontrolled structure.

$$\mathrm{M} = \mathrm{diag}[m_1 \quad m_2 \ldots \quad m_N \quad m_d] \tag{14.14}$$

$$\mathbf{C} = \begin{bmatrix} (c_1+c_2) & -c_2 & & & & & \\ -c_2 & (c_2+c_3) & -c_3 & & & & \\ & . & . & & & & \\ & . & . & . & & & \\ & & . & . & . & & \\ & & & & -c_N & (c_N+c_d) & -c_d \\ & & & & & -c_d & c_d \end{bmatrix} \tag{14.15}$$

$$\mathbf{K} = \begin{bmatrix} (k_1+k_2) & -k_2 & & & & & \\ -k_2 & (k_2+k_3) & -k_3 & & & & \\ & . & . & & & & \\ & . & . & . & & & \\ & & . & . & . & & \\ & & & & -k_N & (k_N+k_d) & -k_d \\ & & & & & -k_d & k_d \end{bmatrix} \tag{14.16}$$

$$\mathbf{x}(t) = [x_1 \quad x_2 \ldots \quad x_N \quad x_d]^{\mathrm{T}} \tag{14.17}$$

Because of the nondiagonal damping and stiffness matrices, the equations of the structure are coupled. In that case, each differential equation of motion depends on the solution of the other equations. In order to obtain the uncoupled form of these equations, the equations can be separated into vibration modes in normal coordinates. In Eq. (14.18), the mode-shape matrix $\mathbf{\Phi}$ is used to transform the generalized coordinates $\mathbf{Y}(t)$ to the geometric coordinates $\mathbf{x}(t)$.

$$\mathbf{x}(t) = \mathbf{\Phi}\ \mathbf{Y}(t) \tag{14.18}$$

The coupled dynamic equations can be written with this transformation by premultiplying the transpose of the ith mode-shape vector ϕ_i^{T} as seen in Eq. (14.19).

$$\phi_i^{\mathrm{T}}\mathrm{M}\mathbf{\Phi}\ddot{\mathrm{Y}}(t) + \phi_i^{\mathrm{T}}\mathrm{C}\mathbf{\Phi}\dot{\mathrm{Y}}(t) + \phi_i^{\mathrm{T}}\mathbf{K}\mathbf{\Phi}\mathrm{Y}(t) = \phi_i^{\mathrm{T}}\mathrm{P}(t) \tag{14.19}$$

In order to separate the equations into the vibration modes, the special property of mode shapes called orthogonality conditions can be used. The orthogonality conditions can be explained by the free vibration of the structural systems (Clough and Penzien, 1993; Chopra, 2001; and Hart and Wong, 1999). The equations of motion forü an undamped system in free vibration are written as:

$$\mathbf{M}\ddot{\mathbf{x}}(t) + \mathbf{K}\mathbf{x}(t) = \mathbf{0} \tag{14.20}$$

in which $\mathbf{0}$ is a zero vector. The free-vibration motion can be assumed as simple harmonic as given in Eq. (14.21).

$$\mathbf{x}(t) = \hat{\mathbf{x}}\ \sin(wt + \theta) \tag{14.21}$$

$\hat{\mathrm{x}}$ represents the shape of the system and θ is a phase angle. The second time derivations of Eq. (14.21) is given as Eq. (14.22).

$$\ddot{\mathbf{x}}(t) = -w^2\hat{\mathbf{x}}\sin(wt + \theta) = -w^2\mathbf{x}(t) \tag{14.22}$$

By substituting Eqs. (14.21) and (14.22) into Eq. (14.20),

$$-w^2\mathbf{M}\hat{\mathbf{x}}\sin(wt + \theta) + \mathbf{K}\hat{\mathbf{x}}\sin(wt + \theta) = 0 \tag{14.23}$$

can be written. Since the sine term is arbitrary, this term can be eliminated. Thus, the equation can be given as Eq. (14.24).

$$(\mathbf{K} - w^2\mathbf{M})\hat{\mathbf{x}} = \mathbf{0} \tag{14.24}$$

According to Eq. (14.24), $\hat{\mathbf{x}}$ is equal to zero for a stationary solution. The frequency equation given in Eq. (14.25) must be solved in order to find the nonzero solution of $\hat{\mathbf{x}}$.

$$|\mathbf{K} - w^2\mathbf{M}| = \mathbf{0} \tag{14.25}$$

The roots of the frequency equation are the square of the natural frequencies of the system. By applying the transformations $w = w_i$ and $\hat{\mathbf{x}} = \phi_i$, the Eq. (14.24) can be written as:

$$(\mathbf{K} - w_i^2\mathbf{M})\phi_i = 0 \tag{14.26}$$

For frequencies of structural vibration (w_i) and modes shapes (ϕ_i), the subscript represents the corresponding vibration mode. For the mth and nth modes, the equations can be written as:

$$\mathbf{K}\phi_m = w_m^2\mathbf{M}\phi_m \tag{14.27}$$

$$\mathbf{K}\phi_n = w_n^2\mathbf{M}\phi_n \tag{14.28}$$

and Eqs. (14.29) and (14.30) can be obtained by premultiplying these equations with ϕ_n^{T} and ϕ_m^{T}, respectively.

$$\phi_n^{\mathrm{T}}\mathbf{K}\phi_m = w_m^2\phi_n^{\mathrm{T}}\mathbf{M}\phi_m \tag{14.29}$$

$$\phi_m^{\mathrm{T}}\mathbf{K}\phi_n = w_n^2\phi_m^{\mathrm{T}}\mathbf{M}\phi_n \tag{14.30}$$

The transpose of $\mathbf{M}$ and $\mathbf{K}$ are equal to themselves because these matrices are symmetric. The usage of this property is demonstrated in Eqs. (14.31) and (14.32).

$$\phi_m^{\mathrm{T}}\mathbf{K}\phi_n = (\phi_n^{\mathrm{T}}\mathbf{K}\phi_m)^{\mathrm{T}} = \phi_n^{\mathrm{T}}\mathbf{K}\phi_m \tag{14.31}$$

$$\phi_m^{\mathrm{T}}\mathbf{M}\phi_n = (\phi_n^{\mathrm{T}}\mathbf{M}\phi_m)^{\mathrm{T}} = \phi_n^{\mathrm{T}}\mathbf{M}\phi_m \tag{14.32}$$

In that case, it is possible to write the Eqs. (14.33) and (14.34) as

$$(w_n^2 - w_m^2)\phi_n^{\mathrm{T}}\mathbf{M}\phi_m = 0 \tag{14.33}$$

$$\left(\frac{1}{w_n^2} - \frac{1}{w_m^2}\right)\phi_n^{\mathrm{T}}\mathbf{K}\phi_m = 0 \tag{14.34}$$

If $w_n \neq w_m$, the orthogonality conditions of mode shape can be written as Eqs. (14.35) and (14.36).

$$\phi_n^{\mathrm{T}}\mathbf{M}\phi_m = 0 \quad n \neq m \tag{14.35}$$

$$\phi_n^{\mathrm{T}}\mathbf{K}\phi_m = 0 \quad n \neq m \tag{14.36}$$

If $n = m$, the nonzero parts can be shown with modal mass (M_i) and modal stiffness (K_i) as Eqs. (14.37) and (14.38), respectively.

$$M_i = \phi_i^{\mathrm{T}} \mathbf{M} \phi_i \tag{14.37}$$

$$K_i = \phi_i^{\mathrm{T}} \mathbf{K} \phi_i \tag{14.38}$$

The orthogonality condition can be also written for damping as seen in Eq. (14.39). The modal damping (C_i) is given in Eq. (14.40).

$$\phi_n^{\mathrm{T}} \mathbf{C} \phi_m = \mathbf{0} \quad n \neq m \tag{14.39}$$

$$C_i = \phi_i^{\mathrm{T}} \mathbf{C} \phi_i \tag{14.40}$$

By using the ortogonality conditions, all components except the ith mode term in the mass, damping, and stiffness expressions of Eq. (14.19) is zero. So Eq. (14.19) can be written as:

$$\mathrm{M}_i \ddot{Y}_i(t) + C_i \dot{Y}_i(t) + K_i Y_i(t) = P_i(t) \tag{14.41}$$

in which M_i, C_i, K_i, $P_i(t)$, and $Y_i(t)$ are generalized mass, damping, stiffness, load, and displacement of ith normal mode, respectively. Natural frequency and damping ratio of ith normal mode are represented by ω_i and ξ_i, respectively. Alternatively, Eq. (14.41) can be written as:

$$\ddot{Y}_i(t) + 2\xi_i \omega_i \dot{Y}_i(t) + \omega_i^2 Y_i(t) = P_i(t)/\mathrm{M}_i \tag{14.42}$$

by substituting

$$w_i^2 = \frac{K_{\mathbf{i}}}{M_{\mathbf{i}}} \tag{14.43}$$

$$2\xi_i w_i = \frac{C_i}{M_i} \tag{14.44}$$

and dividing the Eq. (14.41) with M_i. External loading $\mathbf{P}(t)$ is given in Eq. (14.45) for harmonic sinus loading and in Eq. (14.46) for earthquake loading. In Eqs. (14.45) and (14.46), the amplitude, circular frequency of harmonic loading, and ground acceleration are represented by p_0, $\overline{\omega}$, and $\ddot{x}_{\mathrm{g}}(t)$, respectively. The vector **{1}** represents a column of ones with $N + 1$ element. When the circular frequency of the applied load is equal to the natural frequency of structure, generalized displacement of corresponding normal mode becomes unsteady in the undesired situation called resonance.

$$\mathbf{P}(t) = p_0\{1\} \sin \overline{\omega} t \tag{14.45}$$

$$\mathbf{P}(t) = -\mathbf{M}\{1\}\ddot{x}_g(t) \tag{14.46}$$

By using the modal participation factor for earthquake loading:

$$\Gamma_i = \frac{\phi_i^{\mathrm{T}}\mathbf{M}\{1\}}{M_i} \tag{14.47}$$

the equation of the ith mode can be written shortly as seen in Eq. (14.48),

$$\ddot{Y}_i(t) + 2\xi_i w_i \dot{Y}_i(t) + w_i^2 Y_i(t) = -\Gamma_i \ddot{x}_g(t) \tag{14.48}$$

The response of the system can be described by the Duhamel integral given as Eq. (14.49)

$$\ddot{Y}_i(t) = \frac{-1}{M_i w_i \sqrt{1-\xi_i^2}} \int_0^t \Gamma_i \ddot{x}_g(\tau) \mathrm{e}^{-\xi_i w_i (t-\tau)} \sin\left[w_i \sqrt{1-\xi_i^2}(t-\tau)\right] \mathrm{d}\tau \tag{14.49}$$

By solving the responses of all the modes, the solution at the geometric coordinates is obtained by using the transformation as given in Eq. (14.18).

For the computer analysis, it will be better to analyze the uncoupled equations rather than the coupled ones. Thus, by solving N number of a SDOF structure, the results for a MDOF structure can be obtained by using a loop to combine the modes. In that case, the same program can be used for different structures with different degrees of freedom (Bekdaş and Nigdeli, 2011a,b). By separating the vibration modes, a general structural model can be also analyzed. The formulation of a shear building is given in order to express the formulation more clearly and basically.

14.3 Optimization of TMDs with HS

The optimization of TMD parameters by using HS was studied by Bekdaş and Nigdeli (2011a) under general excitations and specific excitations like the forward-directivity effect (Bekdaş and Nigdeli, 2011b). Also, idealization of the structures as a SDOF structure using the HS optimization process was investigated (Nigdeli and Bekdaş, 2011).

For the optimization of TMDs with HS, a computer code is needed for the dynamic analyses of the structure and optimization process. The dynamic analysis of structure is done for every iteration inside a loop in the optimization code. The objective functions are related with the results of the dynamic analyses. In that case, the equations of motion are not needed to be written in discrete-time form. Matlab with Simulink (The MathWorks Inc, 2010) is suitable for solving the continuous time equations and optimization coding .

At the beginning of the optimization process, structural data, external excitations, HS parameters, and a possible solution range must be defined. For solving a shear MDOF building, structural data contain story masses, stiffness, and damping coefficients. External excitation or excitations can be a harmonic sinus loading for a rapid and general optimization (Bekdaş and Nigdeli, 2011a) or various earthquake data for a longer but specific optimization (Bekdaş and Nigdeli, 2011b; Nigdeli and Bekdaş, 2011).

Application of TMDs can cover some physical and economical restrictions as per the structure. These restrictions can be taken into account by limiting the range of searched parameters. Also, limiting of the range is effective on reaching the optimum solution quickly.

After the structural data and parameters of HS are defined, the dynamic analysis of the uncontrolled structure must be done for the future comparison of the TMD controlled structure. For MDOF structures, the analysis can be performed by transforming equations of motion into generalized coordinates from geometric coordinates. The equation of motion in geometric coordinates is coupled. With this transformation, uncoupled equations as well as degree of freedom can be obtained in generalized coordinates. After the analysis for all modes are obtained, the solution can be transformed back into geometric coordinates. The time history analysis of a mode can be obtained by modeling the equation of motion of the corresponding mode in Matlab Simulink. The block diagram of Eq. (14.42) can be seen in Figure 14.4.

In the block diagram, the displacement of ith normal mode ($Y_i(t)$) is computed according to the defined values at Matlab Workspace. By using the "Integrator" block, the velocity and the displacement of ith mode are obtained by the integration of the acceleration of the mode. "From Workspace" block calls defined values in Workspace. The external excitation data vectors are written in this block. The time and data vectors are entered as the first and second columns, respectively. By using "Gain" blocks, $P_i(t)$, $Y_i(t)$, and $\dot{Y}_i(t)$ are multiplied with the corresponding terms. Then these multiplications are directed to "Sum" block and the acceleration of the ith mode is obtained. The "Clock" block produces the time vector of the solution with a defined time step in Simulink. This time step can be different from the time step of external excitation.

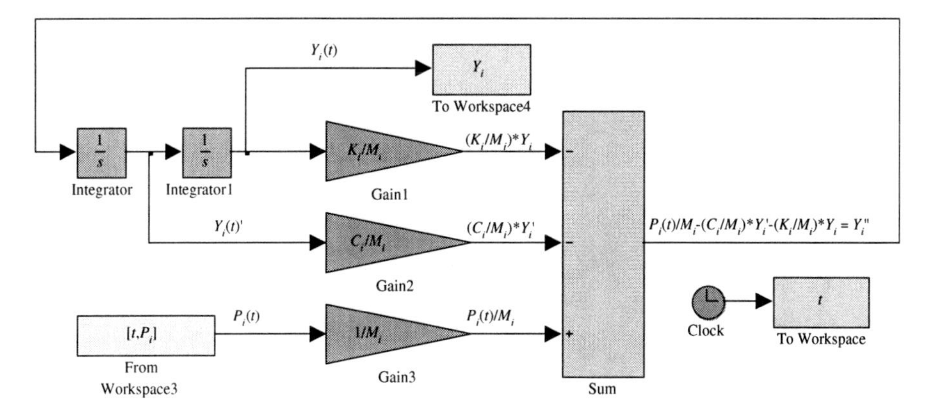

Figure 14.4 Simulink block diagram of the equation of a mode.

Then, the initial HM matrix is generalized with harmony vectors as many as harmony memory size (HMS). In Matlab, the "rand(1)" function can be used to generate random numbers between one and zero. Harmony vectors contain possible values for optimum parameters of TMD according to the range. The size of these vectors depends on the number of the parameters to be optimized. For a whole parameter optimization, mass (or mass ratio), stiffness (or frequency), and damping coefficient (or damping ratio) must be optimized. These TMD parameters are assigned with random numbers within the defined range. Iteratively, each set of vectors (possible TMD parameters) are used for the analysis of the TMD-controlled structure with the same procedure for an uncontrolled structure.

After the initial HM matrix is generalized, stopping criterion or criteria must be checked. The optimization process may contain one or more criterion or criteria. If the criterion or criteria are not satisfied, a new harmony vector is generalized by using the special rules of the HS. A new vector can be generated either from the elements of an existing harmony vector in HM or randomly from the whole range. In this part of the optimization, a modification is made for the adaptation to the TMD problem. If a vector in the HM was chosen for the source of generation, elements of the new vector are assigned with a value around the chosen vector by using a smaller range which is a percentage of the whole solution range. This percentage is also guessed as PAR in classical HS. The possibility to generate a new vector from the existing ones is defined by the value of HMCR. Also, the possibility of choosing the existing vector with best solution may have more of a chance in order to keep the optimization process short. This selection is very suitable to the nature of the musical performances because the musicians tend to play their and their audience's favorite parts of music. When generating a new vector from the HM, the probability of selecting the best vector can be defined as best harmony memory considering rate (BHMCR). This program chooses the other existing harmony vectors randomly with equal probability. If this new vector is better than the existing worst one in the solution, it replaces the worst one. If not, another vector must be generated. The process continues until the criterion or criteria are satisfied. After the optimization process, peak responses and graphics in the time and frequency domain can be obtained.

14.4 Numerical Examples

The HS method for TMD optimization was demonstrated by analyzing two numerical examples. Each numerical example has different possible range and optimization objectives in order to show that the approach is a main source for general and specific problems.

14.4.1 Example 1

In the first example, a ten-story shear building with a TMD attached on the top floor was investigated. The structure is a typical medium-sized civil building with the same story properties. This structure was also analyzed in several research papers (Hadi and Arfiadi, 1998; Lee et al., 2006; Singh et al., 1997). A story of the building has a 360 t mass, 650 MN/m stiffness, and 6.2 MN s/m damping coefficient. The program employing HS seeks TMD parameters randomly within the defined range. The mass was searched between 0.1% and 5% of the total mass of the structure. The stiffness was searched between 100 kN/m and 3000 kN/m. The range of the damping coefficient is 10–200 kN s/m. The Runge–Kutta method was used for solving the problem in Matlab Simulink with 10^{-3} s time step. The HS algorithm parameters, namely, HMS, HMCR, and PAR are 20, 0.5, and 0.1, respectively, in the first numerical example. If the HMS is larger than the selected value, the optimization process will be long. With a 0.5 HMCR, the program has the same probability of generating a harmony from the HM and whole range. A small PAR value is chosen to scan small ranges around the values of existing harmonies in HM. BHMCR is taken in the same way as HMCR. The optimization process was conducted under sine wave ground acceleration with 1g amplitude. The frequency of this excitation is the same with the critical frequency of the structure. The critical frequency of the structure is 6.35 rad/s. At the optimization process, 10 modes at horizontal direction were taken into account. The application has two different stopping criteria. For the first stopping criterion, the maximum first story displacement (x_1) must be reduced with a desired percentage (p). The equation of first criterion is given in Eq. (14.50).

$$\frac{\max|x_1 \text{ without TMD}| - \max|x_1 \text{ with TMD}|}{\max|x_1 \text{ without TMD}|} \leq p \tag{14.50}$$

The second stopping criterion is about the acceleration transfer function (TF) of the first story. TF is the ratio between Laplace transforms of a response and an excitation. The acceleration TF under earthquake loading is given in Eq. (14.51). The TF is unitless, but it can be given in decibels by multiplying the base-10 logarithm of the TF with 20. Equation (14.51) is for s-domain, which is equal to ω_i. The frequency response of the system is given as Eq. (14.52). In order to obtain the amplitude of the TF, the imaginary parts are neglected for TF plots. The TF equations are not dependent on external excitations. By using TF as an optimization objective, a general solution can be also investigated.

$$TF(s) = 20\log_{10}\left[\frac{\ddot{x}(s)}{\ddot{x}_g(s)}\right] = 20\log_{10}\left[\frac{-\mathbf{M}s^2\{1\}}{\mathbf{M}s^2 + \mathbf{C}s + \mathbf{K}}\right] \tag{14.51}$$

$$TF(w) = 20\log_{10}\left[\frac{\mathbf{M}s^2\{1\}}{-\mathbf{M}s^2 + \mathbf{C}si + \mathbf{K}}\right] \tag{14.52}$$

The maximum value at the first resonant peak of frequency response must be reduced under the value of uncontrolled case for the controlled ones. The worst vector is accepted according to Eq. (14.53) by considering both TF and x_1 values. The vector with the highest elimination factor (EF) was eliminated.

$$EF = \frac{\max\ (\text{TF with TMD})}{\max\ (\text{TF without TMD})} + \frac{\max|x_1 \text{ with TMD}|}{\max|x_1 \text{ without TMD}|} \tag{14.53}$$

At the optimization process, the percentage of the reduction (p) was iteratively increased. The desired percentage can be changed after the criterion is satisfied in order to find better values with more reduction. The optimum TMD parameters are found as $m_d = 65$ t, $c_d = 115$ kN s/m, and $k_d = 2570$ kN/m at the value of $p = 65\%$. If the desired percentage is more than 65%, the value of TMD parameters will rapidly increase (Bekdaş and Nigdeli, 2011a,b). The optimum values were compared with the optimum values of Hadi and Arfiadi (GA) ($m_d = 108$ t, $c_d = 151.5$ kN s/m, and $k_d = 3750$ kN/m) and Lee et al. ($m_d = 108$ t, $c_d = 271.79$ kN s/m, and $k_d = 4126.93$ kN/m) for the first structure. The optimum mass and the damping coefficient of HS approach for TMD parameters are lower than the compared methods. Although the optimization process was conducted under a sinus loading, the main aim was to reduce structural responses under earthquakes. The maximum displacements under El Centro (1940) North South (NS) ground acceleration (NGA) record can be seen in Table 14.1 for the three methods.

In the HS optimization, the reductions of displacements are between 45.74% and 49.43%. Although the compared methods use a 66% heavier mass, the reductions are between 32.98% and 38.71%. Also, the dampers are cheaper in the HS

Table 14.1 Maximum Displacements with Respect to Ground Under El Centro (1940) NS Earthquake (Example 1) (Bekdaş and Nigdeli, 2011a,b)

Story	Maximum Absolute Displacement Respect to Ground (m)				% of Reduction		
	Without TMD	With TMD (GA)	With TMD (Lee et al.)	With TMD (HS)	GA	Lee et al.	HS
1	0.031	0.019	0.020	0.016	38.71	35.48	48.39
2	0.060	0.037	0.039	0.031	38.33	35.00	48.33
3	0.087	0.058	0.057	0.044	33.33	34.48	49.43
4	0.112	0.068	0.073	0.057	39.29	34.82	49.11
5	0.133	0.082	0.087	0.068	38.35	34.59	48.87
6	0.151	0.094	0.099	0.078	37.75	34.44	48.34
7	0.166	0.104	0.108	0.087	37.35	34.94	47.59
8	0.177	0.113	0.117	0.094	36.16	33.90	46.89
9	0.184	0.119	0.123	0.099	35.33	33.15	46.20
10	0.188	0.122	0.126	0.102	35.11	32.98	45.74
TMD	–	0.358	0.282	0.395	–	–	–

optimization because of lower damping coefficient values. The effectiveness of a control system can be validated by checking frequency domain responses. In Figure 14.5, the acceleration TF response with respect to ground acceleration is given for the first story of structure. This plot has peak values at natural frequencies as well as at number of degree of freedom. The maximum value at the first resonant state was reduced significantly by the help of the optimum TMD.

The optimum values are also tested by using earthquake excitation records from different places around the world. The records used in analysis are taken from Pacific Earthquake Engineering Research Center (PEER, NGA database). In Table 14.2, the maximum values of structural displacements can be seen. The maximum reduction occurred at the third story under Gazli earthquake (46%) and the minimum reduction occurred at the top story under Northridge earthquake Sylmar record (13%). The TMD is a true optimum because it is useful for all excitations. According to the results, the optimum TMD is significantly effective on reducing peak displacements. The maximum reduction at the first story is approximately 46% under Gazli earthquake, while the minimum reduction is approximately 16% under Northridge earthquake Sylmar excitation. The reductions of top story displacements are approximately between 13% and 38%. The optimum TMD parameters are tested on the top story absolute accelerations ($\ddot{x}_g + \ddot{x}_{10}$) under benchmark earthquakes. As seen in Figure 14.6, the results are effective on the top story absolute accelerations (TS.Abs.Acce.). In addition to the reduction of peak accelerations, the accelerations rapidly decrease under Northridge-Rinaldi excitation, the ratio of the reduction of the top story absolute acceleration is 39.30%.

14.4.2 Example 2

The second example is also a ten-story building, but this structure has different properties in each story (Table 14.3). In the optimization process, six different

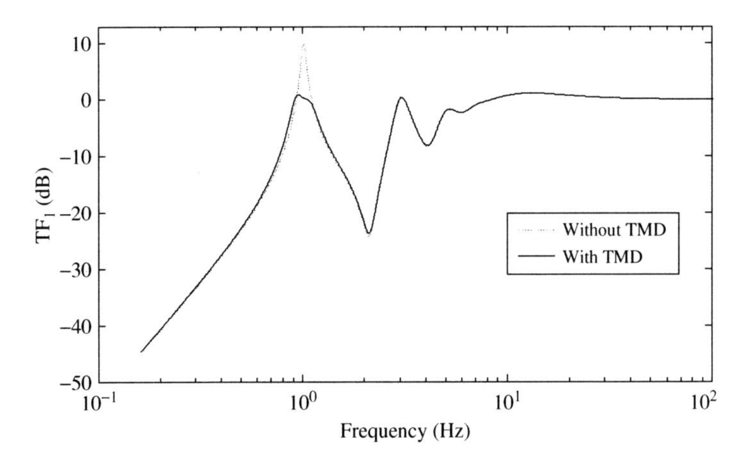

Figure 14.5 Transfer function of the first story (Example 1) (Bekdaş and Nigdeli, 2011a,b).

Table 14.2 Maximum Displacements with Respect to Ground Under Different Earthquake Record (Example 1)

Earthquake Record	Story	1	2	3	4	5	6	7	8	9	10	TMD
Chi Chi, 1999 (CHY080 E)	Without TMD	0.099	0.195	0.288	0.374	0.452	0.520	0.576	0.620	0.650	0.664	–
	With TMD	0.075	0.149	0.220	0.287	0.348	0.402	0.447	0.483	0.507	0.520	1.774
Duzce, 1999 (Bolu 90)	Without TMD	0.059	0.116	0.172	0.224	0.272	0.314	0.351	0.380	0.400	0.410	–
	With TMD	0.040	0.079	0.116	0.151	0.181	0.206	0.229	0.249	0.263	0.270	0.868
Erzincan, 1992 (Erzincan NS)	Without TMD	0.045	0.088	0.128	0.164	0.197	0.224	0.247	0.265	0.277	0.284	–
	With TMD	0.033	0.065	0.094	0.121	0.144	0.164	0.181	0.194	0.204	0.208	0.516
Gazli, 1976 (Karakyr 90)	Without TMD	0.028	0.055	0.080	0.103	0.122	0.138	0.152	0.162	0.171	0.176	–
	With TMD	0.015	0.030	0.043	0.057	0.069	0.081	0.092	0.100	0.106	0.109	0.277
Kobe, 1995 (KJM 000)	Without TMD	0.072	0.144	0.214	0.281	0.342	0.397	0.444	0.481	0.506	0.519	–
	With TMD	0.053	0.106	0.158	0.207	0.254	0.295	0.330	0.358	0.377	0.387	0.942
Tabas, 1978 (Tabas TR)	Without TMD	0.036	0.071	0.105	0.138	0.167	0.193	0.215	0.231	0.243	0.248	–
	With TMD	0.029	0.056	0.080	0.102	0.122	0.138	0.151	0.161	0.168	0.171	0.545
Northridge, 1994 (Sylmar 360)	Without TMD	0.055	0.105	0.148	0.184	0.211	0.231	0.245	0.257	0.266	0.272	–
	With TMD	0.046	0.087	0.122	0.150	0.171	0.187	0.202	0.217	0.229	0.237	0.699
Northridge, 1994 (Rinaldi 228)	Without TMD	0.097	0.190	0.278	0.361	0.437	0.504	0.559	0.602	0.631	0.646	–
	With TMD	0.070	0.136	0.200	0.260	0.314	0.360	0.398	0.427	0.447	0.457	1.584
San Fernando, 1971 (Pacoima Dam 164)	Without TMD	0.064	0.125	0.182	0.235	0.283	0.324	0.359	0.386	0.404	0.413	–
	With TMD	0.047	0.093	0.137	0.177	0.213	0.245	0.270	0.290	0.304	0.312	1.057
Landers, 1992 (Lucerne 000)	Without TMD	0.017	0.033	0.048	0.060	0.072	0.082	0.091	0.098	0.103	0.106	–
	With TMD	0.011	0.021	0.031	0.039	0.046	0.052	0.057	0.061	0.065	0.067	0.252

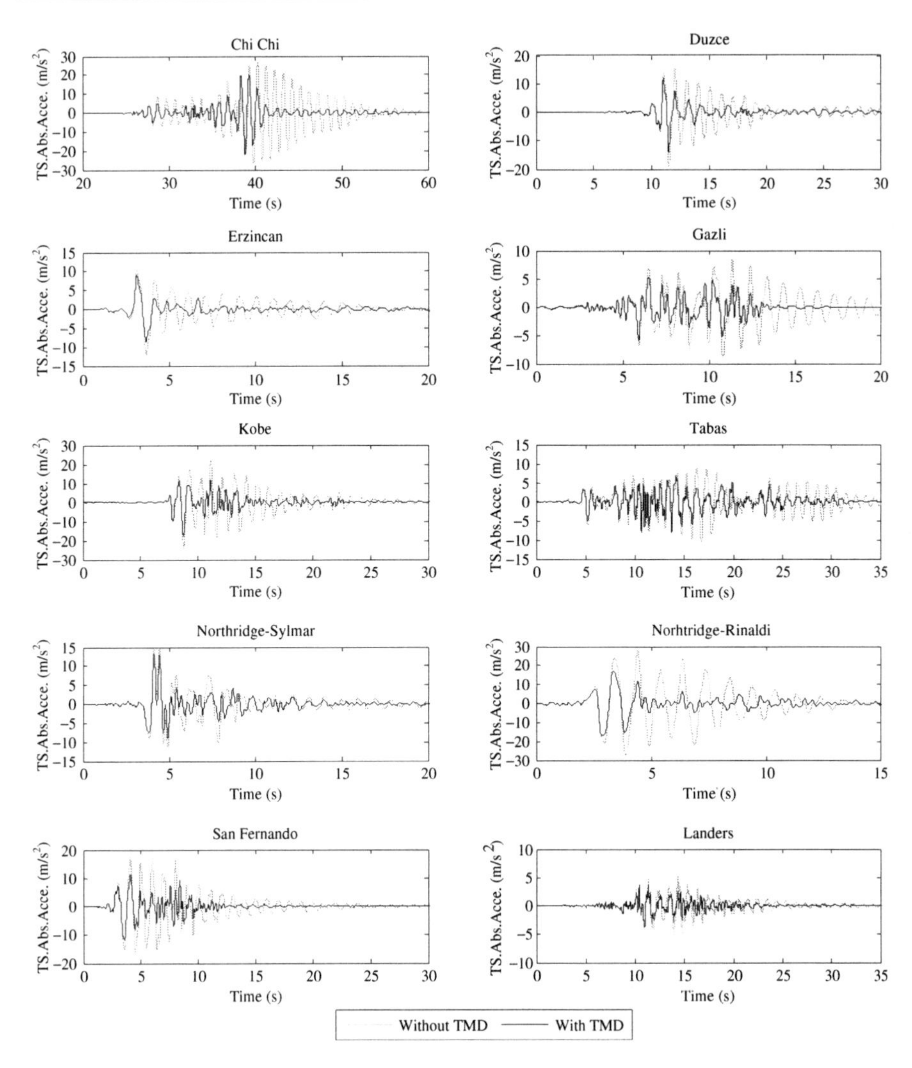

Figure 14.6 The top story acceleration plots under earthquakes (Example 1).

near-fault earthquake excitations with forward directivity were used. The information about these earthquake records is given in Table 14.4. This information contains date, station, component, peak ground acceleration (PGA), peak ground velocity (PGV), and peak ground displacement (PGD) of optimization earthquakes. This example has different HS parameters, range, stopping criteria, and elimination factors. The employed program searches mass, period, and damping ratio of a TMD implemented on top of the structure. The range of the mass is between 1% and 5% of the superstructure mass, while the range for damping ratio is between 5% and 40%. Also, the program searches the period of TMD

Table 14.3 Properties of the Second Example (Sadek et al., 1997)

Story	Mass (*t*)	Stiffness (kN/m)	Damping Coefficient (kN s/m)
10	98	34,310	442.60
9	107	37,430	482.85
8	116	40,550	523.10
7	125	43,670	563.34
6	134	46,790	603.59
5	143	49,910	643.84
4	152	53,020	683.96
3	161	56,140	724.21
2	170	52,260	674.15
1	179	62,470	805.86

Table 14.4 Earthquake Records Used in the HS Optimization (Example 2)

Earthquake	Date	Station	Component	PGA (g)	PGV (cm/s)	PGD (cm)
Loma Prieta	1989	16 LGPC	LGP000	0.563	94.8	41.18
Cape Mendocino	1992	89,156 Petrolia	PET090	0.662	89.7	29.55
Erzincan	1992	95 Erzincan	ERZ-NS	0.515	83.9	27.35
Northridge	1994	77 Rinaldi	RRS228	0.838	166.1	28.78
Northridge	1994	24,514 Sylmar	SYL360	0.843	129.6	32.68
Kobe	1995	0 KJMA	KJM000	0.821	81.3	17.68

between 0.8 and 1.2 times of the superstructure period that is obtained for the first mode. The best and the worst harmony vectors are chosen according to the ratio between the first story displacement (RFSD) of the controlled and uncontrolled structure for the most critical earthquake. RFSD is given in Eq. (14.54) as:

$$\text{RFSD} = \frac{\max|x_1 \text{ without TMD}|}{\max|x_1 \text{ with TMD}|} \tag{14.54}$$

For the stopping criteria, the acceleration TF of the first story must be smaller than the uncontrolled one and RFSD must be smaller than a desired value for all earthquake excitation. If the desired value is smaller than a physical possible value, the program iteratively increases it after several trials assigned by the user. The HS algorithm parameters, namely, HMS, HMCR, and PAR are 5, 0.5, and 0.2, respectively, in the second numerical example. BHMCR is also taken as HMCR.

HMS is smaller than Example 1 because the process of this example is longer. The program checks the results under six earthquake records. Also, PAR is 0.2 to reach the solution more quickly. The value of RFSD is 0.717 for the structure with optimum TMD. The optimum parameters are 53.18 t ($\mu = 3.84\%$), 2.11 s ($k_d = 471.6$ kN/m), and 39.55% ($c_d = 125.26$ kN s/m) for mass, period, and damping ratio of the TMD, respectively. In addition to that, the acceleration TF of the first story is reduced to 2.71 dB from 12.68 dB. Although the optimization was conducted under six different earthquake data, the optimum TMD is effective under benchmark earthquakes as seen in Table 14.5. The value of RFSD is between 0.34 and 0.83 under the optimization and benchmark earthquakes. The maximum reduction rate is at the fifth floor under the Düzce earthquake and the minimum one is at the seventh floor under the Landers earthquake.

The HS approach was also compared with other methods under the El Centro earthquake record that represents a far fault characteristic. Although the optimization process was conducted under near-fault earthquakes, the structural displacements are lower than the compared methods (Table 14.6). The percentage of the reduction at the first floor is 44% for the HS approach, while the best reduction is 17% for the compared methods. The first story displacement graphics can be seen in Figure 14.7. The best reduction at the first story is under the Düzce earthquake (RFSD = 0.46) although this excitation was not used in the optimization process. In addition, the least reduction occurred under the Landers earthquake (RFSD = 0.8) is not low. This shows that the TMD parameters are also effective under benchmark earthquakes that were not considered during optimization.

The best reduction at the top story is under the Tabas earthquake (RFSD = 0.40) and the least reduction occurs under the Landers earthquake (RFSD = 0.74). Although the optimization criterion is about the first story displacement, the optimum TMD is effective on reducing the responses of all floors under optimization and benchmark earthquakes. As observed in the graphics given in Figure 14.7, the structure without TMD cannot obtain steady-state responses. The structural vibrations cannot rapidly be damped after the effective period of earthquakes. This situation can be seen extremely under both optimization (Loma Prieta, Erzincan, and Northridge-Rinaldi) and benchmark earthquakes (Chi Chi and Gazli). This trouble is prevented by the usage of the optimum TMD.

14.5 Conclusion

The HS approach for TMD optimization was validated by analyzing two examples. For the first example, the TMD was optimized for a global solution. The optimum results obtained for a sinus loading are effective on reducing vibrations resulting from earthquakes that were not considered in the optimization process. The other example was conducted under six earthquake records during the optimization process. The TMD can be optimized for the earthquake records that are specific to

Table 14.5 Maximum Displacements with Respect to Ground Under Optimization and Benchmark Earthquakes (Example 2)

Earthquake Record	Story	1	2	3	4	5	6	7	8	9	10	TMD
Loma Prieta, 1989 (LGP000)	Without TMD	0.120	0.249	0.371	0.491	0.607	0.718	0.821	0.908	0.973	1.008	–
	With TMD	0.074	0.156	0.217	0.260	0.285	0.338	0.391	0.439	0.475	0.496	0.932
Cape Mendocino, 1992 (PET090)	Without TMD	0.094	0.198	0.283	0.352	0.406	0.447	0.497	0.552	0.610	0.643	–
	With TMD	0.053	0.112	0.162	0.207	0.249	0.287	0.321	0.350	0.373	0.386	0.694
Erzincan, 1992 (Erzincan NS)	Without TMD	0.142	0.305	0.445	0.575	0.689	0.786	0.863	0.920	0.957	0.976	–
	With TMD	0.068	0.144	0.206	0.262	0.310	0.352	0.388	0.417	0.440	0.455	1.003
Northridge, 1994 (Rinaldi 228)	Without TMD	0.110	0.234	0.336	0.427	0.505	0.568	0.675	0.795	0.889	0.942	–
	With TMD	0.079	0.162	0.228	0.285	0.334	0.375	0.409	0.487	0.556	0.593	0.678
Northridge, 1994 (Sylmar 360)	Without TMD	0.124	0.270	0.397	0.520	0.635	0.737	0.826	0.895	0.941	0.964	–
	With TMD	0.070	0.145	0.209	0.268	0.320	0.373	0.418	0.448	0.475	0.499	0.956
Kobe, 1995 (KJM 000)	Without TMD	0.086	0.184	0.264	0.331	0.384	0.431	0.476	0.515	0.547	0.567	–
	With TMD	0.059	0.119	0.160	0.186	0.197	0.229	0.253	0.269	0.296	0.314	0.482
Chi Chi, 1999 (CHY080 E)	Without TMD	0.105	0.209	0.276	0.335	0.443	0.557	0.666	0.759	0.827	0.864	–
	With TMD	0.061	0.124	0.173	0.218	0.273	0.332	0.390	0.439	0.475	0.492	0.573
Duzce, 1999 (Bolu 90)	Without TMD	0.079	0.170	0.248	0.321	0.386	0.441	0.486	0.521	0.545	0.557	–
	With TMD	0.037	0.076	0.103	0.122	0.132	0.154	0.178	0.205	0.232	0.249	0.289
San Fernando, 1971 (Pacoima Dam 164)	Without TMD	0.077	0.173	0.262	0.355	0.445	0.528	0.599	0.660	0.706	0.732	–
	With TMD	0.054	0.109	0.155	0.201	0.248	0.293	0.337	0.377	0.411	0.427	0.629
Landers, 1992 (Lucerne 000)	Without TMD	0.027	0.054	0.072	0.090	0.105	0.119	0.129	0.138	0.154	0.163	–
	With TMD	0.022	0.042	0.055	0.069	0.082	0.094	0.106	0.114	0.118	0.121	0.249
Gazli, 1976 (Karakyr 90)	Without TMD	0.060	0.129	0.187	0.244	0.310	0.379	0.446	0.506	0.555	0.584	–
	With TMD	0.036	0.075	0.110	0.145	0.177	0.208	0.241	0.277	0.310	0.331	0.559
Tabas, 1978 (Tabas TR)	Without TMD	0.121	0.254	0.362	0.458	0.546	0.634	0.721	0.799	0.860	0.895	–
	With TMD	0.068	0.137	0.184	0.213	0.231	0.258	0.293	0.323	0.345	0.360	0.699

the site of the structure. Although the optimization was conducted under the records that represent the characteristic of the region, earthquakes with unexpected characteristics may occur. For that reason, the optimum solutions were tested with benchmark earthquakes. The performance of TMD under benchmark earthquakes is also effective as observed under optimization earthquakes.

The percentage of the reduction of the first story displacements are between 16.36% and 48.39% for the first example under benchmark earthquakes. Although the optimization process are carried out under sinus loading, the TMD is effective on the excitation of the earthquakes. The reduction percentages of the first story are between 18.52% and 53.16% for the second example. Also, the HS approach is feasible for obtaining specific results according to the physical and economic condition of the structure by defining the limits of the range for the parameters including the mass, stiffness, and damping of the TMD. By using classical closed form formulas, it is not possible to find a satisfied solution for a big mass with low damping or a small mass with high damping.

The optimization objectives are related to the first story displacement and acceleration TF. In addition to that objective, top story absolute acceleration and the displacement of all stories are investigated after the optimization. The optimum TMD is capable of reducing all inspected responses. The number of optimization objectives can be increased but this will also increase the computational effort. In addition to the responses, the limitations in structural design codes can be added as stopping criteria for the optimization objectives.

As a conclusion, the usage of the HS algorithm approach is effective for obtaining the best optimum TMD parameters, and the feasibility of the method was supported by the numerical examples.

Table 14.6 Maximum Displacements Respect to Ground for Different Optimization Methods Under El Centro Earthquake (Example 2)

Story	Maximum Absolute Displacement with Respect to Ground (m)					
	Without TMD	**Den Hartog**	**Warburton**	**Sadek et al.**	**Hadi and Arfiadi (GA)**	**HS Approach**
1	0.041	0.034	0.036	0.036	0.034	0.023
2	0.088	0.074	0.079	0.077	0.072	0.049
3	0.129	0.106	0.114	0.113	0.105	0.070
4	0.166	0.136	0.147	0.145	0.134	0.088
5	0.197	0.163	0.177	0.172	0.160	0.102
6	0.222	0.187	0.206	0.194	0.184	0.115
7	0.252	0.213	0.236	0.219	0.210	0.132
8	0.286	0.239	0.267	0.245	0.236	0.151
9	0.313	0.261	0.292	0.266	0.258	0.167
10	0.327	0.276	0.310	0.281	0.272	0.177
TMD	–	0.602	0.751	0.456	0.635	0.323

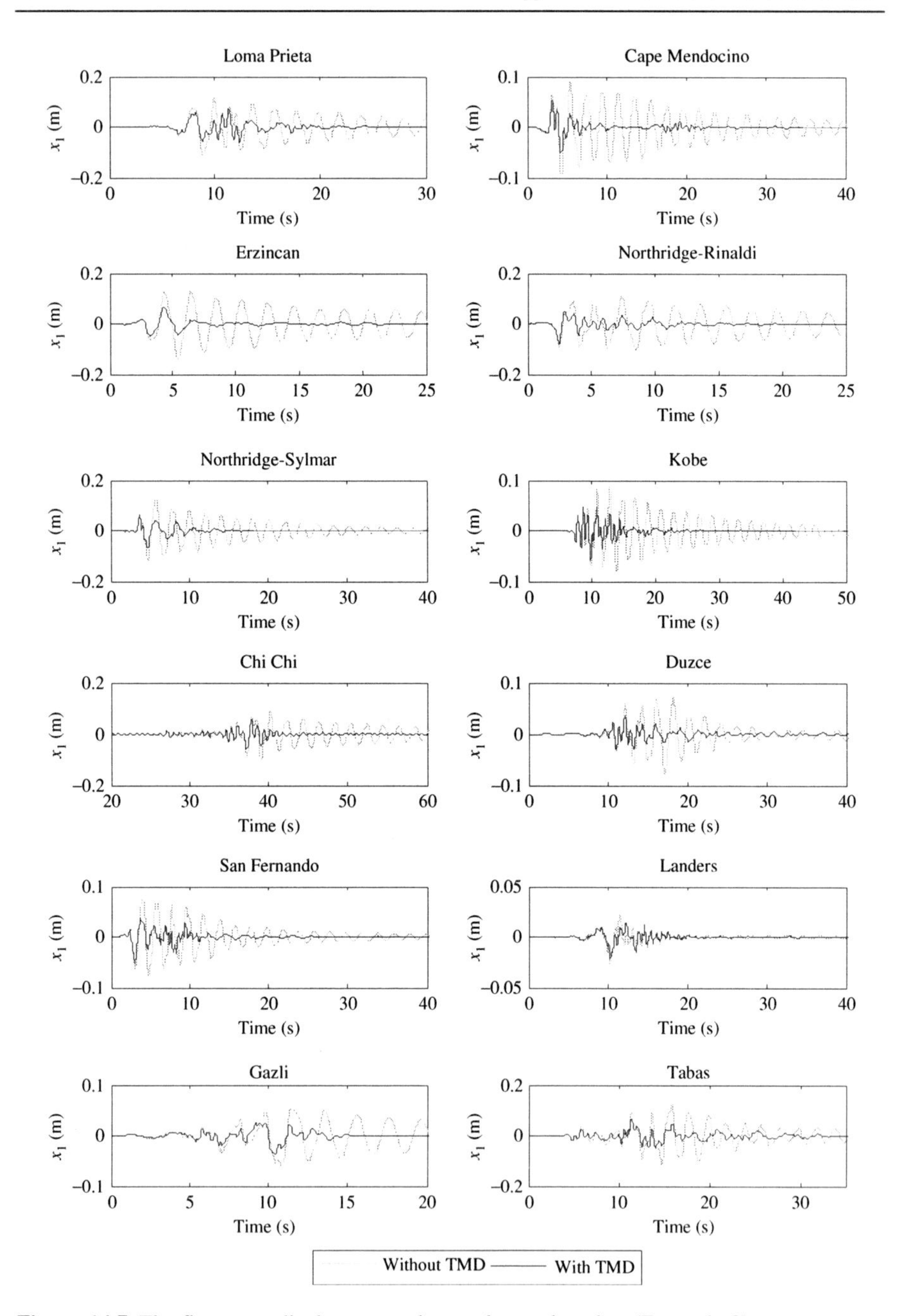

Figure 14.7 The first story displacement plots under earthquakes (Example 2).

Acknowledgments

The authors would like to express their appreciation to Miyamoto International and Yusuf Zahit Gündoğdu for photos of the LAX Theme Building.

References

Afshar, A., Haddad, O.B., Mario, M.A., Adams, B.J., 2007. Honey-bee mating optimization (HBMO) algorithm for optimal reservoir operation. J. Franklin Inst. 344 (5), 452–462.

Aldemir, U., 2006. Optimal control of structures with semiactive-tuned mass dampers. J. Sound Vib. 266, 847–874.

Bakre, S.V., Jangid, R.S., 2007. Optimal parameters of tuned mass damper for damped main system. Struct. Control Health Monit. 14, 448–470.

Bekdaş, G., Nigdeli, S.M., 2011a. Estimating optimum parameters of tuned mass dampers using harmony search. Eng. Struct. 33, 2716–2723.

Bekdaş, G., Nigdeli, S.M., 2011b. Investigation of SDOF idealization for structures with optimum tuned mass dampers. In: Natural Cataclysms and Global Problems of the Modern Civilization Geocataclysms 2011, September 19–21, Istanbul, Turkey.

Bishop, R.E.D., Welbourn, D.B., 1952. The problem of the dynamic vibration absorber. Engineering (London).174–769.

Bozer, A., Altay, G., 2011. Hybrid tracking controller with attached tuned mass damper. Struct. Control Health Monit. doi: 10.1002/stc.497.

Carotti, A., Turci, E., 1999. A tuning criterion for the inertial tuned damper. Design using phasors in the Argand–Gauss plane. Appl. Math. Model. 23, 199–217.

Chang, C.C., 1999. Mass dampers and their optimal designs for building vibration control. Eng. Struct. 21, 454–463.

Chopra, A.K., 2001. Dynamics of Structures: Theory and Applications to Earthquake Engineering. Prentice Hall, Englewood Cliffs, NJ.

Clough, R.W., Penzien, J., 1993. Dynamics of Structures. McGraw-Hill, New York, NY.

Den Hartog, J.P., 1947. Mechanical Vibrations. McGraw-Hill, New York, NY.

Desu, N.B., Deb, S.K., Dutta, A., 2006. Coupled tuned mass dampers for control of coupled vibrations in asymmetric buildings. Struct. Control Health Monit. 13, 897–916.

Dorigo, M., Maniezzo, V., Colorni, A., 1996. The ant system: optimization by a colony of cooperating agents. IEEE Trans. Syst. Man Cybern. Part B. 26, 29–41.

Erol, O.K., Eksin, I., 2006. A new optimization method: big bang–big crunch. Adv. Eng. Softw. 37, 106–111.

Falcon, K.C., Stone, B.J., Simcock, W.D., Andrew, C., 1967. Optimization of vibration absorbers: a graphical method for use on idealized systems with restricted damping. J. Mech. Eng. Sci. 9, 374–381.

Frahm, H., 1911. Device for damping of bodies. US Patent No: 989958.

Gandomi, A.H., Alavi, A.H., 2012. Krill Herd: a new bio-inspired optimization algorithm. Commun. Nonlinear Sci. Numer. Simul. 17 (12), 4831–4845.

Gandomi, A.H., Yang, X.-S., Alavi, A.H., 2012a. Cuckoo search algorithm: a metaheuristic approach to solve structural optimization problems. Eng. Comput. doi: 10.1007/s00366-011-0241-y.

Gandomi, A.H., Yang, X.S., Talatahari, S., Alavi, A.H., 2012b. Metaheuristic algorithm in modeling and optimization. In: Gandomi, et al., (Ed.), Metaheuristic Applications in Structures and Infrastructures. Elsevier, Waltham, MA (Chapter 1).

Geem, Z.W., 2008. Novel derivative of harmony search algorithm for discrete design variables. Appl. Math. Comput. 199, 223–230.
Geem, Z.W., Sim, K.-B., 2010. Parameter setting free harmony search algorithm. Appl. Math. Comput. 217, 3881–3889.
Geem, Z.W., Kim, J.H., Loganathan, G.V., 2001. A new heuristic optimization algorithm: harmony search. Simulation. 76, 60–68.
Glover, F., 1977. Heuristic for integer programming using surrogate constraints. Decis. Sci. 8, 156–166.
Goldberg, D.E., 1989. Genetic Algorithms in Search, Optimization and Machine Learning. Addison Wesley, Boston, MA.
Hadi, M.N.S., Arfiadi, Y., 1998. Optimum design of absorber for MDOF structures. J. Struct. Eng. ASCE. 124, 1272–1280.
Hart, G.C., Wong, K., 1999. Structural Dynamics for Structural Engineering. John Wiley & Sons Inc., New York, NY.
Hasancebi, O., Erdal, F., Saka, M.P., 2010. Adaptive harmony search method for structural optimization. J. Struct. Eng. 136, 419–431.
Hoang, N., Fujino, Y., Warnitchai, P., 2008. Optimum tuned mass damper for seismic applications and practical design formulas. Eng. Struct. 30, 707–715.
Holland, J.H., 1975. Adaptation in Natural and Artificial Systems. University of Michigan Press, Ann Arbor, MI.
Ioi, T., Ikeda, K., 1978. On the dynamic vibration damped absorber of the vibration system. Bull. JSME. 21, 64–71.
Karaboga, D., 2005. An Idea Based on Honey Bee Swarm for Numerical Optimization. (Technical Report TR06). Computer Engineering Department, Engineering Faculty, Erciyes University, Turkey.
Kareem, A., 1997. Modelling of base-isolated buildings with passive dampers under winds. J. Wind Eng. Ind. Aerodyn. 72, 323–333.
Kennedy, J., Eberhart, R.C., 1995. Particle swarm optimization. In: Proceedings of IEEE International Conference on Neural Networks No. IV, Perth, Australia, 27 November—1 December, pp. 1942–1948.
Kirkpatrick, S., Gelatt, C., Vecchi, M., 1983. Optimization by simulated annealing. Science. 220, 671–680.
Lee, K.S., Geem, Z.W., 2005. A new meta-heuristic algorithm for continuous engineering optimization: harmony search theory and practice. Comput. Methods Appl. Mech. Eng. 194, 3902–3933.
Lee, K.S., Geem, Z.W., Lee, S.H., Bae, K.W., 2005. The harmony search heuristic algorithm for discrete structural optimization. Eng. Optim. 37, 663–684.
Lee, C.-L., Chen, Y.-T., Chung, L.-L., Wang, Y.-P., 2006. Optimal design theories and applications of tuned mass dampers. Eng. Struct. 28, 43–53.
Leung, A.Y.T., Zhang, H., 2009. Particle swarm optimization of tuned mass dampers. Eng. Struct. 31, 715–728.
Leung, A.Y.T., Zhang, H., Cheng, C.C., Lee, Y.Y., 2008. Particle swarm optimization of TMD by non-stationary base excitation during earthquake. Earthquake Eng. Struct. Dyn. 37, 1223–1246.
Lin, C.C., Wang, J.F., Ueng, J.M., 2001. Vibration control identification of seismically excited MDOF structure-PTMD systems. J. Sound Vib. 240, 87–115.
Marano, G.C., Greco, R., Chiaia, B., 2010. A comparison between different optimization criteria for tuned mass dampers design. J. Sound Vib. 329, 4880–4890.

Miyamoto, H.K., Gilani, A.S.J., Gündoğdu, Y.Z., 2011. Innovative seismic retrofit of an iconic building. In: Seventh National Conference on Earthquake Engineering, 30 May–3 June, Istanbul, Turkey.

Nakrani, S., Tovey, C., 2004. On honey bees and dynamic allocation in an internet server colony. Adapt. Behav. 12 (3–4), 223–240.

Nigdeli, S.M., Bekdaş, G., 2011. Optimization of tuned mass damper parameters for structures subjected to earthquakes with forward directivity. In: Natural Cataclysms and Global Problems of the Modern Civilization Geocataclysms, 19–21 September 2011, Istanbul, Turkey.

Ormondroyd, J., Den Hartog, J.P., 1928. The theory of dynamic vibration absorber. Trans. Am. Soc. Mech. Eng. 50, 9–22.

Pacific Earthquake Engineering Research Center: NGA database (PEER NGA Database). <http://peer.berkeley.edu/nga/>.

Pourzeynali, S., Lavasani, H.H., Modarayi, A.H., 2007. Active control of high rise building structures using fuzzy logic and genetic algorithms. Eng. Struct. 29, 346–357.

Rana, R., Soong, T.T., 1998. Parametric study and simplified design of tuned mass dampers. Eng. Struct. 20, 193–204.

Rüdinger, F., 2007. Tuned mass damper with nonlinear viscous damping. J. Sound Vib. 300, 932–948.

Sadek, F., Mohraz, B., Taylor, A.W., Chung, R.M., 1997. A method of estimating the parameters of tuned mass dampers for seismic applications. Earthquake Eng. Struct. Dyn. 26, 617–635.

Sahab, M.G., Toropov, V.V., Gandomi, A.H., 2012. A review on traditional and modern structural optimization: problems and techniques (Chapter 2, this book) In: Gandomi, et al., (Ed.), Metaheuristic Applications in Structures and Infrastructures. Elsevier.

Singh, M.P., Matheu, E.E., Suarez, L.E., 1997. Active and semi-active control of structures under seismic excitation. Earthquake Eng. Struct. Dyn. 26, 193–213.

Singh, M.P., Singh, S., Moreschi, L.M., 2002. Tuned mass dampers for response control of torsional buildings. Earthquake Eng. Struct. Dyn. 31, 749–769.

Snowdon, J.C., 1959. Steady-state behavior of the dynamic absorber. J. Acoust. Soc. Am. 31, 1096–1103.

Steinbuch, R., 2011. Bionic optimization of the earthquake resistance of high buildings by tuned mass dampers. J. Bionic Eng. 8, 335–344.

The MathWorks Inc. 2010. MATLAB R2010a. Natick, MA, USA.

Thompson, A.G., 1981. Optimum damping and tuning of a dynamic vibration absorber applied to a force excited and damped primary system. J. Sound Vib. 77, 403–415.

Warburton, G.B., 1982. Optimum absorber parameters for various combinations of response and excitation parameters. Earthquake Eng. Struct. Dyn. 10, 381–401.

Warburton, G.B., Ayorinde, E.O., 1980. Optimum absorber parameters for simple systems. Earthquake Eng. Struct. Dyn. 8, 197–217.

Weber, B., Feltrin, G., 2010. Assessment of long-term behavior of tuned mass dampers by system identification. Eng. Struct. 32, 3670–3682.

Yadav, P., Kumar, R., Panda, S.K., Chang, C.S., 2012. An intelligent tuned harmony search algorithm for optimisation. Inf. Sci. 196, 47–72.

Yang, X.-S., 2005. Engineering optimizations via nature-inspired virtual bee algorithms. In: Mira, J., Alvarez J., (Ed.), Artificial Intelligence and Knowledge Engineering Applications: A Bioinspired Approach. LNCS, 3562, 317–323.

Yang, X.-S., 2008. Nature-Inspired Metaheuristic Algorithms. Luniver Press, United Kingdom.

Yang, X.-S., Gandomi, A.H., 2012. Bat algorithm: a novel approach for global engineering optimization. Eng. Comput. 29 (5), 464–483.

15 Identification of Passive Devices for Vibration Control by Evolutionary Algorithms

Giuseppe Carlo Marano[1]*, Giuseppe Quaranta*[2]*, Jennifer Avakian*[1] *and Alessandro Palmeri*[3]

[1]Department of Civil Engineering and Architecture, Technical University of Bari, Bari, Italy, [2]Department of Civil and Environmental Engineering, University of California Davis, Davis, CA, USA, [3]School of Civil and Building Engineering, Loughborough University, Loughborough, UK

15.1 Introduction

Several devices have been proposed in the past decades in order to reduce the effects of dynamic loads on structures and infrastructures (Marano et al., 2007; Palmeri, 2006; Palmeri and Ricciardelli, 2006; Terenzi, 1999). Within this framework, the present chapter is concerned with the mechanical identification of fluid viscous dampers, a class of passive devices with many civil engineering applications, most of them aimed at increasing the structural protection level against wind and earthquake loadings. Among the most interesting features of viscous dampers are: (i) low maintenance costs; (ii) usability for several earthquakes without damage; and (iii) viscous forces out-of-phase with the elastic forces, so that the stress in the main structural system does not increase. From a manufacturing standpoint, a fluid viscous damper typically consists of a piston within a damper housing filled with a compound of silicone or similar type of oil. The energy is dissipated as the fluid passes through small orifices from one side of the piston to the other. Viscous dampers are usually arranged in one of the following configurations: (i) as a diagonal or chevron bracing element within a steel or concrete frame; (ii) as a part of the cable stays of long-span bridges; (iii) as a part of tuned mass dampers; (iv) as a part of a base isolation system to increase the energy dissipation; or (v) as a device for allowing free thermal movement. Viscous dampers can be effectively used in the construction of new buildings, or in retrofitting existing structures. Their importance in vibration control has increased, thanks to effective energy dissipation capabilities and a wide range of applications. Because of their inherent nonlinear behavior, the mechanical

Metaheuristic Applications in Structures and Infrastructures. DOI: http://dx.doi.org/10.1016/B978-0-12-398364-0.00015-2

characteristics of viscous dampers must be carefully investigated in order to provide a reliable support for designing an efficient protection strategy. As a consequence, appropriate identification techniques are needed.

Numerous parametric identification methodologies for dynamic systems were proposed in the past years, and each of them had some intrinsic drawbacks and limitations. Among the available numerical techniques, nonclassical approaches based on soft computing methods are attracting increasing attention. These identification methodologies rely on metaheuristic paradigms frequently adopted in structural engineering applications (Gandomi et al., 2011, 2012; Yang and Gandomi, 2012), but so far, are basically limited to structural optimization problems. In this context, the parametric identification problem of multi-degree-of-freedom linear systems was resolved using genetic algorithms (Marano et al., 2011), particle swarm optimization, and differential evolution (Tang et al., 2008). An overview of the most recent applications in this field (Marano et al., 2009) also revealed that the genetic algorithms are frequently employed in the parametric identification of nonlinear models; see for instance Monti et al. (2010). Particle swarm optimization and differential evolution techniques were also considered in the parametric identification of hysteresis models (Kao and Fung, 2009) and Van der Pol–Duffing oscillators (Quaranta et al., 2010). Kwok et al. (2007) addressed the genetic algorithm–based identification of a magnetorheological fluid damper using a nonsymmetric version of the original Bouc–Wen hysteretic model. Yun et al. (2008) considered several (parametric and nonparametric) approaches in order to identify a full-scale viscous damper for large flexible bridge structures. About the parametric techniques, the capability of the adaptive random search is explored by Yun et al. (2008).

In this chapter, the mechanical system identification of nonlinear viscous dampers is framed as an optimization problem in which the difference between the experimental and the analytical value of the force experienced by the device is minimized. To this end, a convenient measure of the "distance" between experimental and analytical results is introduced. The optimal set of parameters is thus derived by minimizing this distance using evolutionary-based strategies. Final results demonstrate the effectiveness of the proposed identification strategy.

15.2 Parametric Identification of Fluid Viscous Dampers

15.2.1 Mechanical Model

The application of nonclassical methods for the parametric identification of viscous dampers requires (i) the definition of an appropriate single-degree-of-freedom mechanical model and (ii) the formulation of the objective (or cost) function to be minimized. The first source of concern is, therefore, the definition of an appropriate law for damping model. Because of the relevant difficulties in selecting a reliable damping model, several laws are taken into account in this study.

A nonlinear viscous power law damping is considered in Terenzi (1999) to model the dynamic behavior of some protection systems. This kind of model was used to represent some available devices (i.e., Jarret elastomeric spring dampers) and has recently been adopted in the analysis (Rüdinger, 2007) and optimum design (Rüdinger, 2006) of tuned mass dampers subjected to random vibrations. Therefore, the parametric identification of this nonlinear model can be of interest for some real engineering applications. The simplest fractional model (hereafter indicated as "*fractional viscous model*") is:

$$F = m\ddot{y} + c_\alpha\,\mathrm{sgn}(\dot{y})|\dot{y}|^\alpha \tag{15.1}$$

where y is the displacement (upper dots indicate the time derivative), m is the mass, and c_α is the damping coefficient. The time variable t is here omitted for the sake of conciseness. Moreover, α is the damping law exponent, whose value lies between 0 and 1. If $\alpha = 1$, then the linear viscous damping law is retrieved. On the other hand, if $\alpha = 0$, then the model reproduces a dry friction-based behavior. To better approximate the experimental response, more complex models can be considered. The "*fractional viscous-linear and quadratic elastic model*," for instance, is obtained by introducing the terms k_1y and $k_2\mathrm{sgn}(y)y^2$ in Eq. (15.1):

$$F = m\ddot{y} + c_\alpha\,\mathrm{sgn}(\dot{y})|\dot{y}|^\alpha + k_1y + k_2\,\mathrm{sgn}(y)y^2 \tag{15.2}$$

Finally, the following model is considered:

$$F = m\ddot{y} + c_\alpha\,\mathrm{sgn}(\dot{y})|\dot{y}|^\alpha + k_\beta\,\mathrm{sgn}(y)|y|^\beta \tag{15.3}$$

and it is named "*fractional viscous-fractional elastic model*," where β is the elastic law exponent.

15.2.2 Problem Formulation

On denoting $\mathbf{x}$, a vector of model parameters, the parametric estimation problem is formulated as follows:

$$\begin{aligned} &\min\ \{f(\mathbf{x})\} \\ &\quad \text{s.t.} \\ &\quad \mathbf{x}^l \le \mathbf{x} \le \mathbf{x}^u \end{aligned} \tag{15.4}$$

The global minimum of the problem in Eq. (15.4) is denoted as $\mathbf{x}^*$ and it is the best estimation of the model parameters. In Eq. (15.4), $\mathbf{x}^l$ and $\mathbf{x}^u$ are the lower bound and the upper bound of $\mathbf{x}$, respectively. The objective function $f(\mathbf{x})$ is the cumulative discrepancy between the time histories of the experimental and analytical forces over the entire experiment, that is:

$$f(\mathbf{x}) = \frac{1}{t_{\mathrm{end}} - t_{\mathrm{start}}}\int_{t_{\mathrm{start}}}^{t_{\mathrm{end}}} \left|F_{\mathrm{exp}}(t) - F(t|\mathbf{x})\right| dt \tag{15.5}$$

where t_{start} and t_{end} are the initial and final instant time of the experimental records, respectively.

15.3 Differential Evolution Algorithms

15.3.1 Mutation Operator

The main idea behind differential evolution algorithms (DEAs) is the use of the differences between randomly selected individuals in order to perturb a third individual, and the named target vector (Storn and Price, 1997). In so doing, the algorithm estimates the gradient in a zone of the search space rather than in a precise point of it. Therefore, the main goal of this process (called mutation) is to enable diversity in the current population as well as to move the possible solutions toward the global optimum of the optimization problem. In order to describe the inner work of DEAs, let ${}^{k}\mathbf{x}_i = \{{}^{k}x_{i1},\ldots,{}^{k}x_{ij},\ldots,{}^{k}x_{in}\}$ be the ith individual (with $i = 1,\ldots,N$) at iteration k. The initial population ${}^{0}\mathbf{x}_i$ for $i = 1,\ldots, N$ is defined by generating randomly the collection of N solutions within the specified search space. In this study, the Latin hypercube sampling technique has been used to generate the best initial population with minimum correlation between samples (Monti et al., 2010). At iteration $k \geq 1$, a mutation vector ${}^{k}\mathbf{z}_i$ is computed using one of the following alternatives:

$$ {}^{k}\mathbf{z}_i = {}^{(k-1)}\mathbf{x}_{r1} + \varphi_1\left({}^{(k-1)}\mathbf{x}_{r2} - {}^{(k-1)}\mathbf{x}_{r3}\right) \tag{15.6} $$

$$ {}^{k}\mathbf{z}_i = {}^{(k-1)}\mathbf{x}_{best} + \varphi_1\left({}^{(k-1)}\mathbf{x}_{r1} - {}^{(k-1)}\mathbf{x}_{r2}\right) \tag{15.7} $$

$$ {}^{k}\mathbf{z}_i = {}^{(k-1)}\mathbf{x}_{i} + \varphi_1\left({}^{(k-1)}\mathbf{x}_{r1} - {}^{(k-1)}\mathbf{x}_{r2}\right) + \varphi_2\left({}^{(k-1)}\mathbf{x}_{best} - {}^{(k-1)}\mathbf{x}_{i}\right) \tag{15.8} $$

$$ {}^{k}\mathbf{z}_i = {}^{(k-1)}\mathbf{x}_{best} + \varphi_1\left({}^{(k-1)}\mathbf{x}_{r3} - {}^{(k-1)}\mathbf{x}_{r4}\right) + \varphi_2\left({}^{(k-1)}\mathbf{x}_{r1} - {}^{(k-1)}\mathbf{x}_{r2}\right) \tag{15.9} $$

$$ {}^{k}\mathbf{z}_i = {}^{(k-1)}\mathbf{x}_{r1} + \varphi_1\left({}^{(k-1)}\mathbf{x}_{r4} - {}^{(k-1)}\mathbf{x}_{r5}\right) + \varphi_2\left({}^{(k-1)}\mathbf{x}_{r2} - {}^{(k-1)}\mathbf{x}_{r3}\right) \tag{15.10} $$

From Eqs. (15.6) to (15.10), $r1$, $r2$, $r3$, and $r4$ denote integer numbers randomly selected within the set $\{1,\ldots,i-1,i+1,\ldots, N\}$, and $r1 \neq r2 \neq r3 \neq r4$. The candidate solution ${}^{(k-1)}\mathbf{x}_{best}$ is the current best performer in the population. The coefficients φ_1 and φ_2 are the mutation coefficients. They are real positive constants usually ranging from 0.40 to 1.00 (0.50 in this study). These parameters control the amplification level due to the mutation operator. Any alternative mutation operator leads to different versions (Storn and Price, 1997): rand/1/bin, best/1/bin, current-to-best/1/bin, best/2/bin, rand/2/bin, respectively. A projection scheme is used in order to ensure the feasibility of ${}^{k}\mathbf{z}_i$. This consists of considering the projection on the prescribed interval of each out-of-bound trial vector ${}^{k}\mathbf{z}_i$.

15.3.2 Crossover Operator

The possible solutions ${}^{k}\mathbf{z}_i$ and the current population ${}^{(k-1)}\mathbf{x}_i$ are subjected to binomial crossover:

$$ {}^{k}u_{ij} = \begin{cases} {}^{k}z_{ij} & \text{if } r_{\text{xover}} \leq P_{\text{xover}} \text{ or } j = \text{randint}(1, n) \\ {}^{(k-1)}x_{ij} & \text{otherwise} \end{cases} \tag{15.11} $$

where r_{xover} is a random number generated by using the uniform probability density function in the range [0,1]. The parameter P_{xover} is the probability of crossover, and it takes values between 0 and 1, in accordance with the user's preference (a common assumption is $P_{\text{xover}} = 0.50$). Moreover, randint($1,n$) is an integer number randomly selected within the set $\{1,\ldots,n\}$. This condition aims at ensuring that at least one parameter from ${}^{k}\mathbf{z}_i$ is taken into account in order to build the vector ${}^{k}\mathbf{u}_i$.

15.3.3 Selection Operator

The selection operator employs a very simple one-to-one competition scheme between ${}^{k}\mathbf{u}_i$ and ${}^{(k-1)}\mathbf{x}_i$:

$$ {}^{k}\mathbf{x}_i = \begin{cases} {}^{k}\mathbf{u}_i & \text{if } f\left({}^{k}\mathbf{u}_i\right) < f\left({}^{(k-1)}\mathbf{x}_i\right) \\ {}^{(k-1)}\mathbf{x}_i & \text{otherwise} \end{cases} \tag{15.12} $$

which means that the winner is the best performer between ${}^{k}\mathbf{u}_i$ and ${}^{(k-1)}\mathbf{x}_i$. The output of this operator is a new population for the next generation. In this study, the evolutionary search is stopped once a maximum number of iterations L is reached.

15.4 Particle Swarm Optimization Algorithms

15.4.1 General Model

Two categories of optimizers can be formulated within the framework of swarm intelligence-based techniques. Classical particle swarm optimization algorithms (PSOAs) rely on the Newtonian dynamic to regulate the movement of the particles. As a consequence, position and velocity of each particle can be determined simultaneously. On the contrary, quantum-behaved particle swarm optimization algorithms (Q-PSOAs) are formulated in accordance with quantum theories, where the term "trajectory" is meaningless. The present study is limited to the application of PSOAs.

The ith particle (with $i = 1,\ldots,N$) at iteration k has two attributes, a velocity ${}^k\mathbf{v}_i = \{{}^k\mathbf{v}_{i1},\ldots,{}^k\mathbf{v}_{ij},\ldots,{}^k\mathbf{v}_{in}\}$ and a position ${}^k\mathbf{x}_i = \{{}^k\mathbf{x}_{i1},\ldots,{}^k\mathbf{x}_{ij},\ldots,{}^k\mathbf{x}_{in}\}$. In order to protect the cohesion of the swarm, the velocity ${}^k\mathbf{v}_{ij}$ is forced to be (in absolute value) less than $v_j^{\max}$, with $\mathbf{v}^{\max} = \{v_1^{\max},\ldots,v_j^{\max},\ldots,v_n^{\max}\}$. Typically, it is assumed $\mathbf{v}^{\max} = \gamma(\mathbf{x}^u - \mathbf{x}^l)/\tau$, where $\gamma = 0.50$ (the time factor $\tau = 1$ is introduced to provide a physically consistent formalism). The initial positions ${}^0\mathbf{x}_i$ are defined by generating randomly the collection of N solutions within the assigned search space. Similarly, ${}^0v_{ij}$ is randomly generated using a uniform distribution between $-v_j^{\max}$ and $+v_j^{\max}$. At iteration $k \geq 1$, the velocity ${}^k\mathbf{v}_i$ and the position ${}^k\mathbf{x}_i$ are evaluated as follows (Shi and Eberhart, 1998):

$$ {}^k\mathbf{v}_i = w\,{}^{(k-1)}\mathbf{v}_i + \lambda_1\,{}^k\mathbf{r}_{i1} \times ({}^k\mathbf{x}_i^{\mathrm{Pb}} - {}^{(k-1)}\mathbf{x}_i) + \lambda_2\,{}^k\mathbf{r}_{i2} \times ({}^k\mathbf{x}^{\mathrm{Gb}} - {}^{(k-1)}\mathbf{x}_i) \tag{15.13} $$

$$ {}^k\mathbf{x}_i = {}^{(k-1)}\mathbf{x}_i + \tau\,{}^k\mathbf{v}_i \tag{15.14} $$

where w is the inertia weight whereas λ_1 and λ_2 are the acceleration factors (so-called cognitive and social parameter, respectively). In Eq. (15.13), ${}^k\mathbf{r}_{i1}$ and ${}^k\mathbf{r}_{i2}$ are vectors whose n terms are random numbers uniformly distributed between 0 and 1, and the symbol $\times$ denotes the term-by-term multiplication. The symbol ${}^k\mathbf{x}_i^{\mathrm{Pb}}$ denotes the best previous position of the ith particle (known also as "pbest"):

$$ {}^k\mathbf{x}_i^{\mathrm{Pb}} = \begin{cases} {}^{(k-1)}\mathbf{x}_i^{\mathrm{Pb}} & \text{if } f({}^{(k-1)}\mathbf{x}_i) > f({}^{(k-1)}\mathbf{x}_i^{\mathrm{Pb}}) \\ {}^{(k-1)}\mathbf{x}_i & \text{otherwise} \end{cases} \tag{15.15} $$

and ${}^0\mathbf{x}_i^{\mathrm{Pb}} = {}^0\mathbf{x}_i$ by assumption. If ${}^k\mathbf{x}^{\mathrm{Gb}}$ is the best position among all the particles in the swarm (known as "Gbest"), the swarm is said to be fully informed or fully connected (this version of PSOAs is called global PSOA). The best performer ${}^k\mathbf{x}^{\mathrm{Gb}}$ is:

$$ {}^k\mathbf{x}^{\mathrm{Gb}} = \arg\min_{i=1,\ldots,N}\{f({}^k\mathbf{x}_i^{\mathrm{Pb}})\} \tag{15.16} $$

The magnitude of the updated particle's velocity must be less than the maximum admissible velocity in order to preserve the swarm cohesion, and thus:

$$ {}^k v_{ij} = \begin{cases} \mathrm{sgn}[{}^k v_{ij}]v_j^{\max} & \text{if } |{}^k v_{ij}| > |v_j^{\max}| \\ {}^k v_{ij} & \text{otherwise} \end{cases} \tag{15.17} $$

where $\mathrm{sgn}[\cdot]$ is the sign operator. Furthermore, each particle is forced to move within the feasible search space according to the following rules:

$$ ({}^k x_{ij}, {}^k v_{ij}) = \begin{cases} ({}^k x_{ij}, {}^k v_{ij}) & \text{if } x_j^l \leq {}^k x_j \leq x_j^u \\ ({}^k x_{ij} = x_j^l, {}^k v_{ij} = 0) & \text{if } {}^k x_j < x_j^l \\ ({}^k x_{ij} = x_j^u, {}^k v_{ij} = 0) & \text{otherwise} \end{cases} \tag{15.18} $$

In Eq. (15.18), the unfeasible particle's velocity is fixed at zero for the next iteration in order to avoid considering any points outside the search space.

A typical operator, called "craziness operator," is also performed to increase the direction diversity in the swarm. Let P_{cr} be the probability of craziness, and the particle's velocity component is replaced as follows:

$$\text{if} \quad r_{\text{cr}} \leq P_{\text{cr}} \rightarrow {}^{k}v_{ij} \sim U(-v_j^{\max}, v_j^{\max}) \tag{15.19}$$

where r_{cr} is a uniform random number between 0 and 1, and $U(\cdot)$ is the uniform probability density function. Typical values for P_{cr} are less than 0.10–0.20.

Following iteratively this simple set of instructions, the swarm is expected to "fly" toward the global optimum $\mathbf{x}^*$ of the problem. The evolutionary search is stopped once a maximum number of iterations L is achieved.

15.4.2 Inertia Weight and Acceleration Factors

The constriction factor χ is an alternative to the use of the static inertia weight, and it was initially proposed by Clerc (1999) to replace the $\mathbf{v}^{\max}$ clamping. The constriction model describes the way of choosing w, λ_1, and λ_2 as follows:

$${}^{k}\mathbf{v}_i = \chi\left[{}^{(k-1)}\mathbf{v}_i + \lambda_1\,{}^{k}\mathbf{r}_{i1} \times ({}^{k}\mathbf{x}_i^{\text{Pb}} - {}^{(k-1)}\mathbf{x}_i) + \lambda_2\,{}^{k}\mathbf{r}_{i2} \times ({}^{k}\mathbf{x}^{\text{Gb}} - {}^{(k-1)}\mathbf{x}_i)\right] \tag{15.20}$$

$$\chi = \frac{2}{\left|2 - \psi - \sqrt{\psi^2 - 4\psi}\right|} \qquad \psi = \lambda_1 + \lambda_2 \geq 4 \tag{15.21}$$

The new particle's position is computed as indicated in Eq. (15.14). Typical values for this model are $\lambda_1 = \lambda_2 = 2.05$.

The model given by Eq. (15.13) can be modified by assuming inertia weight and acceleration factors as dynamic parameters (i.e., by placing ${}^{k}w$, ${}^{k}\lambda_1$, ${}^{k}\lambda_2$ instead of w, λ_1, λ_2, respectively). A linear dynamic inertia weight was proposed by Shi and Eberhart (1999):

$${}^{k}w = ({}^{0}w - {}^{L}w)\frac{L-k}{L} + {}^{L}w \tag{15.22}$$

where ${}^{0}w$ and ${}^{L}w$ are the initial and final value of the inertia weight. Usual values for ${}^{0}w$ and ${}^{L}w$ are 0.9 and 0.4, respectively. Similarly,

Ratnaweera et al. (2004) proposed linearly iteration-dependent models for the acceleration factors:

$$
\begin{aligned}
{}^{k}\lambda_1 &= ({}^{0}\lambda_1 - {}^{L}\lambda_1)\frac{L-k}{L} + {}^{L}\lambda_1 \\
{}^{k}\lambda_2 &= ({}^{0}\lambda_2 - {}^{L}\lambda_2)\frac{L-k}{L} + {}^{L}\lambda_2
\end{aligned}
\tag{15.23}
$$

In several papers, ${}^{k}\lambda_1$ changes from 2.5 to 0.5 and ${}^{k}\lambda_2$ from 0.5 to 2.5.

15.4.3 *Chaotic Particle Swarm Optimization*

One of the major drawbacks of the PSOA is its premature convergence, especially for search spaces with several local optima. In order to overcome this problem, some researches proposed the introduction of chaotic maps with certainty, ergodicity, and pseudorandomness properties into PSOA so as to improve the global convergence (Chuanwen and Bompard, 2005). Chaos-based optimization algorithms can carry out overall searches at higher speeds than stochastic ergodic searches that depend on probabilities due to the nonrepetition of chaos (Coelho and Herrera, 2007). In this study, chaotic maps are adopted in order to select the numerical values for the parameters of the particle's velocity. The inertia weight is updated by adopting the logistic map (Chuanwen and Bompard, 2005):

$$
{}^{k}\tilde{w} = \sigma\,{}^{(k-1)}\tilde{w}(1 - {}^{(k-1)}\tilde{w}) \tag{15.24}
$$

where σ is assumed equal to 4 in order to obtain ergodicity in (0,1). The adopted value of the chaotic inertia weight is scaled within the interval [0.40, 0.90]. For what concerns the acceleration factors, the Zaslavskii map is proposed in Coelho and Herrera (2007):

$$
{}^{k}\zeta_1 = \mathrm{mod}[{}^{(k-1)}\zeta_1 + \nu + \delta\,{}^{k}\zeta_2, 1] \tag{15.25}
$$

$$
{}^{k}\zeta_2 = \cos(2\pi\,{}^{(k-1)}\zeta_1) + e^{-\rho}\,{}^{(k-1)}\zeta_2 \tag{15.26}
$$

where mod[·] is the modulus operator (signed remainder after division). The use of the Zaslavskii map has been theoretically justified in Coelho and Herrera (2007) due to its ergodicity and unpredictability, given the presence of a strange attractor with a large Lyapunov exponent for $\nu = 400$, $\rho = 3$, and $\delta = 12.6695$ (in this case, ${}^{k}\zeta_2 \in [-1.0512, 1.0512]$). According to Coelho and Herrera (2007), the Zaslavskii map-based chaotic PSOA should be more capable of escaping from local optima than could random search. In the present applications, both chaotic acceleration factors are functions depending on the results of Eq. (15.26), but final values are scaled within the interval [0.50, 2.50].

15.4.4 Passive Congregation

The dynamic of natural swarms can be modeled by taking into account two forms of grouping (He et al., 2004). One form of grouping is the aggregation, which refers to a grouping by environmental forces. The passive aggregation is a passive grouping due to physical phenomena (i.e., transport caused by water currents) whereas the active aggregation is a grouping by attractive resources (i.e., food or space within the environment). The second form of grouping is the congregation, which refers to a grouping by social forces. It depends on the group itself and not on the environment. The passive congregation is the attraction between members of the group without the existence of observable social behaviors and therefore resembles a random phenomenon. On the other hand, the social congregation may occur under an active information transfer, so that the behavior of the group depends on the relationship among the members of the group. On accounting for a passive congregation, He et al. (2004) proposed the following rule to update the particle's velocity:

$$
\begin{aligned}
{}^{k}\mathbf{v}_i = w\,{}^{(k-1)}\mathbf{v}_i + \lambda_1\,{}^{k}\mathbf{r}_{i1} \times ({}^{k}\mathbf{x}_i^{\mathrm{Pb}} - {}^{(k-1)}\mathbf{x}_i) + \lambda_2\,{}^{k}\mathbf{r}_{i2} \times ({}^{k}\mathbf{x}^{\mathrm{Gb}} - {}^{(k-1)}\mathbf{x}_i) \\
+ \lambda_3\,{}^{k}\mathbf{r}_{i3} \times ({}^{k}\mathbf{x}_{\mathrm{rand}} - {}^{(k-1)}\mathbf{x}_i)
\end{aligned}
\tag{15.27}
$$

in which ${}^{k}\mathbf{x}_{\mathrm{rand}}$ is a particle selected randomly from the swarm, ${}^{k}\mathbf{r}_{i3}$ is a vector whose terms are random numbers uniformly distributed between 0 and 1, and λ_3 is another acceleration factor (called passive congregation coefficient, a constant positive real value). The adopted (constant) numerical values for the control parameters in Eq. (15.27) are taken from He et al. (2004).

15.5 Viscous Damper Identification Using Experimental Data

15.5.1 Experimental Setup

The parametric identification of a nonlinear viscous damper is performed by means of the presented PSOAs and DEAs. Experimental data from a standardized experimental test are considered in order to compare their performances as well as to verify the reliability of the methodology. To this end, a full-scale fluid viscous damper was subjected to dynamic loading in order to measure its response. The damper testing machine basically consists of a high resistance steel frame to withstand loads of tension and compression up to 2200 kN (Figure 15.1). The prescribed force-controlled or displacement-controlled loading condition is generated by means of a servant cylinder. The maximum resultant force at the load cell is equal to 1400 kN. The displacement of the viscous damper is measured by means of a transducer directly placed on the device. As usual in dynamic testing, the viscous damper was subjected to a harmonic-type excitation. The applied loading history and the viscous damper's response are shown in Figure 15.2. Velocity and acceleration records are

Figure 15.1 Viscous fluid damper installed on a damper testing machine.

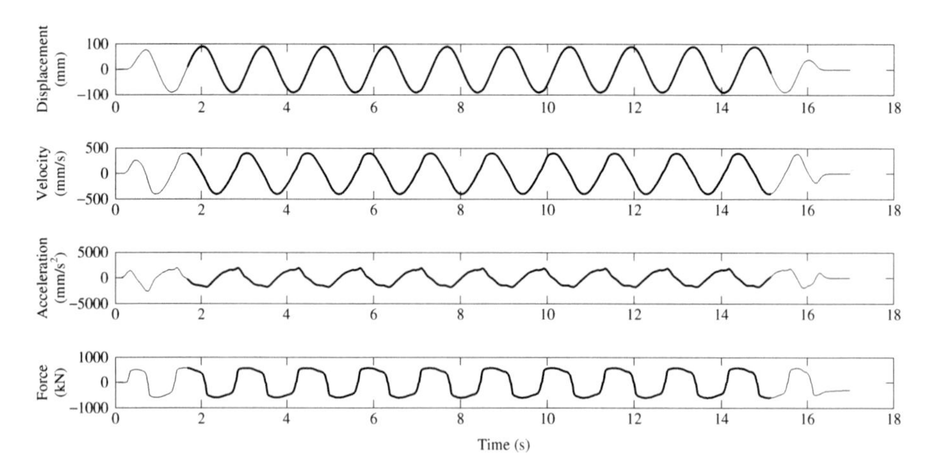

Figure 15.2 Results of the experimental test.

obtained by numerical differentiation of the measured displacement time history using a third-order algorithm. The experimental records of displacements and dynamic load are subjected to wavelet-based denoising.

15.5.2 Nonclassical Parametric Identification Methods

The investigated nonclassical identification methods are listed in Table 15.1. The algorithms listed in Table 15.1 were compared by assuming $N = 100$ and $L = 300$. Each algorithm was performed 100 times in order to allow for a statistical comparison. The mean value, the worst value, the best value, and the standard deviation of the objective function were carried out from this set of simulations and are denoted as "Mean," "Max," "Min," and "Std," respectively.

15.5.3 Results

Statistical results about the objective function are listed in Tables 15.2–15.4. The experimental response of the viscous damper and that obtained using the best parameters estimation are compared in Figures 15.3–15.5. The fractional viscous

Table 15.1 Nonclassical Parametric Identification Techniques

Nomenclature	Operators
DEA01	DEA with mutation operator as in Eq. (15.6)
DEA02	DEA with mutation operator as in Eq. (15.7)
DEA03	DEA with mutation operator as in Eq. (15.8)
DEA04	DEA with mutation operator as in Eq. (15.9)
DEA05	DEA with mutation operator as in Eq. (15.10)
PSOA01	PSOA with dynamic parameters, see Eqs. (15.22)–(15.23)
PSOA02	PSOA with constriction factor, see Eq. (15.20)
PSOA03	Chaotic-PSOA based on Eqs. (15.24)–(15.26)
PSOA04	PSOA with passive congregation, see Eq. (15.27)

Table 15.2 Statistical Results of the Objective Function for *Fractional Viscous Model*

Algorithm	Mean	Max	Min	Std
DEA01	0.164191	0.164191	0.164191	8.41e-17
DEA02	0.164191	0.164191	0.164191	1.08e-13
DEA03	0.164191	0.164191	0.164191	1.19e-15
DEA04	0.164191	0.164191	0.164191	5e-11
DEA05	0.16434	0.170427	0.164191	0.000881
PSOA01	0.164191	0.164191	0.164191	2.23e-10
PSOA02	0.164191	0.164191	0.164191	1.05e-10
PSOA03	0.164191	0.164191	0.164191	1.54e-10
PSOA04	0.1642	0.164274	0.164192	1.3e-05

Table 15.3 Statistical Results of the Objective Function for *Fractional Viscous-Linear and Quadratic Elastic Model*

Algorithm	Mean	Max	Min	Std
DEA01	0.155021	0.164191	0.102115	0.019806
DEA02	0.092584	0.101791	0.082498	0.005109
DEA03	0.138002	0.164191	0.087112	0.025274
DEA04	0.140153	0.164191	0.101658	0.022115
DEA05	0.408732	0.765939	0.20799	0.127324
PSOA01	0.146748	0.164191	0.087326	0.018556
PSOA02	0.159732	0.164191	0.133735	0.007633
PSOA03	0.15842	0.164191	0.097282	0.012153
PSOA04	0.120046	0.150552	0.084546	0.016257

Table 15.4 Statistical Results of the Objective Function for *Fractional Viscous-Fractional Elastic Model*

Algorithm	Mean	Max	Min	Std
DEA01	0.116981	0.164191	0.094823	0.025933
DEA02	0.118978	0.134621	0.103289	0.005985
DEA03	0.144597	0.164191	0.124318	0.007748
DEA04	0.149775	0.154804	0.132414	0.004135
DEA05	0.239024	0.485202	0.161835	0.088663
PSOA01	0.11641	0.164191	0.091191	0.020292
PSOA02	0.144547	0.164191	0.109297	0.019807
PSOA03	0.127939	0.164191	0.098549	0.019287
PSOA04	0.151494	0.163625	0.105089	0.017173

model provides the worst objective function values (Table 15.2). Although the identification of the displacement–force relationship is satisfactory, the model has evident limitations in simulating the force–velocity characteristic of the damper (Figure 15.3). The introduction of the elastic force improves the objective function values (see Tables 15.3 and 15.4). The minimum objective function value is obtained by assuming the fractional viscous-linear and quadratic elastic model (Table 15.3). This model leads to a very good approximation of both displacement–force and velocity–force relationships (Figure 15.4).

The investigated nonclassical identification techniques showed comparable performances. Moreover, the standard deviation values are small, meaning that these methods are very robust. It seems that DEA05 exhibits the worst performances. The PSOA01 is the best performer among the particle swarm optimization techniques. Overall, the best performances are obtained by means of DEA02.

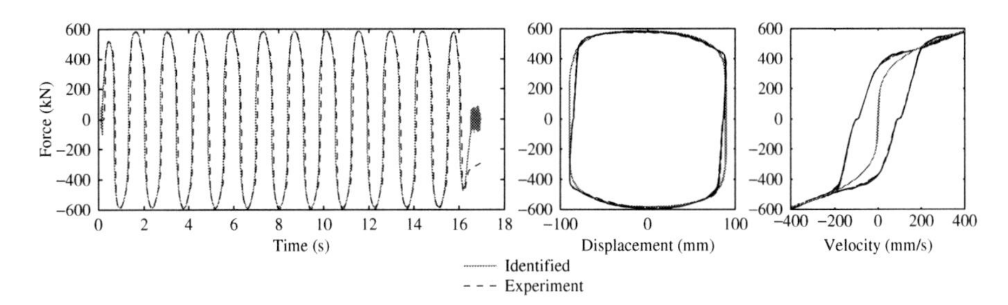

Figure 15.3 Parametric identification using the *fractional viscous model.*

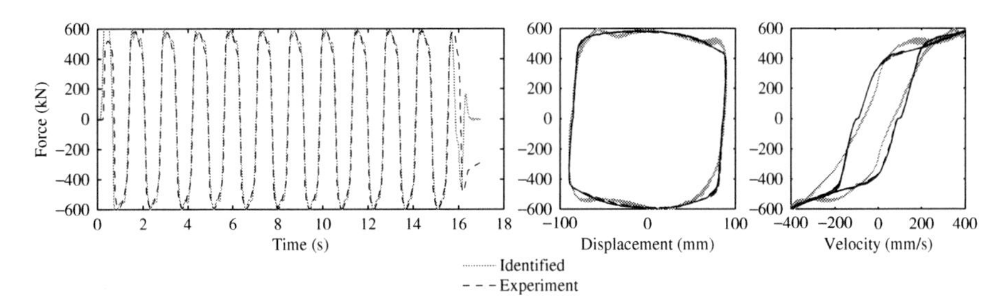

Figure 15.4 Parametric identification using the *fractional viscous-linear and quadratic elastic model.*

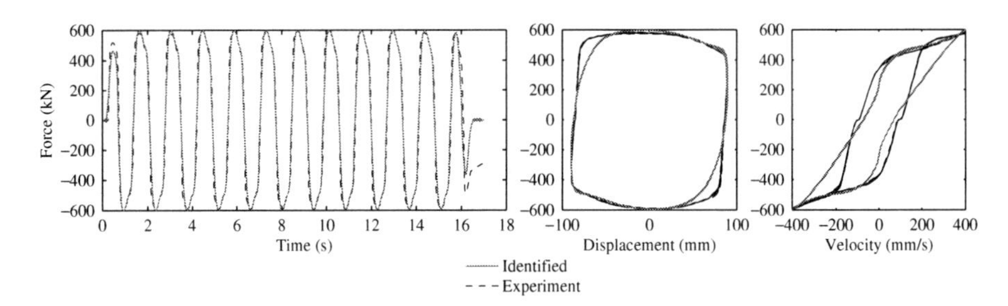

Figure 15.5 Parametric identification using the *fractional viscous-fractional elastic model.*

On considering a pure fractional viscous model, the technical datasheet of the viscous damper indicates a damping law exponent equal to 0.20. This result is in good agreement with the best parameter estimation obtained in this study by assuming the fractional viscous model, since it is found $\alpha = 0.2902$. As it is possible to infer from Table 15.2, this value is approximately the same for all simulations. When the fractional viscous-linear and quadratic elastic model is considered, it is found that the damping law exponent is equal to 0.6762, whereas $k_1 = 5.0190$ kN/mm^2 and $k_2 = 0.0445$ kN/mm^2, with no significant variations from run to run.

15.6 Conclusions

An efficient and robust strategy for the parametric identification of nonlinear viscous dampers from dynamic tests has been presented. The parametric identification is framed as a minimization problem, which has been solved by means of five DEAs and four particle swarm optimization techniques. The investigated nonclassical parametric identification techniques performed very well and demonstrated a satisfactory stability.

Final results confirm that mechanical models with fractional viscous damping provide good approximations of the experimental displacement–force response of fluid viscous dampers. A pure fractional viscous damping model, however, does not provide a correct simulation of the velocity–force relationship. On the contrary, the fractional viscous model incorporating both linear and quadratic elastic terms leads to very good performances. This model is capable of simulating the displacement–force relationship as well as the velocity–force response of the fluid viscous damper. Therefore, it is concluded that a pure fractional viscous model—which is regularly implemented in the current state-of-practice for the simulation of structural systems equipped with fluid viscous dampers—is not reliable for accurate predictions of the velocity–force relationship. Although several technical codes and guidelines suggest this mathematical law, more refined models should be explored. In this perspective, the considered fractional viscous model with linear and quadratic elastic terms is a good compromise between simplicity and accuracy.

Acknowledgment

The authors wish to thank Eng. Ciro Caramia and Eng. Oronzo Patronella (SISMALAB srl) for having performed the experiment.

References

Chuanwen, J., Bompard, E., 2005. A hybrid method of chaotic particle swarm optimization and linear interior for reactive power optimization. Math. Comput. Simul. 68 (1), 57–65.

Clerc, M., 1999. The swarm and the queen: towards a deterministic and adaptive particle swarm optimization. Proceedings of the Congress on Evolutionary Computation. IEEE Service Center, Piscataway, Washington DC, USA, pp. 1951–1957.

Coelho, L.S., Herrera, B.M., 2007. Fuzzy identification based on a chaotic particle swarm optimization approach applied to a nonlinear yo-yo motion system. IEEE Trans. Ind. Electron. 54 (6), 3234–3245.

Gandomi, A.H., Yang, X.S., Alavi, A.H., 2011. Mixed variable structural optimization using firefly algorithm. Comput. Struct. 89 (23–24), 2325–2336.

Gandomi, A.H., Yang, X.S., Alavi, A.H., 2012. Cuckoo search algorithm: A metaheuristic approach to solve structural optimization problems. Eng. Comput. doi: 10.1007/s00366-011-0241-y.

He, S., Wu, Q.H., Wen, J.Y., Saunders, J.R., Paton, R.C., 2004. A particle swarm optimizer with passive congregation. Biosystems. 78 (1−3), 135−147.

Kao, C.C., Fung, R.F., 2009. Using the modified PSO method to identify a Scott-Russell mechanism actuated by a piezoelectric element. Mech. Syst. Signal Process. 23 (5), 1652−1661.

Kwok, N.M, Ha, Q.P, Nguyen, M.T, Li, J., Samali, B., 2007. Bouc−Wen model parameter identification for a MR fluid damper using computationally efficient GA. ISA Trans. 46 (2), 167−179.

Marano, G.C, Trentadue, F., Greco, R., 2007. Stochastic optimum design criterion for linear damper devices for seismic protection of buildings. Struct. Multi. Optim. 33 (6), 441−455.

Marano, G.C., Quaranta, G., Monti, G., 2009. Genetic algorithms in mechanical systems identification: state-of-the-art review. In: Topping, B.H.V., Tsompanakis, Y. (Eds.), Soft Computing in Civil and Structural Engineering. Saxe-Coburg Publications, Stirlingshire, UK, pp. 43−72. (Chapter 2).

Marano, G.C., Quaranta, G., Monti, G., 2011. Modified genetic algorithm for the dynamic identification of structural systems using incomplete measurements. Comput. Aided Civil Infrastruct. Eng. 26 (2), 92−110.

Monti, G., Quaranta, G., Marano, G.C., 2010. Genetic-algorithm-based strategies for dynamic identification of nonlinear systems with noise-corrupted response. ASCE J. Comput. Civil Eng. 24 (2), 173−187.

Palmeri, A., 2006. Correlation coefficients for structures with viscoelastic dampers. Eng. Struct. 28 (8), 1197−1208.

Palmeri, A., Ricciardelli, F., 2006. Fatigue analyses of buildings with viscoelastic dampers. J. Wind Eng. Ind. Aerodyn. 94 (5), 377−395.

Quaranta, G., Monti, G., Marano, G.C., 2010. Parameters identification of Van der Pol−Duffing oscillators via particle swarm optimization and differential evolution. Mech. Syst. Signal Process. 24 (7), 2076−2095.

Ratnaweera, A., Halgamuge, S.K., Watson, H.C., 2004. Self-organizing hierarchical particle swarm optimizer with time-varying acceleration coefficients. IEEE Trans. Evol. Comput. 8 (3), 240−255.

Rüdinger, F., 2006. Optimal vibration absorber with nonlinear viscous power law damping and white noise excitation. J. Eng. Mech. 132 (1), 46−53.

Rüdinger, F., 2007. Tuned mass damper with nonlinear viscous damping. J. Sound Vib. 300 (3−5), 932−948.

Shi, Y., Eberhart, R., 1998. A modified particle swarm optimizer. In: Proceedings of the IEEE World Congress on Computational Intelligence. Anchorage, AK, USA, pp. 69−73.

Shi, Y., Eberhart, R., 1999. Empirical study of particle swarm optimization. In: Proceedings of the Congress on Evolutionary Computation. Washington DC, USA, pp. 1945−1949.

Storn, R., Price, K., 1997. Differential evolution—a simple and efficient heuristic for global optimization over continuous spaces. J. Global Optim. 11 (4), 359−431.

Tang, H., Xue, S., Fan, C., 2008. Differential evolution strategy for structural system identification. Comput. Struct. 86 (21−22), 2004−2012.

Terenzi, G., 1999. Dynamics of SDOF systems with nonlinear viscous damping. J. Eng. Mech. 125 (8), 956−963.

Yang, X.S., Gandomi, A.H., 2012. Bat algorithm: a novel approach for global engineering optimization. Eng. Comput. 29 (5), 464−483.

Yun, H.-B., Tasbighoo, F., Masri, S.F., Caffrey, J.P., Wolfe, R.W., Makris, N., et al., 2008. Comparison of modeling approaches for full-scale nonlinear viscous dampers. J. Vib. Control. 14 (1−2), 51−76.

16 Structural Optimization for Frequency Constraints

Saeed Gholizadeh

Department of Civil Engineering, Urmia University, Urmia, Iran

16.1 Introduction

Up to now, a great deal of research has been done in the field of structural optimization covering a variety of optimization techniques such as algorithms, constraint approximations, sensitivity analysis, size optimization, shape optimization, and topology optimization. This chapter deals with structural optimization with dynamic frequency constraints because of its importance in structural design. Structural optimization for frequency constraints is impressively useful for manipulating the dynamic characteristics of structures. For example, in most low-frequency vibration problems, fundamental frequency and mode shape of the structure considerably affect the dynamic response of the structure to dynamic excitation. In such cases, the ability to manipulate the selected frequency and mode shapes can significantly improve the performance of the structure. Thus, it can be concluded that the natural frequencies are fundamental parameters affecting the dynamic behavior of structures. Therefore, some limitations should be imposed on the natural frequency range to reduce the domain of vibration and also to prevent the resonance phenomenon in the dynamic response of structures (Gholizadeh et al., 2008). Also, to reduce the construction costs of structures, they should be as lightweight as possible. To achieve these purposes, optimization techniques can be effectively employed.

The optimization of structures with frequency constraints is a nonlinear dynamic optimization problem. Frequency constraints are highly nonlinear, nonconvex, and implicit with respect to the design variables. In this case, the application of gradient-based optimization techniques may not produce appropriate results. Because these methods usually employ derivative calculations and their success in comprehensive exploration of the design space is highly dependent on selection of a good starting point, they may converge to the local optima of the handled problems (Gomes, 2011; Kaveh and Zolghadr, 2011). A great deal of research in the field of structural optimization for constrained frequencies employing mathematical programming approaches has been reviewed by Gomes (2011).

Furthermore, natural frequencies of structures are sensitive to shape changes and linking shape and sizing variables increase the number of design variables. Also,

Metaheuristic Applications in Structures and Infrastructures. DOI: http://dx.doi.org/10.1016/B978-0-12-398364-0.00016-4

shape and sizing variables have essentially different physical demonstrations and their changes are of widely different significance. These cause significantly great mathematical difficulties and lead to difficult convergence. In order to overcome these numerical difficulties, global search optimization techniques should be utilized.

In the recent years, metaheuristic algorithms have emerged as the global search approaches that are best able to tackle complex optimization problems. By taking a glance at the literature, it can be found that the most popular metaheuristics are genetic algorithm (GA), particle swarm optimization (PSO), ant colony optimization (ACO), and harmony search (HS) techniques. A comprehensive review of metaheuristics and their applications in the field of structural optimization may be found in Lamberti and Pappalettere (2011).

Metaheuristics have also been employed for structural optimization with frequency constraints. Lingyun et al. (2005) proposed an enhanced GA for optimization of truss structures with frequency constraints. They handled the frequency constraints by a penalty function method. They hybridized simplex search and GA developed following a nature-based scheme of niche. Three truss structures were analyzed and compared with literature results; in most of the cases, their methodology produced better results than those reported in the references.

Salajegheh et al. (2007) employed an improved GA and neural networks to find the optimal weight of structures subjected to multiple natural frequency constraints. The evolutionary algorithm used was the virtual subpopulation (VSP) method. To decrease the computational burden of the optimization process, the structural modal analysis was replaced by back propagation (BP) and wavelet back propagation (WBP) neural networks. It was demonstrated that the best results were found using the VSP with WBP network.

Gholizadeh et al. (2008) combined VSP and other classes of neural networks to find the optimal structures for multiple constrained natural frequencies. In their work, the desired frequencies were predicted by a radial basis function (RBF) and a wavelet radial basis function (WRBF) neural networks. Two examples were presented: the first one is a simple 10-bar planar truss, and the second one is a more complex problem, a 200-bar double-layer grid space structure. In their work, VSP incorporating the WRBF found the best results.

Torkzadeh et al. (2008) presented some gradient-based methods for the optimum design of structures subjected to multiple frequency constraints. The main contribution of their work was to create the higher order approximate eigenvalues to enhance the evaluation of the frequencies and thus enhance the quality of the optimization. In their work, a number of 3D steel-framed structures were optimized, and the results were compared in terms of accuracy and computational time.

Zuo et al. (2011) presented a paper where structural optimization for frequency constraints had been implemented by GA using adaptive eigenvalue reanalysis methods. They reported that the reanalysis technique, which was derived primarily from the Kirsch's combined approximations method, is highly accurate for a problem with repeated eigenvalues. The required number of basis vectors at every generation was adaptively determined and the rules for selecting the initial number of basis vectors were given. Numerical examples of truss design demonstrated the efficiency of the proposed methodology in reducing the computational time involved in the design process of large-scale structures with high accuracy.

Gomes (2011) employed PSO for structural truss mass optimization on size and shape considering frequency constraints. In this work, the PSO algorithm was briefly revised highlighting its most important features. Four benchmark examples regarding the optimization of trusses on shape and size with frequency constraints were presented. The results showed that the PSO algorithm performed similar to other methods and even better in some cases.

Kaveh and Zolghadr (2011) utilized the charged system search (CSS) algorithm and its enhanced version to optimize various truss structures with multiple frequency constraints. The CSS algorithm was introduced by Kaveh and Talatahari (2010) for design of structural problems. This method utilizes the governing laws of Coulomb and Gauss from electrostatics and the Newtonian laws of mechanics. Inspired by these laws, a model is created to formulate the structural optimization method. The results showed that the CSS algorithms performed better than other optimization techniques for most of the benchmark examples.

In the field of structural optimization with frequency constraints, most research relates to skeletal structures, as noted above. In the case of continuous structures, there are only a few published papers.

Gholizadeh and Seyedpoor (2011a) proposed a methodology to find the optimal shape of arch dams subjected to constrained natural frequencies. The optimization was carried out by the VSP. In order to reduce the computational cost of the optimization process, the arch dam natural frequencies were predicted by BP and WBP neural networks. Their numerical results demonstrated the computational merits of the WBP-based VSP procedure.

In the most recent paper, Gholizadeh and Seyedpoor (2011b) proposed an efficient soft computing-based methodology to achieve optimal shape design of arch dams subjected to natural frequency constraints. They employed VSP and PSO as two powerful metaheuristics to effectively perform the optimization task. Because their work considered fluid–structure interaction, computing the natural frequency of the dam by finite element analysis (FEA) during the optimization process was computationally expensive. To reduce the computational burden, two well-known neural network techniques, BP and RBF, were utilized to efficiently predict the frequencies of the dam–water system during the optimization process. Their numerical results showed that PSO incorporating a BP neural network provided the best results.

The optimization problem of skeletal structures with frequency constraints using the gradient-based approaches and advanced metaheuristics has been addressed by many researchers. However, in the case of continuous structures with constrained frequencies, the optimization problem was not sufficiently addressed. In this chapter, two types of structural optimization problems with frequency constraints are addressed. The first type includes size optimization problems of the well-known 10-bar 2D and 72-bar 3D trusses and size and configuration optimization problems of the well-known 37-bar 2D truss and 52-bar 3D lattice dome structure. The second problem focuses on one continuous structure. Therefore, the main aim of presenting such an example is to deal with shape optimization of arch dams with limited natural frequencies while considering fluid–structure interaction. To tackle this problem, VSP, PSO, and HS metaheuristic algorithms are utilized as optimizers. As this is a large-scale, complex problem, the computational effort is high; evaluating these parameters by the FEM leads to a very long optimization time. In this case, neural computing tools are used to obviate this computational rigor. The employed neural computing tools are standard BP and RBF neural networks and an adaptive neuro-fuzzy inference system (ANFIS). The ANFIS is an efficient computational tool for approximation purposes. Generally, fuzzy inference is the process of formulating the mapping from a given input to an output using fuzzy logic.

Numerical results show that in all examples the computational performance of the PSO is better than those of the VSP and HS. Also it is observed that HS reveals better performance compared with VSP. For arch dam optimization, the performance generality of the BP is the best, while the accuracy of the ANFIS is better than that of the RBF. Employing the neural network models highly speeds up the optimization process.

16.2 Formulation of a Structural Optimization Problem with Frequency Constraints

In structural optimization, the aim is usually to minimize the structural weight subject to some constraints. The problem may be formulated as follows:

$$\text{Minimize} \quad f(X) \tag{16.1}$$

$$\text{Subject to} \quad g_i(X) \leq 0, \quad i = 1, 2, \ldots, n_g \tag{16.2}$$

$$X_j \in \Delta, \quad j = 1, 2, \ldots, n_d \tag{16.3}$$

where X is a vector of design variables, $f(X)$ represents objective function, $g(X)$ is the behavioral constraint, and n_g and n_d are the number of constraints and design variables, respectively. A given set of values is expressed by Δ and components of design variable vector, X_j, can take values only from this set.

The above-mentioned constrained optimization problem can be converted into an unconstrained problem using the exterior penalty function method (EPFM) (Vanderplaats, 1984) by constructing a function of the following form:

$$\Phi(X, r_p) = f(X)(1 + p(X)) \tag{16.4}$$

$$p(X) = r_p \sum_{i=1}^{n_g} [\max\{0, \; g_i(X)\}]^2 \tag{16.5}$$

where Φ, p, and r_p are the pseudo-objective function, penalty function, and penalty parameter, respectively.

The main aim of the structural optimization problems with frequency constraints is to minimize the structural weight subject to a number of constraints on natural frequencies of the structure. This problem for purely size optimization and simultaneously size and configuration optimization may be formulated.

A pure size optimization problem with frequency constraints can be formulated as follows:

$$\text{Minimize: } w(X_S) = \sum_{k=1}^{ne} \rho_k l_k A_k \tag{16.6}$$

$$\text{Subject to: } g_i(X_S) = \frac{F_i}{F_{all}} - 1 \le 0, \quad i = 1, 2, \ldots, nf \tag{16.7}$$

$$X_S = [A_1, A_2, \ldots, A_{n1}]^T \in \Delta_S \tag{16.8}$$

where w is the weight of the structure; ρ_k, l_k, and A_k are the kth element material density, length, and cross-sectional area, respectively; F_i and F_{all} are the ith natural frequency and its corresponding allowable value, respectively, and nf is the number of constrained frequencies. Also, X_S is the vector of size variables, Δ_S is the given set of variables, and n_1 is the number of size design variables.

A simultaneous size and configuration optimization problem with frequency constraints can be stated as follows:

$$\text{Minimize: } w(X_S, X_C) = \sum_{k=1}^{ne} \rho_k l_k A_k \tag{16.9}$$

$$\text{Subject to: } g_i(X_S, X_C) = \frac{F_i}{F_{all}} - 1 \le 0, \quad i = 1, 2, \ldots, nf \tag{16.10}$$

$$X_S = [A_1, A_2, \ldots, A_{n1}]^T \in \Delta_S, \quad X_C = [C_1, C_2, \ldots, C_{n1}]^T \in \Delta_C \tag{16.11}$$

where X_C are the vectors of nodal coordinates, Δ_C is the given set of configuration variables, and n_2 is the number of configuration design variables.

Using the EPFM, these constrained optimization problems are converted to the following unconstrained ones:

For pure size optimization:

$$\Phi(X_S, r_p) = w(X_S)\left(1 + r_p \sum_{i=1}^{nf} \left[\max\left\{0, \left(\frac{F_i}{F_{all}} - 1\right)\right\}\right]^2\right) \tag{16.12}$$

For simultaneous size and configuration optimization:

$$\Phi(X_S, X_C, r_p) = w(X_S, X_C)\left(1 + r_p \sum_{i=1}^{nf} \left[\max\left\{0, \left(\frac{F_i}{F_{all}} - 1\right)\right\}\right]^2\right) \tag{16.13}$$

Formulation of the optimization problem of an arch dam considering dam–water interaction for frequency constraints is presented in the next section.

16.3 Formulation of Optimization Problem of an Arch Dam with Frequency Constraints

At first, the geometrical model of arch dam is explained. For the central vertical section of double-curvature arch dam one polynomial of nth order can be used to determine the curve of upstream boundary and another one can be used to determine the thickness (Gholizadeh and Seyedpoor, 2011b). If h and s are the height and the slope of the dam at crest, respectively, and the point where the slope of the upstream face equals to zero that is $z = \beta h$ then for the curve of upstream face, a polynomial of second order may be considered as follows:

$$y(z) = b(z) = -sz + sz^2/(2\beta h) \tag{16.14}$$

In order to formulate the thickness of the central vertical section, the height of the dam is divided into n segments. If t_{ci} is the thickness of the central vertical section and z_i denotes the z coordinate at ith level, the thickness can be expressed as:

$$t_c(z) = \sum_{i=1}^{n+1} \left(\frac{\prod_{k=1}^{n+1}(z - z_k)}{\prod_{k=1}^{n+1}(z_i - z_k)} t_{ci} \right), \quad k \neq i \tag{16.15}$$

The shape of the horizontal section of a parabolic dam is determined by the following two parabolic surfaces, respectively, for upstream and downstream faces (Gholizadeh and Seyedpoor, 2011b):

$$y_u(x, z) = 0.5x^2 \left(\sum_{i=1}^{n+1} \left(\frac{\prod_{k=1}^{n+1}(z - z_k)}{\prod_{k=1}^{n+1}(z_i - z_k)} r_{ui} \right) \right)^{-1} + b(z) \tag{16.16}$$

$$y_d(x, z) = 0.5x^2 \left(\sum_{i=1}^{n+1} \left(\frac{\prod_{k=1}^{n+1}(z - z_k)}{\prod_{k=1}^{n+1}(z_i - z_k)} r_{di} \right) \right)^{-1} + b(z) + t_c(z) \tag{16.17}$$

where r_{ui} and r_{di} are the values of r_u and r_d at ith level, respectively.

In fluid–structure interaction problems, the discretized structural dynamic equation and fluid equation need to be considered simultaneously (Aftabi and Lotfi, 2007; Kucukarslan et al., 2005). The matrix form of wave equation can be presented as follows (Gholizadeh and Seyedpoor, 2011b):

$$\mathbf{M}_f \ddot{P}_e + \mathbf{C}_f \dot{P}_e + \mathbf{K}_f P_e + \rho_w \mathbf{Q}^T (\ddot{U}_e + \ddot{U}_g) = 0 \tag{16.18}$$

where $\mathbf{M}_f$, $\mathbf{C}_f$, and $\mathbf{K}_f$ are fluid mass, damping, and stiffness matrices, respectively, and P_e, $\ddot{U}_e$, and $\ddot{U}_g$ are nodal pressure, acceleration, and ground acceleration vectors, respectively. Also, $\rho_w \mathbf{Q}^T$ in the above relation is often referred to as a coupling matrix.

The discretized structural dynamics equation for ground motion can be formulated using the finite elements as follows:

$$\mathbf{M}_s \ddot{U}_e + \mathbf{C}_s \dot{U}_e + \mathbf{K}_s U_e = -\mathbf{M}_s \ddot{U}_g + \mathbf{Q} P_e \tag{16.19}$$

where $\mathbf{M}_s$, $\mathbf{C}_s$, and $\mathbf{K}_s$ are structural mass, damping, and stiffness matrices, respectively, and U_e is nodal relative displacement vector. Also, $\mathbf{Q}P_e$ term in Eq. (16.19) represents nodal force vector associated with hydrodynamic pressure produced by the reservoir.

Equations (16.18) and (16.19) describe the complete finite element discretized equations for the fluid–structure interaction problem and are written in assembled form as:

$$\begin{aligned} &\begin{bmatrix} \mathbf{M}_s & 0 \\ \mathbf{M}_{fs} & \mathbf{M}_f \end{bmatrix} \begin{Bmatrix} \ddot{U}_e \\ \ddot{P}_e \end{Bmatrix} + \begin{bmatrix} \mathbf{C}_s & 0 \\ 0 & \mathbf{C}_f \end{bmatrix} \begin{Bmatrix} \dot{U}_e \\ \dot{P}_e \end{Bmatrix} \\ &+ \begin{bmatrix} \mathbf{K}_s & \mathbf{K}_{fs} \\ 0 & \mathbf{K}_f \end{bmatrix} \begin{Bmatrix} U_e \\ P_e \end{Bmatrix} = \begin{Bmatrix} -\mathbf{M}_s \ddot{U}_g \\ -\mathbf{M}_{fs} \ddot{U}_g \end{Bmatrix} \end{aligned} \tag{16.20}$$

where $\mathbf{M}_{fs} = \rho_w \mathbf{Q}^T$ and $\mathbf{K}_{fs} = -\mathbf{Q}$.

Equation (16.20) can also be written alternatively in a more compact form as:

$$\mathbf{M}\ddot{U} + \mathbf{C}\dot{U} + \mathbf{K}U = F(t) \tag{16.21}$$

where $\mathbf{M}$, $\mathbf{C}$, and $\mathbf{K}$ are mass, damping, and stiffness matrices of the dam–reservoir system, respectively. Obviously, $\mathbf{M}$ and $\mathbf{K}$ are not symmetric matrices. Since the system damping matrix needs to be included in modal analysis, the eigenproblem becomes a quadratic eigenvalue problem as:

$$(\lambda_i^2 \mathbf{M} + \lambda_i \mathbf{C} + \mathbf{K})\,\phi_i = 0, \quad i = 1, \ldots, n_{df} \tag{16.22}$$

The above equation needs to be solved to get the complex eigenvalues λ_i given by:

$$\lambda_i = \sigma_i \pm \omega_i j, \quad i = 1, \ldots, n_{df} \tag{16.23}$$

where σ_i and ω_i are real and imaginary part of the eigenvalue and $j = \sqrt{-1}$. In this case, natural frequency is calculated as:

$$F_i = \frac{\sqrt{\sigma_i^2 + \omega_i^2}}{2\pi}, \quad i = 1, \ldots, n_{df} \tag{16.24}$$

The concrete volume of the arch dam body is considered as an objective function. Therefore, the optimization problem of the arch dam is formally stated as follows:

$$\text{minimize: } w(X) = \iint_{Area} |y_d(x,z) - y_u(x,z)|\, dx\, dz \tag{16.25}$$

$$\text{Subject to: } g_i(X) = \frac{F_i}{F_{all}} - 1 \le 0, \quad i = 1, 2, \ldots, nf \tag{16.26}$$

$$g_{gj}(X) = \frac{r_{dj}}{r_{uj}} - 1 \leq 0, \quad j = 1, \ldots, n+1 \tag{16.27}$$

$$g_c(X) = \frac{s}{s_{all}} - 1 \leq 0 \tag{16.28}$$

$$\begin{cases} g^u_{sj}(X) = \dfrac{\varphi_j}{\varphi^u} - 1 \leq 0 \\ g^l_{sj}(X) = 1 - \dfrac{\varphi_j}{\varphi^l} \leq 0 \end{cases}, \quad j = 1, \ldots, n+1 \tag{16.29}$$

$$X^{\mathrm{T}} = \{s\beta t_{c1} \ldots t_{cn+1}\ r_{u1} \ldots r_{un+1}\ r_{d1} \ldots r_{dn+1}\} \tag{16.30}$$

where $g_c(X)$ is the geometric constraint that is applied for facile construction; s is the slope of overhang at the upstream face of dam and s_{all} is its allowable value; $g^u_{si}(X)$ and $g^l_{si}(X)$ are the constraints ensuring the sliding stability of the dam, and φ_i is the ith central angle of arch dam and usually $90 \leq \varphi_i \leq 130$; X may have $3n + 5$ components involving shape parameters of arch dam.

Using the EPFM, the above-mentioned constrained optimization problem of a dam can be easily converted to the following unconstrained ones:

$$\begin{aligned} \Phi(X, r_p) = w(X)\Bigg(1 + r_p \sum_{i=1}^{nf} \left[\max\left\{0, \left(\frac{F_i}{F_{all}} - 1\right)\right\}\right]^2 \\ + r_p \sum_{j=1}^{n+1} \left[\max\left\{0, \left(\frac{r_{dj}}{r_{uj}} - 1\right)\right\}\right]^2 + r_p \left[\max\left\{0, \left(\frac{s}{s_{all}} - 1\right)\right\}\right]^2 \\ + r_p \sum_{j=1}^{n+1} \left[\max\left\{0, \left(\frac{\varphi_j}{\varphi^u} - 1\right)\right\}\right]^2 + r_p \sum_{j=1}^{n+1} \left[\max\left\{0, \left(1 - \frac{\varphi_j}{\varphi^l}\right)\right\}\right]^2\Bigg) \end{aligned} \tag{16.31}$$

In this chapter, VSP, PSO, and HS metaheuristics are selected for implementing the optimization task. In the subsequent section, these algorithms are described.

16.4 Metaheuristics

In order to comprehensively explore a larger fraction of the design space, stochastic search techniques reveal their promising abilities in comparison with gradient-based optimization methods. In recent years, a variety of stochastic metaheuristic optimization methods inspired by nature were developed. The metaheuristics, due to their high potential for simple computer implementation, have now emerged as the most practical approaches for solving many complex problems. However, there are several newly developed metaheuristics (Gandomi and Alavi, 2012; Gandomi et al., 2011, 2012a; Yang and Gandomi, 2012) based on a huge number of publications in the field; GAs, PSO, and HS are the most popular metaheuristics and many successful applications of them have been reported. In this chapter, GA, PSO, and HS are considered. As metaheuristics have already been reviewed and described by Gandomi et al., 2012b and Sahab et al., 2012, the basic concepts of GA, PSO, and HS are briefly described below.

16.4.1 Genetic Algorithm

GA is a stochastic search algorithm based on principles of natural competition between individuals for appropriating limited natural sources. Success of the winner normally depends on their genes, and reproduction by such individuals causes the spread of their genes. By successive selection of superior individuals and reproducing them, the population will be led to obtain more natural resources. The GA simulates this process and calculates the optimum of objective functions. In general, the standard GA is not convenient for finding the solutions to complex problems. The VSP method (Gholizadeh and Salajegheh, 2010; Gholizadeh and Samavati, 2011; Salajegheh and Gholizadeh, 2005) is an alternative to overcome this shortcoming of GA.

In this modified GA, an initial population with a small number of individuals is selected; this population is much smaller than that in standard GA. Then, all the necessary operations of the standard GA are carried out and the optimal solution is achieved. As the size of the population is small, the method converges to a premature solution. In each generation, the best individual is saved. Then, the best solution is repeatedly copied to create a new population and the remaining members of the population are randomly selected. Thereafter, the optimization process is repeated using standard GA with a reduced population to achieve a new solution.

16.4.2 Particle Swarm Optimization

The PSO is based on the social behavior of animals such as fish schooling, insect swarming, and bird flocking. The PSO has been proposed by Kennedy (1977) to simulate the graceful motion of bird swarms as a part of a sociocognitive study.

The PSO involves a number of particles, which are randomly initialized in the search space. These particles are referred to as a swarm. The particles fly through the search space and their positions are updated based on the best positions of individual particles and the best of the swarm in each iteration. The objective function is evaluated for each particle at each grid point, and the fitness values of particles are obtained to determine the best position in the search space (Kennedy and Eberhart, 1995). In iteration k, the swarm is updated using the following equations:

$$V_i^{k+1} = \omega^k V_i^k + c_1 r_1 (P_i^k - X_i^k) + c_2 r_2 (P_g^k - X_i^k) \tag{16.32}$$

$$X_i^{k+1} = X_i^k + V_i^{k+1} \tag{16.33}$$

where X_i and V_i represent the current position and the velocity of the ith particle, respectively; P_i is the best previous position of the ith particle (pbest) and P_g is the best global position among all the particles in the swarm (gbest); r_1 and r_2 are two uniform random sequences generated from interval [0, 1]; c_1 and c_2 are the cognitive and social scaling parameters, respectively. The inertia weight used to discount the previous velocity of particle preserved is expressed by ω.

16.4.3 HS Algorithm

The HS algorithm was proposed by Geem et al. (2001) and Lee and Geem (2004, 2005). This approach is based on the musical performance process that occurs when a musician searches for a better harmony. The HS algorithm consists of five basic steps that can be summarized as follows:

A harmony memory (HM), the harmony memory considering rate (HMCR), the pitch adjusting rate (PAR), and the maximum number of searches should be specified. To improvise a new HM, a new harmony vector is generated. Thus, the new value of the ith design variable can be chosen from the possible range of ith column of the HM or from the entire possible range as follows:

$$x_i^{\text{new}} = \begin{cases} x_i^j \in \{x_i^1, x_i^2, \ldots, x_i^{\text{HMS}}\}^{\text{T}} & \text{with the probability of HMCR} \\ x_i \in \Delta_i & \text{with the probability of } (1 - \text{HMCR}) \end{cases} \tag{16.34}$$

where Δ_i is the set of the potential range of values for ith design variable. The HMCR is the probability of choosing one value from the significant values stored in the HM. Pitch adjusting is performed only after a value has been chosen from the HM. If the pitch-adjustment decision for x_i^{new} is "Yes," then a neighboring value with the probability of PAR% $\times$ HMCR is taken for it as follows:

$$x_i^{\text{new}} \leftarrow \begin{cases} x_i^{\text{new}} \pm u(-1, +1) \times \text{bw} & \text{with the probability of PAR} \times \text{HMCR} \\ x_i^{\text{new}} & \text{with the probability of PAR} \times (1 - \text{HMCR}) \end{cases} \tag{16.35}$$

where $u(-1, +1)$ is a uniform distribution between -1 and $+1$; also bw is an arbitrary distance bandwidth for the continuous design variables.

If x_i^{new} is better than the worst vector in the HM, the new harmony is substituted for the existing worst harmony. The optimization process of HS is repeated by continuing improvising new harmonies until a termination criterion is satisfied.

The computational burden of evaluating the natural frequencies of large-scale structures by FEM during the optimization process usually is high. Therefore, it is necessary to adopt a computational strategy for decreasing the computational time. In this chapter, neural network techniques are used to achieve this purpose.

16.5 Neural Networks

Nowadays, neural networks are considered as more appropriate techniques for simplification of complex and time-consuming problems due to their ability to process and map external data and information based on past experiences (Gholizadeh et al., 2009). In this section, the following three well-known and popular neural network models are employed: the RBF neural network, the BP neural network, and the ANFIS.

16.5.1 RBF Model

RBF neural networks are two-layer feed forward networks. The hidden layer consists of RBF neurons with Gaussian activation functions. The outputs of RBF neurons have significant responses to the inputs only over a range of values called the receptive field. During the training, the receptive field radius of RBF neurons is determined such that the neurons can cover the input space properly. To train the hidden layer, no training is accomplished and the transpose of training input matrix is taken as the layer weight matrix (Wasserman, 1993). In order to adjust output layer weights, a supervised training algorithm is employed. The output layer weight matrix is calculated from the following equation:

$$W_2^{\mathrm{RBF}} = \Delta^{-1}\mathrm{T} \tag{16.36}$$

where T is the target matrix, Δ is the output of the hidden layer, and W_2^{RBF} is the output layer weight matrix.

16.5.2 BP Model

Standard BP (Hagan et al., 1996) is a gradient descent optimization algorithm, which adjusts the weights in the steepest descent direction according to the following equation:

$$W_{k+1}^{\mathrm{BP}} = W_k^{\mathrm{BP}} - \eta_k G_k \tag{16.37}$$

where W_k^{BP} and G_k are the weight and the current gradient matrices, respectively, and η_k is the learning rate.

The Levenberg–Marquardt (LM) (Hagan and Menhaj, 1999) algorithm was designed to approach second-order training speed without having to compute the Hessian matrix. In the LM algorithm, the weights updating is achieved as follows:

$$W_{k+1}^{\mathrm{BP}} = W_k^{\mathrm{BP}} - [J^{\mathrm{T}}J + \mu I]^{-1}J^{\mathrm{T}}E \tag{16.38}$$

where J is the Jacobian matrix that contains first derivatives of the network errors with respect to the weights, and E is a vector of network errors. Also, μ is a correction factor (Hagan et al., 1996).

16.5.3 ANFIS Model

There are two types of fuzzy inference systems (Mamdani and Assilian, 1975) that can be implemented: Mamdani-type and Sugeno-type (Sugeno, 1985; Paiva and Dourado, 2001). In this chapter, the Sugeno-type method of fuzzy inference based on an adaptive network, namely, the ANFIS, is employed. In this chapter, an ANFIS model containing n_m inputs with two associated membership functions (MFs) per dimension and the value of a desired natural frequency as output is explained. So, it has 2^{n_m} fuzzy if−then rules where the kth rule can be expressed as:

$$\text{Rule } k\text{: if } (tp_1 \text{ is } M_1^{j_1}) \text{ and } (tp_2 \text{ is } M_2^{j_2}) \ldots \text{ and } (tp_{n_m} \text{ is } M_{n_m}^{j_{n_m}}) \text{ then } F_k = c_{k_0} + \sum_{i=1}^{n_m} c_{k_i}\, tp_i \tag{16.39}$$

where $\{tp_1, tp_2, \ldots, tp_{n_m}\}^{\mathrm{T}}$ is input vector; $M_1^{j_1}, M_2^{j_2}, \ldots, M_{n_m}^{j_{n_m}}$ $(j_i = 1, 2)$ are labels for classifying MFs of the inputs; F_k is the linear output of kth rule and $c_{k_0}, c_{k_1}, \ldots, c_{k_{n_m}}$ are parameters of output membership function.

Typical topology of the ANFIS model is shown in Figure 16.1. The ANFIS model consists of five layers briefly explained below:

Layer 1: All the nodes in this layer are adaptive nodes. They generate membership grades of the inputs. The outputs of this layer $O^1_{M_i^{j_i}}$ are given by

$$O^1_{M_i^{j_i}} = \mu_{M_i^{j_i}}(tp_i), \quad i = 1, \ldots, n_m \tag{16.40}$$

where $M_i^{j_i}$ $(j_i = 1, 2)$ represent two MFs corresponding to ith input tp_i which can be triangular, trapezoidal, Gaussian functions, or other shapes. In this study, the generalized bell-shaped MFs defined below are utilized.

$$\mu_{M_i^{j_i}}(tp_i, a_i^{j_i}, b_i^{j_i}, c_i^{j_i}) = \frac{1}{1 + \left|\left(\frac{tp_i - c_i^{j_i}}{a_i^{j_i}}\right)\right|^{2b_i^{j_i}}}, \quad i = 1, 2, \ldots, n_m \tag{16.41}$$

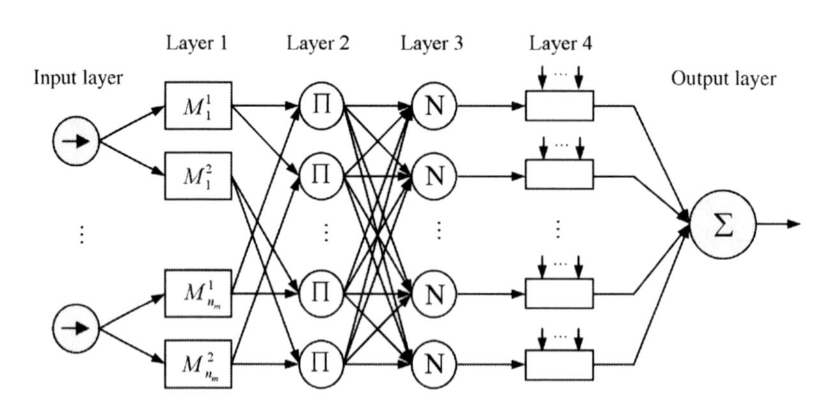

Figure 16.1 Typical topology of ANFIS model.

where $a_i^{j_i}, b_i^{j_i}, c_i^{j_i}$ are the parameters of the MFs, governing the bell-shaped functions.

Layer 2: The nodes in this layer are fixed nodes and they perform a simple multiplier. The outputs of this layer O_k^2 are represented as:

$$O_k^2 = w_k = \mu_{M_1^{j_1}}(tp_1)\mu_{M_2^{j_2}}(tp_2)\ldots\mu_{M_{nm}^{j_{nm}}}(tp_{n_m}), \quad k = 1, 2, \ldots, 2^{n_m} \tag{16.42}$$

Layer 3: The nodes in this layer are also fixed nodes and they play a normalization role in the network. The outputs of this layer O_k^3 can be represented as:

$$O_k^3 = \overline{w}_k = \frac{w_k}{w_1 + w_2 + \cdots + w_{2^{n_m}}}, \quad k = 1, 2, \ldots, 2^{n_m} \tag{16.43}$$

Layer 4: All the nodes in this layer are adaptive, whose output is simply the product of the normalized firing strength and a first-order polynomial (for a first-order Sugeno model). Thus, the outputs of this layer O_k^4 are:

$$O_k^4 = \overline{w}_k F_k = \overline{w}_k(c_{k0} + c_{k1}tp_1 + \cdots + c_{k_{nm}}tp_{n_m}), \quad k = 1, 2, \ldots, 2^{n_m} \tag{16.44}$$

Layer 5: The single node in this layer is a fixed node and computes the overall output O^5 as the summation of all incoming signals, that is,

$$\begin{aligned} O^5 &= \sum_{k=1}^{2^{nm}} \overline{w}_k F_k. \\ &= \sum_{k=1}^{2^{nm}} \overline{w}_k(c_{k_0} + c_{k1}tp_1 + \cdots + c_{k_{nm}}tp_{n_m}) \\ &= \sum_{k=1}^{2^{nm}} ((\overline{w}_k c_{k_0}) + (\overline{w}_k c_{k1})tp_1 + \cdots + (\overline{w}_k c_{k_{nm}})tp_{n_m}) \end{aligned} \tag{16.45}$$

The task of the learning algorithm for this architecture is to tune all the modifiable parameters to make the ANFIS output match the training data. The detailed algorithm and mathematical background of the training algorithm of the ANFIS can be found in Jang (1993).

16.6 Numerical Examples

The numerical examples are presented in two parts. In the first part, four optimization problems of skeletal structures with frequency constraints are addressed. In the second part, the shape optimization of arch dams with limited natural frequencies considering fluid–structure interaction is solved. A set of benchmark structural optimization problems can be found in Gandomi and Yang (2011). In this section, a personal Pentium IV 3.0 GHz is used and all of the algorithms are coded in MATLAB (2009). For all the presented examples, the parameters of the metaheuristics are as follows:

- For VSP: the number of individuals = 30, crossover rate = 0.9, mutation rate = 0.001.
- For PSO: the number of particles = 30, $c_1 = c_2 = 2.0$, $\omega_{\min} = 0.4$, $\omega_{\max} = 0.9$.
- For HS: the size of harmony memory = 30, HMCR = 0.9, PAR = 0.25.

The maximum number of generations for VSP and PSO is chosen to be 125. This means that the maximum number of structural analyses is limited to 3750. In order that the results of the HS can be compared with the other employed algorithms, the maximum number of analyses for HS is also 3750.

16.6.1 First Example: 10-Bar Truss

The 10-bar planar truss is shown in the Figure 16.2. A nonstructural mass of 454 kg is attached to the free nodes. This problem includes 10 design variables including the cross-sectional area of each member. The Young's modulus and material density are 6.98×10^{10} Pa and 2770 kg/m^3, respectively. The minimum value of the cross-sectional area is 0.645×10^{-4} m^2 for all elements. This well-known benchmark problem has been investigated by many researchers such as Gomes (2011), Lingyun et al. (2005), Kaveh and Zolghadr (2011), Sedaghati et al. (2002), and Wang et al. (2004). The frequency constraints are as follows:

$$fr_1 \geq 7 \text{ Hz}, \quad fr_2 \geq 15 \text{ Hz}, \quad \text{and } fr_3 \geq 20 \text{ Hz} \tag{16.46}$$

The results of VSP, PSO, and HS metaheuristics, including optimal designs and their corresponding natural frequencies, are compared with the other methods in Table 16.1. These results show that among the above-mentioned three metaheuristics, PSO has the best computational performance while HS is better than VSP. The convergence history of the VSP, PSO, and HS is shown in Figure 16.3.

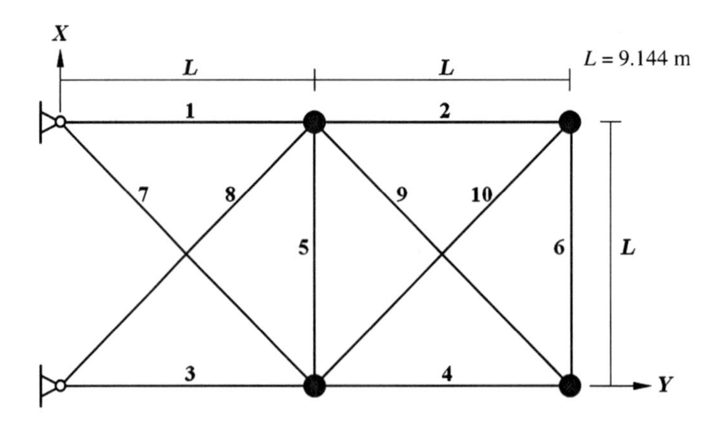

Figure 16.2 10-Bar truss.

Table 16.1 Results of VSP, PSO, and HS Compared with Other Methods for 10-Bar Truss

Element No.	Sedaghati et al. (2002)	Wang et al. (2004)	Lingyun et al. (2005)	Gomes (2011)	Kaveh and Zolghadr (2011)	Present Work		
						VSP	PSO	HS
1	38.245	32.456	42.23	37.712	39.569	31.843	32.361	37.172
2	9.916	16.577	18.555	9.959	16.740	11.228	16.111	16.479
3	38.619	32.456	38.851	40.265	34.361	38.559	38.678	37.988
4	18.232	16.577	11.222	16.788	12.994	16.708	13.045	9.335
5	4.419	2.115	4.783	11.576	0.645	5.910	0.645	4.853
6	4.419	4.467	4.451	3.955	4.802	4.154	5.105	4.378
7	20.097	22.810	21.049	25.308	26.182	24.538	28.059	23.117
8	24.097	22.810	20.949	21.613	21.260	20.742	20.581	24.719
9	13.890	17.490	10.257	11.576	11.766	15.800	9.514	9.999
10	11.452	17.490	14.342	11.186	11.392	12.851	14.953	13.718
Weight (kg)	537.01	553.8	542.75	537.98	529.25	539.40	530.22	535.45
Analyses	–	–	–	2000	4000	3750	3750	3750
fr_1 (Hz)	6.992	7.011	7.008	7.000	7.000	7.006	7.003	7.001
fr_2 (Hz)	17.599	17.302	18.148	17.786	16.238	18.115	16.155	17.642
fr_3 (Hz)	19.973	20.001	20.000	20.000	20.000	20.034	20.006	20.001

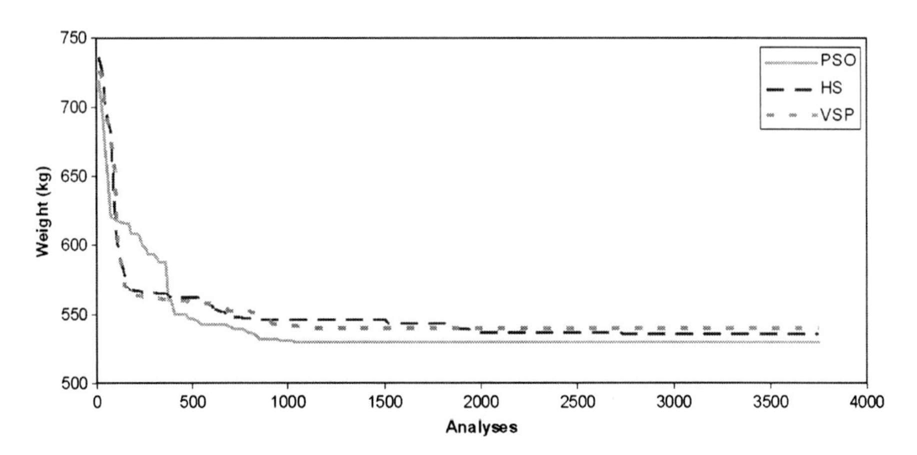

Figure 16.3 Convergence history of the VSP, PSO, and HS algorithms in the case of 10-bar truss.

16.6.2 Second Example: 72-Bar Truss

The configuration of the 72-bar space truss is shown in Figure 16.4. Four nonstructural masses of 2270 kg are attached to the nodes 1 through 4. This problem also has been investigated by Gomes (2011), Kaveh and Zolghadr (2011), and Sedaghati et al. (2002). The elements are classified in 16 design groups as given in Table 16.2. The Young's modulus and material density are 6.98×10^{10} Pa and 2770 kg/m^3, respectively. The minimum cross-sectional area is 0.645×10^{-4} m^2. The frequency constraints are as follows:

$$fr_1 = 4\ \text{Hz and}\ fr_3 \geq 6\ \text{Hz} \tag{16.47}$$

The results of VSP, PSO, and HS metaheuristics, including optimal designs and their corresponding natural frequencies, are compared with the other methods in Table 16.2. These given results demonstrate that among the employed metaheuristics, PSO has the best computational performance and HS is better than VSP. The convergence history of the VSP, PSO, and HS is shown in Figure 16.5.

16.6.3 Third Example: 37-Bar Truss

This example is one of the sizing and shape optimization benchmark problems investigated by Gomes (2011), Kaveh and Zolghadr (2011), Lingyun et al. (2005), and Wang et al. (2004). The 37-bar truss is shown in Figure 16.6. The elements of the lower chord are modeled as bar elements with constant cross-sectional areas of 4×10^{-3} m^2, and the other bars are modeled as simple bar elements. The structural members are grouped in a symmetrical manner to design the sizing variables. Young's modulus is 2.1×10^{-11} N/m^2 and the material density is $\rho = 7800$ kg/m^3.

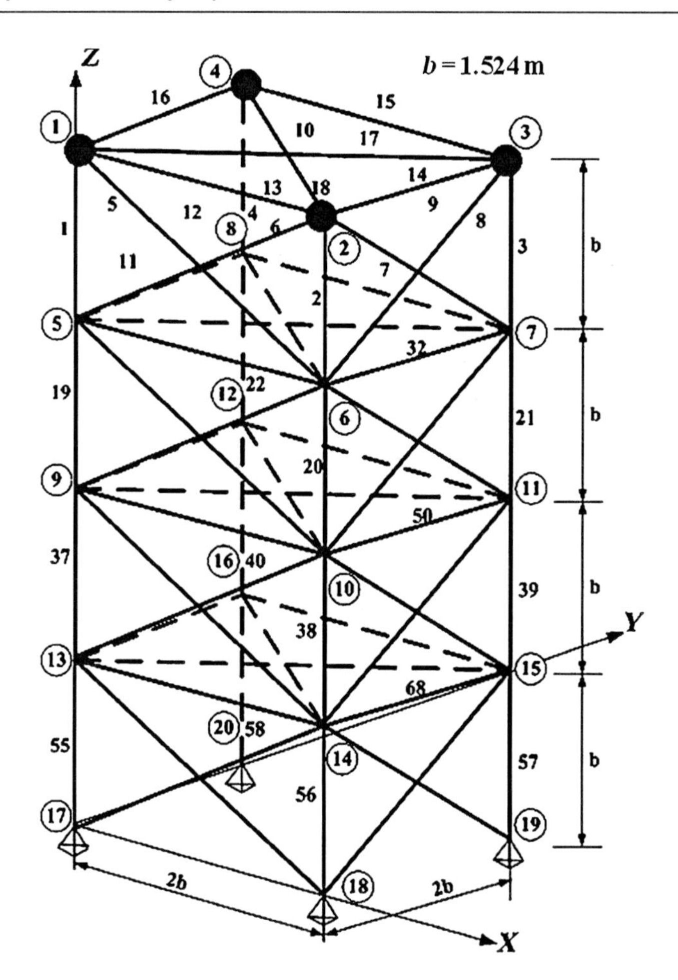

Figure 16.4 72-Bar truss.

Also a nonstructural mass of 10 kg is attached to each node on the lower chord, which stays fixed during the optimization process. The *y*-coordinate of all the nodes on the upper chord can vary in a symmetrical way. Thus, there are 19 design variables including 5 shape and 14 size variables. The minimum cross-sectional area is 10^{-4} m^2 and the frequency constraints are as follows:

$$fr_1 \geq 20 \text{ Hz},\ fr_2 \geq 40 \text{ Hz},\ fr_3 \geq 60 \text{ Hz} \tag{16.48}$$

The results of VSP, PSO, and HS metaheuristics, including optimal designs and their corresponding natural frequencies, are compared with the other methods in Table 16.3. These given results demonstrate that among the employed metaheuristics, PSO has the best computational performance and HS is better than VSP. The convergence history of the VSP, PSO, and HS is shown in Figure 16.7.

Table 16.2 Results of VSP, PSO, and HS Compared with Other Methods for 72-Bar Truss

No.	Element Groups	Sedaghati et al. (2002)	Gomes (2011)	Kaveh and Zolghadr (2011)	Present Work		
					VSP	PSO	HS
1	1–4	3.499	2.987	2.252	3.748	3.650	3.349
2	5–12	7.932	7.849	9.109	9.528	6.529	8.159
3	13–16	0.645	0.645	0.648	0.645	0.645	0.645
4	17–18	0.645	0.645	0.645	0.645	0.645	0.645
5	19–22	8.056	8.765	7.946	7.426	8.687	10.434
6	23–30	8.011	8.153	7.703	7.074	8.562	7.798
7	31–34	0.645	0.645	0.647	0.645	0.645	0.645
8	35–36	0.645	0.645	0.646	0.645	0.645	0.645
9	37–40	12.812	13.450	13.46	12.844	11.722	12.631
10	41–48	8.061	8.073	8.250	6.839	8.558	9.376
11	49–52	0.645	0.645	0.645	0.645	0.645	0.645
12	53–54	0.645	0.645	0.646	0.645	0.645	0.645
13	55–58	17.279	16.684	18.36	17.256	17.240	15.514
14	59–66	8.088	8.159	7.053	9.027	8.492	6.834
15	67–70	0.645	0.645	0.645	0.645	0.645	0.645
16	71–72	0.645	0.645	0.646	0.645	0.645	0.645
Weight (kg)		327.605	328.823	328.393	329.76	327.33	328.61
Analysis		–	42,840	4000	3750	3750	3750
fr_1 (Hz)		4.000	4.000	4.000	4.0002	4.0001	4.0000
fr_3 (Hz)		6.000	6.000	6.004	6.0024	6.0003	6.0001

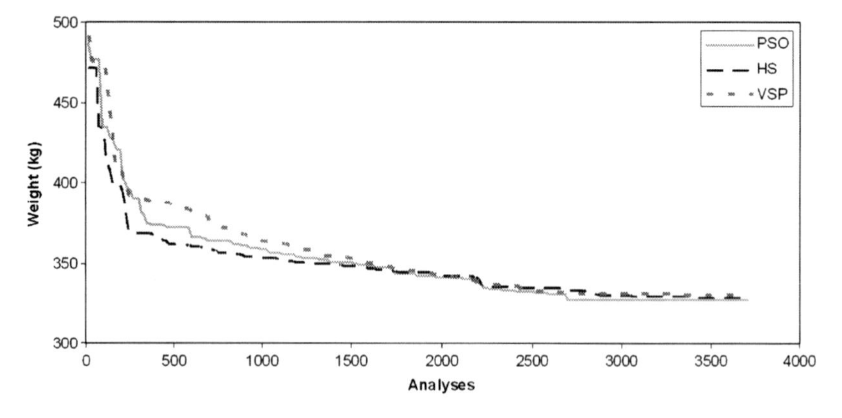

Figure 16.5 Convergence history of the VSP, PSO, and HS algorithms in the case of 72-bar truss.

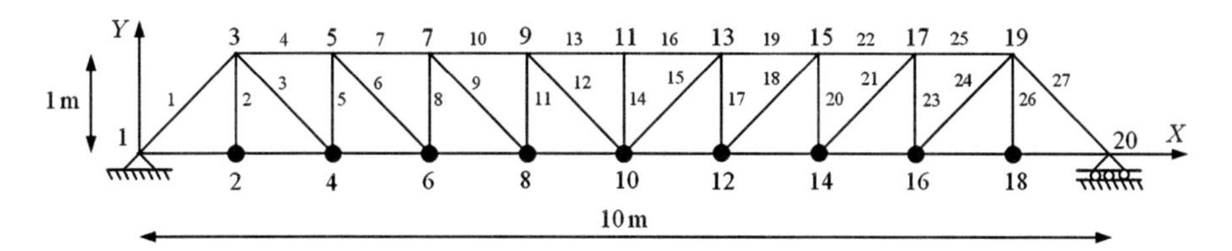

Figure 16.6 37-Bar truss.

16.6.4 Fourth Example: 52-Bar Truss

The optimization problem of 52-bar space truss, shown in Figure 16.8, has been inspected by Gomes (2011), Lin et al. (1982), Kaveh and Zolghadr (2011), and Lingyun et al. (2005). A nonstructural mass of 50 kg is added to free nodes. All free nodes are permitted to move in a symmetrical manner in the range of ± 2 m from their initial position. In this example, there are eight size variables corresponding to cross-sectional areas of elements 1–8 shown in Figure 16.8 and five shape variables associated with Z_A, X_B, Z_B, X_F, and Z_F. Young's modulus is 2.1×10^{-11} N/m^2 and the material density is $\rho = 7800$ kg/m^3. The lower and upper bounds on cross-sectional areas are 10^{-4} and 10^{-3} m^2, respectively. The frequency constraints are as follows:

$$fr_1 \leq 15.916 \text{ Hz}, \; fr_2 \geq 28.648 \text{ Hz} \tag{16.49}$$

The results of VSP, PSO, and HS metaheuristics, including optimal designs and their corresponding natural frequencies, are compared with the other methods in Table 16.4. These given results demonstrate that among the employed metaheuristics, PSO has the best computational performance and HS is better than VSP. The convergence history of the VSP, PSO, and HS is shown in Figure 16.9.

16.6.5 Fifth Example: Arch Dam

This example is the optimization of the Morrow Point arch dam for frequency constraints, which was solved previously by Gholizadeh and Seyedpoor (2011b). The dam structure is 143 m high, with a crest length of 221 m. The thin arch structure ranges in thickness from 3.7 m at the crest to 16 m at the base. The dam construction required 273,600 m^3 of concrete.

In the present study, the finite element model of the Morrow Point arch dam considering fluid–structure interaction is employed. The arch dam is treated as a 3D linear structure. An eight-noded solid element is utilized to discretize of the dam body. The reservoir is assumed to be uniform shape and an eight-noded fluid element is used to discretize the fluid medium and the interface of the fluid–structure interaction problem (Aftabi and Lotfi, 2007). The element has four degrees of freedom per node: translations in the nodal x, y, and z directions, and pressure. The translations, however, are applicable only at nodes that are on the interface. In this study, interaction between dam and the foundation rock is not considered and it is

Table 16.3 Results of VSP, PSO, and HS Compared with Other Methods for 37-Bar Truss

No.	Design Variables	Wang et al. (2004)	Lingyun et al. (2005)	Gomes (2011)	Kaveh and Zolghadr (2011)	Present Work		
						VSP	PSO	HS
1	Y_3, Y_{19} (m)	1.2086	1.1998	0.9637	1.0289	0.9854	1.0796	1.2842
2	Y_5, Y_{17} (m)	1.5788	1.6553	1.3978	1.3868	1.6386	1.5789	1.7268
3	Y_7, Y_{15} (m)	1.6719	1.9652	1.5929	1.5893	2.0060	1.8985	2.1154
4	Y_9, Y_{13} (m)	1.7703	2.0737	1.8812	1.6405	2.1518	2.0920	2.2533
5	Y_{11} (m)	1.8502	2.3050	2.0856	1.6835	2.4273	2.1979	2.3400
6	A_1, A_{27} (mc^2)	3.2508	2.8932	2.6797	3.4484	2.8475	3.1246	2.2122
7	A_2, A_{26} (mc^2)	1.2364	1.1201	1.1568	1.5045	1.0000	1.2063	1.1647
8	A_3, A_{24} (mc^2)	1.0000	1.0000	2.3476	1.0039	1.6919	1.0000	1.0000
9	A_4, A_{25} (mc^2)	2.5386	1.8655	1.7182	2.5533	1.8903	2.2022	1.8950
10	A_5, A_{23} (mc^2)	1.3714	1.5962	1.2751	1.0868	1.4607	1.4374	1.4738
11	A_6, A_{21} (mc^2)	1.3681	1.2642	1.4819	1.3382	1.6608	1.5261	1.4303
12	A_7, A_{22} (mc^2)	2.4290	1.8254	4.6850	3.1626	1.8377	2.1247	1.6667
13	A_8, A_{20} (mc^2)	1.6522	2.0009	1.1246	2.2664	2.0282	1.6451	2.0240
14	A_9, A_{18} (mc^2)	1.8257	1.9526	2.1214	1.2668	1.8983	1.5320	2.0069
15	A_{10}, A_{19} (mc^2)	2.3022	1.9705	3.8600	1.7518	1.6902	1.3645	1.3382
16	A_{11}, A_{17} (mc^2)	1.3103	1.8294	2.9817	2.7789	1.4101	1.4120	1.6521
17	A_{12}, A_{15} (mc^2)	1.4067	1.2358	1.2021	1.4209	2.1382	2.0114	1.6760
18	A_{13}, A_{16} (mc^2)	2.1896	1.4049	1.2563	1.0100	1.4276	1.2560	1.5066
19	A_{14} (mc^2)	1.0000	1.0000	3.3276	2.2919	1.7679	1.0806	1.0000
	Weight (kg)	366.50	368.84	377.20	367.59	372.72	366.69	369.49
	Analyses	–	–	12,500	4000	3750	3750	3750
	fr_1 (Hz)	20.0850	20.0013	20.0001	20.0028	20.2160	20.1170	20.2760
	fr_2 (Hz)	42.0743	40.0305	40.0003	40.0155	40.1189	40.6049	40.0540
	fr_3 (Hz)	62.9383	60.0000	60.0001	61.2798	60.0278	60.1951	60.2692

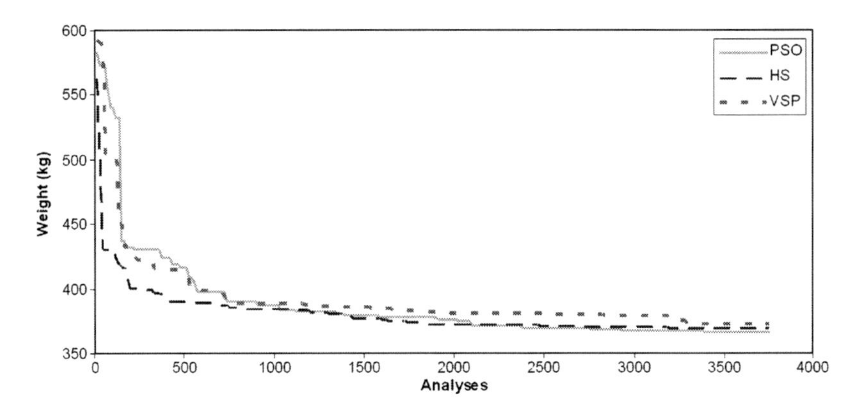

Figure 16.7 Convergence history of the VSP, PSO, and HS algorithms in the case of 37-bar truss.

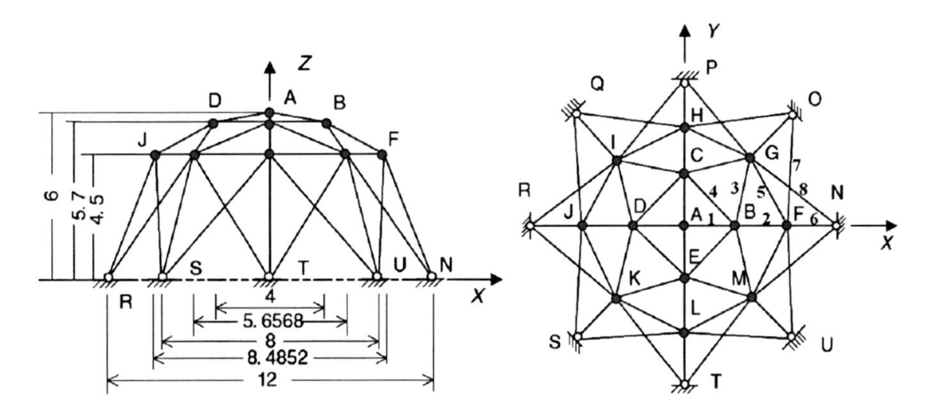

Figure 16.8 52-Bar truss.

assumed to be rigid to avoid the extra complexities that would otherwise arise. Interaction between the fluid and foundation rock is approximately considered through a damping boundary condition applied along the bottom and sides of the reservoir (Gholizadeh and Seyedpoor, 2011b). The finite element model of the arch dam–water system is shown in Figure 16.10. Details of finite element model verification based on an idealized symmetric model of the Morrow Point arch dam is given by Seyedpoor et al. (2009).

To create the dam geometry, three fifth-order functions are considered for $t_c(z)$, $r_u(z)$, and $r_d(z)$. So, by accounting for two shape parameters needed to define the curve of upstream face $b(z)$, the dam can be modeled by 20 shape design variables as:

$$X^T = \{s\ \beta\ t_{c1}\ t_{c2}\ t_{c3}\ t_{c4}\ t_{c5}\ t_{c6}\ r_{u1}\ r_{u2}\ r_{u3}\ r_{u4}\ r_{u5}\ r_{u6}\ r_{d1}\ r_{d2}\ r_{d3}\ r_{d4}\ r_{d5}\ r_{d6}\} \tag{16.50}$$

Table 16.4 Results of VSP, PSO, and HS Compared with Other Methods for 52-Bar Truss

No.	Design Variables	Lin et al. (1982)	Lingyun et al. (2005)	Gomes (2011)	Kaveh and Zolghadr (2011)	Present Work		
						VSP	PSO	HS
1	Z_A (m)	4.3201	5.8851	5.5344	6.1590	5.6997	6.1065	5.7540
2	X_B (m)	1.3153	1.7623	2.0885	2.2609	1.6422	2.3288	1.6944
3	Z_B (m)	4.1740	4.4091	3.9283	3.9154	4.1334	3.9444	4.0191
4	X_F (m)	2.9169	3.4406	4.0255	4.0836	3.0124	4.0738	3.3239
5	Z_F (m)	3.2676	3.1874	2.4575	2.5106	3.3119	2.5830	3.0199
6	A_1 (cm^2)	1.0000	1.0000	0.3696	1.0335	1.0000	1.0000	1.0000
7	A_2 (cm^2)	1.3300	2.1417	4.1912	1.0960	1.0000	1.0000	1.3326
8	A_3 (cm^2)	1.5800	1.4858	1.5123	1.2449	1.3480	1.2395	1.3510
9	A_4 (cm^2)	1.0000	1.4018	1.5620	1.2358	1.6921	1.1870	1.4499
10	A_5 (cm^2)	1.7100	1.9111	1.9154	1.4078	1.7199	1.3143	1.5515
11	A_6 (cm^2)	1.5400	1.0109	1.1315	1.0022	1.0000	1.0000	1.0000
12	A_7 (cm^2)	2.6500	1.4693	1.8233	1.6024	1.6697	1.5015	1.4257
13	A_8 (cm^2)	2.8700	2.1411	1.0904	1.4596	1.9915	1.7300	1.9729
	Weight (kg)	298.02	236.05	228.38	197.34	222.67	199.92	210.54
	Analyses	–	–	12,500	4000	3750	3750	3750
	fr_1 (Hz)	15.22	12.81	12.751	11.849	10.1419	12.9790	10.1451
	fr_2 (Hz)	29.28	28.65	28.649	28.649	28.711	28.657	28.6850

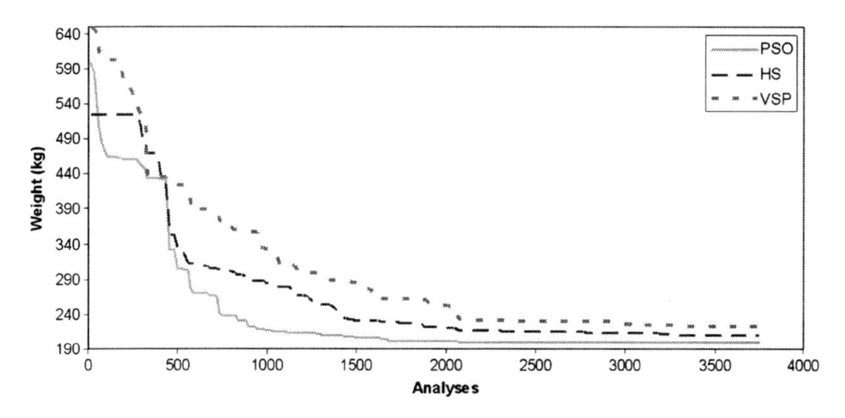

Figure 16.9 Convergence history of the VSP, PSO, and HS algorithms in the case of 52-bar truss.

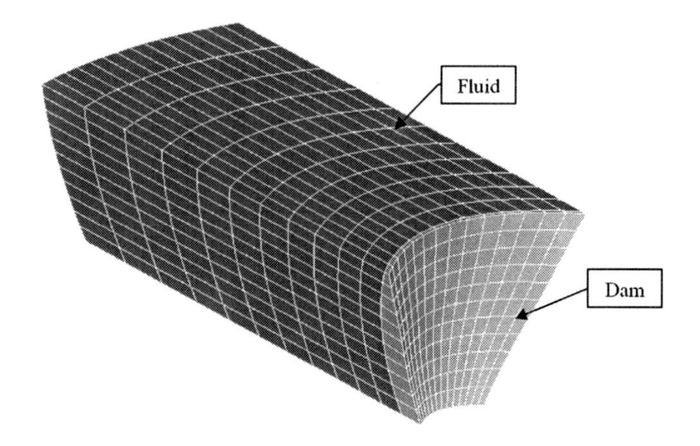

Figure 16.10 Finite element model of arch dam—water system.

The lower and upper bounds of design variables required in the optimization process can be determined using preliminary design methods (Seyedpoor et al., 2009):

$$\begin{array}{llll}
0. \leq s \leq 0.3 & 3\text{ m} \leq t_{c1} \leq 10\text{ m} & 100\text{ m} \leq r_{u1} \leq 135\text{ m} & 100\text{ m} \leq r_{d1} \leq 135\text{ m} \\
0.5 < \beta \leq 1 & 5\text{ m} \leq t_{c2} \leq 15\text{ m} & 85\text{ m} \leq r_{u2} \leq 115\text{ m} & 85\text{ m} \leq r_{d2} \leq 115\text{ m} \\
 & 10\text{ m} \leq t_{c3} \leq 20\text{ m} & 70\text{ m} \leq r_{u3} \leq 100\text{ m} & 70\text{ m} \leq r_{d3} \leq 100\text{ m} \\
 & 15\text{ m} \leq t_{c4} \leq 25\text{ m} & 60\text{ m} \leq r_{u4} \leq 80\text{m} & 60\text{ m} \leq r_{d4} \leq 80\text{ m} \\
 & 20\text{ m} \leq t_{c5} \leq 30\text{ m} & 45\text{ m} \leq r_{u5} \leq 60\text{ m} & 45\text{ m} \leq r_{d5} \leq 60\text{ m} \\
 & 25\text{ m} \leq t_{c6} \leq 35\text{ m} & 30\text{ m} \leq r_{u6} \leq 45\text{ m} & 30\text{ m} \leq r_{d6} \leq 45\text{ m}
\end{array} \quad (16.51)$$

According to the results presented by Varshney (1982), constraints on natural frequency are taken as follows:

$$fr_1 \geq 2.0\,\text{Hz},\quad fr_3 \geq 2.5\,\text{Hz},\quad fr_5 \geq 3.0\,\text{Hz},\quad fr_7 \geq 3.5\,\text{Hz},\quad fr_9 \geq 4.0\,\text{Hz} \tag{16.52}$$

The computational effort of design optimization of the arch dam with frequency constraints is usually high due to the fact that a huge number of eigenproblems should be solved. To reduce the computational effort, RBF, BP, and ANFIS models are trained to predict the required natural frequencies during the optimization process. The errors between exact and approximate frequencies are calculated as follows:

$$\text{err} = \left| \frac{fr_{\text{ap}} - fr_{\text{ex}}}{fr_{\text{ex}}} \right| \times 100 \tag{16.53}$$

where fr_{ap} and fr_{ex} represent the approximate and exact frequencies, respectively.

Inputs and outputs of the employed neural network models are the design variables of the arch dams (X, defined by Eq. (16.50)) and their corresponding natural frequencies, respectively, as follows:

$$I_{\text{NN}} = X \tag{16.54}$$

$$O_{\text{NN}} = \{fr_1\, fr_3\, fr_5\, fr_7\, fr_9\}^{\text{T}} \tag{16.55}$$

where I_{NN} and O_{NN} represent neural network models' input and output vectors, respectively.

In the neural networks' training and testing phases, 200 training pairs are randomly generated, and 160 and 40 samples are used for training and testing the generality of the networks, respectively. As the spent time to each sample analysis is 0.125 min, the total time spent on the data generation phase is equal to 25 min. Using the aforementioned data, RBF and BP neural networks are trained. The spent times to train the RBF and BP are 0.005 and 1.5 min, respectively. The numbers of hidden layer neurons in RBF and BP networks are 160 and 8, respectively. It should be noted that the number of RBF neurons is equal to the number of training samples, but the number of BP neurons are determined by trial and error. In this case, the number of BP neurons is changed and the testing errors are monitored. The best results are observed in the case of eight hidden layer neurons. As the ANFIS model provides only one output, in this work five ANFIS models are trained to predict the desired frequencies. The time spent to training the ANFIS models equals 2.5 min. Training results are given in Table 16.5. It is observed that all the neural network models possess appropriate generalization and can be employed in the optimization process. These results demonstrate also that the best performance is associated with the BP model, while the ANFIS model is better than the RBF model.

Table 16.5 Maximum and Mean Errors of RBF, BP, and ANFIS Models

Metric	Neural Network Model	Error (%)					Average
		fr_1	fr_3	fr_5	fr_7	fr_9	
Mean Error	RBF	2.0999	1.9295	0.6551	2.3570	2.6959	1.9475
	BP	0.9506	0.9194	0.5120	0.8540	1.4166	0.9305
	ANFIS	1.3420	1.1123	0.5756	1.0098	1.8902	1.1860
Maximum Error	RBF	7.0498	5.6769	1.8198	9.9876	8.8495	6.6767
	BP	3.6059	3.5939	1.3714	4.2425	4.7331	3.5094
	ANFIS	5.0897	4.9834	1.4561	5.4509	6.0073	4.5975

In this example, the maximum number of structural analyses is limited to 30,000. Results of optimization by the VSP, PSO, and HS using the FEM, RBF, BP, and ANFIS models for evaluating the natural frequencies during the optimization process are given in Table 16.6. Also, the natural frequencies of the optimum designs evaluated by direct FEM are given in this table. It is clear that all of the presented optimum designs are feasible.

The errors of approximate frequencies of optimum designs predicted by the RBF, BP, and ANFIS models are compared with their corresponding accurate ones in Table 16.7. It can be observed that the accuracy of approximate frequencies obtained by all the methods is good. The results reveal that the solutions found by PSO are better than those of the VSP and HS and the results of HS is also better than that of VSP. In the case of all employed metaheuristics, the best results are obtained using the BP model. Also, the overall time of optimization by a neural network is 0.0071 times that of optimization by FEM. This demonstrates that the overall time of optimization can be significantly reduced using neural networks.

16.7 Conclusions

The natural frequencies are important parameters that affect the dynamic behavior of structures. By imposing some constraints on the natural frequencies, the dynamic behavior of structures may be improved. This aim can be reliably implemented by employing optimization techniques. In this work, the computational performance of the most popular metaheuristics such as GA, PSO, and HS are studied in the framework of constrained frequency-based structural optimization processes.

The numerical simulation problems presented in this chapter include two parts. In the first part, size and shape optimization of truss structures is achieved. These examples include size optimization problems of the well-known 10-bar 2D and 72-bar 3D trusses and size and configuration optimization problems of the well-known 37-bar 2D truss and 52-bar 3D lattice dome structure. Numerical results of optimization demonstrate that in these examples the computational performance of

Table 16.6 Optimum Designs of the Arch Dam Obtained by Several Methods

Design Variable No.	VSP				PSO				HS			
	FEM	RBF	BP	ANFIS	FEM	RBF	BP	ANFIS	FEM	RBF	BP	ANFIS
1	0.2999	0.2858	0.2750	0.2502	0.2999	0.2991	0.2998	0.2999	0.2955	0.299257	0.2406	0.2879
2	0.5219	0.7843	0.5298	0.7172	0.6947	0.5160	0.5155	0.5068	0.6174	0.506097	0.5865	0.7473
3	4.6574	6.0612	6.3963	7.5143	6.1672	5.4086	5.9755	5.5549	6.3139	5.720229	6.7593	5.7972
4	7.0115	7.7083	6.5279	7.6423	6.1673	5.7099	5.9807	5.7125	6.6345	5.835631	6.9235	6.1040
5	13.6145	12.7407	12.5317	11.8680	12.4629	12.8612	12.0880	13.2865	12.1251	13.29861	12.1464	11.9218
6	17.6295	16.5000	17.8409	16.0450	17.0227	17.8769	16.7978	18.3147	16.4266	18.36948	16.2372	16.3714
7	20.1103	20.0413	21.3141	20.1294	20.0000	20.2559	20.0029	20.5241	20.2174	20.66206	20.0461	20.0098
8	34.9258	34.9096	30.4450	34.8157	31.8335	25.4234	26.5262	25.9123	28.4478	25.3271	34.8836	34.1474
9	117.3292	111.1720	113.9780	110.3799	104.4904	114.2867	111.4166	110.3740	115.8557	120.9144	119.1608	118.6377
10	98.6615	101.0291	104.2424	104.0493	100.9819	103.0090	105.7608	103.3122	102.1618	111.2764	106.0466	102.8373
11	83.9818	89.2355	91.0641	90.9459	93.9109	89.4789	94.9474	91.4513	86.7537	89.59603	90.1217	87.0345
12	71.4023	76.6612	75.6576	75.2504	77.5043	73.2069	77.3438	75.0629	71.3829	66.86798	71.0155	70.2838
13	48.2445	48.7862	58.1460	45.9726	45.5906	55.7948	57.7257	56.9693	56.5231	48.16405	53.4024	54.1019
14	41.4219	41.4688	41.6742	42.0064	41.9360	41.8266	38.7295	41.7370	33.6215	37.90583	41.8920	39.1186
15	100.2810	111.1639	103.1934	104.0517	104.0740	113.4527	102.3784	110.2701	111.5385	120.5546	103.8380	115.0908
16	96.9862	100.8533	102.2870	103.9243	100.9819	97.1300	102.3218	99.9164	99.7891	104.5402	103.2140	94.2742
17	83.9565	89.1002	85.9613	90.9164	83.8124	82.4762	84.7974	85.8551	82.7624	84.69364	89.2499	80.2347
18	67.2637	76.5022	72.8285	75.0088	65.6551	69.0673	69.4996	71.5565	67.4016	64.70181	70.8249	66.4524
19	48.2199	48.6498	58.0564	45.9649	45.5901	55.7311	57.6994	56.9532	55.8643	45.92391	53.3852	53.0988
20	31.5909	31.4888	35.9020	31.7827	36.4202	40.6737	38.7172	38.2861	32.6709	35.06673	31.3475	36.1170
Volume (m^3)	242,466.30	247,572.50	244,711.65	245,834.49	229,668.34	234,277.55	231,115.72	232,708.14	239,202.85	240,891.75	240,103.45	240,189.44
Optimization time (min)	3750.0	0.3	0.3	0.3	3750.0	0.3	0.3	0.3	3750.0	0.3	0.3	0.3
Training time (min)	–	25.0	26.5	27.5	–	25.0	26.5	27.5	–	25.0	26.5	27.5
Overall time (min)	3750.0	25.3	26.8	27.8	3750.0	25.3	26.8	27.8	3750.0	25.3	26.8	27.8
fr_1 (Hz)	2.2292	2.2667	2.2822	2.2388	2.2702	2.2642	2.2815	2.2672	2.1702	2.2757	2.1997	2.2393
fr_3 (Hz)	2.5001	2.5793	2.5372	2.5272	2.5710	2.5672	2.5150	2.5308	2.5002	2.5168	2.5017	2.6480
fr_5 (Hz)	3.3055	3.3322	3.2865	3.3149	3.1522	3.2663	3.2227	3.1996	3.2673	3.2701	3.2869	3.3535
fr_7 (Hz)	3.5873	3.6362	3.6801	3.8049	3.5004	3.5435	3.6031	3.5047	3.5753	3.6634	3.7357	3.6309
fr_9 (Hz)	4.0006	4.1729	4.0884	4.1067	4.1422	4.1173	4.0659	4.0960	4.1206	4.0830	4.0658	4.1791

Table 16.7 Error Percentage of Approximate Frequencies of Optimum Dams

Frequency No.	VSP			PSO			HS		
	RBF	**BP**	**ANFIS**	**RBF**	**BP**	**ANFIS**	**RBF**	**BP**	**ANFIS**
1	8.2603	0.3124	0.3444	9.6612	2.4363	9.3871	6.4388	3.2835	4.7865
2	2.8511	2.2988	0.8958	2.1821	1.2751	1.1741	3.1473	2.6797	3.0076
3	0.9941	0.8580	1.0180	4.1756	2.8496	4.5893	3.0365	1.1221	1.5413
4	3.9694	3.7180	5.1795	8.4388	2.8577	5.4063	4.0279	6.2440	6.3305
5	0.5727	2.4738	0.1370	2.8062	0.5843	2.2736	1.8067	1.4092	1.5938
Average	3.3295	1.9322	1.5149	5.4528	2.0006	4.5661	3.6914	2.9477	3.4519

the PSO is better than those of the VSP and HS. Also, it is observed that HS reveals better performance compared with VSP.

In the second part, shape optimization of arch dams is implemented considering fluid–structure interaction. As evaluation of the natural frequencies of such large-scale problem is time-consuming, these parameters are predicted using properly trained BP, RBF, and ANFIS neural network models. Training results demonstrate that all the neural network models possess appropriate generalization and can be employed in the optimization process. These results also reveal that the BP model generally produces the best performance, while the ANFIS is better than RBF.

The optimization results show that the best solution is found by PSO and the results of HS are also better than those of the VSP. In the case of all employed metaheuristics, the best results are obtained using the BP model; however, the solutions found using ANFIS are better than those of RBF. The important point is that overall time of optimization by using neural network models is 0.0071 times that of optimization by FEM. This reveals the high efficiency of the neural networks to significantly reduce the overall time of optimization.

References

Aftabi, A., Lotfi, V., 2007. Linear dynamic analysis of arch dams utilizing modified efficient fluid hyper-element. Eng. Struct. 29 (10), 2654–2661.

Gandomi, A.H., Alavi, A.H., 2012. Krill Herd: a new bio-inspired optimization algorithm. Commun. Nonlin. Sci. Numer. Simul. 17 (12), 4831–4845.

Gandomi, A.H., Yang, X.S., 2011. Benchmark problems in structural optimization. In: Koziel, S., Yang, X.S. (Eds.), Computational Optimization, Methods and Algorithms. Springer, pp. 259–281. Studies in Computational Intelligence 356.

Gandomi, A.H., Yang, X.S., Alavi, A.H., 2011. Mixed variable structural optimization using firefly algorithm. Comput. Struct. 89 (23–24), 2325–2336.

Gandomi, A.H., Yang, X.S., Talatahari, S., Alavi, A.H., 2012a. Metaheuristics in modeling and optimization. In: Gandomi, et al., (Eds.), Metaheuristic Applications in Structures and Infrastructures. Elsevier, Waltham, MA (Chapter 1).

Gandomi, A.H., Yang, X.S., Alavi, A.H., 2012b. Cuckoo search algorithm: a metaheuristic approach to solve structural optimization problems. Eng. Comput. doi: 10.1007/s00366-011-0241-y.

Geem, Z.W., Kim, J.H., Loganathan, G.V., 2001. A new heuristic optimization algorithm: harmony search. Simulations. 76 (2), 60–68.

Gholizadeh, S., Salajegheh, E., 2010. Optimal design of structures for earthquake loading by self-organizing radial basis function neural networks. Adv. Struct. Eng. 13 (2), 339–356.

Gholizadeh, S., Samavati, O.A., 2011. Structural optimization by wavelet transforms and neural networks. Appl. Math. Model. 35 (2), 915–929.

Gholizadeh, S., Seyedpoor, S.M., 2011a. Optimum design of arch dams for frequency limitations. Inter. J. Optimiz. Civil Eng. 1 (1), 1–14.

Gholizadeh, S., Seyedpoor, S.M., 2011b. Shape optimization of arch dams by metaheuristics and neural networks for frequency constraints. Sci. Iran. 18 (5), 1020–1027.

Gholizadeh, S., Salajegheh, E., Torkazadeh, P., 2008. Structural optimization with frequency constraints by genetic algorithm using wavelet radial basis function neural network. J. Sound Vib. 312 (1–2), 316–331.

Gholizadeh, S., Salajegheh, J., Salajegheh, E., 2009. An intelligent neural system for predicting structural response subject to earthquakes. Adv. Eng. Softw. 40 (8), 630–639.

Gomes, H.M., 2011. Truss optimization with dynamic constraints using a particle swarm algorithm. Expert Syst. Appl. 38 (1), 957–968.

Hagan, M.T., Menhaj, M., 1999. Training feed-forward networks with the Marquardt algorithm. IEEE Trans. Neural Netw. 5 (6), 989–993.

Hagan, M.T., Demuth, H.B., Beal, M.H., 1996. Neural Network Design. PWS Publishing Company, Boston, MA.

Jang, J.S.R., 1993. ANFIS: Adaptive-network-based fuzzy inference systems. IEEE Trans. Syst. Man Cybern. 23 (3), 665–685.

Kaveh, A., Talatahari, S., 2010. A novel heuristic optimization method: charged system search. Acta Mech. 213 (3–4), 267–289.

Kaveh, A., Zolghadr, A., 2011. Shape and size optimization of truss structures with frequency constraints using enhanced charged system search algorithm. Asian J. Civil Eng. 12 (4), 487–509.

Kennedy, J., 1977. The particle swarm: social adaptation of knowledge. In: International Conference on Evolutionary Computation. IEEE, Piscataway, NJ, pp. 303–308.

Kennedy, J., Eberhart, R.C., 1995. Particle swarm optimization. In: International Conference on Neural Networks. Perth, Australia, 1942–1945.

Kucukarslan, S., Coskun, B., Taskin, B., 2005. Transient analysis of dam–reservoir interaction including the reservoir bottom effects. J. Fluids Struct. 20 (8), 1073–1084.

Lamberti, L., Pappalettere, C., 2011. Metaheuristic design optimization of skeletal structures: a review. Comput. Technol. Rev. 4 (1), 1–32.

Lee, K.S., Geem, Z.W., 2004. A new structural optimization method based on the harmony search algorithm. Comput. Struct. 82 (9–10), 781–798.

Lee, K.S., Geem, Z.W., 2005. A new meta-heuristic algorithm for continuous engineering optimization: harmony search theory and practice. Comput. Methods Appl. Mech. Eng. 194 (36–38), 3902–3933.

Lin, J.H., Chen, W.Y., Yu, Y.S., 1982. Structural optimization on geometrical configuration and element sizing with static and dynamic constraints. Comput. Struct. 15 (5), 507–515.

Lingyun, W., Mei, Z., Guangming, W., Guang, M., 2005. Truss optimization on shape and sizing with frequency constraints based on genetic algorithm. J. Comput. Mech. 35 (5), 361–368.

Mamdani, E.H., Assilian, S., 1975. An experiment in linguistic synthesis with a fuzzy logic controller. Inter. J. Man-Mach. Stud. 7 (1), 1–13.

Paiva, R.P., Dourado, A., 2001. Structure and parameter learning of neuro-fuzzy systems: a methodology and a comparative study. J. Intell. Fuzzy Syst. 11 (3), 147–161.

Sahab, M.G., Toropov, V.V., Gandomi, A.H., 2012. A review on traditional and modern structural optimization: problems and techniques. In: Gandomi, et al., (Eds.), Metaheuristic Applications in Structures and Infrastructures. Elsevier, Chapter 2.

Salajegheh, E., Gholizadeh, S., 2005. Optimum design of structures by an improved genetic algorithm using neural networks. Adv. Eng. Softw. 36 (11–12), 757–767.

Salajegheh, E., Gholizadeh, S., Torkzadeh, P., 2007. Optimal design of structures with frequency constraints using wavelet back propagation neural network. Asian J. Civil Eng. 8 (1), 97–111.

Sedaghati, R., Suleman, A., Tabarrok, B., 2002. Structural optimization with frequency constraints using finite element force method. AIAA J. 40 (2), 382–388.

Seyedpoor, S.M., Salajegheh, J., Salajegheh, E., Gholizadeh, S., 2009. Optimum shape design of arch dams for earthquake loading using fuzzy inference system and wavelet neural networks. Eng. Optimiz. 41 (5), 473–493.

Sugeno, M., 1985. Industrial Applications of Fuzzy Control. Elsevier Science.

The Language of Technical Computing, 2009. MATLAB. Math Works Inc.

Torkzadeh, P., Salajegheh, J., Salajegheh, E., 2008. Efficient methods for structural optimization with frequency constraints using higher order approximations. Inter. J. Struct. Stability Dyn. 8 (3), 439–450.

Vanderplaats, G.N., 1984. Numerical Optimization Techniques for Engineering Design: with Applications. McGraw-Hill, New York, NY.

Varshney, R.S., 1982. Concrete Dams. Oxford and IBH Publishing, New Delhi.

Wang, D., Zhang, W.H., Jiang, J.S., 2004. Truss optimization on shape and sizing with frequency constraints. AIAA J. 42 (3), 622–630.

Wasserman, P.D., 1993. Advanced Methods in Neural Computing. Prentice Hall, New York, NY.

Yang, X.S., Gandomi, A.M., 2012. Bat algorithm: a novel approach for global engineering optimization. Eng. Comput. 29 (5), 464–483.

Zuo, W., Xu, T., Zhang, H., Xu, T., 2011. Fast structural optimization with frequency constraints by genetic algorithm using adaptive eigenvalue reanalysis methods. Struct. Multi. Optimiz. 43 (6), 799–810.

17 Optimum Performance-Based Seismic Design of Frames Using Metaheuristic Optimization Algorithms

Siamak Talatahari

Marand Faculty of Engineering, University of Tabriz, Tabriz, Iran

17.1 Introduction

Traditionally, the main objective of seismic design is to protect human life with some expectation for structural damage that could be repaired after the earthquake event. The design of structures can be formulated as problems of optimization in which a measure of performance is to be optimized while satisfying all constraints. Determining the practical optimum design of steel frames by reducing the material cost as one of the major factors in the construction of a building is the aim of structural optimization. The problem is often to minimize the weight or volume of the structural system considering the required constraints specified by design codes. Chapter 8 (Talatahari and Kaveh, 2013) reviews the optimization of steel frames; however, in this chapter we will focus on performance-based seismic optimum design of frames. Finding performance-based optimum design of frames is more difficult and therefore choosing an appropriate optimization method is an important issue. Metaheuristic algorithms seem more suitable tools than conventional methods for this kind of problems due to their capability of exploring and finding promising regions in the search space in an affordable time (Kaveh and Talatahari, 2010a). Metaheuristic algorithms tend to perform well for most engineering optimization problems and this is because these methods refrain from simplifying or making assumptions about the original form (Kaveh and Talatahari, 2010b).

This chapter presents the seismic optimum design of steel frames considering four performance levels using the metaheuristic algorithms containing the genetic algorithm (GA), ant colony optimization (ACO), particle swarm optimization (PSO), as well as particle swarm ant colony optimization (PSACO) as an advanced hybrid metaheuristic (Kaveh and Talatahari, 2009a). The

Metaheuristic Applications in Structures and Infrastructures. DOI: http://dx.doi.org/10.1016/B978-0-12-398364-0.00017-6

nonlinear analysis is required to reach to the structural response at various performance levels. Therefore, the refined plastic hinge analysis method is developed to estimate the nonlinear behavior of the entire structural system and members effectively. In the refined plastic hinge analysis method, the geometric nonlinearity of a steel frame structure, the gradual plastification of member sections, and the geometric imperfection of column members are included. Metaheuristics can obtain the optimum sections among the set of candidate solutions. The weight of the structure is considered as the target function, and the candidate solutions are discrete sectional sizes that are selected from a set of W-shape sections. The constraints are the roof drift at various performance levels.

17.2 A Brief Review of Metaheuristic Algorithms

Many of metaheuristic algorithms are simulated from natural processes (Kaveh and Talatahari, 2010c). The most well-known and frequently used evolutionary optimization technique, GAs, inspired from Darwin's natural selection theorem, is based on the idea of the survival of the fittest (Holland, 1975). Differential evolution (DE), developed by Storn and Price (1997), is a vector-based evolutionary algorithm that carries out operations over each component (Gandomi et al., 2012a). PSO is another stochastic optimization method inspired by the social behavior of bird flocking and fish schooling (Kennedy and Eberhart, 1995). A bat-inspired algorithm was recently proposed by Yang (2010a) based on the echolocation of microbats (Altringham, 1996; Richardson, 2008). In the bat algorithm, the echolocation characteristics of microbats are idealized to establish an optimization algorithm (Gandomi et al., 2012b). ACO is a cooperative search technique that mimics the foraging behavior of real-life ant colonies (Dorigo, 1992). The artificial bee colony algorithm, developed by Karaboga (2005), is based on the foraging behavior of honey bees. Krill herd, a novel biologically inspired algorithm proposed by Gandomi and Alavi (2012), is based on simulating the herding behavior of krill individuals. The firefly algorithm (FA) mimics the social behavior of fireflies. Fireflies communicate, search for prey, and find mates using bioluminescence with varied flashing patterns. By mimicking the nature of their behavior, some of the flashing characteristics of fireflies as the base of the FA were idealized (Yang, 2010b). Harmony search algorithm is another metaheuristic algorithm that is based on musical performance processes that occur when a musician searches for a better state of harmony, such as during jazz improvization (Geem and Kim, 2001). The charged system search is based on the Coulomb and Gauss laws from electrical physics and the governing laws of motion from the Newtonian mechanics (Kaveh and Talatahari, 2010d). Cuckoo search (CS) is one of the latest nature-inspired metaheuristic algorithms, developed by Yang and Deb (2009) and Gandomi et al., 2012c.

17.3 Statement of Seismic Design of Frames

The most popular optimization objective for design of steel frames is the cost, which is a function of the weight, handling, fabrication, erection, leakage of resistance, maintenance, and the type of the connections (Kaveh et al., 2010). Unfortunately, a complete description of the real cost of a building before its construction is not often possible since accurate cost data require information from many factors that are unpredictable and not precisely defined. The mathematical formulation of the structural optimization problems can be expressed as minimizing the weight of structures as the cost function without taking into consideration other influencing tributary parameters:

$$\text{Minimize: } W(X) = \sum_{j=1}^{\text{ne}} \rho\, L_j \cdot A_j \tag{17.1}$$

where $W(X)$ is the weight of the structure, X is the vector of design variables taken from W-shaped sections found in the AISC design manual (2001), ne is the number of members, ρ is the material mass density, L_j and A_j are the length and the cross-sectional area of the member j, respectively.

The economical structure must have not only the least weight but should also be strong enough to resist earthquake loads to prevent causing damage in the structures. Therefore, the optimum design should also satisfy the design constraints. Lateral deflections of a building may cause human discomfort and minor damage of nonstructural components. Extreme inelastic lateral deflections due to a severe earthquake can cause the failure of mechanical, electrical, and plumbing systems, or suspended ceilings and equipment to fall, thereby posing threats to human life. Such loading can also increase the possibility of building instability, thereby reducing the structural safety. As a result, it is essential to control the lateral drift of building frameworks under seismic loading. The lateral drift constraints at various performance levels can be expressed as (Kaveh et al., 2010):

$$\text{OP Level } \Delta^{\text{OP}}(X) \le \overline{\Delta}^{\text{OP}} \tag{17.2}$$

$$\text{IO Level } \Delta^{\text{IO}}(X) \le \overline{\Delta}^{\text{IO}} \tag{17.3}$$

$$\text{LS Level } \Delta^{\text{LS}}(X) \le \overline{\Delta}^{\text{LS}} \tag{17.4}$$

$$\text{CP Level } \Delta^{\text{CP}}(X) \le \overline{\Delta}^{\text{CP}} \tag{17.5}$$

where Δ^{level} is the lateral drift and $\overline{\Delta}^{\text{level}}$ is the allowable lateral drift. Here, OP, IO, LS, and CP are the different performance levels that are described as follows.

Each building's performance level is made up of a structural performance level that describes the limiting damage state of the structural systems and a nonstructural performance level that describes the limiting damage state of the nonstructural systems. Three structural performance levels and four nonstructural performance levels can be used to form the four basic building performance levels. Thus, a number of building performance levels (or particular damage states) defined in the literature are as operational (OP), immediate occupancy (IO), life safety (LS), and collapse prevention (CP) (FEMA-273, 1997). The operational level is that at which a building has sustained minimal or no damage to its structural and nonstructural components, and the building is suitable for normal occupancy or use; a building at the immediate occupancy level has sustained minimal or no damage to its structural elements and only minor damage to its nonstructural components and is safe to be reoccupied immediately; a building at the life safety level has experienced extensive damage to its structural and nonstructural components and, while the risk to life is low, repairs may be required before re-occupancy can occur; the collapse prevention level is when a building has reached a state of impending partial or total collapse, where the building may have suffered a significant loss of lateral strength and stiffness with some permanent lateral deformation, but the major components of the gravity load-carrying system should still continue to carry gravity load demands.

It is common to adopt the target displacement as the maximum allowable roof drift for a specified performance level, and in this study, roof drifts of 0.4%, 0.7%, 2.5%, and 5% of the height of the building, are taken as the allowable roof drifts for the OP, IO, LS, and CP performance levels in the design optimization process, respectively.

Four acceleration design spectra, which represent four different earthquake levels corresponding to 50%, 20%, 10%, and 2% probability of exceeding in a 50-year period, are taken as the basis for calculating the (equivalent static) seismic loading for the four performance levels OP, IO, LS, and CP, respectively. Without loss of generality, the calculation of spectral acceleration S_a^i for each design spectra $i$can be expressed as (FEMA-273, 1997):

$$S_a^i = \begin{cases} F_a S_s^i (0.4 + 3T/T_0) & \text{if} \quad 0 < T \leq 0.2T_0^i \\ F_a S_s^i & \text{if} \quad 0.2T_0^i < T \leq T_0^i \\ F_v S_1^i / T & \text{if} \quad T > T_0^i \end{cases} \quad (i = \text{OP, IO, LS, CP}) \tag{17.6}$$

where T is the elastic fundamental period of the structure, which is computed from structural analysis at elastic range:

$$T = \frac{2\pi}{\omega} = 2\pi \left[\frac{\sum_{s=1}^{\text{ns}} m_s \nu_s^2}{g \sum_{s=1}^{\text{ns}} m_s \nu_s} \right]^{1/2} \tag{17.7}$$

where m_s,ν_s are mass and lateral drift of story s at elastic range, respectively, and T_0^i is the period at which the constant acceleration and constant velocity regions of the response spectrum intersect.

$$T_0^i = \frac{F_v S_1^i}{F_a S_s^i} \tag{17.8}$$

where S_s^i, S_1^i correspond to the short-period and the 1 s period response acceleration parameters, respectively, and F_a,F_v are the site coefficient determined respectively from tables 2.13 and 2.14 in FEMA-273 (1997).

Having S_a^i from Eq. (17.6), the corresponding base shear V_b for the four performance levels is then found as:

$$V_b^i = W_s \cdot S_a^i / g \ \ (i = \text{OP, IO, LS, CP}) \tag{17.9}$$

Then, the load pattern is defined as follows:

$$P_s = V_b \left(G_s H_s^k / \sum_{m=1}^{ns} G_m H_m^k \right) \quad (s = 1, 2, \ldots..., \text{ns}) \tag{17.10}$$

where W_S is the seismic weight of the structure, g is the gravitational acceleration, P_s is the lateral load applied at story level s, H_s and H_m are the heights from the base of the building to story level s and m, respectively, G_s and G_m are the seismic weights for story s and m, respectively, and k is the exponent whose value depends on the period of the building (here, $k = 2$).

17.4 Pushover Analysis for Performance-Based Design

With the emergence of the performance-based design methodology, there is a need to develop the corresponding analytical methods (Kaveh et al., 2010). The expectation is that the pushover analysis will provide adequate information on seismic demands imposed by the design ground motion on the structural system and its components. The static pushover analysis is known as a practical approach due to the simplicity and ability to estimate component and system deformation demands with acceptable accuracy without the intensive computational and modeling effort of a dynamic analysis. The static pushover analysis was first presented by Saiidi and Sozen (1981).

There are various methods of static pushover analyses to predict the seismic demands on building frameworks under equivalent static earthquake loading (Biddah and Naumoski, 1995; Bracci et al., 1997; Ferhi and Truman, 1996a, 1996b; Kilar and Fajfar, 1997; Lawson et al., 1994; Moghadam and Tso, 1995). However, this study developed a computer-based pushover analysis procedure that is adopted; it was originally conceived for the elastic analysis of steel frameworks with semi-rigid connections (Xu, 1992 ; Xu and Grierson, 1993).

Many of the previous studies suggest obtaining the main characteristics of the seismic behavior with a nonlinear static analysis under monotonically increasing loads. The suggested pushover analysis in this paper is inspired by second-order inelastic analysis of semi-rigid framed structures that rigidity-factor is replaced with plasticity factor in stiffness matrix. According to this approach, fictitious plastic-hinge connections are introduced at the two ends of beam-column elements and semi-rigid analysis techniques were modified for the nonlinear load-deformation analysis of building frameworks under increasing seismic loads. Here, the elastic stiffness matrix is comprised of both the first-order and the second-order geometric properties

$$K = S_e C_e + S_g C_g \tag{17.11}$$

The matrix K consists of two parts; the first part is conceived from the method of Monfortoon and Wu (1963) that employs the rigidity-factor concept to develop a first-order elastic analysis technique for semi-rigid frames (i.e., $S_e \times C_e$), and the second part is conceived from the method of Xu (1992) that considers the rigidity-factor concept to develop a second-order elastic analysis technique for semi-rigid frames (i.e., $S_g \times C_g$). Here S_e and S_g are the standard first-order elastic and the second-order geometric stiffness matrices, respectively, when the member has rigid moment-connections; C_e and C_g are the corresponding correction matrices that account for the reduced rotational stiffness of the semi-rigid moment-connections.

The displacement-control pushover analysis procedure is not suitable for the performance-based seismic design because the displacement of the roof is considered as the constraint for the optimization problem; therefore, structural designs are based on the loads specified by governing codes. When the magnitude of an earthquake loading is determined from the corresponding acceleration spectrum provided by a specific design code (FEMA 273, 1997), a load-control pushover analysis procedure is utilized to evaluate the seismic demand at a specified earthquake loading level in which monotonically increasing lateral loads along with constant gravity loads are applied to a framework until a control node (referred to the building roof) sways to a predefined "target" lateral displacement, or to a "target" base shear, which corresponds to performance levels that were described in Section 17.3. In each loading level, the structural stiffness matrix (Eq. (17.11)) is updated using plasticity factor values at the two ends of each element, $p_i(i = 1,2)$, and then the analysis is repeated for the next load increment. The value of plasticity factor p is conceived from the rigidity factor used in semi-rigid analysis. This factor r_i defines the rotational stiffness of the connection and can be interpreted as the ratio of the end-rotation α_i of the member to the combined rotation θ_i of the member as:

$$r_i = \frac{\alpha_i}{\theta_i} = \frac{1}{1 + (3\mathrm{EI}/RL)} \quad (i = 1, 2) \tag{17.12}$$

where R is the rotational stiffness of connection i and EIand $L$are the bending stiffness and length of the connected member, respectively. In fact, upon replacing connection rotational stiffness R with section postelastic flexural stiffness in Eq. (17.12), the degradation of the flexural stiffness of a member section experiencing postelastic behavior can be characterized by the plasticity factor:

$$p = \frac{1}{1 + (3\mathrm{EI}/R^p L)} \tag{17.13}$$

where $R^p = \mathrm{d}M/\mathrm{d}\phi$ is the section postelastic flexural stiffness and p is the plasticity factor which varies in the range $0 \le p \le 1$ as the postelastic flexural stiffness varies between that for an ideal elastic section, ($\mathrm{d}M/\mathrm{d}\phi = \infty$), and fully plastic section, ($\mathrm{d}M/\mathrm{d}\phi = 0$), respectively (see Figure 17.1). Besides, the factor p can be used to estimate the corresponding percentage extent of section plastic behavior through the following expression:

$$\%\text{plasticity} = 100(1 - p) \tag{17.14}$$

For utilization of the variation in postelastic flexural of plastic hinge section under increasing moment, it is necessary to adopt a nonlinear moment-rotation ($M - \phi$) relationship as:

$$M(\phi) = M_y + \sqrt{(M_p - M_y)^2 + [(M_p - M_y)(\phi_p - \phi)/\phi]^2} \tag{17.15}$$

where $M_y = S\sigma_{ye}$ and $M_p = Z\sigma_{ye}$ are the known first-yield and fully plastic moment capacities of the member section, respectively (S and Z are the elastic and plastic section module, respectively, and σ_{ye} is the expected yield stress of the material) and ϕ is the extent of postelastic rotation occurring somewhere between first yielding ($\phi = 0$) and full plastification ($\phi = \phi_p$) of the cross section. From Eq. (17.15), the postelastic moment varies in the range $M_y \le M(\phi) \le M_p$ as the plastic rotation varies in the range $0 \le \phi \le \phi_p$, as shown in Figure 17.1.

From Figure 17.1, for moment levels less than M_y, no plastic rotation occurs and the fictitious connection has a flexural stiffness $\mathrm{d}M/\mathrm{d}\phi = \infty$. Conversely, beyond the point where the moment level reaches M_p, the change in postelastic moment $\mathrm{d}M = 0$ and the postelastic section flexural stiffness $\mathrm{d}M/\mathrm{d}\phi = 0$ (zero stiffness implies that the member section has fully formed a plastic hinge and freely allows rotational discontinuity). That is, the postelastic flexural stiffness R^p varies

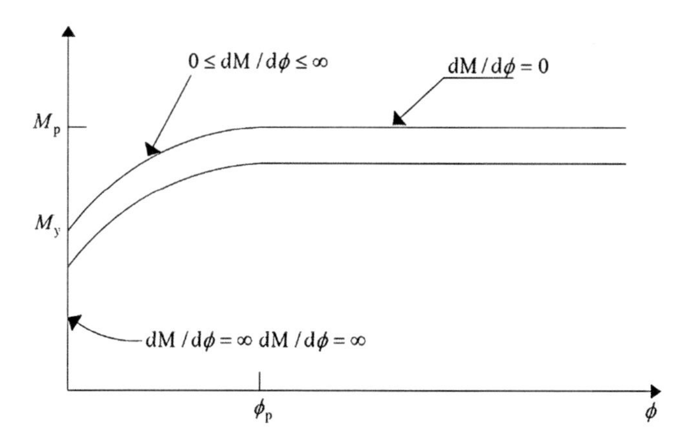

Figure 17.1 Postelastic moment-curvature relation.

in the range $0 \leq dM/d\phi \leq \infty$ as the cross-sectional behavior progresses from initial yielding at the M_y moment level ($\phi = 0$) to full plastification at the M_p moment level ($\phi = \phi_p$), and upon differentiating Eq. (17.15) with respect to ϕ, R^p is defined by the following function:

$$R^p = \frac{dM(\phi)}{d\phi} = \frac{(M_p - M_y)^2(\phi_p - \phi)}{\phi_p^2\sqrt{(M_p - M_y)^2 - [(M_p - M_y)(\phi_p - \phi)/\phi_p]^2}} \quad (\phi \leq \phi_p) \qquad (17.16)$$

According to Figure 17.1, if $M \leq M_y$, then $\phi = 0$; otherwise, if $M > M_y$, then Eq. (17.15) is solved for $M(\phi) = M$ to find the corresponding value of the postelastic curvature for the section as:

$$\phi = \phi_p \left[1 - \sqrt{1 - \left(\frac{M/M_y - 1}{f_s - 1} \right)^2} \right] \quad (\phi \leq \phi_p) \qquad (17.17)$$

Since the postelastic analysis procedure must be extended to account for combined stress states (combined bending moment M and axial force N for members of planner frameworks), a reduction in the moment capacity of a member cross section is incorporated through the interaction constraint equation (Figure 17.1). The value of ϕ is substituted into Eq. (17.16) to update the postelastic flexural stiffness $dM/d\phi$ of the member section to its value prevailing at the beginning of the next load step, then the updated $dM/d\phi$ value is used to revise the section plasticity factor p at the two end sections of each member through Eq. (17.12). Then the structure stiffness matrix is updated and the analysis procedure is repeated for the next load increment. Load increments extend until the structure base shear reaches to a specific performance level, as shown in Figure 17.2.

17.5 Utilized Metaheuristic Algorithms

A review of utilized metaheuristic algorithms including GA, ACO, PSO, and heuristic PSACO is presented in the following subsections.

17.5.1 Genetic Algorithms

The GA developed by Holland (1975) is a stochastic search method that mimics natural evolution and is based on the concept of "survival of the fittest." The GA is inspired by the principles of genetics and evolution and mimics the reproduction behavior observed in biological populations. The GA has an advantage over traditional optimization techniques since it does not require derivatives of the objective function and can hence be applied to solve complex and discontinuous optimization problems.

The GA is inspired by biological evolution, cross-breeding, and trial solutions and allows only the best solutions to survive and propagate successive generations.

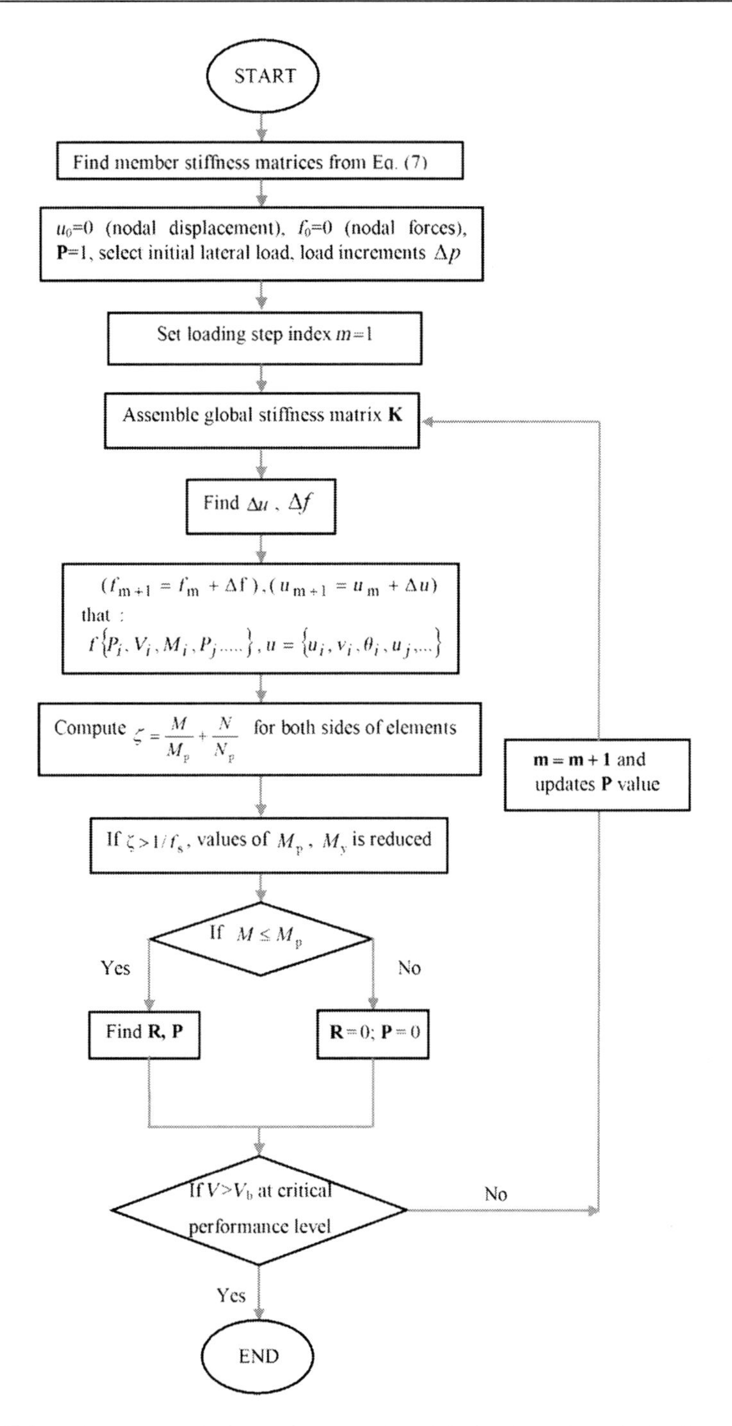

Figure 17.2 Pushover analysis for performance-based design (Kaveh et al., 2010).

It deals with a population of individual (or chromosome) solutions, which undergo constant changes by means of genetic operations of reproduction, crossover, and mutation. GA begins its search from a randomly generated population of designs that evolve over successive generations (iterations), eliminating the need for a user-supplied starting point. The solutions are ranked according to their fitness with respect to the objective function where the fitted individuals are more likely to reproduce and propagate the next generation. Based on the value of the objective function, individuals (parents) are selected for reproduction of the next generation by exchanging genetic information to form children (crossover). The parents are then removed and replaced in the population by the children to keep a stable population size. The result is a new generation with (normally) better fitness. Occasionally, mutation is introduced into the population to prevent the convergence to a local optimum and help generate unexpected directions in the parameter space.

The more GA iterates, the better the chance would be to generate an optimal solution. After a number of generations, the population is expected to evolve artificially, and the (near) optimal solution will be reached. The measure of success is the convergence to a population with identical members. In this study, the population size is set to 50; the crossover and migration fraction are taken as 0.8 and 0.2, respectively.

In recent years, several studies regarding the application of the second-order analysis methods to the optimum design of steel frame structures by using of GAs have been conducted. Pezeshk et al. (2000) utilized GAs for the geometrically nonlinear optimal design of plane steel frames. The results were compared to those obtained by linear elastic analysis and the AISC-LRFD elastic analysis methods. Finding the optimum design of plane steel frames with semi-rigid connections using a GA and a geometrically nonlinear analysis was performed by Saka and Kameshki (1997). An automated design algorithm for partially restrained and fully restrained steel frames was implemented using an object-oriented evolutionary algorithm and the distributed plasticity (plastic zone) model for consideration of both member material nonlinearity as well as connection nonlinearity in the frame analysis by Foley et al. (2004). Xu et al. (2006) introduced a multicriteria optimization method for performance-based seismic design of steel building frameworks under equivalent static seismic loading.

17.5.2 Ant Colony Optimization

ACO was introduced as a novel nature-inspired method for the solution of difficult combinatorial optimization problems (Dorigo, 1992). The inspiring source of ACO is the foraging behavior of real ants (Kaveh et al., 2008). When searching for food, ants initially explore the area surrounding their nests in a random manner. As soon as an ant finds a food source, it evaluates it and carries some food back to the nest. During the return trip, the ant deposits a pheromone trail on the ground. The pheromone deposited, the amount of which may depend on the quantity and quality of the food, guides other ants to the food source. This indirect communication among ants via pheromone trails enables them to find the shortest paths between their nest and food sources. This capability of real ant colonies has inspired the definition of artificial ant colonies that can find approximate solutions to optimization problems.

The main characteristic of the ACO is that after each iteration, the pheromone values are updated (global updating rule). The pheromone associated with the edge joining cities i and j is updated by the following relationship:

$$\tau_{ij}(t+n) = (1-\rho).\tau_{ij}(t) + \rho.\Delta\tau_{ij}(t) \tag{17.18}$$

where $\rho \in (0,1)$ is the pheromone decay coefficient and $\Delta\tau_{ij}(t)$ is computed for the best ant as follows:

$$\Delta\tau_{ij}(t) = 1/L^{+} \tag{17.19}$$

where L^{+} is the length of navigated tour for the best ant.

According to the local updating rule, after each selection, the pheromone value related to the selected path is updated by the local updating rule as:

$$\tau_{ij}(t+1) = \xi.\tau_{ij}(t) \tag{17.20}$$

where ξ is an adjustable parameter between 0 and 1, representing the persistence of the trail.

In the construction of a solution, ants select the following city to be visited through a stochastic mechanism. When ant m is in city i, the probability of going to city j is given by:

$$p_{ij}^{m} = \begin{cases} \dfrac{[\tau_{ij}]^{\alpha}[\eta_{ij}]^{\beta}}{\sum_{j \in N_i}[\tau_{ij}]^{\alpha}[\eta_{ij}]^{\beta}} & \text{if} \quad j \in N_i \\ 0 & \text{otherwise} \end{cases} \tag{17.21}$$

where N_i is the set all the neighbor nodes of node i, α and β are two parameters that control the relative weight of the pheromone trail versus the heuristic information, η_{ij} is called visibility for cross-sectional area j, in member i, is equal to the reverse of the considering cross-sectional area:

$$\eta_{ij} = \frac{1}{A_{ij}} \tag{17.22}$$

where A_{ij} is the jth cross-sectional area of member i.

ACO has widely been applied to various engineering problems in the recent years such as water distribution system optimization (Maier et al., 2003), optimal soil hydraulic parameters (Abbaspour et al., 2001), optimum design of channels (Nourani et al., 2009), and structural optimization problems (Kaveh et al., 2008) among many others. In addition, Kaveh and Jahanshahi (2008) utilized ant colony systems to optimize the process of finding the collapse load factor of frame structures. Performance-based design of steel frames was also performed by Kaveh et al. (2010).

17.5.3 Particle Swarm Optimization

The PSO algorithm, inspired by social behavior simulation, was initially proposed by Kennedy and Eberhart, 1995. The PSO is a population-based optimization algorithm that involves a number of particles that are randomly initialized in the search space. The particles move through the search space and their positions are updated based on the best positions of individual particles (called pbest) and the best of the swarm (called gbest) in each iteration. This matter is shown mathematically as the following equations:

$$v_i^d(k+1) = w \times v_i^d(k) + c_1 \times \text{rand}_{1i}^d \times (\text{pbest}_i^d(k) - x_i^d) + c_2 \times \text{rand}_{2i}^d \times (\text{gbest}^d(k) - x_i^d)) \tag{17.23}$$

$$x_i^d(k+1) = x_i^d(k) + v_i^d(k+1) \tag{17.24}$$

where x_i and v_i represent the current position and the velocity of the ith particle, respectively, the superscript d stands for dth design variable, rand_{1i}^d and rand_{2i}^d represent random numbers between 0 and 1, $\text{pbest}_i^d(k)$ is the best position visited by each particle itself, $\text{gbest}^d(k)$ corresponds to the global best position in the swarm up to iteration k, and c_1 and c_2 represent cognitive and social parameters, respectively. A larger inertia weight makes the global exploration easier while a smaller inertia weight tends to facilitate local exploitation. Similar to the other studies (Talatahari et al., 2012), here the inertia weight is reduced linearly from 0.9 to 0.4 during the optimization process.

17.5.4 PSACO Algorithm

The PSACO, a hybridized approach based on PSO and ACO, is described in this section. PSACO utilizes a PSO algorithm as a global search, and the ACO approach worked as a local search. PSACO was utilized to solve different civil engineering examples such as a benchmark engineering problem (Kaveh and Talatahari, 2009b), truss design with continuous variables (Kaveh and Talatahari, 2009a), truss design with discrete variables (Kaveh and Talatahari, 2009c), frame optimization (Kaveh and Talatahari, 2008; Kaveh and Talatahari, 2009d), as well as many others.

Compared to other evolutionary algorithms based on heuristics, the advantages of PSO consist of easy implementation and a smaller number of parameters to be adjusted. However, it is known that the original PSO had difficulties in controlling the balance between exploration (global investigation of the search place) and exploitation (the fine search around a local optimum). In order to improve this character of PSO, one method is to hybridize PSO with other approaches such as ACO.

The implementation of PSACO algorithm consists of two stages. In the first stage, it applies PSO, while ACO is implemented in the second stage. ACO works as a local search, wherein, ants apply pheromone-guided mechanism to refine the positions found by particles in the PSO stage. In PSACO, a simple pheromone-guided mechanism of ACO is proposed to be applied for the local search. The proposed ACO algorithm handles M ants equal to the number of particles in PSO.

In the ACO stage, each ant generates a solution around gbest_i^d which can be expressed as:

$$z_i^d = N(\text{gbest}_i^d, \sigma_i) \tag{17.25}$$

In the above equation, $N(\text{gbest}_i^d, \sigma)$ denotes a random number normally distributed with mean value gbest_i^d and variance σ_i, where:

$$\sigma_i = (x_{i,\max} - x_{i,\min}) \times \eta \tag{17.26}$$

η is used to control the step size. The normal distribution with mean gbest_i^d can be considered as a continuous pheromone that has the maximum value in gbest_i^d and decreases with going away from it. In ACO algorithms, the probability of selecting a path with more pheromone is greater than other paths. Similarly, in the normal distribution, the probability of selecting a solution in the neighborhood of gbest_i^d is greater than the others. This principle is used in the PSACO algorithm as a helping factor to guide the exploration and to increase the controlling in exploitation.

In the present method, the objective function value of the new solution in the ACO stage is computed and if it is better, the current position of ant i, is replaced by the current position of particle i in the swarm.

17.6 Design Examples

Two building frameworks are selected for seismic optimum design using the meta-heuristic algorithm (Kaveh et al., 2010). These frames have previously been used to illustrate the pushover analysis technique by Hasan et al. (2002).

The expected yield strength of steel material used for column members is $\sigma_{\text{ye}} = 397$ MPa, while $\sigma_{\text{ye}} = 339$ MPa is considered for beam members. The properties of the different W-shaped sections are available in the manual of the AISC (2001), including the elastic modulus E and the area A, moment of inertia I, elastic modulus S, and plastic modulus Z of the cross section. The constant gravity load w is accounted for a tributary-area width of 4.57 m and dead-load and live-load factors of 1.2 and 1.6, respectively. The values for the site coefficients F_a,F_v is presented in Table 17.1, which are obtained based on the site class and the values of the response acceleration parameters S_s^i and S_1^i.

For pushover analysis, the postelastic curvature increment beyond the first-yield curvature ϕ_{y} when plasticity first penetrates through the full depth of a member cross section is taken to be $\phi_{\text{p}} = 0.09$ radians, while the value of the ultimate

Table 17.1 Performance Level Site Parameters

Frame	Site Class	Performance Level	Earthquake Level	$S_s(g)$	$S_l(g)$	F_a	F_v
Three-story	D	OP	50%/50	0.365	0.11	1.50	2.36
		IO	20%/50	0.658	0.198	1.27	2.00
		LS	10%/50	0.794	0.237	1.18	1.92
		CP	2%/50	1.15	0.346	1.04	1.70
Nine-story	D	OP	50%/50	0.109	0.035	1.60	2.40
		IO	20%/50	0.180	0.058	1.60	2.40
		LS	10%/50	0.250	0.080	1.60	2.40
		CP	2%/50	1.100	0.410	1.06	1.59

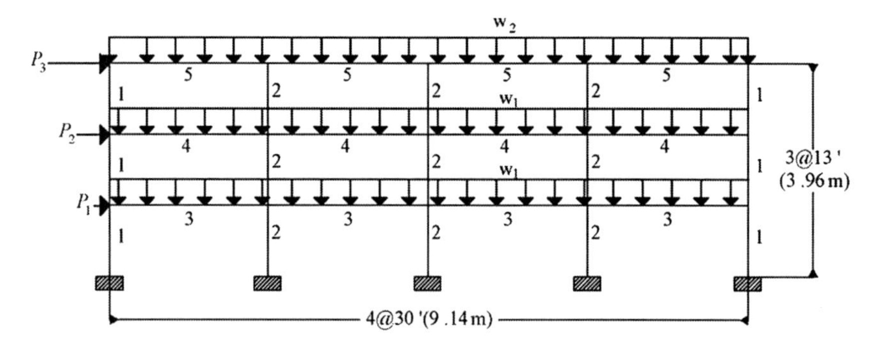

Figure 17.3 Three-story steel moment-frame.

postelastic curvature increment ϕ_u is set arbitrarily large so that member sections do not experience abrupt local failure.

17.6.1 Four-Bay Three-Story Steel Frame

The configuration, grouping of the members, and applied loads of the four-bay three-story framed structure are shown in Figure 17.3 (Kaveh et al., 2010). The 27 members of the structure are categorized into five groups, as indicated in the figure. The modulus of elasticity is taken as $E = 200$ GPa. The constant gravity load of $w_1 = 32$ kN/m is applied to the first- and second-story beams, while the gravity load of $w_2 = 28.7$ kN/m is applied to the roof beams. The seismic weight is 4688 kN for each of the first and second stories, and 5071 kN for the roof story.

The performance-based optimum results for the metaheuristic algorithm are summarized in Table 17.2. The PSACO, ACO, and GA need 4500, 3900, and 6800 analyses to reach a convergence, while PSO needs 8500 analyses. The best PSACO design results in a frame that weighs 279.2 kN, which is 8.8%, 2.5%, and 1.5%

Table 17.2 The Statistical Information of Performance-Based Optimum Designs for the Four-Bay Three-Story Frame

Algorithm	PSACO	PSO	ACO (Kaveh et al., 2010)	GA (Kaveh et al., 2010)	A Conventional Design (Hassan et al., 2005)
Best weight (kN)	279.2	286.3	283.4	303.9	412.9
Average weight (kN)	290.4	302.4	294.3	321.5	–
Worst weight (kN)	298.5	310.7	303.2	339.7	–
Std. dev. (kN)	6.453	10.453	7.566	14.332	–
Average number of analyses	4,500	8,500	3,900	6,800	–

lighter than the design of GA, PSO, and ACO, respectively, while the result of conventional design (Hassan et al., 2005) is approximately 40% more than the result of metaheuristic algorithms. In a series of 20 different design runs, the average weight of the PSACO designs is 290.4 kN, with a standard deviation of 6.45 kN, while the average weight of the PSO and ACO designs is 302.4 and 294.3 kN, respectively. The standard deviation values are 10.45 and 7.56 kN for the PSO and ACO, respectively.

17.6.2 Five-Bay Nine-Story Steel Frame

A five-bay nine-story steel frame is considered as shown in Figure 17.4. The dimensions, the applied loading system, and the member groupings are also provided in this figure. The material has a modulus of elasticity equal to $E = 200$ GPa. The 108 members of the structure are categorized into 15 groups, as indicated in the Figure 17.4. The constant gravity load of $w_1 = 32$ kN/m is applied to the beams in the first to the eighth story, while $w_2 = 28.7$ kN/m is applied to the roof beams. The seismic weights are 4942 kN for the first story, 4857 kN for each of the second to eighth stories, and 5231 kN for the roof story. In this example, each of the five beam element groups is chosen from all 267 W-shapes, while the eight column element groups are limited to W14 sections (37 W-shapes).

Table 17.3 presents the statistical results obtained by the metaheuristic algorithms. The best PSACO design results in a frame weighing 1601.32 N which is 5.1%, 1.9%, and 7.6% lighter than the PSO, ACO, and GA. In order to converge to a solution for the PSACO algorithm, approximately 6000 frame analyses are required, which are less than the 12,500 and 9700 analyses necessary for the PSO and GA. The ACO needs only 5600 analyses to find an optimum result.

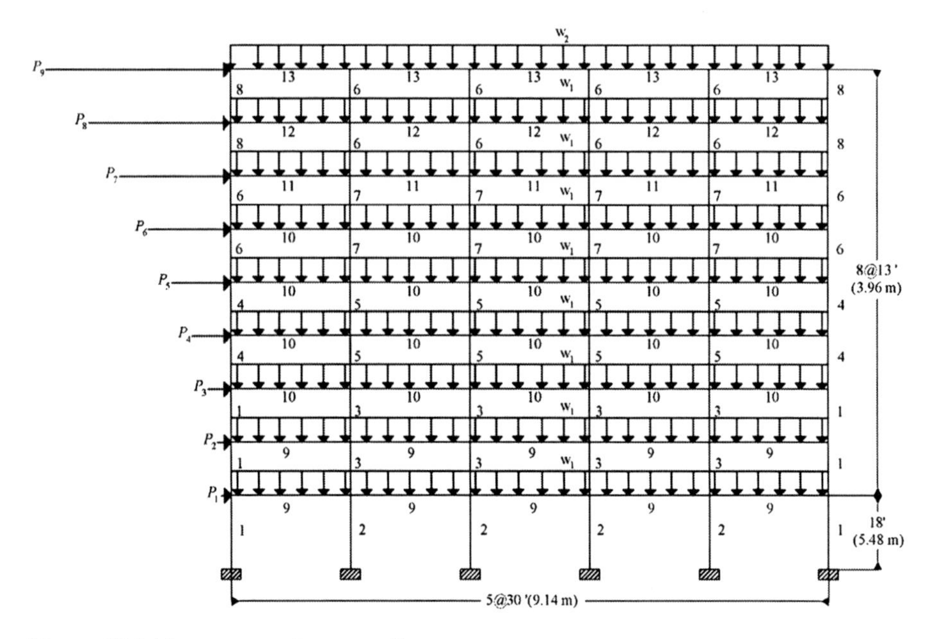

Figure 17.4 Nine-story steel moment-frame.

Table 17.3 The Statistical Information of Performance-Based Optimum Designs for the 5-bay 9-Story Frame

Algorithm	PSACO	PSO	ACO (Kaveh et al., 2010)	GA (Kaveh et al., 2010)
Best weight (kN)	1601.32	1682.63	1631.83	1723.1
Average weight (kN)	1650.55	1725.36	1696.2	1791.4
Worst weight (kN)	1759.65	1813.25	1786.94	1943.2
Std. dev. (kN)	38.52	66.35	49.33	78.33
Average number of analyses	6000	12,500	5600	9700

17.7 Concluding Remarks

The main objective of conventional seismic design of structures is to protect human life with some expectation for structural damage that could be repaired after the earthquake event. On the other hand, the construction cost is one of the major factors. Therefore, we have two objective functions: structural costs and structural damages. In this chapter, metaheuristic optimization methods are used to find optimum seismic designs of steel frames.

The problem is formulated to minimize the weight of the structure considering the required constraints specified by design codes. Therefore, considering multiple performance levels and optimum design of structures to reduce both the structural cost and the ductility demand becomes necessary. As a result, it is essential to control the lateral drift of building frameworks under seismic loading at various performance levels.

To find lateral drift of a building at various performance levels, a nonlinear analysis must be done. To fulfill this aim, a simple computer-based method for pushover analysis of steel building frameworks subject to equivalent-static earthquake loading is utilized. The method accounts for first-order elastic and second-order geometric stiffness properties, and the influence that combined stresses have on plastic behavior, and employs a conventional elastic analysis procedure modified by a plasticity factor to trace elastic–plastic behavior over the range of performance levels for a structure (Kaveh et al., 2010). The plasticity factor is shown analogous to a similar rigidity factor for elastic analysis of semi-rigid frames, and the stiffness properties for semi-rigid analysis are directly adopted for pushover analysis.

Two examples are considered to investigate the capability of metaheuristic algorithms. The GA, ACO, PSO, as well as an advanced hybrid method (i.e., PSACO) are utilized to find the optimum seismic design of examples. Compared to conventional design, all of the selected algorithms can find a better result with approximately 40% lighter designs. Performing a vast exploring of the search space and testing different designs during the optimization process improves the results of the metaheuristic algorithms. The obtained results indicate that the PSACO algorithm compared to GA, ACO, and PSO can find a better optimum seismic design of structures.

References

Abbaspour, K.C., Schulin, R., van Genuchten, M.T.H., 2001. Estimating unsaturated soil hydraulic parameters using ant colony optimization. Adv. Water Res. 24, 827–841.

Altringham, J.D., 1996. Bats: Biology and Behavior. Oxford Univesity Press, Oxford.

American Institute of Steel Construction (AISC), 2001. Manual of Steel Construction: Load and Resistance Factor Design. American Institute of Steel Construction (AISC) 2001, Chicago, IL.

Biddah, A.C., Naumoski, N., 1995. Use of pushover test to evaluate damage of reinforced concrete frame structures subjected to strong seismic ground motions. Proceedings of the Seventh Canadian Conference on Earthquake Engineering. Montreal, Canada.

Bracci, J.M., Kunnath, S.K., Reinhorn, A.M., 1997. Seismic performance and retrofit evaluation of reinforced concrete structures. J. Struct. Eng. 123 (1), 3–10.

Dorigo, M., 1992. Optimization, Learning and Natural Algorithms (in Italian). Dipartimento di Elettronica, Politecnico di Milano, IT (Ph.D. Thesis).

Federal Emergency Management Agency, FEMA-273, 1997. NEHRP Guideline for the Seismic Rehabilitation of Buildings. Building Seismic Safety Council, Washington, DC.

Ferhi, A., Truman, K.Z., 1996a. Behaviour of asymmetric building systems under a monotonic load—I. Eng. Struct. 18 (2), 133–141.

Ferhi, A., Truman, K.Z., 1996b. Behaviour of asymmetric building systems under a monotonic load—II. Eng. Struct. 18 (2), 142–153.

Foley, C.M., Schinler, D., Voss, M.S., 2004. Optimized design of fully and partially restrained steel frames using advanced analysis and object-oriented evolutionary computation. Technical Report to the National Science Foundation. CMS9813216.

Gandomi, A.H., Alavi, A.H., 2012. Krill herd: a new bio-inspired optimization algorithm. Commun. Nonlin. Sci. Numer. Simul. 17 (12), 4831–4845.

Gandomi, A.H., Yang, X.S., Talatahari, S., Deb, S., 2012a. Coupled eagle strategy and differential evolution for unconstrained and constrained global optimization. Comput. Math. Appl. 63 (1), 191–200.

Gandomi, A.H., Yang, X.S., Alavi, A.H., Talatahari, S., 2012b. Bat algorithm for constrained optimization tasks. Neural Comput. Appl. doi: 10.1007/s00521-012-1028-9.

Gandomi, A.H., Yang, X.S., Talatahari, S., Deb, S., 2012c. Design optimization of truss structures using cuckoo search algorithm. Struct. Des. Tall Spec. Buildings. doi: 10.1002/tal.1033.

Geem, Z.W., Kim, J.H., 2001. A new heuristic optimization algorithm: harmony search. Simulation. 76, 60–68.

Hasan, R., Xu, L., Grierson, D.E., 2002. Pushover analysis for performance-based seismic design. Comput. Struct. 80 (31), 2483–2493.

Hassan, R., Cohanim, B., Weck, O., 2005. A comparison of particle swarm optimization and the genetic algorithm. Proceedings of Forty-sixth AIAA/ASME/ASCE/AHS/ASC Structures. Structural Dynamics & Materials Conference, Austin, Texas, pp. 18–21.

Holland, J., 1975. Adaptation in Natural and Artificial Systems. The University of Michigan Press, Ann Arbor, MI.

Karaboga, D., 2005. An Idea Based on Honey Bee Swarm for Numerical Optimization. Technical Report-TR06. Computer Engineering Department, Erciyes University Kayseri, Turkey.

Kaveh, A., Jahanshahi, M., 2008. Plastic limit analysis of frames using ant colony systems. Comput. Struct. 86, 1152–1163.

Kaveh, A., Talatahari, S., 2008. A discrete particle swarm ant colony optimization for design of steel frames. Asian J. Civil Eng. 9 (6), 563–575.

Kaveh, A., Talatahari, S., 2009a. Particle swarm optimizer, ant colony strategy and harmony search scheme hybridized for optimization of truss structures. Comput. Struct. 87 (5–6), 267–283.

Kaveh, A., Talatahari, S., 2009b. Engineering optimization with hybrid particle swarm and ant colony optimization. Asian J. Civil Eng. 10 (6), 611–628.

Kaveh, A., Talatahari, S., 2009c. A particle swarm ant colony optimization algorithm for truss structures with discrete variables. J. Construct. Steel Res. 65 (8–9), 1558–1568.

Kaveh, A., Talatahari, S., 2009d. Hybrid algorithm of harmony search, particle swarm and ant colony for structural design optimization. Stud. Comput. Intell. 239, 159–198.

Kaveh, A., Talatahari, S., 2010a. An improved ant colony optimization for constrained engineering design problems. Eng. Comput. 27 (1), 155–182.

Kaveh, A., Talatahari, S., 2010b. Charged system search for optimum grillage systems design using the LRFD-AISC code. J. Construct. Steel Res. 66 (6), 767–771.

Kaveh, A., Talatahari, S., 2010c. Optimum design of skeletal structures using imperialist competitive algorithm. Comput. Struct. 88 (21–22), 1220–1229.

Kaveh, A., Talatahari, S., 2010d. A novel heuristic optimization method: charged system search. Acta Mech. 213 (3–4), 267–289.

Kaveh, A., Farhmand Azar, B., Talatahari, S., 2008. Ant colony optimization for design of space trusses. Int. J. Space Struct. 23 (3), 167–181.

Kaveh, A., Farahmand Azar, B., Hadidi, A., Rezazadeh Sorochi, F., Talatahari, S., 2010. Performance-based seismic design of steel frames using ant colony optimization. J. Construct. Steel Res. 66 (4), 566–574.

Kennedy, J., Eberhart, R.C., 1995. A new optimizer using particle swarm theory. Proceedings of the Sixth International Symposium on Micro Machine and Human Science. Nagoya, Japan.

Kilar, V., Fajfar, P., 1997. Simple pushover analysis of asymmetric buildings. Earthquake Eng. Struct. Dyn. 26, 233–249.

Lawson, R.S., Vance, V., Krawinkler, H., 1994. Nonlinear static push-over analysis why, when, and how? Proceedings of Fifth U.S. National Conference on Earthquake Engineering. Chicago, IL, EERI, 1, pp. 283–292.

Maier, H.R., Simpson, A.R., Zecchin, A.C., Foong, W.K., Phang, K.Y., Seah, H.Y., et al., 2003. Ant colony optimization for design of water distribution systems. J. Water Res. Plan. Manag. 129 (3), 200–209.

Moghadam, A.S., Tso, W.K., 1995. 3-D pushover analysis for eccentric buildings. Proceedings of Seventh Canadian Conference on Earthquake Engineering. Montreal, Canada.

Monfortoon, G.R., Wu, T.S., 1963. Matrix analysis of semi-rigidly connected steel frames. J. Struct. Div. 89 (6), 13–42.

Nourani, V., Talatahari, S., Monadjemi, P., Shahradfar, S., 2009. Application of ant colony optimization to optimal design of open channels. J. Hydr. Res. 47 (5), 656–665.

Pezeshk, S., Camp, C.V., Chen, D., 2000. Design of nonlinear framed structures using genetic optimization. J. Struct. Eng. 126 (3), 352–358.

Richardson, P., 2008. Bats. Natural History Museum, London.

Saiidi, M., Sozen, M.A., 1981. Simple nonlinear seismic analysis of R/C structures. J. Struct. Div. 107 (ST5), 937–952.

Saka, M.P., Kameshki, E.S., 1997. Optimum design of nonlinear steel frames with semi-rigid connections using a genetic algorithm. Comput. Struct. 79, 1593–1604.

Storn, R., Price, K.V., 1997. Differential evolution: a simple and efficient heuristic for global optimization over continuous spaces. J. Global Opt. 11 (4), 341–359.

Talatahari, S., Kaveh, A., 2013. Optimum design of skeletal structures via big bang-big crunch algorithm. In: Gandomi, A.H., Yang, X.S., Talatahari, S. Alavi, A.H. (Eds.), Metaheuristic Applications in Structures and Infrastructures, Elsevier.

Talatahari, S., Kheirollahi, M., Farahmandpour, C., Gandomi, A.H., 2012. Optimum design of truss structures using multi stage particle swarm optimization. Neural Comput. Appl. doi: 10.1007/s00521-012-1072-5.

Xu, L., 1992. Geometrical stiffness and sensitivity matrices for optimization of semirigid steel frameworks. Struct. Optim. 5 (1–2), 95–99.

Xu, L., Grierson, D.E., 1993. Computer-automated design of semirigid steel frameworks. J. Struct. Eng. 119 (6), 1740–1760.

Xu, L., Gong, Y., Grierson, D.E., 2006. Seismic design optimization of steel building frameworks. J. Struct. Eng. 132 (2), 277–286.

Yang, X.S., 2010a. Firefly algorithm, stochastic test functions and design optimisation. Inter. J. Bio-Inspired Comput. 2 (2), 78–84.

Yang, X.S., 2010b. A new metaheuristic bat-inspired algorithm. In: Gonzalez, J R, et al., (Eds.), Nature Inspired Cooperative Strategies for Optimization (NISCO 2010), vol. 284. Springer, Berlin, pp. 65–74. , Studies in Computational Intelligence.

Yang, X.S., Deb, S., 2009. Cuckoo search via Lévy flights. Proceeings of World Congress on Nature & Biologically Inspired Computing. IEEE Publications, Coimbatore, India, pp. 210–214.

18 Expression Programming Techniques for Formulation of Structural Engineering Systems

Amir Hossein Gandomi[1] *and Amir Hossein Alavi*[2]

[1]Department of Civil Engineering, The University of Akron, Akron, OH, USA, [2]School of Civil Engineering, Iran University of Science and Technology, Tehran, Iran

18.1 Introduction

In contrast to other civil engineering problems, many structural engineering systems lack a precise analytical theory or model for their solutions. In order to cope with the complexity of structural engineering problems and the spatial variability of the involved materials, traditional forms of engineering design solutions have widely been developed. The information has usually been collected, synthesized, and presented in the form of design charts, tables, or empirical formulae (Alavi and Gandomi, 2011a; Shahin et al., 2001).

Recent technological progress has generated extremely accurate and reliable computer-aided pattern recognition and data-classification methods. Pattern-recognition systems, as an example, learn adaptively from experience and extract various discriminators. Artificial neural networks (ANNs) are the most widely used pattern recognition procedures. ANNs have been successfully employed to capture nonlinear interactions between various parameters in complex structural engineering systems (Alavi and Gandomi, 2011b; Alavi et al., 2009; Cabalar and Cevik, 2009; Gandomi and Alavi, 2011b; Javadi, 2006; Juang et al., 2001). Despite the acceptable performance of ANNs in most cases, they have some fundamental disadvantages that limit their use by several researchers. There is no definite function in ANN methodology to be used for the calculation of output. ANN only has final synaptic weights to obtain outcome in a parallel manner. However, due to nonlinearity and complexity, more robust tools are required to assess the behavior of many structural engineering problems.

Genetic programming (GP) (Koza, 1992) is a new alternative approach to overcome the limitations of ANNs. GP is a specialization of genetic algorithms (GA) where solutions are computer programs rather than binary strings. One of the main features of GP over other soft computing tools (e.g., ANNs) is its ability to

Metaheuristic Applications in Structures and Infrastructures. DOI: http://dx.doi.org/10.1016/B978-0-12-398364-0.00018-8

generate simplified prediction equations without assuming prior form of the existing relationship (Alavi et al., 2011a). For the last decade, GP and its variants have been pronounced as powerful methods for simulating the behavior of civil engineering problems (Alavi et al., 2011c, 2012; Gandomi and Alavi, 2011a, 2012a,b; Gandomi et al., 2010a,b,c, 2011d; Javadi et al., 2006; Narendra et al., 2006). Gene expression programming (GEP) (Ferreira, 2001) is another recent extension to GP that evolves computer programs of different sizes and shapes encoded in linear chromosomes of fixed length. The GEP chromosomes are composed of multiple genes, each gene encoding a smaller subprogram. Multiexpression programming (MEP) (Oltean and Dumitrescu, 2002) is another linear variant of GP that uses a linear representation of chromosomes. MEP has a special ability to encode multiple computer programs of a problem in a single chromosome. Based on numerical experiments, the GEP and MEP approaches are able to significantly outperform similar techniques (Oltean and Groşan, 2003a,b). In contrast with classical GP and ANNs, application of GEP and MEP in the field of structural engineering is new and original (Baykasoglu et al., 2009; Cevik, 2007; Cevik et al., 2010; Gandomi et al., 2011c,d; Gesoglu et al., 2009).

This study investigates the potential of GEP and MEP in simulating the nonlinear complex behavior of structural engineering systems. In order to demonstrate the formulation capabilities of these EP techniques, they are applied to two practical examples of structural engineering. A comparative study between the proposed formulation results and the existing models in the literature is conducted. The GEP and MEP models are developed based on reliable experimental results collected through an extensive literature review.

18.2 Genetic Programming

GP is a symbolic optimization technique that creates computer programs to solve a problem through simulating the biological evolution of living organisms (Koza, 1992). The main difference between GP and GA is the representation of the solution. GA creates a string of numbers that represent the solution. The classical GP solutions are computer programs represented as tree structures and expressed in a functional programming language (such as List Processing [LISP]) (Alavi et al., 2011a; Koza, 1992). The fitness of each program generated by GP is evaluated using a fitness function. Thus, the fitness function is the objective function that GP aims to optimize (Gandomi et al., 2011b). In addition to classical tree-based GP, there are other types of GP where programs are represented in different ways. These are linear and graph-based GP (Alavi and Gandomi, 2011a; Banzhaf et al., 1998). The emphasis of the present study is placed on the linear-based GP techniques.

18.2.1 Expression Programming

Computers do not naturally run tree-shaped programs. Therefore, slow interpreters have to be used as a part of classical tree-based GP. Conversely, by evolving the binary bit patterns, the use of an expensive interpreter is avoided.

Consequently, a linear GP system can run several orders of magnitude faster than comparable interpreting systems. The enhanced speed of the linear variants of GP (e.g., linear genetic programming (LGP) and MEP) permits conducting many runs in realistic timeframes. This leads to deriving consistent and high-precision models with little customization (Francone and Deschaine, 2004; Gandomi et al., 2011b).

Several linear variants of GP such as GEP (Ferreira, 2001, 2006) and MEP (Oltean and Dumitrescu, 2002) have recently been proposed. EP techniques are the most common linear-based GP methods. These variants make a clear distinction between the genotype and the phenotype of an individual (Oltean and Groşan, 2003a).

GEP is a natural development of GP first proposed by Ferreira (2001). GEP consists of five main components (Alavi and Gandomi, 2011a): function set, terminal set, fitness function, control parameters, and termination condition. GEP uses a fixed length of character strings to represent solutions to the problems, which are afterwards expressed as parse trees of different sizes and shapes. These trees are called GEP expression trees (Alavi and Gandomi, 2011a). One advantage of the GEP technique is that the creation of genetic diversity is extremely simplified as genetic operators work at the chromosome level. The multigenic nature of GEP allows the evolution of more complex programs composed of several subprograms. Each GEP gene contains a list of symbols with a fixed length that can be any element from a function set like $\{+, -, \times, /, \text{Log}\}$ and the terminal set like $\{a, b, c, 1\}$. The function set and terminal set must have the closure property: each function must able to take any value of data type which can be returned by a function or assumed by a terminal (Alavi and Gandomi, 2011a; Gandomi et al., 2011a).

MEP is another subarea of GP. It was first introduced by Oltean and Dumitrescu (2002). Linear chromosomes are used by MEP for solution encoding. This technique encodes multiple computer programs in a single chromosome. A program with the best fitness represents the chromosome. The MEP decoding process is not more complicated than other GP variants storing a single program in a chromosome (Alavi et al., 2010a). The steady-state algorithm of MEP starts by the creation of a random population of computer programs (Alavi and Gandomi, 2012; Oltean and Groşan, 2003a,b).

The representation of the MEP solutions is similar to the procedure followed by C and Pascal to convert expressions into machine code. Functions and terminals are a part of a population member created by MEP. The terminal and function symbols are elements in the terminal and function sets, respectively. A function set can contain the basic arithmetic operations or any other mathematical functions. The terminal set can contain numerical constants, logical constants, and variables. Each gene encodes a terminal or a function symbol. The first symbol in a chromosome is a terminal symbol. The best expression is selected after controlling the fitness of all expression in an MEP chromosome using a fitness function (Oltean and Groşan, 2003a).

Comprehensive details about GEP and MEP can be found in Alavi and Gandomi (2011a).

18.3 Application to Structural Engineering Problems

18.3.1 Review of State of the Art

Applications of the GEP and MEP techniques to solving problems in structural engineering are quite new. Cevik (2007) and Pala (2008) used GEP for characteristic modeling of cold-formed steel. GEP is utilized by Gesoglu et al. (2009) for mechanical modeling of concrete. Baykasoglu et al. (2009) proposed prediction models for high-strength concrete parameters via GEP. Cevik et al. (2010) proposed a GEP-based formula for predicting torsional strength of reinforced concrete (RC) beams. Gandomi et al. (2011e) utilized GEP to develop a new prediction model for the load capacity of castellated steel beams. Gandomi et al. (2012) recently used GEP to relate the concrete triaxial strength to mix design parameters. Although Oltean and Groşan (2003a) have shown that the MEP algorithm can outperform LGP and GEP, it has been rarely used for engineering modeling (e.g., Alavi et al., 2010a). Based on an extensive literature review, the only application of MEP to structural engineering problems is done by Gandomi et al. (2011c) for the modeling of uplift capacity of suction caisson. The uplift capacity of suction caisson is the only structural system that has been modeled using different metaheuristic algorithms. Therefore, it can provide a sound basis for conducting a brief comparative study. The uplift capacity problem is solved using the finite element method (FEM) (Deng and Carter, 1999), ANN (Rahman et al., 2001), neurogenetic network (NGN) (Pai, 2005), evolutionary polynomial regression (EPR) (Rezania et al., 2008), hybrid GP and simulated annealing (GP/SA), least squares regression (LSR) (Alavi et al., 2010b), and classical tree-based GP, LGP, and GEP (Alavi et al., 2011b). All of these studies have used the same database. Table 18.1 summarizes the results

Table 18.1 Performance Statistics of Different Methods for the Prediction of the Uplift Capacity of Suction Caisson

Methods	Experimental Versus Predicted			Experimental/Predicted		
	R	MAE	RMSE	Mean	SD	Covariance
FEM	0.996	8.49	11.88	0.910	0.206	0.226
LSR	0.871	33.55	51.36	0.826	0.299	0.363
Metaheuristic Algorithms						
ANN	0.986	11.01	17.77	0.980	0.261	0.266
NGN	0.922	34.66	48.76	0.844	0.492	0.583
EPR	0.995	10.81	15.95	0.970	0.207	0.213
GP/SA	0.998	8.723	11.122	0.931	0.130	0.140
Tree-based GP	0.988	18.19	13.90	1.369	1.804	1.317
LGP	0.997	11.60	14.92	0.927	0.199	0.215
GEP	0.991	17.23	24.87	0.922	0.256	0.278
MEP	0.996	9.78	16.19	0.892	0.166	0.187

obtained by different methods for the prediction of the uplift capacity. Correlation coefficient (R), root mean squared error (RMSE), mean absolute error (MAE), mean of experimental to predicted ratio, standard deviation (SD), and covariance are used to evaluate the performance of the models.

As can be observed from this table, the FEM and LSR have the best and worst performance, respectively. Among eight different metaheuristic algorithms, MEP has the best performance. The performance of MEP is comparable with FEM. Furthermore, the GEP algorithm has a very good performance and outperforms LSR, NGN, and tree-based GP.

18.3.2 Numerical Experiments

This chapter investigates the applicability of using the GEP and MEP approaches to formulate structural engineering problems. The investigated problems are as follows:

1. Prediction of load capacity of RC columns.
2. Prediction of hysteretic energy demand in steel moment resisting frames.

The GEP and MEP models were developed based on experimental results obtained from the literature. Various parameters involved in the GEP and MEP algorithms are presented in Table 18.2. The major task is to define the hidden function connecting the input variables and output variables. There are 13 parameters for GEP and eight parameters for MEP. The former five parameters are similar for each algorithm. The parameter selection will affect the generalization capability of the GEP and MEP models. Several runs are conducted to come up with a parameterization of GEP and MEP that provided enough robustness and generalization to solve the problems. The effective training time specifies the number of generations in GEP and MEP. For all the cases, three levels are set for the number of generations. A fairly large number of generations are tested on each run to find models with minimum error. For each case, the program is run until there is no longer significant improvement in the performance of the models or the run is terminated automatically. Three levels are set for the population size. Large populations are used with the runs to guarantee sufficient diversity. Note that a run will take longer with a larger population size. Two levels are considered for the crossover and mutation rates. The success of the algorithms usually increases with increasing the head size and number of genes in GEP and chromosome length in MEP. In this case, the complexity of the evolved functions increases and the speeds of the algorithms decrease. Different optimal levels are considered for these parameters as tradeoffs between the running time and the complexity of the evolved solutions. Basic arithmetic operators and mathematical functions are utilized to get the optimum models. The values considered for the other parameters are based on some suggested values (e.g., Alavi and Gandomi, 2011a; Baykasoglu et al., 2008). Also, several preliminary runs are conducted and the performance behavior is observed to choose the final parameters. All of the combinations of the parameters are tested and 10 replications are carried out for each combination.

Table 18.2 Parameter Settings for the GEP and MEP Algorithms

Algorithms		Parameters	Parameter Setting
Common parameters	GEP, MEP	Number of generation	100, 250, 500
		Population size	500, 2500, 5000
		Function set	+, −, ×, /, √, power, exp, log, ln
		Mutation rate (%)	10, 90
		Fitness function	Linear error function
Algorithm-specific parameters	MEP	Crossover rate (%)	50, 95
		Crossover type	Uniform
		Chromosome length	50−80 genes
	GEP	Number of genes	1−3
		Head size	3, 5, 8
		Linking function	+
		One-point recombination rate (%)	30, 50
		Two-points recombination rate (%)	30
		Gene recombination rate	10
		Gene transposition rate (%)	10
		Numerical constants	Integer, Floating-point

The GEP algorithm is implemented by GeneXproTools (GEPSOFT, 2006) software. The source code of MEP (Oltean, 2004) in C++ is modified by the authors to be utilizable for the available problems.

Overfitting is one of the essential problems in machine-learning generalization. An efficient approach to prevent overfitting is to test the derived models on a validation set to find a better generalization (Banzhaf et al., 1998). This strategy is considered in this study for improving the generalization of the models. For this aim, the available datasets are randomly divided into learning, validation, and testing subsets. The learning data are taken for training (genetic evolution). The validation data are used to specify the generalization capability of the models on data they did not train on (model selection). Thus, both the learning and validation data are involved in the modeling process and are categorized into one group referred to as "training data." The model with the best performance on both of the learning and validation datasets is finally selected as the outcome of the run. The testing datasets are further employed to measure the performance of the optimal models obtained by GEP and MEP on data without a role in building the models. To obtain a consistent data division, several combinations of the training and testing sets are considered. The selection was in a way that the maximum, minimum, mean, and SD of parameters are consistent in the training and testing datasets. Out of the available data for each problem, approximately 85% are used for the training purpose (70% for learning and 15% for validation) and 15% for the testing of the generalization capability of the GEP and MEP models.

18.3.3 Model Selection

To obtain the best prediction models, the best GEP- and MEP-based formulas are chosen on the basis of a multiobjective strategy as below:

1. The simplicity of the model, although this is not a predominant factor.
2. Providing the best fitness value on the learning set of data.
3. Providing the best fitness value on a validation set of data.

The first objective must be controlled by the user. For the other objectives, the following objective function (OBJ) is used as a measure of how well the model-predicted output agrees with the experimentally measured output. The selections of the best models are deduced by the minimization of the following function:

$$\mathrm{OBJ} = \left(\frac{\mathrm{No.}_{\mathrm{Learning}} - \mathrm{No.}_{\mathrm{Validation}}}{\mathrm{No.}_{\mathrm{Training}}}\right) \frac{\mathrm{RMSE}_{\mathrm{Learning}} + \mathrm{MAE}_{\mathrm{Learning}}}{R^2_{\mathrm{Learning}}} + \frac{2\mathrm{No.}_{\mathrm{Validation}}}{\mathrm{No.}_{\mathrm{Training}}} \frac{\mathrm{RMSE}_{\mathrm{Validation}} + \mathrm{MAE}_{\mathrm{Validation}}}{R^2_{\mathrm{Validation}}} \tag{18.1}$$

where $\mathrm{No.}_{\mathrm{Train}}$, $\mathrm{No.}_{\mathrm{Learning}}$, and $\mathrm{No.}_{\mathrm{Validation}}$ are the number of training, learning, and validation data, respectively. It is well known that only R is not a good indicator of prediction accuracy of a model. This is because by shifting the output values of a model equally, the R value will not change. The constructed objective function takes into account the changes of R, RMSE, and MAE together. Higher R values and lower RMSE and MAE values result in lowering OBJ and, consequently, indicate a more precise model. In addition, the above function considers the effects of different data divisions for the training and testing data.

18.3.4 Prediction Problems

18.3.4.1 A Case Study of Concrete Structures: Shear Strength of RC Columns

Columns are the most important element of RC structures as their failure leads the building to collapse. The significant effect of shear forces on RC columns subjected to lateral load is well understood. Shear failure of RC columns reduces the lateral strength of the building and may lead to rapid strength degradation. Hence, providing precise estimations of the shear strength of RC columns is an essential consideration in structural design (Choe, 2006). Several analytical solutions are presented by design codes to determine the shear strength of circular columns. The contributions of concrete and transverse reinforcement to the shear strength of circular columns are two key components included in these design codes (Caglar, 2009).

There is some published information on the behavioral modeling of RC columns. In this context, ANNs have been applied to the prediction of the shear strength of circular RC columns (Caglar, 2009). As mentioned above, ANNs have some fundamental disadvantages that limit their usage in practical calculations. To overcome these limitations and provide a more robust method, the GEP and MEP approaches are used as alternative solutions to simulate the characteristics of RC columns. These methods have an advantage that once the model is trained, it can be used as a quick and accurate tool for evaluating the shear strength of circular RC columns without the need for any manual testing.

18.3.4.1.1 Model Construction and Analysis

The GEP- and MEP-based correlations for predicting the shear strength of circular RC columns were developed using 47 datasets gathered by Choe (2006). The data are from 12 different experimental studies of RC columns. The database includes the measurements of several mechanical and geometrical variables. Herein, four influencing parameters are used as the predictor variables based on a literature review (Caglar, 2009). The statistics of different input and output parameters involved in the model development are given in Table 18.3. $\rho_l f_{yl}$, $\rho_w f_{yh}$, $A_e\sqrt{f_c}/1000$, and P/A_g are the considered input variables. ρ_l and ρ_w are, respectively, longitudinal and transverse reinforcement ratios in percentage. f_{yl}, f_{yh}, and f_c' are, respectively, yield stress of longitudinal reinforcement, hoops, and concrete compressive strength in MPa. A_e and A_g are effective and gross cross-sectional areas, respectively. P is the axial load in kN and the output is the shear force (V) in kN. Of the available data, 38 datasets are used as the training data and the rest are used for the testing of the proposed models. Furthermore, the proposed correlations are compared with different codes such as ACI-318 (2005), ASCE-ACI (1973), ATC-32 (1996), and CALTRANS (1996).

Table 18.3 The Variables Used in Model Development

	$\rho_l f_{yl}$ (MPa)	$\rho_w f_{yh}$ (MPa)	$A_e\sqrt{f_c}$ (in 1000s)	P/A_g (MPa)	V (kN)
Mean	1056.1	197.48	756.86	2.8173	321.08
Standard error	60.818	20.414	50.240	0.6521	17.334
Median	1155	166.77	650.47	1.4236	321.50
Mode	1395.2	280.5	654.02	0	230
SD	374.91	125.84	309.70	4.0200	106.86
Sample variance	140557	15836	95915	16.160	11418
Kurtosis	− 1.5969	0.4517	0.4562	2.1352	− 0.1795
Skewness	− 0.2959	0.7849	1.2164	1.6602	0.1572
Range	1080.8	531.26	1198.1	14.427	486
Minimum	439.2	26	267.55	0	93
Maximum	1520	557.26	1465.6	14.427	579
Confidence level (95.0%)	123.23	41.362	101.80	1.3213	35.123

The formulations of the shear strength of circular RC columns, V (kN), for the best result by the GEP and MEP algorithms are as given below:

$$V_{\mathrm{GEP}} = 9 + \frac{1}{7}\left(\rho_{\mathrm{l}} f_{\mathrm{yl}} + \rho_{\mathrm{w}} f_{\mathrm{yh}} + \frac{A_{\mathrm{e}}\sqrt{f_c'}}{875}\right) + 7\frac{P}{A_{\mathrm{g}}}(\rho_{\mathrm{l}} f_{\mathrm{yl}} + \rho_{\mathrm{w}} f_{\mathrm{yh}})\left(\rho_{\mathrm{w}} f_{\mathrm{yh}} + \frac{A_{\mathrm{e}}\sqrt{f_{\mathrm{c}}'}}{1000} + \frac{P}{A_{\mathrm{g}}}\right) \tag{18.2}$$

$$V_{\mathrm{MEP}} = 10 - \left(2\frac{A_{\mathrm{e}}\sqrt{f_{\mathrm{c}}'}}{1000} - \rho_{\mathrm{l}} f_{\mathrm{yl}}\left(\frac{A_{\mathrm{e}}\sqrt{f_{\mathrm{c}}'}}{1000} + \rho_{\mathrm{w}} f_{\mathrm{yh}}\right)\right.$$
$$\left. + (\rho_{\mathrm{w}} f_{\mathrm{yh}})^2 - \rho_{\mathrm{l}} f_{\mathrm{yl}}\frac{P}{A_{\mathrm{g}}}\left(18 + \frac{P}{A_{\mathrm{g}}}\right)\right) \Big/ \left(2\rho_{\mathrm{l}} f_{\mathrm{yl}} + \frac{A_{\mathrm{e}}\sqrt{f_{\mathrm{c}}'}}{1000}\right) \tag{18.3}$$

Figure 18.1 shows a comparison between the experimental values and the values predicted by GEP and MEP. Table 18.4 presents the performance of the proposed models on the training and testing datasets. Performance indices of different models on the entire database are summarized in Table 18.5. Based on the values of the performance measures, it can be observed that the GEP and MEP models are able to predict the target values with a high degree of accuracy. Comparing the performance of the proposed models, it can be seen that the GEP model has produced better results than the MEP-based model. The results clearly demonstrate that the ACI-318, ATC-32, ASCE-ACI, and CALTRANS methods are not efficient in estimating the shear strength of circular RC columns.

18.3.4.2 A Case Study of Steel Structures: Hysteretic Energy Demand in Steel Moment Frames

Energy-based parameters have widely been used as seismic design parameters (Housner, 1956). An efficient way to analyze damage of a structure is to evaluate the amount of energy imparted to the structure during an earthquake. This energy is called the total energy input (EI). Hysteretic energy or hysteretic energy demand

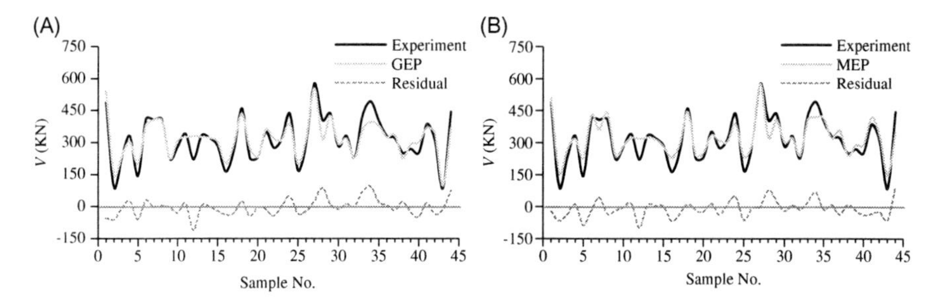

Figure 18.1 Experimental versus predicted shear strength of circular RC columns by (A) GEP and (B) MEP.

Table 18.4 Performance Statistics of the GEP and MEP Models for the Assessment of the Shear Strength of RC Columns

Method	Experimental Versus Predicted			Experimental/Predicted		
	R	**MAE**	**RMSE**	**Mean**	**SD**	**Covariance**
All datasets						
GEP	0.9346	30.3284	40.2143	0.9765	0.1373	0.1406
MEP	0.9391	31.6158	40.7709	0.9550	0.1509	0.1580
Training datasets						
GEP	0.9330	29.5420	40.0555	0.9786	0.1391	0.1421
MEP	0.9432	29.5311	38.7068	0.9606	0.1406	0.1464
Testing datasets						
GEP	0.9397	35.3094	41.2053	0.9630	0.1367	0.1419
MEP	0.9143	44.8189	51.9736	0.9196	0.2185	0.2377

Table 18.5 Overall Performance of Different Models for the Assessment of the Shear Strength of RC Columns

Method	Experimental Versus Predicted			Experimental/Predicted		
	R	**MAE**	**RMSE**	**Mean**	**SD**	**Covariance**
Standard codes						
ACI-318	0.6950	70.0659	100.3677	0.9477	0.2274	0.2399
ATC-32	0.6314	74.1205	109.3941	1.0152	0.2811	0.2769
ASCE-ACI	0.7320	79.9432	114.0106	0.9334	0.2529	0.2709
CALTRANS	0.6186	75.5386	105.1456	1.1725	0.6032	0.5144
Expression programming						
GEP	0.9346	30.3284	40.2143	0.9765	0.1373	0.1406
MEP	0.9391	31.6158	40.7709	0.9550	0.1509	0.1580

is a part of the EI which is dissipated through the hysteretic behavior. The hysteretic energy demand is the source of damage to the structural component. The structural failure happens when the earthquake-induced hysteretic energy demand for a structure is larger than the hysteretic energy dissipation capacity of the structure. The hysteretic energy and its distribution throughout the structure are dependent on both the structural systems and the ground motion. Therefore, the hysteretic energy parameter can be regarded as a design parameter, in particular, when the damage is expected not to exceed some specified limits (Bertero and Teran-Gilmore, 1994). One of the major concerns for assessing the response of structures for low-performance levels in performance-based earthquake resistant design is to

Table 18.6 The Variables Used in Model Development

	I	N_s	Z	T (s)	η	E_H/E_I	E_H/m (cm/s)^2
Mean	2	10.667	2	2.3611	0.1293	0.8330	6728.8
Standard error	0.1601	1.3806	0.1601	0.2225	0.0132	0.0114	1258.5
Median	2	9	2	2.2862	0.11	0.85	4387
SD	0.8321	7.1737	0.8321	1.1559	0.0688	0.0593	6539.2
Sample variance	0.6923	51.462	0.6923	1.3361	0.0047	0.0035	4.3×10^7
Kurtosis	−1.56	−1.56	−1.56	−1.56	−1.56	0.7545	1.7077
Skewness	0	0.3623	0	0.1047	0.4302	−0.9968	1.5391
Range	2	17	2	2.7754	0.162	0.23	24198
Minimum	1	3	1	1.0109	0.058	0.68	415
Maximum	3	20	3	3.7863	0.22	0.91	24613
Confidence level (95.0%)	0.3291	2.8378	0.3291	0.4573	0.0272	0.0234	2586.8

determine the hysteretic energy demand and dissipation capacity and level of damage of the structure to a predefined earthquake ground motion (Riddell and Garcia, 2001).

18.3.4.2.1 Model Construction and Analysis

In the proposed GEP and MEP models, earthquake intensity (I), number of stories (N_s), soil type (Z), fundamental period (T), strength index (η), and hysteretic energy to energy imparted to the structure ratio (E_H/E_I) are the variables used to predict the hysteretic energy demand (E_H/m). The descriptive statistics of different input and output parameters involved in the model development are given in Table 18.6. The GEP- and MEP-based correlations are developed based on 27 datasets presented by Akbas (2006). Out of the available data, 22 datasets are used as the training data and the rest are used for the testing of the proposed models. No rational prediction model for the hysteretic energy demand has been yet developed that would encompass the influencing variables considered in this study. Therefore, it is not possible to conduct a comparative study between the proposed models and the existing solutions.

18.3.4.2.2 GEP- and MEP-Based Formulations of Hysteretic Energy

The optimal hysteretic energy prediction equations derived by means of the GEP and MEP algorithms are as follows:

$$\left(\frac{E_H}{m}\right)_{GEP} = 4096\frac{IN_s}{\eta} - 12IN_sT(4-T)(T+3) - \frac{ZITE_I}{2E_H} \tag{18.4}$$

$$\left(\frac{E_\mathrm{H}}{m}\right)_{\mathrm{MEP}} = \frac{36I}{T}\left(Z\eta\frac{E_\mathrm{I}}{E_\mathrm{H}} - 5\frac{E_\mathrm{H}}{E_\mathrm{I}}\right)\left(\frac{E_\mathrm{H}}{E_\mathrm{I}}\left(\frac{\eta}{Z} - N_s + 4\right) - 5\right) \times \left(1 - \frac{T}{N_s - 4}\right)\left(4 + I - \frac{E_H}{E_\mathrm{I}}\right) \quad (18.5)$$

A comparison of the experimental versus predicted E_H/m by GEP and MEP is shown in Figure 18.2. Performance statistics of the GEP and MEP models are summarized in Table 18.7. The results indicate that the GEP and MEP models are able to predict hysteretic energy with a high degree of accuracy. As it is seen, GEP has produced generally better outcomes than MEP on the testing and entire data.

18.4 Model Validity

Smith (1986) suggested the following criteria for evaluating the performance of a model:

- If a model gives correlation coefficient $(R) > 0.8$, a strong correlation exists between the predicted and measured values.

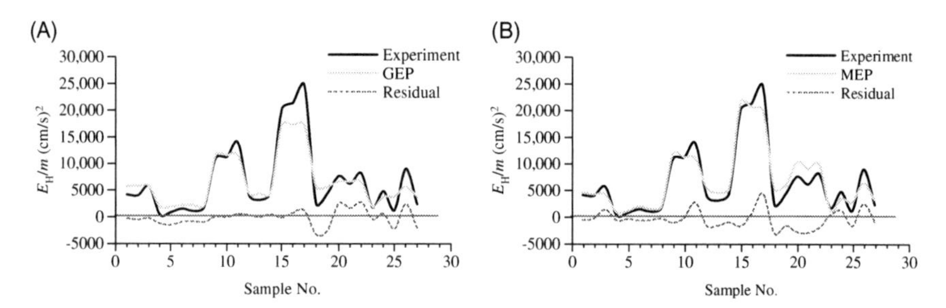

Figure 18.2 Measured versus predicted hysteretic energy by (A) GEP and (B) MEP.

Table 18.7 Overall Performance of EP Models for Hysteretic Energy Assessment

Method	Experimental Versus Predicted			Experimental/ Predicted		
	R	MAE	RMSE	Mean	SD	Covariance
All datasets						
GEP	0.9698	1602.6	2224.3	0.8902	0.3404	0.3824
MEP	0.9650	1.4021	1.7616	0.8842	0.2816	0.3184
Training datasets						
GEP	0.9601	1564.5	2213.9	0.8482	0.3497	0.4123
MEP	0.9619	1381.0	1772.8	0.8500	0.2799	0.3293
Testing datasets						
GEP	0.9978	1770.5	2269.3	1.0752	0.2421	0.2252
MEP	0.9725	1494.6	1711.3	1.0346	0.2630	0.2542

Table 18.8 Statistical Parameters of the GEP and MEP Models

Item	Formula	Condition	Case I: RC Columns		Case II: Hysteretic Energy	
			GEP	MEP	GEP	MEP
1	Eq. (18.5)	$0.8 < R$	0.9397	0.9143	0.9978	0.9598
2	$k = \frac{\sum_{i=1}^{n}(h_i \times t_i)}{h_i^2}$	$0.85 < k < 1.15$	0.9982	0.9776	1.2101	1.0494
3	$k' = \frac{\sum_{i=1}^{n}(h_i \times t_i)}{t_i^2}$	$0.85 < k' < 1.15$	0.9845	0.9953	0.8230	0.9365
4	$R_m = R^2 \times \left(1 - \sqrt{\lvert R^2 - Ro^2 \rvert}\right)$	$0.5 < R_m$	0.9397	0.9143	0.6296	0.7415
where	$Ro^2 = 1 - \frac{\sum_{i=1}^{n}(t_i - h_i^o)^2}{\sum_{i=1}^{n}(t_i - \bar{t}_i)^2}$,	$h_i^o = k \times t_i$	1.0000	0.9962	0.8605	0.9923
	$Ro'^2 = 1 - \frac{\sum_{i=1}^{n}(h_i - t_i^o)^2}{\sum_{i=1}^{n}(h_i - \bar{h}_i)^2}$,	$t_i^o = k' \times h_i$	0.9973	0.9997	0.8829	0.9870

In all cases, the error values (e.g., RMSE and MAE) should be at the minimum. The model can therefore be judged as very good. It can be observed from Tables 18.4, 18.5, and 18.7 that the GEP and MEP models with high R and low RMSE and MAE values are able to predict the target values to an acceptable degree of accuracy. Furthermore, new criteria recommended by Golbraikh and Tropsha (2002) are checked for the external validation of the models on the test datasets. It is suggested that at least one slope of regression lines (k or k') through the origin should be close to 1. Recently, Roy and Roy (2008) introduced a confirm indicator (R_m) of the external predictability of models. For $R_m > 0.5$, the condition is satisfied. Either the squared correlation coefficient (through the origin) between predicted and experimental values (Ro^2) or the coefficient between experimental and predicted values (Ro'^2) should be close to R^2 and to 1. The considered validation criteria and the relevant results obtained by the models are presented in Table 18.8. The EP models are considered valid, if they satisfy the required conditions. The validation phase ensures that the derived models are strongly valid and it is not established by chance.

18.5 Conclusions

In this paper, two promising expression programming techniques, namely GEP and MEP, are employed for the analysis of complex structural engineering systems. The capabilities of the GEP and MEP methodologies are illustrated by application to two practical structural engineering problems: (1) shear strength of RC columns and (2) hysteretic energy demand in steel moment resisting frames. Reliable databases from the previously published experimental results are used to develop the models. The following conclusions can be derived from the results presented in this research:

1. Despite high nonlinearity in the behavior of the investigated structural systems, the proposed GEP and MEP models give reasonable estimations of the target values. Furthermore, the proposed models efficiently satisfy the conditions of different criteria considered for their external validation.
2. The proposed GEP and MEP models efficiently take into consideration the effects of several parameters representing the engineering behavior of the structural problems.
3. In general, the prediction capabilities of the GEP and MEP models are found to be very close to each other. The results clearly demonstrate that the GEP and MEP techniques can reliably be applied to formulate the structural engineering problems.
4. GEP and MEP provide prediction equations that are relatively short and simple and can be used for routine design practice via hand calculations.
5. Unlike the traditional methods, GEP and MEP do not require any simplifying assumptions in developing the models.

The GEP and MEP methods are particularly practical for the situations where good experimental data are available, the behavior is too complex, or the conventional constitutive models are unable to effectively describe various aspects of the behavior.

References

ASCE-ACI, 1973. Shear strength of reinforced concrete members ASCE-ACI joint task committee 426. J. Struct. Eng. 99, 1091–1187.

Akbas, B., 2006. A neural network model to assess the hysteretic energy demand in steel moment resisting frames. Struct. Eng. Mech. 23, 2–10.

Alavi, A.H., Gandomi, A.H., 2011a. A robust data mining approach for formulation of geotechnical engineering systems. Eng. Comput. 28 (3), 242–274.

Alavi, A.H., Gandomi, A.H., 2011b. Prediction of principal ground-motion parameters using a hybrid method coupling artificial neural networks and simulated annealing. Comput. Struct. 89 (23-24), 2176–2194.

Alavi, A.H., Gandomi, A.H., 2012. Energy-based models for assessment of soil liquefaction. Geosci. Front. 3 (4), 541–555.

Alavi, A.H., Gandomi, A.H., Gandomi, M., Sadat Hosseini, S.S., 2009. Prediction of maximum dry density and optimum moisture content of stabilized soil using RBF neural networks. IES J Part A: Civil Struct. Eng. 2 (2), 98–106.

Alavi, A.H., Gandomi, A.H., Sahab, M.G., Gandomi, M., 2010a. Multi expression programming: a new approach to formulation of soil classification. Eng. Comput. 26 (2), 111–118.

Alavi, A.H., Gandomi, A.H., Mousavi, M., Mollahasani, A., 2010b. High-precision modeling of uplift capacity of suction caissons using a hybrid computational method. Geomech. Eng. 2 (4), 253–280.

Alavi, A.H., Ameri, M., Gandomi, A.H., Mirzahosseini, M.R., 2011a. Formulation of flow number of asphalt mixes using a hybrid computational method. Constr. Build. Mater. 25 (3), 1338–1355.

Alavi, A.H., Aminian, P., Gandomi, A.H., Arab Esmaeili, M., 2011b. Genetic-based modeling of uplift capacity of suction caissons. Expert Syst. Appl. 38 (10), 12608–12618.

Alavi, A.H., Gandomi, A.H., Mollahasani, A., 2001c. A genetic programming-based approach for performance characteristics assessment of stabilized soil. Variants of Evolutionary Algorithms for Real-World Applications. In: Chiong, R., Weise, T., Michalewicz, Z. (Eds.), Springler-Verlag, Berlin., pp. 343–375 (Chapter 9).

Alavi, A.H., Gandomi, A.H., Bolury, J., Mollahasani, A., 2012. Linear and tree-based genetic programming for solving geotechnical engineering problems. In: Yang, X.S., et al., (Eds.), Metaheuristics in Water Resources, Geotechnical and Transportation Engineering. Elsevier, Waltham, MA, USA.

American Concrete Institute (ACI), 2005. Building Code Requirements for Structural Concrete. ACI Committee 318. Farmington Hills, MI.

Applied Technology Council, 1996. Improved Seismic Design Criteria for California Bridges: Provisional Recommendation. Report No. ATC-32. Redwood City, CA.

Banzhaf, W., Nordin, P., Keller, R., Francone, F., 1998. Genetic Programming - —An Introduction. On the Automatic Evolution of Computer Programs and its Application. dpunkt/Morgan Kaufmann, San Francisco, CA, USA.

Baykasoglu, A., Gullub, H., Canakci, H., Ozbakir, L., 2008. Prediction of compressive and tensile strength of limestone via genetic programming. Expert Syst. Appl. 35 (1–2), 111–123.

Baykasoglu, A., Oztas, A., Ozbay, E., 2009. Prediction and multi-objective optimization of high-strength concrete parameters via soft computing approaches. Expert Syst. Appl. 36 (3), 6145–6155.

Bertero, V.V., Teran-Gilmore, A., 1994. Use of energy concepts in earthquake-resistant analysis and design: issues and future directions. Advances in Earthquake Engineering Practice, Short Course in Structural Engineering:Architectural and Economic Issues. University of California, Berkeley, CA.

CALTRANS Memo to Designers 20-4, 1996. Attachment B, earthquake retrofit analysis for single column bents.

Cabalar, A.F., Cevik, A., 2009. Modelling damping ratio and shear modulus of sand-mica mixtures using neural networks. Eng. Geol. 104, 31–40.

Caglar, N., 2009. Neural network-based approach for determining the shear strength of circular reinforced concrete columns. Constr. Build. Mater. 23, 3225–3232.

Cevik, A., 2007. A new formulation for web crippling strength of cold-formed steel sheeting using genetic programming. J. Constr. Steel Res. 63, 867–883.

Cevik, A., Arslan, M.H., Köroğlu, M.H., 2010. Genetic-programming-based modeling of RC beam torsional strength. KSCE J. Civil Eng. 14 (3), 371–384.

Choe, L.Y., 2006. Shear Strength of Circular Reinforced Concrete Columns. M.Sc. Thesis. The Ohio State University, Columbus, OH, USA.

Deng, W., Carter, J.P., 1999. Analysis of Suction Caissons Subjected to Inclined Uplift Loading. Centre for Geotechnical Research, The University of Sydney, Sydney, Australia.

Ferreira, C., 2001. Gene expression programming: a new adaptive algorithm for solving problems. Complex Syst. 13 (2), 87–129.

Ferreira, C., 2006. Gene Expression Programming: Mathematical Modeling by an Artificial Intelligence. second ed. Springer-Verlag, Germany.

Francone, F.D., Deschaine, L.M., 2004. Extending the boundaries of design optimization by integrating fast optimization techniques with machine–code-based, linear genetic programming. Inf. Sci. 161, 99–120.

GEPSOFT, 2006. GeneXproTools Owner's Manual. Version 4.0. Available from: <http://www.gepsoft.com/>. (last accessed on March 12, 2011).

Gandomi, A.H., Alavi, A.H., 2011a. Multi-stage genetic programming: a new strategy to nonlinear system modeling. Inf. Sci. 181 (23), 5227–5239.

Gandomi, A.H., Alavi, A.H., 2011b. Applications of Computational Intelligence in Behavior Simulation of Concrete Materials. In: Yang, X.S., Koziel, S. (Eds.), Computational Optimization and Applications in Engineering and Industry, vol. 359. Springer SCI, pp. 221–243. (Chapter 9), Brooklyn, NY, USA.

Gandomi, A.H., Alavi, A.H., 2012a. A new multi-gene genetic programming approach to nonlinear system modeling. Part I: materials and structural engineering problems. Neural Comput. Appl. 21 (1), 171–187.

Gandomi, A.H., Alavi, A.H., 2012b. A new multi-gene genetic programming approach to nonlinear system modeling. Part II: geotechnical and earthquake engineering problems. Neural Comput. Appl. 21 (1), 189–201.

Gandomi, A.H., Alavi, A.H., Sahab, M.G., 2010a. New formulation for compressive strength of CFRP confined concrete cylinders using linear genetic programming. Mater. Struct. 43 (7), 963–983.

Gandomi, A.H., Alavi, A.H., Sahab, M.G., Arjmandi, P., 2010b. Formulation of elastic modulus of concrete using linear genetic programming. J. Mech. Sci. Technol. 24 (6), 1011–1017.

Gandomi, A.H., Alavi, A.H., Arjmandi, P., Aghaeifar, A., Seyednoor, M., 2010c. Genetic programming and orthogonal least squares: a hybrid approach to modeling the compressive strength of CFRP-confined concrete cylinders. J. Mech. Mater. Struct. Math. Sci. 5 (5), 735–753.

Gandomi, A.H., Alavi, A.H., Mirzahosseini, R., Moghdas Nejad, F., 2011a. Nonlinear genetic-based models for prediction of flow number of asphalt mixtures. J. Mater. Civil Eng. 23 (3), 248–263.

Gandomi, A.H., Alavi, A.H., Yun, G.J., 2011b. Nonlinear modeling of shear strength of SFRC beams using linear genetic programming. Struct. Eng. Mech. 38 (1), 1–25.

Gandomi, A.H., Alavi, A.H., Yun, G.J., 2011c. Formulation of uplift capacity of suction caissons using multi expression programming. KSCE J. Civil Eng. 15 (2), 363–373.

Gandomi, A.H., Tabatabaie, S.M., Moradian, M.H., Radfar, A., Alavi, A.H., 2011d. A new prediction model for load capacity of castellated steel beams. J. Constr. Steel Res. 67 (7), 1096–1105.

Gandomi, A.H., Alavi, A.H., Mousavi, M., Tabatabaei, S.M., 2011e. A hybrid computational approach to derive new ground-motion attenuation models. Eng. Appl. Artif. Intell. 24 (4), 717–732.

Gandomi, A.H., Babanajad, S.K., Alavi, A.H., Farnam, Y., 2012. A novel approach to strength modeling of concrete under triaxial compression. J. Mater. Civil Eng. 24 (9), 1132–1143.

Gesoglu, M., Güneyisi, E., Özturan, T., Özbay, E., 2009. Modeling the mechanical properties of rubberized concretes by neural network and genetic programming. Mater. Struct. 43 (1-2), 31–45.

Golbraikh, A., Tropsha, A., 2002. Beware of q^2!. J. Mol. Graphics Model. 20 (4), 269–276.

Housner, G.W., 1956. Limit design of structures to resist earthquakes. Proceedings of the First World Conference on Earthquake Engineering, Berkeley, CA.

Javadi, A.A., 2006. Estimation of air losses in compressed air tunneling using neural network. Tunn. Undergr. Sp. Technol. 21 (1), 9–20.

Javadi, A.A., Rezania, M., Mousavi Nezhad, M., 2006. Evaluation of liquefaction induced lateral displacements using genetic programming. Comput. Geotech. 33 (4-5), 222–233.

Juang, C.H., Jiang, T., Christopher, R.A., 2001. Three-dimensional site characterisation: neural network approach. Geotechnique. 51 (9), 799–809.

Koza, J., 1992. Genetic Programming: On the Programming of Computers by Means of Natural Selection. MIT Press, Cambridge, MA.

Narendra, B.S., Sivapullaiah, P.V., Suresh, S., Omkar, S.N., 2006. Prediction of unconfined compressive strength of soft grounds using computational intelligence techniques: a comparative study. Comput. Geotech. 33 (3), 196–208.

Oltean, M., 2004. Multi Expression Programming Source Code. Available from: <http://www.mep.cs.ubbcluj.ro/>. (last accessed on September 12, 2010).

Oltean, M., Dumitrescu, D., 2002. Multi Expression Programming. Technical Report, UBB-01-2002. Babeş-Bolyai University, Cluj-Napoca, Romania.

Oltean, M., Groşan, C., 2003a. A comparison of several linear genetic programming techniques. Adv. Complex Syst. 14 (4), 1–29.

Oltean, M., Groşan, C., 2003b. Solving classification problems using infix form genetic programming. In: Berthold, M. (Ed.), Intelligent Data Analysis. Springer-Verlag, Berlin, pp. 242–252. (LNCS 2810).

Pai, G.A.V., 2005. Prediction of uplift capacity of suction caissons using a neuro-genetic network. Eng. Comput. 21 (2), 129–139.

Pala, M., 2008. Genetic programming-based formulation for distortional buckling stress of cold-formed steel members. J. Constr. Steel Res. 64 (12), 1495–1504.

Rahman, M.S., Wang, J., Deng, W., Carter, J.P., 2001. A neural network model for the uplift capacity of suction caissons. Comput. Geotech. 28 (4), 269–287.

Rezania, M., Javadi, A.A., Giustolisi, O., 2008. An evolutionary-based data mining technique for assessment of civil engineering systems. Eng. Comput. 25 (5-6), 500–517.

Riddell, R., Garcia, J.E., 2001. Hysteretic energy spectrum and damage control. Earthq. Eng. Struct. Dyn. 30 (12), 1791–1816.

Roy, P.P., Roy, K., 2008. On some aspects of variable selection for partial least squares regression models. QSAR Comb. Sci. 27, 302–313.

Shahin, M.A., Maier, H.R., Jaksa, M.B., 2001. Artificial neural network applications in geotechnical engineering. Aust. Geomech. 36 (1), 49–62.

Smith, G.N., 1986. Probability and Statistics in Civil Engineering. Collins, London.

19 An Evolutionary Divide-and-Conquer Strategy for Structural Identification

Chan Ghee Koh[1] *and Thanh N. Trinh*[2]

[1]Department of Civil and Environmental Engineering, National University of Singapore, Singapore, [2]Applied Computing and Mechanics Laboratory, Structural Engineering Institute, Swiss Federal Institute of Technology in Lausanne (EPFL), Lausanne, Switzerland

19.1 Introduction

Large structural systems such as high-rise buildings, long-span bridges, and offshore platforms often require inspection and maintenance for sustainable and safe usage. The condition of these large structures can be assessed by determining their key parameters using structural identification. Its feasibility for practical implementation has been enhanced greatly due to recent rapid advances in sensor technology, wireless communication, and computational power. To make this work, however, it is essential to have a good numerical strategy to quantify system characteristics accurately and efficiently even with limited and noisy data. Although considerable progress has been made in civil engineering, there remain many challenges in achieving good identification for large systems.

The main goal of this chapter is to present an identification strategy that is able to identify unknown parameters of large structures. This strategy combines two complementary methods, a divide-and-conquer method and an evolutionary algorithm, which work on different principles to enhance the accuracy of identification results. While the former method reduces the identification problem size, the latter focuses on the improvement of the search effectiveness. This strategy is named *evolutionary divide-and-conquer strategy*. It works by dividing a large structure with many unknowns into many smaller parts, called *substructures*, each with a manageable number of unknowns that are identified more accurately using a multi-feature genetic algorithm (GA). The GA search capability is improved through adopting several new features that allows global and local searches to be conducted simultaneously. The application of this strategy into structural identification is referred to as *substructural identification (Sub-SI)*.

Metaheuristic Applications in Structures and Infrastructures. DOI: http://dx.doi.org/10.1016/B978-0-12-398364-0.00019-X

19.2 Recent Studies on Sub-SI

Sub-SI that used gradient search methods such as least squares or the Kalman filter as a search engine has undergone a great deal of development in the last two decades. Koh et al. (1991) proposed a Sub-SI method in a time domain using the extended Kalman filter (EKF) with a weighed global iteration algorithm to estimate the unknown structural parameters of substructures. The performance of the Sub-SI method was much better in terms of accuracy and computational efficiency than that of identifying the whole structure at one go, referred to as *complete structural identification (CSI)*. This Sub-SI method was then extended to the identification of frame structures, particularly focusing on the identification at the element level (Oreta and Tanabe, 1994). The merit of the EKF is the simplicity in the formulation of the state equation for parameter estimation. However, the use of the EKF has inherent difficulties, such as the need for good initial guess values for unknown parameters and the possibility of divergence due to the nonlinear nature of the algorithm. Yun and Lee (1996) presented a Sub-SI method using an autoregressive and moving average with the stochastic input model (Lee and Yun, 1991) and a sequential prediction error method (Goodwin and Sin, 1984). Nevertheless, the maximum noise level considered in that study was limited (5%).

Tee et al. (2005) proposed two Sub-SI methods using observer/Kalman filter identification (Juang et al., 1993) and the eigensystem realization algorithm (Juang and Pappa, 1985) as search engines. The first method was based on the first-order state-space formulation of the substructure, whereas the second method performed identification in the first- and second-order model identification. Numerical and experimental studies were carried out, and a fairly large structural system of 50 degrees of freedom (DOFs) was numerically studied with measurement noise at 5%. These Sub-SI methods improved the accuracy of identification results compared to all the previous Sub-SI methods. Nonetheless, these Sub-SI methods require the measurement of accelerations to be available at all DOFs (i.e., complete measurement). To avoid the need for complete measurement, Tee et al. (2009) further improved Sub-SI by integrating a condensed model identification and recovery (CMIR) method (Koh et al., 2006).

Although these methods have attained a certain degree of success, several drawbacks remain in their practical application. With the increase in available computational speed, nonclassical methods such as artificial neural network (ANN) and GA are becoming popular alternatives to gradient search methods in recent years (Topping and Tsompanakis, 2009). Substructural identification using neural networks as a search engine was first presented by Yun and Bahng (2000). Xu and Du (2006) recently proposed a neural network (NN)–based Sub-SI method in a time domain by directly using acceleration measurements. While all the Sub-SI methods mentioned here can deal with the identification of stiffness and damping parameters, they are not able to identify unknown stiffness, damping, and mass parameters simultaneously. To address this problem, Sub-SI methods using GA as a search engine have been proposed in recent years.

There are several advantageous features of GA, such as relative ease of implementation and desirable characteristics of global convergence. Koh et al. (2003) employed GA as a search engine in the framework of the divide-and-conquer method, namely, GA-based Sub-SI methods. Assuming mass values as unknown parameters, these Sub-SI methods could simultaneously identify all unknown stiffness, damping, and mass parameters, and they yielded good results. While the idea of Sub-SI seems straightforward, the difficulty lies in accounting for the interaction effect at the interface DOFs, where the substructure of interest is separated from the remaining part of a structure. The interaction effects are treated as interface forces that are computed based on all the measurement acceleration, velocity, and displacement at the interface DOFs. For practicability, these methods used acceleration measurements only to compute the interface force by adopting the concept of "quasi-static" displacement vectors. The displacements of internal DOFs are decomposed into quasi-static displacement and "relative" displacement. Nevertheless, inaccuracy could result from (i) using relative acceleration rather than directly using response acceleration and (ii) neglecting the velocity-dependent part in the interface forces. Therefore, to improve the accuracy of identification results, it is necessary to develop a Sub-SI method that uses direct accelerations to compute interface forces (Laory et al., 2012). Moreover, it should be noted that in these Sub-SI methods, only a simple GA is used as the search engine. With the goal of enhancing the identification's robustness, it is worthwhile to improve the GA's search capability for Sub-SI methods. Several GA-based Sub-SI methods using the quasi-static displacement concept and the simple GA can be found in Koh and Thanh (2009), Narayana and Shankar (2006), and Sandesh and Shankar (2006).

A strategy is proposed herein that integrates a multifeature GA into the framework of a divide-and-conquer method to identify the structural parameters of large structures. This strategy uses direct acceleration measurements, without requiring velocity and displacement measurements. The effectiveness of the proposed strategy is illustrated in numerical studies. In addition, the strategy is applied to detect local damage at critical zones of large structures, commonly known as *local damage detection*. The experimental study verifies the applicability of the strategy on a 10-story steel frame.

19.3 Multifeature GA

19.3.1 A Simple GA

GA was introduced in the 1960s is a stochastic search method belonging to the family of metaheuristic methods. It is essentially based on Darwin's theory of natural selection and natural genetics (Goldberg, 1989). It combines survival of the fittest individuals with randomized information exchange. To emulate evolution in the natural world, a GA is composed of three operators: *reproduction* (or selection), *crossover*, and *mutation*. Recognizing that these operators can be modeled in an artificial system, a computational model based on GA was well developed by Holland (Raphael and Smith, 2003). More details about GA can be found in Chapter 1 of this book. In the context of identification, GA has been shown to have several key advantages over several gradient search methods (Koh et al., 2000):

- Enhancement of global convergence by conducting a population-to-population search;
- No requirement of gradient information or other auxiliary knowledge; only the objective function and corresponding fitness values drive the search process;
- Relative ease of implementation and convenient use of any measured response in defining the fitness function;
- A robust self-start feature with random initial guess in a relatively wide search range;
- A high level of concurrence and thus suitable for parallel computing when needed;
- An objective function is defined in terms of any desired response quantity at the user's convenience.

With significant advantages as mentioned previously, the GA-based soft computing approach has been successfully used in civil engineering applications such as construction scheduling and structural optimization (Maeck and De Roeck, 2003; Maeck et al., 2001; Trinh, 2010; Yang and Soh, 1997; Ye et al., 2000). Nevertheless, in the context of structural identification with more challenges (such as a large number of unknowns, presence of I/O noise, and incomplete measurements), the use of simple GA alone does not necessarily work. It is essential to improve GA to make it work more effectively.

19.3.2 A Multifeature GA

The main idea behind the multifeature GA method is to integrate several novel features for improving the search capability of GA, thereby increasing the accuracy and reliability of solutions (Perry et al., 2006a,b). These features are integrated into an improved GA based on migration and artificial selection (iGAMAS). In addition, the convergence of solutions during the search process is enhanced by introducing a search space reduction method (SSRM). From an algorithm point of view, the multifeature GA method is composed of two iteration loops: inner and outer loops that correspond to iGAMAS and SSRM.

The iGAMAS method simultaneously explores the search space as a *global* search and exploits promising individuals as a *local* search. This method integrates several novel features into the GA, including concurrent evolution of several species (subpopulations) with different roles, migration between species, artificial selection, regeneration, reintroduction, and a variable data length procedure. The layout of iGAMAS is illustrated in Figure 19.1. More details can be found in Koh and Perry (2010).

The motivation to develop SSRM (Figure 19.2) comes from the fact that the GA's convergence and accuracy are highly dependent on the search space. By progressively and adaptively reducing the search space, more accurate and efficient identification is possible. The idea is to narrow the search space for parameters (unknowns) that converge quickly, so as to dedicate the search effort to the

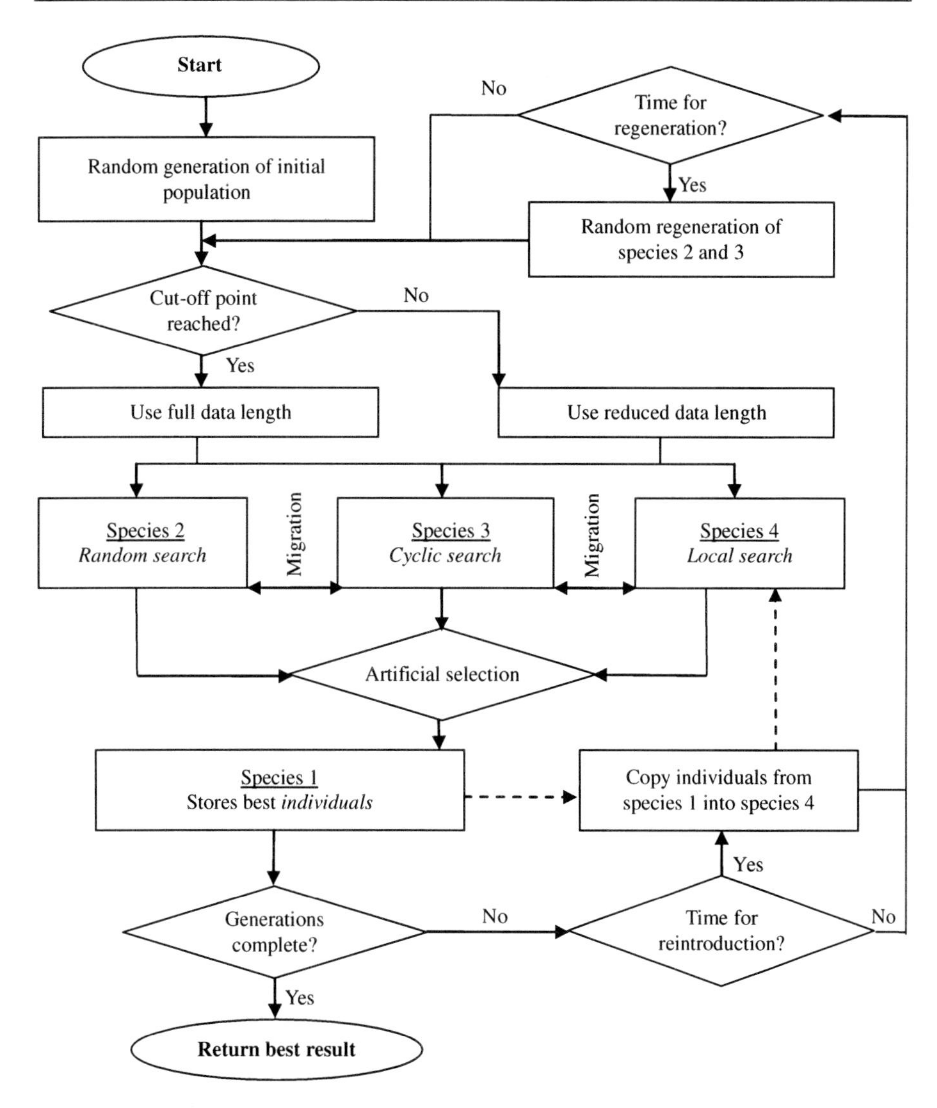

Figure 19.1 An improved GA based on migration and artificial selection (iGAMAS).

remaining parameters. After several runs of iGAMAS, the results are used to compute two statistical parameters (means and standard deviations) for each parameter. The standard deviation indicates the variation of each parameter, and thus its search space can be adjusted accordingly. The multi-feature GA method has shown advantages and robustness over the simple GA. The numerical study indicated a significant improvement in the reliability and accuracy of the identified parameters compared to a simple GA. The significance of this method was demonstrated through successfully identifying a 20-DOF unknown-mass system, involving

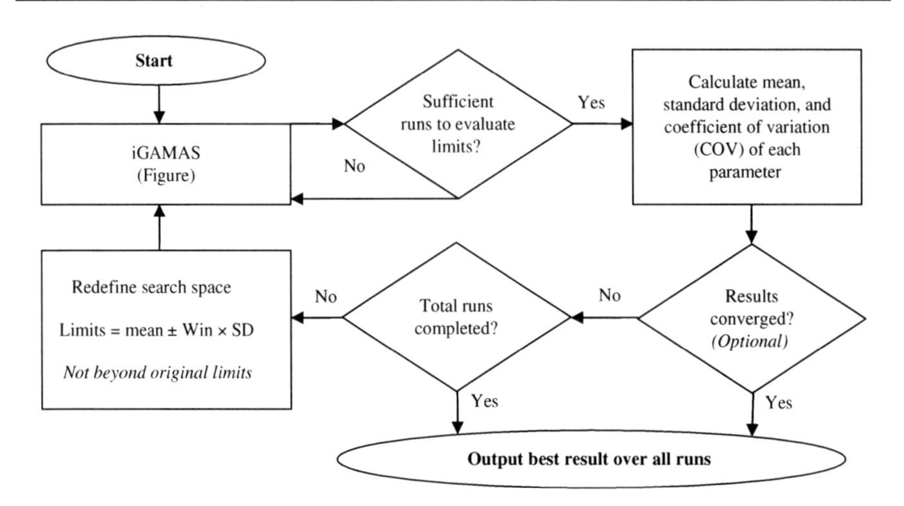

Figure 19.2 Search space reduction method (SSRM).

42 unknowns (Perry et al., 2006b). This approach also has been applied for the identification of various problems (Koh and Perry, 2010; Perry and Koh, 2008; Perry et al., 2007; Thanh et al., 2009; Wang et al., 2010). It is important to note that in these problems, all the structures were identified at one go, a process usually known as CSI.

Although this method has been successfully applied to structural systems, its applicability is limited to structures with typically not more than 50 unknowns. For large structural systems, the modeling of such systems often involves a large number of degrees of freedom, involving a large number of unknowns. Therefore, this method would face three challenges: (i) numerical difficulties to obtain accurate identification results, (ii) the need for a large number of sensors, and (iii) expensive computations required for managing and processing the large sets of data collected.

19.4 Divide-and-Conquer-Based Structural Identification

To address the above-mentioned challenges, the divide-and-conquer method provides a good solution by dividing the whole structure into manageable smaller parts, known as "substructures," on which the identification is carried out independently. This procedure is referred to as Sub-SI. There are six advantages when employing the divide-and-conquer method to identify large structures:

- Since each substructure involves a reasonable number of degrees of freedom and unknown parameters compared to that of the whole structure, the speed and capability of converging to an accurate solution is improved.

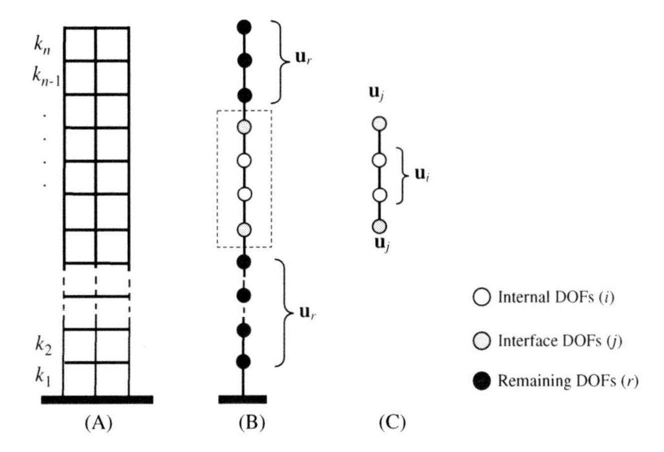

Figure 19.3 (A) Shear building; (B) lumped-mass system; and (C) a substructure.

- It is not necessary to monitor all the DOFs of a structure simultaneously. Instead, only the critical zones (where damage is likely to occur) need to be monitored, allowing for a significant reduction in the number of sensors required and for efficient management of data.
- The effect of measurement errors on the accuracy of identification results can be greatly minimized due to fewer sensors being used.
- Modeling errors induced due to mathematical modeling and inappropriate assumption of uncertain boundary conditions in structural identification are inevitable (Catbas et al., 2007; Saitta et al., 2005; Smith and Saitta, 2008). Sub-SI minimizes the error by identifying substructures that are completely independent of the boundary of a structure.
- If there is no force to be applied within a substructure, its unknown parameters can be identified without the need of force information.

Since each substructure is independently identified, the identification of many substructures can be conducted concurrently using parallel computing techniques. In particular, for structural health monitoring (SHM) applications using smart sensing technology (Spencer et al., 2007), Sub-SI can be employed as a distributed computing strategy (DCS) in a hierarchical SHM system. Communication and data processing in DCS mainly take place in the group of local sensors, thus reducing the transmission of large amounts of data.

Sub-SI is conceptually demonstrated through a shear building, as shown in Figure 19.3. For substructuring, the equations of motion of the entire structure can be written in partition form as

$$\begin{bmatrix} \mathbf{M}_{rr} & \mathbf{M}_{rj} & \mathbf{0} \\ \mathbf{M}_{jr} & \mathbf{M}_{jj} & \mathbf{M}_{ji} \\ \mathbf{0} & \mathbf{M}_{ij} & \mathbf{M}_{ii} \end{bmatrix} \begin{Bmatrix} \ddot{\mathbf{u}}_r \\ \ddot{\mathbf{u}}_j \\ \ddot{\mathbf{u}}_i \end{Bmatrix} + \begin{bmatrix} \mathbf{C}_{rr} & \mathbf{C}_{rj} & \mathbf{0} \\ \mathbf{C}_{jr} & \mathbf{C}_{jj} & \mathbf{C}_{ji} \\ \mathbf{0} & \mathbf{C}_{ij} & \mathbf{C}_{ii} \end{bmatrix} \begin{Bmatrix} \dot{\mathbf{u}}_r \\ \dot{\mathbf{u}}_j \\ \dot{\mathbf{u}}_i \end{Bmatrix}$$
$$+ \begin{bmatrix} \mathbf{K}_{rr} & \mathbf{K}_{rj} & \mathbf{0} \\ \mathbf{K}_{jr} & \mathbf{K}_{jj} & \mathbf{K}_{ji} \\ \mathbf{0} & \mathbf{K}_{ij} & \mathbf{K}_{ii} \end{bmatrix} \begin{Bmatrix} \mathbf{u}_r \\ \mathbf{u}_j \\ \mathbf{u}_i \end{Bmatrix} = \begin{Bmatrix} \mathbf{P}_r \\ \mathbf{P}_j \\ \mathbf{P}_i \end{Bmatrix} \tag{19.1}$$

where **M**, **C**, and **K** are the mass, damping, and stiffness matrices, respectively; **u**, **u̇**, and **ü** are the displacement, velocity, and acceleration vectors, respectively; **P** is the input force vector; subscript i denotes the internal DOFs of the substructure of interest, and subscript j denotes the interface DOFs of the substructure with the remaining structures r. The equation of motion for the substructure of interest which is extracted from the above equation is written as

$$\begin{bmatrix} \mathbf{M}_{ij} & \mathbf{M}_{ii} \end{bmatrix} \begin{Bmatrix} \ddot{\mathbf{u}}_j \\ \ddot{\mathbf{u}}_i \end{Bmatrix} + \begin{bmatrix} \mathbf{C}_{ij} & \mathbf{C}_{ii} \end{bmatrix} \begin{Bmatrix} \dot{\mathbf{u}}_j \dot{\mathbf{u}}_i \end{Bmatrix} + \begin{bmatrix} \mathbf{K}_{ij} & \mathbf{K}_{ii} \end{bmatrix} \begin{Bmatrix} \mathbf{u}_j \\ \mathbf{u}_i \end{Bmatrix} = \mathbf{P}_i \tag{19.2}$$

Treating interaction effects at the interface DOFs, where the substructure of interest is separated from the remaining parts of a structure, as interface forces to the substructure (Koh et al., 1991, 2003), this equation can be rewritten as

$$\mathbf{M}_{ii}\ddot{\mathbf{u}}_i + \mathbf{C}_{ii}\dot{\mathbf{u}}_i + \mathbf{K}_{ii}\mathbf{u}_i = \mathbf{P}_i + \mathbf{P}_j \tag{19.3}$$

where interface force $\mathbf{P}_j$ is treated as an input force to the substructure and is computed from interface measurements as

$$\mathbf{P}_j = -\left(\mathbf{M}_{ij}\ddot{\mathbf{u}}_j + \mathbf{C}_{ij}\dot{\mathbf{u}}_j + \mathbf{K}_{ij}\mathbf{u}_j\right) \tag{19.4}$$

Thus far, the Sub-SI procedure seems relatively straightforward if the accelerations, velocities, and displacements at the interface DOFs are measurable to compute interface force using this approach. In reality, nevertheless, measuring accelerations is often preferred over velocity and displacement quantities. Therefore, the difficulty of using Sub-SI lies in appropriately accounting for the interaction effects at the interface DOFs using only acceleration measurements. To eliminate the requirement of velocity and displacement, a numerical integration is adopted to estimate velocity and displacement values from the measured interface acceleration as follows:

$$\dot{\mathbf{u}}_j^{k+1} = \dot{\mathbf{u}}_j^k + \frac{\Delta t}{2}\left(\ddot{\mathbf{u}}_j^k + \ddot{\mathbf{u}}_j^{k+1}\right); \quad \mathbf{u}_j^{k+1} = \mathbf{u}_j^k + \frac{\Delta t}{2}\left(\dot{\mathbf{u}}_j^k + \dot{\mathbf{u}}_j^{k+}1\right) \tag{19.5}$$

where Δt is the time step.

Note that using Eq. (19.5), interface velocity and displacement quantities are reasonably achieved if acceleration measurements are noise free. In reality, the noise contamination in measurement signals is inevitable, and thus it may result in a small drift in the integrated velocity and displacement time histories due to accumulating noise errors on numerical integration, leading to a small drift in the interface forces, $\mathbf{P}_j$. From a frequency analysis point of view, this drift may be regarded as low-frequency force components. It is also noted that the natural frequencies of substructures are usually higher than those of the whole structure. The dynamic responses of substructures are predominantly excited by force components with their frequencies close to substructures' frequencies. Therefore, the responses from the low-frequency components of the interface force would be negligible.

The layout of the Sub-SI strategy is shown in Figure 19.4. Sub-SI employs the multifeature GA method as a search engine in the framework of the divide-and-conquer method. Therefore, this strategy is an effective combination of two complementary advantages that are based on two different principles. While the divide-and-conquer method reduces the structure size to be identified, the multifeature GA enhances search effectiveness. For each trial set of unknown parameters, the forward (or dynamic) analysis is carried out by numerically solving the dynamic equation given in Eq. (19.3) for each substructure. Then, the simulated acceleration time histories obtained from each trial set is compared to the measured acceleration time histories using an objective function. The goal of GA is to adjust the trial parameters in order to best match the simulated *internal* acceleration time histories with the measured internal ones. Consequently, the square of errors summed over N_m measured internal DOFs and the data length L_t of acceleration time histories is minimized. The objective function is computed as the inverse of the squared error as

$$f_e = \frac{1}{\varepsilon + \sum_{i=1}^{N_m} \sum_{j=1}^{L_t} (\ddot{u}_{i,j}^{m} - \ddot{u}_{i,j}^{s})^2} \quad (19.6)$$

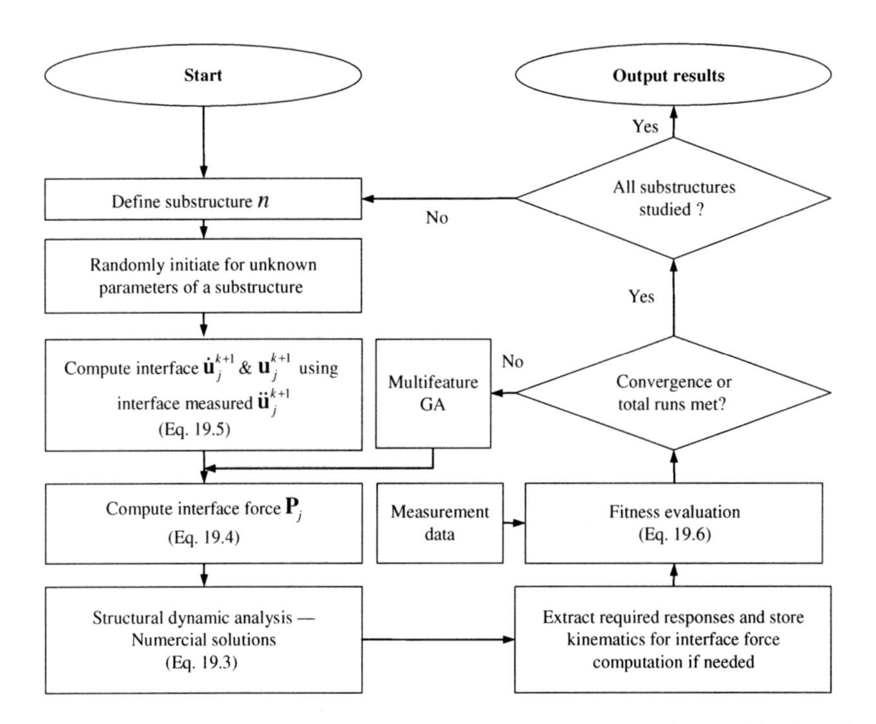

Figure 19.4 A layout of the evolutionary divide-and-conquer strategy for identification of large structures.

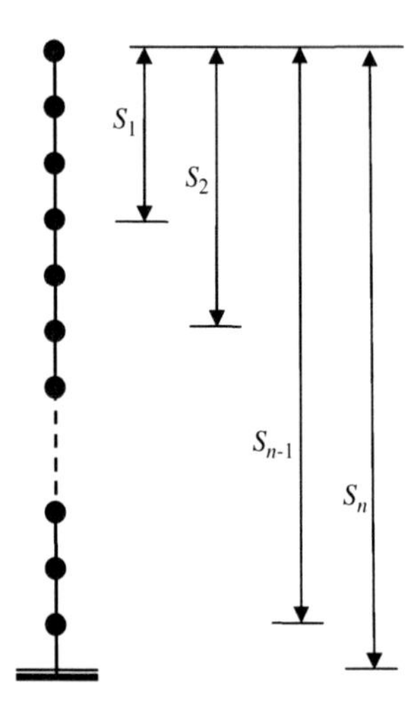

Figure 19.5 Progressive substructural identification (Pro-SI).

where superscripts m and s denote measured and simulated quantities, respectively. A small value, ε, is used to avoid singularity when the simulated and measured responses happen to match exactly. Its value is chosen to be 0.001 for this study.

Structural identification using more measurements is generally more reliable than identification using few measurements. Therefore, given a limited number of available sensors, the accuracy of identification can be enhanced if the measurement system allows the shifting of sensors into the substructure under investigation. If this is not possible, an alternative is to use as many response measurements as possible by progressively expanding the domain of substructure identification as shown in Figure 19.5, referred to as *progressive substructural identification* (Pro-SI) (Koh et al., 2003). Pro-SI can be seen as a variation of Sub-SI, with the same idea of dividing the whole structure into many substructures to improve the identification performance. The primary difference is that the substructures progressively enlarge while still keeping the number of unknowns manageable. This is done by carrying forward the identified parameters of the current substructure to the identification of the next substructure and treating them as known values. Hence, Pro-SI conducts identification of substructures in sequence.

19.5 Numerical Study

The performance of the evolutionary divide-and-conquer strategy is illustrated and assessed through identification of a 100-DOF shear building that involves 202 unknowns (100 mass parameters, 100 stiffness parameters, and 2 damping ratio parameters). For comparison, CSI is also used to identify this structure as a whole without substructuring.

The initial search range of half to double the exact values is assigned for all unknown parameters of the structure. To consider the stability of GA's random search, all identifications are repeated five times and the average of identified

Table 19.1 GA Parameters Used for the Known-Mass and Unknown-Mass Systems in the Numerical Simulation

Parameter	Unknown-Mass System
Population size	65×3
Runs	5/25
Generations	200
Crossover rate	0.4
Mutation rate	0.2
Window width	4.0
Migration	0.05
Regeneration	3
Reintroduction	100

parameters is presented. The GA parameters are shown in Table 19.1. The identifications are carried out in the presence of 5% and 10% noise. To simulate a noise-contaminated signal, $\mathbf{x}_{con}$, noise is added to a clean signal $\mathbf{x}_{cle}$ as

$$\mathbf{x}_{con} = \mathbf{x}_{cle} + E_{lev} \times \text{RMS}(\mathbf{x}_{cle}) \times \mathbf{N}_{oise} \tag{19.7}$$

where E_{lev} is a given noise level, RMS($\mathbf{x}_{cle}$) is the root-mean-square value of the clean signal vector, and $\mathbf{N}_{oise}$ is a randomly generated noise vector of Gaussian distribution with zero mean and unit standard deviation.

While existing Sub-SI studies deal with identification of stiffness and/or damping parameters only, the greater challenge here is to identify not only stiffness and damping but also mass parameters simultaneously. In this regard, many time domain methods based on state-space formulation and frequency domain methods are not applicable (Koh et al., 1991; Yun and Bahng, 2000). Therefore, this study focuses on identifying a large structure with unknown-mass information. Its structural properties are $k_1 - k_{30} = 16 \times 10^5$ kN/m, $k_{31} - k_{60} = 12 \times 10^5$ kN/m, $k_{61} - k_{100} = 9 \times 10^5$ kN/m, $m_1 - m_{30} = 15 \times 10^5$ kg, $m_{31} - m_{60} = 12 \times 10^5$ kg, and $m_{61} - m_{100} = 8 \times 10^5$ kg. The natural periods of the first two modes are 10.5 and 4.1 s. Rayleigh damping is assumed with a damping ratio of 2% for the first two modes. Two excitation forces are applied at the third and eighth nodes of every 10 levels. Responses of the building is simulated using a time step of 0.001 s. Six accelerations at the first, third, fifth, sixth, eighth, and tenth nodes of every 10 stories (60% of DOFs) are extracted and used as "measured" data in the identification. For Sub-SI, the complete structure is divided into 20 substructures: $S_1 = [100 - 96]$, $S_2 = [96 - 91]$,. . .,$S_{20} = [6 - 1]$. For Pro-SI, the identification is executed in 20 steps: $S_1 = [100 - 96]$, $S_2 = [100 - 91]$,. . .,$S_{20} = [100 - 1]$.

Tables 19.2 and 19.3 present the mean and maximum absolute errors of identified stiffness and mass values. The accurate identification results can be achieved in the noise-free case. To be able to identify such a large structure with incomplete measurement and to achieve a mean error of less than 3% in a situation with less than 10% noise is remarkable. Indeed, compared to the best results obtained from the previous Sub-SI methods that could identify a structure of 50-DOFs under 5%

Table 19.2 Absolute Error in Identified Stiffness of a 100-DOF Structure

Noise Level (%)	Mean Error (%)			Maximum Error (%)		
	CSI	Sub-SI	Pro-SI	CSI	Sub-SI	Pro-SI
0	4.80	0.00	0.02	13.15	0.07	0.13
5	5.76	0.95	0.89 (5.1)	15.59	2.99	2.81
10	6.91	2.97	2.48	17.99	9.75	7.36

Note: The value in parentheses is the mean absolute error of the identification result of a 50-DOF unknown-mass system under 5% noise in a recent study (Koh et al., 2003).

Table 19.3 Absolute Error in Identified Mass of a 100-DOF Structure

Noise Level (%)	Mean Error (%)			Maximum Error (%)		
	CSI	Sub-SI	Pro-SI	CSI	Sub-SI	Pro-SI
0	3.36	0.00	0.01	14.67	0.07	0.07
5	3.79	0.81	0.86 (5)	13.69	3.91	2.71
10	4.71	2.50	2.55	14.45	8.24	6.73

Note: The value in parentheses is the mean absolute error of the identification result of a 50-DOF unknown-mass system under 5% noise in a recent study (Koh et al., 2003).

noise as shown by Koh et al. (2003), the results achieved from this study are much better, despite a larger system of 100 DOFs to be identified. Good results also confirm that the effect of the low-frequency component of interface forces on the identification of a substructure could be negligible.

Note that for a large structure, Sub-SI and Pro-SI yield much more accurate results than CSI. For stiffness identification under 10% noise, the mean error for CSI is about 6.91%, but it reduces to 2.97% and 2.48% for Sub-SI and Pro-SI, respectively. Correspondingly, the maximum absolute error also reduces from 17.99% to 9.75% and 7.36%, respectively. Likewise, for mass identification, there is an improvement, with reduction in the mean error from 4.71% for CSI to 2.55% for Pro-SI, and correspondingly in the maximum error from 14.45% to 6.73%. Furthermore, the efficiency of the evolutionary divide-and-conquer strategy is demonstrated in the saving of computational time of 27% for Sub-SI and 2% for Pro-SI compared to CSI (391 min on a Core 2 Duo, 3 GHz PC).

19.6 Applications to Local Damage Detection

In the context of structural health monitoring, structural damage usually takes place at some critical zones whose members may likely reach their maximum strength prior to the members in other zones, and whose members are significantly affected by environmental factors such as corrosion. Therefore, it is more efficient and economical if we can monitor critical zones locally instead of the whole large structure.

Damage is often manifested through changes in physical properties of structural members. Therefore, by applying the Sub-SI strategy to identify changes in stiffness parameters before and after damage occurrence, damage in each member can be detected, localized, and quantified within a substructure. Here, the goal is not only to identify damage locations but also to quantify damage extents (or magnitudes) using the measured acceleration responses before and after damage takes place. A damage extent D_i in member i is quantified as a ratio of the loss in stiffness to the original undamaged stiffness:

$$D_i = \frac{k_i^{\mathrm{u}} - k_i^{\mathrm{d}}}{k_i^{\mathrm{u}}} \tag{19.8}$$

where k_i^{u} and k_i^{d} respectively denote the undamaged and damaged stiffness values for member i.

The identified results of the undamaged structure are used as a starting point for identification of the damage structure. This is implemented by assigning the identified parameters of the undamaged substructure to some initial values in the identification of the damaged substructure. Only changes need to be identified, and therefore the identification is to be carried out faster than that for the undamaged structure.

This example assumes that damage takes place at stories 62, 63, and 66 in the 100-story shear building as mentioned in Section 19.5 (Figure 19.3), with damage extents of 10%, 10%, and 20%, respectively. A substructure including stories 61−67 is chosen to identify independently of the remaining parts of the structure. The input force is measured at story 63, and six acceleration measurements are taken at stories 60, 61, 63, 65, 66, and 67. The I/O noise level is set at 10% for both undamaged and damaged states. Note that the responses at other stories outside this substructure are not needed. The identification is carried out five times for the substructure in both undamaged and damaged states, and the average results are used to compute the reduction in stiffness for each story. The GA parameters and search ranges are set exactly the same as those used in Section 19.5 (Table 19.1).

Damage detection typically considers two key aspects: true damage and false damage. True damage is the extent that is identified at the damaged location, whereas false damage is the extent at the undamaged location. Both these aspects are presented in Table 19.4 and graphically shown in Figure 19.6. The first aspect is the true damage identified at the damaged stories 62, 63, and 66, along with the corresponding absolute errors compared to "real" damage. The results indicate that even when 10% noise is present in the measurements of undamaged and damaged structures, the strategy is able to localize and quantify damage accurately, with a small error of 3%.

The second aspect, which is just as important, is the reduction in the damage extent that is falsely reported at the undamaged story. As an indication of this error, the maximum false damage identified at the undamaged stories is also examined. The results highlight that the damage is successfully detected since true damage extent is significantly larger than maximum false damage. Indeed, the true damage extent of 11.59%

Table 19.4 Local Damage Quantification Results Using a Substructure (Stories 60–67) of a 100-story Shear Building (Real Damage 10% at Stories 62 and 63 and 20% at Story 66)

Story	Identified Damage Extent (%)	Absolute Error (%)
62	11.59	1.59
63	13.00	3.00
66	21.26	1.26
Maximum false damage	5.36	

Note: The noise level is 10% in both cases of undamaged and damaged states.

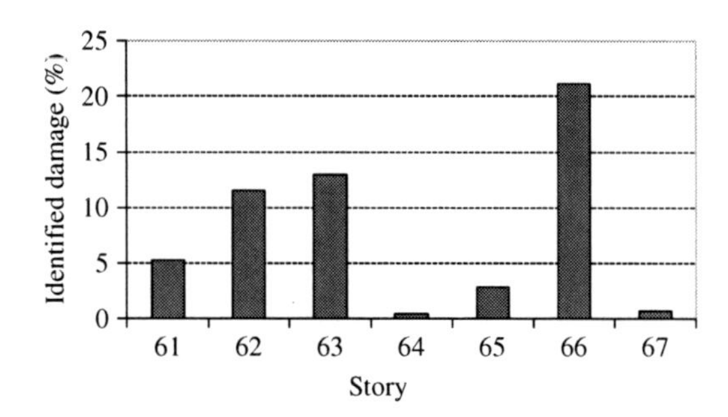

Figure 19.6 Damage quantification results in a substructure (stories 60 to 67) within a 100-story shear building using incomplete acceleration responses and a known input forces under 10% noise (damage 10% at stories 62 and 63 and 20% at story 66).

is twice as much as the maximum false damage of 5.36%. In addition, Figure 19.6 shows that the strategy is able to identify damage that takes place in locations that are close to one another (e.g., damage at stories 62 and 63). Also, it successfully quantifies damage at different extents (e.g., damage extents of 10% at stories 62 and 63 and 20% at story 66). These inferences are verified further in experimental study.

19.7 Experimental Verification

The strategy has successfully identified damage at a specific zone in a large structure in the numerical study. Thus, it is highly desirable to examine the performance of this strategy on a real model. The aim of this experimental study on a 10-story steel frame is to assess the proposed strategy in terms of the following aspects:

- The capability of identifying a single-damage and multiple-damage events with some closeness to one another;
- The effect of different damage extents on the accuracy of identification results;
- The effect of damage outside a substructure on the accuracy of identification results;
- Verification of the accuracy of the damage detection through comparing the results that are identified in various substructures;
- The margin of true damages compared to maximum false damages.

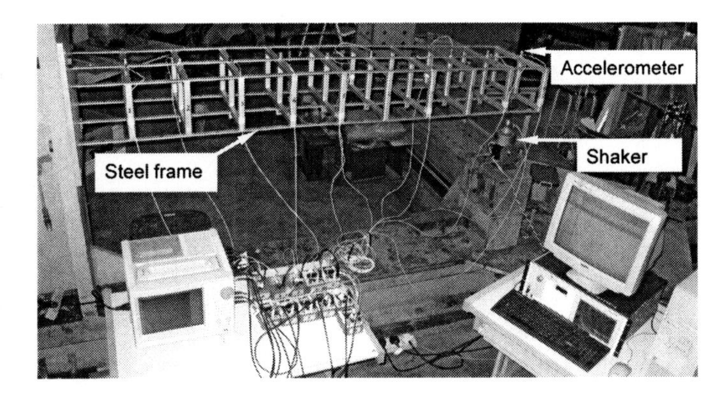

Figure 19.7 Test setup in the laboratory.

Figure 19.7 shows the experimental setup, where the frame is fixed horizontally to a strong frame for ease of force application. A vertical force was generated by an electromagnetic shaker (Labworks ET-126B) on the steel frame via a connection rod at level 10 (free end). To study the consistency of identification results, four different forces (A, B, C, and D) were generated, each with 1000 data points at a time step of 0.1 s.

The overall dimension of the frame was 2 m in height and 0.2 m $\times$ 0.4 m in the plan. It was fabricated by six columns of rectangular cross section 25×4.5 (in millimeters) and five beams of square hollow section $25 \times 25 \times 3$ at each of these 10 stories. As the beams were relatively stiff, the frame behaved as a shear building that is schematically represented as a lumped-mass system. By symmetry of the structure and loading, the significant motion at each level was a single-direction horizontal translation. The lumped mass value was 3.25 kg for each of levels 1−9 and 3 kg for level 10. The natural frequencies of this frame ranged from 8.3 Hz for the first mode to 115.2 Hz for the tenth mode.

A total of 10 accelerometers were mounted on the upper plane of each story to measure accelerations of the frame at a sampling rate of 10 kHz. The signals from the force sensor were also recorded using a digital oscilloscope. Although the highest frequency of the frame corresponding to the tenth mode was 115.2 Hz, the high sampling frequency allows for a more accurate simulation of the response during identification since the frequencies of a substructure were often higher than those of the complete structure. It was also noted that the accuracy of interface acceleration affects the identification results of a substructure. Signals of duration 0.1 s starting immediately before the application of shaker force on the frame were used as input in the identification strategy.

In the study, 10 damage scenarios were generated by progressively cutting one of the center columns (Figure 19.8) at five stories (8, 5, 4, 2, and 9 in this order). The damage scenarios were named D1−D10, as summarized in Table 19.5. Since each story has six columns, cutting one column resulted in a reduction of approximately 16.67% in the corresponding story stiffness. Therefore, cutting one and two columns corresponds to theoretical damage extents of 16.7% and 33.3%.

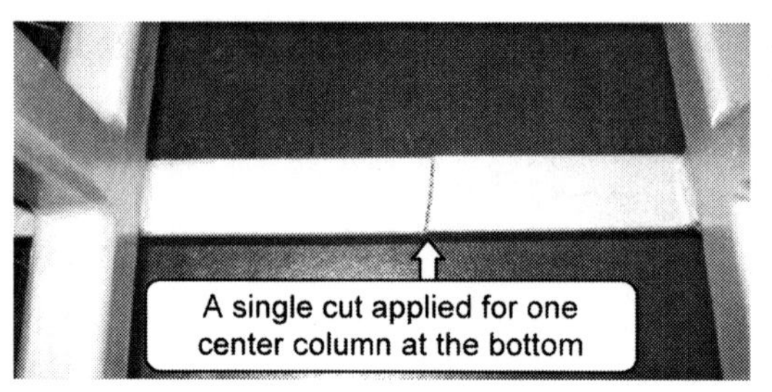

Figure 19.8 A single cut at one column corresponding to damage 16.67%.

Table 19.5 Damage Scenarios Considered

Damage Scenario	Damaged Stories	Corresponding Damage Indices
D1	8th	16.7%
D2	8th	33.3%
D3	5th + 8th	16.7% + 33.3%
D4	4th + 5th + 8th	16.7% + 16.7% + 33.3%
D5	4th + 5th + 8th	33.3% + 16.7% + 33.3%
D6	4th + 5th + 8th	33.3% + 33.3% + 33.3%
D7	2nd + 4th + 5th + 8th	16.7% + 33.3% + 33.3% + 33.3%
D8	2nd + 4th + 5th + 8th	33.3% + 33.3% + 33.3% + 33.3%
D9	2nd + 4th + 5th + 8th + 9th	33.3% + 33.3% + 33.3% + 33.3% + 16.7%
D10	2nd + 4th + 5th + 8th + 9th	33.3% + 33.3% + 33.3% + 33.3% + 33.3%

Damage at stories 4, 5, 8, and 9 can be detected by considering a 7-story substructure from story 3 to story 9. The search range for stiffness is 200–600 kN/m for all stiffness parameters, 2.5–3.5 kg for all mass parameters, and 0–4 and 0–0.0002 for the two damping parameters. Note that all stiffness, mass, and damping parameters of the substructure are treated as unknowns. The acceleration measurements at levels 2–9 are used in damage identification.

Figure 19.9 presents typical damage identification results using complete measurements. For two single-damage scenarios D1 and D2, the damage identification results are in very good agreement with the theoretical damage extents of 16.7% and 33.3%, respectively. The maximum absolute error in these scenarios is 2.6%. For multiple damages (e.g., D10 with five damaged stories), all damaged stories (four in the substructure considered) are well identified with good mean absolute error of about 2.5%. Good results demonstrate that the strategy is capable of quantifying two adjacent damaged stories, such as stories 4 and 5 (D4, D7, and D8), and stories 8 and 9 (D10).

One of the concerns for local damage detection is that damage occurring outside the substructure considered may affect the accuracy of the identification results. This concern is addressed through examining scenarios D7 and D8 in Figure 19.9. Comparing the identified damage extents in these scenarios, it is seen that although two damage extents of 16.67% and 33.33% subsequently occurred at story 2 outside this substructure, the identified damage magnitudes at stories 4, 5, and 8 of the

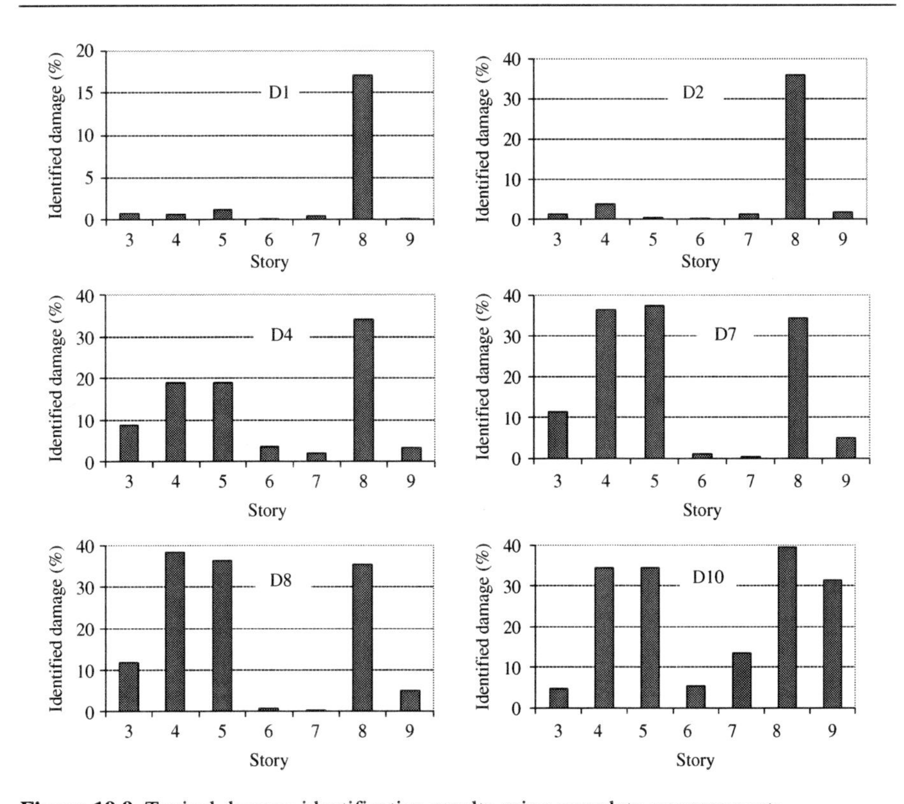

Figure 19.9 Typical damage identification results using complete measurements.

substructure in both scenarios were very similar. This implies that damage within the substructure considered can be identified independently without the effects of any damage outside it.

Measurement and model errors may cause the damage index at the undamaged story to be larger than zero; this is known as *false damage.* For practical applications, it is important that the true damage extent identified exceeds the maximum false damage extent by a sufficient margin. Figure 19.9 shows that the identified magnitude of the true damage is usually at least two times higher than that of maximum false damage, except for scenario D9, where the identified true damage at story 9 is only 37% higher than the maximum false damage magnitude (8.70% at story 3).

To examine the strategy in the case of incomplete measurements, the same identification procedure is applied for the same substructure mentioned above, but the measurements at stories 5 and 7 are no longer used. Typical identification results for the substructure are shown in Figure 19.10.

For a single-damage scenario D1, the proposed strategy still achieves good results using incomplete measurement. A maximum absolute error is less than 2%. Similarly, for a multiple-damage scenario D10, the identification results are also

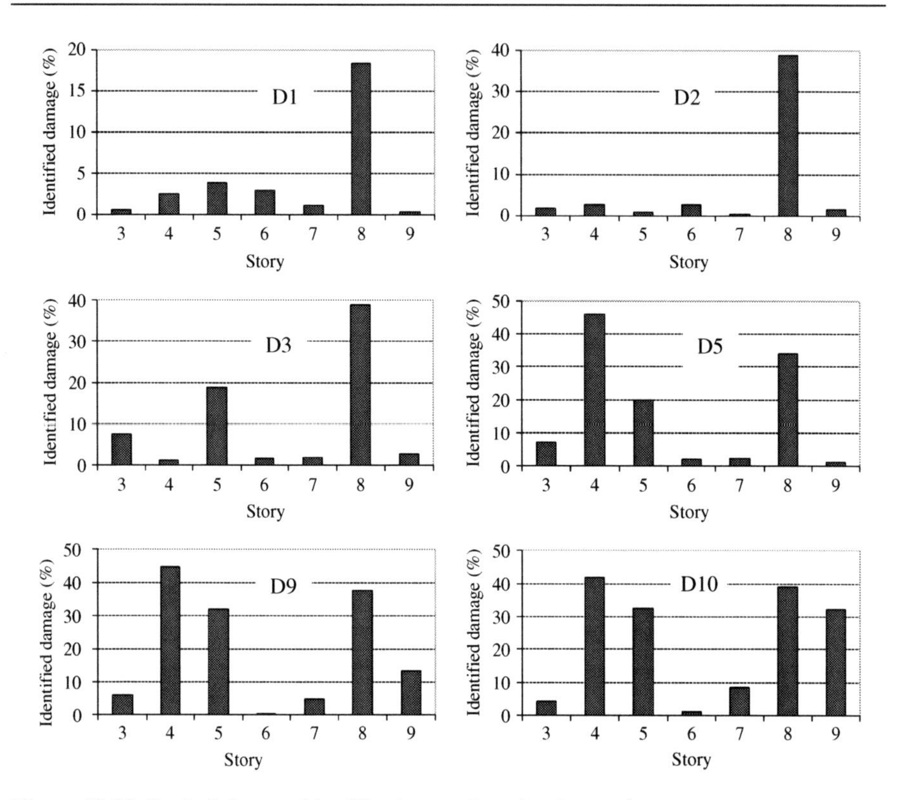

Figure 19.10 Typical damage identification results using incomplete measurements.

relatively good, with a maximum error of 8.63%. The identification results using incomplete measurements are not as good as than those using complete measurements [e.g., D5 (maximum error 12.62%) and D9 (maximum error 11.47%)] due to the coupling effect of damage at two adjacent stories.

19.8 Conclusions

By combining two complementary advantages of a multifeature GA and a divide-and-conquer method, the proposed strategy is useful for the identification of large structures with unknown stiffness, mass, and damping parameters. The study shows that it is able to identify stiffness parameters of a large structure up to 100 DOFs, involving 202 unknowns, yielding results with the mean absolute error of less than 2% even when using responses contaminated by 10% noise. In addition, this strategy is capable of locally identifying damage at critical zones where damage likely takes place. Multiple damage events at critical zones, with some closeness to one another, can be identified using only acceleration responses and input excitation forces within the selected substructures. The extent of a single damage is identified more accurately than those of multiple damage events. When several damage events with different magnitudes occur at adjacent locations, the accuracy of identification results can be improved by utilizing complete measurements. The extent of true damage exceeds that of maximum false damage by a reasonable margin.

References

Catbas, F.N., Ciloglu, S.K., Hasancebi, O., Grimmelsman, K., Aktan, A.E., 2007. Limitations in structural identification of large constructed structures. J. Struct. Eng. 133, 1051–1066.

Goldberg, D.E., 1989. Genetic Algorithms in Search, Optimization, and Machine Learning. Addison-Wesley, Boston, MA.

Goodwin, G.C., Sin, K.S., 1984. Adaptive Filtering Prediction and Control. Prentice Hall, Englewood Cliffs, NJ.

Juang, J.N., Pappa, R.S., 1985. An eigensystem realization algorithm for modal parameter identification and model reduction. J. Guidance Control Dyn. 8, 620–627.

Juang, J.N., Phan, M., Horta, L.G., Longman, R.W., 1993. Identification of observer/Kalman filter Markov parameters: theory and experiments. J. Guidance Control Dyn. 16, 320–329.

Koh, C.G., Perry, M.J., 2010. Structural Identification and Damage Detection Using Genetic Algorithms. Taylor & Francis, London.

Koh, C.G., Thanh, T.N., 2009. Challenges and strategies in using genetic algorithms for structural identification. In: Topping, B.H.V., Tsompanakis, Y. (Eds.), Soft Computing in Civil and Structural Engineering. Saxe-Coburg Publications, Stirlingshire.

Koh, C.G., See, L.M., Balendra, T., 1991. Estimation of structural parameters in time domain. A substructure approach. Earthquake Eng. Struct. Dyn. 20, 787–801.

Koh, C.G., Hong, B., Liaw, C.Y., 2000. Parameter identification of large structural systems in time domain. J. Struct. Eng. 126, 957–963.

Koh, C.G., Hong, B., Liaw, C.Y., 2003. Substructural and progressive structural identification methods. Eng. Struct. 25, 1551–1563.

Koh, C.G., Tee, K.F., Quek, S.T., 2006. Condensed model identification and recovery for structural damage assessment. J. Struct. Eng. 132, 2018–2026.

Laory, I., Nizar, B.H.A., Trinh, T.N., Smith, I.F.C., 2012. Measurement system configuration for damage identification of continuously monitored structures. J. Bridge Eng. 17 (6), 857–866.

Lee, C.-G., Yun, C.-B., 1991. Parameter identification of linear structural dynamic systems. Comput. Struct. 40, 1475–1487.

Maeck, J., De Roeck, G., 2003. Damage assessment using vibration analysis on the Z24-bridge. Mech. Syst. Signal Process. 17, 133–142.

Maeck, J., Peeters, B., De Roeck, G., 2001. Damage identification on the Z24 bridge using vibration monitoring. European COST F3 Conference on System Identification and Structural Health Monitoring, June 2000. IOP Publishing, 512–517.

Narayana, M.K., Shankar, K., 2006. Time domain identification of structural parameters without interface measurement using substructural analysis. Second International Congress on Computational Mechanics and Simulation, Guwahati, India.

Oreta, A.W.C., Tanabe, T.-A., 1994. Element identification of member properties of framed structures. J. Struct. Eng. 120, 1961–1975.

Perry, M.J., Koh, C.G., 2008. Output-only structural identification in time domain: Numerical and experimental studies. Earthquake Eng. Struct. Dyn. 37, 517–533.

Perry, M.J., Koh, C.G., Choo, Y.S., 2006a. Identification of damage in a steel frame using a modified genetic algorithm. Fourth World Conference on Structural Control and Monitoring. San Diego, CA, United States.

Perry, M.J., Koh, C.G., Choo, Y.S., 2006b. Modified genetic algorithm strategy for structural identification. Comput. Struct. 84, 529–540.

Perry, M.J., Halkyard, J.E., Koh, C.G., 2007. Rapid preliminary design of floating offshore structures using a modified genetic algorithm. Am. Soc. Mech. Eng.777–784.

Raphael, B., Smith, I.F.C., 2003. Fundamentals of Computer-Aided Engineering. Wiley, West Sussex.

Saitta, S., Raphael, B., Smith, I.F.C., 2005. Data mining techniques for improving the reliability of system identification. Adv. Eng. Inform. 19, 289–298.

Sandesh, S., Shankar, K., 2006. Time domain parametric identification of plate bending rigidity coefficients using substructural approach. Second International Congress on Computational Mechanics and Simulation (ICCMS-06), Guwahati, India.

Smith, I.F.C., Saitta, S., 2008. Improving knowledge of structural system behavior through multiple models. J. Struct. Eng. 134, 553–561.

Spencer, J.B.F., Nagayama, T., Rice, J.A., Agha, G.A., 2007. Smart Sensor Technology: A New Paradigm for Structural Health Monitoring, Yokohama, Japan.

Tee, K.F., Koh, C.G., Quek, S.T., 2005. Substructural first- and second-order model identification for structural damage assessment. Earthquake Eng. Struct. Dyn. 34, 1755–1775.

Tee, K.F., Koh, C.G., Quek, S.T., 2009. Numerical and experimental studies of a substructural identification strategy. Struct. Health Monit. 8, 397–410.

Thanh, T.N., Koh, C.G., Choo, Y.S., 2009. Identification of spudcan fixity for a jack-up rig. The 28th International Conference on Ocean, Offshore, and Arctic Engineering. Honolulu, HI, USA, ASME 43413, 47–53.

Topping, B.H.V., Tsompanakis, Y., 2009. Soft Computing in Civil and Structural Engineering. Saxe-Coburg Publications, Stirlingshire.

Trinh, T.N., 2010. Evolutionary Divide-And-Conquer Strategy for Identification of Structural Systems and Moving Forces. National University of Singapore, Ph.D. Thesis.

Wang, X.M., Koh, C.G., Thanh, T.N., Zhang, J., 2010. System identification of jack-up platform by spectral analysis. The 29th International Conference on Ocean, Offshore, and Arctic Engineering. Shanghai, China, ASME 49095, 411–420.

Xu, B., Du, T., 2006. Direct substructural identification methodology using acceleration measurements with neural networks. Proceedings of SPIE—The International Society for Optical Engineering, San Diego, CA, United States. SPIE 6178.

Yang, J., Soh, C.K., 1997. Structural optimization by genetic algorithms with tournament selection. J. Comput. Civil Eng. 11, 195–200.

Ye, Z.P., Li, X.L., Dang, C.Y., 2000. Optimization of the main parts of hydroelectric sets using hybrid genetic algorithm. J. Mater. Process. Technol.105.

Yun, C.-B., Bahng, E.Y., 2000. Substructural identification using neural networks. Comput. Struct. 77, 41–52.

Yun, C.-B., Lee, H.-J., 1996. Damage Estimation Using Substructural Identification in Time Domain. ASCE, pp. 846–849.

Part Three

Construction Management and Maintenance

20 Swarm Intelligence for Large-Scale Optimization in Construction Management

Emad E. Elbeltagi

Structural Engineering Department, Mansoura University, Mansoura, Egypt

20.1 Introduction

The difficulties associated with using mathematical optimization on large-scale engineering problems have contributed to the development of alternative solutions. Construction project managers have to make optimization decisions frequently during the life of the project in order to meet the project deadline, the limited budget, and/or the limited resources. Optimization problems often are solved either by using mathematical approaches or using approximate methods. Mathematical approaches employ linear, integer, or dynamic programming techniques to arrive at optimum solutions. While mathematical optimization models can reach the best solution, optimizing real-life problems is challenging because of their big domain. Finding exact solutions to these problems turns out to be Non-deterministic Polynomial time (NP)-hard complex with a large number of variables and nonlinear objective functions (Lovbjerg, 2002). To overcome these problems, researchers have proposed swarm intelligence (SI)—based algorithms for finding near-optimum solutions to problems.

SI is the collective behavior of decentralized, self-organized systems, whether natural or artificial. Swarm intelligence algorithms (SAs) are stochastic search methods that mimic the natural biological evolution and/or the social behavior of species. Among their advantages are that they work on a set of solutions at a time instead of a single solution and thus they can search the whole problem space, and they can provide intermediate results at any time during the computation (Eiben, 2002; Lovbjerg, 2002). SAs are typically made up of a population of simple agents interacting locally with one another and with their environment. The inspiration often comes from nature, especially biological systems. The agents follow very simple rules, and although there is no centralized control structure dictating how individual agents should behave, interactions between such agents lead to the emergence of "intelligent" global behavior that is unknown to the individual agents.

Metaheuristic Applications in Structures and Infrastructures. DOI: http://dx.doi.org/10.1016/B978-0-12-398364-0.00020-6

The first evolutionary-based technique introduced was genetic algorithms (GAs; Holland, 1975). GAs were developed based on the Darwinian principle of "survival of the fittest" and the natural process of evolution through reproduction. During the past 10 years, SAs have been introduced as well. Recent developments in SAs include techniques inspired by different natural processes, such as memetic algorithms (MAs), particle swarm optimization (PSO), ant colony optimization (ACO), modified shuffled frog leaping (MSFL), firefly algorithms, krill herd algorithms, cuckoo search algorithms, and bat algorithms, among others (Dorigo et al., 1996; Eberhart and Shi, 1998; Eusuff and Lansey, 2003; Gandomi and Alavi, 2012; Gandomi et al., 2011, 2012a; Kennedy and Eberhart, 1995; Yang and Gandomi, 2012).

Four SAs are introduced here: MAs, PSO, ACO, and MSFL. A brief description of each algorithm is presented, along with its application, to solve the construction time–cost trade-off (TCT) problem and to decide the repair decisions for bridge decks. Based on these applications, a comparison among the four algorithms is then presented in terms of processing the speed and quality of the solution obtained. Also, the results of applying the four algorithms are compared to the well-known GAs. Finally, comments on the applicability of SAs in construction management are presented.

20.2 SI-Based Optimization Algorithms

The following subsections present a brief description of four SAs: MAs, PSO, ACO, and MSFL. More information on these algorithms can be found in Gandomi et al. (2012b).

20.2.1 Memetic Algorithms

MAs are inspired by Dawkins's notion of a meme (Dawkins, 1976). MAs are similar to GAs, but the elements that form a chromosome are called *memes*. The unique aspect of MAs is that all chromosomes and their offspring are allowed to gain some experience, through a local search, before being involved in the evolutionary process (Merz and Freisleben, 1997). A pseudocode for MA is shown in Figure 20.1.

Similar to the GAs, an initial population is created at random. Afterward, a local search is performed on each population member to improve its experience and thus obtain a population of local optimum solutions. Then, crossover and mutation operators are applied, similar to GAs, to produce offspring chromosomes. Next, these offspring chromosomes are subjected to a local search so that local optimality is maintained.

Merz and Freisleben (1997) proposed an approach to perform local search through a pairwise interchange. In this method, the local search neighborhood is defined as the set of all solutions that can be reached from the current solution by

```
Begin;
    Generate random population of N solutions (chromosomes);
    For each individual i ∈ N: calculate fitness (i);
    For each individual i ∈ N: do local-search (i);
        For i = 1 to number of generations;
            Randomly select an operation (crossover or mutation);
            If crossover;
                Select two chromosomes at random i_a and i_b;
                Generate on offspring chromosome i_c = crossover (i_a and i_b);
                i_c = local-search (i_c);
            Else If mutation;
                Select one chromosome i at random;
                Generate an offspring chromosome i_c = mutate (i);
                i_c = local-search (i_c);
            End if;
            Calculate the fitness of the offspring chromosome;
            If i_c is better than the worst chromosome then replace the worst chromosome by i_c;
        Next i;
    Check if termination = true;
End;
```

Figure 20.1 Pseudocode of the MAs.

swapping two memes in the chromosome. For a chromosome of length n, the neighborhood size for the local search i is

$$N = 1/2 \cdot n(n-1) \tag{20.1}$$

The local search algorithm, however, can be designed to suit the nature of each particular problem. For example, local search can be conducted by adding or subtracting an incremental value from every gene and testing the chromosome's performance. The change is kept if the chromosome's performance improves; otherwise, the change is ignored. The parameters involved in MAs are population size, number of generations, crossover rate, mutation rate, and a local search mechanism.

20.2.2 Particle Swarm Optimization

PSO is inspired by the social behavior of a flock of migrating birds trying to reach a destination. In PSO, each solution is a "bird" and is referred to as a "particle" (Kennedy and Eberhart, 1995). The evolutionary process is initialized with a group of random particles (solutions). The ith particle (solution) is represented by its position as a point in an s-dimensional space, where s is the number of variables. Throughout the process, each particle i monitors three values: its current position ($X_i = x_{i1}, x_{i2},\ldots,x_{is}$), the best position that it reached in previous cycles ($P_i = p_{i1}, p_{i2},\ldots,p_{is}$), and its flying velocity ($V_i = v_{i1}, v_{i2},\ldots,v_{is}$). In each cycle, the position (P_g) of the best particle (g) is calculated as the best fitness of all particles.

Accordingly, each particle updates its velocity V_i to catch up with the best particle g, as follows:

$$\text{New } V_i = \omega \cdot \text{current } V_i + c_1 \cdot \text{rand}(\,) \cdot (P_i - X_i) + c_2 \cdot \text{Rand}(\,)\, x(P_g - X_i) \qquad (20.2)$$

As such, using the new velocity V_i, the particle's updated position becomes

$$\text{New position } X_i = \text{current position } X_i + \text{New } V_i; \quad V_{max} \geq V_i \geq -V_{max} \qquad (20.3)$$

where c_1 and c_2 are two positive constants named as learning factors ($c_1 = c_2 = 2$); rand(·) and Rand(·) are two random functions in the range [0, 1], and V_{max} is an upper limit on the maximum change of particle velocity. The operator ω (decreases linearly with time from a value of 1.4 to 0.5) is an *inertia weight* employed as an improvement proposed by Shi and Eberhart (1998) to control the impact of the previous history of velocities on the current velocity and plays the role of balancing the global and the local search.

20.2.3 Ant Colony Optimization

ACO was based on the observation of real ant colonies. When ants are traveling, they deposit a substance called pheromone, forming a pheromone trail. The trail is used by other ants, and over time, there is a higher probability that the trail with higher pheromone concentration will be chosen. Due to this positive feedback (autocatalytic) process, all ants will choose the shorter path (Dorigo et al., 1996).

Implementing the ACO requires a representation of S variables for each ant k ($k = 1,2,\ldots,N$); each variable i has a set of n_i options with their values l_{ij}, and their associated pheromone concentrations $\{\tau_{ij}\}$, where $i = 1,2,\ldots,s$, and $j = 1,2,\ldots n$ (Figure 20.2). Each ant is evaluated according to an objective function. Then, pheromone concentration associated with each variable is changed in a way to reinforce good solutions, as follows:

$$\tau_{ij}(t) = \rho\tau_{ij}(t-1) + \Delta\tau_{ij}; \quad t = 1,2,\ldots,T \qquad (20.4)$$

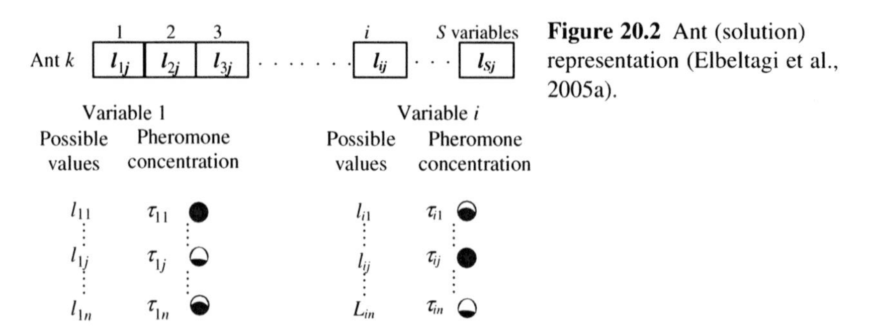

Figure 20.2 Ant (solution) representation (Elbeltagi et al., 2005a).

where T is the number of iterations; τ_{ij} (t) is the pheromone concentration associated with option l_{ij} at iteration t; τ_{ij} $(t-1)$ is the concentration of pheromone at the previous iteration $(t-1)$; $\Delta\tau_{ij}$ is the change in pheromone concentration; and ρ = pheromone evaporation rate (0 to 1). The change in pheromone concentration $\Delta\tau_{ij}$ is calculated as (Dorigo et al., 1996)

$$\Delta\tau_{ij} = \sum_{k=1}^{m} \begin{cases} R/\text{fitness}_k & \text{if option } l_{ij} \text{ is chosen by ant } k \\ 0 & \text{otherwise} \end{cases} \tag{21.5}$$

where R is a constant called the *pheromone reward factor*; and fitness_k is the value of the objective function (solution performance) calculated for ant k.

Once the pheromone is updated, the next iteration starts by changing the ants' paths (i.e., associated variable values) in a manner that respects pheromone concentration and some heuristic preference. As such, an ant k at iteration t will change the value for each variable according to the following probability (Dorigo et al., 1996):

$$P_{ij}(k,t) = \frac{[\tau_{ij}(t)]^{\alpha} \times [\eta_{ij}]^{\beta}}{\sum_{l_{ij}} [\tau_{ij}(t)]^{\alpha} \times [\eta_{ij}]^{\beta}} \tag{20.6}$$

where $P_{ij}(k,t)$ is the probability that option l_{ij} is chosen by ant k for variable i at iteration t; $\tau_{ij}(t)$ is the pheromone concentration associated with option l_{ij} at iteration t; η_{ij} is the heuristic factor for preferring among available options and is an indicator of how good it is for ant k to select option l_{ij}; and α and β are parameters that control the relative importance of pheromone concentration versus the heuristic factor (both α and β are greater than zero).

20.2.4 MSFL Algorithm

The MSFL algorithm combines the benefits of the genetic-based MAs and the social behavior−based PSO. In the SFL, the population consists of a set of frogs (solutions) that is partitioned into subsets referred to as *memeplexes*. Within each memeplex, the individual frogs evolve through a process of memetic evolution. After defined number of memetic evolution steps, ideas are passed among memeplexes in a shuffling process (Liong and Atiquzzaman, 2004). The local search and the shuffling processes continue until defined convergence criteria are satisfied (Eusuff and Lansey, 2003).

In the SFL, an initial population of N frogs is created randomly. For s-dimensional problems (s variables), a frog i is represented as $X_i = (x_{i1}, x_{i2}, \ldots, x_{is})$. Afterward, the frogs are sorted in a descending order according to their fitness. Then, the entire population is divided into m memeplexes, each containing n frogs (i.e., $N = m \times n$). In this process, the first frog goes to the first memeplex, the second frog goes to the second memeplex, frog n goes to the m memeplex, and frog $n + 1$ goes to the first memeplex, and so on.

Within each memeplex, the individuals with the best and the worst fitness are identified as X_b and X_w, respectively. Also, the individual with the global best fitness is identified as X_g. Then, the position of the individual with the worst fitness is adjusted as follows (Elbeltagi et al., 2007; Eusuff and Lansey, 2003):

$$\text{Change in frog position } (D_i) = \text{rand}() \cdot C(X_b - X_w) \tag{20.7}$$

$$\text{New position, } X_w = \text{current position } X_w + D_i, \quad D_{max} \geq D_i \geq -D_{max} \tag{20.8}$$

where rand(·) is a random number between 0 and 1; D_{max} is the maximum allowed change in the frog's position; C is the *search-acceleration factor* proposed by Elbeltagi et al. (2007) and it takes a positive value ranging from 1.3 to 2, and is used to make a balance between the local and global searches. If this process produces a better solution, it replaces the worst frog. Otherwise, the calculations in Eqs. (20.7) and (20.8) are repeated but with respect to the global best frog (i.e., X_g replaces X_b). The calculations then continue for a specific number of iterations. The evolution process in each memeplex is shown in the flowchart in Figure 20.3.

20.3 Experiments and Discussion

In order to experiment with the previously mentioned algorithms on construction management, the four SAs have been coded using the Visual Basic for Applications (VBA) macro-programming language, and all the experiments took place on a 2-GHz AMD laptop. Table 20.1 presents a summary of the main characteristics/differences of the four presented algorithms and the GAs.

Each algorithm has its own parameters that affect its performance in terms of solution quality and processing time. To obtain the most suitable parameter values that suit the test problems, a large number of experiments were conducted. For each algorithm, an initial setting of the parameters was established using values previously reported in the literature (Elbeltagi et al., 2005a; Eusuff and Lansey, 2003; Hegazy and Elbeltagi, 1999; Kennedy and Eberhart, 1995; Shi and Eberhart, 1998). Then, the parameters were fine-tuned along with the experimentation. The final parameter values adopted for each of the four SAs and the GAs are given in Table 20.2.

20.3.1 Project TCT Problem

The objective of the traditional critical path method (CPM) analysis is to find the duration required to perform a specific project. However, in real construction projects, activities are scheduled under available resources. The activity duration is a function of resource availability where different resource combinations have their own costs. Accordingly, the scheduler needs to take into account of the trade-off between direct cost and duration (Leu et al., 2001). For example, using more workers may save time, but the project direct cost could increase. In general, the less

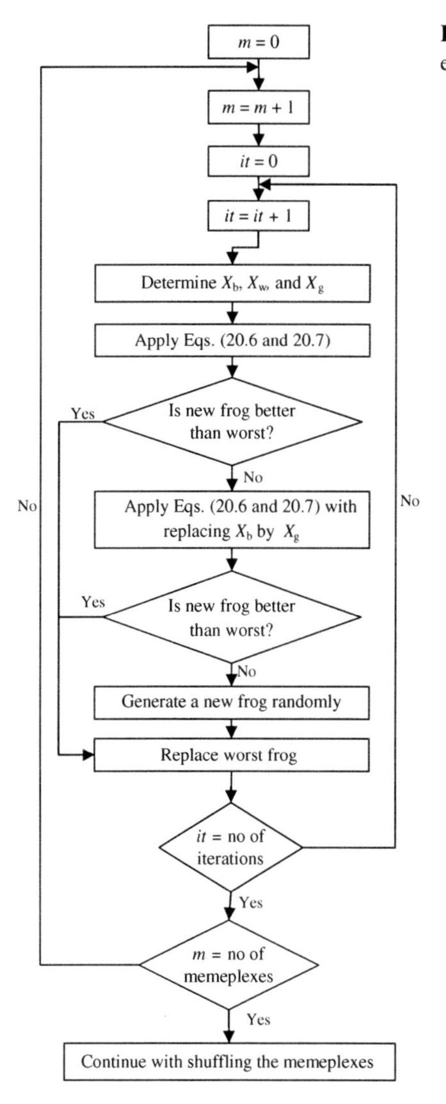

Figure 20.3 Flowchart of a memeplex evolution of the MSFL.

expensive the resources used, the longer it takes to complete an activity, and this relationship is named the *TCT*. TCT is one of the difficult decisions to optimize, especially with real-life projects involving hundreds of activities (Liu et al., 1995). The objective of the TCT, therefore, is to reduce the original CPM project duration to meet a specific deadline at minimum cost.

The performance of the four different SAs is compared using a TCT problem. This problem is a case study for an 18-activity project and was described by Feng et al. (1997). The activities, preceding activities, and the duration and cost data associated with the different methods of construction of the activities are presented in Table 20.3. The 18 activities were initially set to their least-cost option (all-normal

Table 20.1 Comparison Among the Five Algorithms

GAs	MAs	PSO	MSFL	ACO
• Starts with a group of a randomly generated population. • An individual solution (chromosome) consists of a set of genes equal to the number of variables. • Can be used for optimizing continuous or discrete functions. • Uses either binary coding or real numbers to represent variables. • Can be used for solving combinatorial problems. • Genes can be transmitted only from parent to offspring. • Uses the principle of survival of the fittest through selection, crossover, and mutation. • Crossover probability usually set from 0.5 to 0.95, mutation probability usually set 0.1–0.01. • Number of chromosomes in order of hundreds.	• Starts with a group of a randomly generated population. • An individual solution (chromosome) consists of a set of memes equal to the number of variables. • Can be used for optimizing continuous or discrete functions. • Uses either binary coding or real numbers to represent variables. • Can be used for solving combinatorial problems. • Memes can be transmitted between any two individuals. • Uses the principle of survival of the fittest through selection, crossover, and mutation. • Improved ideas can be incorporated immediately rather than waiting for full generation as in GAs. • MAs = GAs + local search. • Designed to search for local optimal solution. • Number of chromosomes ranges from 20 to 40.	• Starts with a group of randomly generated particles (swarm). • An individual solution (particle) consists of a set of values representing the problem variables. • Usually used for optimizing continuous nonlinear functions. • Uses real numbers to represent variables. • Each particle flies toward its best local position (local search) and the global best particle (global search). • Learning factors c_1 and c_2 are usually equal and ranges from 0 to 4. • Number of particles ranges from 20 to 40.	• Starts with a group of randomly generated frogs. • An individual solution (frog) consists of a set of memes equal to the number of variables. • Mainly used for optimizing discrete functions. • Uses real numbers to represent variables. • Can be used for solving combinatorial problems. • It follows the idea of MAs by applying local search through dividing the population into a set of subgroups (memeplexes). • Each memeplex is evolved as an independent culture by following the best frog. • It is an extension of the process used in PSO. • Information is passed between memeplexes in a shuffling process among all memeplexes. • Number of frogs ranges from 100 to 200.	• Starts with a group of ants generated randomly, with all decisions having equal probabilities. • An individual solution (ant or cycle) is equivalent to an individual member of GAs. • Mainly used for optimizing discrete functions. • Uses real numbers to represent variables. • Can be used for solving combinatorial problems. • A solution is constructed incrementally as ants move from one decision point to the next until all points are covered. • One iteration in ACO is equivalent to one generation in GAs. • In each iteration, the pheromone trail is updated. • Newly generated solutions will follow the trails that have higher amount of pheromone. • Pheromone evaporation is used as a local search tool. • Number of ants ranges from 10 to 20.

Table 20.2 Parameters for the Different SAs and GAs

Algorithm	Parameters
GA	Population = 200; Crossover = 0.8; Mutation = 0.08 Number of generations = 500 or stopping criteria satisfied
MA	Same as Gas
PSO	Population = 40; Iterations = 1000 Maximum velocity = 2; ω set as a time variant (1.2–0.4) decreasing with the increase of number of generations, at any generation i, $\omega = 0.4 + 0.8 \times$ *(number of generations−i)/ (number of generations−1)*
ACO	Population = 30; $\alpha = 0.5$; $\beta = 2.5$; $\rho = 0.4$; $R = 10$ Iterations = 100
MSFL	Population = 200; Number of memeplexes = 20 Number of frogs per memeplex = 10 Iteration per memeplex = 10; $C = 2.1 - 1.3$

schedule). The total direct cost of the project in this case is $99,740, with project duration being 169 days. With the all-normal schedule exceeding the desirable 110-day duration, it is required to search for a least-cost approach to minimizing the 169-day duration.

In the TCT problem formulation, each activity in the project can have up to five discrete methods of construction. Each method of construction is a combination of resources such as labor, equipment, and material. Three data elements are required for each method of construction: its index, its estimated duration, and its direct cost. The decision variables, therefore, are determining the optimum set of activities' methods of construction that minimizes the total project cost while not exceeding the target completion time. The objective is to minimize the total project cost and could be formulated as follows:

$$\text{Min} \sum_{i=1}^{p} C_{ij} + T \cdot I \tag{20.9}$$

where p is the number of activities; C_{ij} is the direct cost of activity i using its method of construction j; T is the total project duration; and I is the daily indirect cost (assumed to be $500/day). The solution of the TCT problem is represented in the form of a string, as shown in Figure 20.4.

In order to experiment with each algorithm, 20 trial runs were performed using each algorithm. The performance of the three algorithms was compared using three criteria: (1) the percentage of success, as represented by the number of trials required for the objective function to reach its known target value (110 days); (2) the average project duration of the solution obtained in all trials; and (3) the processing time to reach the optimum target value. The results of applying the four algorithms are reported in Table 20.4, along with the GA results as reported in Elbeltagi et al. (2005a).

Table 20.3 TCT Problem Data

Activity No.	Preceding Activities	Optional Methods of Construction									
		Option 1		Option 2		Option 3		Option 4		Option 5	
		Duration (Days)	Cost ($)	Duration (Days)	Cost ($)	Duration (Days)	Cost ($)	Duration (Days)	Cost ($)	Duration (Days)	Cost ($)
1	–	14	2,400	15	2,150	16	1,900	21	1,500	24	1,200
2	–	15	3,000	18	2,400	20	1,800	23	1,500	25	1,000
3	–	15	4,500	22	4,000	33	3,200	–	–	–	–
4	–	12	45,000	16	35,000	20	30,000	–	–	–	–
5	1	22	20,000	24	17,500	28	15,000	30	10,000	–	–
6	1	14	40,000	18	32,000	24	18,000	–	–	–	–
7	5	9	30,000	15	24,000	18	22,000	–	–	–	–
8	6	14	220	15	215	16	200	21	208	24	120
9	6	15	300	18	240	20	180	23	150	25	100
10	2, 6	15	450	22	400	33	320	–	–	–	–
11	7, 8	12	450	16	350	20	300	–	–	–	–
12	5, 9, 10	22	2,000	24	1,750	28	1,500	30	1,000	–	–
13	3	14	4,000	18	3,200	24	1,800	–	–	–	–
14	4, 10	9	3,000	15	2,400	18	2,200	–	–	–	–
15	12	12	4,500	16	3,500	–	–	–	–	–	–
16	13, 14	20	3,000	22	2,000	24	1,750	28	1,500	30	1,000
17	11, 14, 15	14	4,000	18	3,200	24	1,800	–	–	–	–
18	16, 17	9	3,000	15	2,400	18	2,200	–	–	–	–

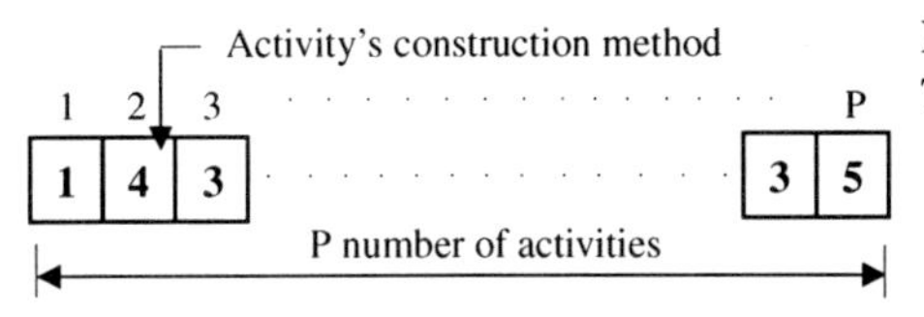

Figure 20.4 Solution representation of the TCT problem.

Table 20.4 Results of the TCT Problem

Algorithm	Minimum Project Duration (Days)	Average Project Duration (Days)	Minimum Cost ($)	Average Cost ($)	% Success Rate	Processing Time (in seconds)
PSO	110	112	161,270	161,940	60	15
ACO	110	122	161,270	166,675	20	10
Modified SFL	110	110	161,270	161,270	100	8
MAs	110	114	161,270	162,495	20	21
GAs (Elbeltagi et al., 2005a)	113	120	162,270	164,772	0	16

Using the GA to solve the TCT problem, the minimum solution obtained was 113 days with a minimum total cost of $162,270; but the success rate for reaching the optimum solution was zero, as shown in Table 20.4 (Elbeltagi et al., 2005a). When applying the MA to the TCT problem, it was able to reach the optimum project duration of 110 days and a total cost of $161,270, with a 20% success rate and an average cost that improved on that of the GA (Table 20.4). The PSO algorithm outperformed all other algorithms when used to solve the TCT problem except for the MSFL, with a success rate of 60% and average total cost of $161,940, as shown in Table 20.4. While the ACO algorithm was able to achieve a success rate of 20%, the average total cost of the 20 runs was greater than that of all other algorithms (Table 20.4). The results obtained using the MSFL algorithm outperformed all other algorithms, with a success rate of 100% (i.e., the MSFL was able to reach the optimum solution in all 20 trial runs) compared to 0% for the original GAs. The project cost reached its minimum value of $161,270. Also, the average processing time was 8 s, which was less than the other algorithms.

Further experiments were conducted to examine the performance of the four SAs in large-scale projects. Projects with 90, 180, and 270 activities were constructed by copying the 18-activity project several times in the Microsoft Project software as subprojects. Every subproject was given a finish-to-start relationship with the previous one. Accordingly, the overall project duration becomes multiples of the original project duration (169 days using all-normal durations). In all experiments, the deadline duration was set to 110 days times the numbers of subprojects (e.g., 110×5, 110×10, and 110×15 days, respectively), and indirect cost of $500/day was used. While the ACO, PSO, and MA were able to reach the optimum or near-optimum solution, the processing times increased exponentially with the increase in the number of activities. On the other hand, the MSFL algorithm was able to reach the optimum solution in most experiments (e.g., in the case of the 90-activity project, 550-day project duration was achieved with a minimum total cost of $810,550) for a recorded processing time of only 9 min. These results suggest that there is the potential for solving large-scale optimization problems using the MSFL algorithm.

20.3.2 Bridge-Deck Repair-Strategy Problem

The components of a typical asset management system include the time-dependent deterioration model, repair cost model, and improvement model; constraints; and the decision support module (Hegazy et al., 2004; Figure 20.5). In this study, a bridge-deck management system is used for applying the presented SAs. In this problem, the different SAs are used to determine the optimum or near-optimum repair decision for each bridge-deck so that the total life cycle cost (TLCC) is minimized. A detailed description of this problem is presented in Hegazy et al. (2004).

The objective is to minimize the present value of the TLCC of repairing selected bridges throughout the planning horizon (five years), while maintaining an acceptable condition for each bridge, as follows:

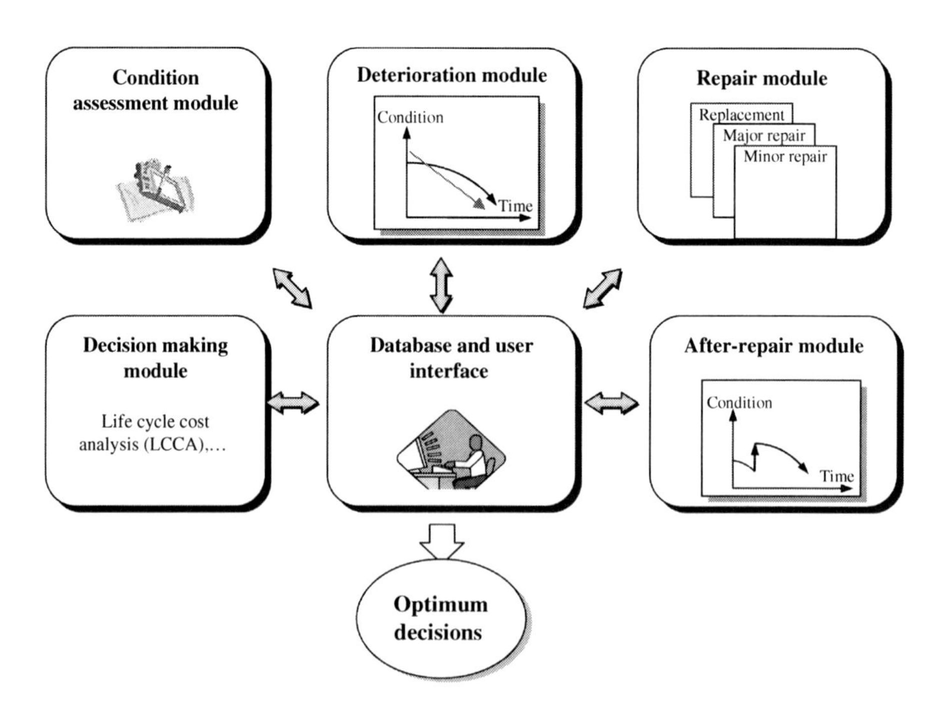

Figure 20.5 Components of a typical management system.

$$\text{Min TLCC} = \sum_{t=1}^{T} \sum_{i=1}^{N} \frac{C_{ti}}{(1+r)^t} \tag{20.10}$$

where C_{ti} is the repair cost of bridge-deck i at time t; r is the discount rate; T is the number of years; and N is the number of bridges. The objective function is subjected to the following constraints: yearly life cycle cost (LCC) should be less than or equal to yearly budget limits; condition rating of individual facilities greater than or equal to minimum acceptable level; and the overall condition rating for the whole network is greater than or equal to predefined value (Elbeltagi et al., 2005b). The number of variables equal $N \times T$, and the solution is structured as a string of elements equal to the number of variables, as shown in Figure 20.6. Each element can take an integer value from 0 to 3 corresponding to one of the repair options (i.e., 0 = do nothing; 1 = light repair; 2 = medium repair; and 3 = extensive repair). Experiments were carried out using different number of bridges, 10, 50, and 100, to represent different sizes of bridge networks. A total of 10 trial runs were performed for each number of bridges using all five algorithms.

The performance of the four algorithms is compared using four criteria: (1) the percentage of success (i.e., how many trials out of 10 were able to provide a solution without violating the condition constraints on both individual bridges and the whole network); (2) the best solution obtained (least LCC); (3) the average solution (i.e., the average value of the LCC for all success trials); and (4) the average processing time for all successful trials. In all experiments, the solution stopped when the value of the objective function (TLCC) did not improve after 10 consecutive iterations (Elbeltagi et al., 2005b).

The results of applying the five algorithms to a different number of bridges (10, 50, and 100) are given in Table 20.5. For the experiments with large networks of bridges (50 and 100 bridges), the networks were constructed by copying the 10-bridge network several times. As such, the solution obtained from the 10-bridge network (which reached a near-optimum solution) was used to measure the success in the larger networks (Elbeltagi et al., 2005b).

The results presented in Table 20.5 show that in the case of 10 bridges, GAs, MAs, and MSFL could obtain solutions that satisfy all the condition constraints in all trial runs, while the PSO achieved only 80% success, whereas the ACO achieved only 60% success. However, the best TLCC cost obtained with the SFL, \$5,733,000, is less than that obtained with all other algorithms. Also, the average

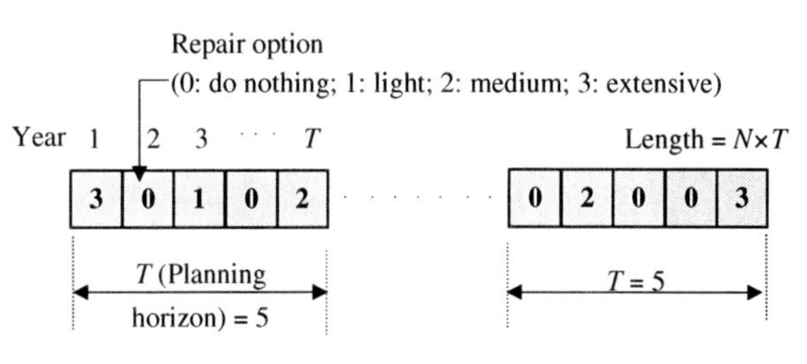

Figure 20.6 Bridge-deck repair solution representation.

Table 20.5 Bridge-Deck Repair Experiments

Algorithm	**Comparison Criterion**	**Number of Bridges**		
		10	**50**	**100**
PSO	% Success	80	40	10
	Best cost ($)	6,533,000	45,426,000	95,717,000
	Average cost ($)	7,720,000	47,655,000	95,717,000
	Time (h:min:s)	00:05:46	01:19:55	02:16:54
ACO	% Success	60	40	0
	Best cost ($)	6,873,000	44,866,000	–
	Average cost ($)	7,725,000	47,345,000	–
	Time (h:min:s)	00:02:26	01:30:24	–
MSFL	% Success	100	100	80
	Best cost ($)	5,733,000	30,126,000	67,206,333
	Average cost ($)	5,956,000	31,638,000	67,903,111
	Time (h:min:s)	00:01:37	00:26:06	01:28:34
MAs	% Success	100	100	50
	Best cost ($)	6,013,000	44,220,000	97,543,667
	Average cost ($)	6,756,000	46,715,000	98,031,111
	Time (h:min:s)	00:26:46	10:54:41	24:13:22
GAs	% Success	100	80	20
	Best cost ($)	6,073,333	44,866,000	98,346,667
	Average cost ($)	6,833,333	46,91,833	99,031,111
	Time (h:min:s)	00:02:16	00:49:02	02:17:40

TLCC in the case of the MSFL is much less than that obtained with all other algorithms (as shown in Table 20.5).

As the number of bridges increased to 50, the GAs was able to achieve only 80% success compared to 100% using the MSFL and MAs. Also, the success rate for the PSO and the ACO was 50% and 40%, respectively. Despite the high success rate of the MAs, its processing time was very large as the local search consumes a large amount of time as the chromosome length increased. In addition, in this case, the MSFL algorithms achieved the least processing time of 8 min, 6 s, which is very small comparable to other algorithms. The results still have the same trend, with 100 bridges where the MSFL outperformed all algorithms. In the case of 100 bridges, while the MAs achieved a 50% success rate, its processing time became very long. In this case, the ACO was unable to achieve any feasible solution, and the trials were prematurely converged. As shown in Table 20.5, as the number of bridges increased, the processing times increased exponentially, with the MSFL having the least processing time. In the case of 50 bridges, for example, the average processing time of the GAs was 49 min, 2 s, compared to 26 min, 6 s using the MSFL (Elbeltagi et al., 2005b).

Based on the previous discussion, the complexity of the problem substantially increases as the number of bridges increases. The variables, constraints, and the whole solution space become too large. For example, in the case of 100 bridges, the number of possible solutions is 4^{500}, which is extremely large. From this experimentation, problem size still represents a huge challenge for optimizing bridge maintenance and repair decisions.

20.4 Conclusions

In this chapter, four SI-based search algorithms were presented: MAs, PSO, ACO, and MSFL. A brief description of each algorithm was presented. The well-known TCT problem (as an application from the construction management domain) was used to compare the performance of the five algorithms. In this test problem, the SAs were used to search for the least-cost combination of construction methods so that the total project cost is minimized and the project deadline is met. The same algorithms were used to find the optimum (or near-optimum) repair options for bridge-deck considering LCC. The comparative results of the four SAs were also presented, along with a comparison with the well-known GAs.

A total of 20 trial runs were studied to evaluate the performance of the different algorithms. The results showed that the MSFL algorithm outperformed all other algorithms, with the highest success rate (number of runs that reached the optimum solution out of the 20 trial runs). Also, the processing time with MSFL was the least among all other algorithms. The PSO method was generally found to perform better than other algorithms in terms of success rate and solution quality, while being second-best in terms of processing time.

The advantage of SAs steams from their ability to search a population of points and improving this set of points rather than a single initial point. Therefore, it can quickly find optimal or near-optimal solutions. In addition, SAs can work with large problem sizes; as the size of the problem increases, the computational time increases to reach the optimal or near-optimal solution. On the other hand, problems with fewer decision variables can be solved using mathematical optimization methods. While SAs do not guarantee finding the global optimal solution within a finite time, they generally prove their validity in finding near-optimal solutions to complicated problems. SAs also can be used to generate initial near-optimal solutions quickly, which can be used as a good starting point for mathematical optimization techniques.

While the SAs have been successfully applied, large projects still require long processing times. Various experiments have showed that the key issue is to determine the set of parameters that optimize the performance of a SA. As shown by the increase in processing time as the problem size increases, problem size still represents a challenge for optimization. The complexity of the problem substantially increases as the number of variables increase and the solution space becomes too large. For example, in the case of a 400-activity project, with three possible construction methods for each activity, the number of possible solutions is 3^{400}, which is extremely large. The problem is also expected to become even much larger if the model is expanded to the case of multiple bridge components (e.g., deck, substructure, and superstructure). As such, some strategies may need to be applied to reduce the number of bridges (i.e., decision variables) to include in the optimization. One approach is to divide the project into smaller subprojects or sort the bridges according to their condition and optimize the decisions only for the set of bridges with a condition below a threshold value. This is expected to reduce the number of decision variables and accordingly improve the quality of solutions.

Finally, this chapter discusses the implementation of new, nontraditional tools for large-scale optimization that have great potential for application in the construction management domain.

References

Dawkins, R., 1976. The Selfish Gene. Oxford University Press, Oxford.

Dorigo, M., Maniezzo, V., Colorni, A., 1996. Ant system: optimization by a colony of cooperating agents. IEEE Trans. Syst., Man Cybern. 26 (1), 29–41.

Eberhart, R., Shi, Y., 1998. Comparison between genetic algorithms and particle swarm optimization. Proceedings of the Seventh Annual Conference on Evolutionary Programming. Springer Verlag, San Diego, CA, pp. 611–618.

Eiben, A.E., 2002. Evolutionary computing: the most powerful problem solver in the universe. *Dutch Math. Archive* (Nederlands Archief voor Wiskunde). 5/3 (2), 126–131.

Elbeltagi, E., Hegazy, T., Grierson, D., 2005a. Comparison among five evolutionary-based optimization algorithms. J. Adv. Eng. Inf. 19 (1), 43–53.

Elbeltagi, E., Elbehairy, H., Hegazy, T., Grierson, D., 2005b. Evolutionary algorithms for optimizing bridge deck rehabilitation. International Conference on Computing in Civil Engineering (CD-ROM), July 12–15, Cancun, Mexico, Paper # 8542.

Elbeltagi, E., Hegazy, T., Grierson, D., 2007. A modified shuffled-frog-leaping optimization algorithm: applications to project management. J. Struct. Infrastruct. Eng. 3 (1), 53–60.

Eusuff, M.M, Lansey, K.E., 2003. Optimization of water distribution network design using the shuffled frog leaping algorithm. J. Water Res. Planning Manage. 129 (3), 210–225.

Feng, C., Liu, L., Burns, S., 1997. Using genetic algorithms to solve construction time–cost trade-off problems. J. Comput. Civil Eng. 11 (3), 184–189.

Gandomi, A.H., Alavi, A.H., 2012. Krill herd algorithm: a new bio-inspired optimization algorithm. Commun. Nonlinear Sci. Numer. Simul. 17 (12), 4831–4845.

Gandomi, A.H., Yang, X.S., Alavi, A.H., 2011. Mixed variable structural optimization using firefly algorithm. Comput. Struct. 89 (23–24), 2325–2336.

Gandomi, A.H., Yang, X.S., Alavi, A.H., 2012a. Cuckoo search algorithm: a metaheuristic approach to solve structural optimization problems. Eng. Comput. doi: 10.1007/s00366-011-0241-y.

Gandomi, A.H., Yang, X.S., Talatahari, S., Alavi, A.H., 2012b. Metaheuristics in modeling and optimization. In: Gandomi, A., Yang, X-S., Talatahari, S., Alavi, A. (Eds.), Metaheuristic Applications in Structures and Infrastructures (1st Edition, Chapter 1). Elsevier, Waltham, MA.

Hegazy, T., Elbeltagi, E., 1999. EvoSite: an evolution-based model for site layout planning. J. Comput. Civil Eng. 13 (3), 198–206.

Hegazy, T., Elbeltagi, E., Elbehairy, H., 2004. Bridge deck management system with integrated life cycle cost optimization. Trans. Res. Record: J. Trans. Res. Board. 44–50 (No. 1866, TRB, National Research Council, Washington, DC).

Holland, J., 1975. Adaptation in Natural and Artificial Systems. University of Michigan Press, Ann Arbor, MI.

Kennedy, J., Eberhart, R., 1995. Particle swarm optimization. Proceedings of the IEEE International Conference on Neural Networks (Perth, Australia). IEEE Service Center, Piscataway, NJ, pp. 1942–1948.

Leu, S.S., Chen, A.T., Yang, C.H., 2001. A GA-based fuzzy optimal model for construction time–cost trade-off. Inter. J. Project Manage. 19, 47–58.

Liong, S.-Y., Atiquzzaman, M.D., 2004. Optimal design of water distribution network using shuffled complex evolution. J. Inst. Eng. 44 (1), 93–107.

Liu, L., Burns, S.A., Feng, C.-W., 1995. Construction time–cost trade-off analysis using LP/IP hybrid method. J. Constr. Eng. Manage. 121 (4), 446–454.

Lovbjerg, M., 2002. Improving Particle Swarm Optimization by Hybridization of Stochastic Search Heuristics and Self-Organized Criticality. Masters Thesis, Aarhus Universitet, Denmark.

Merz, P., Freisleben, B., 1997. A genetic local search approach to the quadratic assignment problem. In: Bäck, C.T. (Ed.), Proceedings of the Seventh International Conference on Genetic Algorithms. Morgan Kaufmann, pp. 465–472.

Shi, Y., Eberhart, R., 1998. A modified particle swarm optimizer. *Proceedings of the IEEE International Conference on Evolutionary Computation*. IEEE Press, Piscataway, NJ, pp. 69–73.

Yang, X.S., Gandomi, A.H., 2012. Bat algorithm: a novel approach for global engineering optimization. Eng. Comput. 29 (5), 464–483.

21 Network-Level Infrastructure Management Based on Metaheuristics

Katharina C. Lukas and André Borrmann

Chair for Computational Modelling and Simulation, Technische Universität München

21.1 Introduction

Most industrialized countries today face the problem of aging infrastructure, since many infrastructural assets such as roads and bridges, were built during the 1960s and 1970s. To keep the infrastructure safe, maintenance measures have to be undertaken.

These maintenance measures have costs not just for the asset manager but also for the public. The costs for the manager can be measured as the money spent on maintenance. This is influenced not only by the kind of measures undertaken but also by possible savings resulting from synergies with third-party construction sites. The costs to the public are harder to measure. The public is affected by noise and pollution from construction sites, but the main impact can be seen in interruptions to traffic flow, and, in the worst case, traffic jams. The impact on traffic is especially high if different construction sites in an urban area are not coordinated, resulting in one set of works blocking the detour of the other.

Traditionally, the planning of maintenance measures for a large stock of infrastructural assets is done manually. As a consequence, some of the considerations just mentioned can be taken into account only marginally, if at all.

For example, steadying the budget flow, which reduces the administrative effort significantly, is not considered in this approach. Also, the impact on traffic flow that occurs due to capacity reductions on roads with construction sites can be considered in only a very simplified way. The benefit of postponing one maintenance site to avoid the simultaneous closure of two streets also cannot be evaluated.

In this chapter, different metaheuristic approaches to this problem are evaluated to ascertain their usefulness in creating feasible schedules with minimal impact on traffic flow.

Metaheuristic Applications in Structures and Infrastructures. DOI: http://dx.doi.org/10.1016/B978-0-12-398364-0.00021-8

21.2 Problem Description

Buildings are subject to deterioration that affects not only their serviceability but also their safety. Maintenance measures are required to ensure that both are upheld. Traditionally, a deterministic prognosis to identify the latest time at which maintenance work has to be undertaken is done based on regular inspections. Recent research has focused on the idea of a probabilistic prognosis. In these new approaches, the prognosis is performed using Markov chains (Vesikari, 2008) or different probabilistic deterioration functions (Borrmann et al., 2012; Hammad et al., 2006; Okasha and Frangopol, 2010). But whichever method is used, the manager is provided with a deadline for the maintenance of the structure.

However, for managers of a large number of infrastructural assets, whether in the public or private sector, it is not always desirable to maintain such structures at the latest possible time. The decision as to which structure to maintain and when is influenced by a number of considerations.

Annual budget limits and the limited availability of project managers and working crews restrict the number of assets that can be maintained each year. In addition, the influence of construction sites on traffic makes some combinations of parallel maintenance works more attractive than others. In some cases, it may be possible to exploit synergies resulting from maintenance work undertaken by third parties, such as streetcar operators.

Given the importance of safety of infrastructure facilities (i.e., the constraint that structures have to be maintained before they reach a critical condition), it is not sufficient to plan an optimal schedule for only 1 year in advance. Optimizing a schedule of works for only the next year can result in problems in planning a feasible schedule for the year after that, or, if it is possible, a schedule that has a very high impact on traffic. Such an approach is far too shortsighted; therefore, a schedule for the next 5–10 years has to be generated.

Constructing a schedule is a highly complex combinatorial problem. It may, for example, be beneficial to bring forward the maintenance of a structure with a repair-by deadline outside the considered time frame so that it falls within the time frame. As such, it can be necessary to consider more assets than those that will be part of the solution. Additionally, the impact on traffic as an objective function is nonlinear, as the simultaneous (partial) closure of a number of assets will have a highly different effect on traffic than the sum of the effects of individual closures.

In this chapter, a single problem is considered, with the impact on traffic as the only objective. This objective is evaluated by simulating the scenarios for all the years of a schedule using a traffic simulator (VISUM by PTV AG (2009a)). Later extensions to more objectives are possible. The following constraints are considered: a limited budget per year, a fixed number of maintenance projects per year, and safety considerations (i.e., the maintenance of all assets before they reach a critical condition).

A solution to this problem is a schedule for the next y years, where for each year, a fixed number o of assets will be scheduled for maintenance. Choosing a

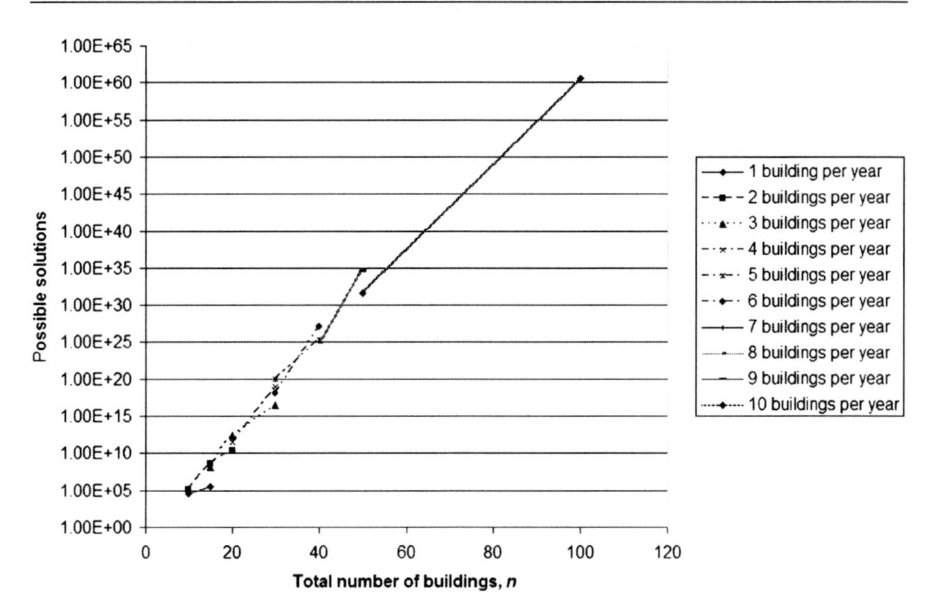

Figure 21.1 A combinatorial explosion. The number of possible solutions increases massively with the problem size. All values are given for a 5-year schedule.

fixed number of maintenance actions is suitable as it reflects the aforementioned constraint of available working crews. A schedule is feasible if all assets are scheduled before their respective repair-by deadlines so that they do not reach a critical condition, and if the costs for all the years are within the given budget. The quality measure (objective function) for such a feasible schedule is the time that the vehicles spend inside the network in the most critical year of the schedule. For example, a schedule over 5 years (year 1: 470 veh × h[1], year 2: 469 veh × h, year 3: 474 veh × h, year 4: 457 veh × h, and year 5: 440 veh × h) has a quality measure of 474 veh × h. The objective is to minimize this value.

As with most combinatorial problems, the solution space of the described problem increases massively with the problem size. In Figure 21.1, the size of the solution space for different combinations of the total number n of assets considered and the number o of possible maintenances per year for a 5-year schedule is shown. While there are 30,240 different schedules for $n = 10$ and $o = 1$, for $n = 20$ and $o = 2$, there are already 2.095×10^{10} possibilities. A typical problem size derived from real-world practice is $n = 100$ and $o = 10$, which has a solution space of 4.877×10^{60}. Even if infeasible solutions are excluded, the search space is still too large to consider exhaustive enumeration. As direct (mathematically closed) approaches for solving this kind of strongly constrained, nonlinear optimization

[1] The variable veh × h stands for "vehicle hours."

problems are not available, the application of metaheuristics is a good choice for finding good and feasible schedules.

The metaheuristics described in this chapter are ant colony optimization (ACO) and genetic algorithms (GAs). Alternative approaches are, for example, simulated annealing (SA; Kirkpatrick et al., 1983), particle swarm optimization (PSO; Kennedy and Eberhart, 1995), bee colony optimization (BCO; Teodorović and Dell'Orco, 2005), the firefly algorithm (FA; Gandomi et al., 2011; Yang, 2009), or the cuckoo search algorithm (Gandomi et al., 2012).

The algorithms were chosen for their good applicability to combinatorial optimization problems. Still, both algorithms need tuning to the individual problem types. While this is relatively easy for GAs, in ACO, a high number of parameters have to be set. But with the right set of parameters ACO can demonstrate higher intelligence in the solving process. One type of ACO, the ant colony system, has another advantage over GAs: it is able to find good solutions with far fewer calls to the objective function, which is the main factor that influences the run time.

In the following sections, both types of heuristics and the required adaptations to the infrastructure management system are described. These algorithms were tested using a simplified problem to compare their applicability to the infrastructure management problem.

21.3 Ant Colony Optimization

Since its first formulation in the Ph.D. thesis of Marco Dorigo (Dorigo, 1992), ACO has been used to solve a number of different problems. These include theoretical problems such as the travelling salesman problem (Dorigo, 1992; Dorigo and Gambardella, 1997; Dorigo et al., 1996; Gambardella and Dorigo, 1996), the vehicle routing problem (Bullnheimer et al., 1997, 1999), the quadratic assignment problem (Maniezzo, 1999; Maniezzo and Colorni, 1999; Stützle and Dorigo, 1999; Stützle and Hoos, 1996; Taillard and Gambardella, 1997) or the job shop scheduling problem (Colorni et al., 1994). But ACO has also been used on real-world problems such as project scheduling (Christodoulou, 2010; Merkle et al., 2002), creating university timetables (Socha et al., 2002), construction site planning (Ning et al., 2009), and pipe routing (Christodoulou and Ellinas, 2010).

The idea is based on the behavior of natural ants (Bonabeau et al., 1997; Goss et al., 1989). These animals are able to find the shortest route between their nest and a food source by communicating via trails of pheromones. When searching for food, ants leave behind trails that serve as a guide for the ants yet to come. Trails that lead to food sources in near proximity have a higher density of pheromones because the ants using them return earlier. As a consequence, other ants use these trails too, reinforcing the pheromone density even more.

In ACO, artificial ants search their way through a graph describing the optimization problem in such a way that a route through the graph corresponds with a solution to the problem.

This chapter describes two types of ACOs: the rank-based ant system (Bullnheimer et al., 1999) and the ant colony system (Dorigo and Gambardella, 1997), and the necessary adaptations applied to them to solve the infrastructure management problem as described in Section 21.2.

21.3.1 The ACO Algorithm in General

The basic algorithm is the same for all types of ACO. In each iteration step, all ants of the colony construct individual solutions to the problem by searching for a route through the problem graph. The choice of which node is taken next is decided probabilistically: For each decision, an ant has to take the probability p_j to choose node j as follows:

$$p_j = \frac{\tau_j^{\alpha}\eta_j^{\beta}}{\sum_{l \in N}\tau_l^{\alpha}\eta_l^{\beta}} \tag{21.1}$$

where α and β are fixed constants; τ_j and τ_l describe the amount of pheromone at node j and l, respectively; and η provides additional heuristic information on the desirability of the node. N is the set of all selectable nodes, that is, all nodes of the graph that are allowed according to the set rules for the choice.

After all the ants are finished constructing routes, the quality of the individual solution is evaluated. According to this evaluation, the pheromone trails are updated before the next iteration. The diverse types of ACO differ significantly in the way this pheromone update is performed.

21.3.2 Rank-Based Ant System

In the rank-based ant system (Bullnheimer et al., 1999), the pheromone trails are altered by two different updates. The first one, the so-called evaporation, affects all nodes of the graph: the amount of pheromone on each node i is reduced by a factor ρ:

$$\tau_i \leftarrow (1-\rho) \times \tau_i \tag{21.2}$$

This is done to avoid unlimited accumulation of the pheromone trails and to eliminate old nonpromising solutions from the memory of the algorithm, thus encouraging further exploration of the search space (Dorigo and Stützle, 2004).

Following this update, a second update is performed on selected nodes: the w best ants from the colony are selected and given a rank r according to their quality (the best ant has rank 1). After this, on all nodes i belonging to the ant of rank r, τ_i is increased by

$$(w - r + 1) \times \Delta\tau_i^r \tag{21.3}$$

$\Delta\tau_i^r$ is computed according to the quality of the solution. For a minimization problem, $\Delta\tau_j^r$ can be chosen as the inverse of the objective function value, and for a maximization problem, the value should be directly proportional to the objective function value.

In addition, a separate ant, the so-called Elitist Ant, is used to store the best solution found so far. This ant deposits in each iteration an amount of pheromone of

$$(w+1) \times \Delta\tau_i^e \tag{21.4}$$

on all nodes i that belong to its schedule, so long as no better solution has been found. $\Delta\tau_j^e$ is thereby computed according to $\Delta\tau_j^r$.

21.3.3 Ant Colony System

The pheromone update in the ant colony system (Dorigo and Gambardella, 1997) differs markedly from that in the rank-based ant system. Only one ant, the elitist ant (defined the same way as in the rank-based ant system), is used for the global pheromone update. On all nodes i belonging to its solution, the amount of pheromone is modified by

$$\tau_i \leftarrow (1-\rho)\tau_i + \rho\Delta\tau_i^e \tag{21.5}$$

$\Delta\tau_j^e$ is computed the same way as in the rank-based ant system, so the nodes of the elitist solution are not only the only nodes where pheromone is deposited but also the only ones where evaporation takes place.

To encourage exploration of the search space away from the elitist solution, a local pheromone update is performed by all ants by eliminating pheromone during the construction of their routes. If an ant chooses node i, the amount of pheromone τ_i on i is modified according to

$$\tau_i \leftarrow (1-\xi)\tau_i + \xi\tau_0 \tag{21.6}$$

where ξ is a parameter $0 < \xi < 1$ and τ_0 is the initial amount of pheromone for all nodes. Therefore, a node chosen by many ants becomes less attractive, thus pushing the search to different regions of the search space.

In addition to the different modes of pheromone update, there is another difference between the ant colony system and other types of ACO: the ant colony system uses a much stricter choice rule. With the probability q_0, an ant chooses its next node not with a probabilistic choice based on the values computed with Eq. (21.1) but with the highest value of $\tau_i^\alpha \times \eta_i^\beta$. This aggressive choice rule pushes the search again in the neighborhood of the elitist solution.

If the parameters ρ, ξ, and q_0 are chosen well, the ant colony system shows a very good balance between exploitation of knowledge from previous iterations and exploration of the search space.

Another advantage of the ant colony system, especially for problems with highly time-consuming fitness evaluations, is that it needs only a few ants, as opposed to other types of ACO, which perform better the more ants there are in the colony. For the ant colony system, the optimal number of ants depends on ρ and q_0 (Dorigo and Gambardella, 1997). For the typical values $\rho = 0.1$ and $q_0 = 0.9$, the optimal number of ants is 10.

21.3.4 ACO for the Infrastructure Management Problem

21.3.4.1 Problem Graph

A possible formulation for the infrastructure management problem is a layered graph (Figure 21.2) with the layers representing the years under consideration and the nodes per layer representing the infrastructural assets subject to maintenance plans in the network. A solution to the combinatorial problem is then described by a path with o selected nodes per layer.

To model the parallel choice of assets, we do not use single ants; rather, we use ant teams to construct the maintenance schedules based on an idea by Lee (2009). The number of ants in the team corresponds to the number of parallel maintenance measures per year. For every year i (layer in the graph), each ant from a team k chooses a different node j following Eq. (21.1), the stricter choice rule of the ant colony system. The set N hereby contains all nodes representing assets that were not selected by any ant from k in year i or any earlier year.

21.3.4.2 Guidance to Feasible Solutions

The feasible region for the infrastructure management problem represents only a small part of the search space. Furthermore, there is no known problem representation to describe it as a joined set where every feasible solution can be reached

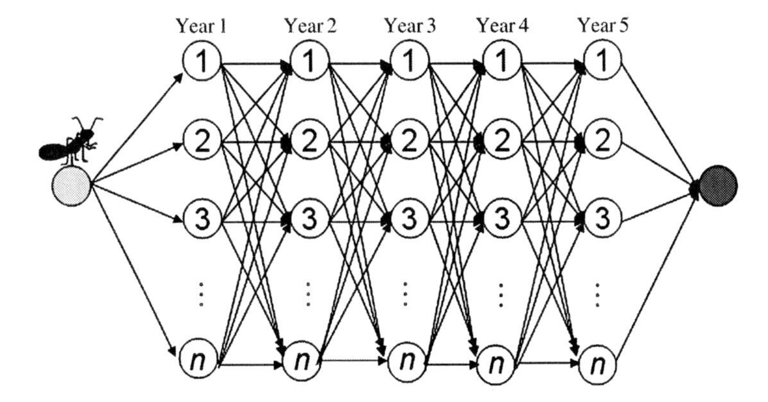

Figure 21.2 Problem graph for the infrastructure management problem: the layers in the graph represent the years. Per year, n nodes exist representing the assets that are considered. A solution to the combinatorial problem consists of o selected nodes per layer.

directly from every other without crossing infeasible regions. For this reason, further attention has to be given to the question of how to handle infeasible solutions.

A first, but rather naive, idea would be not to allow infeasible solutions to manipulate the pheromone trails at all. But this approach would cause some major problems. Given the relatively small size of the feasible region with regard to the overall search space, the probability of finding a feasible solution is very low. A number of iterations will be required to find any feasible solution at all. Without further guidance, this solution will be found only by chance. A possible way to guide the search in this step of the algorithm is described in Section 21.3.4.3.

If the first feasible solution is found only by random search, it is not possible to draw further conclusions about its quality for the objective function. There is a high probability that this solution will be bad. However, as this solution is the only one on which pheromones are deposited, the search will now concentrate on its neighborhood. If there is no information on infeasible solutions to learn from, the search remains trapped in this region, causing the search to converge prematurely on suboptimal solutions.

This suggests that infeasible solutions should also be allowed to deposit pheromones. To prevent the algorithm from converging on an infeasible solution, whose objective function value may be better than the global (feasible) optimum, we make use of the elitist ant: by allowing only feasible solutions to be stored in the elitist ant, the search is encouraged to exploit the best feasible solution found so far. By simultaneously allowing infeasible solutions to deposit pheromone by becoming part of the elitist group of the w iteration best ants, the search is still able to cross infeasible regions and, in the next iteration, could then find possible good feasible solutions in the neighborhood of one of the depositing infeasible solutions.

Of course, these considerations apply only to the rank-based ant system, not to the ant colony system. For the ant colony system, where only the elitist ant deposits pheromones, there is no difference whether or not infeasible solutions are excluded from pheromone depositing or from becoming the elitist solution. Nevertheless, the rule of eliminating pheromones as the ants walk supports such a high diversity of solutions in one iteration that the previously described problems do not appear as frequently. The mechanisms described in the next section ensure that the ant colony system finds not only a high diversity of solutions but also a high diversity of feasible solutions in each iteration.

21.3.4.3 Reinterpretation of η

In the infrastructure management problem, defining η is difficult because the benefit of choosing one node over another node j is not known. This is due to the fact that the quality of a schedule does not depend on a single choice, but only on combinations of nodes chosen for the different years. So, unlike in other problem types where ACO has been used, using η to measure the impact of a choice on the objective function does not make sense.

Instead of omitting η completely, we propose to use it for a different goal: by setting η_{ij} higher based on how urgent the maintenance of an asset j is in year i, the search is guided from the outset more toward feasible regions in the search space. This is because the critical assets are given preference, making them more likely to be scheduled before their repair-by deadline.

This definition of η_{ij} is sufficient for the rank-based ant system, as its probabilistic choice rule means it still has enough freedom in its search. The choice rule here is so open that in the first few iterations, almost all of the ants fulfill only some of the security constraints (i.e., some of the critical assets are not scheduled). Only over a number of iterations will the pheromone accumulation result in a preference for the critical nodes, because even if the single ants have not chosen all of them, it is highly probable that over the total colony (and over a number of iterations), all of them have been chosen at least once. This freedom allows the ants to not only reduce the constraint violations to meet the security constraint over many iterations but also to meet the budget constraint without further guidance.

This is different in the ant colony system: here, the stricter choice rule leads the ants in the first iteration to solutions in the immediate neighborhood of greedy solutions, i.e. the solution found by a Greedy Algorithm. In this case, where the only guidance for the ants is the grade of criticalness of the nodes, this means that the solutions found in the first iteration almost certainly fulfill the security constraint. For some problem instances, these near-to-greedy solutions will not fulfill the budget constraints, though (e.g., when all assets with deadline 1 have very high costs and all assets with low costs, which would be needed to stay within the budget constraint, have very high deadline values). Therefore, a number of iterations are wasted on finding solutions that satisfy both constraints. As soon as such a solution is found, even if it is far from the optimum, the search concentrates rapidly around it as it is the only solution on which pheromones are deposited. This limits further exploration.

To avoid this, η_{ij} has to be chosen in such a way that it will contain not only information about the safety but also about the budget. The fulfillment of the budget constraint as the objective function depends on the combination of assets rather than on the single assets (if the costs also have to be between a lower and a higher limit, a combination of only low-cost assets as well as of only high-cost assets must be avoided: only a combination of assets with low costs and assets with high costs will satisfy the constraint). As a consequence, η_{ij} has to be adapted with every choice.

The idea is to compute, after each choice of an asset j for the year i, the estimated costs for the year i by multiplying the average cost of all structural maintenance measures chosen up to that point in i by the total number of assets to be maintained in i. If these estimated costs lie within the limits, η_{il} is computed for all assets l the same way as in the Rank-based Ant System, by setting it to be reverse-proportional to the distance between its deadline dl and the year i. If the estimated costs lie below the lower limit, the choice of assets with high maintenance costs has to be encouraged. Therefore, for all assets l, η_{il} has to be set to

$$\eta_{il} = \frac{1}{(\mathrm{d}l - i)} \times \frac{c_l}{\mathrm{scal}} \tag{21.7}$$

with c_l the costs of maintaining asset l and scal an apt scaling factor.

If the estimated costs are higher than the higher budget limit, η_{il} for all assets l has to be set to the reverse of Eq. (21.7) so that the choice of assets with low maintenance costs is encouraged:

$$\eta_{il} = \frac{1}{(\mathrm{d}l - i)} \times \frac{\mathrm{scal}}{c_l} \tag{21.8}$$

21.3.4.4 Dynamic Elitist Ant

In Merkle et al. (2002), an approach to preventing the rank-based ant system from running into local optima is described. When the elitist ant has not changed over a given number of iterations (e.g., 5), a new elitist ant is selected, even if no better feasible solution has been found. For this new elitist ant, the best feasible solution that is not the same as that favored by the old elitist ant is selected. This prevents the search from concentrating too heavily around the elitist solution. If the new elitist ant is far enough from the old one, there is a possibility that the search can be guided toward new regions.

A similar procedure is used when the elitist ant has taken over the whole elitist group, that is, all w iteration best ants have the same value. If the elitist ant is not changed, the search would run into a dead end, as pheromone is deposited on one solution only. There is almost no chance for it now to find new solutions, let alone better ones. By switching the elitist ant to the next best feasible solution outside the elitist group, the algorithm is given the opportunity to escape this dominant solution.

21.4 Genetic Algorithms

The general methodology of GAs was developed by Holland (1975). It has been successfully applied to a number of real-life problems, as well as different maintenance scheduling issues, such as the maintenance of generators in power plants (Burke and Smith, 1997; Lapa et al., 2005; Muñoz et al., 1997; Negnevitsky and Kelareva, 1999), oil storage tanks (Li et al., 2002), water pipes (Dridi et al., 2008), and railways (Budai et al., 2009). It has also been applied to optimizing the maintenance of highway bridges, but so far under the sole consideration of the maintenance costs (Morcous and Lounis, 2004).

The underlying idea is based on Darwin's theory of evolution: the solutions to an optimization problem are treated as individuals of a species. The objective function describes the environment of this species. The better an individual fulfills the objective function, the better it is adapted to this environment. By crossover, mutation, and selection over a number of generations, the population of solutions gradually evolves into a population of better-adapted individuals (i.e., individuals that obtain better values for the objective function).

In general, the algorithm takes the following steps. First, the solutions have to be encoded as genes. This can be done in many different ways. The most common representations are binary strings, integer strings, real-value strings, and permutations. The choice of which representation to take depends on the problem. Together with the operators for crossover and mutation, the representation defines the fitness landscape (Merz and Freisleben, 1999). A good representation describes a fitness landscape that is almost nonrugged (i.e., there is only little difference in the fitness values between neighboring solutions) and one where every feasible solution can be reached from any other feasible solution without crossing infeasible regions.

Regardless of the chosen encoding, the following actions are performed as part of each GA. First, a suitable initial population is generated. Subsequently, an iterative procedure is executed consisting of the selection, crossover, and mutation in each cycle. One single cycle consisting of these steps is called a *generation*.

The first step involves the selection of a subset of the population, called the *mating pool*, which can be performed either by a roulette wheel selection with the probabilities defined either as fitness proportional, rank-based, or by a tournament. From this mating pool, two or more individuals are selected randomly to create new offspring in a crossover operation. The crossover operators are designed with the goal of preserving in the offspring as many qualities of both parents as possible. The new individual is also randomly changed using a mutation operator to promote exploration of the search space.

The population of the next generation is selected from all the individuals from the last generation and the newly created offspring. As the population's size stays the same for all generations, some of the individuals have to be dropped. Possible selection mechanisms for the survivors are age-based replacement, fitness-proportional selection, and tournament selection.

With this new population, the next iteration starts with the selection of a new mating pool.

21.4.1 GAs for the Infrastructure Management Problem

For the infrastructure management problem, several different representations exist for GAs. The two most promising are described in this section.

One possibility is to describe an individual solution by an integer matrix of the size $y \times o$ (Figure 21.3). The value of an entry represents the identification number (ID) of an asset, and the row represents the year; for example, an entry 57 in the third row of the matrix means that asset number 57 is maintained in the third year of the schedule. The column in which the ID is located does not have a particular meaning.

For crossover, the offspring inherits $y/2$ rows from each of the parents. In doing so, one has to ensure that an ID does not appear more than once in the schedule. If this is the case, the algorithm first tries to replace it with the element at the same place from the other parent. If this element is also already part of the schedule, it is replaced by a random unscheduled element. This crossover operator is designed in a way that it preserves as many of both parents' characteristics in the combinations

27	65	62	88	17	75	82	55	87	37
66	68	53	25	56	67	60	43	6	52
95	74	2	3	14	57	11	36	91	41
70	44	32	13	16	84	24	76	90	5
80	47	85	48	8	83	89	46	50	34

Figure 21.3 Matrix representation for the infrastructure management problem. The rows represent the years, and the value of the entry represents the IDs of the assets. The shown gene describes a schedule over the next 5 years, where, each year, 10 assets can be maintained. In the first year of this schedule, the assets with the IDs 17, 27, 37, 55, 62, 65, 75, 82, 87, and 88 are scheduled to be maintained.

of bridges as possible, as the crucial gene information lies in its rows. Unfortunately, no crossover operator exists for this representation that works without resetting random values that are part of neither of the parents if double occurrences of any asset are to be avoided.

Mutation changes every entry of the matrix with a mutation probability p_m. On average, one entry should be changed [i.e., $p_m \approx 1/(y \times o)$]. The entry is replaced by a random unscheduled element.

An alternative representation is a permutation of the length n. Here, the place in the string stands for the ID of the asset and the entry encodes the year for which its maintenance is scheduled: 0 to $o-1$ denotes maintenance in the first year of the schedule, o to $2o-1$ means maintenance in the second year, and so on. A number greater than $(y \times o)--1$ means that the asset is not scheduled (Figure 21.4).

The crossover operation is realized as a partially mapped crossover (Goldberg and Lingel, 1985). The steps of this operation are as follows:

1. Choose two random crossover points and copy the segment in between from the first parent to the first offspring (see first part of Figure 21.5).
2. Starting from the first crossover point, search for elements in the second parent that have not been copied. If an element i is found, look in the offspring to see which element j has been copied to its place. Set i in the offspring to the place that j has in the second parent. If this place is already taken by an element k, set i in the place that k has in the second parent, and so on (see second part of Figure 21.5).
3. Fill the remaining elements of the offspring by copying from the second parent (see third part of Figure 21.5).

Like the crossover mechanism for the matrix representation, this operator preserves most of the characteristics of the parents when generating offspring for the permutation representation.

1	2	3	4	5	6	7	8	9	10	11	12	13	14	15	16	17	18	19	20	21	22	23	24	25
90	22	23	58	39	18	69	44	77	73	26	76	33	24	80	34	4	95	67	70	97	85	54	36	13
26	27	28	29	30	31	32	33	34	35	36	37	38	39	40	41	42	43	44	45	46	47	48	49	50
93	0	96	56	84	89	32	99	49	88	27	9	98	87	71	29	91	17	31	94	47	41	43	63	48
51	52	53	54	55	56	57	58	59	60	61	62	63	64	65	66	67	68	69	70	71	72	73	74	75
82	19	12	92	7	14	25	50	75	16	78	2	55	51	1	10	15	11	60	30	53	62	86	21	5
76	77	78	79	80	81	82	83	84	85	86	87	88	89	90	91	92	93	94	95	96	97	98	99	100
37	64	61	79	40	66	6	45	35	42	57	8	3	46	38	28	81	68	74	20	52	59	83	72	65

Figure 21.4 Permutation representation for the infrastructure management problem. The actual gene consists only of the elements shown in white; the gray cells are shown only for better orientation. The schedule here is the same as the one shown in Figure 21.3. As the described schedule consists from $y = 5$ years and $o = 10$ maintained assets per year, entries of $y \times o = 50$ or greater (e.g., the entry 90 for asset 1) mean that the corresponding assets are not scheduled. The entry 22 for asset 2 represents maintenance in the third year of the schedule.

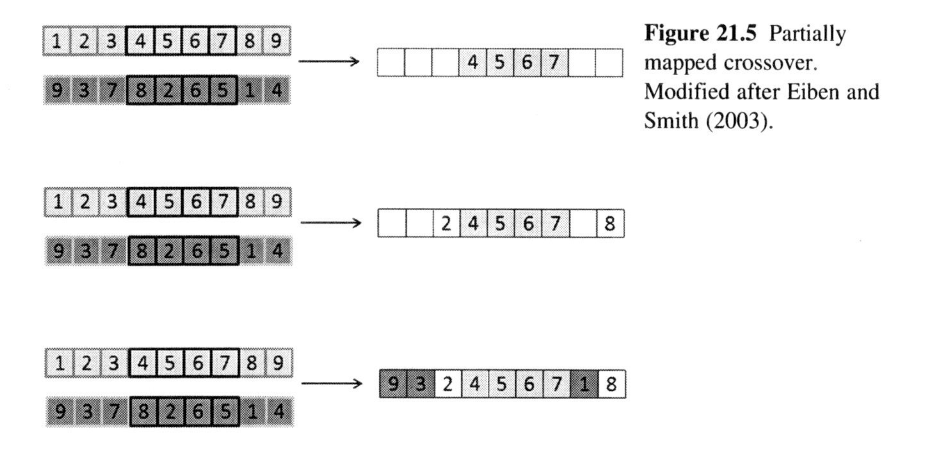

Figure 21.5 Partially mapped crossover. Modified after Eiben and Smith (2003).

For this representation, mutation is performed by exchanging two random entries of the chromosome. In concrete terms, this means that the years of maintenance are exchanged between two corresponding assets. This mutation is applied to the offspring with probability 1, so every new individual is mutated.

21.4.2 Repair Functions

Both representations of the infrastructure management problem exhibit a common issue: in both cases, the region of feasible solutions is disjoined, so it is possible that to get from one feasible solution to another, it will be necessary to cross a region of infeasible solutions.

At the same time, the feasible region represents only a very small part of the solution space. Accordingly, the start generation is most unlikely to contain feasible solutions.

This combination can lead to the search becoming trapped in an "island" in which the first feasible solution is found. As a consequence, the algorithm converges to the local optimum in this island, which may lie very far from the global optimum (Figure 21.6).

To overcome this problem, a repair function is introduced. After a new infeasible solution is created (either in the start generation or by crossover and mutation), an attempt is made to repair it.

In the matrix representation, for all assets whose maintenance deadline is within the considered period but have been scheduled too late or not at all, the algorithm attempts to exchange it. If the asset was scheduled too late, the function looks for other assets in all the years before its deadline that have a maintenance deadline outside the considered time span. The first such asset changes places with the

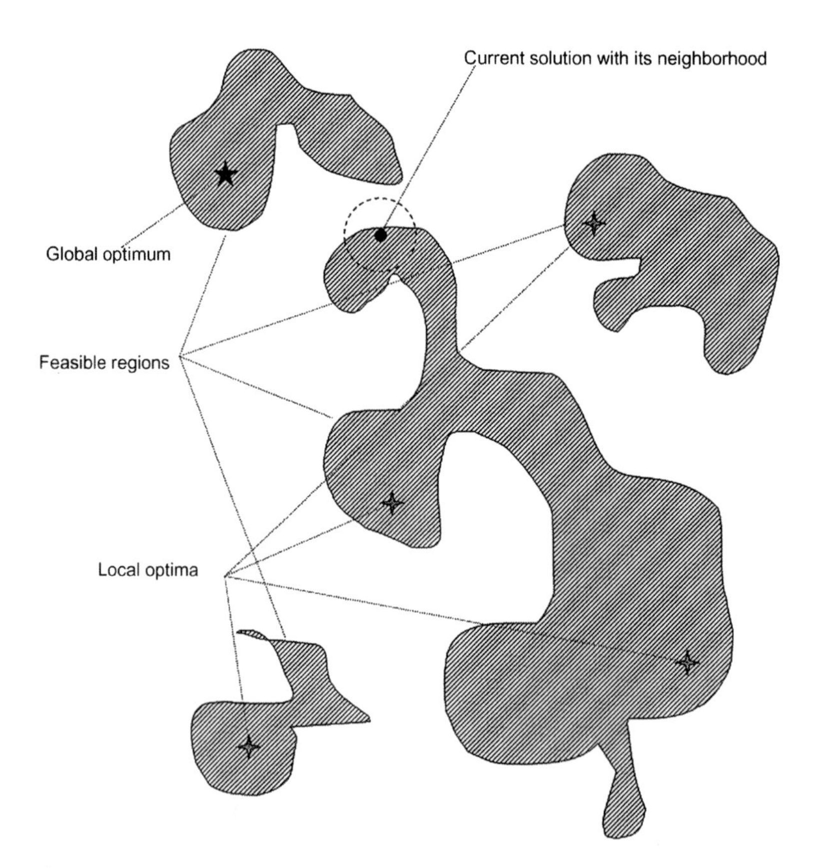

Figure 21.6 Schematic view of the search space. Feasible regions are shaded. If the first feasible solution is found at the place of the black dot without repair functions, the probability of the search reaching the global optimum is very low.

critical asset. If no matching asset (i.e., with a deadline outside the schedule) is found in the years before the deadline of the critical element, the asset with the most distant deadline in these years is chosen for exchange, and in turn, a new place is searched for this element inside its deadline in the same way. If the asset was not scheduled for maintenance at all, the same procedure is undertaken, but the asset with a noncritical deadline is eliminated from the schedule. To avoid biasing some solutions using this repair function, the critical assets are processed in random order.

To repair an infeasible individual in the permutation representation, for each critical asset that was scheduled too late or never, a noncritical asset is searched for that was scheduled within the deadline of the critical one, and the two entries are exchanged. In this representation, there is no danger of bias and the critical elements can be processed in the order they appear (i.e., ordered by their IDs).

Both these repair functions permit the search to cross infeasible regions between feasible islands without running too far into infeasible regions and so allow finding feasible solutions in earlier generations (Figure 21.6).

21.4.3 Additional Ways of Handling Infeasibility

These repair functions alone are not sufficient to handle the infeasible solutions. The algorithms described in Section 21.4.2 are able to reduce the number of constraint violations but cannot guarantee the creation of a feasible solution from an infeasible one, for example, when feasible exchanges cannot be found for some unscheduled or late-scheduled critical assets. In addition, the budget constraint is not considered at all.

Therefore, even with repairs to infeasible individuals, it is very unlikely that a feasible solution will be found in the first few generations. To avoid unnecessary calls to the traffic simulator, so long as no feasible individual has been found, the objective function is replaced by a minimization of constraint violations. Using this objective function, evolution is performed in the usual way until a feasible individual is created by crossover, mutation, and repair.

As soon as such a feasible individual is found, all currently existing individuals (the parent population as well as the offspring produced in the current generation before the feasible one) are evaluated with the traffic simulator. The objective function is reset to the minimization of the impact on traffic. Infeasible solutions are penalized with a static penalty. The penalized fitness $\hat{f}(x)$ for an individual x is computed as

$$\hat{f}(x) = f(x) + \lambda \times c(x) \tag{21.9}$$

with $f(x)$ as the value of the objective function x, $c(x)$ the number of constraint violations (i.e., unscheduled or late-scheduled critical assets) in x, and λ a penalty factor. This λ is chosen so that if half of all possible constraint violations occur, the fitness value is roughly doubled.

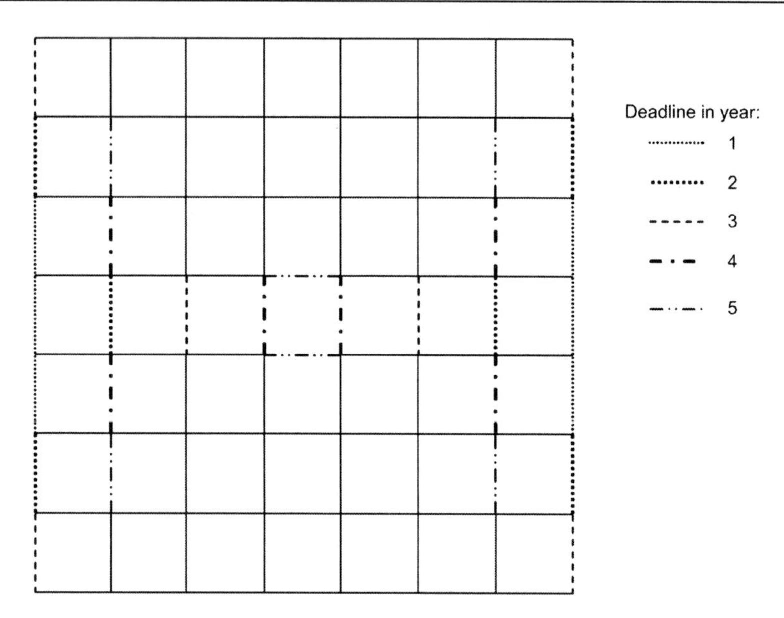

Figure 21.7 Road net for the test runs. All roads between the crossroads have a length of 100 m and a capacity of 1000 vehicles/day. The sources and targets of all the traffic are in the four corners: a flow of 1000 vehicles per day flows from each corner to each of the other corners.

The combination of minimization of constraint violations, repair functions, and static penalties allows the algorithm to find feasible solutions quickly, while at the same time preserving a certain level of diversity inside the population.

21.5 Test Case

To test and compare the different approaches, an artificial scenario (depicted in Figure 21.7) was designed. It is a chessboard layout of 7×7 roads in both the horizontal and vertical directions. Each of these roads has a length of 100 m and a capacity of 1000 vehicles per day. The maintenance deadlines of the roads[2] are marked in Figure 21.7; all the roads not marked have a deadline greater than five. The sources and targets of traffic are in the four corners of the network. From each corner, there is a flow of 1000 vehicles per day to each of the other corners.

The scenario has been designed this way so that a human observer can intuitively find good feasible solutions for a better comparability of the algorithms' results.

[2] In contrast to the rest of the chapter, for the test case, we speak of "roads" rather than of "assets." As the described algorithms can be applied to the maintenance of bridges or tunnels as well as to road pavement, "road" is used here to denote a road containing any type of infrastructural asset (bridge, tunnel, pavement, etc.) that has to be maintained.

For the individual roads, maintenance costs of between 100,000 and 400,000 units have been assigned. The annual maintenance budget is between 2,000,000 and 3,000,000 units.

A schedule is created for the next 5 years. Each year, 10 roads can be maintained, resulting in a search space size of 9.97×10^{63}. Maintenance works reduce the capacity of a road by 50%. The best manually found feasible solution to this scenario has an objective value of 6.08×10^8 veh $\times$ s.

The scenarios described by the schedules produced by the heuristic optimization algorithms are evaluated using VISUM (PTV AG, 2009a). The assignment of the traffic is realized as equilibrium assignment (Beckmann et al., 1955; PTV AG, 2009b). VISUM Standard is used for the daily course of traffic.

21.6 Test Results

Table 21.1 summarizes the results of each of the 10 runs with the different settings. The GAs with both representations, matrix and permutation, were run over 50 generations with a population size of 50 and a mating pool size of 50. Parent and survivor selections were both performed as tournament selection. The GA without a repair function was run with the same parameters with the matrix representation over 50 generations.

Table 21.1 Results of 10 Test Runs, Each with Different Settings (in veh $\times$ s)

	Mean Value Over 10 Runs	**Best Found Value**	**Standard Deviation Over 10 Runs**	**Standard Deviation (%)**
Rank-based ant system	6.24×10^8	6.15×10^8	13.5×10^6	2.1
Rank-based ant system with dynamic elitist ant	6.18×10^8	6.11×10^8	3.87×10^6	0.6
Ant colony system (η only security)	6.31×10^8	6.17×10^8	14.7×10^6	2.3
Ant colony system (η cost and security)	6.11×10^8	6.10×10^8	0.65×10^6	0.1
GAs matrix representation	6.06×10^8	6.05×10^8	0.81×10^6	0.1
GAs permutation representation	6.07×10^8	6.06×10^8	1.13×10^6	0.2
GAs matrix representation without repair	6.23×10^8	6.09×10^8	11.3×10^6	1.8

All tested algorithms are able to find feasible solutions to the problem. The rank-based ant system and ant colony system, with η only considering the safety constraint, both get stuck into local optima. A dynamic elitist ant shows only a slight improvement for the rank-based ant system. Both representations for GAs with repair and ant colony system with η considering safety and costs are able to find solutions in the range of the manually found solution reliably. The best-known solution was found by the GA with matrix representation. Without a repair function, the convergence of GAs is much slower as much time is wasted on infeasible solutions. As a consequence, after the same number of generations, its solutions are worse than those found using GAs with a repair function.

The rank-based ant system was run for 50 generations with 50 ants, and $w = 5$. ρ was set to 0.4. The ant colony system was run with 10 ants and $\rho = 0.1$ for 100 generations.

The run time for the GAs and the rank-based ant system was almost the same—about 12 h—as it is mainly influenced by the number of callings to the traffic simulator. This number is determined by the number of the considered years multiplied by the population size. For this reason, the ant colony system is much faster, even with twice the number of generations, as it needs only a fifth of the evaluations per generation due to the smaller population.

All the tested algorithms were able to find feasible solutions during their run time in all the test runs, but the quality of those found solutions differs substantially. Only the GAs with a repair function and the ant colony system with η considering safety and costs were able to find solutions in the range of the manually found solutions. The best-known solution to the problem was also found by GAs. All three named approaches found those solutions with high confidence. The standard deviation between the results from different runs is only 0.1% (0.2% for GAs with permutation representation). All of them exhibit good convergence behavior (see Figure 21.8 for one run of the GA with matrix representation). In the GA with matrix representation, as well as in the ant colony system, with η considering both cost and safety after each generation, a feasible solution exists. For the GA with permutation representation, the same applies, but in most runs, this occurred only after the first few iterations.

Even though the ant colony system shows slightly inferior results to both the GA representations in this test case, it may prove to be more attractive for use in practice, as it needs a much shorter run time to produce high-quality results.

Without the repair functions in GAs, a lot of time is spent on the creation and evaluation of infeasible solutions. Consequently, the algorithm needs many more generations between the changes of the best individual in the population, and thus for convergence. As a result, even after 50 generations, the search is far from the good results and the results of different runs vary considerably. However, after

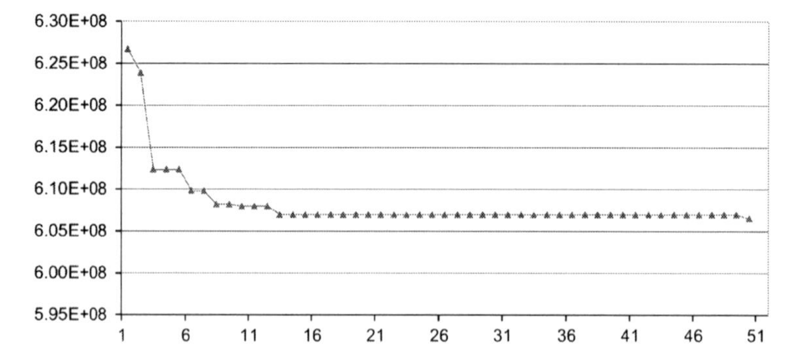

Figure 21.8 Fifty generations in one run using the GA with matrix representation. The results shown are the best feasible solutions in the population after each generation.

many more generations, the GAs without repair will also converge on solutions in the range of the best-known solution.

The rank-based ant system and ant colony system with η only considering the safety constraint both get stuck in local optima far from the best-known solution. The reason for this can be seen in Figure 21.9. The algorithms need a number of iterations to find a feasible solution at all. As this first-found feasible solution now becomes the elitist ant (or, in the ant colony system, the only pheromone-depositing ant), the search will concentrate on its neighborhood. For a few iterations, the algorithms may be able to find better feasible solutions in this region until it reaches a local optimum. The influence of this will be even higher than that of the first-found feasible solution, and it will rapidly take over the whole elitist group and thereafter the whole colony. The probability that solutions apart from it will be found becomes lower with each iteration, and the probability that one of these solutions is better tends toward zero. As the algorithms always converge to local optima in the neighborhood of the first feasible solution that was found by chance, the results in different test runs differ considerably.

By dynamically changing the elitist ant in the rank-based ant system as proposed by Merkle et al. (2002), the search is able to leave the neighborhood of the first-found feasible solution. But this will only be of benefit to the search if the new elitist ant is in the neighborhood of another local optimum and better than or nearer the global optimum than the old one, which is not always the case. Otherwise, the search will stay trapped in the same region of the search space. If the first-found

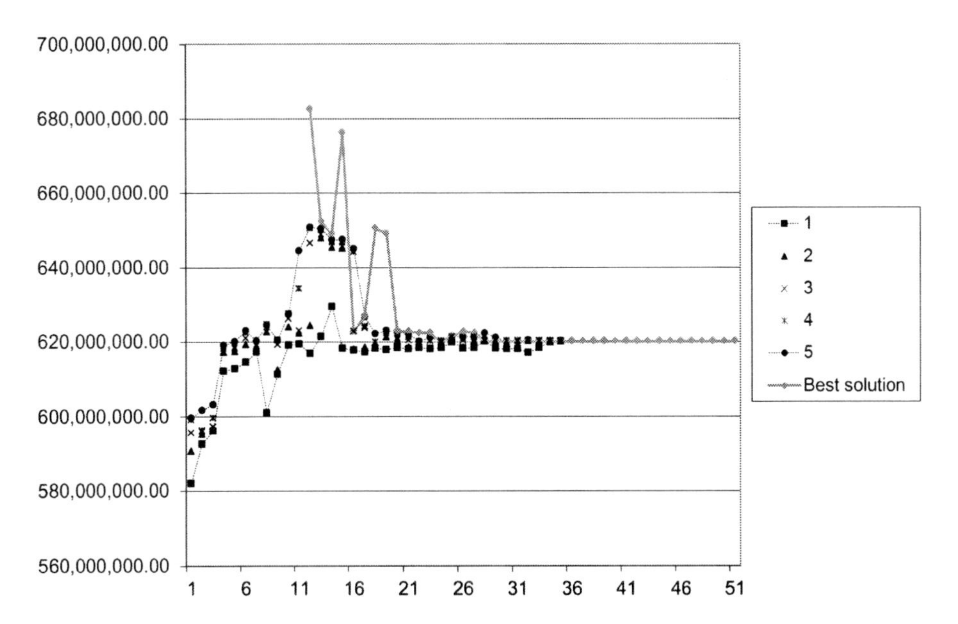

Figure 21.9 The elitist group and the best feasible solution over 50 iterations of a run of the rank-based ant colony system. It can be seen that a relatively good feasible solution takes over the elitist group in only a few iterations, thus leading to premature convergence.

feasible solution lies very far from the global optimum, it will become almost impossible to reach it by always exchanging the elitist ant with another solution that lies in the right direction. Therefore, this approach shows only a slight improvement to the rank-based ant system with a fixed elitist ant. The algorithm again runs into local optima, though different ones from those found with a static elitist ant, as it is now able to "hop" from one local optimum to another and thus to reach ones that lie farther inside the feasible regions.

21.7 Summary and Outlook

In this chapter, we introduced the problem of finding an optimal schedule for the maintenance of infrastructural buildings on network level with regard to the impact on traffic. As it represents a complex combinatorial problem, the use of metaheuristics for solving it is proposed. ACO and GAs were adapted to this problem and tested using a simplified test instance.

Extensive test runs show that all proposed algorithms are able to find feasible schedules, but there are differences in the quality of the found solutions. Using a combination of repair functions and static penalties, both problem representations for GAs can be made to find high-quality solutions to the test problem reliably. In contrast, the tested dialects of ACO, the rank-based ant system and the ant colony system, get stuck in local optima without further modifications. By using the greedy heuristic part in the choice function such that the choices that are advantageous to producing feasible solutions are made more attractive, ant colony system can be adapted to produce solutions of similar quality to those achieved with the GAs.

Even though the solutions found by the modified ant colony system are still slightly inferior to the ones found with GAs, the shorter run time needed to produce them may make the ant colony system the more promising choice for practical applications. Additional tests on a more realistic problem scenario will follow.

A further enhancement to a multiobjective optimization so that additional wishes of the infrastructure manager, such as synergies with third-party construction sites, should also be considered to make the approach more applicable in practice. Both GAs and ACO could possibly be adapted to this multiobjective optimization. Which approach proves to be more suitable in this application will be explored in future research.

Acknowledgments

This project was funded by the German Federal Ministry of Economics and Technology.

References

Beckmann, M.J., McGuire, C.B., Winsten, C.B., 1955. Studies in the economics of transportation, Yale University Press, New Haven, CT.

Bonabeau, E., Theraulaz, G., Deneubourg, J.L., 1997. Self-organization in social insects. Trends Ecol. Evol. 12 (5), 188–193.

Borrmann, A., Lukas, K., Zintel, M., Schießl, P., Kluth, M., 2012. BIM-based life-cycle management for reinforced concrete buildings. Int. J. 3D Inf. Model. 1 (1), 1–24.

Budai, G., Dekker, R., Kaymak, U., 2009. Genetic and Memetic Algorithms for Scheduling Railway Maintenance Activities. Rotterdam, The Netherlands.

Bullnheimer, B., Hartl, R.F., Strauss, C., 1997. Applying the ant system to the vehicle routing problem. Proceedings of the Second International Conference on Metaheuristics, 285–296.

Bullnheimer, B., Hartl, R.F., Strauss, C., 1999. A new rank-based version of the ant system: a computational study. Cent. Eur. J. Oper. Res. Econ. 7, 25–38.

Burke, E.K., Smith, A.J., 1997. A memetic algorithm for the maintenance scheduling problem. Proceedings of the ICONIP, pp. 469–473.

Christodoulou, S., 2010. Scheduling resource constrained projects with ant colony optimization artificial agents. J. Comput. Civ. Eng. 24 (1), 45–55.

Christodoulou, S.E., Ellinas, G., 2010. Pipe routing through ant colony optimization. J. Infrastruct. Syst. 16 (2), 149–159.

Colorni, A., Dorigo, M., Maniezzo, V., Trubian, M., 1994. Ant system for job-shop scheduling. Belgian J. Oper. Res. Stat. Compu. Sci. 34 (1), 39–53.

Dorigo, M., 1992. Optimization, Learning and Natural Algorithms. Politecnico di Milano, Milan.

Dorigo, M., Gambardella, L.M., 1997. Ant colony system: a cooperative learning approach to the traveling salesman problem. IEEE Trans. Evol. Comput. 1 (1), 53–66.

Dorigo, M., Stützle, T., 2004. Ant Colony Optimization. Cambridge, MA, Cambridge.

Dorigo, M., Maniezzo, V., Colorni, A., 1996. Ant system: optimization by a colony of cooperating agents. IEEE Trans. Syst. Man Cybern. Part B—Cybern. 26 (1), 29–41.

Dridi, L., Parizeau, M., Mailhot, A., Villeneuve, J.-P., 2008. Using evolutionary optimization techniques for scheduling water pipe renewal considering a short planning horizon. Comput. Aided Civ. Infrastruct. Eng. 23, 625–635.

Eiben, A.E., Smith, J.E., 2003. Introduction to Evolutionary Computing. Springer-Verlag, Berlin.

Gambardella, L., Dorigo, M., 1996. Solving symmetric and asymmetric TSPs by ant colonies. Proceedings of IEEE International Conference on Evolutionary Computation, pp. 622–627.

Gandomi, A.H., Yang, X.-S., Alavi, A.H., 2011. Mixed variable structural optimization using firefly algorithm. Comput. Struct. 89 (23–24), 2325–2336.

Gandomi, A.H., Yang, X.-S., Alavi, A.H., 2011. Cuckoo search algorithm: a metaheuristic approach to solve structural optimization problems. Eng. Comput. doi: 10.1007/s00366-011-0241-y, 1–19.

Goldberg, D.E., Lingel, R., 1985. Alleles, loci, and the travelling salesman problem. Proceedings of the First International Conference on Genetic Algorithms and Their Applications, L. Erlbaum Associates Inc., pp. 41–49.

Goss, S., Aron, S., Deneubourg, J.L., 1989. Self-organized shortcuts in the Argentine ant. Naturwissenschaften. 76 (12), 579–581.

Hammad, A., Zhang, C., Hu, Y., 2006. Mobile model-based lifecycle management systems. Comput. Aided Civ. Infrastruct. Eng. 21, 530–547.

Holland, J.H., 1975. Adaption in Natural and Artificial Systems. The University of Michigan, Ann Arbor, MI.

Kennedy, J., Eberhart, R., 1995. Particle swarm optimization. Proceedings of the IEEE International Conference on Neural Networks, (4), 1942–1948.

Kirkpatrick, S., Gelatt, C.D., Vecchi, M.P., 1983. Optimization by simulated annealing. Science. 220, 671–680.

Lapa, C.M.F., Pereira, C.M.N.A., de Barros, M.P., 2005. A model for preventive maintenance planning by genetic algorithms based in cost and reliability. Reliab. Eng. Syst. Saf. 91, 233–240.

Lee, H.-Y., 2009. Optimizing schedule for improving the traffic impact of work zone on roads. Autom. Constr. 18, 1034–1044.

Li, S.-T., Ting, C.-K., Lee, C., Chen, S.-C., 2002. Maintenance scheduling of oil storage tanks using Tabu-based genetic algorithm. Proceedings of the Fourteenth International Conference on Tools with Artificial Intelligence, IEEE, 209–215.

Maniezzo, V., 1999. Exact and approximate nondeterministic tree-search procedures for the quadratic assignment problem. INFORMS J. Comput. 11 (4), 358–369.

Maniezzo, V., Colorni, A., 1999. The ant system applied to the quadratic assignment problem. IEEE Trans. Knowl. Data Eng. 11 (5), 769–778.

Merkle, D., Middendorf, M., Schmeck, H., 2002. Ant colony optimization for resource-constrained project scheduling. IEEE Trans. Evol. Comput. 6 (4), 333–346.

Merz, P., Freisleben, B., 1999. Fitness landscape and memetic algorithm design. New Ideas Optimization, D. Corne, M. Dorigo, F. Glover (Eds.), McGraw-Hill, London. 245–260.

Morcous, G., Lounis, Z., 2004. Maintenance optimization of infrastructure networks using genetic algorithms. Autom. Constr. 14, 129–142.

Muñoz, A., Martorell, S., Serradell, V., 1997. Genetic algorithms in optimizing surveillance and maintenance of components. Reliab. Eng. Syst. Saf. 57, 107–120.

Negnevitsky, M., Kelareva, G.. 1999. Genetic algorithms for maintenance scheduling in power systems. Sixth International Conference on Neural Information Processing, IEEE, (2) 94–103.

Ning, X., Lam, K.-C., Lam, M.C.-K., 2009. Dynamic construction site layout planning using max–min ant system. Autom. Constr. 19, 55–65.

Okasha, N.M., Frangopol, D.M., 2010. Advanced modeling for efficient computation of life-cycle performance prediction and service-life estimation of bridges. J. Comput. Civ. Eng. 24 (4), 548–556.

PTV AG. 2009a. VISUM 11.0—User Manual.

PTV AG. 2009b. VISUM 11.0—Fundamentals.

Socha, K., Knowles, J., Sampels, M., 2002. A max–min ant system for the university course timetabling problem. Ant Algorithms: Proceedings of the Third International Workshop (ANTS 2002), Lecture Notes in Computer Science. 2463, 1–13.

Stützle, T., Dorigo, M., 1999. ACO algorithms for the quadratic assignment problem. New Ideas Optim. 33–50.

Stützle, T., Hoos, H., 1996. Improving the ant system: a detailed report on the max–min ant system, Technical Report AIDA–96–12, FG Intellektik, FB Informatik, TU Darmstadt, Germany.

Taillard, E., Gambardella, L., 1997. Adaptive Memories for the Quadratic Assignment Problem. Research Report. IDSIA Lugano, Switzerland.

Teodorović, D., Dell'Orco, M., 2005. Bee colony optimization—a cooperative learning approach to complex transportation problems. Adv. OR AI Methods Transp.51–60.

Vesikari, E., 2008. Life-cycle management tools for civil infrastructure. Proceedings of the Third International Workshop on Lifetime Engineering of Civil Infrastructure, Ube, Japan.

Yang, X.-S., 2009. Firefly algorithms for multimodal optimization. Stoch. Algorithms Found. App. 169–178.

22 Large-Scale Maintenance Optimization Problems for Civil Infrastructure Systems

Sehyun Tak, Sunghoon Kim and Hwasoo Yeo

Department of Civil and Environmental Engineering, Korea Advanced Institute of Science and Technology, Daejeon, Republic of Korea

22.1 Introduction

Maintenance optimization based on life cycle cost (LCC) assessment is applied to many civil infrastructure systems such as pavements and bridges to reduce operating and socioeconomic cost. However, due to the large size of infrastructure networks, maintaining such networks is challenging, especially when many constraints exist and the budget is limited. So, it is important to develop new strategies for managing public infrastructures in a way that ensures long-term sustainability under constrained budgets. This chapter reviews metaheuristic methods for infrastructure maintenance optimization, followed by current critical issues such as deterministic and stochastic problems, single- and multi-facility problems, and infrastructure interdependencies.

22.2 Large-Scale Maintenance Optimization Problem

22.2.1 Maintenance Optimization Formulation

The goal of the infrastructure maintenance management is to maintain the level of service of the entire infrastructure with available funds and resources. Infrastructure maintenance management system contains various maintenance objectives such as maximizing network performance, maximizing the cost-effectiveness of maintenance activities, minimizing the user cost. The objectives also include minimizing the present worth of the total maintenance cost with a certain set of constraints such as budget, condition state, manpower, and equipment (Chootinan et al., 2006). Among these various objectives, three goals are commonly used in the infrastructure maintenance optimization problem, as shown in Table 22.1. The cost-based model and bi-objective model are commonly used because the budget for infrastructure maintenance is generally limited.

Metaheuristic Applications in Structures and Infrastructures. DOI: http://dx.doi.org/10.1016/B978-0-12-398364-0.00022-X

Table 22.1 Problem Statement of Infrastructure Maintenance Optimization

	Objective	Constraints
Cost-based formulation	Minimize the present value of life-cycle maintenance cost	Condition index Safety index
Performance-based formulation	Maximize the performance of the entire pavement network	Budget limitations User costs
Bi-objective formulation	Maximize the weighted function of cost and performance	Condition index Budget limitations

22.2.2 Deterministic Versus Stochastic Problem

Information on infrastructure condition plays an important role in the infrastructure maintenance optimization problem because the selection of optimal maintenance policy is determined depending on the information on the infrastructure's current and future conditions. Using such information, there are already many studies related to performance indicators, deterioration modeling, and their effects on the maintenance optimization problem.

In general, deterioration models are classified into two categories: deterministic and stochastic models. Deterministic models, such as straight-line extrapolation, regression models, and curve-fitting techniques, depend on mathematical or statistical formulations for explaining the relationship between the factors affecting infrastructure deterioration and measures of infrastructure condition (Morcous, 2000). Unfortunately, deterministic models cannot predict the future condition of infrastructure accurately because the models do not consider all variables related to the deterioration process due to the existence of unobserved variables. Furthermore, while specifying a deterioration model, the methods do not capture the uncertainties that occur during the condition-generating process of an infrastructure element.

The existing uncertainties in the infrastructure deterioration model can be attributed to the two types of factors: exogenous and endogenous factors (Durango and Madanat, 2002). Exogenous factors are mainly caused by the environment and level of utilization. For example, exogenous factors such as weather and user demands are difficult to predict. Therefore, the effects of exogenous factors foster uncertainty in the deterioration model. Endogenous factors such as design characteristics and materials stem from the different reactions to such exogenous conditions. For example, two segments on the same road often show different deterioration rates, despite the fact that they consist of the same materials and design characteristics. These kinds of unknown variability are considered as uncertainties of endogenous factors in the deterioration model. Due to the uncertainties in the deterioration model, the deterministic model may overestimate the effect of repair action or underestimate the deterioration process. An underestimation of the maintenance cost or overestimation of the overall condition may be caused by the uncertainties that are not considered in deterministic models (Chootinan et al., 2006). Due to the error generated from the deterministic model, safety hazards can be caused, and moreover, the errors and safety hazards are propagated over the process.

To overcome the limitations of the deterministic model, the stochastic model is now increasingly used in the infrastructure deterioration model to deal with a high degree of uncertainties. The Markovian decision process (MDP), which can take into account the inherent randomness in deterioration, is one of the most popular stochastic techniques. In the MDP model, a finite set of the facility conditions, which are measured by a discrete state and the deterioration process, is represented by transition probabilities.

Although the MDP model has been used in many studies, it is still based on assumptions and has some limitations. The transition probability matrix, which is obtained by Markov process modeling, is based on the assumption that the transition probability matrix does not depend on the age of the infrastructure elements. This assumption differs from the mechanistic knowledge of material behavior and deterioration rate of the actual situation because the deterioration rate of the infrastructure is significantly affected by its age (Guignier and Madanat, 1999). For example, corrosion occurring in the early stage spreads into surrounding areas as time passes. Therefore, the corrosion dramatically accelerates the deterioration process. This model does not consider any uncertainties in the process of measuring facility conditions. Unfortunately, the inspection error may significantly affect the accuracy of the results such as cost and final condition state (Smilowitz and Madanat, 2000). So, new methods such as joint MDP, latent MDP, and adaptive control have recently been developed to overcome the limitations of the MDP model (Guillaumot et al., 2003).

22.2.3 Single-Facility Versus Multi-Facility Problem

The infrastructure maintenance management system supports decision making for maintaining and repairing infrastructure facilities experiencing deterioration. The goal of infrastructure maintenance optimization is to find the cost-effective way to increase the infrastructure facilities' performances by properly deciding when and how to repair. Infrastructure maintenance optimization problems can be classified into single-facility and multi-facility problems. The selection of repair methods and timing to repairing each infrastructure are considered at a single-facility level, while resource allocation and prioritization for repairing the infrastructure network are considered at a multi-facility level.

Single-facility-level maintenance optimization has been developed to achieve desirable solutions from the long-term economic point of view. LCC analysis is one of the typical ways to approach the single-facility-level maintenance optimization problem because LCC can optimally balance the lifetime structure performance and the whole-life maintenance cost for the single-facility-level infrastructure maintenance optimization. This analysis mainly focuses on selecting the repair methods, estimated costs, expected improvements, and timing for the repair.

Figure 22.1A shows an example of performance profiles of essential maintenance strategy and optimal maintenance strategy. The "no-maintenance strategy" is not a cost-effective way due to the high LCC caused by the replacement costs, which are generally much higher than the repair costs. Optimal maintenance is achieved by adequate balancing between maintenance costs and performance. If the repair action is applied infrequently, operation costs and maintenance costs increase, and then the LCCs can be very high. On the other hand, if the repair action is applied frequently, it is uneconomical because the excessive expenditure increases the average infrastructure condition more than necessary. Proper maintenance yields a gently sloping performance curve that indicates enhanced average conditions over pavement life cycle and longer service life. To solve the single-facility-level maintenance optimization problem, many optimization techniques have been adopted, such as optimal control theory, linear and nonlinear programming, dynamic programming, integer programming, and calculus of variations (Friesz and Fernandez, 1979; Li and Madanat, 2002; Madanat and Benakiva, 1994; Ouyang and Madanat, 2006; Tsunokawa and Schofer, 1994; Tzafestas and Botsaris, 1986).

Single-facility-level maintenance optimization focuses on achieving desirable solutions by choosing the optimal action and time without accounting for budget constraints. However, in practice, budget limitations require a different method than the single-facility-level maintenance optimization. Multi-facility-level maintenance optimization is needed to allocate the resources to all infrastructures to be repaired due to budget limits. Therefore, for the multi-facility problem, the objective is to find the optimal set of repair actions to allocate the resources

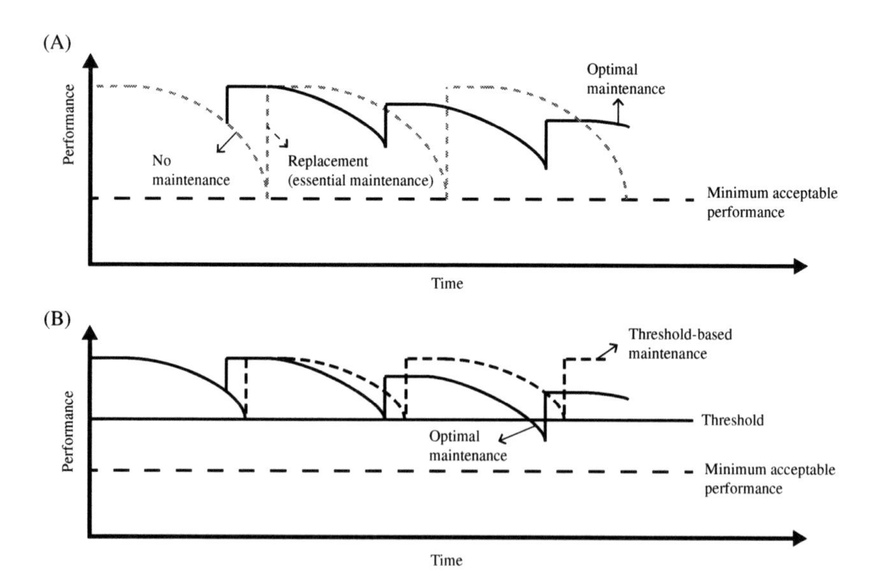

Figure 22.1 Performance profiles of two different maintenance strategies: (A) essential maintenance and optimal maintenance (or preventive maintenance); (B) threshold-based maintenance and optimal maintenance.

appropriately based on the complex optimization techniques. This problem can be solved by priority ranking and mathematical optimization. Priority ranking techniques sort all infrastructures in descending order based on the calculated values of service level, benefit/cost ratio, and sufficiency rating (Mohamed et al., 1995). Starting with the infrastructure with highest values, financial resources are allocated to that infrastructure until the available funds are exhausted. Multi-facility-level maintenance optimization can also be solved by linear programming based on the assumption that inspections are performed at predetermined and fixed time intervals (Smilowitz and Madanat, 2000). This model selects an infrastructure to minimize the total expected cost for the system provider and users.

The output of the multi-facility-level maintenance model is a set of candidate infrastructures for maintenance, repair, and rehabilitation (MR&R) activities (Johnston and Zia, 1984). Therefore, it can effectively overcome the computational difficulties encountered when attempting to apply an individual model to each facility. However, in this model, the individual segments of infrastructures are assumed to be homogeneous.

Therefore, the results at the multi-facility level cannot be the optimal budget, and the long-term costs become greater than needed at the view of the single-facility-level. This is because they do not consider each individual infrastructure's needs at the multi-facility level.

Most of the current infrastructure maintenance management systems are developed to support either the single-facility-level decision or the multi-facility-level decision. Dealing with single-facility and multi-facility levels separately may not be the optimal way of finding the solution. Therefore, the single-facility- and multi-facility-level decisions should be integrated because both level decisions are very interrelated (Thompson et al., 2003). Several attempts have been made to integrate the advantages of the single-facility level and multi-facility level. Most of these attempts used the genetic algorithm (GA) technique because this technique can effectively solve multi-objective optimization problems (Chootinan et al., 2006; Frangopol and Liu, 2007; Furuta et al., 2006; Neves et al., 2006).

The threshold maintenance strategy has been used recently for the multi-facility maintenance optimization problems (Sathaye and Madanat, 2011). This technique simplifies the LCC analysis by using an optimal trigger roughness value, as shown in Figure 22.1(b). All infrastructures whose condition states are lower than the trigger value are resurfaced with a fixed intensity. Therefore, this model can efficiently produce the optimal solutions without any complex information such as initial condition state or deterioration rate. However, this model has some restrictions because the results sensitively respond to the trigger roughness.

22.2.4 Interdependency Issues for the Multi-Facility Problem (Network Level)

Infrastructures are highly interconnected and mutually dependent in complex ways. If one infrastructure is repaired to maintain its condition, the repaired infrastructure can directly or indirectly affect other infrastructures in a system. For example, a temporary interruption of a subway system for maintenance causes the increase in use of alternative ground transportation such as buses and private vehicles, and this increase may cause congestion on roads. Therefore, interdependencies of infrastructures are to be identified, understood, and analyzed when dealing with the multi-facility maintenance optimization.

Figure 22.2 shows a hypothetical transportation network to explain the infrastructure interdependencies. Interdependencies in infrastructure maintenance optimization problems can be classified into economic interdependencies and demand interdependencies (Sarutipand, 2008). Economic interdependencies reflect costs associated with throughput loss and benefits associated with coordinating intervention schedules of adjacent facilities. These interdependencies can be explained on the basis of Figure 22.2. If infrastructures B and D are disrupted at the same time for maintenance, the user cost increases much more than twice the cost of disrupting each infrastructure individually because users who demand to move from node 1 to node 2 have to move via node 3. Therefore, infrastructures B and D are repaired at different times in order to minimize the disruption to the system. This negative effect is also seen when infrastructures A and F are repaired at the same time. This kind of relation is called a *substitutable relationship*. On the other hand, if infrastructures B and C are repaired at the same time, the agency cost can be decreased much more than the cost of repairing each infrastructure individually because personnel and equipment delivery costs are saved when adjacent facilities are being repaired during the same period. Therefore, infrastructures B and C are repaired at same time in order to minimize the repair costs. This positive effect is also seen when infrastructures D and E are repaired together. This kind of relation is called a *complementary relationship*.

Demand interdependencies are strongly related to the effect of demand on infrastructure condition. If infrastructure F is disrupted for maintenance, the demand for infrastructure A increases. Consequently, maintenance costs, vehicle operating costs, and congestion costs increase due to the accelerated deterioration process. However, despite the importance of interdependencies of infrastructures, there are only a few experiments dealing with interdependency issues for the multi-facility problem (Dekker et al., 1996; Durango-Cohen and Sarutipand, 2006).

Interdependencies of infrastructures should be significantly considered on MR&R policies for multi-facility transportation systems because they can lead to

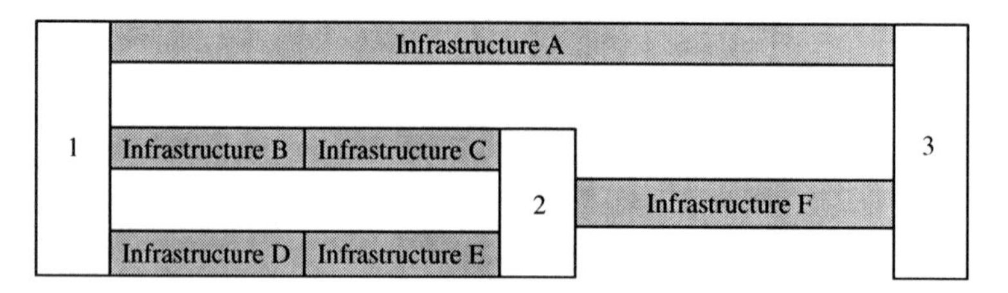

Figure 22.2 Hypothetical transportation network with three paths between nodes 1 and 4.

optimal coordination of maintenance policies, which is more appealing in practice. Therefore, more research about interdependencies of infrastructures is needed.

22.3 Metaheuristic Solution Approaches

Traditional mathematical optimization methods such as linear programming and dynamic programming techniques often fail in solving Nondeterministic Polynomial (NP) hard problems with large numbers of variables and nonlinear objective functions (Lvbjerg, 2002). Furthermore, infrastructure maintenance management problems deal with constrained, nonlinear, and nondifferentiable structures that cannot be solved efficiently by classic optimization techniques. Metaheuristic solution approaches were introduced by many researchers to handle these kinds of optimization problems while overcoming the drawback of using a traditional optimization tool for the infrastructure maintenance management system.

Metaheuristics are high-level strategies for exploring search spaces. Metaheuristics are divided into trajectory methods and population-based methods. Population-based methods are used more frequently because they do a better job identifying promising areas in a certain search space, whereas trajectory methods are better at exploring promising areas in the search space (Blum and Roli, 2003). Generally, population-based methods perform better than trajectory-based methods such as simulated annealing (SA) and tabu search for infrastructure maintenance management in terms of computational efficiency and simplicity. This is mainly due to the difference in the concept of recombining solutions to obtain new ones. So, in this section, population-based algorithms such as genetic algorithm (GA), ant colony optimization (ACO), the shuffled frog leaping (SFL) algorithm, and the hybrid algorithm for maintenance management are reviewed.

22.3.1 Genetic Algorithms

GAs are evolutionary methods motivated by the principles of natural selection and "survival of the fittest" (Beasley et al., 1993). GAs have been used to solve infrastructure maintenance management problems by many researchers (Avijit Maji, 2007; Cheu et al., 2004; Chootinan et al., 2006; Frangopol and Liu, 2007; Jha and Abdullah, 2006; Liu and Frangopol, 2005; Morcous and Lounis, 2005; Okasha and Frangopol, 2009). Although there are many other metaheuristics such as SA, tabu search, ACO, and SFL, GAs are highly recognized for the following three reasons. First, the strength of GAs, which is the population-based metaheuristic, is certainly based on the concept of recombining solutions to obtain new ones. GAs generate a population of multiple solutions (genes) and use probabilistic rules to generate a new and better population. Second, GAs can find the optimal solutions more efficiently and accurately in a timely fashion because their search spaces are usually larger than those of the trajectory method (Goldberg, 1989). Finally, GAs evaluate only the objective function and do not require auxiliary information such as derivatives, which can make the mathematics of the problem very simple.

Figure 22.3 shows the basic steps of performing the GA process, and Figure 22.4 illustrates the concept of GA. In step 1, a set of possible solutions to the problem is established, and each individual (gene) in the population is encoded into genetic representation (i.e., chromosomes). Each individual (gene) in the population represents a single infrastructure that must be considered in the maintenance management system. Each code in the gene is a type of repair action. In step 2, each individual in the initial population is evaluated based on the fitness value computed from the fitness function, which is determined based on the objective function and the constraints. In infrastructure maintenance management, the fitness

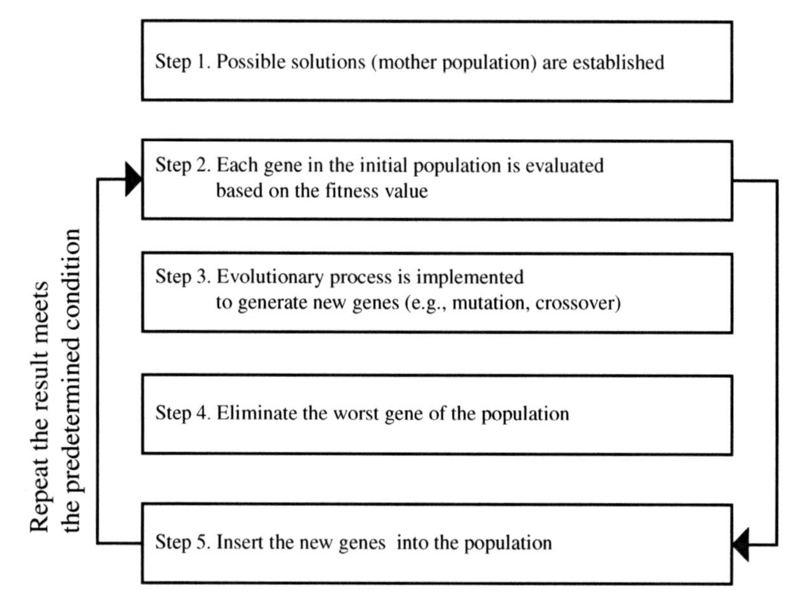

Figure 22.3 The GA process.

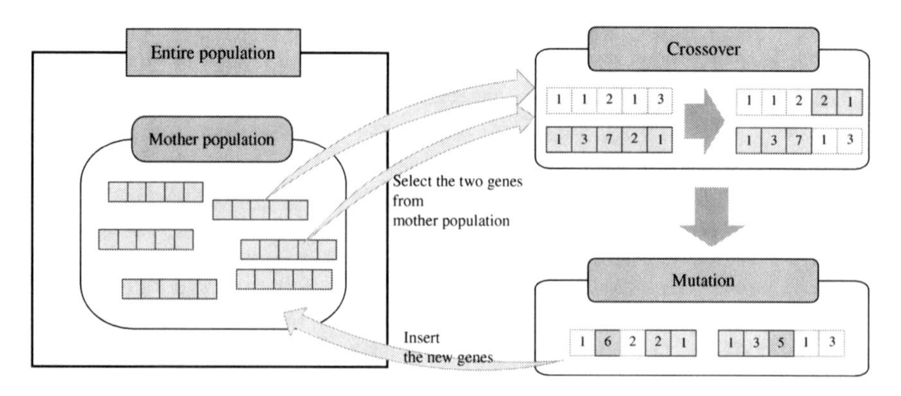

Figure 22.4 An illustration of GA.

value of a solution is generally defined as total cost, the sum of the condition state, or both. In step 3, an evolutionary process such as selection, mutation, or crossover is implemented on the parent pool to generate an offspring pool. In steps 4 and 5, the offspring solutions are evaluated based on the fitness value and the bad solution is replaced by the better solution.

In the GA algorithm, the crossover, mutation, and pool size are the important factors because both the ratio and the numbers of crossover and mutation can significantly affect the optimal value and computation time (Avijit Maji, 2007). The crossover operation is highly related with intensification. Higher crossover probability means that it can generate the promising solution more quickly. Mutation operation is highly related with diversification (Blum and Roli, 2003). Higher mutation probability means that it has a high possibility of generating the new solution, which has never been selected. On the other hand, a very low mutation probability may cause the local optima because it reduces the possibility of exploring a new solution. To avoid stagnation in the evolutionary process, a mutation rate (<0.1) is usually used (Goldberg, 1989). Therefore, balancing between the diversification and intensification strategy is important.

22.3.2 Ant Colony Optimization

ACO, developed by Dorigo (Dorigo and Stützle, 2004), is an iterative algorithm and belongs to the class of metaheuristics. It is based on the foraging behavior of ants in nature, which are capable of finding the shortest path between their nest and a source of food by *stigmergy*, which is an indirect form of communication.

The ACO algorithm is composed of three parts: ant-based solution construction, pheromone update, and iteration (Blum, 2005). In the ant-based solution construction phase, the sequences of solution components represent the series of artificial ants' position changes that are determined by the stochastic-mechanism based pheromone. One solution component refers to an ant's position, which represents the repair action in the infrastructure maintenance management system. In the ACO algorithm, pheromone update is the important phase because the differences between ACO algorithms such as ant system (AS), max−min ant system (MMAS), and ant colony system (ACS) are caused by the difference of pheromone rules (Dorigo et al., 2006). In AS, pheromone values are updated by all ants that pass the node. In AS, even though a path is insufficient, if the number of ants that have passed through the path is great, too much pheromone can be accumulated on the insufficient path. In other words, due to the amount of excessively accumulated pheromone on the path, the insufficient path may be recognized and selected incorrectly as the good path. MMAS is an algorithm that tries to overcome this shortcoming of AS. Unlike AS, in which pheromone is updated time to time during the construction process, only the pheromone value of the best ant is updated at the end of the construction process in MMAS. ACS is similar to MMAS in terms of updating pheromone at the end of the process, but the difference is that in ACS, the processes of local pheromone updates are included during iterations to diversify the search.

ACO shows better performance than GA in terms of computation time when optimizing maintenance dates because, in such cases, a priori information can be included in the information matrix (Samrout et al., 2005). Due to the strength of this algorithm, ACO has been applied to several optimization problems such as scheduling problems, vehicle routing problems, the set packing problem, transportation networks, and preventive maintenance (Blum, 2005; Gandibleux et al., 2005; Merkle et al., 2002; Samrout et al., 2005; Vitins and Axhausen, 2009). Generally, ACO algorithms are competitive with other optimization techniques when applied to the problems that are not overly constrained. However, ACO has limitations in determining the optimal repair action in situations where many constraints exist and there is less a priori information about the problem. Especially for the infrastructure maintenance management system, ACO has some drawbacks related to accuracy and processing time.

22.3.3 Shuffled Frog Leaping

The SFL algorithm is a memetic metaheuristic method for combinatorial optimization. The meme is the fundamental unit of information, which is similar to the gene. Both the meme and gene spread through the population in terms of evolution. The difference between memes and genes is that a meme is mind related. It is the unit for carrying cultural ideas from one element to another. It can self-replicate and mutate by selective pressures.

A memetic algorithm is a population-based approach for heuristic search. Comparing memetic and genetic evolution, they have the same basic principle: generation of possible solutions and then selection of the best one based on measured fitness, combination of solution, and mutation. However, memetic evolution is much more flexible than genetic evolution. In genetic evolution, the offspring can be produced only by selecting two parent chromosomes and exchanging their information. Higher organisms may take several iterations to propagate because genes are transmitted between generations. However, memes can be transmitted between any two individuals. Therefore, memetic evolution can produce more diverse offspring than genetic evolution, and the spreading speed of memetic evolution is much faster than that of genetic evolution.

The SFL algorithm, developed by Eusuff and Lansey (2003), can be used to solve many complex optimization problems, which are nonlinear, nondifferentiable, and multi-modal (Zhang et al., 2008). The strength of SFL is the balance between a wide scan of large solution space and a deep search of promising locations for global optimum. SFL consists of two main approaches: global exploration and local exploration. Figure 22.5A illustrates memetic evolution in a submemeplex in SFL, and Figure 22.5B shows the shuffling process. The exploration approach of SFL is summarized as follows: (1) SFL randomly generates a virtual population within the feasible search space, and each frog (solution) has the same structure as in the GA; (2) The frogs are ranked in descending order according to their fitness; (3) The whole population of frogs is then partitioned into subsets called *memeplexes* according to the frogs' ranked priority; (4) Each memeplex evolves based on local

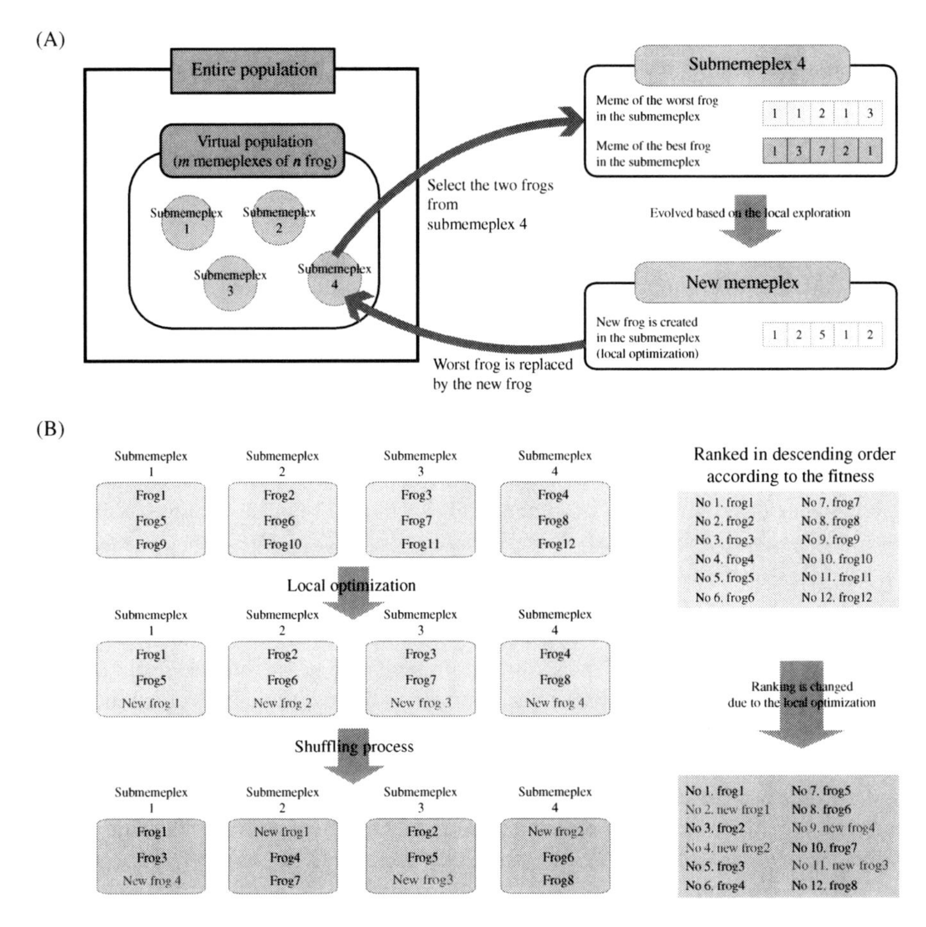

Figure 22.5 (A) An illustration of memetic evolution in a submemeplex in SFL. (B) An illustration of the shuffling process.

exploration, which refers to the concept of local optimization; and (5) The solutions of evolved memeplexes are mixed and formed through a shuffling process, which rearranges the frogs based on the fitness value in descending order. Steps 4 and 5 are continuously repeated until the predetermined stopping criterion is achieved. The shuffling process promotes a global information exchange among the frogs. In the local exploration approach, each memeplex is evolved independently using a process similar to the particle swarm optimization (PSO) algorithm.

This algorithm shows better performance than other metaheuristics such as GA, evolutionary algorithm (EA), and ACO (Elbehairy, 2008; Elbehairy et al., 2006). However, SFL has limitations when the optimal solution lies at the boundaries of the feasible region because, in this case, SFL needs more time and larger number of memeplexes (Eusuff et al., 2006). Due to the strength of this algorithm, SFL has been applied to several optimization problems such as project management

(Elbeltagi et al., 2007), the unit commitment problem (Ebrahimi et al., 2011), water distribution network design (Eusuff and Lansey, 2003; Eusuff et al., 2006), and bridge deck repairs (Elbehairy et al., 2006).

22.3.4 Hybridization of Metaheuristics

Single metaheuristic solution approaches have some limitations as mentioned previously. So, hybridization of metaheuristics is proposed in order to reduce these shortcomings while enhancing the advantages of the single metaheuristics. In this chapter, a hybrid algorithm of ACO and GA is reviewed (Samrout et al., 2007). As mentioned above, ACO shows good performance in terms of computation time. However, ACO has limitations in determining the optimal repair action in the situation where many constraints exist and there is not a great deal of a priori information about the problem. In this system, ACO is only used to search the best maintenance dates, as shown in Figure 22.6. Each ant i builds a solution, which has information about timing to apply repair action, and all solutions are evaluated and the most feasible solution is recorded.

GA is used in the second part of the process to find the best combination of actions that minimize the maintenance costs while satisfying all the constraints. New possible combinations of actions are generated based on the most feasible solution, which was discovered in the first part. An evolutionary process is implemented on the solutions, which have information about both the times and actions. Then, it stops the process when certain conditions are satisfied.

This hybrid algorithm shows better performance than general ACO in terms of computation time. The time consumption of ACO increases faster than the hybrid algorithm as the number of infrastructures increases. So, the hybrid approach becomes more and more efficient as the numbers of infrastructures and ants increase. However, this model still has some limitations. This model does not consider the multi-repair action for one infrastructure. For example, even though

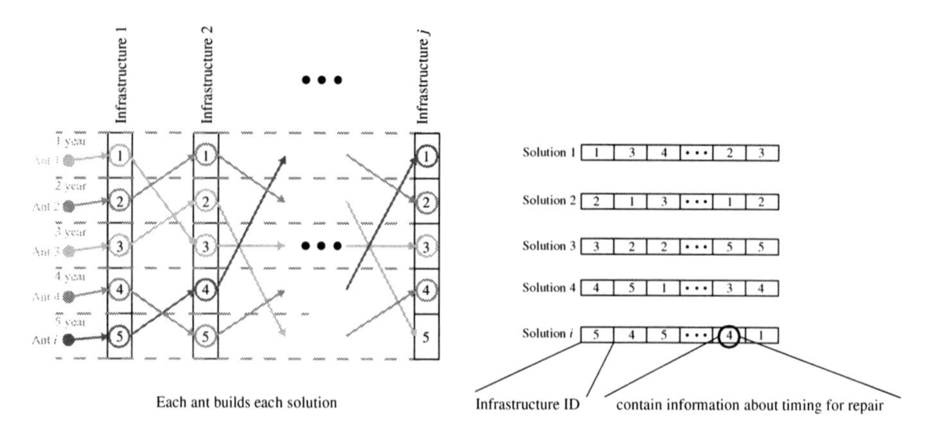

Figure 22.6 Example of solution construction by ACO.

several iterations of repair actions with a low price are more cost-effective than one repair action with a high price, it is difficult to deal with these aspects in this hybrid algorithm.

22.3.5 Summary

In this section, various metaheuristic algorithms such as GA, ACO, SFL, and the hybrid algorithm for maintenance management were reviewed.

The metaheuristics show good performance in the area of infrastructure maintenance optimization. However, the metaheuristics have to carefully determine the proper parameters because the results from such algorithms fluctuate greatly depending on the main parameters such as crossover, mutation, pheromone evaporation, and memeplex. Even though many researchers have studied the determination of proper parameters for varying sizes and numbers of actions, a more precise way of determining each parameter can be the subject of further studies (Frangopol and Liu, 2007).

In addition to the metaheuristics algorithms explained previously, metaheuristics have been newly developed such as the krill herd algorithm (KHA), (Gandomi and Alavi, 2012) and the Paddy Field Algorithm (PFA), (Premaratne et al., 2009). The purpose of the KHA, which is based on the simulation of the herding of krill swarms in response to specific biological and environmental processes, is to find the best solution throughout the continuous process of balancing between globally and locally optimum solutions. The FFA is inspired by the plant population to reduce the possibility of falling into the local optimum. However, since these algorithms are newly developed, only a few cases have been applied to the infrastructure maintenance management system. Continuous attempts are needed to overcome the limitations of the current metaheuristics that apply to infrastructure maintenance management.

22.4 Case Studies

The infrastructure maintenance optimization field has been a fertile ground for the application of metaheuristics techniques, as demonstrated by the many applications mentioned previously. So, many metaheuristics techniques (mainly GA, ACO, SFL, and EA) have been used in infrastructure maintenance management with various degrees of success. Table 22.2 presents the summary of the application of metaheuristics in the field of infrastructure maintenance management.

Some of the characteristics of metaheuristics algorithms can be identified from the case studies listed in Table 22.2. First, the success rate generally decreases as the number of infrastructures increases. The success rate is one of the important criteria, and it represents how many times, out of n trials, the system is able to provide a solution while satisfying the constraints on the single-facility and multi-facility levels. Among various metaheuristics algorithms, SFL shows the best performance,

Table 22.2 Case Studies of Metaheuristics Applications in Infrastructure Maintenance Optimization

Metaheuristics Technique	Reference	Service Life (Years)	Objective Function	Solution Results	Case Study	Size	Success Rate (%)		Computation Time (h:min:s)	Number of Repair Actions
							Performance	Optimality		
EA	Yeo et al. (2012)	40	Max performance	Individual	PV	20	–	96.6		3
						200	–		00:00:12	3
						2000			01:49:36	3
GA	Elbehairy (2008)	5	Min cost	Individual	BM	10	100	–	00:02:16	4
						100	20	–	02:17:40	4
	Elbeltagi et al. (2007)	5	Min cost	Individual	BM	50	60	–		5
	Morcous and Lounis (2005)	15	Max Performance	Group	BM		–	–	10,000 iteration	3
	Frangopol and Liu (2007)	30	Multi-objectives	Individual	BM	13				5
	Liu and Frangopol (2005)	50	Multi-objectives	Individual	BM	Single facility				4
	Avijit Maji (2007)	5	Min cost	Individual	IM	12			150 iteration	2
	Chootinan et al. (2006)	10	Multi-objectives	Group	PV	53				

	Jha and Abdullah (2006)	10	Min cost	Group	IM	32				2
PSO	Elbehairy (2008)	5	Min cost	Individual	BM	10	80	–	00:05:46	4
						100	10	–	02:16:54	4
PSH	Yeo et al. (2012)	40	Max performance	Individual	PV	20	–	97.8		3
						200	–		00:15:25	3
ACO	Elbehairy (2008)	5	Min cost	Individual	BM	10	60	–	00:02:26	4
						100	0	–		4
SFL	Elbehairy (2008)	5	Min cost	Individual	BM	10	100	–	00:01:37	4
						100	80	–	01:28:34	4
	Elbeltagi et al. (2007)	5	Min cost	Individual	BM	50	10	–		5
MSFL	Elbeltagi et al. (2007)	5	Min cost	Individual	BM	50	100	–		5

Multi-objectives: Represent that more than two objective optimization are applied, for example, consider boat maximize the performance and minimize the cost.
BR, bridge maintenance; IN, infrastructure maintenance; PM, pavement maintenance.
Group: Represent that maintenance optimization is applied to groups of facilities and not to individual facilities.
Individual: Represent that maintenance optimization is applied to individual facilities and not to groups of facilities.
Performance: Represent the percentage of success, that is, how many trials out of 10 were able to provide a solution without violating the constraints on both individual bridges and the whole network.
Optimality: Represent the optimal total expected cost ratio (total system expected cost to go/real optimal costs).

with an 80% success rate when the number of infrastructures is 100 (Elbehairy, 2008; Elbeltagi et al., 2007). ACO shows the worst performance when the number of infrastructures increases to 100. Therefore, solving the infrastructure maintenance optimization problem with only ACO is not recommended. Second, the computation times for all successful trials dramatically increase as the number of infrastructures increase. EA and SFL show better performance than the other algorithms. As suggested by Yeo et al. (2012), EA shows the fastest speed due to its simple and efficient problem formulation structure adopting a bi-level, bottom-up approach. Third, four kinds of repair decisions (i.e., 0 = do nothing, 1 = light repair, 2 = medium repair, and 3 = replacement) are most frequently used. The number of types of repair decisions is the important criterion in the infrastructure optimization problem because the computation time, success rate, and quality of results depend greatly on this number. If the number of types of repair decisions is too small, the results from the optimization do not accurately reflect the real situation.

Even though the metaheuristics algorithms used in these case studies show good performance, some drawbacks still remain. Generally, the success rate significantly decreases as the number of infrastructure increases and various constraints are generated. So, it is essential to develop an efficient preprocessing or other method to avoid violating the constraints with less computation time and high success rate. Figure 22.7 shows a good example for this problem. Many algorithms for maintenance optimization are based on the "one *i*-year optimization method," which determines consecutive *i*-year maintenance policy by one optimization. However, "*i* year-by-year optimization method," which considers each year individually in *i* times consecutive optimizations, is more effective in terms of computation time and the satisfaction of constraints (Elbehairy et al., 2006). In addition, the "*i* year-by-year optimization method" is more realistic to apply to the infrastructure maintenance management because the budget for the maintenance policy is determined for each year. Moreover, there exist many uncertainties in the process of predicting the future condition of the infrastructures, and these uncertainties grow as the number of predicted years increases. Therefore, the "*i* year-by-year consecutive optimization" method is more suitable.

22.5 Summary

In this chapter, we presented and compared the metaheuristic methods for infrastructure maintenance optimization. Generally, metaheuristics algorithms have huge potential to solve the NP-hard problem. So, many cases have been studied for supporting infrastructure maintenance, as mentioned previously. However, there are some limitations in applying the cases directly to the infrastructure maintenance system. First, most of the current researches deal only with small problem sizes. As shown in Table 22.2, all the problem sizes are smaller than 100 except the case studied by Yeo et al. (2012). Therefore, it is insufficient to say that these methods can be directly applied to problems with realistic sizes. Second, the results from

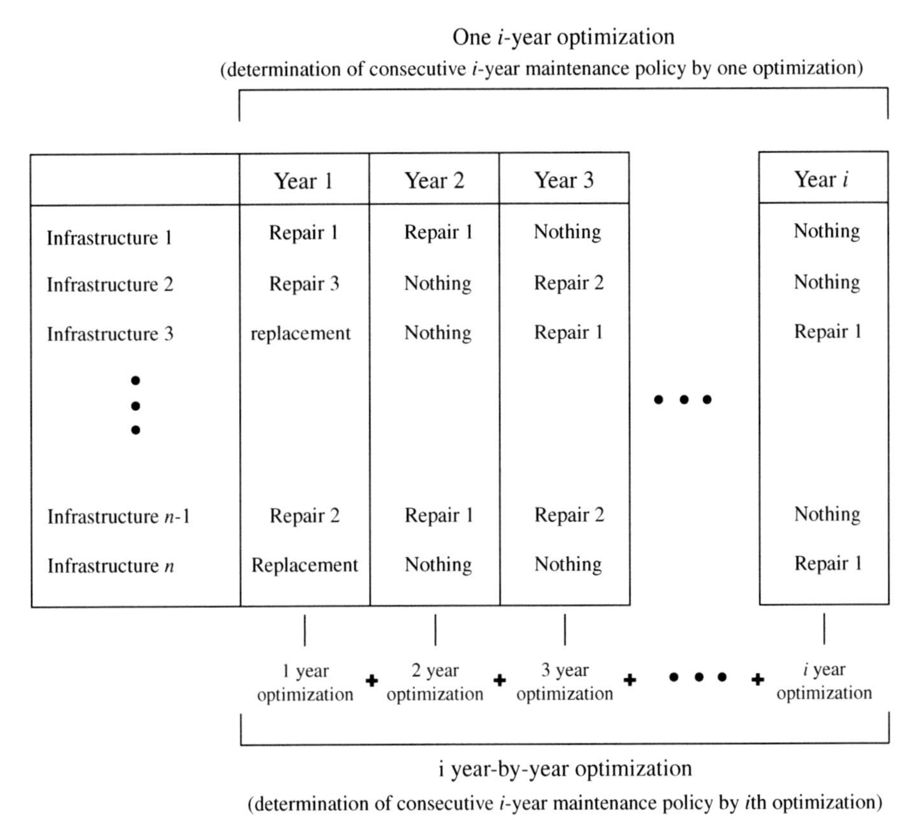

Figure 22.7 Year-by-year optimization method.

metaheuristics algorithms have rarely been verified using the success rate. The results from the other combinatorial problem, such as finding the shortest path, are relatively easy to verify because most of these problems are based on the standard problem. However, infrastructure needs a new method to verify the results because the problems in reality do not always have the standard form to be compared. Finally, the efficiency, in terms of computation time, has to be enhanced because the infrastructure maintenance addresses the trade-off between the success rate and computation time. Future research is recommended to overcome these limitations.

References

Avijit Maji, M.K.J., 2007. Modeling highway infrastructure maintenance schedules with budget constraints. Trans. Res. Rec. J. Trans. Res. Board. 19–26.

Beasley, D., Bull, D.R., Martin, R.R., 1993. An overview of genetic algorithms. 1. Fundamentals. Univer. Comput. 15, 58–69.

Blum, C., 2005. Ant colony optimization: introduction and recent trends. Phys. Life Rev. 2, 353–373.

Blum, C., Roli, A., 2003. Metaheuristics in combinatorial optimization: overview and conceptual comparison. ACM Comput. Surv. 35, 268–308.

Cheu, R.L., Wang, Y., Fwa, T.F., 2004. Genetic algorithm–simulation methodology for pavement maintenance scheduling. Comput. Aided Civil Infrastruct. Eng. 19, 446–455.

Chootinan, P., Chen, A., Horrocks, M.R., Bolling, D., 2006. A multi-year pavement maintenance program using a stochastic simulation-based genetic algorithm approach. Trans. Res. Part A Policy Pract. 40, 725–743.

Dekker, R., Schouten, F.V.D., Wildeman, R.E., 1996. A review of multicomponent maintenance models with economic dependence. Math. Methods Oper. Res. 45, 24.

Dorigo, M., Stützle, T., 2004. Ant Colony Optimization. MIT Press, Cambridge, MA.

Dorigo, M., Birattari, M., Stützle, T., 2006. Ant colony optimization. Comput. Intell. Mag. IEEE. 1, 28–39.

Durango, P.L., Madanat, S.M., 2002. Optimal maintenance and repair policies in infrastructure management under uncertain facility deterioration rates: an adaptive control approach. Trans. Res. Part A Policy Pract. 36, 763–778.

Durango-Cohen, P.L., Sarutipand, P., 2006. Capturing interdependencies and heterogeneity in the management of multifacility transportation infrastructure systems. J. Infrastruct. Syst. 13, 9.

Ebrahimi, J., Hosseinian, S.H., Gharehpetian, G.B., 2011. Unit commitment problem solution using shuffled frog leaping algorithm. IEEE Trans. Power Syst. 26, 573–581.

Elbehairy, H., 2008. Bridge Management System with Integrated Life Cycle Cost Optimization. Ph.D. Thesis, University of Waterloo.

Elbehairy, H., Elbeltagi, E., Hegazy, T., Soudki, K., 2006. Comparison of two evolutionary algorithms for optimization of bridge deck repairs. Comput. Aided Civil Infrastruct. Eng. 21, 561–572.

Elbeltagi, E., Hegazy, T., Grierson, D., 2007. A modified shuffled frog-leaping optimization algorithm: applications to project management. Struct. Infrastruct. Eng. 3, 53–60.

Eusuff, M.M., Lansey, K.E., 2003. Optimization of water distribution network design using the shuffled frog leaping algorithm. J. Water Resour. Plan. Manage. 129, 210–225.

Eusuff, M.M., Lansey, K., Pasha, F., 2006. Shuffled frog-leaping algorithm: a memetic meta-heuristic for discrete optimization. Eng. Optimiz. 38, 129–154.

Frangopol, D.M., Liu, M., 2007. Maintenance and management of civil infrastructure based on condition, safety, optimization, and life-cycle cost. Struct. Infrastruct. Eng. 3, 29–41.

Friesz, T.L., Fernandez, J.E., 1979. Model of optimal transport maintenance with demand responsiveness. Trans. Res. Part B Method. 13, 317–339.

Furuta, H., Kameda, T., Nakahara, K., Takahashi, Y., Frangopol, D.M., 2006. Optimal bridge maintenance planning using improved multi-objective genetic algorithm. Struct. Infrastruct. Eng. 2, 33–41.

Gandibleux, X., Jorge, J., Angibaud, S.E., Delorme, X., Rodriguez, J., 2005. An ant colony optimization inspired algorithm for the set packing problem with application to railway infrastructure. The Sixth Metaheuristics International Conference (MIC2005), Vienna, Austria, 390–396.

Gandomi, A.H., Alavi, A.H., 2012. Krill herd: a new bio-inspired optimization algorithm. Commun. Nonlinear Sci. Numer. Simul. 17 (12), 4831–4845.

Goldberg, D.E., 1989. Genetic Algorithms in Search, Optimization, and Machine Learning. Addison-Wesley, Reading, MA.

Guignier, F., Madanat, S., 1999. Optimization of infrastructure systems maintenance and improvement policies. J. Infrastruct. Syst. 5, 124–134.

Guillaumot, V.M., Durango-Cohen, P.L., Madanat, S.M., 2003. Adaptive optimization of infrastructure maintenance and inspenction decisions under performance model uncertainty. J. Infrastruct. Syst. 9, 7.

Jha, M.K., Abdullah, J., 2006. A Markovian approach for optimizing highway life-cycle with genetic algorithms by considering maintenance of roadside appurtenances. J. Franklin Inst. Eng. Appl. Math. 343, 404–419.

Johnston, D.W., Zia, P., 1984. Level of service system for bridge evaluation. Trans. Res. Board. 1–8.

Li, Y.W., Madanat, S., 2002. A steady-state solution for the optimal pavement resurfacing problem. Trans. Res. Part A Policy Pract. 36, 525–535.

Liu, M., Frangopol, D.M., 2005. Multiobjective maintenance planning optimization for deteriorating bridges considering condition, safety, and life-cycle cost. J. Struct. Eng. 131, 833–842.

Lvbjerg, M., 2002. Improving Particle Swarm Optimization by Hybridization of Stochastic Search Heuristics and Self-Organized Criticality. Masters Thesis, Aarhus Universitet.

Madanat, S., Benakiva, M., 1994. Optimal inspection and repair policies for infrastructure facilities. Trans. Sci. 28, 55–62.

Merkle, D., Middendorf, M., Schmeck, H., 2002. Ant colony optimization for resource-constrained project scheduling. IEEE Trans. Evol. Comput. 6, 333–346.

Mohamed, H.A.H., Engineering, C.E.C.U.D., Co, P., 1995. Development of Optimal Strategies for Bridge Management Systems. Carleton University, Ottawa, ON.

Morcous, G. 2000. Case-Based Reasoning for Modeling Bridge Deterioration. Ph.D. Thesis, Concordia University.

Morcous, G., Lounis, Z., 2005. Maintenance optimization of infrastructure networks using genetic algorithms. Autom. Constr. 14, 129–142.

Neves, L.A.C., Frangopol, D.M., Petcherdchoo, A., 2006. Probabilistic lifetime-oriented multiobjective optimization of bridge maintenance: combination of maintenance types. J. Struct. Eng. 132, 1821–1834.

Okasha, N.M., Frangopol, D.M., 2009. Lifetime-oriented multi-objective optimization of structural maintenance considering system reliability, redundancy and life-cycle cost using GA. Struct. Saf. 31, 460–474.

Ouyang, Y., Madanat, S., 2006. An analytical solution for the finite-horizon pavement resurfacing planning problem. Trans. Res. Part B Method. 40, 767–778.

Premaratne, U., Samarabandu, J., Sidhu, T., 2009. A new biologically inspired optimization algorithm. Fourth International Conference on Industrial and Information Systems. Sri Lanka, 279–284.

Samrout, M., Yalaoui, F., Chatelet, E., Chebbo, N., 2005. New methods to minimize the preventive maintenance cost of series–parallel systems using ant colony optimization. Reliab. Eng. Syst. Saf. 89, 346–354.

Samrout, M., Châtelet, E., Kouta, R., Chebbo, N., 2007. Decisions making optimization in a complex system's maintenance policy using metaheuristics, UGF and proportional hazard model. Inter. J. Computat. Sci. 1, 17.

Sarutipand, P., 2008. An Optimization Framework for Management of Transportation Infrastructure Systems with Interdependent Facilities. Ph.D. Thesis, Northwestern University.

Sathaye, N., Madanat, S., 2011. A bottom-up solution for the multi-facility optimal pavement resurfacing problem. Trans. Res. Part B Method. 45, 1004–1017.

Smilowitz, K., Madanat, S., 2000. Optimal inspection and maintenance policies for infrastructure networks. Comput. Aided Civil Infrastruct. Eng. 15, 5−13.

Thompson, P.D., Sobanjo, J.O., Kerr, R., 2003. Florida DOT project-level bridge management models. J. Bridge Eng. 8, 345−352.

Tsunokawa, K., Schofer, J.L., 1994. Trend curve optimal-control model for highway pavement maintenance—case-study and evaluation. Trans. Res. Part A Policy Pract. 28, 151−166.

Tzafestas, S.G., Botsaris, C.A., 1986. Reliability-measures and optimal maintenance policies for a standby system with repair facilities subject to breakdown. J. Franklin Inst. Eng. Appl. Math. 321, 309−323.

Vitins, B.J., Axhausen, K.W., 2009. Optimization of large transport networks using the ant colony heuristic. Comput. Aided Civil Infrastruct. Eng. 24, 1−14.

Yeo, H., Yoon, Y., Madanat, S., 2012. Algorithms for bottom-up maintenance optimisation for heterogeneous infrastructure systems. Structural and Infrastructure Engineering: Maintenance, Management, Life-cycle Design and Performance. 9 (4), 317−328.

Zhang, X., Hu, X., Cui, G., Wang, Y., Niu, Y., 2008. An improved shuffled frog leaping algorithm with cognitive behavior. Proceedings of the 7th World Congress on Intelligent Control and Automation, pp. 6197−6202.

23 Metaheuristic Applications in Bridge Infrastructure Maintenance Scheduling Considering Stochastic Aspects of Deterioration

Manoj K. Jha[1], *Monique Head*[1] *and Shobeir Pirayeh Gar*[2]

[1]Department of Civil Engineering, Morgan State University, Baltimore, MD, USA, [2]Zachry Department of Civil Engineering, Texas A&M University, College Station, TX, USA

23.1 Introduction

Highway infrastructure consists of pavement, bridges, tunnels, and other minor assets, such as signs, guardrails, and luminaries. Their upkeep and timely maintenance is essential for motorists' mobility and safety. Due to heavy traffic volume on urban roads, infrastructure elements deteriorate rapidly. In order to ensure adequate service, a maintenance optimization scheduling model over a specified planning horizon is generally desired (Jha and Abdullah, 2006). Such a model should consider the deterioration of highway features in order to realistically predict the maintenance schedule over a planning horizon.

The objective of this chapter is to develop a maintenance scheduling optimization model for bridges considering stochastic aspects of deterioration and to discuss an innovative experimental technique for examining the deterioration of bridges.

23.2 Deterioration Modeling

Since the deterioration of the highway infrastructure over time is a result of both routine wear and tear and random (stochastic) events, such as accidents and adverse weather, we develop a highway infrastructure optimization model to reflect the deterministic and stochastic aspects of deterioration caused by both routine and unexpected

Metaheuristic Applications in Structures and Infrastructures. DOI: http://dx.doi.org/10.1016/B978-0-12-398364-0.00023-1

conditions warranting routine maintenance, repair, and rehabilitation (MR&R) and reactive activities. The model is solved using a genetic algorithm (GA), and a case study from the City of Baltimore Department of Transportation is presented.

23.2.1 Model Formulation

23.2.1.1 Routine Maintenance (Deterministic Approach)

The formulation for routine maintenance (Jha, 2010) assumes maintenance inspection and implementation of maintenance actions (also referred to as "policies" in much of the literature) at prespecified, discrete intervals (typically, every year). We classify the maintainable highway infrastructure elements (also referred to as *appurtenances*) in two categories: those lasting perpetually, such as the pavement, and those lasting nonperpetually, such as signs, guardrails, and luminaries.

Let $k = 1,2,\ldots,\mathrm{KK}$ represent the total number of nonperpetual and perpetual appurtenances ($\mathrm{KK} = K_{\mathrm{np}} + K_{\mathrm{p}}$; K_{np} = number of nonperpetual appurtenances and K_{p} = number of perpetual appurtenances) within the analysis highway section. For this analysis, let $j = 1,2,\ldots, J$ represent possible maintenance actions (policies) to be undertaken, $i = 1,2,\ldots, I$ possible condition states of the appurtenances, and $t = 1,2,\ldots, T$ possible time periods. Then the objective function for the routine maintenance case can be expressed as:

$$\text{Min} \sum_{t=1}^{T} \sum_{k=1}^{\mathrm{KK}} \sum_{i=1}^{I} \sum_{j=1}^{J} \alpha^t w_{kij}^t \, c(k,i,j) \tag{23.1}$$

where α^t is the discount factor at time t, w_{kij}^t is the probability that appurtenance k will be in condition state i if action j is applied in time t, and $c(k,i,j)$ is the maintenance cost of appurtenance k for applying action j, resulting in its condition state i. The constraints are given as:

$$w_{kij}^t \geq 0 \quad \forall k,i,j,t \tag{23.2}$$

$$\sum_i \sum_j w_{kij}^t = 1 \quad \forall k,t \tag{23.3}$$

$$\sum_j w_{kij}^1 = q_{ki}^1 \quad \forall k,i \tag{23.4}$$

$$\sum_j w_{k(i+1)j}^t = \sum_i \sum_j w_{kij}^{t-1} p_{ki(i+1)}(j) \quad \forall k,(i+1) \tag{23.5}$$

$$\sum_k \sum_i \sum_j \alpha^t c(k,i,j) \leq B^t \quad \forall t \tag{23.6}$$

Equation (23.2) ensures that condition probability is always non-negative. Equation (23.3) ensures that total condition probability is 1. Equation (23.4) implies that condition state at the beginning of the analysis (year 1) is known, and Eq. (23.5) specifies the likelihood that the condition state will move from i to $(i+1)$ in year t if action j is applied in the previous year $(t-1)$. Equation (23.6) implies that the total maintenance cost in a given year cannot exceed the available budget for that year. B^t is the available budget in year t.

23.2.1.2 Reactive Maintenance (Stochastic Approach)

Due to random incidents, such as inclement weather and accidents, roadside features may be damaged earlier than their expected life spans, which leads to reactive maintenance. The conditions and actions to be undertaken in this case are governed by a reactive approach rather than inspections at fixed-time (discrete) intervals. We assume that while accidents and inclement weather are stochastic (unpredictable) in nature, and damage caused by such incidents is also unpredictable, the likelihood of the inspection of roadside features will depend on the probability of an accident or inclement weather. The deterministic formulation is modified as follows to reflect the stochastic case:

$$\text{Min} \sum_{t=1}^{T} \sum_{k=1}^{\text{KK}} \sum_{i=1}^{I} \sum_{j=1}^{J} \alpha^t w_{kij}^t \, c(k,i,j) \min\{\gamma(P^t), 1\} \tag{23.7}$$

where

$$\begin{cases} \gamma(P^t) = 1 & \text{if } P^t \geq t_c \\ 0 & \text{otherwise} \end{cases} \tag{23.8}$$

In Eq. (23.8), P^t is the probability of an MR&R action being applied in year t as a result of an accident/inclement weather. It means that if the deterioration caused by an accident/inclement weather reaches a threshold value t_c, then the full cost of the appropriate maintenance action should be taken into consideration. If the probable damage is less than t_c, then only a minor repair may be necessary. The threshold value can be adjusted by user feedback and experience in practical situations. The other constraints under this case are same as Eqs. (23.2)–(23.6).

The resulting optimization problem considering the effects of both deterministic and stochastic aspects of deterioration can be expressed by combining Eqs. (23.1) and (23.7), which can be expressed as:

$$\text{Min} \sum_{t=1}^{T} \sum_{k=1}^{\text{KK}} \sum_{i=1}^{I} \sum_{j=1}^{J} \alpha^t w_{kij}^t \, c(k,i,j)[1 + \min\{\gamma(P^t), 1\}] \tag{23.9}$$

The constraints are given as Eqs. (23.1)–(23.5).

23.3 Solution Algorithm

Several metaheuristic algorithms can be found in the literature (Gandomi and Alavi, 2012; Gandomi et al., 2011, 2012) that may be suitable to solve the aforementioned optimization problem. However, here a GA is developed to solve the optimization problem represented by Eq. (23.9). For the maintenance optimization problem, the objective is to come up with the best sequence of actions over a planning horizon that will minimize the discounted maintenance cost resulting in an enhanced highway life cycle. Let Λ_k be the chromosome vector for the kth infrastructure element or appurtenance (assuming a time horizon of T years and an annual maintenance plan). Thus, the chromosome matrix can be given as:

$$\boldsymbol{\Lambda} = \begin{bmatrix} \lambda_{11} & \lambda_{12} & . & . & \lambda_{1T} \\ \lambda_{21} & \lambda_{22} & . & . & \lambda_{2T} \\ . & . & . & . & . \\ . & . & . & . & . \\ \lambda_{k1} & \lambda_{k2} & . & . & \lambda_{kT} \end{bmatrix} \tag{23.10}$$

where λ is a gene (or annual maintenance action to be chosen over a T-year planning horizon).

Assuming two actions ($j = 1, 2$ in Eq. (23.10)), the solution can be encoded using binary digits (0, 1), where 0 = do nothing and 1 = an intermediate MR&R action. Thus, a possible encoded solution over a 10-year planning horizon might look like this:

$$\Lambda_k = [0010000010]_k \tag{23.11}$$

Equation (23.11) implies that MR&R action is needed in years 3 and 9. It can be seen that with the binary coding option, the solution space for each appurtenance consists of 2^T members. Thus, for KK appurtenances, the solution space consists of $\text{KK} \times 2^T$ members. In general, if there are J actions available for each of the appurtenances, then the solution space will consist of $\text{KK} \times J^T$ members. It can be seen that the problem can easily be Non deterministic Polynomial time-hard (NP-hard) as KK, J, and T grow.

The GA allows building a relatively smaller initial population and uses customized operators to improve solution quality over successive search generations by developing a selection/replacement scheme. Usually, initial population size is much smaller (less than 100), but through a proper selection/replacement scheme (which use probabilistic rules), a GA can still explore the entire solution space without getting stuck in local optima.

23.3.1 A Numerical Example

The major highway infrastructure network of the city of Baltimore primarily consists of pavements and bridges. Baltimore has the responsibility of maintenance and rehabilitation of 299 bridges within the city limits and watershed bridges in Baltimore County. All bridges have been inspected biannually since 1987, and a small percentage of the bridges whose condition warrants rerating based on the conditions are inspected annually. The bridges are included in the Structural Inventory and Appraisal System (SAIS); conditions are noted on a plan and recommendations are prepared for repair and maintenance of each bridge. All these data are loaded in a bridge management software program called PONTIS, owned by the Maryland State Highway Administration. After inspection, a Bridge Sufficiency Rate (BSR) is calculated from 0 to 100, with 100 being the best.

Under federal highway policy, bridges with a BSR of greater than 80 do not require any maintenance; bridges with BSR between 50 and 80 need MR&R; and bridges with BSR less than 50 need to be replaced. Any query to the PONTIS data can provide the elements of bridge to be rehabilitated/replaced, type of repairs recommended, and the cost of repair for the particular element or other elements of the bridge. Most bridge repair work is performed under contract. Minor repair and maintenance work is assigned to the Maintenance Division. A preventive maintenance program is currently being developed using the SAIS program.

We perform a numerical example to obtain optimal actions to be undertaken for 16 bridges labeled 1001 through 1509, as shown in the southwest quadrant of Figure 23.1. The conditions based on the Year 1 BSR is shown in Table 23.1. The following three possible actions are assumed:

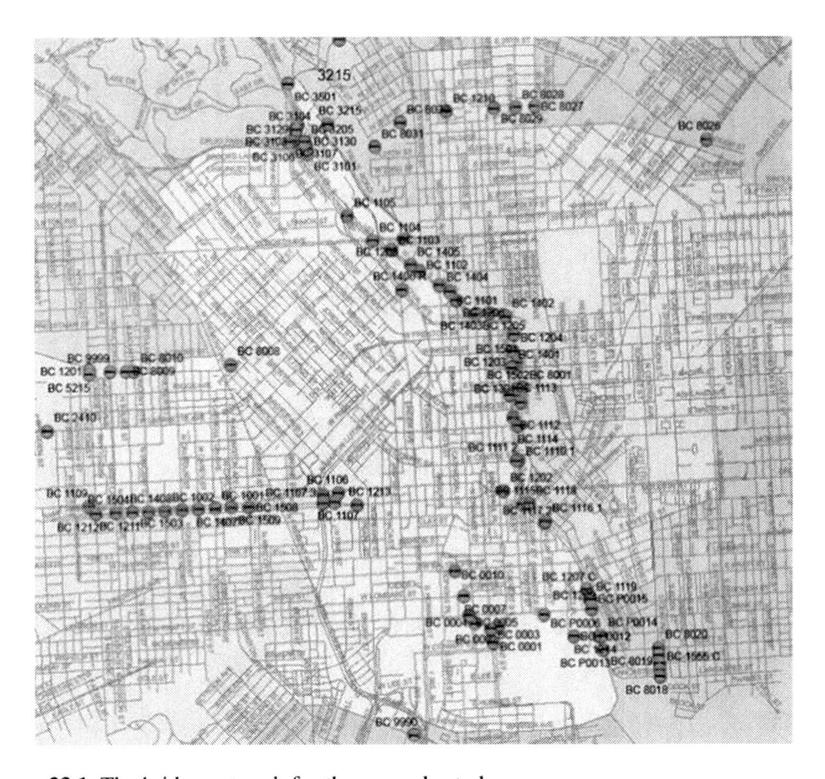

Figure 23.1 The bridge network for the example study.

Table 23.1 Condition of the Bridges for the Example Study

Bridge Number	Year									
	1	**2**	**3**	**4**	**5**	**6**	**7**	**8**	**9**	**10**
1001	0.8									
1002	0.8									
1106	0.7									
1107	0.9									
1108	0.6									
1109	0.6									
1211	0.6									
1212	0.6				$\text{pc}_i = f(c_i, \text{tr}_i) = c_i + \text{tr}_i$					
1213	0.7									
1407	0.7									
1408	0.8									
1503	0.6									
1504	0.6									
1507	0.8									
1508	0.8									
1509	0.7									

0 = Do nothing for $\text{BSR} \geq 0.8$
1 = Intermediate MR&R for $50 \leq \text{BSR} \leq 80$
2 = Full rehabilitation for $\text{BSR} < 0.5$

The genetic encoding takes up tertiary values instead of binary values as shown in Eq. (23.11).

Table 23.2 shows normal or adverse (due to random incidents) conditions warranting routing or reactive maintenance, respectively. Table 23.3 shows the present cost of undertaking MR&R actions and full rehabilitation of the 16 bridges.

23.3.2 Results and Discussion

The optimal results shown in Table 23.4 indicate the appropriate actions (i.e., do nothing, intermediate MR&R, or full rehabilitation) to be undertaken over the 10-year planning horizon. The total present cost for undertaking the bridge maintenance program over a 10-year period can then be calculated using the figures shown in Table 23.3. Please note that the adverse incidents in years 2 and 6 have caused full replacements of bridges 1212 and 1211, respectively, in those years.

The developed model is very useful for obtaining optimal infrastructure maintenance rehabilitation plan over any specified planning horizon while considering the stochasticity of incidents. Although the example is performed on a bridge network, pavement and other infrastructure networks can be analyzed in a similar fashion, and appropriate optimal maintenance plans can be obtained for them.

Table 23.2 Normal (N) and Adverse (A) Conditions of the Bridges Warranting Routine or Reactive Maintenance

Bridge Number	Year									
	1	**2**	**3**	**4**	**5**	**6**	**7**	**8**	**9**	**10**
1001	N	N	N	A	N	N	N	N	N	N
1002	N	N	A	N	N	N	N	A	N	N
1106	N	N	N	A	N	N	N	N	N	N
1107	A	N	N	N	N	N	N	N	N	N
1108	N	N	N	N	A	N	N	N	N	N
1109	N	N	N	N	A	N	N	N	A	N
1211	N	N	N	N	N	A	N	N	N	N
1212	A	N	N	N	N	N	N	N	N	N
1213	N	N	N	N	N	N	N	N	N	N
1407	N	N	N	N	N	N	N	N	N	N
1408	A	N	N	N	A	N	N	N	N	N
1503	A	N	N	N	N	N	N	N	N	N
1504	N	N	N	N	A	N	N	N	N	N
1507	A	N	N	N	N	N	N	N	N	N
1508	N	N	N	N	N	N	A	N	N	N
1509	N	N	N	A	N	N	N	N	N	N

Table 23.3 Cost of MR&R and Full Rehabilitation in the Present Year

Bridge Number	Cost of Routine MR&R in US Dollars in the Present Year (Year 1)	Cost of Full Rehabilitation in US Dollars in the Present Year (Year 1)
1001	200,000	10 M
1002	220,000	12.5 M
1106	150,000	8 M
1107	300,000	15 M
1108	130,000	7 M
1109	140,000	9 M
1211	320,000	16 M
1212	360,000	20 M
1213	120,000	6 M
1407	155,000	8 M
1408	160,000	9 M
1503	340,000	18 M
1504	220,000	11 M
1507	120,000	7 M
1508	135,000	9 M
1509	300,000	14 M

An interest rate of 6% is used to calculate future costs.

Table 23.4 Optimal Maintenance Schedule

Bridge Number	Year									
	1	2	3	4	5	6	7	8	9	10
1001	0	0	0	1	0	0	0	0	1	0
1002	0	0	0	0	0	1	0	1	0	0
1106	1	0	0	1	0	0	0	0	1	0
1107	0	1	0	0	0	0	1	0	0	0
1108	1	0	0	0	1	0	0	0	0	1
1109	1	0	0	0	1	0	0	0	0	1
1211	1	0	0	0	0	2	0	0	1	0
1212	1	2	0	0	0	0	0	1	0	0
1213	1	0	0	0	0	1	0	0	0	0
1407	1	0	0	0	0	0	1	0	0	0
1408	0	1	0	0	1	0	0	0	0	1
1503	1	0	0	0	0	1	0	0	0	0
1504	1	0	0	0	1	0	0	0	0	1
1507	0	1	0	0	0	0	1	0	0	0
1508	0	0	0	0	0	1	1	0	0	0
1509	1	0	0	1	0	0	0	0	1	0

0, do nothing; 1, intermediate MR&R; 2, full replacement.

23.4 Experimental Procedure

Deterioration of bridges can exert serious, widespread, and prolonged impacts on various societal sectors and commerce. Whenever maintenance has not been managed considering preventive and life-cycle costs, the results have caused extraordinary casualties and economic losses that could have been prevented if a risk analysis assessment to derive mitigation strategies had been implemented. Current maintenance practices in the United States do not include risk criteria to aid in optimal decision making for bridge maintenance and monitoring, where uncertainty lies in determining which bridge to repair first and which mitigation strategy should be deployed.

Structural deterioration, particularly due to corrosion and aging, is of paramount importance when considering metaheuristic applications in bridge infrastructure maintenance scheduling. Aside from the natural, time-dependent process of aging bridge infrastructure, structural deterioration can be triggered and even accelerated by a number of factors, such as fatigue, spalling or cracking of concrete, buildup of debris in joints and steel bearings, and exposure to and ingress of chlorides from deicing salts applied for snow or ice removal, chemical spills, and coastal or marine environments (Melchers and Frangopol, 2008). This, in turn, can lead to corrosion of structural steel elements like rebars, anchor bolts, keeper plates, bearings, and/or girders. Chloride-induced corrosion from deicing salts has been identified as resulting in significantly higher degradation than chloride-induced corrosion from marine environments (Stewart and Rosowsky, 1998). Chloride attacks can reduce the strength of members due to a decrease in cross-sectional areas due to reduction of the diameter of corroded reinforcement, increase the bearing coefficient of friction given the accumulation of debris, and increase the probability of structural failure of the bridge-bearing system due to corrosion of the anchor bolts and keeper plates, if present (Ghosh and Padgett, 2010). Therefore, the way in which structural systems are modeled to capture structural deterioration due to corrosion depends on the overall influence of the variability of initial reinforcement diameter, rate of corrosion, and corrosion initiation time. Moreover, heavy, frequent traffic demands and unpredictable events like earthquakes and other natural phenomena can further amplify the associated risk and vulnerability of structurally deteriorated bridges.

In the United States, particularly, there is a need to seek alternative solutions to better address transportation infrastructure needs, in light of the woefully poor statistics cited by the American Society of Civil Engineers (ASCE) in its Infrastructure Report Card (Giroux, 2010). While the concept of sustainability is not a new one, ASCE's Infrastructure Report Card noted that there is a major concern for addressing life-cycle costs and maintainability of the nation's bridges. In fact, over half of the approximately 600,000 bridges in the United States are approaching the end of their design lives, and nearly a quarter of these bridges are in need of major retrofit or replacement to eliminate deficiencies and improve load ratings (ASCE, 2009). Therefore, there is a need for systematic and sustainable procedures to provide the required and specific maintenance intervention to the bridges according to their importance, deterioration level, and risk level involved for society. Risk and reliability-based criteria can be applied to set cost-effective strategies to mitigate the risk of bridges according to their deterioration level, importance to society, and age, among other important aspects; and to identify appropriate retrofit strategies for the bridges based on service conditions and cost-reliability relationships. These criteria can be used for the prioritization needed in the decision-making process to promote optimum safety and sustainability. Moreover, systematic procedures to facilitate bridge managers' decision-making process by anticipating the effect of preventive maintenance actions and retrofit schemes based on risk- and reliability-based criteria are needed. The government and society will benefit, as the national bridge system may be better protected and the limited financial resources may be better allocated for bridge maintenance and retrofit.

23.5 Evaluation of FRP Composite Materials in Bridge Applications

Fiber-reinforced polymer (FRP) composites are not new; they were developed within the last 70 years. For example, the defense industry has used these materials for the development of stealth aircraft (Tang and Podolny, 1998). Other researchers have explored the potential for FRP composites for various applications (Christopoulous et al., 2008; Holden et al., 2003). However, bridge application for these composites is growing in popularity, given confidence in and familiarity with their properties, where FRP composite materials consist of many forms and can contain various elements to achieve different performance measures. FRP composite materials generally have high strength, high fatigue resistance, corrosion resistance, and are even lightweight. These characteristics are especially desirable for bridge applications, and as a result these materials have been employed in both the aerospace industry and military applications. High-strength fibers can be contained in nonmetallic prestressing tendons by which the fibers can consist of glass, carbon, or aramids. These fibers can be used like rope or as a composite rod containing a matrix of plastic or epoxy resin.

23.5.1 State-of-the-Art and State-of-the-Practice

The state-of-the-art and state-of-the-practice of these materials in bridge applications have grown from the first pedestrian FRP bridge built by Israelis in 1975 to other pedestrian bridges with pultruded shapes consisting of hybrid glass and carbon FRP composites that have an increase in stiffness with relatively little additional cost. Bridge deck systems were constructed and tested in the early 1990s, where the first US all-composite vehicular public bridge was opened for service on December 4, 1996, in Russell, KS. The construction process for this bridge consisted of FRP decks being shop-fabricated with composite honeycomb cells sandwiched between two face sheets (Tang and Podolny, 1998). The bridge was installed in 1 day, which is a definitely accelerated construction time that helps to minimize the impact of construction on commuters and commerce, to the benefit which bridges primarily serve within the transportation network. The cost savings of the accelerated construction offset the higher initial costs for the FRP materials. In addition to accelerating construction time, the lightweight nature of FRP composites offer the unique advantage of removing the problem of excessive dead load on long-span bridges (Tang and Podolny, 1998). Furthermore, FRP composites have been used for bridge applications varying from retrofit and repair techniques such as that of Seible et al. (1994) to new construction of bridge decks containing FRP tendons. More than 80 bridge projects using FRP composites have been constructed to date, with less than 40% being in the United States (SPI, 2008). While FRP materials had been deployed in various structures, standard test methods and material characterization did not exist in the United States until the Federal Highway Administration funded research conducted at the University of Wyoming and Penn State University to examine the development of performance specifications (Gilstrap, 1997).

However, there are some noted shortcomings for FRP bridge decks aside from material characterization and long-term durability. FRP decks that are sealed and enclosed make it difficult to access for field inspection and evaluation. Consequently, nondestructive evaluation and nondestructive testing are necessary for these structures to verify structural integrity and monitor in-service conditions. This, in turn, has led to structural health-monitoring systems for these structures. While success of these materials has been demonstrated, there is still a need to evaluate further the performance of FRP composites, particularly of aramid fiber reinforced polymers

(AFRPs), which show promise in bridge applications given their unique properties. FRP materials have been used in hybrid bridge construction employing FRPs as reinforcing steel, cable, and tendon systems, as well as laminates for highway bridges. In light of this previous research, evaluation, and validation, studies still need to be carried out to discern the specific advantages, characteristics, shortcomings, and even future needs for concrete bridge systems that are prestressed with AFRP tendons. As mentioned previously, the design of FRP composite bridge structures has largely focused on stiffness rather than strength requirements (Tang and Podolny, 1998).

23.5.2 Advantages and Challenges of AFRP Composite Materials

AFRP composite materials can be considered for nonmetallic prestressing tendons, where *aramid* is a generic name for aromatic polyamide (Nanni, 1993). They are commonly referred to as "aromatic fibers" since they are comprised of linked benzene rings and amide bonds. AFRPs consist of organic fibers that are manufactured by extruding a polymer solution through a spinneret. The aramid fibers are very strong in longitudinal tension but have low transverse stiffness and poor flexural and longitudinal compressive properties (Gilstrap, 1997; Nanni, 1993). The fibers are anisotropic resulting in a maximum strength in the direction of the fiber axis (Gilstrap, 1997). Aramid fibers include Kevlar fibers manufactured by DuPont, Twaron fibers manufactured by Enka in Holland, and Technora produced by the Teijin Corporation in Japan (Dolan, 1990).

One reason for using AFRP tendons as a high-performance material over prestressing steel is due to its high corrosive resistance, high tensile strength, high impact resistance, light weight, high thermal stability, and insensitivity to electric fields. Corrosion-resistant materials assist in preventing premature spalling or corrosion-induced cracking, where longitudinal cracks and leakage between deck slabs and beams are of major concern. While corrosion resistance is an added benefit for rebars and tendons, a high-modulus aramid tendon like Technora, produced by Teijin Corporation in Japan, has the ability to undergo a reasonable amount of strain before it fails. Stress–strain distributions of aramid tendons have been shown to be nearly linear to failure (Dolan et al. (2001)). Although the failure of AFRP tendons is brittle, they have high tensile strengths up to 300 ksi and have a specific gravity more than six times less than that of steel. Having a lower modulus compared to steel is beneficial in that a longer extension of the tendon is required during initial stressing and losses from creep and shrinkage of the concrete are less (Dolan, 1990). Tests on pre- and post-tensioned slabs using Kevlar-reinforced tendons have been conducted by Dolan (1989). Dolan's research consisted of four or six rods, each with a diameter of 3 mm and 5000 Kevlar fibers. A vinylester resin was applied for the fabrication of the rod. Dolan also conducted one-quarter scale tests on two 134-mm-long, 102-mm-deep, and 153-mm-wide beams. The results from his quarter-scale beam tests using the Kevlar aramid tendons showed considerable ductility and virtually complete recovery after the removal of the load. In fact, the AFRP tendons may exhibit good recovery similar to carbon fiber reinforced polymer (CFRP) tendons, as revealed by Grace and Sayed (1997), where small residual displacements can be beneficial, especially during an earthquake. Previous work conducted by Grace and Sayed (1997) from the Lawrence Technological University in Southfield, Michigan, has demonstrated that the use of CFRP-composite prestressing tendons can be designed to perform in a ductile manner. More research conducted by Dolan et al. (2001) focused on statically loaded single-span prestressed beams with AFRP tendons; however, little attention has been paid to the evaluation of a bridge system consisting of full-scale beams and slab prestressed with AFRP tendons and subjected to cyclic (fatigue) loads. This proposed research aims to examine full-scale concrete beams and slabs prestressed with Technora aramid fibers produced by Teijin Corporation in Japan. Technora fibers are reported to be less sensitive to moisture than other aramids and are proposed as the material of choice for testing concrete slabs and bulb T-beams that are prestressed with AFRP tendons and subjected to service, cyclic, and ultimate loads. Proper characterization of materials for the selection of fibers and anchorage are critical to the advancement of FRPs in prestressing applications. Consideration should be given to these materials, given their subjectivity to fatigue and creep under sustained loads for long-term durability.

23.6 Application of AFRP Bars in a Full-Scale Bridge Deck Slab

A bridge deck slab is typically under higher risk of corrosion compared to other bridge components because of direct and broader exposure to the environment and traffic load, as well as deicing chemicals. This is a major concern as the bridge infrastructure is an integral part of the network that connects people and businesses. However, the significant cost of repair and rehabilitation programs in terms of budget and induced downtime considerably affect the economy. Over the past couple of decades, application of AFRP-composite materials as an alternative for steel reinforcement has been increasingly noticed, as they have a much higher strength, lighter weight, and noncorrosive nature compared to conventional reinforcing steel. The initial cost of FRP bars is larger than steel; however, it is a small fraction of immense budget spent in future for replacement and repair of bridge decks deteriorated due to corrosion of reinforcing steel.

While the application of FRP bars in concrete bridges, particularly bridge deck slabs, is on a fast development track, most of the research programs are focused on individual concrete members like beams, slabs, and piles (Dolan et al., 2001). In fact, there is a lack of experimental data on full-scale laboratory specimens where the dimensions, boundary conditions, and structural connections are all modeled to realistically resemble a bridge deck system. The results of such full-scale tests are more informative regarding the overall structural behavior where the performance of each component can be directly monitored and also more conclusive in providing design recommendations and filling the engineering knowledge gaps. In this section, the results of an experimental investigation on a full-scale bridge deck slab reinforced and prestressed with AFRP bars are presented. The load-deflection diagram, cracking pattern, and failure mechanism are the main outcomes discussed herein.

23.6.1 Bridge Deck Specimen Layout and Test Setup

A full-scale bridge deck slab consisting of two full-depth precast panels connected through a cast-in-place seam (wet joint) was tested in the High-Bay Structural and Materials Testing Laboratory of Texas A&M University. The precast panels are reinforced in parallel with traffic direction (y axis) and prestressed perpendicular to the traffic direction (x axis) with AFRP bars. Three concrete beams spaced 1830 mm apart center-to-center support the bridge deck slab and are connected to it via six shear pockets per beam. The composite action between the slab and support beams is provided through two high-strength bolts embedded in each pocket, which act as shear connectors. There is a 63-mm hunch between the bridge deck slab and support beams filled with a high performance grout. As shown in Figures 23.2A and B, the

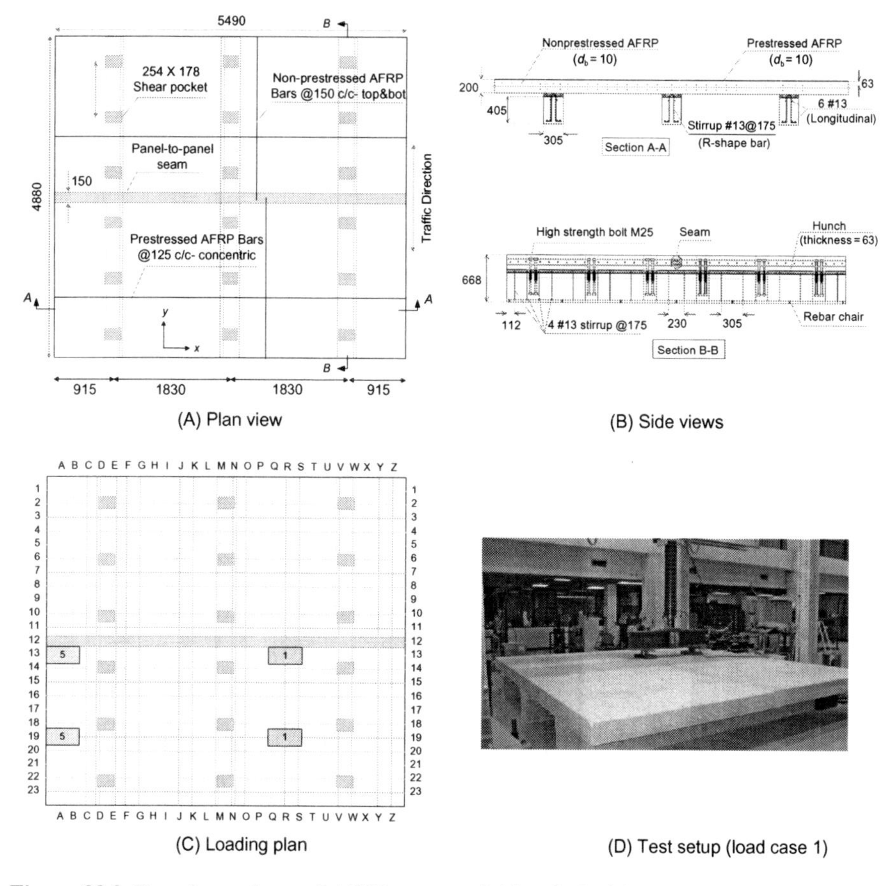

(A) Plan view

(B) Side views

(C) Loading plan

(D) Test setup (load case 1)

Figure 23.2 Experimental test of AFRP concrete bridge deck slab.

bridge deck slab covers two 1800-mm interior spans and two 915-mm overhangs at both sides. The slab is tested under a monotonically increased concentrated load until failure. A 2200-kN actuator is used to apply the load. Load cases 1 and 5 of this experimental research are discussed herein, which represent the axle load of a truck on the interior span and overhang, respectively. The loads are located right at the edge of the panel-to-panel seam to induce the critical load case. Figure 23.2C shows the loading plan and test setup. The applied load is measured through an in-series load cell attached to the tip of the actuator and 34 string pots are used underneath the bridge deck to measure the deflections in both the x- and y-directions below the applied load. To map the cracking pattern, a 200×200 mm grid is depicted on top and beneath the bridge deck from A to Z in the x-direction and 1–23 in the y-direction.

23.6.2 Material Properties

The AFRP bars used in this research are Arapree®, with a 10-mm diameter and sand blast coating. Uniaxial tension tests revealed the rupture stress, strain, and modulus of elasticity of the bars equal to 1380 MPa, 0.02 GPa, and 69 GPa, respectively. As a comparison with conventional steel rebars, the strength and modulus of elasticity of AFRP bars are approximately 3.5 times and one-third that of steel, respectively. The specified cylinder compressive strength of concrete for the first panel, second panel, and seam were found equal to 41, 35, and 38 MPa, respectively. The specified cubes' compressive strength of Sika grout was found equal to 35 MPa. Acceptable consistency in the strength of material was achieved between the precast panels, panel-to-panel seam, and haunch.

23.6.3 AASHTO LRFD Criteria

According to AASHTO LRFD (2010), the wheel load of HS20 truck is equal to 71 kN. If the lane factor of 1.2, the impact factor of 1.33, and the live factor of 1.75 are applied, the maximum factored load will be equal to 200 kN. The load-bearing capacity of the slab is expected to be larger than the specified maximum factored load. The deflection limit of span/800 has been stipulated in AASHTO LRFD for vehicular loads. It should be noted that this deflection limit is to control the vibration, which is a serviceability issue rather than a safety issue. The center of load on the overhang is only required to be 300 mm from the barrier face, per AASHTO LRFD (2010); however, in this test, the load is applied on the edge of the overhang to induce larger moments at support and the more critical case, accordingly.

23.7 Experimental Results

For the interior span (load case 1), flexural cracks were first observed close to the support beams on top of the deck at about the 250-kN wheel load (axle load/2) and also underneath the deck at the location of loading plates. As the load increased, the top cracks propagated around the loading plate and the bottom cracks propagated in a fan shape. At approximately 700-kN load and 19-mm deflection, extensive flexural cracks were eventually followed by failure in a punching manner due to restrained edges. The cracking pattern and load deflection response are shown in Figure 23.3A. In this figure, the black lines represent the cracks on the top surface, and the gray lines indicate the cracks beneath the slab. The load deflection diagram is compared to the control specimen, reinforced with steel, tested by Mander et al. (2010) under the same load cases. As seen in the figure, the AFRP specimen exhibited larger strength and similar deformability compared to the control specimen. Likewise, for the overhang (load case 5), flexural cracks were first observed at about 125 kN on top of the deck and close to the exterior support beam. As load increased, flexural cracks were formed

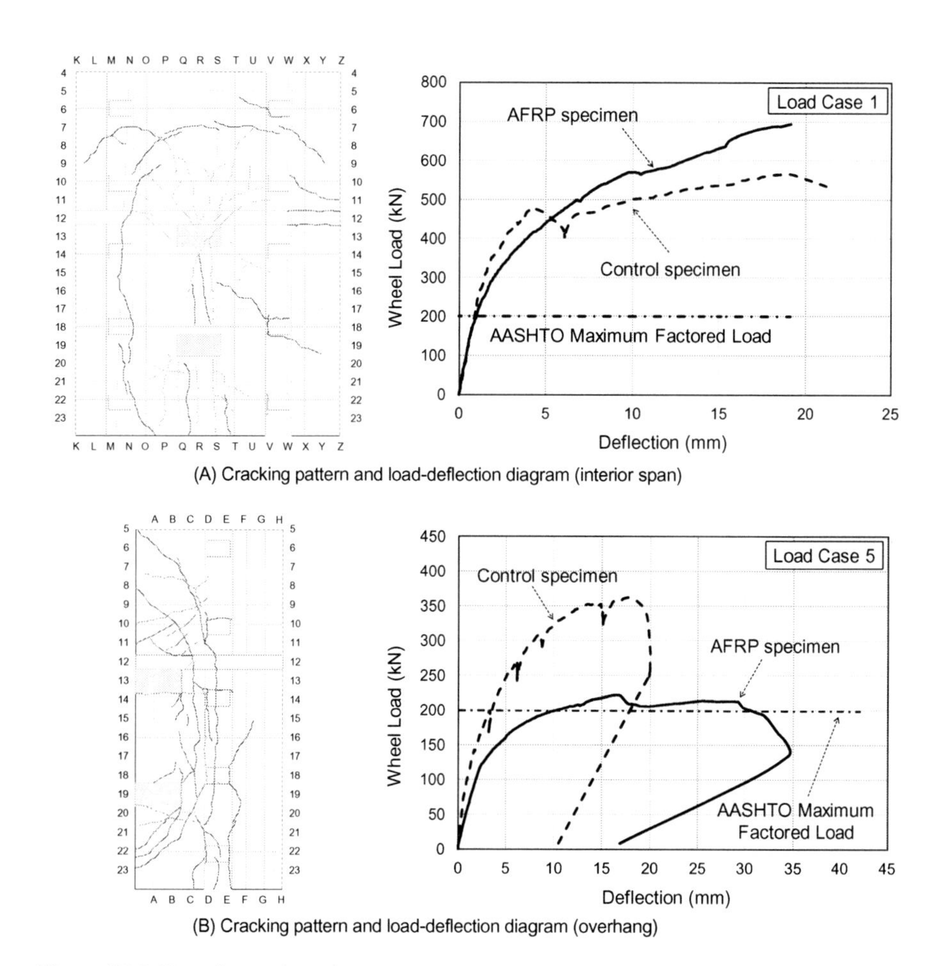

Figure 23.3 Experimental results.

underneath the deck below the loading plates and the top flexural cracks propagated toward the slab edge in a triangle shape. The cracks started to widen, and the slab experienced a large deformation (about 35 mm) and finally failed in a flexural mechanism.

The load-bearing capacity of the overhang was observed to be 220 kN (Figure 23.3B). Comparison with the control specimen shows the ultimate strength of the AFRP specimen is less than that of the steel specimen; however, the induced deformability is considerably larger.

23.8 Discussion

At both the interior span and the overhang, the load-bearing capacity of the slab was larger than the required strength specified by AASHTO LRFD (2010). For the interior span, the deflection at the service load was about 1 mm, which is less than the allowable value (span/800 = 2.25 mm). While AFRP bars behave linearly until rupture and do not have a yielding point, considerable deformability was observed in both the interior and the overhang. This is basically attributed to the low modulus of elasticity of AFRP bars (say, one-third that of steel). In other words, when flexural cracks appear on top of the deck and close to the support due to large negative moments, the flexural stiffness of the bridge deck in the x-direction substantially drops due to the low modulus of elasticity of the bars, and hence the moment is largely redistributed and the y-direction undergoes a larger flexural demand. A drop in the flexural stiffness of the bridge deck section and subsequent moment redistribution continue until the failure mechanism completes. Flexural failure mechanism was evident for both the interior span and the overhang from cracking pattern and load deflection response. However, due to the restrained edges of the interior span, the slab eventually failed in a punching manner. For both load cases, no sign of local failure at a panel-to-panel seam was observed, which shows the sufficient flexural and shear connectivity between the panels. Such structural integrity was implied by the complete transfer of flexural cracks between the panels. Experimental results conclusively showed a satisfactory structural performance of the AFRP concrete bridge deck slab from both strength and serviceability viewpoints.

Acknowledgments

The authors appreciate the assistance provided by Dr. Jawad Abdullah of the Maryland State Highway Administration and Mr. Bimal Devkota of the Baltimore City Department of Transportation in helping out with portions of the research related to infrastructure maintenance inspection and scheduling. This work was jointly carried out at the Morgan State University's Center for Advanced Transportation and Infrastructure Engineering Research and at Texas A&M University with funding from the National Science Foundation through grant CMMI-0927333.

References

AASHTO, 2010. LRFD Bridge Design Specifications, American Association of State Highway and Transportation Officials, Washington, DC.

ASCE, 2009. ASCE: infrastructure fact sheet. <http://www.infrastructurereportcard.org/sites/default/files/RC2009_bridges.pdf/> (accessed 22.05.09.).

Christopoulous, C., Tremblay, R., Kim, H.-J., Lacerte, M., 2008. Self-centering energy dissipative bracing system for the seismic resistance of structures: development and validation. J. Struct. Eng. 134 (1), 96–107.

Dolan, C., 1990. Developments in non-metallic prestressing tendons. PCI J. 35 (5), 80–88.

Dolan, C.W., 1989. Prestressed Concrete Using Kevlar Reinforced Tendons. Cornell University (Ph.D. Dissertation), Ithaca, NY.

Dolan, C.W., Hamilton III, H.R., Nanni, A., 2001. Design Recommendations for Concrete Structures Prestressed with FRP Tendons. Prepared for the Federal Highway Administration. Report No. DTFH61-96-0019. University of Wyoming, Pennsylvania State University, and University of Missouri-Rolla, FHWA, Washington, DC.

Gandomi, A.H., Alavi, A.H., 2012. Krill herd: a new bio-inspired optimization algorithm. Commun. Nonlin. Sci. Numer. Simul. 17 (12), 4831–4845.

Gandomi, A.H., Yang, X.S., Alavi, A.H., 2011. Mixed variable structural optimization using firefly algorithm. Comput. Struct. 89 (23–24), 2325–2336.

Gandomi, A.H., Yang, X.S., Alavi, A.H., 2012. Cuckoo search algorithm: a metaheuristic approach to solve structural optimization problems. Eng. Comput. doi: 10.1007/s00366-011-0241-y.

Ghosh, J., Padgett, J.E., 2010. Aging considerations in the development of time-dependent seismic fragility curves. J. Struct. Eng. 136 (12), 1497–1511.

Gilstrap, J.M., 1997. Characterization of Fiber Reinforced Polymer Prestressing Tendons. M. S. Thesis, Department of Civil and Architectural Engineering, Laramie, WY.

Giroux, R.P., 2010. Sustainable Bridges. Web Aspire (Supplement to Perspective, Spring 2010) Chicago, IL.

Grace, N., Sayed, G.,1997. Ductility of prestressed concrete bridges using internal/external CFRP strands. Presented at the Scotland Conference on Composites. Scotland.

Holden, T., Restrepo, J., Mander, J.B., 2003. Seismic performance of precast reinforced and prestressed concrete walls. J. Struct. Eng. 129 (3), 286–296.

Jha, M.K., 2010. Optimal highway infrastructure maintenance scheduling considering deterministic and stochastic aspects of deterioration. In: Gopalakrishnan, K., Peeta, S. (Eds.), Sustainable and Resilient Critical Infrastructure Systems. Springer, Springer-Verlag, Berlin, Heidelberg, pp. 231–248.

Jha, M.K., Abdullah, J., 2006. A Markovian approach for optimizing highway life-cycle with genetic algorithms by considering maintenance of roadside appurtenances. J. Franklin Inst. 343, 404–419.

Mander, T.J., Henley, M.D., Scott, R.M., Head, M.H., Mander, J.B., 2010. Experimental performance of full-depth precast, prestressed concrete overhang, bridge deck panels. J. Bridge Eng. 15 (5), 503–510.

Melchers, R.E., Frangopol, D.M., 2008. Probabilistic modeling of structural degradation. Reliab. Eng. Syst. Saf. 93 (3), 363–500.

Nanni, A., 1993. Fiber-reinforced-plastic (FRP) reinforcement for concrete structures: properties and applications. Dev. Civ. Eng. 42, pp. 1–450.

Seible, F., Hegemier, G.A., Priestly, M.J.N., Innamorato, D., et al., 1994. Seismic Retrofitting of Squat Circular Bridge Piers with Carbon Fiber Jackets. Advanced Composites Technology Transfer Consortium, Report No. ACTT-94/04, UCSD, 55p.

SPI, 2008. A Look at the World's FRP Composites Bridges. SPI Composites Institute, New York, NY (A Publication of the Market Development Alliance).

Stewart, M.G., Rosowsky, D.V., 1998. Time-dependent reliability of deteriorating reinforced concrete bridge decks. Struct. Saf. 20 (1), 91–109.

Tang, B., Podolny, Jr. W., 1998. A successful beginning for FRP composite materials in bridge applications. Proceedings of the FHWA International Conference on Corrosion and Rehabilitation of Reinforced Concrete Structures. Orlando, FL, December.

CPSIA information can be obtained at www.ICGtesting.com
Printed in the USA
BVOW01*0618310713

327116BV00006B/110/P

9 780123 983640